List of the Elements with their Symbols and Atomic Masses*

Element	Symbol	Atomic Number	Atomic Mass†	Element	Symbol	Atomic Number	Atomic Mass†
Actinium	Ac	89	(227)	Mendelevium	Md	101	(256)
Aluminum	Al	13	26.98	Mercury	Hg	80	200.6
Americium	Am	95	(243)	Molybdenum	Mo	42	95.94
Antimony	Sb	51	121.8	Neodymium	Nd	60	144.2
Argon	Ar	18	39.95	Neon	Ne	10	20.18
Arsenic	As	33	74.92	Neptumium	Np	93	(237)
Astatine	At	85	(210)	Nickel	Ni	28	58.69
Barium	Ba	56	137.3	Niobium	Nb	41	92.91
Berkelium	Bk	97	(247)	Nitrogen	N	7	14.01
Beryllium	Be	4	9.012	Nobelium	No	102	(253)
Bismuth	Bi	83	209.0	Osmium	Os	76	190.2
Bohrium	Bh	107	(262)	Oxygen	O	8	16.00
Boron	B	5	10.81	Palladium	Pd	46	106.4
Bromine	Br	35	79.90	Phosphorus	P	15	30.97
Cadmium	Cd	48	112.4	Platinum	Pt	78	195.1
Calcium	Ca	20	40.08	Plutonium	Pu	94	(242)
Californium	Cf	98	(249)	Polonium	Po	84	(210)
Carbon	C	6	12.01	Potassium	K	19	39.10
Cerium	Ce	58	140.1	Praseodymium	Pr	59	140.9
Cesium	Cs	55	132.9	Promethium	Pm	61	(147)
Chlorine	Cl	17	35.45	Protactinium	Pa	91	(231)
Chromium	Cr	24	52.00	Radium	Ra	88	(226)
Cobalt	Co	27	58.93	Radon	Rn	86	(222)
Copper	Cu	29	63.55	Rhenium	Re	75	186.2
Curium	Cm	96	(247)	Rhodium	Rh	45	102.9
Darmstadtium	Ds	110	(281)	Roentgenium	Rg	111	272
Dysprosium	Dy	66	162.5	Rubidium	Rb	37	85.47
Einsteinium	Es	99	(254)	Ruthenium	Ru	44	101.1
Erbium	Er	68	167.3	Rutherfordium	Rf	104	(257)
Europium	Eu	63	152.0	Samarium	Sm	62	150.4
Fermium	Fm	100	(253)	Scandium	Sc	21	44.96
Fluorine	F	9	19.00	Seaborgium	Sg	106	(263)
Francium	Fr	87	(223)	Selenium	Se	34	78.96
Gadolinium	Gd	64	157.3	Silicon	Si	14	28.09
Gallium	Ga	31	69.72	Silver	Ag	47	107.9
Germanium	Ge	32	72.59	Sodium	Na	11	22.99
Gold	Au	79	197.0	Strontium	Sr	38	87.62
Hafnium	Hf	72	178.5	Sulfur	S	16	32.07
Hahnium	Ha	105	(260)	Tantalum	Ta	73	180.9
Hassium	Hs	108	(265)	Technetium	Tc	43	(99)
Helium	He	2	4.003	Tellurium	Te	52	127.6
Holmium	Ho	67	164.9	Terbium	Tb	65	158.9
Hydrogen	H	1	1.008	Thallium	Tl	81	204.4
Indium	In	49	114.8	Thorium	Th	90	232.0
Iodine	I	53	126.9	Thulium	Tm	69	168.9
Iridium	Ir	77	192.2	Tin	Sn	50	118.7
Iron	Fe	26	55.85	Titanium	Ti	22	47.88
Krypton	Kr	36	83.80	Tungsten	W	74	183.9
Lanthanum	La	57	138.9	Uranium	U	92	238.0
Lawrencium	Lr	103	(257)	Vanadium	V	23	50.94
Lead	Pb	82	207.2	Xenon	Xe	54	131.3
Lithium	Li	3	6.941	Ytterbium	Yb	70	173.0
Lutetium	Lu	71	175.0	Yttrium	Y	39	88.91
Magnesium	Mg	12	24.31	Zinc	Zn	30	65.39
Manganese	Mn	25	54.94	Zirconium	Zr	40	91.22
Meitnerium	Mt	109	(266)				

*All atomic masses have four significant figures. These values are recommended by the Committee on Teaching of Chemistry, International Union of Pure and Applied Chemistry.

†Masses of the longest-lived isotope for radioactive elements are given in parentheses.

Foundations *of*
General, Organic,
and Biochemistry

Katherine J. Denniston
Towson University

Joseph J. Topping
Towson University

Boston   Burr Ridge, IL   Dubuque, IA   New York   San Francisco   St. Louis
Bangkok   Bogotá   Caracas   Kuala Lumpur   Lisbon   London   Madrid   Mexico City
Milan   Montreal   New Delhi   Santiago   Seoul   Singapore   Sydney   Taipei   Toronto

Higher Education

FOUNDATIONS OF GENERAL, ORGANIC, AND BIOCHEMISTRY

Published by McGraw-Hill, a business unit of The McGraw-Hill Companies, Inc., 1221 Avenue of the Americas, New York, NY 10020. Copyright © 2008 by The McGraw-Hill Companies, Inc. All rights reserved. No part of this publication may be reproduced or distributed in any form or by any means, or stored in a database or retrieval system, without the prior written consent of The McGraw-Hill Companies, Inc., including, but not limited to, in any network or other electronic storage or transmission, or broadcast for distance learning.

Some ancillaries, including electronic and print components, may not be available to customers outside the United States.

♻ This book is printed on recycled, acid-free paper containing 10% postconsumer waste.

1 2 3 4 5 6 7 8 9 0 QPD/QPD 0 9 8 7

ISBN 978–0–07–351106–1
MHID 0–07–351106–4

Publisher: *Thomas D. Timp*
Senior Sponsoring Editor: *Tamara Hodge*
Senior Developmental Editor: *Donna Nemmers*
Marketing Manager: *Todd Turner*
Senior Project Manager: *Gloria G. Schiesl*
Lead Production Supervisor: *Sandy Ludovissy*
Lead Media Project Manager: *Judi David*
Executive Media Producer: *Linda Meehan Avenarius*
Media Producer: *Daryl Bruflodt*
Senior Coordinator of Freelance Design: *Michelle D. Whitaker*
Cover/Interior Designer: *Elise Lansdon*
Senior Photo Research Coordinator: *Lori Hancock*
Photo Research: *Connie Mueller*
Compositor: *Techbooks*
Typeface: 10/12 *Palatino*
Printer: *Quebecor World Dubuque, IA*

(USE) Cover Image: *Human Fingerprint: Digital Vision / Getty Images; DNA Double Helix Model: ©Fredrik Skold / Getty Images; Technician Examining CAT Scans: ©Adam Crowley / Getty Images; Washing Oil from Bird: ©Benelux Press / Getty Images*

The credits section for this book begins on page C-1 and is considered an extension of the copyright page.

Library of Congress Cataloging-in-Publication Data

Denniston, K. J. (Katherine J.)
    Foundations of general, organic, and biochemistry / Katherine J. Denniston, Joseph J. Topping. — 1st ed.
        p. cm.
    Includes index.
    ISBN 978-0-07-351106-1 — ISBN 0-07-351106-4 (acid-free paper)
        1. Chemistry—Textbooks. 2. Organic chemistry—Textbooks. 3. Biochemistry—Textbooks. I. Topping, Joseph J. II. Title.

QD31.3.D46 2008
540--dc22

                                           2006047036

www.mhhe.com

We are thankful to our families, whose patience and support made it possible for us to embark on this new adventure.

—Katherine J. Denniston

—Joseph J. Topping

About the Authors

Katherine J. Denniston is the Associate Dean of the Jess and Mildred Fisher College of Science and Mathematics, Director of Premedical and Predental Programs, and Professor in the Department of Biological Sciences.

Formerly the Director of the Center for Science and Mathematics Education, Dr. Denniston has a long-standing interest in reform of undergraduate and K-12 education. She was the Project Director of the Maryland Collaborative for Teacher Preparation II, a National Science Foundation-funded statewide Collaborative for the preparation of science and mathematics specialists to teach in middle school. In addition, she was Director of the Maryland Educators' Summer Research Program, which facilitated research experiences for inservice and preservice teachers, and co-Director of the Maryland Governor's Academy for Mathematics and Science, a summer residence professional development program for Maryland teachers. Dr. Denniston was co-Principal Investigator on an NSF grant to introduce inquiry-based laboratories into the first semester biology course required of all biology majors.

From 2002–2004, Dr. Denniston served as a Program Officer in the Division of Undergraduate Education at the National Science Foundation, working with a number of programs, including the Course, Curriculum, and Laboratory Improvement; Science, Technology, Engineering, and Mathematics Talent Expansion; Advanced Technological Education Teacher Preparation; and the Robert Noyce Scholarship Programs.

Before coming to Towson University in 1985, Denniston earned her Ph.D. in Microbiology from the Pennsylvania State University. She was a post-doctoral fellow in the Department of Genetics at the University of Wisconsin, Madison, a Senior Staff Fellow at the National Cancer Institute, and a Research Assistant Professor at the Division of Molecular Virology and Immunology, Georgetown University. At Towson, Dr. Denniston has taught a wide variety of courses from the introductory to the graduate level.

Dr. Denniston has published extensively on various aspects of molecular biology and virology. She has also published articles and presented workshops on science education and was co-editor of *Recombinant DNA*, published by Dowden, Hutchinson, and Ross, Inc.

Joseph Topping was born in Amsterdam, New York. He received his B.S. degree in chemistry from Le Moyne College in Syracuse, and his M.S. and Ph.D. from the University of New Hampshire. After doing postdoctoral research at the Ames Laboratory of Iowa State University, he joined the chemistry department at Towson University, where he has taught since 1970.

Professor Topping has been involved with a number of initiatives designed to improve the quality of middle and secondary school instruction, most notably the NSF-sponsored Maryland Collaborative for Teacher Preparation. He has a number of papers and presentations in the literature, including baseline studies of contamination of the Chesapeake Bay and its tributaries. He is actively involved with the Maryland Section of the American Chemical Society and is a member of the Board of Governors of the Eastern Analytical Symposium.

Professor Topping's hobbies include golf, fly fishing, softball, and collecting vintage baseball cards. He is an avid reader of history, particularly, the history of baseball.

Brief Contents

Contents

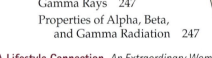

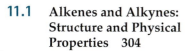

15 Lipids and Their Functions in Biochemical Systems 440

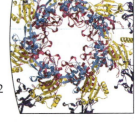

16 Protein Structure and Enzymes 470

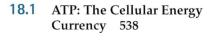

Applications in Chemistry

A LIFESTYLE
Connection

A MEDICAL
Connection

AN ENVIRONMENTAL
Perspective

CHEMISTRY
at the Crime Scene

Preface

This first edition of *Foundations of General, Organic, and Biochemistry* is the answer to a long-standing dream of ours to write a text that would condense the critical topics of chemistry into a book that could serve a one-semester, as well as a two-semester, chemistry course. The key to doing this well is to put the "general" back into general chemistry, providing students with a view of chemistry as an integrated discipline. Our strategy to accomplish this goal has been to tighten the inorganic chapters to emphasize those topics essential to understanding basic concepts in organic chemistry and biochemistry, as well as to highlight critical applications in related careers. This approach allows us to provide sufficient opportunity for students to develop an understanding of organic and biochemistry, areas often neglected for lack of time.

Reaching the Student Audience

Like two-semester books, this text is designed for majors in health-related fields. However, our target audience for this book is somewhat broader. We recognize the increased need for two-year programs to teach biotechnologists, medical technicians, forensic technologists, and agriculture and environmental managers, in addition to the nursing programs that are the typical target audience for such texts. Our response to this expanded set of needs has been to write a book fostering student understanding of key concepts in chemistry that are needed to successfully pursue these varied career paths.

With this target audience in mind, we have used a writing style and level intended to stimulate student interest and maintain that interest throughout the course. We have also integrated concepts and applications to produce scientifically literate students—those who are able to recognize the connections between the chemistry they are studying and the world in which they live. Such students are better able to make well-informed decisions on the scientific and technological issues facing the world today. Both practical understanding and scientific literacy are promoted through inclusion of special boxed topics connecting chemistry to medical, environmental, and forensic applications, and the often open-ended "For Further Understanding" questions that follow, which require students to think "beyond the box."

Key Features of *Foundations of General, Organic, and Biochemistry*

Throughout the project, we have been true to one goal: to write a book that is student-oriented and readable. The ultimate goal is to promote student learning and to facilitate teaching. We want to engage students, appeal to visual learners, and provide a variety of pedagogical tools to support the needs of different types of learners. We have utilized a variety of strategies to accomplish these goals.

Engaging Students

Students learn better when they can see a clear relationship between the subject material they are studying and real life. We have written the text to help students make connections between the principles of chemistry and their life experiences, as well as their future professional experiences. This is accomplished through the inclusion of numerous boxed topics called *Connections*. These short stories present real-world situations involving one or more topics that students will encounter in the chapter.

- *Medical Connections* relate chemistry to a health concern or a diagnostic application.
- *Environmental Perspectives* deal with issues, including the impact of chemistry on the ecosystem and the way in which these environmental changes affect human health.
- *Lifestyle Connections* delve into chemistry and society, and include such topics as gender issues in science and historical viewpoints.

- *Chemistry at the Crime Scene* essays explore the chemistry behind the emergent field of forensic science.

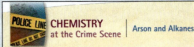

CHEMISTRY at the Crime Scene | Arson and Alkanes

In September of 2005, Thomas Sweatt was sentenced to life in prison for setting 45 fires in the Washington, D.C. area. Aside from millions of dollars in property damage, two people died as a result of these fires. Mr. Sweatt confessed to the fires, which terrorized the Washington metropolitan area over a two-year period, stating that he was "addicted to setting fires."

Authorities estimate that one-third of all fires are arson and the Federal Bureau of Investigation reports that arson is more common in the United States than anywhere else in the world. Although arsonists may set fires as terrorist acts, to defraud insurance companies, to gain revenge, or to cover up another crime, other arsonists are mentally ill pyromaniacs like Mr. Sweatt.

As soon as a fire is extinguished and the scene is secure, investigators immediately gather evidence to determine the cause of the fire. They study the pattern of the fire to determine the point of origin. This is critical, because this is where they must sample for the presence of accelerants, flammable substances that cause fires to burn hotter and spread more quickly. The most common accelerants are mixtures of hydrocarbons, including gasoline, kerosene, or diesel fuel.

Because all of these accelerants contain molecules that evaporate, they may be detected at the point of origin by trained technicians or "sniffer dogs." However, a much more advanced technology is also available; it is called *headspace gas chromatography*. Gas chromatography separates and identifies components of a sample based on differences in their boiling points. Each gas mixture produces its own unique "chemical fingerprint" or chromatogram. Crime scene technicians collect debris from the point of origin and seal it in an airtight vial. In the laboratory, they heat the vial so the hydrocarbons evaporate and are trapped in the headspace of the vial. These gases are then collected with a needle and syringe and injected into the gas chromatogram for analysis and identification.

To be absolutely certain that an accelerant has been used, the crime scene technicians also collect debris from control sites away from the point of origin. The reason for this is that pyrolysis, the decomposition or transformation of a compound caused by heat, may produce products that simulate accelerants. If an accelerant is found at the point of origin and not among the other pyrolysis products, it can be concluded that arson was the cause of the blaze.

Of course, the next priority is to catch the arsonist. Crime scene technicians collect and analyze physical evidence, including fingerprints, footprints, and other artifacts, found at the crime scene. In the case of Mr. Sweatt, it was DNA fingerprint evidence from articles of clothing left at the crime scenes that led to his capture and conviction and ended his two-year arson spree.

FOR FURTHER UNDERSTANDING

Investigate the composition of gasoline, diesel fuel, and kerosene and explain how they can be distinguished from one another using gas chromatography.

Other accelerants that have been used by arsonists include nail polish remover (acetone), grain alcohol (ethanol), and rubbing alcohol (2-propanol or isopropyl alcohol.) What properties do these substances share that make them useful as accelerants?

Learning Tools

In designing the original learning system, we asked ourselves the question: "If we were students, what would help us organize and understand the material covered in this chapter?" With valuable suggestions from reviewers, we have established a set of pedagogical tools to support student learning:

- **Learning Goals:** A set of objectives at the beginning of each chapter previews concepts that will be covered in the chapter. Icons within the margin locate text material that supports the learning goals.
- **Detailed Chapter Outline:** A listing of topic headings is provided for each chapter. Topics are arranged in outline form to help students organize the material in their own minds.
- **Chapter Cross-References:** To help students locate the pertinent background material, references to previous chapters, sections, and perspectives are noted in the margins of the text. These marginal cross references also alert students to upcoming topics related to the information currently being studied.

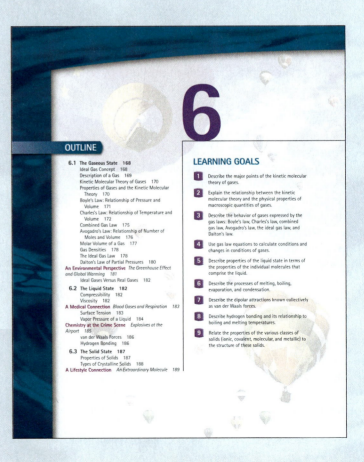

6

OUTLINE

LEARNING GOALS

1. Describe the major points of the kinetic molecular theory of gases.

2. Explain the relationship between the kinetic molecular theory and the physical properties of macroscopic quantities of gases.

3. Describe the behavior of gases expressed by the gas laws: Boyle's law, Charles's law, combined gas law, Avogadro's law, the ideal gas law, and Dalton's law.

4. Use gas law equations to calculate conditions and changes in conditions of gases.

5. Describe properties of the liquid state in terms of the properties of the individual molecules that comprise the liquid.

6. Describe the processes of melting, boiling, evaporation, and condensation.

7. Describe the dipolar attractions known collectively as van der Waals forces.

8. Describe hydrogen bonding and its relationship to boiling and melting temperatures.

9. Relate the properties of the various classes of solids (ionic, covalent, molecular, and metallic) to the structure of these solids.

---

182 | CHAPTER 6 States of Matter

Ideal Gases Versus Real Gases

To this point, we have assumed, in both theory and calculations, that all gases behave as ideal gases. However, in reality there is no such thing as an ideal gas. As we noted at the beginning of this section, the ideal gas is a model (a very useful one) that describes the behavior of individual atoms and molecules; this behavior translates to the collective properties of measurable quantities of these atoms and molecules. Limitations of the model arise from the fact that interactive forces, even between the widely spaced particles of gas, are not totally absent in any sample of gas.

Attractive forces are present in gases composed of polar molecules. Nonuniform charge distribution on polar molecules creates positive and negative regions, resulting in electrostatic attraction and deviation from ideality.

See Sections 3.5 and 6.2 for a discussion of interactions of polar molecules.

Calculations involving polar gases such as HF, NO, and SO_2 based on ideal gas equations (which presume no such interactions) are approximations. However, at low pressures, such approximations certainly provide useful information. Nonpolar molecules, on the other hand, are only weakly attracted to each other and behave much more ideally in the gas phase.

- **Summary of Key Reactions:** In the organic chemistry chapters, each major reaction type is highlighted on a blue background. Major equations are summarized at the end of the chapter, facilitating review.
- **Chapter Summary:** Each major topic of the chapter is briefly reviewed in paragraph form in the end-of-chapter summary. These summaries serve as a mini-study guide, covering the major concepts in the chapter.
- **Key Terms:** Key terms are printed in boldface in the text and defined immediately. Each key term is also listed at the end of the chapter and is accompanied by a section number for easy reference.
- **Glossary of Key Terms:** In addition to being listed at the end of the chapter, each key term from the text is defined in the alphabetical glossary at the end of the book.
- **Further Information Online:** The ARIS website for this textbook provides important readings, equations, and tables of formula weights, mathematics reviews, among numerous other topics, and can be accessed at www.mhhe.com/denniston.
- **Online Animations:** An animation icon alerts the reader to animations that are available online to help bring chemistry to life.

The Art Program

Today's students are much more visually oriented than students of any previous generation. Television and the computer represent alternate modes of learning. We have built upon this observation through use of color, figures, and three-dimensional computer-generated models. This art program enhances the readability of the text and provides alternative pathways to learning.

- **Dynamic Illustrations:** Each chapter is amply illustrated using figures, tables, and chemical formulas. All of these illustrations are carefully annotated for clarity.

SUMMARY OF REACTIONS

Reactions of Alkanes

Combustion:

$$C_nH_{2n+2} + O_2 \longrightarrow CO_2 + H_2O + \text{heat energy}$$

Alkane · · · · Oxygen · · · · Carbon dioxide · · Water

Halogenation:

$$\underset{\text{Alkane}}{R-\overset{\overset{\displaystyle H}{|}}{\underset{\underset{\displaystyle H}{|}}{C}}-H} + \underset{\text{Halogen}}{X_2} \xrightarrow{\text{light or heat}} \underset{\text{Alkyl halide}}{R-\overset{\overset{\displaystyle H}{|}}{\underset{\underset{\displaystyle H}{|}}{C}}-X} + \underset{\text{Hydrogen halide}}{H-X}$$

SUMMARY

10.1 The Chemistry of Carbon

The modern science of organic chemistry began with Wöhler's synthesis of urea in 1828. At that time, people believed that it was impossible to synthesize an organic molecule outside a living system. We now define organic chemistry as the study of carbon-containing compounds. The differences between the ionic bond, which is characteristic of many inorganic substances, and the covalent bond in organic compounds are responsible for the great contrast in properties and reactivity between organic and inorganic compounds. All organic compounds are classified as either *hydrocarbons* or *substituted hydrocarbons*. In substituted hydrocarbons, a hydrogen atom is replaced by a functional group. A *functional group* is an atom or group of atoms arranged in a particular way that imparts specific chemical or physical properties to a molecule. The major families of organic molecules are defined by the specific functional groups that they contain.

10.2 Alkanes

Alkanes are *saturated hydrocarbons*, that is, hydrocarbons that have only carbon and hydrogen atoms that are bonded by carbon–carbon and carbon–hydrogen single bonds. They have the general molecular formula C_nH_{2n+2} and are nonpolar, water-insoluble compounds with low melting and boiling points. In the *I.U.P.A.C. Nomenclature System* the alkanes are named by determining the number of carbon atoms in the parent compound

and numbering the carbon chain to provide the lowest possible number for all substituents. The substituent names and numbers are used as prefixes before the name of the parent compound.

Constitutional or *structural isomers* are molecules that have the same molecular formula but different structures. They have different physical and chemical properties because the atoms are bonded to one another in different patterns.

10.3 Cycloalkanes

Cycloalkanes are a family of organic molecules having C—C single bonds in a ring structure. They are named by adding the prefix *cyclo-* to the name of the alkane parent compound.

10.4 Reactions of Alkanes and Cycloalkanes

Alkanes can participate in *combustion* reactions. In complete combustion reactions, they are oxidized to produce carbon dioxide, water, and heat energy. They can also undergo *halogenation* reactions to produce *alkyl halides*.

KEY TERMS

aliphatic hydrocarbon (10.1)
alkane (10.2)
alkyl group (10.2)
alkyl halide (10.4)
aromatic hydrocarbon (10.1)
combustion (10.4)
condensed formula (10.2)
constitutional isomers (10.2)
cycloalkane (10.3)
functional group (10.1)
halogenation (10.4)
hydrocarbon (10.1)
I.U.P.A.C. Nomenclature System (10.2)

line formula (10.2)
molecular formula (10.2)
parent compound (10.2)
primary (1°) carbon (10.2)
quaternary (4°) carbon (10.2)
saturated hydrocarbon (10.1)
secondary (2°) carbon (10.2)
structural formula (10.2)
structural isomer (10.2)
substituted hydrocarbon (10.1)
substitution reaction (10.4)
tertiary (3°) carbon (10.2)
unsaturated hydrocarbon (10.1)

QUESTIONS AND PROBLEMS

The Chemistry of Carbon

Foundations

10.17 Why is the number of organic compounds nearly limitless?
10.18 What are allotropes?
10.19 What are the three allotropic forms of carbon?
10.20 Describe the three allotropes of carbon.
10.21 Why do ionic substances generally have higher melting and boiling points than covalent substances?

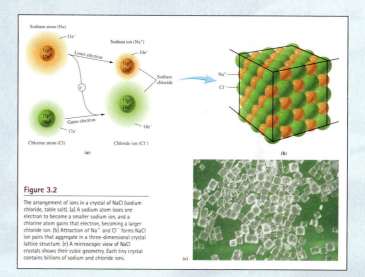

Figure 3.2

The arrangement of ions in a crystal of NaCl (sodium chloride, table salt). (a) A sodium atom loses one electron to become a smaller sodium ion, and a chlorine atom gains that electron, becoming a larger chloride ion. (b) Attraction of Na^+ and Cl^- forms NaCl ion pairs that aggregate in a three-dimensional crystal lattice structure. (c) A microscopic view of NaCl crystals shows their cubic geometry. Each tiny crystal contains billions of sodium and chloride ions.

124 | **CHAPTER 4** Calculations and the Chemical Equation

Figure 4.6

Balancing the equation HCl + Ca → $CaCl_2$ + H_2. (a) Neither product is the correct chemical species. (b) The reactant, HCl, is incorrectly represented as H_2Cl_2. (c) This equation is correct; all species are correct, and the law of conservation of mass is obeyed.

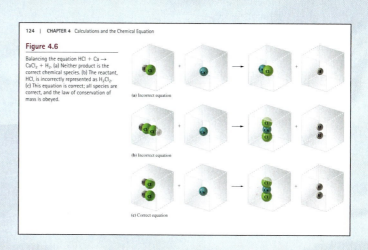

(a) Incorrect equation

(b) Incorrect equation

(c) Correct equation

- **Color-Coding Scheme:** We have color-coded the reactions so that chemical groups being added or removed in a reaction can be quickly recognized. This is done by using red print in chemical equations or formulas to draw the reader's eye to key elements or properties in a reaction or structure. Blue print is used when additional features must be highlighted.

Blue background screens denote generalized chemical and mathematical equations. In the organic chemistry chapters, we use a blue background screen to designate key reactions within the text and in the Summary of Reactions section at the end of the chapter.

Yellow background screens are used to illustrate energy—either as energy stored in electrons or groups of atoms—in the general and biochemistry sections of the text. In the organic chemistry section of the text, we use yellow background screens to show the parent chain of organic compounds.

Certain situations make it necessary to adopt a unique color convention tailored to the material in a particular chapter. For example, in Chapter 16, the structures of amino acids require three colors to draw students' attention to key features of these molecules. For consistency, red is used to denote the acid portion of an amino acid, and blue is used to denote the basic portion of an amino acid. Green print is used to denote the R groups, and a yellow background screen directs the eye to the α-carbon.

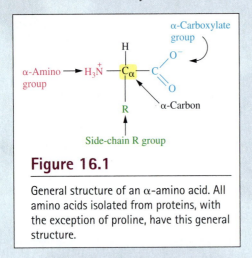

Figure 16.1

General structure of an α-amino acid. All amino acids isolated from proteins, with the exception of proline, have this general structure.

- **Computer-Generated Models:** The students' ability to understand the geometry and three-dimensional structure of molecules is essential to the understanding of organic and biochemical reactions. Computer generated models are used throughout the text because they are both accurate and easily visualized.

Problem Solving and Critical Thinking

The best way to learn chemistry and apply it to practical situations is to develop problem-solving and critical thinking skills. To help students accomplish this, we have created a variety of problems that require recall, fundamental calculations, and complex reasoning.

- **In-Chapter Examples, Solutions, and Problems:** Each chapter includes a number of examples that show the student, step-by-step, how to properly reach the correct solution to model problems. Whenever possible, they are followed by in-text problems that allow the students to test their mastery of information and to build self-confidence.

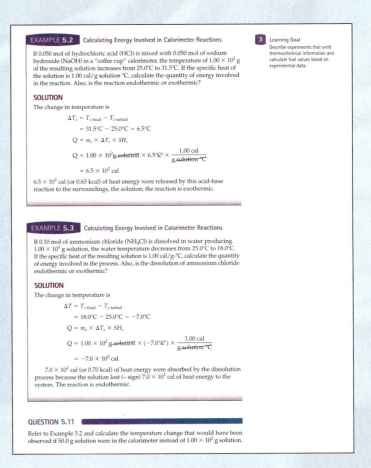

- **End-of-Chapter Problems:** We have created a wide variety of paired concept problems. The answers to the odd-numbered questions are found in the back of the book as reinforcement for the students as they develop problem-solving skills. The students must then apply the same principles to the related even-numbered problems.

volume of any ideal gas is 22.4 L. STP conditions are defined as 273 K (or 0°C) and 1 atm pressure.

Boyle's law, Charles's law, and Avogadro's law may be combined into a single expression relating all four terms, the *ideal gas law:* $PV = nRT$. R is the ideal gas constant (0.0821 L-atm K^{-1} mol^{-1}) if the units P (atmospheres), V (liters), n (number of moles), and T (Kelvin) are used.

The *combined gas law* provides a convenient expression for gas law calculations involving the most common variables: pressure, volume, and temperature.

Dalton's law of partial pressures states that a mixture of gases exerts a pressure that is the sum of the pressures that each gas would exert if it were present alone under similar conditions ($P_t = p_1 + p_2 + p_3 + \cdots$).

6.2 The Liquid State

Liquids are practically incompressible because of the closeness of the molecules. The *viscosity* of a liquid is a measure of its resistance to flow. Viscosity generally decreases with increasing temperature. The *surface tension* of a liquid is a measure of the attractive forces at the surface of a liquid. *Surfactants* decrease surface tension.

The conversion of liquid to vapor at a temperature below the boiling point of the liquid is *evaporation*. Conversion of the gas to the liquid state is *condensation*. The *vapor pressure of the liquid* is defined as the pressure exerted by the vapor at equilibrium at a specified temperature. The *normal boiling point* of a liquid is the temperature at which the vapor pressure of the liquid is equal to 1 atm.

Molecules in which a hydrogen atom is bonded to a small, highly electronegative atom such as nitrogen, oxygen, or fluorine exhibit *hydrogen bonding*. Hydrogen bonding in liquids is responsible for lower than expected vapor pressures and higher than expected boiling points. The presence of *van der Waals forces* and hydrogen bonds significantly affects the boiling points of liquids as well as the melting points of solids.

6.3 The Solid State

Solids have fixed shapes and volumes. They are *incompressible*, owing to the closeness of the particles. Solids may be *crystalline*, having a regular, repeating structure, or *amorphous*, having no organized structure.

Crystalline solids may exist as *ionic solids, covalent solids, molecular solids,* or *metallic solids*. Electrons in metallic solids are extremely mobile, resulting in the high *conductivity* (ability to carry electrical current) exhibited by many metallic solids.

KEY TERMS

amorphous solid (6.3)
Avogadro's law (6.1)
barometer (6.1)
Boyle's law (6.1)
Charles's law (6.1)
combined gas law (6.1)
condensation (6.2)
covalent solid (6.3)
crystalline solid (6.3)
Dalton's law (6.1)
dipole–dipole interactions (6.2)
evaporation (6.2)
hydrogen bonding (6.2)
ideal gas (6.1)
ideal gas law (6.1)
ionic solid (6.3)
kinetic molecular theory (6.1)
London forces (6.2)
melting point (6.3)
metallic bond (6.3)
metallic solid (6.3)
molar volume (6.1)
molecular solid (6.3)
normal boiling point (6.2)
partial pressure (6.1)
pressure (6.1)
standard temperature and pressure (STP) (6.1)
surface tension (6.2)
surfactant (6.2)
van der Waals forces (6.2)
vapor pressure of a liquid (6.2)
viscosity (6.2)

QUESTIONS AND PROBLEMS

Kinetic Molecular Theory

Foundations

6.21 Compare and contrast the gas, liquid, and solid states with regard to the average distance of particle separation.
6.22 Compare and contrast the gas, liquid, and solid states with regard to the nature of the interactions among the particles.
6.23 Describe the molecular/atomic basis of gas pressure.
6.24 Describe the measurement of gas pressure.

Applications

6.25 Why are gases easily compressible?
6.26 Why are gas densities much lower than those of liquids or solids?
6.27 Do gases exhibit more ideal behavior at low or high pressures? Why?
6.28 Do gases exhibit more ideal behavior at low or high temperatures? Why?
6.29 Use the kinetic molecular theory to explain why dissimilar gases mix more rapidly at high temperatures than at low temperatures.
6.30 Use the kinetic molecular theory to explain why aerosol cans carry instructions warning against heating or disposing of the container in a fire.
6.31 Predict and explain any observed changes taking place when an inflated balloon is cooled (perhaps refrigerated).
6.32 Predict and explain any observed changes taking place when an inflated balloon is heated (perhaps microwaved).

Gas Laws

Foundations

6.33 The pressure on a fixed mass of a gas is tripled at constant temperature. Will the volume increase, decrease, or remain the same?
6.34 By what factor will the volume of the gas in Question 6.33 change?

Applications

6.35 Explain why the Kelvin scale is used for gas law calculations.
6.36 The temperature on a summer day may be 90°F. Convert this value to Kelvins.

3.62 Would H_2O or CCl_4 be expected to have a higher melting point? Why?

Drawing Lewis Structures of Molecules

Foundations

3.63 Draw the appropriate Lewis structure for each of the following atoms:
a. H
b. He
c. C
d. N
3.64 Draw the appropriate Lewis structure for each of the following atoms:
a. Be
b. B
c. F
d. S
3.65 Draw the appropriate Lewis structure for each of the following ions:
a. Li^+
b. Mg^{2+}
c. Cl^-
d. P^{3-}
3.66 Draw the appropriate Lewis structure for each of the following ions:
a. Be^{2+}
b. Al^{3+}
c. O^{2-}
d. S^{2-}

Applications

3.67 Give the Lewis structure for each of the following compounds:
a. NCl_3
b. CH_3OH
c. CS_2
3.68 Give the Lewis structure for each of the following compounds:
a. HNO_3
b. CCl_4
c. PBr_3
3.69 Using the VSEPR theory, predict the geometry, polarity, and water solubility of each compound in Question 3.67.
3.70 Using the VSEPR theory, predict the geometry, polarity, and water solubility of each compound in Question 3.68.
3.71 Ethanol (ethyl alcohol or grain alcohol) has a molecular formula of C_2H_5OH. Represent the structure of ethanol using the Lewis electron dot approach.
3.72 Formaldehyde, H_2CO, in water solution has been used as a preservative for biological specimens. Represent the Lewis structure of formaldehyde.
3.73 Acetone, C_3H_6O, is a common solvent. It is found in such diverse materials as nail polish remover and industrial solvents. Draw its Lewis structure if its skeletal structure is

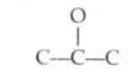

3.74 Ethylamine is an example of an important class of organic compounds. The molecular formula of ethylamine is $CH_3CH_2NH_2$. Draw its Lewis structure.
3.75 Predict whether the bond formed between each of the following pairs of atoms would be ionic, nonpolar, or polar covalent:
a. S and O
b. Si and P
c. Na and Cl
d. Na and O
e. Ca and Br
3.76 Predict whether the bond formed between each of the following pairs of atoms would be ionic, nonpolar, or polar covalent:
a. Cl and Cl
b. H and H
c. C and H
d. Li and F
e. O and O
3.77 Draw an appropriate covalent Lewis structure formed by the simplest combination of atoms in Problem 3.75 for each molecule that involves a nonpolar or polar covalent bond.
3.78 Draw an appropriate covalent Lewis structure formed by the simplest combination of atoms in Problem 3.76 for each molecule that involves a nonpolar or polar covalent bond.

Properties Based on Electronic Structure and Molecular Geometry

3.79 What is the relationship between the polarity of a bond and the polarity of the molecule?
3.80 What effect does polarity have on the solubility of a compound in water?
3.81 What effect does polarity have on the melting point of a pure compound?
3.82 What effect does polarity have on the boiling point of a pure compound?
3.83 Would you expect KCl to dissolve in water?
3.84 Would you expect ethylamine (Question 3.74) to dissolve in water?

FOR FURTHER UNDERSTANDING

1. Predict differences in our global environment that may have arisen if the freezing point and boiling point of water were 20°C higher than they are.
2. Would you expect the compound $C_2S_2H_4$ to exist? Why or why not?
3. Write a Lewis structure for the ammonium ion. Explain why its charge must be +1.
4. Which of the following compounds would be predicted to have the higher boiling point? Explain your reasoning.

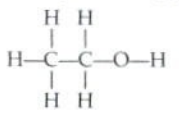

Ethanol

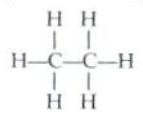

Ethane

5. Why does the octet rule not work well for compounds of lanthanide and actinide elements? Suggest a number other than eight that may be more suitable.

- **For Further Understanding Problems:** Each end-of-chapter Questions and Problems section includes a set of critical thinking problems referred to as "For Further Understanding." These problems are intended to challenge students to integrate concepts to solve more complex problems. They make a perfect complement to the classroom lecture because they provide an opportunity for in-class discussion of complex problems dealing with daily life and the health care sciences. Each of the boxed essays in the text also includes open-ended For Further Understanding problems that challenge the student to investigate the topic further.

CHEMISTRY at the Crime Scene | **Explosives at the Airport**

The images flash across our television screens: a "bomb sniffing" dog being led through an airport or train station, pausing to sniff packages or passengers, looking for anything of a suspicious nature. Or, perhaps, we see a long line of people waiting to pass through a scanning device surrounded by what appears to be hundreds of thousands of dollars worth of electronic gadgetry.

At one level, we certainly know what is happening. These steps are taken to increase the likelihood that our trip, as well as everyone else's, will be as safe and worry-free as possible. From a scientific standpoint, we may wonder how these steps actually detect explosive materials. What do the dog and some electronic devices have in common? How can a dog sniff a solid or a liquid? Surely everyone knows that the nose can only sense gases, and explosive devices are solids or liquids, or a combination of the two.

One potential strategy is based on the concept of vapor pressure, which you have just studied. We now know that liquids, such as water, have a measurable vapor pressure at room temperature. In fact, most liquids and many solids have vapor pressures large enough to allow detection of the molecules in the gas phase. The challenge is finding devices that are sufficiently sensitive and selective, enabling them to detect low concentrations of molecules characteristic of explosives, without becoming confused by thousands of other compounds routinely present in the air.

Each explosive device has its own "signature," a unique mix of chemicals used in its manufacture and assembly. If only one, or perhaps a few, of these compounds has a measurable vapor pressure, it may be detected with a sensitive measuring device.

Dogs are renowned for their keen sense of smell, and some breeds are better than others. Dogs can be trained to signal the presence of certain scents by barking or exhibiting unusual agitation. A qualified handler can recognize these cues and alert appropriate authorities.

Scientific instruments are designed to mimic the scenario described here. A device, the mass spectrometer, can detect very low concentrations of molecules in the air. Additionally, it can distinguish certain "target" molecules, because each different compound has its own unique molar mass. Detection of molecules of interest generates an electrical signal, and an alarm is sounded.

Compounds with high vapor pressures are most easily detected. Active areas of forensic research involve designing a new generation of instruments that are even more sensitive and selective than those currently available. Decreased cost and increased portability and reliability will enable many sites, not currently being monitored, to have the same level of protection as major transit facilities.

FOR FURTHER UNDERSTANDING

Would you expect nonpolar or polar molecules of similar mass to be more easily detected? Why?

Why must an explosives detection device be highly selective?

Supplements for the Instructor

The following items may accompany this text. Please consult your McGraw-Hill representative for policies, prices, and availability as some restrictions may apply.

- McGraw-Hill's *Foundations of General, Organic, and Biochemistry* ARIS website (Assessment, Review, and Instruction System) makes homework meaningful—and manageable—for instructors and students. Instructors can assign and grade chapter-specific homework within the industry's most robust and versatile homework management system. These homework questions can be imported into a variety of course management solutions such as WebCT, Blackboard, and WebAssign. These course cartridges also provide online testing and powerful student tracking features. From the ARIS website, students can access these chapter-specific self-study tools:

  - Self-quizzes
  - Animations
  - Key Terms

Go to www.aris.mhhe.com to learn more, or go directly to this book's ARIS site at www.mhhe.com/denniston.

- An *Instructor's Testing and Resource CD-ROM* includes the Instructor's Solutions Manual, prepared by the authors and by Timothy Dwyer of Villa Julie College. The Instructor's Solutions Manual contains suggestions for organizing lectures, instructional objectives, perspectives on boxed readings from the text, a list of each chapter's key problems and concepts, and contains all answers for the textbook's even-numbered end-of-chapter problems. A Computerized Test Bank of questions for each chapter, prepared by Ann T. Eakes, is also included on this cross-platform CD-ROM.
- Laboratory Resource Guide: Written by Charles H. Henrickson, Larry C. Byrd, and Norman W. Hunter of Western Kentucky University, this helpful prep guide contains the hints that the authors have learned over the years to ensure students' success in the laboratory. This Resource Guide is available through the ARIS course website for this text.
- Over 300 animations are available within McGraw-Hill's *Chemistry Animations DVD*. Instructors can easily view the animations and import them into PowerPoint to create multimedia presentations.
- A set of 100 *printed transparencies* feature key color images and tables from the text to assist instructors with classroom projection needs.

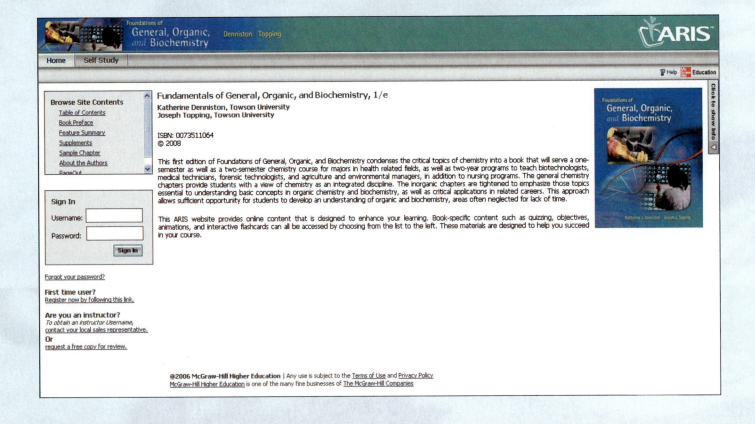

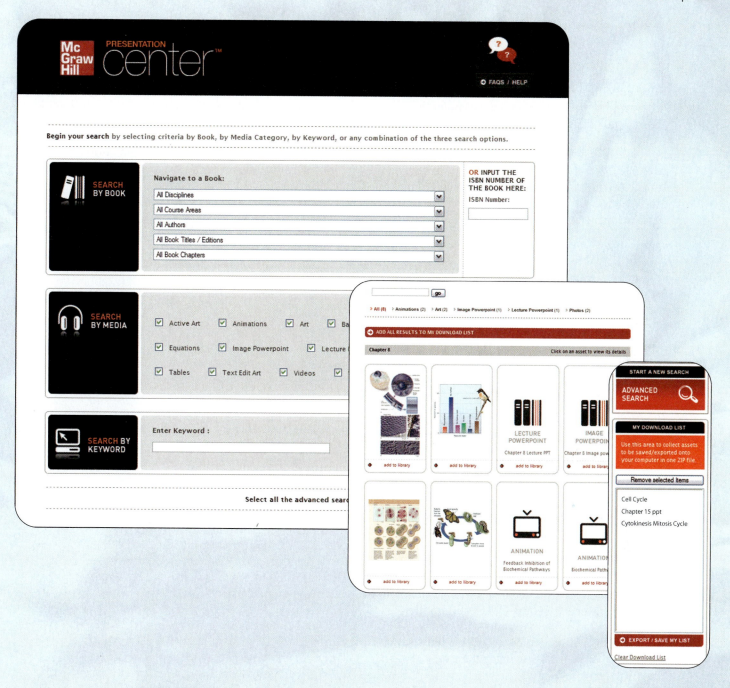

- Build instructional materials wherever, whenever, and however you want! McGraw-Hill Presentation Center is an online digital library containing assets such as photos, artwork, PowerPoint® presentations, and other media types that can be used to create customized lectures, visually enhanced tests and quizzes, compelling course websites, or attractive printed support materials. The McGraw-Hill Presentation Center library includes thousands of assets from many McGraw-Hill titles. This ever-growing resource gives instructors the power to utilize assets specific to an adopted textbook as well as content from all other books in the library. The Presentation Center can be accessed from the instructor side of your textbook's ARIS website, and the Presentation Center's dynamic search engine allows you to explore by discipline, course, textbook chapter, asset type, or keyword. Simply browse, select, and download the files you need to build engaging course materials. All assets are copyrighted by McGraw-Hill Higher Education but can be used by instructors for classroom purposes.

- McGraw-Hill has partnered with eInstruction to provide the revolutionary *Classroom Performance System* (CPS), to bring interactivity into the classroom. CPS is a wireless response system that gives the instructor and students immediate feedback from the entire class. The wireless response pads are essentially remotes that are easy to use and engage students. CPS allows you to motivate student preparation, interactivity, and active learning so you can receive immediate feedback and know what students understand. A text-specific set of questions, formatted for both CPS and PowerPoint, is available via download from the Instructor area of the Online Learning Center.
- *A Laboratory Manual for General, Organic, and Biochemistry* by Henrickson/Byrd/Hunter provides clear and concise laboratory experiments to reinforce students' understanding of concepts. Prelaboratory exercises, questions, and report sheets are coordinated with each experiment to ensure active student involvement and comprehension.

Supplements for the Student

- The *Student Solutions Manual,* prepared by Timothy Dwyer of Villa Julie College, contains chapter-sorted, detailed solutions and explanations for all odd-numbered problems in the text.
- McGraw-Hill's *Foundations of General, Organic, and Biochemistry* ARIS Website (www.mhhe.com/denniston) contains useful study and review tools for students, including chapter-specific quizzing and animations specific to the content in each chapter. Animations have quiz questions associated with them, to check your comprehension of the material.
- *Schaum's Outline of General, Organic, and Biological Chemistry* is written by George Odian and Ira Blei. This supplement provides students with over 1400 solved problems with complete solutions. It also teaches effective problem-solving techniques.

Acknowledgments

We are grateful to our many colleagues at McGraw-Hill for their support, direction, and assistance on this project. In particular, we wish to thank Gloria Schiesl, Senior Project Manager, Donna Nemmers, Senior Developmental Editor, and Thomas Timp, Publisher. We also wish to acknowledge the assistance of Ann Eakes for her work writing the PowerPoint Lecture Outlines and Test Bank ancillaries to support this textbook. Timothy Dwyer, in conjunction with the authors, carefully prepared the Instructor's Solutions Manual and Student Solutions Manual to accompany this text.

This text has been the result of feedback received from many professors who are teaching the course. These reviewers have our appreciation and assurance that their comments received serious consideration.

John R. Amend, *Montana State University*
Maher Atteya, *Georgia Perimeter College*
Mary H. Bailey, *The Ohio State University*
Mark Champagne, *Macomb Community College*
Derald Chriss, *Southern University*
Rajeev B. Dabke, *Columbus State University*
Brent Feske, *Armstrong Atlantic State University*
John W. Francis, *Columbus State Community College*
Karen Frindell, *Santa Rosa Junior College*
Zewdu Gebeyehu, *Columbus State University*
Steven M. Graham, *St. John's University*
Byron E. Howell, *Tyler Junior College*
Michael O. Hurst, *Georgia Southern University*
Bret Johnson, *The College of St. Scholastica*
Mark D. Lord, *Columbus State Community College*
Charles A. Lovelette, *Columbus State University*
Frank R. Milio, *Towson University*
Li-June Ming, *University of South Florida*
John T. Moore, *Stephen F. Austin State University*
Jessica N. Orvis, *Georgia Southern University*
John Paparelli, *San Antonio College*
Neal Phillp, *Bronx Community College/CUNY*
Jerry L. Poteat, *Georgia Perimeter College*
Douglas Raynie, *South Dakota State University*
Christine Rich, *University of Louisville*
Gillian E. A. Rudd, *Northwestern State University*
Susan M. Sawyer, *Kellogg Community College*
Shirish Shah, *Towson University*
David B. Shaw, *Madison Area Technical College*
Howard T. Silverstein, *Georgia Perimeter College*
Robert E. Smith, *Longview Community College*
Luise E. Strange de Soria, *Georgia Perimeter College*
Kim Woodrum, *University of Kentucky*
John Woolcock, *Indiana University of Pennsylvania*
Burl Yearwood, *LaGuardia Community College/CUNY*
Paulos Yohannes, *Georgia Perimeter College*
W. C. Zipperer, *Armstrong Atlantic State University*

Foundations *of*
General, Organic,
and Biochemistry

1

LEARNING GOALS

1 Discuss the approach to science, the scientific method, and distinguish among the terms *hypothesis, theory,* and *scientific law.*

2 Describe the properties of the solid, liquid, and gaseous states.

3 Provide specific examples of physical and chemical properties and physical and chemical change.

4 Distinguish between intensive and extensive properties.

5 Classify matter as element, compound, or mixture.

6 Distinguish between data and results.

7 Learn the major units of measure in the English and metric systems, and be able to convert from one system to another.

8 Report data and results using scientific notation and the proper number of significant figures.

9 Use appropriate units in problem solving.

10 Use density, mass, and volume in problem solving, and calculate the specific gravity of a substance from its density.

Chemistry
Methods and Measurement

Measurement is an integral part of athletic competition. What measurements are important in the game of basketball?

Students choose a career in medicine because they want to help others. In medicine, helping others means easing pain and suffering by treating or curing diseases. One important part of the practice of medicine involves observation. The physician must carefully observe the patient and listen to his or her description of symptoms to arrive at a preliminary diagnosis. Then appropriate tests must be done to determine whether the diagnosis is correct. During recovery the patient must be carefully observed for changes in behavior or symptoms. These changes are clues that the treatment or medication needs to be modified.

These practices are also important in science. The scientist makes an observation and develops a preliminary hypothesis or explanation for the observed phenomenon. Experiments are then carried out to determine whether the hypothesis is reasonable. When performing the experiment and analyzing the data, the scientist must look for any unexpected results that indicate that the original hypothesis must be modified.

Several important discoveries in medicine and the sciences have arisen from accidental observations. A health care worker or scientist may see something quite unexpected. Whether this results in an important discovery or is ignored depends on the training and preparedness of the observer.

It was Louis Pasteur, a chemist and microbiologist, who said, "Chance favors the prepared mind." In the history of science and medicine there are many examples of individuals who have made important discoveries because they recognized the value of an unexpected observation.

One such example is the use of ultraviolet (UV) light to treat infant jaundice. Infant jaundice is a condition in which the skin and the whites of the eyes appear yellow because of high levels of the bile pigment bilirubin in the blood. Bilirubin is a breakdown product of the oxygen-carrying blood protein hemoglobin. If bilirubin accumulates in the body, it can cause brain damage and death. The immature liver of the baby cannot remove the bilirubin.

An observant nurse in England noticed that when jaundiced babies were exposed to sunlight, the jaundice faded. Research based on her observation showed that the UV light changes the bilirubin into another substance that can be excreted. To this day, jaundiced newborns are treated with UV light.

The Pap smear test for the early detection of cervical and uterine cancer was also developed because of an accidental observation. Dr. George Papanicolaou, affectionately called Dr. Pap, was studying changes in the cells of the vagina during the stages of the menstrual cycle. He recognized cells in one

—Continued next page

Continued—

sample that looked like cancer cells. Within five years, Dr. Pap had perfected a technique for staining cells from vaginal fluid and observing them microscopically for the presence of any abnormal cells. The lives of countless women have been saved because a routine Pap smear showed early stages of cancer.

In this first chapter of your study of chemistry, you will learn more about the importance of observation and accurate, precise measurement in medical practice and scientific study. You will also study the scientific method, the process of developing hypotheses to explain observations, and the design of experiments to test those hypotheses.

1.1 The Discovery Process

Chemistry

Chemistry is the study of *matter,* its chemical and physical properties, the chemical and physical changes it undergoes, and the *energy* changes that accompany those processes. **Matter** is anything that has mass and occupies space. The changes that matter undergoes always involve either gain or loss of energy. **Energy** is the ability to do work to accomplish some change. The study of chemistry involves matter, energy, and their interrelationship. Matter and energy are at the heart of chemistry.

The Scientific Method

The **scientific method** is a systematic approach to the discovery of new information. How do we learn about the properties of matter, the way it behaves in nature, and how it can be modified to make useful products? Chemists do this by using the scientific method to study the way in which matter changes under carefully controlled conditions.

The scientific method is not a "cookbook recipe" that, if followed faithfully, will yield new discoveries; rather, it is an organized approach to solving scientific problems. Every scientist brings his or her own curiosity, creativity, and imagination to scientific study. But scientific inquiry still involves some of the "cookbook approach."

Characteristics of the scientific process include the following:

- *Observation.* The description of, for example, the color, taste, or odor of a substance is a result of observation. The measurement of the temperature of a liquid or the size or mass of a solid results from observation.
- *Formulation of a question.* Humankind's fundamental curiosity motivates questions of why and how things work.
- *Pattern recognition.* If a scientist finds a cause-and-effect relationship, it may be the basis of a generalized explanation of substances and their behavior.
- *Developing theories.* When scientists observe a phenomenon, they want to explain it. The process of explaining observed behavior begins with a hypothesis. A **hypothesis** is simply an attempt to explain an observation, or series of observations, in a commonsense way. If many experiments support a hypothesis, it may attain the status of a theory. A **theory** is a hypothesis supported by extensive testing (experimentation) that explains scientific facts and can predict new facts.
- *Experimentation.* Demonstrating the correctness of hypotheses and theories is at the heart of the scientific method. This is done by carrying out carefully designed experiments that will either support or disprove the theory or hypothesis.

A LIFESTYLE Connection | The Scientific Method

The discovery of penicillin by Alexander Fleming is an example of the scientific method at work. Fleming was studying the growth of bacteria. One day, his experiment was ruined because colonies of mold were growing on his plates. From this failed experiment, Fleming made an observation that would change the practice of medicine: Bacterial colonies could not grow in the area around the mold colonies. Fleming hypothesized that the mold was making a chemical compound that inhibited the growth of the bacteria. He performed a series of experiments designed to test this hypothesis.

The key to the scientific method is the design of carefully controlled experiments that will either support or disprove the hypothesis. This is exactly what Fleming did.

In one experiment he used two sets of tubes containing sterile nutrient broth. To one set he added mold cells. The second set (the control tubes) remained sterile. The mold was allowed to grow for several days. Then the broth from each of the tubes (experimental and control) was passed through a filter to remove any mold cells. Next, bacteria were placed in each tube. If Fleming's hypothesis was correct, the tubes in which the mold had grown would contain the chemical that inhibits growth, and the bacteria would not grow. On the other hand, the control tubes (which were never used to grow mold) would allow bacterial growth. This is exactly what Fleming observed.

Within a few years this *antibiotic*, penicillin, was being used to treat bacterial infections in patients.

FOR FURTHER UNDERSTANDING

What is the purpose of the control tubes used in this experiment?

What common characteristics do you find in this story and the Medical Connection on page 6?

- *Summarizing information.* A scientific **law** is nothing more than the summary of a large quantity of information. For example, the law of conservation of matter states that matter cannot be created or destroyed, only converted from one form to another. This statement represents a massive body of chemical information gathered from experiments.

The scientific method involves the interactive use of hypotheses, development of theories, and thorough testing of theories using well-designed experiments and is summarized in Figure 1.1.

QUESTION 1.1

Discuss the meaning of the term *scientific method*.

QUESTION 1.2

Describe an application of reasoning involving the scientific method that has occurred in your day-to-day life.

Models in Chemistry

Hypotheses, theories, and laws are frequently expressed using mathematical equations. These equations may confuse all but the best mathematicians. For this reason, a *model* of a chemical unit or system is often used to make ideas clearer. A good model based on everyday experience, although imperfect, gives a great deal of information in a simple fashion. Consider the fundamental unit of methane, the major component of natural gas, which is composed of one carbon atom (symbolized by C) and four hydrogen atoms (symbolized by H).

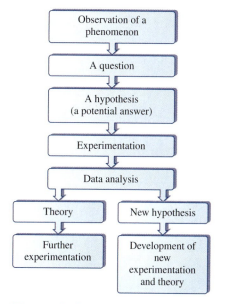

Figure 1.1

The scientific method, an organized way of doing science. A degree of trial and error is apparent here. If experimentation does not support the hypothesis, one must begin the cycle again.

A MEDICAL Connection | Curiosity, Science, and Medicine

Curiosity is one of the most important human traits. Small children constantly ask "why?". As we get older, our questions become more complex, but the curiosity remains.

Curiosity is also the basis of the scientific method. A scientist observes an event, wonders why it happens, and sets out to answer the question. Dr. Michael Zasloff's curiosity may lead to the development of an entirely new class of antibiotics. When he was a geneticist at the National Institutes of Health, his experiments involved the surgical removal of the ovaries of African clawed frogs. After surgery, he sutured (sewed up) the incision and put the frogs back in their tanks. These water-filled tanks were teeming with bacteria, but the frogs healed quickly, and the incisions did not become infected!

Of all the scientists to observe this remarkable healing, only Zasloff was curious enough to ask whether there were chemicals in the frogs' skin that defended the frogs against bacterial infections—a new type of antibiotic. All currently used antibiotics are produced by fungi or are synthesized in the laboratory. One big problem in medicine today is that more and more pathogenic (disease-causing) bacteria are becoming resistant to these antibiotics. Zasloff hoped to find an antibiotic that worked in an entirely new way so that the current problems with antibiotic resistance might be overcome.

Dr. Zasloff found two molecules in frog skin that can kill bacteria. Both are small proteins. Zasloff named them *magainins*, from the Hebrew word for shield. Most of the antibiotics that we now use enter bacteria and kill them by stopping some biochemical process inside the cell. Magainins are more direct; they simply punch holes in the bacterial membrane, and the bacteria explode.

One of the magainins, now chemically synthesized in the laboratory so that no frogs are harmed, may be available to the public in the near future. This magainin can kill a wide variety of bacteria (broad-spectrum antibiotic), and it has passed Phase I human trials. If this compound passes all remaining tests, it will be used in treating deep, infected wounds and ulcers, providing an alternative to traditional therapy.

FOR FURTHER UNDERSTANDING

Why is it important for researchers to continually design and develop new antibacterial substances?

What common characteristics do you find in this work and the discoveries discussed in the introduction to this chapter?

A geometrically correct model of methane can be constructed from balls and sticks. The balls represent the individual units (atoms) of hydrogen and carbon, and the sticks correspond to the attractive forces that hold the hydrogen and carbon together. The model consists of four balls representing hydrogen symmetrically arranged around a center ball representing carbon. The "carbon" ball is attached to each "hydrogen" ball by sticks, as shown:

Color-coding the balls distinguishes one type of matter from another; the geometrical form of the model, all of the angles and dimensions of a tetrahedron, are the same for each methane unit found in nature. Methane is certainly not a collection of balls and sticks, but such models are valuable because they help us understand the chemical behavior of methane and other, more complex substances.

(a) (b) (c) (d)

Figure 1.2

Examples of technology originating from scientific inquiry: (a) synthesis of a new drug, (b) solar energy cells, (c) preparation of solid-state electronics, (d) use of a gypsy moth sex attractant for insect control.

Chemists and physicists have used the observed properties of matter to develop models of the individual units of matter. These models collectively make up what we now know as the atomic theory of matter.

The modeling concept has advanced so far in large part due to the development of sophisticated computer software enabling construction of remarkably accurate three-dimensional molecular models. Compounds are designed and synthesized in the laboratory with the hope that they will perform very specific functions, such as curing diseases that have been resistant to other forms of treatment. Figure 1.2 shows some of the variety of modern technology that has its roots in the understanding of the atom.

1.2 Matter and Properties

Properties are characteristics of matter and are classified as either physical or chemical. In this section, we will learn the meaning of physical and chemical properties and how they are used to characterize matter.

Physical Properties

There are three *states of matter:* the **gaseous state,** the **liquid state,** and the **solid state.** A gas is made up of particles that are widely separated. In fact, a gas will expand to fill any container; it has no definite shape or volume. In contrast, particles of a liquid are closer together; a liquid has a definite volume but no definite shape; it takes on the shape of its container. A solid consists of particles that are close together and that often have a regular and predictable pattern of particle arrangement (crystalline). A solid has both fixed volume and fixed shape. Attractive forces, which exist between all particles, are very pronounced in solids and much less so in gases.

2 Learning Goal

Describe the properties of the solid, liquid, and gaseous states.

3 Learning Goal

Provide specific examples of physical and chemical properties and physical and chemical change.

 **Animation**

The Three States of Matter

(a)

(b)

(c)

Figure 1.3

The three states of matter exhibited by water: (a) solid, as ice; (b) liquid, as ocean water; (c) gas, as humidity in the air.

Light is the energy needed to make the reaction happen. Chlorophyll is the energy absorber that converts light energy to chemical energy.

Chapter 5 discusses the role of energy in chemical reactions.

Figure 1.4

An example of separation based on differences in physical properties. Magnetic iron is separated from other nonmagnetic substances. A large-scale version of this process is important in the recycling industry.

Water is the most common example of a substance that can exist in all three states over a reasonable temperature range (Figure 1.3). Conversion of water from one state to another constitutes a *physical change*. A **physical change** produces a recognizable difference in the appearance of a substance without causing any change in its composition or identity. For example, we can warm an ice cube and it will melt, forming liquid water. Clearly its appearance has changed; it has been transformed from the solid to the liquid state. It is, however, still water; its composition and identity remain unchanged. A physical change has occurred. We could demonstrate the constancy of composition and identity by refreezing the liquid water, re-forming the ice cube. This melting and freezing cycle could be repeated over and over. This very process is a hallmark of our global weather changes. The continual interconversion of the three states of water in the environment (snow, rain, and humidity) clearly demonstrates the retention of the identity of water particles or *molecules*.

A **physical property** can be observed or measured without changing the composition or identity of a substance. As we have seen, melting ice is a physical change. We can measure the temperature when melting occurs; this is the *melting point* of water. We can also measure the *boiling point* of water, when liquid water becomes a gas. Both the melting and boiling points of water, and of any other substance, are physical properties.

A practical application of separation of materials based upon their differences in physical properties is shown in Figure 1.4.

Chemical Properties

We have noted that physical properties can be exhibited, measured, or observed without any change in identity or composition. In contrast, **chemical properties** do result in a change in composition and can be observed only through chemical reactions. A **chemical reaction** is a process of rearranging, removing, replacing, or adding atoms to produce new substances. For example, the process of photosynthesis can be shown as

$$\text{carbon dioxide} + \text{water} \xrightarrow[\text{chlorophyll}]{\text{light}} \text{sugar} + \text{oxygen}$$

This chemical reaction involves the conversion of carbon dioxide and water (the *reactants*) to a sugar and oxygen (the *products*). The products and reactants are clearly different. We know that carbon dioxide and oxygen are gases at room temperature and water is a liquid at this temperature; the sugar is a solid white powder. A chemical property of carbon dioxide is its ability to form sugar under certain conditions. The process of formation of this sugar is the *chemical change*.

EXAMPLE 1.1   Identifying Properties

Can the process that takes place when an egg is fried be described as a physical or chemical change?

SOLUTION

Examine the characteristics of the egg before and after frying. Clearly, some significant change has occurred. Furthermore, the change appears irreversible. More than a simple physical change has taken place. A chemical reaction (actually, several) must be responsible; hence chemical change.

3 Learning Goal
Provide specific examples of physical and chemical properties and physical and chemical change.

QUESTION 1.3

Classify each of the following as either a chemical property or a physical property:

a. color
b. flammability
c. hardness
d. odor
e. taste

QUESTION 1.4

Classify each of the following as either a chemical change or a physical change:

a. water boiling to become steam
b. butter becoming rancid
c. combustion of wood
d. melting of ice in spring
e. decay of leaves in winter

Intensive and Extensive Properties

It is important to recognize that properties can also be classified according to whether they depend on the size of the sample. Consequently, there is a fundamental difference between properties such as density and specific gravity and properties such as mass and volume.

An **intensive property** is a property of matter that is *independent* of the *quantity* of the substance. Density, boiling and melting points, and specific gravity are intensive properties. For example, the boiling point of one single drop of water is exactly the same as the boiling point of a liter of water.

An **extensive property** *depends* on the *quantity* of a substance. Mass and volume are extensive properties. There is an obvious difference between 1 g of silver and 1 kg of silver; the quantities and, incidentally, the value, differ substantially.

See page 28 for a discussion of density and specific gravity.

4 Learning Goal
Distinguish between intensive and extensive properties.

EXAMPLE 1.2   Differentiating Between Intensive and Extensive Properties

Is temperature an extensive or intensive property?

SOLUTION

Imagine two glasses each containing 100 g of water, and each at 25°C. Now pour the contents of the two glasses into a larger glass. You would predict that the mass of the water in the larger glass would be 200 g (100 g + 100 g)

Continued—

EXAMPLE 1.2 —Continued

because mass is an extensive property, dependent on quantity. However, we would expect the temperature of the water to remain the same (not 25°C + 25°C); hence temperature is *an intensive property* . . . independent of quantity.

QUESTION 1.5

Water freezes at 0°C. Is the freezing temperature of water an intensive or extensive property?

QUESTION 1.6

Explain your reasoning in arriving at your answer to Question 1.5.

Classification of Matter

5 Learning Goal
Classify matter as element, compound, or mixture.

Chemists look for similarities in properties among various types of materials. Recognizing these likenesses simplifies learning the subject and allows us to predict the behavior of new substances on the basis of their relationship to substances already known and characterized.

Many classification systems exist. The most useful system, based on composition, is described in the following paragraphs (see also Figure 1.5).

All matter is either a *pure substance* or a *mixture*. A **pure substance** has only one component. Pure water is a pure substance. It is made up only of particles containing two hydrogen atoms and one oxygen atom, that is, water molecules (H_2O).

There are different types of pure substances. Elements and compounds are both pure substances. An **element** is a pure substance that cannot be changed into a simpler form of matter by any chemical reaction. Hydrogen and oxygen, for example, are elements. Alternatively, a **compound** is a substance resulting from the combination of two or more elements in a definite, reproducible way. The elements hydrogen and oxygen may combine to form the compound water, H_2O.

A **mixture** is a combination of two or more pure substances in which each substance retains its own identity. Alcohol and water can be combined in a mixture. They coexist as pure substances because they do not undergo a chemical reaction; they exist as thoroughly mixed discrete molecules. This collection of dissimilar particles is the mixture. A mixture has variable composition; there are an infinite number of combinations of quantities of alcohol and water that can be mixed. For example, the mixture may contain a small amount of alcohol and a large amount of water or vice versa. Each is, however, an alcohol–water mixture.

At present, more than one hundred elements have been characterized. A complete listing of the elements and their symbols is found on the inside front cover of this textbook.

Figure 1.5

Classification of matter. All matter is either a pure substance or a mixture of pure substances. Pure substances are either elements or compounds, and mixtures may be either homogeneous (uniform composition) or heterogeneous (nonuniform composition).

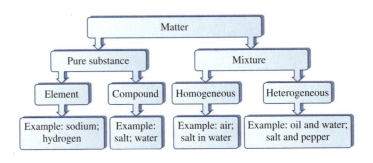

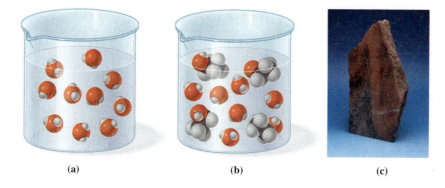

Figure 1.6

Schematic representation of some classes of matter. (a) A pure substance, water, consists of a single component. (b) A homogeneous mixture, ethanol and water, has a uniform distribution of components. (c) A heterogeneous mixture, marble, has a nonuniform distribution of components. The lack of homogeneity is readily apparent.

A detailed discussion of solutions (homogeneous mixtures) and their properties is presented in Chapter 7.

A mixture may be either *homogeneous* or *heterogeneous* (Figure 1.6). A **homogeneous mixture** has uniform composition. Its particles are well mixed, or thoroughly intermingled. A homogeneous mixture, such as alcohol and water, is described as a *solution*. Air, a mixture of gases, is an example of a gaseous solution. A **heterogeneous mixture** has a nonuniform composition. A mixture of salt and pepper is a good example of a heterogeneous mixture. Concrete is also composed of a heterogeneous mixture of materials (various types and sizes of stone and sand present with cement in a nonuniform mixture).

EXAMPLE 1.3   Categorizing Matter

Is seawater a pure substance, a homogeneous mixture, or a heterogeneous mixture?

SOLUTION

Imagine yourself at the beach, filling a container with a sample of water from the ocean. Examine it. You would see a variety of solid particles suspended in the water: sand, green vegetation, perhaps even a small fish! Clearly, it is a mixture, and one in which the particles are not uniformly distributed throughout the water; hence a heterogeneous mixture.

5   Learning Goal
Classify matter as element, compound, or mixture.

QUESTION 1.7

Is each of the following materials a pure substance, a homogeneous mixture, or a heterogeneous mixture?

a. ethyl alcohol
b. blood
c. Alka-Seltzer dissolved in water
d. oxygen in a hospital oxygen tank

QUESTION 1.8

Is each of the following materials a pure substance, a homogeneous mixture, or a heterogeneous mixture?

a. air
b. paint
c. perfume
d. carbon monoxide

1.3 Measurement in Chemistry

Data, Results, and Units

A scientific experiment produces **data.** Each piece of data is the individual result of a single measurement or observation. Examples include the *mass* of a sample and the *time* required for a chemical reaction to occur. Mass, length, volume, time, temperature, and energy are common types of data obtained from chemical experiments.

Results are the outcome of an experiment. Data and results may be identical, but more often several related pieces of data are combined, and logic is used to produce a result.

6 Learning Goal
Distinguish between data and results.

EXAMPLE 1.4 Distinguishing Between Data and Results

In many cases, a drug is less stable if moisture is present, and excess moisture can hasten the breakdown of the active ingredient, leading to loss of potency. Therefore we may wish to know how much water a certain quantity of a drug gains when exposed to air. To do this experiment, we must first weigh the drug sample, then expose it to the air for a period and reweigh it. The change in weight,

$$[\text{weight}_{\text{final}} - \text{weight}_{\text{initial}}] = \text{weight difference}$$

indicates the weight of water taken up by the drug formulation. The initial and final weights are individual bits of *data;* by themselves they do not answer the question, but they do provide the information necessary to calculate the answer: the results. The difference in weight and the conclusions based on the observed change in weight are the *results* of the experiment.

QUESTION 1.9

Classify each piece of information in Example 1.11 (p. 29) as *data* or *results.*

QUESTION 1.10

Classify each piece of information in Example 1.12 (p. 29) as *data* or *results.*

The experiment described in Example 1.4 was really not a very good experiment because many other environmental conditions were not measured. Measurement of the temperature and humidity of the atmosphere and the length of time that the drug was exposed to the air (the creation of a more complete set of data) would make the results less ambiguous.

Any measurement made in the experiment must also specify the units of that measurement. An initial weight of three *ounces* is clearly quite different from three *pounds*. A **unit** defines the basic quantity of mass, volume, time, or whatever quantity is being measured. A number that is not followed by the correct unit usually conveys no useful information.

Proper use of units is central to all aspects of science. The following sections are designed to develop a fundamental understanding of this vital topic.

English and Metric Units

The *English system* is a collection of functionally unrelated units. In the *English system of measurement*, the standard *pound* (lb) is the basic unit of weight. The fundamental unit of *length* is the standard *yard* (yd), and the basic unit of *volume* is the

7 Learning Goal
Learn the major units of measure in the English and metric systems, and be able to convert from one system to another.

standard *gallon* (gal). The English system is used in the United States in business and industry. However, it is not used in scientific work, primarily because it is difficult to convert from one unit to another. For example,

$$1 \text{ foot} = 12 \text{ inches} = 0.33 \text{ yard} = \frac{1}{5280} \text{ mile} = \frac{1}{6} \text{ fathom}$$

Clearly, operations such as the conversion of 1.62 yards to units of miles are not straightforward. In fact, the English "system" is not really a system at all. It is simply a collection of measures accumulated throughout English history. Because they have no common origin, it is not surprising that conversion from one unit to another is not straightforward.

The United States, the last major industrial country to retain the English system, has begun efforts to convert to the metric system. The *metric system* is truly "systematic." It is composed of a set of units that are related to each other decimally, in other words, as powers of ten. Because the *metric system* is a decimal-based system, it is inherently simpler to use and less ambiguous. For example, the length of an object may be represented as

$$1 \text{ meter} = 10 \text{ decimeters} = 100 \text{ centimeters} = 1000 \text{ millimeters}$$

The metric system was originally developed in France just before the French Revolution in 1789. In the metric system, there are three basic units. Mass is represented as the *gram*, length as the *meter*, and volume as the *liter*. Any subunit or multiple unit contains one of these units preceded by a prefix indicating the power of ten by which the base unit is to be multiplied to form the subunit or multiple unit. The most common metric prefixes are shown in Table 1.1.

The same prefix may be used for volume, mass, length, time, and so forth. Consider the following examples:

Other metric units, for time and temperature, will be treated in Section 1.5.

$$1 \text{ milliliter (mL)} = \frac{1}{1000} \text{ liter} = 0.001 \text{ liter} = 10^{-3} \text{ liter}$$

A volume unit is indicated by the base unit, liter, and the prefix *milli-*, which indicates that the unit is one thousandth of the base unit. In the same way,

$$1 \text{ milligram (mg)} = \frac{1}{1000} \text{ gram} = 0.001 \text{ gram} = 10^{-3} \text{ gram}$$

and

$$1 \text{ millimeter (mm)} = \frac{1}{1000} \text{ meter} = 0.001 \text{ meter} = 10^{-3} \text{ meter}$$

The representation of numbers as powers of ten may be unfamiliar to you. This useful notation is discussed in Section 1.4.

TABLE 1.1 Some Common Prefixes Used in the Metric System		
Prefix	Multiple	Decimal Equivalent
mega (M)	10^6	1,000,000.
kilo (k)	10^3	1,000.
deka (da)	10^1	10.
deci (d)	10^{-1}	0.1
centi (c)	10^{-2}	0.01
milli (m)	10^{-3}	0.001
micro (μ)	10^{-6}	0.000001
nano (n)	10^{-9}	0.000000001

Unit Conversion: English and Metric Systems

To convert from one unit to another, we must have a *conversion factor* or series of conversion factors that relate two units. The proper use of these conversion factors is called the *factor-label method*. This method is also termed *dimensional analysis*.

This method is used for two kinds of conversions: to convert from one unit to another within the *same system* or to convert units from *one system to another*.

Conversion of Units Within the Same System

We know, for example, that in the English system,

$$1 \text{ gallon} = 4 \text{ quarts}$$

Because dividing both sides of the equation by the same term does not change its identity,

$$\frac{1 \text{ gallon}}{1 \text{ gallon}} = \frac{4 \text{ quarts}}{1 \text{ gallon}}$$

The expression on the left is equal to unity (1); therefore

$$1 = \frac{4 \text{ quarts}}{1 \text{ gallon}} \qquad \text{or} \qquad 1 = \frac{1 \text{ gallon}}{4 \text{ quarts}}$$

Now, multiplying any other expression by the ratio 4 quarts/1 gallon or 1 gallon/4 quarts will not change the value of the term, because multiplication of any number by 1 produces the original value. However, there is one important difference: The units will have changed.

EXAMPLE 1.5 Using Conversion Factors

Convert 12 gallons to units of quarts.

SOLUTION

$$12 \text{ gal} \times \frac{4 \text{ qt}}{1 \text{ gal}} = 48 \text{ qt}$$

The conversion factor, 4 qt/1 gal, serves as a bridge, or linkage, between the unit that was given (gallons) and the unit that was sought (quarts).

The conversion factor in Example 1.5 may be written as 4 qt/1 gal or 1 gal/4 qt, because both are equal to 1. However, only the first factor, 4 qt/1 gal, will give us the units we need to solve the problem. If we had set up the problem incorrectly, we would obtain

$$12 \text{ gal} \times \frac{1 \text{ gal}}{4 \text{ qt}} = 3 \frac{\text{gal}^2}{\text{qt}}$$

Incorrect units

Clearly, units of gal^2/qt are not those asked for in the problem, nor are they reasonable units. The factor-label method is therefore a self-indicating system; the correct units (those required by the problem) will result only if the factor is set up properly.

Table 1.2 lists a variety of commonly used English system relationships that may serve as the basis for useful conversion factors.

TABLE 1.2 Some Common Relationships Used in the English System

A. Weight	1 pound = 16 ounces
	1 ton = 2000 pounds
B. Length	1 foot = 12 inches
	1 yard = 3 feet
	1 mile = 5280 feet
C. Volume	1 gallon = 4 quarts
	1 quart = 2 pints
	1 quart = 32 fluid ounces

QUESTION 1.11

Convert 155 pounds to tons.

QUESTION 1.12

Convert 1500 feet to miles.

Units within the metric system can be converted by using the factor-label method as well. Unit prefixes that dictate the conversion factor facilitate unit conversion (refer to Table 1.1).

EXAMPLE 1.6 Using Conversion Factors

Convert 10.0 centimeters to meters.

7 Learning Goal
Learn the major units of measure in the English and metric systems, and be able to convert from one system to another.

SOLUTION

First, recognize that the prefix *centi-* means $\frac{1}{100}$ of the base unit, the meter (m), just as one cent is $\frac{1}{100}$ of a dollar. There are 100 cents in a dollar and there are 100 cm in one meter. Thus, our conversion factor is either

$$\frac{1 \text{ m}}{100 \text{ cm}} \quad \text{or} \quad \frac{100 \text{ cm}}{1 \text{ m}}$$

each being equal to 1. Only one, however, will result in proper cancellation of units, producing the correct answer to the problem. If we proceed as follows:

$$10.0 \text{ cm} \times \frac{1 \text{ m}}{100 \text{ cm}} = 0.100 \text{ m}$$

$$\begin{array}{ccc} \text{Data} & \text{Conversion} & \text{Desired} \\ \text{given} & \text{factor} & \text{result} \end{array}$$

we obtain the desired units, meters. If we had used the conversion factor 100 cm/1 m, the resulting units would be meaningless and the answer would have been incorrect:

$$10.0 \text{ cm} \times \frac{100 \text{ cm}}{1 \text{ m}} = 1000 \frac{\text{cm}^2}{\text{m}}$$

Incorrect units

TABLE 1.3 Commonly Used "Bridging" Units for Intersystem Conversions			
Quantity	**English**		**Metric**
Mass	1 pound	=	454 grams
	2.2 pounds	=	1 kilogram
Length	1 inch	=	2.54 centimeters
	1 yard	=	0.91 meter
Volume	1 quart	=	0.946 liter
	1 gallon	=	3.78 liters

QUESTION 1.13

Convert 1.0 liter to each of the following units, using the factor-label method:

a. milliliters   d. centiliters
b. microliters   e. dekaliters
c. kiloliters

QUESTION 1.14

Convert 1.0 gram to each of the following units:

a. micrograms   d. centigrams
b. milligrams   e. decigrams
c. kilograms

Conversion of Units from One System to Another

English and metric conversions are shown in Tables 1.1 and 1.2.

The conversion of a quantity expressed in units of one system to an equivalent quantity in the other system (English to metric or metric to English) requires a *bridging* conversion unit. Examples are shown in Table 1.3.

The conversion may be represented as a three-step process:

1. Conversion from the units given in the problem to a bridging unit.
2. Conversion to the other system using the bridge.
3. Conversion within the desired system to units required by the problem.

EXAMPLE 1.7   Using Conversion Factors Between Systems

7   **Learning Goal**

Learn the major units of measure in the English and metric systems, and be able to convert from one system to another.

Convert 4.00 ounces to kilograms.

SOLUTION

Step 1. A convenient bridging unit for mass is 1 pound = 454 grams. To use this conversion factor, we relate ounces (given in the problem) to pounds:

$$4.00 \ \cancel{\text{ounces}} \times \frac{1 \text{ pound}}{16 \ \cancel{\text{ounces}}} = 0.250 \text{ pound}$$

Continued—

Step 2. Using the bridging unit conversion, we get

$$0.250 \; \cancel{\text{pound}} \times \frac{454 \text{ grams}}{1 \; \cancel{\text{pound}}} = 114 \text{ grams}$$

Step 3. Grams may then be directly converted to kilograms, the desired unit:

$$114 \; \cancel{\text{grams}} \times \frac{1 \text{ kilogram}}{1000 \; \cancel{\text{grams}}} = 0.114 \text{ kilogram}$$

The calculation may also be done in a single step by arranging the factors in a chain:

$$4.00 \; \cancel{oz} \times \frac{1 \; \cancel{lb}}{16 \; \cancel{oz}} \times \frac{454 \; \cancel{g}}{1 \; \cancel{lb}} \times \frac{1 \text{ kg}}{1000 \; \cancel{g}} = 0.114 \text{ kg}$$

Helpful Hint: Refer to the discussion of rounding off numbers on page 24.

EXAMPLE 1.8 **Using Conversion Factors**

Convert 1.5 meters^2 to centimeters^2.

SOLUTION

The problem is similar to the conversion performed in Example 1.6. However, we must remember to include the exponent in the units. Thus

$$1.5 \text{ m}^2 \times \left(\frac{10^2 \text{ cm}}{1 \text{ m}}\right)^2 = 1.5 \; \cancel{\text{m}^2} \times \frac{10^4 \text{ cm}^2}{1 \; \cancel{\text{m}^2}} = 1.5 \times 10^4 \text{ cm}^2$$

Note: The exponent affects both the number *and* unit within the parentheses.

> **7** **Learning Goal**
>
> Learn the major units of measure in the English and metric systems, and be able to convert from one system to another.

QUESTION 1.15

a. Convert 0.50 inch to meters.
b. Convert 0.75 quart to liters.
c. Convert 56.8 grams to ounces.
d. Convert 1.5 cm^2 to m^2.

QUESTION 1.16

a. Convert 0.50 inch to centimeters.
b. Convert 0.75 quart to milliliters.
c. Convert 56.8 milligrams to ounces.
d. Convert 3.6 m^2 to cm^2.

1.4 Significant Figures and Scientific Notation

Information-bearing figures in a number are termed *significant figures*. Data and results arising from a scientific experiment convey information about the way in which the experiment was conducted. The degree of uncertainty or doubt associated with a measurement or series of measurements is indicated by the number of figures used to represent the information.

Significant Figures

Consider the following situation: A student was asked to obtain the length of a section of wire. In the chemistry laboratory, several different types of measuring devices are usually available. Not knowing which was most appropriate, the student decided to measure the object using each device that was available in the laboratory. The following data were obtained:

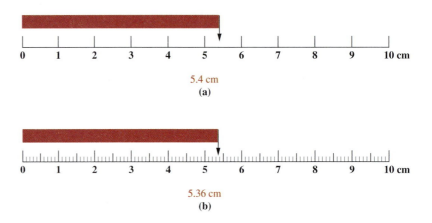

5.4 cm

(a)

5.36 cm

(b)

Two questions should immediately come to mind:

Are the two answers equivalent?

If not, which answer is correct?

In fact, the two answers are *not* equivalent, but *both* are correct. How do we explain this apparent contradiction?

The data are not equivalent because each is known to a different degree of certainty. The answer 5.36 cm, containing three significant figures, specifies the length of the wire more exactly than 5.4 cm, which contains only two significant figures. The term **significant figures** is defined as all digits in a number representing data or results that are known with certainty *plus one uncertain digit*.

The uncertain digit represents the degree of doubt in a single measurement.

In case (a), we are certain that the wire is at least 5 cm long and equally certain that it is *not* 6 cm long because the end of the wire falls between the calibration lines 5 and 6. We can only estimate between 5 and 6, because there are no calibration indicators between 5 and 6. The end of the wire appears to be approximately four-tenths of the way between 5 and 6, hence 5.4 cm. The 5 is known with certainty, and 4 is estimated; there are two significant figures.

The uncertain digit results from an estimation.

In case (b), the ruler is calibrated in tenths of centimeters. The end of the wire is at least 5.3 cm and not 5.4 cm. Estimation of the second decimal place between the two closest calibration marks leads to 5.36 cm. In this case, 5.3 is certain, and the 6 is estimated (or uncertain), leading to three significant digits.

Both answers are correct because each is consistent with the measuring device used to generate the data. An answer of 5.36 cm obtained from a measurement using ruler (a) would be *incorrect* because the measuring device is not capable of that exact specification. On the other hand, a value of 5.4 cm obtained from ruler (b) would be erroneous as well; in that case the measuring device is capable of generating a higher level of certainty (more significant digits) than is actually reported.

In summary, the number of significant figures in a measurement is determined by the measuring device. Conversely, the number of significant figures reported is an indication of the sophistication of the measurement itself.

Recognition of Significant Figures

Only *significant* digits should be reported as data or results. However, are all digits, as written, significant digits? Let's look at a few examples illustrating the guidelines that are used to represent data and results with the proper number of significant digits.

- All nonzero digits are significant.
  7.314 has *four* significant digits.
- The number of significant digits is independent of the position of the decimal point.
  73.14 has *four* significant digits, as does 7.314.
- Zeros located between nonzero digits are significant.
  60.052 has *five* significant figures.
- Zeros at the end of a number (often referred to as trailing zeros) are significant if the number contains a decimal point.
  4.70 has *three* significant figures.
  Helpful Hint: Trailing zeros are ambiguous; the next section offers a solution for this ambiguity.
- Trailing zeros are insignificant if the number does not contain a decimal point and are significant if a decimal point is indicated.
  100 has *one* significant figure; 100. has three significant figures.
- Zeros to the left of the first nonzero integer are not significant; they serve only to locate the position of the decimal point.
  0.0032 has *two* significant figures.

QUESTION 1.17

How many significant figures are contained in each of the following numbers?

a. 7.26
b. 726
c. 700.2
d. 7.0
e. 0.0720

QUESTION 1.18

How many significant figures are contained in each of the following numbers?

a. 0.042
b. 4.20
c. 24.0
d. 240
e. 204

Scientific Notation

It is often difficult to express very large numbers in the proper number of significant figures using conventional notation. The solution to this problem lies in the

8 **Learning Goal**

Report data and results using scientific notation and the proper number of significant figures.

use of **scientific notation,** also referred to as *exponential notation*, which involves the representation of a number as a power of ten.

The speed of light is 299,792,458 m/s. For many calculations, two or three significant figures are sufficient. Using scientific notation, two significant figures, and rounding (p. 24), the speed of light is 3.0×10^8 m/s. The conversion is illustrated using simpler numbers:

$$6200 = 6.2 \times 1000 = 6.2 \times 10^3$$

or

$$5340 = 5.34 \times 1000 = 5.34 \times 10^3$$

You can find further information on the use of exponential notation online at www.mhhe.com/denniston in "A Review of Mathematics."

RULE: To convert a number greater than 1 to scientific notation, the original decimal point is moved x places to the left, and the resulting number is multiplied by 10^x. The exponent (x) is a *positive* number equal to the number of places the original decimal point was moved. ■

Scientific notation is also useful in representing numbers less than 1. For example, the mass of a single helium atom is

0.0000000000000000000000006692 gram

a rather cumbersome number as written. Scientific notation would represent the mass of a single helium atom as 6.692×10^{-24} gram. The conversion is illustrated by using simpler numbers:

$$0.0062 = 6.2 \times \frac{1}{1000} = 6.2 \times \frac{1}{10^3} = 6.2 \times 10^{-3}$$

or

$$0.0534 = 5.34 \times \frac{1}{100} = 5.34 \times \frac{1}{10^2} = 5.34 \times 10^{-2}$$

RULE: To convert a number less than 1 to scientific notation, the original decimal point is moved x places to the right, and the resulting number is multiplied by 10^{-x}. The exponent ($-x$) is a *negative* number equal to the number of places the original decimal point was moved. ■

QUESTION 1.19

Represent each of the following numbers in scientific notation, showing only significant digits:

a. 0.0024
b. 0.0180
c. 224

QUESTION 1.20

Represent each of the following numbers in scientific notation, showing only significant digits:

a. 48.20
b. 480.0
c. 0.126

Error, Accuracy, Precision, and Uncertainty

Error is the difference between the true value and our estimation, or measurement, of the value. Some degree of error is associated with any measurement. Two types of error exist: random error and systematic error. *Random error* causes data from multiple measurements of the same quantity to be scattered in a more or less uniform way around some average value. *Systematic error* causes data to be either smaller or larger than the accepted value. Random error is inherent in the experimental approach to the study of matter and its behavior; systematic error can be found and, in many cases, removed or corrected.

Examples of systematic error include such situations as

- Dust on the balance pan, which causes all objects weighed to appear heavier than they really are.
- Impurities in chemicals used for the analysis of materials, which may interfere with (or block) the desired process.

Accuracy is the degree of agreement between the true value and the measured value. **Uncertainty** is the degree of doubt in a single measurement.

When measuring quantities that show continuous variation, for example, the weight of this page or the volume of one of your quarters, some doubt or uncertainty is present because the answer cannot be expressed with an infinite number of meaningful digits. The number of meaningful digits is determined by the measuring device. The presence of some error is a natural consequence of any measurement.

The simple process of converting the fraction ⅔ to its decimal equivalent can produce a variety of answers that depend on the device used to perform the calculation: pencil and paper, calculator, computer. The answer might be

$$0.67$$

$$0.667$$

$$0.6667$$

and so forth. All are correct, but each value has a different level of uncertainty. The first number listed, 0.67, has the greatest uncertainty.

It is always best to measure a quantity several times. Modern scientific instruments are designed to perform measurements rapidly; this allows many more measurements to be completed in a reasonable period. Replicate measurements of the same quantity minimize the uncertainty of the result. **Precision** is a measure of the agreement of replicate measurements.

It is important to recognize that accuracy and precision are not the same thing. It is possible to have one without the other. However, when scientific measurements are carefully made, the two most often go hand in hand; high-quality data are characterized by high levels of precision and accuracy.

In Figure 1.7, bull's-eye (a) shows the goal of all experimentation: accuracy *and* precision. Bull's-eye (b) shows that the results are repeatable (good precision); however, some error in the experimental procedure has caused the results to center on an incorrect value. This error is systematic, occurring in each replicate measurement. Occasionally, an experiment may show "accidental" accuracy. The precision is poor, but the average of these replicate measurements leads to a correct value. We don't want to rely on accidental success; the experiment should be repeated until the precision inspires faith in the accuracy of the method. Modern measuring devices in chemistry, equipped with powerful computers with immense storage capacity, are capable of making literally thousands of individual replicate measurements to enhance the quality of the result. Bull's-eye (c) describes the most common situation. A low level of precision is all too often associated with poor accuracy.

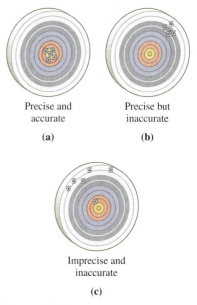

Precise and accurate

(a)

Precise but inaccurate

(b)

Imprecise and inaccurate

(c)

Figure 1.7

An illustration of precision and accuracy in replicate experiments.

Significant Figures in Calculation of Results

Addition and Subtraction

If we combine the following numbers:

$$37.68 \quad \text{liters}$$

$$108.428 \quad \text{liters}$$

$$6.71862 \quad \text{liters}$$

our calculator will show a final result of

$$152.82662 \quad \text{liters}$$

> **Remember the distinction between the words *zero* and *nothing*. Zero is one of the ten digits and conveys as much information as 1, 2, and so forth. *Nothing* implies no information; the digits in the positions indicated by *x*'s could be 0, 1, 2, or any other.**

Clearly, the answer, with eight digits, defines the volume of total material much more accurately than *any* of the individual quantities being combined. This cannot be correct; *the answer cannot have greater significance than any of the quantities that produced the answer.* We rewrite the problem:

$$
\begin{array}{ll}
37.68xxx & \text{liters} \\
108.428xx & \text{liters} \\
+ \quad 6.71862 & \text{liters} \\
\hline
152.82662 & \text{(should be 152.83) liters}
\end{array}
$$

See rules for rounding off discussed on page 24.

where x = no information; x may be any integer from 0 to 9. Adding 2 to two unknown numbers (in the right column) produces no information. Similar logic prevails for the next two columns. Thus, five digits remain, all of which are significant. Conventional rules for rounding off would dictate a final answer of 152.83.

QUESTION 1.21

Report the result of each of the following to the proper number of significant figures:

a. $4.26 + 3.831 =$
b. $8.321 - 2.4 =$
c. $16.262 + 4.33 - 0.40 =$

QUESTION 1.22

Report the result of each of the following to the proper number of significant figures:

a. $7.939 + 6.26 =$
b. $2.4 - 8.321 =$
c. $2.333 + 1.56 - 0.29 =$

Multiplication and Division

In the preceding discussion of addition and subtraction, the position of the decimal point in the quantities being combined has a bearing on the number of significant figures in the answer. In multiplication and division this is not the case. The decimal point position is irrelevant when determining the number of significant figures in the answer. It is the number of significant figures in the data that is important. Consider

$$\frac{4.237 \times 1.21 \times 10^{-3} \times 0.00273}{11.125} = 1.26 \times 10^{-6}$$

The answer is limited to three significant figures; the answer can have *only* three significant figures because two numbers in the calculation, 1.21×10^{-3} and 0.00273, have three significant figures and "limit" the answer. Remember, *the answer can be no more precise than the least precise number from which the answer is derived.* The *least precise number* is the number with the fewest significant figures.

QUESTION 1.23

Report the results of each of the following operations using the proper number of significant figures:

a. $63.8 \times 0.80 =$

b. $\dfrac{63.8}{0.80} =$

c. $\dfrac{53.8 \times 0.90}{0.3025} =$

QUESTION 1.24

Report the results of each of the following operations using the proper number of significant figures:

a. $\dfrac{27.2 \times 15.63}{1.84} =$

b. $\dfrac{13.6}{18.02 \times 1.6} =$

c. $\dfrac{12.24 \times 6.2}{18.02 \times 1.6} =$

Exponents

Now consider the determination of the proper number of significant digits in the results when a value is multiplied by any power of ten. In each case the number of significant figures in the answer is identical to the number contained in the original term. Therefore

You can find further information on calculations involving exponents online at www.mhhe.com/denniston in "A Review of Mathematics."

$$(8.314 \times 10^2)^3 = 574.7 \times 10^6 = 5.747 \times 10^8$$

and

$$(8.314 \times 10^2)^{1/2} = 2.883 \times 10^1$$

Each answer contains four significant figures.

Exact (Counted) and Inexact Numbers

Inexact numbers, by definition, have uncertainty (the degree of doubt in the final significant digit). *Exact numbers*, on the other hand, have no uncertainty. Exact numbers may arise from a definition; there are *exactly* 60 minutes in 1 hour, or there are exactly 1000 mL in 1 liter.

Exact numbers are a consequence of counting. Counting the number of dimes in your pocket or the number of letters in the alphabet are common examples. The fact that exact numbers have no uncertainty means that they do not limit the number of significant figures in the result of a calculation.

For example,

$$4.00 \text{ oz} \times 1 \text{ lb}/16 \text{ oz} = 0.250 \text{ lb (3 significant figures)}$$

or

$$2568 \text{ oz} \times 1 \text{ lb}/16 \text{ oz} = 160.5 \text{ lb} \text{ (4 significant figures)}$$

In both examples, the number of significant figures in the result is governed by the data (the number of ounces) not the conversion factor, which is *exact*, because it is defined.

A good rule of thumb to follow is: In the metric system, the quantity being converted, not the conversion factor, generally determines the number of significant figures.

Rounding Off Numbers

The use of an electronic calculator generally produces more digits for a result than are justified by the rules of significant figures on the basis of the data input. For example, on your calculator,

$$3.84 \times 6.72 = 25.8048$$

The most correct answer would be 25.8, dropping 048.

A number of acceptable conventions for rounding exist. Throughout this book we will use the following:

RULE: When the number to be dropped is less than 5, the preceding number is not changed. When the number to be dropped is 5 or larger, the preceding number is increased by one unit. ■

8 Learning Goal

Report data and results using scientific notation and the proper number of significant figures.

EXAMPLE 1.9 Rounding Numbers

Round off each of the following to three significant figures.

SOLUTION

a. 63.669 becomes 63.7. *Rationale:* 6 > 5.
b. 8.7715 becomes 8.77. *Rationale:* 1 < 5.
c. 2.2245 becomes 2.22. *Rationale:* 4 < 5.
d. 0.0004109 becomes 0.000411. *Rationale:* 9 > 5.

Helpful Hint: Symbol $x > y$ implies "x greater than y." Symbol $x < y$ implies "x less than y."

QUESTION 1.25

Round off each of the following numbers to three significant figures.

a. 61.40
b. 6.171
c. 0.066494

QUESTION 1.26

Round off each of the following numbers to three significant figures.

a. 6.2262
b. 3895
c. 6.885

1.5 Experimental Quantities

Thus far we have discussed the scientific method and its role in acquiring data and converting the data to obtain the results of the experiment. We have seen that such data must be reported in the proper units with the appropriate number of significant figures. The quantities that are most often determined include mass, length, volume, time, temperature, and energy. Now let's look at each of these quantities in more detail.

Mass

Mass describes the quantity of matter in an object. The terms *weight* and *mass*, in common usage, are often considered synonymous. They are not, in fact. **Weight** is the force of gravity on an object:

$$\text{weight} = \text{mass} \times \text{acceleration due to gravity}$$

When gravity is constant, mass and weight are directly proportional. But gravity is not constant; it varies as a function of the distance from the center of the earth. Therefore weight cannot be used for scientific measurement because the weight of an object may vary from one place on the earth to the next.

Mass, on the other hand, is independent of gravity; it is a result of a comparison of an unknown mass with a known mass called a *standard mass*. Balances are instruments used to measure the mass of materials.

Examples of common balances used for the determination of mass are shown in Figure 1.8.

The common conversion units for mass are as follows:

$$1 \text{ gram (g)} = 10^{-3} \text{ kilogram (kg)} = \frac{1}{454} \text{ pound (lb)}$$

In chemistry, when we talk about incredibly small bits of matter such as individual atoms or molecules, units such as grams and even micrograms are much too large. We don't say that a 100-pound individual weighs 0.0500 ton; the unit does not fit the quantity being described. Similarly, an atom of a substance such as hydrogen is very tiny. Its mass is only 1.661×10^{-24} gram.

9 Learning Goal

Use appropriate units in problem solving.

Energy will be considered in Chapter 5.

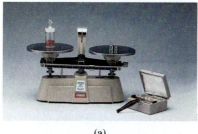

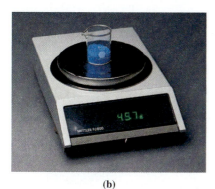

(a)

(b)

(c)

Figure 1.8

Three common balances that are useful for the measurement of mass. (a) A two-pan comparison balance for approximate mass measurement suitable for routine work requiring accuracy to 0.1 g (or perhaps 0.01 g). (b) A top-loading single-pan electronic balance that is similar in accuracy to (a) but has the advantages of speed and ease of operation. The revolution in electronics over the past twenty years has resulted in electronic balances largely supplanting the two-pan comparison balance in routine laboratory usage. (c) An analytical balance that is capable of precise mass measurement (three to five significant figures beyond the decimal point). A balance of this type is used when the highest level of precision and accuracy is required.

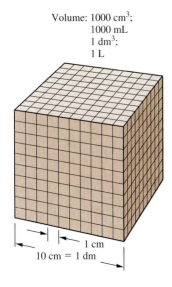

Volume: 1000 cm³;
1000 mL
1 dm³;
1 L

1 cm
10 cm = 1 dm

Volume: 1 cm³;
1 mL

1 cm

Figure 1.9

The relationships among various volume units.

One *atomic mass unit* (amu) is a more convenient way to represent the mass of one hydrogen atom, rather than 1.661×10^{-24} gram:

$$1 \text{ amu} = 1.661 \times 10^{-24} \text{ g}$$

Units should be chosen to suit the quantity being described. This can easily be done by choosing a unit that gives an exponential term closest to 10^0.

Length

The standard metric unit of *length,* the distance between two points, is the meter. Large distances are measured in kilometers; smaller distances are measured in millimeters or centimeters. Very small distances such as the distances between atoms on a surface are measured in *nanometers* (nm):

$$1 \text{ nm} = 10^{-7} \text{ cm} = 10^{-9} \text{ m}$$

Common conversions for length are as follows:

$$1 \text{ meter (m)} = 10^2 \text{ centimeters (cm)} = 3.94 \times 10^1 \text{ inch (in.)}$$

Volume

The standard metric unit of *volume,* the space occupied by an object, is the liter. A liter is the volume occupied by 1000 grams of water at 4 degrees Celsius (°C). The volume, 1 liter, also corresponds to

$$1 \text{ liter (L)} = 10^3 \text{ milliliters (mL)} = 1.06 \text{ quarts (qt)}$$

The relationship between the liter and the milliliter is shown in Figure 1.9.

Typical laboratory glassware used for volume measurement is shown in Figure 1.10. The volumetric flask is designed to *contain* a specified volume, and the graduated cylinder, pipet, and buret *dispense* a desired volume of liquid.

Time

The standard metric unit of time is the second. The need for accurate measurement of time by chemists may not be as apparent as that associated with mass, length,

Figure 1.10

Common laboratory equipment used for the measurement of volume. Graduated (a) cylinders, (b) pipets, and (c) burets are used for the delivery of liquids. (d) Volumetric flasks are used to contain a specific volume. A graduated cylinder is usually used for measurement of approximate volume; it is less accurate and precise than either pipets or burets.

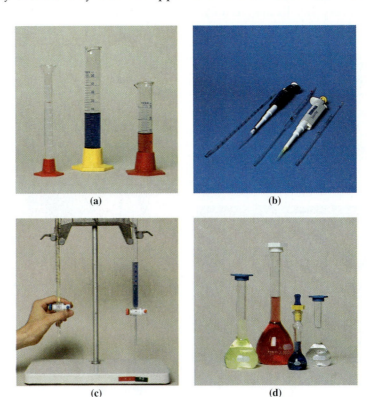

(a) (b)

(c) (d)

and volume. It is necessary, however, in many applications. In fact, matter may be characterized by measuring the time required for a certain process to occur. The rate of a chemical reaction is a measure of change as a function of time.

Temperature

Temperature is the degree of "hotness" of an object. This may not sound like a very "scientific" definition, and, in a sense, it is not. We know intuitively the difference between a "hot" and a "cold" object, but developing a precise definition to explain this is not easy. We may think of the temperature of an object as a measure of the amount of heat in the object. However, this is not strictly true. An object increases in temperature because its heat content has increased and vice versa; however, the relationship between heat content and temperature depends on the quantity and composition of the material.

Many substances, such as mercury, expand as their temperature increases, and this expansion provides us with a way to measure temperature and temperature changes. If the mercury is contained within a sealed tube, as it is in a thermometer, the height of the mercury is proportional to the temperature. A mercury thermometer may be calibrated, or scaled, in different units, just as a ruler can be. Three common temperature scales are *Fahrenheit (°F), Celsius (°C),* and *Kelvin (K).* Two convenient reference temperatures that are used to calibrate a thermometer are the freezing and boiling temperatures of water. Figure 1.11 shows the relationship between the scales and these reference temperatures.

Although Fahrenheit temperature is most familiar to us, Celsius and Kelvin temperatures are used exclusively in scientific measurements. It is often necessary to convert a temperature reading from one scale to another. To convert from Fahrenheit to Celsius, we use the following formula:

$$°C = \frac{°F - 32}{1.8}$$

To convert from Celsius to Fahrenheit, we solve the formula (above) for °F, resulting in

$$°F = 1.8°C + 32$$

To convert from Celsius to Kelvin, we use the formula

$$K = °C + 273.15$$

The Kelvin scale is of particular importance because it is directly related to molecular motion. As molecular speed increases, the Kelvin temperature proportionately increases.

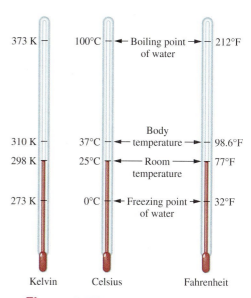

Figure 1.11

The freezing point and boiling point of water, body temperature, and room temperature expressed in the three common units of temperature.

 You can find further information on temperature conversions online at www.mhhe.com/denniston in "A Review of Mathematics."

The Kelvin symbol does not have a degree sign. The degree sign implies a value that is *relative* to some standard. Kelvin is an *absolute* scale.

| **EXAMPLE 1.10** | **Converting from Fahrenheit to Celsius and Kelvin** |

Normal body temperature is 98.6°F. Calculate the corresponding temperature in degrees Celsius:

SOLUTION

Using the expression relating °C and °F,

$$°C = \frac{°F - 32}{1.8}$$

Substituting the information provided,

$$= \frac{98.6 - 32}{1.8} = \frac{66.6}{1.8}$$

Continued—

9 Learning Goal
Use appropriate units in problem solving.

EXAMPLE 1.10 —*Continued*

results in:

$$= 37.0°C$$

Calculate the corresponding temperature in Kelvin units:

SOLUTION

Using the expression relating K and °C,

$$K = °C + 273.15$$

substituting the value obtained in the first part,

$$= 37.0 + 273.15$$

results in

$$= 310.2 \text{ K}$$

QUESTION 1.27

The freezing temperature of water is 32°F. Calculate the freezing temperature of water in

a. Celsius units   b. Kelvin units

QUESTION 1.28

When a patient is ill, his or her temperature may increase to 104°F. Calculate the temperature of this patient in

a. Celsius units   b. Kelvin units

Density and Specific Gravity

Both mass and volume are a function of the *amount* of material present (extensive property). **Density,** the ratio of mass to volume,

$$d = \frac{\text{mass}}{\text{volume}} = \frac{m}{V}$$

is *independent* of the amount of material (intensive property). Density is a useful way to characterize or identify a substance because each substance has a unique density (Figure 1.12).

One milliliter of air and 1 milliliter of iron do not weigh the same. There is much more mass in 1 milliliter of iron; its density is greater.

Density measurements were used to discriminate between real gold and "fool's gold" during the gold rush era. Today, the measurement of the density of a substance is still a valuable analytical technique. The densities of a number of common substances are shown in Table 1.4.

In density calculations, the mass is usually represented in grams, and volume is given in either milliliters (mL) or cubic centimeters (cm³ or cc):

$$1 \text{ mL} = 1 \text{ cm}^3 = 1 \text{ cc}$$

The unit of density would therefore be g/mL, g/cm³, or g/cc.

Animation
Solid-Liquid Density

10 Learning Goal
Use density, mass, and volume in problem solving, and calculate the specific gravity of a substance from its density.

Intensive and extensive properties are described on page 9.

TABLE 1.4 Densities of Some Common Materials

Substance	Density (g/mL)	Substance	Density (g/mL)
Air	0.00129 (at 0°C)	Methyl alcohol	0.792
Ammonia	0.000771 (at 0°C)	Milk	1.028–1.035
Benzene	0.879	Oxygen	0.00143 (at 0°C)
Bone	1.7–2.0	Rubber	0.9–1.1
Carbon dioxide	0.001963 (at 0°C)	Turpentine	0.87
Ethyl alcohol	0.789	Urine	1.010–1.030
Gasoline	0.66–0.69	Water	1.000 (at 4°C)
Gold	19.3	Water	0.998 (at 20°C)
Hydrogen	0.000090 (at 0°C)	Wood	0.3–0.98
Kerosene	0.82	(balsa, least dense; ebony	
Lead	11.3	and teak, most dense)	
Mercury	13.6		

Figure 1.12

Density (mass/volume) is a unique property of a material. A mixture of wood, water, brass, and mercury is shown, with the cork—the least dense—floating on water. Additionally, brass, with a density greater than water but less than liquid mercury, floats on the interface between these two liquids.

EXAMPLE 1.11 Calculating the Density of a Solid

2.00 cm³ of aluminum weigh 5.40 g. Calculate the density of aluminum in units of g/cm^3.

SOLUTION

The density expression is

$$d = \frac{m}{V} = \frac{g}{cm^3}$$

Substituting the information given in the problem,

$$= \frac{5.40 \text{ g}}{2.00 \text{ cm}^3}$$

results in

$$= 2.70 \text{ g}/cm^3$$

EXAMPLE 1.12 Calculating the Mass of a Gas from Its Density

Air has a density of 0.0013 g/mL. What is the mass of a 6.0-L sample of air?

SOLUTION

$$0.0013 \text{ g}/mL = 1.3 \times 10^{-3} \text{ g}/mL$$

(The decimal point is moved three positions to the right.) This problem can be solved by using conversion factors:

$$6.0 \text{ L air} \times \frac{10^3 \text{ mL air}}{1 \text{ L air}} \times \frac{1.3 \times 10^{-3} \text{ g air}}{\text{mL air}} = 7.8 \text{ g air}$$

10 **Learning Goal**

Use density, mass, and volume in problem solving, and calculate the specific gravity of a substance from its density.

EXAMPLE 1.13 **Using the Density to Calculate the Mass of a Liquid**

Calculate the mass, in grams, of 10.0 mL of mercury (symbolized Hg) if the density of mercury is 13.6 g/mL.

SOLUTION

Using the density as a conversion factor from volume to mass,

$$m = (10.0 \ \cancel{mL \ Hg})\left(13.6 \ \frac{g \ Hg}{\cancel{mL \ Hg}}\right)$$

Cancellation of units results in

$$= 136 \ g \ Hg$$

---

10 **Learning Goal**

Use density, mass, and volume in problem solving, and calculate the specific gravity of a substance from its density.

EXAMPLE 1.14 **Using the Density to Calculate the Volume of a Liquid**

Calculate the volume, in milliliters, of a liquid that has a density of 1.20 g/mL and a mass of 5.00 grams.

SOLUTION

Using the density as a conversion factor from mass to volume,

$$V = (5.00 \ \cancel{g \ liquid})\left(\frac{1 \ mL \ liquid}{1.20 \ \cancel{g \ liquid}}\right)$$

Cancellation of units results in

$$= 4.17 \ mL \ liquid$$

QUESTION 1.29

The density of ethyl alcohol (200 proof, or pure alcohol) is 0.789 g/mL at 20°C. Calculate the mass of a 30.0-mL sample.

QUESTION 1.30

Calculate the volume, in milliliters, of 10.0 g of a saline solution that has a density of 1.05 g/mL.

Specific gravity is frequently referenced to water at 4°C, its temperature of maximum density (1.000 g/mL). Other reference temperatures may be used. However, the temperature must be specified.

For convenience, values of density are often related to a standard, well-known reference, the density of pure water at 4°C. This "referenced" density is called the **specific gravity**, the ratio of the density of the object in question to the density of pure water at 4°C.

$$\text{specific gravity} = \frac{\text{density of object (g/mL)}}{\text{density of water (g/mL)}}$$

Specific gravity is a *unitless* term. Because the density of water at 4.0°C is 1.00 g/mL, the numerical values for the density and specific gravity of a substance are equal. That is, an object with a density of 2.00 g/mL has a specific gravity of 2.00 at 4°C.

A MEDICAL Connection | Diagnosis Based on Waste

Any archaeologist would say that you can learn a great deal about the activities and attitudes of a society by finding the remains of its dump sites and studying its waste.

Similarly, urine, a waste product consisting of a wide variety of metabolites, may be analyzed to indicate abnormalities in various metabolic processes or even unacceptable behavior (recall the steroid tests in Olympic competition).

Many of these tests must be performed by using sophisticated and sensitive instrumentation. However, a very simple test, the measurement of the specific gravity of urine, can be an indicator of diabetes mellitus. Excess sugar molecules in urine "fit" between water molecules. Consequently, the total volume of the urine changes only very slightly; however, the mass changes measurably. The density, the ratio m/V, increases; hence, the specific gravity also increases. The normal range for human urine specific gravity is 1.010–1.030.

A hydrometer, a weighted glass bulb inserted in a liquid, may be used to determine specific gravity. The higher it floats in the liquid, the more dense the liquid. A hydrometer that is calibrated to indicate the specific gravity of urine is called a urinometer.

Although hydrometers have been replaced by more modern measuring devices that use smaller samples, these newer instruments operate on the same principles as the hydrometer.

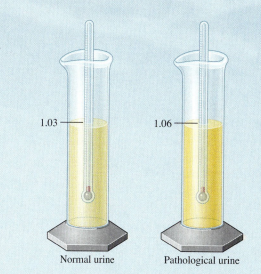

Normal urine   Pathological urine

A hydrometer, used in the measurement of the specific gravity of urine.

FOR FURTHER UNDERSTANDING

Give reasons that may account for such a broad range of "normal" values.

Could results depend on food or medicine consumed prior to the test?

Routine hospital tests involving the measurement of the specific gravity of urine and blood samples are frequently used as diagnostic tools. For example, diseases such as kidney disorders and diabetes change the composition of urine. This compositional change results in a corresponding change in specific gravity. This change is easily measured and provides the basis for a quick preliminary diagnosis. This topic is discussed in greater detail in A Medical Connection: Diagnosis Based on Waste.

SUMMARY

1.1 The Discovery Process

Chemistry is the study of matter and the changes that matter undergoes. *Matter* is anything that has mass and occupies space. The changes that matter undergoes always involve either gain or loss of energy. *Energy* is the ability to do work (to accomplish some change). Thus a study of chemistry involves matter, energy, and their interrelationship.

The *scientific method* consists of six interrelated processes: observation, questioning, pattern recognition, development of *theories* from *hypotheses*, experimentation, and

summarizing information. A *law* summarizes a large quantity of information.

The development of the scientific method has played a major role in civilization's rapid growth during the past two centuries.

1.2 Matter and Properties

Properties (characteristics) of matter may be classified as either physical or chemical. *Physical properties* can be observed without changing the chemical composition of the sample. *Chemical properties* result in a change in composition and can be observed only through *chemical reactions*.

Intensive properties are independent of the quantity of the substance. *Extensive properties* depend on the quantity of a substance.

Three states of matter exist (*solid, liquid,* and *gas*); these states of matter are distinguishable by differences in physical properties.

All matter is classified as either a *pure substance* or a *mixture.* A pure substance is a substance that has only one component. A mixture is a combination of two or more pure substances in which the combined substances retain their identity.

A *homogeneous mixture* has uniform composition. Its particles are well mixed. A *heterogeneous mixture* has a nonuniform composition.

An *element* is a pure substance that cannot be converted into a simpler form of matter by any chemical reaction. A *compound* is a substance produced from the combination of two or more elements in a definite, reproducible fashion.

1.3 Measurement in Chemistry

Science is the study of humans and their environment. Its tool is experimentation. A scientific experiment produces *data.* Each piece of data is the individual result of a single measurement. Mass, length, volume, time, and temperature, are common types of data obtained from chemical experiments.

Results are the outcome of an experiment. Usually, several pieces of data are combined, using a mathematical equation, to produce a result.

A *unit* defines the basic quantity of mass, volume, time, and so on. A number that is not followed by the correct unit usually conveys no useful information.

The metric system is a decimal-based system in contrast to the English system. In the metric system, mass is represented as the gram, length as the meter, and volume as the liter. Any subunit or multiple unit contains one of these units preceded by a prefix indicating the power of ten by which the base unit is to be multiplied to form the subunit or multiple unit. Scientists favor this system over the not-so-systematic English units of measurement.

To convert one unit to another, we must set up a *conversion factor* or series of conversion factors that relate two units. The proper use of these conversion factors is referred to as the factor-label method. This method is used either to convert from one unit to another within the same system or to convert units from one system to another. It is a very useful problem-solving tool.

1.4 Significant Figures and Scientific Notation

Significant figures are all digits in a number representing data or results that are known with certainty plus the first uncertain digit. The number of significant figures associated with a measurement is determined by the measuring device. Results should be rounded off to the proper number of significant figures.

Error is defined as the difference between the true value and our estimation, or measurement, of the value. *Accuracy* is the degree of agreement between the true and measured values. *Uncertainty* is the degree of doubt in a single measurement. The number of meaningful digits in a measurement is determined by the measuring device. *Precision* is a measure of the agreement of replicate measurements.

Very large and very small numbers may be represented with the proper number of significant figures by using *scientific notation.*

1.5 Experimental Quantities

Mass describes the quantity of matter in an object. The terms weight and mass are often used interchangeably, but they are not equivalent. *Weight* is the force of gravity on an object. The fundamental unit of mass in the metric system is the gram. One atomic mass unit (amu) is equal to 1.661×10^{-24} g.

The standard metric unit of length is the meter. Large distances are measured in kilometers; smaller distances are measured in millimeters or centimeters. Very small distances (on the atomic scale) are measured in nanometers (nm). The standard metric unit of volume is the liter. A liter is the volume occupied by 1000 grams of water at 4 degrees Celsius. The standard metric unit of time is the second, a unit that is used in the English system as well.

Temperature is the degree of "hotness" of an object. Many substances, such as liquid mercury, expand as their temperature increases, and this expansion provides us with a way to measure temperature and temperature changes. Three common temperature scales are Fahrenheit (°F), Celsius (°C), and Kelvin (K).

Density is the ratio of mass to volume and is a useful way of characterizing a substance. Values of density are often related to a standard reference, the density of pure water at 4°C. This "referenced" density is the *specific gravity,* the ratio of the density of the object in question to the density of pure water at 4°C.

KEY TERMS

accuracy (1.4)	homogeneous mixture (1.2)
chemical property (1.2)	hypothesis (1.1)
chemical reaction (1.2)	intensive property (1.2)
chemistry (1.1)	law (1.1)
compound (1.2)	liquid state (1.2)
data (1.3)	mass (1.5)
density (1.5)	matter (1.1)
element (1.2)	mixture (1.2)
energy (1.1)	physical change (1.2)
error (1.4)	physical property (1.2)
extensive property (1.2)	precision (1.4)
gaseous state (1.2)	properties (1.2)
heterogeneous mixture (1.2)	pure substance (1.2)

result (1.3)

scientific method (1.1)

scientific notation (1.4)

significant figures (1.4)

solid state (1.2)

specific gravity (1.5)

temperature (1.5)

theory (1.1)

uncertainty (1.4)

unit (1.3)

weight (1.5)

QUESTIONS AND PROBLEMS

The Discovery Process

Foundations

1.31 Define each of the following terms:
  a. chemistry
  b. matter
  c. energy

1.32 Define each of the following terms:
  a. hypothesis
  b. theory
  c. law

1.33 Give the base unit for each of the following in the metric system:
  a. mass
  b. volume
  c. length

1.34 Give the base unit for each of the following in the metric system:
  a. time
  b. temperature
  c. energy

Applications

1.35 Discuss the difference between the terms *mass* and *weight*.

1.36 Discuss the difference between the terms *data* and *results*.

1.37 Distinguish between specific gravity and density.

1.38 Distinguish between a hypothesis and a law.

1.39 Stem-cell research has the potential to provide replacement "parts" for the human body. Is this statement a hypothesis, theory, or law? Explain your reasoning.

1.40 Observed increases in global temperatures are caused by elevated levels of carbon dioxide. Is this statement a hypothesis, theory, or law? Explain your reasoning.

Matter and Properties

Foundations

1.41 Describe what is meant by a physical property.

1.42 Describe what is meant by a physical change.

1.43 Distinguish between a homogeneous mixture and a heterogeneous mixture.

1.44 Distinguish between an intensive property and an extensive property.

Applications

1.45 Label each of the following as either a physical change or a chemical reaction:
  a. An iron nail rusts.
  b. An ice cube melts.
  c. A limb falls from a tree.

1.46 Label each of the following as either a physical change or a chemical reaction:
  a. A puddle of water evaporates.
  b. Food is digested.
  c. Wood is burned.

1.47 Label each of the following properties of sodium as either a physical property or a chemical property:
  a. Sodium is a soft metal (can be cut with a knife).
  b. Sodium reacts violently with water to produce hydrogen gas and sodium hydroxide.

1.48 Label each of the following properties of sodium as either a physical property or a chemical property:
  a. When exposed to air, sodium forms a white oxide.
  b. Sodium melts at 98°C.
  c. The density of sodium metal at 25°C is 0.97 g/cm^3.

1.49 Label each of the following as either a pure substance or a mixture:
  a. water
  b. table salt (sodium chloride)
  c. blood

1.50 Label each of the following as either a pure substance or a mixture:
  a. sucrose (table sugar)
  b. orange juice
  c. urine

1.51 Label each of the following as either a homogeneous mixture or a heterogeneous mixture:
  a. a soft drink
  b. a saline solution
  c. gelatin

1.52 Label each of the following as either a homogeneous mixture or a heterogeneous mixture:
  a. gasoline
  b. vegetable soup
  c. concrete

1.53 Label each of the following as either an intensive property or an extensive property:
  a. mass
  b. volume
  c. density

1.54 Label each of the following as either an intensive property or an extensive property:
  a. specific gravity
  b. temperature
  c. heat content

Measurement in Chemistry

Foundations

1.55 Convert 2.0 pounds to
  a. ounces
  b. tons
  c. grams
  d. milligrams
  e. dekagrams

1.56 Convert 5.0 quarts to
  a. gallons
  b. pints
  c. liters
  d. milliliters
  e. microliters

1.57 Convert 3.0 grams to
  a. pounds
  b. ounces
  c. kilograms
  d. centigrams
  e. milligrams

1.58 Convert 3.0 meters to
  a. yards
  b. inches
  c. feet
  d. centimeters
  e. millimeters

1.59 Convert 50.0°F to
    a. °C
    b. K

1.60 Convert −10.0°F to
    a. °C
    b. K

1.61 Convert 20.0°C to
    a. K
    b. °F

1.62 Convert 300.0 K to
    a. °C
    b. °F

Applications

1.63 A 150-lb adult has approximately 9 pints of blood. How many liters of blood does the individual have?

1.64 If a drop of blood has a volume of 0.05 mL, how many drops of blood are in the adult described in Problem 1.63?

1.65 A patient's temperature is 38.5°C. To what Fahrenheit temperature does this correspond?

1.66 A newborn is 21 inches long and weighs 6 lb 9 oz. Describe the baby in metric units.

1.67 Which distance is shorter, 5.0 cm or 5.0 in.?

1.68 Which volume is smaller, 50.0 mL or 0.500 L?

Significant Figures and Scientific Notation

Foundations

1.69 How many significant figures are contained in each of the following numbers?
    a. 10.0
    b. 0.214
    c. 0.120
    d. 2.062
    e. 10.50
    f. 1050

1.70 How many significant figures are contained in each of the following numbers?
    a. 3.8×10^{-3}
    b. 5.20×10^{2}
    c. 0.00261
    d. 24
    e. 240
    f. 2.40

1.71 Round the following numbers to three significant figures:
    a. 3.873×10^{-3}
    b. 5.202×10^{-2}
    c. 0.002616
    d. 24.3387
    e. 240.1
    f. 2.407

1.72 Round the following numbers to three significant figures:
    a. 123700
    b. 0.00285792
    c. 1.421×10^{-3}
    d. 53.2995
    e. 16.96
    f. 507.5

Applications

1.73 Perform each of the following arithmetic operations, reporting the answer with the proper number of significant figures:
    a. (23)(657)
    b. 0.00521 + 0.236
    c. $\dfrac{18.3}{3.0576}$
    d. 1157.23 − 17.812
    e. $\dfrac{(1.987)(298)}{0.0821}$

1.74 Perform each of the following arithmetic operations, reporting the answer with the proper number of significant figures:
    a. $\dfrac{(16.0)(0.1879)}{45.3}$
    b. $\dfrac{(76.32)(1.53)}{0.052}$
    c. (0.0063)(57.8)
    d. 18 + 52.1
    e. 58.17 − 57.79

1.75 Express the following numbers in scientific notation (use the proper number of significant figures):
    a. 12.3
    b. 0.0569
    c. −1527
    d. 0.000000789
    e. 92,000,000
    f. 0.005280
    g. 1.279
    h. −531.77

1.76 Express each of the following numbers in decimal notation:
    a. 3.24×10^{3}
    b. 1.50×10^{-4}
    c. 4.579×10^{-1}
    d. -6.83×10^{5}
    e. -8.21×10^{-2}
    f. 2.9979×10^{8}
    g. 1.50×10^{0}
    h. 6.02×10^{23}

Experimental Quantities

Foundations

1.77 Calculate the density of a 3.00×10^{2}-g object that has a volume of 50.0 mL.

1.78 What volume, in liters, will 8.00×10^{2} g of air occupy if the density of air is 1.29 g/L?

1.79 What is the mass, in grams, of a piece of iron that has a volume of 1.50×10^{2} mL and a density of 7.20 g/mL?

1.80 What is the mass of a femur (leg bone) having a volume of 118 cm³? The density of bone is 1.8 g/cm³.

Applications

1.81 The specific gravity of a patient's urine sample was measured as 1.008. Given that the density of water is 1.000 g/mL at 4°C, what is the density of the urine sample?

1.82 The density of grain alcohol is 0.789 g/mL. Given that the density of water at 4°C is 1.00 g/mL, what is the specific gravity of grain alcohol?

1.83 You are given three bars of metal. Each is labeled with its identity (lead, uranium, platinum). The lead bar has a mass of 5.0×10^{1} g and a volume of 6.36 cm³. The uranium bar has a mass of 75 g and a volume of 3.97 cm³. The platinum bar has a mass of 2140 g and a volume of 1.00×10^{2} cm³. Which of these metals has the lowest density? Which has the greatest density?

1.84 Refer to Problem 1.83. Suppose that each of the bars had the same mass. How could you determine which bar had the lowest density or highest density?

1.85 The density of methanol at 20°C is 0.791 g/mL. What is the volume of a 10.0-g sample of methanol?

1.86 The density of methanol at 20°C is 0.791 g/mL. What is the mass of a 50.0-mL sample of methanol?

FOR FURTHER UNDERSTANDING

1. An instrument used to detect metals in drinking water can detect as little as one microgram of mercury in one liter of water. Mercury is a toxic metal; it accumulates in the body and is responsible for the deterioration of brain cells. Calculate the number of mercury atoms you would consume if you drank one liter of water that contained only one microgram of mercury. (The mass of one mercury atom is 3.3×10^{-22} grams.)

2. Yesterday's temperature was 40°F. Today it is 80°F. Bill tells Sue that it is twice as hot today. Sue disagrees. Do you think Sue is correct or incorrect? Why or why not?

3. Aspirin has been recommended to minimize the chance of heart attacks in persons who have already had one or more occurrences. If a patient takes one aspirin tablet per day for ten years, how many pounds of aspirin will the patient consume? (Assume that each tablet is approximately 325 mg.)

4. Design an experiment that will allow you to measure the density of your favorite piece of jewelry.

5. The diameter of an aluminum atom is 250 picometers (1 picometer = 10^{-12} meters). How many aluminum atoms must be placed end to end to make a "chain" of aluminum atoms one foot long?

2

OUTLINE

LEARNING GOALS

1 Describe the important properties of protons, neutrons, and electrons.

2 Calculate the number of protons, neutrons, and electrons in any atom.

3 Distinguish among atoms, ions, and isotopes and calculate atomic masses from isotopic abundance.

4 Trace the history of the development of atomic theory, beginning with Dalton.

5 Recognize the important subdivisions of the periodic table: periods, groups (families), metals, and nonmetals.

6 Use the periodic table to obtain information about an element.

7 Describe the relationship between the electronic structure of an element and its position in the periodic table.

8 Write electron configurations for atoms of the most commonly occurring elements.

9 Use the octet rule to predict the charge of common cations and anions.

10 Use the periodic table and its predictive power to estimate the relative sizes of atoms and ions, as well as relative magnitudes of ionization energy and electron affinity.

The Structure of the Atom and the Periodic Table

Recall for a moment the first time that you sat down in front of a computer. Perhaps it was connected to the Internet; somewhere in its memory was a word processor program, a spreadsheet, a few games, and many other features with strange-sounding names. Your challenge, very simply, was to use this device to access and organize information. Several manuals, all containing hundreds of pages of bewilderment, were your only help. How did you overcome this seemingly impossible task?

We are quite sure that you did not succeed without doing some reading and talking to people who had experience with computers. Also, you did not attempt to memorize every single word in each manual.

Success with a computer or any other storehouse of information results from developing an overall understanding of the way in which the system is organized. Certain facts must be memorized, but seeing patterns and using these relationships allows us to accomplish a wide variety of tasks that involve similar logic.

The study of chemistry is much like "real life." Just as it is impossible to memorize every single fact that will allow you to run a computer or drive an automobile in traffic, it is equally impossible to learn every fact in chemistry. Knowing the organization and logic of a process, along with a few key facts, makes a task manageable.

One powerful organizational device in chemistry is the periodic table. Its use in organizing and predicting the behavior of all of the known elements (and many of the compounds formed from these elements) is the subject of this chapter.

Learning is facilitated by organizing information. How can the periodic table make the process of learning chemistry easier?

Figure 2.1

Sophisticated techniques, such as scanning tunneling electron microscopy, provide visual evidence for the structure of atoms and molecules. Each dot represents the image of a single iron atom. Even more amazing, the iron atoms have been arranged on a copper surface in the form of the Chinese characters representing the word *atom*.

1 Learning Goal

Describe the important properties of protons, neutrons, and electrons.

Radioactivity and radioactive decay are discussed in Chapter 9.

2 Learning Goal

Calculate the number of protons, neutrons, and electrons in any atom.

Recall from Chapter 1 (p. 26) that one atomic mass unit (amu) is equivalent to 1.661×10^{-24} g.

2.1 Composition of the Atom

The basic structural unit of an element is the **atom,** which is the smallest unit of an element that retains the chemical properties of that element. A tiny sample of the element copper, too small to be seen by the naked eye, is composed of billions of copper atoms arranged in some orderly fashion. Each atom is incredibly small. Only recently have we been able to "see" atoms using modern instruments such as the scanning tunneling microscope (Figure 2.1).

Electrons, Protons, and Neutrons

We know from experience that certain kinds of atoms can "split" into smaller particles and release large amounts of energy; this process is *radioactive decay*. We also know that the atom is composed of three primary particles: the *electron*, the *proton*, and the *neutron*. Although other subatomic fragments with unusual names (neutrinos, gluons, quarks, and so forth) have also been discovered, we shall concern ourselves only with the primary particles: the protons, neutrons, and electrons.

We can consider that the atom is composed of two distinct regions:

1. The **nucleus** is a small, dense, positively charged region in the center of the atom. The nucleus is composed of positively charged **protons** and uncharged **neutrons.**
2. Surrounding the nucleus is a diffuse region of negative charge populated by **electrons,** the source of the negative charge. Electrons are very low in mass in contrast to protons and neutrons.

The properties of these particles are summarized in Table 2.1.

Atoms of various types differ in their number of protons, neutrons, and electrons. The number of protons determines the identity of the atom. As such, the number of protons is *characteristic* of the element. When the number of protons is equal to the number of electrons, the atom is neutral because the charges are balanced and effectively cancel one another.

We may represent an element symbolically as follows:

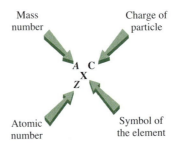

The **atomic number** (Z) is equal to the number of protons in the atom, and the **mass number** (A) is equal to the *sum* of the number of protons and neutrons (the mass of the electrons is so small as to be insignificant in comparison to that of the nucleus).

TABLE 2.1 Selected Properties of the Three Basic Subatomic Particles			
Name	**Charge**	**Mass (amu)**	**Mass (grams)**
Electron (e)	−1	5.4×10^{-4}	9.1095×10^{-28}
Proton (p)	+1	1.00	1.6725×10^{-24}
Neutron (n)	0	1.00	1.6750×10^{-24}

If

$$\text{number of protons} + \text{number of neutrons} = \text{mass number}$$

then, if the number of protons is subtracted from each side,

$$\text{number of neutrons} = \text{mass number} - \text{number of protons}$$

or, because the number of protons equals the atomic number,

$$\text{number of neutrons} = \text{mass number} - \text{atomic number}$$

For an atom, in which positive and negative charges cancel, the number of protons and electrons must be equal and identical to the atomic number.

EXAMPLE 2.1 **Determining the Composition of an Atom**

Calculate the numbers of protons, neutrons, and electrons in an atom of fluorine. The atomic symbol for the fluorine atom is $^{19}_{9}F$.

2 **Learning Goal**
Calculate the number of protons, neutrons, and electrons in any atom.

SOLUTION

The mass number 19 tells us that the total number of protons + neutrons is 19. The atomic number, 9, represents the number of protons. The difference, $19 - 9$, or 10, is the number of neutrons. The number of electrons must be the same as the number of protons, hence 9, for a neutral fluorine atom.

QUESTION 2.1

Calculate the number of protons, neutrons, and electrons in each of the following atoms:

a. $^{32}_{16}S$
b. $^{23}_{11}Na$

QUESTION 2.2

Calculate the number of protons, neutrons, and electrons in each of the following atoms:

a. $^{1}_{1}H$
b. $^{244}_{94}Pu$

Isotopes

Isotopes are atoms of the same element having different masses *because they contain different numbers of neutrons*. In other words, isotopes have different mass numbers. For example, all of the following are isotopes of hydrogen:

3 **Learning Goal**
Distinguish among atoms, ions, and isotopes and calculate atomic masses from isotopic abundance.

$^{1}_{1}H$	$^{2}_{1}H$	$^{3}_{1}H$
Hydrogen	Deuterium	Tritium
(Hydrogen-1)	(Hydrogen-2)	(Hydrogen-3)

Isotopes are often written with the name of the element followed by the mass number. For example, the isotopes $^{12}_{6}C$ and $^{14}_{6}C$ may be written as carbon-12 (or C-12) and carbon-14 (or C-14), respectively.

Certain isotopes (radioactive isotopes) of elements emit particles and energy that can be used to trace the behavior of biochemical systems. These isotopes otherwise behave identically to any other isotope of the same element. Their

A detailed discussion of the use of radioactive isotopes in the diagnosis and treatment of diseases is found in Chapter 9.

CHEMISTRY at the Crime Scene | Microbial Forensics

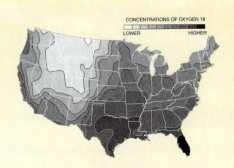

CONCENTRATIONS OF OXYGEN 18

LOWER   HIGHER

We learned that atoms of the same element are not identical; common elements are really a mixture of two or more isotopes, differing in mass because they contain different numbers of neutrons. Furthermore, the atomic mass is the weighted average of the masses of these various isotopes.

It seems to follow from this that the relative amounts of these isotopes would be the same, no matter where in the world we obtain a sample of the element. In reality, this is not true. There exist small, but measurable differences between elemental isotopic ratios; these differences correlate with their locations.

Water, of course, contains oxygen atoms, and a small fraction of these atoms are oxygen-18. We know that lakes and rivers across the United States contain oxygen-18 in different concentrations. Ocean water is richer in oxygen-18. Consequently, freshwater supplies closest to oceans are similarly enriched, perhaps because water containing oxygen-18 is a bit heavier and falls to earth more rapidly, before the clouds are carried inland, far from the coast. Scientists have been able to map the oxygen-18 distribution throughout the United States, creating contour maps, similar to weather maps, showing regions of varying concentrations of the isotope.

How can we use these facts to our advantage? The Federal Bureau of Investigation certainly has an interest in ascertaining the origin of microbes that may be used by terrorists. Anthrax, distributed to several sites shortly after September 11 through the United States Postal Service, is but one familiar example. If we could carefully measure the ratio of oxygen-18 to oxygen-16 in weaponized biological materials, it might be possible to match this ratio to that found in water supplies in a specific location. We would then have an indication of the region of the country where the material was cultured, perhaps leading to the arrest and prosecution of the perpetrators of the terrorist activity.

This is certainly not an easy task. However the F.B.I. has assembled an advisory board of science and forensic experts to study this and other approaches to tracking the origin of biological weapons of mass destruction. Just like the "smoking gun," these strategies, termed *microbial forensics,* could lead to more effective prosecution and a safer society for all of us.

FOR FURTHER UNDERSTANDING

Search the Web for more information on the properties of anthrax.

How might the approach, described above, be applied to determining the origin of an oil spill at sea?

Atomic mass units are convenient for representing the mass of very small particles, such as individual atoms. Refer to the discussion of units in Chapter 1, p. 26.

chemical behavior is identical; it is their nuclear behavior that is unique. As a result, a radioactive isotope can be substituted for the "nonradioactive" isotope, and its biochemical activity can be followed by monitoring the particles or energy emitted by the isotope as it passes through the body.

The existence of isotopes explains why the average masses, measured in atomic mass units (amu), of the various elements are not whole numbers. This is contrary to what we would expect from proton and neutron masses, which are very close to unity.

Consider, for example, the mass of one chlorine atom, containing 17 protons (atomic number) and 18 neutrons:

$$17 \text{ protons} \times \frac{1.00 \text{ amu}}{\text{proton}} = 17.00 \text{ amu}$$

$$18 \text{ neutrons} \times \frac{1.00 \text{ amu}}{\text{neutron}} = 18.00 \text{ amu}$$

$$17.00 \text{ amu} + 18.00 \text{ amu} = 35.00 \text{ amu (mass of chlorine atom)}$$

Inspection of the periodic table reveals that the mass of chlorine is actually 35.45 amu, *not* 35.00 amu. The existence of isotopes accounts for this difference. A natural sample of chlorine is composed principally of two isotopes, chlorine-35 and chlorine-37, in approximately a 3:1 ratio, and the tabulated mass is the *weighted average* of the two isotopes. In our calculation, the chlorine atom referred to is the isotope that has a mass number of 35 amu.

The weighted average of the masses of all of the isotopes of an element is the **atomic mass** and should be distinguished from the mass number, which is the sum of the number of protons and neutrons in a single isotope of the element.

Example 2.2 demonstrates the calculation of the atomic mass of chlorine.

> The weighted average is not a true average but is corrected by the relative amounts (the weighting factor) of each isotope present in nature.

EXAMPLE 2.2 Determining Atomic Mass

Calculate the atomic mass of naturally occurring chlorine if 75.77% of chlorine atoms are $^{35}_{17}Cl$ (chlorine-35) and 24.23% of chlorine atoms are $^{35}_{17}Cl$ (chlorine-37).

3 Learning Goal
Distinguish among atoms, ions, and isotopes and calculate atomic masses from isotopic abundance.

SOLUTION

Step 1. Convert each percentage to a decimal fraction.

$$75.77\% \text{ chlorine-35} \times \frac{1}{100\%} = 0.7577 \text{ chlorine-35}$$

$$24.23\% \text{ chlorine-37} \times \frac{1}{100\%} = 0.2423 \text{ chlorine-37}$$

Step 2. Multiply the decimal fraction of each isotope by the mass of that isotope to determine the isotopic contribution to the average atomic mass.

contribution to atomic mass by chlorine-35	=	fraction of all Cl atoms that are chlorine-35	×	mass of a chlorine-35 atom
	=	0.7577	×	35.00 amu
	=	26.52 amu		

contribution to atomic mass by chlorine-37	=	fraction of all Cl atoms that are chlorine-37	×	mass of a chlorine-37 atom
	=	0.2423	×	37.00 amu
	=	8.965 amu		

Step 3. The weighted average is the sum of the isotopic contributions:

atomic mass of naturally occurring Cl	=	contribution of chlorine-35	+	contribution of chlorine-37
	=	26.52 amu	+	8.965 amu
	=	35.49 amu		

which is very close to the tabulated value of 35.45 amu. An even more exact value would be obtained by using a more exact value of the mass of the proton and neutron (experimentally known to a greater number of significant figures).

Whenever you do calculations such as those in Example 2.2, before even beginning the calculation, you should look for an approximation of the value sought. Then do the calculation and see whether you obtain a reasonable number (similar to your anticipated value). In the preceding problem, if the two isotopes have masses of 35 and 37, the atomic mass must lie somewhere between the two

> A hint for numerical problem solving: Estimate (at least to an order of magnitude) your answer before beginning the calculation using your calculator.

extremes. Furthermore, because the majority of a naturally occurring sample is chlorine-35 (about 75%), the value should be closer to 35 than to 37. An analysis of the results often avoids problems stemming from untimely events such as pushing the wrong button on a calculator.

3 **Learning Goal**

Distinguish among atoms, ions, and isotopes and calculate atomic masses from isotopic abundance.

EXAMPLE 2.3 Determining Atomic Mass

Calculate the atomic mass of naturally occurring carbon if 98.90% of carbon atoms are $^{12}_{6}C$ (carbon-12) with a mass of 12.00 amu and 1.11% are $^{13}_{6}C$ (carbon-13) with a mass of 13.00 amu. (Note that a small amount of $^{14}_{6}C$ is also present but is small enough to ignore in a calculation involving three or four significant figures.)

SOLUTION

Step 1. Convert each percentage to a decimal fraction.

$$98.90\% \text{ carbon-12} \times \frac{1}{100\%} = 0.9890 \text{ carbon-12}$$

$$1.11\% \text{ carbon-13} \times \frac{1}{100\%} = 0.0111 \text{ carbon-13}$$

Step 2.

contribution to atomic mass by carbon-12	=	fraction of all C atoms that are carbon-12	×	(mass of a carbon-12 atom)
	=	0.9890	×	12.00 amu
	=	11.87 amu		

contribution to atomic mass by carbon-13	=	fraction of all C atoms that are carbon-13	×	mass of a carbon-13 atom
	=	0.0111	×	13.00 amu
	=	0.144 amu		

Step 3. The weighted average is

atomic mass of naturally occurring carbon	=	contribution of carbon-12	+	contribution of carbon-13
	=	11.87 amu	+	0.144 amu
	=	12.01 amu		

Helpful Hint: Because most of the carbon is carbon-12, with very little carbon-13 present, the atomic mass should be very close to that of carbon-12. Approximations, before performing the calculation, provide another check on the accuracy of the final result.

QUESTION 2.3

The element neon has three naturally occurring isotopes. One of these has a mass of 19.99 amu and a natural abundance of 90.48%. A second isotope has a mass of 20.99 amu and a natural abundance of 0.27%. A third has a mass of 21.99 amu and a natural abundance of 9.25%. Calculate the atomic mass of neon.

QUESTION 2.4

The element nitrogen has two naturally occurring isotopes. One of these has a mass of 14.003 amu and a natural abundance of 99.63%; the other isotope has a mass of 15.000 amu and a natural abundance of 0.37%. Calculate the atomic mass of nitrogen.

Ions

Ions are electrically charged particles that result from a gain of one or more electrons by the parent atom (forming negative ions, or **anions**) or a loss of one or more electrons from the parent atom (forming positive ions, or **cations**).

Formation of an anion may occur as follows:

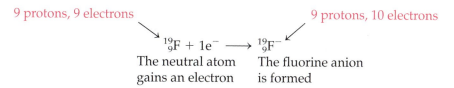

9 protons, 9 electrons   9 protons, 10 electrons

$$^{19}_{9}F + 1e^- \longrightarrow {}^{19}_{9}F^-$$

The neutral atom   The fluorine anion
gains an electron   is formed

> **3 Learning Goal**
> Distinguish among atoms, ions, and isotopes, and calculate atomic masses from isotopic abundance.

Alternatively, formation of a cation of sodium may proceed as follows:

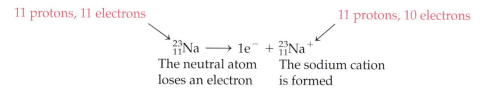

11 protons, 11 electrons   11 protons, 10 electrons

$$^{23}_{11}Na \longrightarrow 1e^- + {}^{23}_{11}Na^+$$

The neutral atom   The sodium cation
loses an electron   is formed

> **Ions are often formed in chemical reactions, when one or more electrons are transferred from one substance to another.**

Note that the electrons gained are written to the left of the reaction arrow (they are reactants), whereas the electrons lost are written as products to the right of the reaction arrow. For simplification, the atomic and mass numbers are often omitted because they do not change during ion formation. For example, the sodium cation would be written as Na^+ and the anion of fluorine as F^-.

2.2 Development of Atomic Theory

With this overview of our current understanding of the structure of the atom, we now look at a few of the most important scientific discoveries that led to modern atomic theory.

> **4 Learning Goal**
> Trace the history of the development of atomic theory, beginning with Dalton.

Dalton's Theory

The first experimentally based theory of atomic structure was proposed by John Dalton, an English schoolteacher, in the early 1800s. Dalton proposed the following description of atoms:

1. All matter consists of tiny particles called atoms.
2. An atom cannot be created, divided, destroyed, or converted to any other type of atom.
3. Atoms of a particular element have identical properties.
4. Atoms of different elements have different properties.
5. Atoms of different elements combine in simple whole-number ratios to produce compounds (stable aggregates of atoms).
6. Chemical change involves joining, separating, or rearranging atoms.

Figure 2.2

An illustration of John Dalton's atomic theory. (a) Atoms of the same element are identical but differ from atoms of any other element. (b) Atoms combine in whole-number ratios to form compounds.

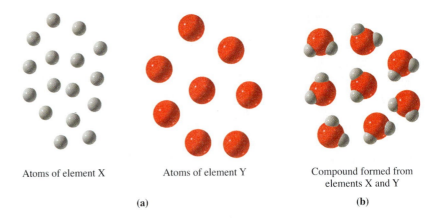

Atoms of element X   Atoms of element Y   Compound formed from elements X and Y

(a)   (b)

Although Dalton's theory was founded on meager and primitive experimental information, we regard much of it as correct today. Postulates 1, 4, 5, and 6 are currently regarded as true. The discovery of the processes of nuclear fusion, fission ("splitting" of atoms), and radioactivity has disproved the postulate that atoms cannot be created or destroyed. Postulate 3, that all the atoms of a particular element are identical, was disproved by the discovery of isotopes.

Fusion, fission, radioactivity, and isotopes are discussed in some detail in Chapter 9. Figure 2.2 uses a simple model to illustrate Dalton's theory.

Evidence for Subatomic Particles: Electrons, Protons, and Neutrons

The next major discoveries occurred almost a century later (1879–1897). Although Dalton pictured atoms as indivisible, various experiments, particularly those of William Crookes and Eugene Goldstein, indicated that the atom is composed of charged (+ and −) particles.

Crookes connected two metal electrodes (metal discs connected to a source of electricity) at opposite ends of a sealed glass vacuum tube. When the electricity was turned on, rays of light were observed traveling between the two electrodes. They were called **cathode rays** because they traveled from the *cathode* (the negative electrode) to the *anode* (the positive electrode).

Later experiments by J. J. Thomson, an English scientist, demonstrated the electrical and magnetic properties of cathode rays (Figure 2.3). The rays were deflected toward the positive electrode of an external electric field. Because opposite charges attract, this indicates the negative character of the rays. Similar experiments with an external magnetic field showed a deflection as well; hence these cathode rays also have magnetic properties.

4 Learning Goal

Trace the history of the development of atomic theory, beginning with Dalton.

Crookes's cathode ray tube was the forerunner of the computer screen and the television.

 **Animation**

Thomson's Cathode Ray Tube

Figure 2.3

Illustration of an experiment demonstrating the charge of cathode rays. The application of an external electric field causes the electron beam to deflect toward a positive charge, implying that the cathode ray is negative.

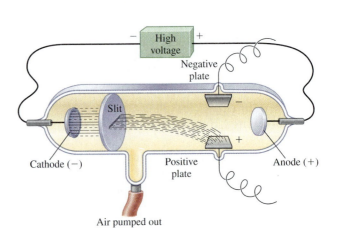

A change in the material used to fabricate the electrode discs brought about no change in the experimental results. This suggested that the ability to produce cathode rays is a characteristic of all materials.

In 1897, Thomson announced that cathode rays are streams of negative particles of energy. These particles are *electrons*. Similar experiments, conducted by Goldstein, led to the discovery of particles that are equal in charge to the electron but opposite in sign. These particles, much heavier than electrons (actually 1837 times as heavy), are called *protons*.

As we have seen, the third fundamental atomic particle is the *neutron*. It has a mass virtually identical (it is less than 1% heavier) to that of the proton and has zero charge. The neutron was first postulated in the early 1920s, but it was not until 1932 that James Chadwick demonstrated its existence with a series of experiments involving the use of small particle bombardment of nuclei.

Evidence for the Nucleus

In the early 1900s, it was believed that protons and electrons were uniformly distributed throughout the atom. However, an experiment by Hans Geiger led Ernest Rutherford (in 1911) to propose that the majority of the mass and positive charge of the atom was actually located in a small, dense region, the *nucleus*, and small, negatively charged electrons occupy a much larger volume outside the nucleus.

To understand how Rutherford's theory resulted from the experimental observations of Geiger, let us examine this experiment in greater detail. Rutherford and others had earlier demonstrated that some atoms spontaneously "decay" to produce three types of radiation: alpha (α), beta (β), and gamma (γ) radiation. This process is known as *natural radioactivity* (Figure 2.4). Geiger used radioactive materials, such as *radium*, as projectile sources, "firing" alpha particles at a thin metal foil target (gold leaf). He then observed the interaction of the metal and alpha particles with a detection screen (Figure 2.5) and found that

a. Most alpha particles pass through the foil without being deflected.
b. A small fraction of the particles were deflected, some even *directly back to the source.*

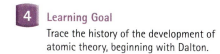

Learning Goal
Trace the history of the development of atomic theory, beginning with Dalton.

Animation
The Rutherford Experiment

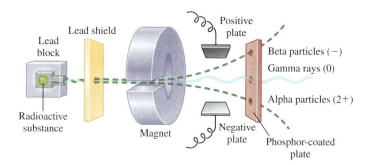

Figure 2.4

Types and characteristics of radioactive emissions. The direction taken by the radioactive emissions indicates the presence of three types of emissions: positive, negative, and neutral components.

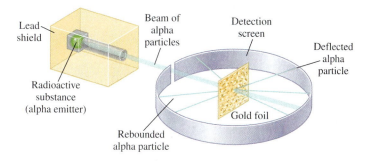

Figure 2.5

The alpha particle scattering experiment. Most alpha particles passed through the foil without being deflected; a few were deflected from their path by nuclei in the gold atoms.

Figure 2.6

(a) A model of the atom (credited to Thomson) prior to the work of Geiger and Rutherford. (b) A model of the atom supported by the alpha-particle scattering experiments of Geiger and Rutherford.

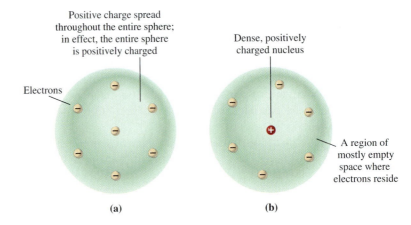

Positive charge spread throughout the entire sphere; in effect, the entire sphere is positively charged

Electrons

Dense, positively charged nucleus

A region of mostly empty space where electrons reside

(a)   (b)

Rutherford interpreted this to mean that most of the atom is empty space, because most alpha particles were not deflected. Further, most of the mass and positive charge must be located in a small, dense region; collision of the heavy and positively charged alpha particle with this small dense and positive region (the nucleus) caused the great deflections. Rutherford summarized his astonishment at observing the deflected particles: "It was almost as incredible as if you fired a 15-inch shell at a piece of tissue and it came back and hit you."

The significance of Rutherford's contribution cannot be overstated. It caused a revolutionary change in the way that scientists pictured the atom (Figure 2.6). His discovery of the nucleus is fundamental to our understanding of chemistry. Chapter 9 will provide much more information on a special branch of chemistry: nuclear chemistry.

Animation

Rutherford's Experiment and a New Atomic Model

QUESTION 2.5

Describe the experiment that provided the basis for our understanding of the nucleus.

QUESTION 2.6

Describe the series of experiments that characterized the electron.

2.3 The Periodic Law and the Periodic Table

5 Learning Goal

Recognize the important subdivisions of the periodic table: periods, groups (families), metals, and nonmetals.

In 1869, Dmitri Mendeleev, a Russian, and Lothar Meyer, a German, working independently, found ways of arranging elements in order of increasing atomic mass such that elements with similar properties were grouped together in a *table of elements*. The **periodic law** is embodied by Mendeleev's statement, "the elements if arranged according to their atomic weights (masses), show a distinct *periodicity* (regular variation) of their properties." The *periodic table* (Figure 2.7) is a visual representation of the periodic law.

Chemical and physical properties of elements correlate with the electronic structure of the atoms that make up these elements. In turn, the electronic structure correlates with position on the periodic table.

A thorough familiarity with the arrangement of the periodic table allows us to predict electronic structure and physical and chemical properties of the various elements. It also serves as the basis for understanding chemical bonding.

The concept of "periodicity" may be illustrated by examining a portion of the modern periodic table. The elements in the second row (beginning with lithium, Li, and proceeding to the right) show a marked difference in properties. However,

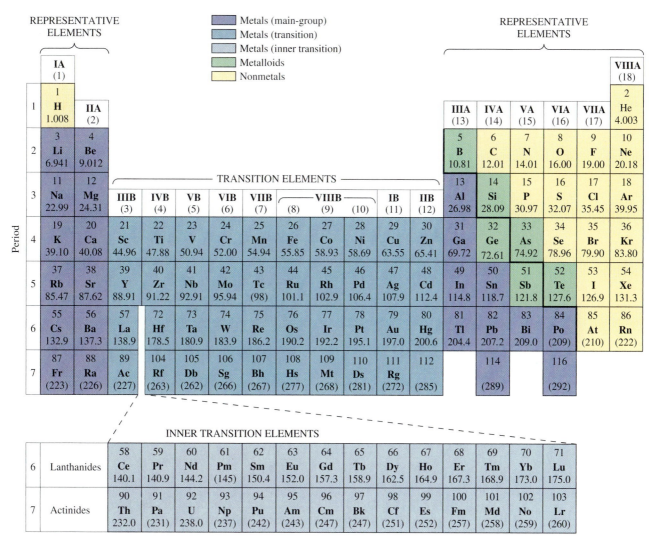

Figure 2.7

Classification of the elements: the periodic table.

---

sodium (Na) has properties similar to those of lithium, and sodium is therefore placed below lithium; once sodium is fixed in this position, the elements Mg through Ar have properties remarkably similar (though not identical) to those of the elements just above them. The same is true throughout the complete periodic table.

Mendeleev arranged the elements in his original periodic table in order of increasing atomic mass. However, as our knowledge of atomic structure increased, atomic numbers became the basis for the organization of the table. Remarkably, his table was able to predict the existence of elements not known at the time.

The modern periodic law states that *the physical and chemical properties of the elements are periodic functions of their atomic numbers.* If we arrange the elements in order of increasing number of protons, the properties of the elements repeat at regular intervals.

Not all elements are equally important to an introductory study of chemistry. Table 2.2 lists twenty of the elements that are most important to biological systems, along with their symbols and a brief description of their functions.

We will use the periodic table as our "map," just as a traveler would use a road map. A short time spent learning how to read the map (and remembering to carry it along on your trip!) is much easier than memorizing every highway and intersection.

TABLE 2.2 Summary of the Most Important Elements in Biological Systems

Element	Symbol	Significance
Hydrogen	H	Components of major biological molecules
Carbon	C	
Oxygen	O	
Nitrogen	N	
Phosphorus	P	
Sulfur	S	
Potassium	K	Produce electrolytes responsible for fluid balance and nerve transmission
Sodium	Na	
Chlorine	Cl	
Calcium	Ca	Bones, nerve function
Magnesium	Mg	
Zinc	Zn	Essential trace metals in human metabolism
Strontium	Sr	
Iron	Fe	
Copper	Cu	
Cobalt	Co	
Manganese	Mn	
Cadmium	Cd	"Heavy metals" toxic to living systems
Mercury	Hg	
Lead	Pb	

The information learned about one element relates to an entire family of elements grouped as a recognizable unit within the table.

Numbering Groups in the Periodic Table

The periodic table created by Mendeleev has undergone numerous changes over the years. These modifications occurred as more was learned about the chemical and physical properties of the elements. The labeling of groups with Roman numerals followed by the letter *A* (representative elements) or *B* (transition elements) was standard, until 1983, in North America and Russia. However, in other parts of the world, the letters *A* and *B* were used in a different way. Consequently, two different periodic tables were in widespread use. This certainly created some confusion.

In 1983, the International Union of Pure and Applied Chemistry (IUPAC), recommended that a third system, using numbers 1–18 to label the groups, replace both of the older systems. Unfortunately, multiple systems now exist and this can cause confusion for both students and experienced chemists.

The periodic tables in this textbook are "double labeled." Both the old (Roman numeral) and new (1–18) systems are used to label the groups. The label that you use is simply a guide to reading the table; the real source of information is in the structure of the table itself. The following sections will show you how to extract useful information from this structure.

> Mendeleev's original periodic table included only the elements known at the time, less than half of the current total.

Periods and Groups

5 Learning Goal
Recognize the important subdivisions of the periodic table: periods, groups (families), metals, and nonmetals.

A **period** is a horizontal row of elements in the periodic table. The periodic table consists of six periods containing 2, 8, 8, 18, 18, and 32 elements. The seventh period is still incomplete but potentially holds 32 elements. Note that the *lanthanide series*, a collection of 14 elements that are chemically and physically similar to the element lanthanum, is part of period six. It is written separately for convenience of presentation and is inserted between lanthanum (La), atomic number 57, and hafnium (Hf), atomic number 72. Similarly, the *actinide series*, consisting of 14 elements similar to the element actinium, is inserted between actinium, atomic number 89, and rutherfordium, atomic number 104.

Groups or *families* are columns of elements in the periodic table. The elements of a particular group or family share many similarities, as in a human family. The similarities extend to physical and chemical properties that are related to similarities in electronic structure (that is, the way in which electrons are arranged in an atom).

Group A elements are called **representative elements,** and Group B elements are **transition elements.** Certain families also have common names. For example, Group IA (or 1) elements are also known as the **alkali metals;** Group IIA (or 2), the **alkaline earth metals;** Group VIIA (or 17), the **halogens;** and Group VIIIA (or 18), the **noble gases.**

> *Representative elements* **are also known as** *main-group elements.* **These terms are synonymous.**

Metals and Nonmetals

A **metal** is a substance whose atoms tend to lose electrons during chemical change, forming positive ions. A **nonmetal,** on the other hand, is a substance whose atoms may gain electrons, forming negative ions.

A closer inspection of the periodic table reveals a bold zigzag line running from top to bottom, beginning to the left of boron (B) and ending between polonium (Po) and astatine (At). This line acts as the boundary between *metals,* to the left, and *nonmetals,* to the right. Elements straddling the boundary have properties intermediate between those of metals and nonmetals. These elements are referred to as **metalloids.** The metalloids include boron (B), silicon (Si), germanium (Ge), arsenic (As), antimony (Sb), tellurium (Te), polonium (Po), and astatine (At).

Metals and nonmetals may be distinguished by differences in their physical properties in addition to their chemical tendency to lose or gain electrons. Metals have a characteristic luster and generally conduct heat and electricity well. Most (except mercury) are solids at room temperature. Nonmetals, on the other hand, are poor conductors, and several are gases at room temperature.

> **5** Learning Goal
> Recognize the important subdivisions of the periodic table: periods, groups (families), metals, and nonmetals.

> **Note that aluminum (Al) is classified as a metal, not a metalloid.**

Atomic Number and Atomic Mass

The atomic number is the number of protons in the nucleus of an atom of an element. It also corresponds to the nuclear charge, the positive charge from the nucleus. Both the atomic number and the average atomic mass of each element are readily available from the periodic table. For example,

> **6** Learning Goal
> Use the periodic table to obtain information about an element.

$$20 \longleftarrow \text{atomic number}$$

$$\text{Ca} \longleftarrow \text{symbol}$$

$$\text{calcium} \longleftarrow \text{name}$$

$$40.08 \longleftarrow \text{atomic mass}$$

More detailed periodic tables may also include such information as the electron arrangement, relative sizes of atoms and ions, and most probable ion charges.

QUESTION 2.7

Refer to the periodic table (Figure 2.7) and find the following information:

a. the symbol of the element with an atomic number of 40
b. the mass of the element sodium (Na)

c. the element whose atoms contain 24 protons
d. the known element that should most resemble the as-yet undiscovered element with an atomic number of 117

QUESTION 2.8

Refer to the periodic table (Figure 2.7) and find the following information:

a. the symbol of the noble gas in period 3
b. the element in Group IVA with the smallest mass
c. the only metalloid in Group IIIA
d. the element whose atoms contain 18 protons

QUESTION 2.9

For each of the following element symbols, give the name of the element, its atomic number, and its atomic mass.

a. He
b. F
c. Mn

QUESTION 2.10

For each of the following element symbols, give the name of the element, its atomic number, and its atomic mass:

a. Mg
b. Ne
c. Se

2.4 Electron Arrangement and the Periodic Table

Animation
Electron Configurations

7 Learning Goal

Describe the relationship between the electronic structure of an element and its position in the periodic table.

A primary objective of studying chemistry is to understand the way in which atoms join together to form chemical compounds. The most important factor in this *bonding process* is the arrangement of the electrons in the atoms that are combining. The periodic table is helpful because it provides us with a great deal of information about the electron arrangement of atoms.

We have seen (p. 49) that elements in the periodic table are classified as either representative or transition. Representative elements consist of all group 1, 2, and 13–18 elements (IA–VIIIA). All others are transition elements. The guidelines that we will develop for writing electron arrangements are intended for representative elements. Electron arrangements for transition elements include several exceptions to the rule.

Valence Electrons

If we picture two spherical objects that we wish to join together, perhaps with glue, the glue can be applied to the surface and the two objects can then be brought into contact. We can extend this analogy to two atoms that are modeled as spherical objects. Although this is not a perfect analogy, it is apparent that the surface interaction is of primary importance. Although the positively charged nucleus and "interior" electrons certainly play a role in bonding, we can most

A MEDICAL Connection

Copper Deficiency and Wilson's Disease

An old adage tells us that we should consume all things in moderation. This is very true of many of the trace minerals, such as copper. Too much copper in the diet causes toxicity and too little copper results in a serious deficiency disease.

Copper is extremely important for the proper functioning of the body. It aids in the absorption of iron from the intestine and facilitates iron metabolism. It is critical for the formation of hemoglobin and red blood cells in the bone marrow. Copper is also necessary for the synthesis of collagen, a protein that is a major component of the connective tissue. It is essential to the central nervous system in two important ways. First, copper is needed for the synthesis of norepinephrine and dopamine, two chemicals that are necessary for the transmission of nerve signals. Second, it is required for the deposition of the myelin sheath (a layer of insulation) around nerve cells. Release of cholesterol from the liver depends on copper, as does bone development and proper function of the immune and blood clotting systems.

The estimated safe and adequate daily dietary intake (ESADDI) for adults is 1.5–3.0 mg. Meats, cocoa, nuts, legumes, and whole grains provide significant amounts of copper. The accompanying table shows the amount of copper in some common foods.

Although getting enough copper in the diet would appear to be relatively simple, it is estimated that Americans often ingest only marginal levels of copper, and we absorb only 25–40% of that dietary copper. Despite these facts, it appears that copper deficiency is not a serious problem in the United States.

Individuals who are at risk for copper deficiency include people who are recovering from abdominal surgery, which causes decreased absorption of copper from the intestine. Others at risk are premature babies and people who are sustained solely by intravenous feedings that are deficient in copper. In addition, people who ingest high doses of antacids or take excessive supplements of zinc, iron, or vitamin C can develop copper deficiency because of reduced copper absorption. Because copper is involved in so many processes in the body, it is not surprising that the symptoms of copper deficiency are many and diverse. They include anemia; decreased red and white blood cell counts; heart disease; increased levels of serum cholesterol; loss of bone; defects in the nervous system, immune system, and connective tissue; and abnormal hair.

Just as too little copper causes serious problems, so does an excess of copper. At doses greater than about 15 mg, copper causes toxicity that results in vomiting. The effects of extended exposure to excess copper are apparent when we look at Wilson's disease. This is a genetic disorder in which excess copper cannot be removed from the body and accumulates in the cornea of the eye, liver, kidneys, and brain. The symptoms include a greenish ring around the cornea, cirrhosis of the liver, copper in the urine, dementia and paranoia, drooling, and progressive tremors. As a result of the condition, the victim generally dies in early adolescence. Wilson's disease can be treated

Copper in One-Cup Portions of Food

Food	Mass of copper (mg)
Sesame seeds	5.88
Cashews	3.04
Oysters	2.88
Sunflower seeds	2.52
Peanuts, roasted	1.85
Crabmeat	1.71
Walnuts	1.28
Almonds	1.22
Cereal, All Bran	0.98
Tuna fish	0.93
Wheat germ	0.70
Prunes	0.69
Kidney beans	0.56
Dried apricots	0.56
Lentils, cooked	0.54
Sweet potato, cooked	0.53
Dates	0.51
Whole milk	0.50
Raisins	0.45
Cereal, C. W. Post, Raisins	0.40
Grape Nuts	0.38
Whole-wheat bread	0.34
Cooked cereal, Roman Meal	0.32

Source: David C. Nieman, Diane E. Butterworth, and Catherine N. Nieman, *Nutrition,* Revised First Edition. Copyright 1992 Wm. C. Brown Communications, Inc., Dubuque, Iowa. All Rights Reserved. Reprinted by permission.

with moderate success if it is recognized early, before permanent damage has occurred to any tissues. The diet is modified to reduce the intake of copper; for instance, such foods as chocolate are avoided. In addition, the drug penicillamine is administered. This compound is related to the antibiotic penicillin but has no antibacterial properties; rather it has the ability to bind to copper in the blood and enhance its excretion by the kidneys into the urine. In this way the brain degeneration and tissue damage that are normally seen with the disease can be lessened.

FOR FURTHER UNDERSTANDING

Why is there an upper limit on the recommended daily amount of copper?

Iron is another essential trace metal in our diet. Go to the Web and find out if upper limits exist for daily iron consumption.

easily understand the process by considering only the outermost electrons. We refer to these as *valence electrons.* **Valence electrons** are the outermost electrons in an atom, which are involved, or have the potential to become involved, in the bonding process.

> Metals tend to have fewer valence electrons, and nonmetals tend to have more valence electrons.

For representative elements the number of valence electrons in an atom corresponds to the number of the *group* or *family* in which the atom is found. For example, elements such as hydrogen and sodium (in fact, all alkali metals, Group IA or 1) have one valence electron. From left to right in period 2, beryllium, Be (Group IIA or 2), has two valence electrons; boron, B (Group IIIA or 3), has three; carbon, C (Group IVA or 4), has four; and so forth.

An atom may have electrons in several different energy levels. These energy levels are symbolized by n; the lowest energy level is assigned a value of $n = 1$. Each energy level may contain up to a fixed maximum number of electrons. For example, the $n = 1$ energy level may contain a maximum of two electrons. Thus hydrogen (atomic number = 1) has one electron and helium (atomic number = 2) has two electrons in the $n = 1$ level. Only these elements have electrons *exclusively* in the first energy level:

$n = 1$ $n = 1$

Hydrogen: one-electron atom   **Helium:** two-electron atom

These two elements make up the first period of the periodic table. Period 1 contains all elements whose *maximum* energy level is $n = 1$. In other words, the $n = 1$ level is the *outermost* electron region for hydrogen and helium. Hydrogen has one electron and helium has two electrons in the $n = 1$ level.

The valence electrons of elements in the second period are in the $n = 2$ energy level. You must fill the $n = 1$ level with two electrons before adding electrons to the next level. The third electron of lithium (Li) and the remaining electrons of the second period elements must be in the $n = 2$ level and are considered the valence electrons for lithium and remaining second period elements.

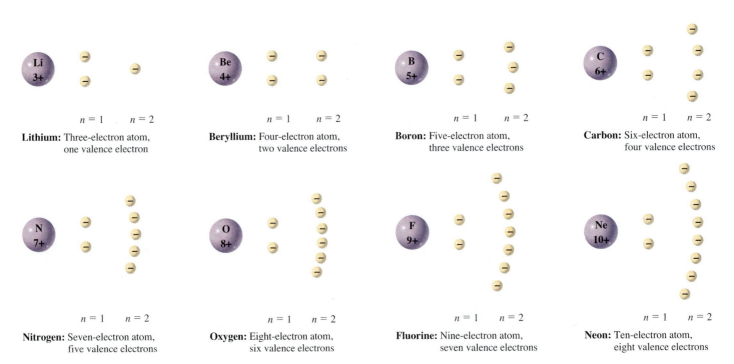

$n = 1$ $n = 2$

Lithium: Three-electron atom, one valence electron

$n = 1$ $n = 2$

Beryllium: Four-electron atom, two valence electrons

$n = 1$ $n = 2$

Boron: Five-electron atom, three valence electrons

$n = 1$ $n = 2$

Carbon: Six-electron atom, four valence electrons

$n = 1$ $n = 2$

Nitrogen: Seven-electron atom, five valence electrons

$n = 1$ $n = 2$

Oxygen: Eight-electron atom, six valence electrons

$n = 1$ $n = 2$

Fluorine: Nine-electron atom, seven valence electrons

$n = 1$ $n = 2$

Neon: Ten-electron atom, eight valence electrons

TABLE 2.3 The Electron Distribution for the First Twenty Elements of the Periodic Table						
Element Symbol and Name	Total Number of Electrons	Total Number of Valence Electrons	Electrons in $n = 1$	Electrons in $n = 2$	Electrons in $n = 3$	Electrons in $n = 4$
H, hydrogen	1	1	1	0	0	0
He, helium	2	2	2	0	0	0
Li, lithium	3	1	2	1	0	0
Be, beryllium	4	2	2	2	0	0
B, boron	5	3	2	3	0	0
C, carbon	6	4	2	4	0	0
N, nitrogen	7	5	2	5	0	0
O, oxygen	8	6	2	6	0	0
F, fluorine	9	7	2	7	0	0
Ne, neon	10	8	2	8	0	0
Na, sodium	11	1	2	8	1	0
Mg, magnesium	12	2	2	8	2	0
Al, aluminum	13	3	2	8	3	0
Si, silicon	14	4	2	8	4	0
P, phosphorus	15	5	2	8	5	0
S, sulfur	16	6	2	8	6	0
Cl, chlorine	17	7	2	8	7	0
Ar, argon	18	8	2	8	8	0
K, potassium	19	1	2	8	8	1
Ca, calcium	20	2	2	8	8	2

Third period elements fill the $n = 3$ level. Argon, the last element in that period has 18 total electrons, 8 in the $n = 3$ level, and, consequently, 8 valence electrons.

The electron distribution (arrangement) of the first twenty elements of the periodic table is given in Table 2.3.

Two general rules of electron distribution are based on the periodic law:

RULE 1: The number of valence electrons in an atom equals the Roman numeral group number for all representative (A group) elements. ■

RULE 2: The energy level ($n = 1, 2$, etc.) in which the valence electrons are located corresponds to the *period* in which the element may be found. ■

For example,

Group IA	*Group IIA*	*Group IIIA*	*Group VIIA*
Li	Ca	Al	Br
one valence electron in $n = 2$ energy level; period 2	two valence electrons in $n = 4$ energy level; period 4	three valence electrons in $n = 3$ energy level; period 3	seven valence electrons in $n = 4$ energy level; period 4

EXAMPLE 2.4 Determining Electron Arrangement

Provide the total number of electrons, total number of valence electrons, and energy level in which the valence electrons are found for the silicon (Si) atom.

7 **Learning Goal**

Describe the relationship between the electronic structure of an element and its position in the periodic table.

———— *Continued—*

EXAMPLE **2.4** *—Continued*

SOLUTION

Step 1. Determine the position of silicon in the periodic table. Silicon is found in Group IVA and period 3 of the table. Silicon has an atomic number of 14.

Step 2. The atomic number provides the number of electrons in an atom. Silicon therefore has 14 electrons.

Step 3. Because silicon is in Group IV, only 4 of the 14 electrons are valence electrons.

Step 4. Silicon has 2 electrons in $n = 1$, 8 electrons in $n = 2$, and 4 electrons in the $n = 3$ level.

QUESTION 2.11

For each of the following elements, provide the *total* number of electrons and *valence* electrons in its atom:

   a. Na
   b. Mg
   c. S
   d. Cl
   e. Ar

QUESTION 2.12

For each of the following elements, provide the *total* number of electrons and *valence* electrons in its atom:

   a. K
   b. F
   c. P
   d. O
   e. Ca

The Quantum Mechanical Atom

Animation
Atom Structure

Niels Bohr hypothesized that certain fixed **energy levels** could be occupied by electrons surrounding each atomic nucleus. He also believed that each level was defined by a spherical *orbit* around the nucleus, located at a specific distance from the nucleus. The concept of certain fixed energy levels is referred to as the **quantization** of energy. The implication is that only these orbits, or *quantum levels*, as described by Max Planck, are allowed locations for electrons.

This has become known as the Bohr theory and Niels Bohr was honored with the Nobel Prize, recognizing his insight and his contribution to our understanding of the arrangement of electrons in the atom. However, the success of Bohr's theory was short-lived. Atoms with more than one electron could not be explained by Bohr's theory.

The Bohr theory, even with its limitations, served as a necessary first step. His ideas were refined and expanded by an Austrian physicist, Erwin Schrödinger. He described electrons in atoms in probability terms, developing equations that emphasize the wavelike character of electrons. Although Schrödinger's approach was founded on complex mathematics, we can readily use models of electron probability regions to enable us to gain a reasonable insight into atomic structure without the need to understand the underlying mathematics.

Schrödinger's theory, often described as quantum mechanics, incorporates Bohr's principal energy levels ($n = 1, 2$, and so forth); however, it is proposed that each of these levels is made up of one or more sublevels. Each sublevel, in turn, contains one or more atomic orbitals. In the following section, we shall look at each of these regions in more detail and learn to predict how electrons are arranged in stable atoms.

Energy Levels and Sublevels

Principal Energy Levels

The principal energy levels are designated $n = 1, 2, 3$, and so forth. The number of possible sublevels in a principal energy level is also equal to n. When $n = 1$, there can be only one sublevel; $n = 2$ allows two sublevels, and so forth.

The total electron capacity of a principal level is $2(n)^2$. For example,

$n = 1$	$2(1)^2$	Capacity $= 2e^-$
$n = 2$	$2(2)^2$	Capacity $= 8e^-$
$n = 3$	$2(3)^2$	Capacity $= 18e^-$

Sublevels

A **sublevel** is a set of equal-energy orbitals within a principal energy level. The sublevels, or subshells, are symbolized as s, p, d, f, and so forth.

We specify both the principal energy level and type of sublevel when describing the location of an electron—for example, $1s, 2s, 2p$. Energy level designations for the first four principal energy levels follow:

- The first principal energy level ($n = 1$) has one possible sublevel: $1s$.
- The second principal energy level ($n = 2$) has two possible sublevels: $2s$ and $2p$.
- The third principal energy level ($n = 3$) has three possible sublevels: $3s, 3p$, and $3d$.
- The fourth principal energy level ($n = 4$) has four possible sublevels: $4s, 4p, 4d$, and $4f$.

These sublevels increase in energy in the following order:

$$s < p < d < f$$

Orbitals

An **atomic orbital** is a specific region of a sublevel containing a maximum of two electrons.

Figure 2.8 depicts a model of an s orbital. It is spherically symmetrical, much like a Ping-Pong ball. Its volume represents a region where there is a high probability of finding electrons of similar energy. This probability decreases as we approach the outer region of the atom. The nucleus is at the center of the s orbital. At that point, the probability of finding the electron is zero; electrons cannot reside in the nucleus. Only one s orbital can be found in any n level. Atoms with many electrons, occupying a number of n levels, have an s orbital in each n level. Consequently, $1s, 2s, 3s$, and so forth are possible orbitals.

Figure 2.9 describes the shapes of the three possible p orbitals within a given level. Each has the same shape, and that shape appears much like a dumbbell;

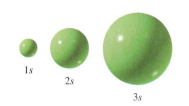

Figure 2.8

Representation of s orbitals.

Figure 2.9

Representation of the three p orbitals, p_x, p_y, and p_z.

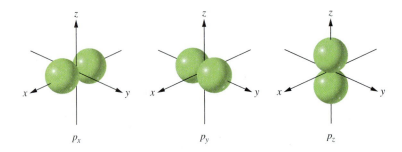

p_x p_y p_z

these three orbitals differ only in the direction they extend into space. Imaginary coordinates x, y, and z are superimposed on these models to emphasize this fact. These three orbitals, termed p_x, p_y, and p_z, may coexist in a single atom.

In a similar fashion, five possible d orbitals and seven possible f orbitals exist. The d orbitals exist only in $n = 3$ and higher principal energy levels; f orbitals exist only in $n = 4$ and higher principal energy levels. Because of their complexity, we will not consider the shapes of d and f orbitals.

Electrons in Sublevels

We can deduce the maximum electron capacity of each sublevel based on the information just given.

For the s sublevel:

$$1 \text{ orbital} \times \frac{2e^- \text{ capacity}}{\text{orbital}} = 2e^- \text{ capacity}$$

For the p sublevel:

$$3 \text{ orbitals} \times \frac{2e^- \text{ capacity}}{\text{orbital}} = 6e^- \text{ capacity}$$

For the d sublevel:

$$5 \text{ orbitals} \times \frac{2e^- \text{ capacity}}{\text{orbital}} = 10e^- \text{ capacity}$$

For the f sublevel:

$$7 \text{ orbitals} \times \frac{2e^- \text{ capacity}}{\text{orbital}} = 14e^- \text{ capacity}$$

Electron Spin

Section 2.2 discusses the properties of electrons demonstrated by Thomson.

As we have noted, each atomic orbital has a maximum capacity of two electrons. The electrons are perceived to *spin* on an imaginary axis, and the two electrons in the same orbital must have opposite spins: clockwise and counterclockwise. Their behavior is analogous to two ends of a magnet. Remember, electrons have magnetic properties. The electrons exhibit sufficient magnetic attraction to hold themselves together despite the natural repulsion that they "feel" for each other, owing to their similar charge (remember, like charges repel). Electrons must therefore have opposite spins to coexist in an orbital. A pair of electrons in one orbital that possess opposite spins are referred to as *paired* electrons.

8 Learning Goal

Write electron configurations for atoms of the most commonly occurring elements.

Electron Configuration and the Aufbau Principle

The arrangement of electrons in atomic orbitals is referred to as the atom's **electron configuration.** The *aufbau*, or building up, *principle* helps us to represent the electron configuration of atoms of various elements. According to this principle, electrons fill the lowest energy orbital that is available first. We should also recall that the maximum capacity of an s level is two, that of a p level is six, that of a d level is ten, and that of an f level is fourteen electrons. Consider the following guidelines for writing electron configurations:

Guidelines for Writing Electron Configurations

- Obtain the total number of electrons in the atoms from the atomic number found in the periodic table.
- Electrons in atoms occupy the lowest energy orbitals that are available, beginning with $1s$.
- Each principal energy level, n, can contain only n subshells.
- Each sublevel is composed of one (s) or more (three p, five d, seven f) orbitals.
- No more than two electrons can be placed in any orbital.
- The maximum number of electrons in any principal energy level is $2(n)^2$.
- The theoretical order of orbital filling is depicted in Figure 2.10.

Now let us look at several elements:

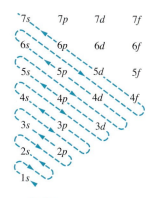

Figure 2.10

A useful way to remember the filling order for electrons in atoms. Begin adding electrons at the bottom (lowest energy) and follow the arrows. Remember: no more than two electrons in each orbital.

Hydrogen

Hydrogen is the simplest atom; it has only one electron. That electron must be in the lowest principal energy level ($n = 1$) and the lowest orbital (s). We indicate the number of electrons in a region with a *superscript,* so we write $1s^1$.

Helium

Helium has two electrons, which will fill the lowest energy level. The ground state (lowest energy) electron configuration for helium is $1s^2$.

Lithium

Lithium has three electrons. The first two are configured as helium. The third must go into the orbital of the lowest energy in the second principal energy level; therefore the configuration is $1s^2\,2s^1$.

Beryllium Through Neon

The second principal energy level can contain eight electrons [$2(2)^2$], two in the s level and six in the p level. The "building up" process results in

Be	$1s^2\,2s^2$
B	$1s^2\,2s^2\,2p^1$
C	$1s^2\,2s^2\,2p^2$
N	$1s^2\,2s^2\,2p^3$
O	$1s^2\,2s^2\,2p^4$
F	$1s^2\,2s^2\,2p^5$
Ne	$1s^2\,2s^2\,2p^6$

Sodium Through Argon

Electrons in these elements retain the basic $1s^2\,2s^2\,2p^6$ arrangement of the preceding element, neon; new electrons enter the third principal energy level:

Na	$1s^2\,2s^2\,2p^6\,3s^1$
Mg	$1s^2\,2s^2\,2p^6\,3s^2$

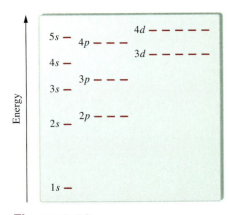

Figure 2.11

An orbital energy-level diagram. Electrons fill orbitals in the order of increasing energy.

8 **Learning Goal**

Write electron configurations for atoms of the most commonly occurring elements.

Al	$1s^2\,2s^2\,2p^6\,3s^2\,3p^1$
Si	$1s^2\,2s^2\,2p^6\,3s^2\,3p^2$
P	$1s^2\,2s^2\,2p^6\,3s^2\,3p^3$
S	$1s^2\,2s^2\,2p^6\,3s^2\,3p^4$
Cl	$1s^2\,2s^2\,2p^6\,3s^2\,3p^5$
Ar	$1s^2\,2s^2\,2p^6\,3s^2\,3p^6$

By knowing the order of filling of atomic orbitals, lowest to highest energy, you may write the electron configuration for any element. The order of orbital filling can be represented by the diagram in Figure 2.10. Such a diagram provides an easy way of predicting the electron configuration of the elements. Remember that the diagram is based on an energy scale, with the lowest energy orbital at the beginning of the "path" and the highest energy orbital at the end of the "path." An alternative way of representing orbital energies is through the use of an energy level diagram, such as the one in Figure 2.11.

EXAMPLE 2.5 **Writing the Electron Configuration of Tin**

Tin, Sn, has an atomic number of 50; thus we must place fifty electrons in atomic orbitals. We must also remember the total electron capacities of orbital types: s, 2; p, 6; d, 10; and f, 14. The first principal energy level has one sublevel, the second has two sublevels, and so on. The order of filling (Figure 2.10 or 2.11) is 1s, 2s, 2p, 3s, 3p, 4s, 3d, 4p. The electron configuration is as follows:

$$1s^2\,2s^2\,2p^6\,3s^2\,3p^6\,4s^2\,3d^{10}\,4p^6\,5s^2\,4d^{10}\,5p^2$$

As a check, count electrons in the electron configuration (add all of the super-scripted numbers) to see that we have accounted for all fifty electrons of the Sn atom.

QUESTION 2.13

Give the electron configuration for an atom of

a. sulfur
b. calcium

QUESTION 2.14

Give the electron configuration for an atom of

a. potassium
b. phosphorus

Shorthand Electron Configurations

As we noted earlier, the electron configuration for the sodium atom (Na, atomic number 11) is

$$1s^2\,2s^2\,2p^6\,3s^1$$

The electron configuration for the preceding noble gas, neon (Ne, atomic number 10), is

$$1s^2\, 2s^2\, 2p^6$$

The electron configuration for sodium is really the electron configuration of Ne, with $3s^1$ added to represent one additional electron. So it is permissible to write

[Ne] $3s^1$ as equivalent to $1s^2\, 2s^2\, 2p^6\, 3s^1$

[Ne] $3s^1$ is the *shorthand electron configuration* for sodium.
Similarly,

[Ne] $3s^2$ representing Mg,

[Ne] $3s^2\, 3p^5$ representing Cl, and

[Ar] $4s^1$ representing K

are valid electron configurations.

The use of abbreviated electron configurations, in addition to being faster and easier to write, serves to highlight the valence electrons, those electrons involved in bonding. The symbol of the noble gas represents the *core*, nonvalence electrons, and the valence electron configuration follows the noble gas symbol.

QUESTION 2.15

Give the *shorthand* electron configuration for each atom in Question 2.13.

QUESTION 2.16

Give the *shorthand* electron configuration for each atom in Question 2.14.

2.5 The Octet Rule

Elements in the last family, the noble gases, have either two valence electrons (helium) or eight valence electrons (neon, argon, krypton, xenon, and radon). These elements are extremely stable and were often termed *inert gases* because they do not readily bond to other elements, although they can be made to do so under extreme experimental conditions. A full $n = 1$ energy level (as in helium) or an outer *octet* of electrons (eight valence electrons, as in all of the other noble gases) is responsible for this unique stability.

Atoms of elements in other groups are more reactive than the noble gases because in the process of chemical reaction, they are trying to achieve a more stable "noble gas" configuration by gaining or losing electrons. This is the basis of the **octet rule** which states that elements usually react to attain the electron configuration of the noble gas closest to them in the periodic table (a stable octet of electrons). In chemical reactions, they will gain, lose, or share the minimum number of electrons necessary to attain this more stable energy state. The octet rule, although simple in concept, is a remarkably reliable predictor of chemical change, especially for representative elements.

> We may think of stability as a type of contentment; a noble gas atom does not need to rearrange its electrons or lose or gain any electrons to get to a more stable, lower energy, or more "contented" configuration.

Ion Formation and the Octet Rule

Metals and nonmetals differ in the way in which they form ions. Metallic elements (located at the left of the periodic table) tend to form positively charged ions

9 Learning Goal
Use the octet rule to predict the charge of common cations and anions.

called *cations*. Positive ions are formed when an atom loses one or more electrons, for example,

$$Na \longrightarrow Na^+ \quad + \quad e^-$$

Sodium atom
$(11e^-, 1 \text{ valence } e^-)$

Sodium ion
$(10e^-)$

$$Mg \longrightarrow Mg^{2+} \quad + \quad 2e^-$$

Magnesium atom
$(12e^-, 2 \text{ valence } e^-)$

Magnesium ion
$(10e^-)$

$$Al \longrightarrow Al^{3+} \quad + \quad 3e^-$$

Aluminum atom
$(13e^-, 3 \text{ valence } e^-)$

Aluminum ion
$(10e^-)$

In each of these cases, the atom has lost *all* of its valence electrons. The resulting ions have the same number of electrons as the nearest noble gas atom.

These ions are particularly stable. Each ion is **isoelectronic** (that is, it has the same number of electrons) with its nearest noble gas neighbor and has an octet of electrons in its outermost energy level.

> Recall that the prefix *iso* (Greek *isos*) means equal.

Sodium is typical of each element in its group. Knowing that sodium forms a 1+ ion leads to the prediction that H, Li, K, Rb, Cs, and Fr also will form 1+ ions. Furthermore, magnesium, which forms a 2+ ion, is typical of each element in its group; Be^{2+}, Ca^{2+}, Sr^{2+}, and so forth are the resulting ions.

Nonmetallic elements, located at the right of the periodic table, tend to gain electrons to become isoelectronic with the nearest noble gas element, forming negative ions called *anions*.

> Section 3.2 discusses the naming of ions.

Consider:

$$F \quad + \quad 1e^- \longrightarrow F^- \quad (\text{isoelectronic with Ne, } 10e^-)$$

Fluorine atom
$(9e^-, 7 \text{ valence } e^-)$

Fluoride ion
$(10e^-)$

> The ion of fluorine is the *fluoride ion;* the ion of oxygen is the *oxide ion;* and the ion of nitrogen is the *nitride ion.*

$$O \quad + \quad 2e^- \longrightarrow O^{2-} \quad (\text{isoelectronic with Ne, } 10e^-)$$

Oxygen atom
$(8e^-, 6 \text{ valence } e^-)$

Oxide ion
$(10e^-)$

$$N \quad + \quad 3e^- \longrightarrow N^{3-} \quad (\text{isoelectronic with Ne, } 10e^-)$$

Nitrogen atom
$(7e^-, 5 \text{ valence } e^-)$

Nitride ion
$(10 e^-)$

As in the case of positive ion formation, each of these negative ions has an octet of electrons in its outermost energy level.

The element fluorine, forming F^-, indicates that the other halogens, Cl, Br, and I, behave as a true family and form Cl^-, Br^-, and I^- ions. Also, oxygen and the other nonmetals in its group form 2^- ions; nitrogen and phosphorus form 3^- ions. It is important to recognize that ions are formed by gain or loss of *electrons*. No change occurs in the nucleus; the number of protons remains the same.

QUESTION 2.17

Give the charge of the most probable ion resulting from each of the following elements. With what element is the ion isoelectronic?

a. Ca   d. Mg
b. Sr   e. P
c. S

A MEDICAL Connection | Dietary Calcium

"Drink your milk!" "Eat all of your vegetables!" These imperatives are almost universal memories from our childhood. Our parents knew that calcium, present in abundance in these foods, was an essential element for the development of strong bones and healthy teeth.

Recent studies, spanning the fields of biology, chemistry, and nutrition science indicate that the benefits of calcium go far beyond bones and teeth. This element has been found to play a role in the prevention of disease throughout our bodies.

Calcium is the most abundant mineral (metal) in the body. It is ingested as the calcium ion (Ca^{2+}) either in its "free" state or "combined," as a part of a larger compound; calcium dietary supplements often contain ions in the form of calcium carbonate. The acid naturally present in the stomach produces the calcium ion:

$$CaCO_3 + 2H^+ \longrightarrow Ca^{2+} + H_2O + CO_2$$

calcium   stomach   calcium water carbon
carbonate acid   ion   dioxide

Vitamin D serves as the body's regulator of calcium ion uptake, release, and transport in the body. (You can find further information online at www.mhhe.com/denniston, in "Lipid-Soluble Vitamins.")

Calcium is responsible for a variety of body functions including:

- transmission of nerve impulses
- release of "messenger compounds" that enable communication among nerves
- blood clotting
- hormone secretion
- growth of living cells throughout the body

The body's storehouse of calcium is bone tissue. When the supply of calcium from external sources, the diet, is insufficient, the body uses a mechanism to compensate for this shortage. With vitamin D in a critical role, this mechanism removes calcium from bone to enable other functions to continue to take place. It is evident then that prolonged dietary calcium deficiency can weaken the bone structure. Unfortunately, current studies show that as many as 75% of the American population may not be consuming sufficient amounts of calcium. Developing an understanding of the role of calcium in premenstrual syndrome, cancer, and blood pressure regulation is the goal of three current research areas.

Calcium and premenstrual syndrome (PMS). Dr. Susan Thys-Jacobs, a gynecologist at St. Luke's-Roosevelt Hospital Center in New York City, and colleagues at eleven other medical centers are conducting a study of calcium's ability to relieve the discomfort of PMS. They believe that women with chronic PMS have calcium blood levels that are normal only because calcium is continually being removed from the bone to maintain an adequate supply in the blood. To complicate the situation, vitamin D levels in many young women are very low (as much as 80% of a person's vitamin D is made in the skin, upon exposure to sunlight; many of us now minimize our exposure to the sun because of concerns about ultraviolet radiation and skin cancer). Because vitamin D plays an essential role in calcium metabolism, even if sufficient calcium is consumed, it may not be used efficiently in the body.

Colon cancer. The colon is lined with a type of cell (epithelial cell) that is similar to those that form the outer layers of skin. Various studies have indicated that by-products of a high-fat diet are irritants to these epithelial cells and produce abnormal cell growth in the colon. Dr. Martin Lipkin, Rockefeller University in New York, and his colleagues have shown that calcium ions may bind with these irritants, reducing their undesirable effects. It is believed that a calcium-rich diet, low in fat, and perhaps use of a calcium supplement can prevent or reverse this abnormal colon cell growth, delaying or preventing the onset of colon cancer.

Blood pressure regulation. Dr. David McCarron, a blood pressure specialist at the Oregon Health Sciences University, believes that dietary calcium levels may have a significant influence on hypertension (high blood pressure). Preliminary studies show that a diet rich in low-fat dairy products, fruits, and vegetables, all high in calcium, may produce a significant lowering of blood pressure in adults with mild hypertension.

The take-home lesson appears clear: a high calcium, low fat diet promotes good health in many ways. Once again, our parents were right!

FOR FURTHER UNDERSTANDING

Distinguish between "free" and "combined" calcium in the diet.

Why might calcium supplements be ineffective in treating all cases of osteoporosis?

QUESTION 2.18

Which of the following pairs of atoms and ions are isoelectronic?

a. Cl^-, Ar   d. Li^+, Ne
b. Na^+, Ne   e. O^{2-}, F^-
c. Mg^{2+}, Na^+

The transition metals tend to form positive ions by losing electrons, just like the representative metals. Metals, whether representative or transition, share this characteristic. However, the transition elements are characterized as "variable valence" elements; depending on the type of substance with which they react, they may form more than one stable ion. For example, iron has two stable ionic forms:

$$Fe^{2+} \text{ and } Fe^{3+}$$

Copper can exist as

$$Cu^{+} \text{ and } Cu^{2+}$$

and elements such as vanadium, V, and manganese, Mn, can each form four different stable ions.

Predicting the charge of an ion or the various possible ions for a given transition metal is not an easy task. Energy differences between valence electrons of transition metals are small and not easily predicted from the position of the element in the periodic table. In fact, in contrast to representative metals, the transition metals show great similarities within a *period* as well as within a *group*.

2.6 Trends in the Periodic Table

Atomic Size

Many atomic properties correlate with electronic structure, hence, with their position in the periodic table. Given the fact that interactions among multiple charged particles are very complex, we would not expect the correlation to be perfect. Nonetheless, the periodic table remains an excellent guide to the prediction of properties.

If our model of the atom is a tiny sphere whose radius is determined by the distance between the center of the nucleus and the boundary of the region where the valence electrons have a probability of being located, the size of the atom will be determined principally by two factors.

1. The energy level (*n* level) in which the outermost electron(s) is (are) found increases as we go *down* a group. (Recall that the outermost *n* level correlates with period number.) Thus the size of atoms should increase from top to bottom of the periodic table as we fill successive energy levels of the atoms with electrons (Figure 2.12).
2. As the magnitude of the positive charge of the nucleus increases, its "pull" on all of the electrons increases, and the electrons are drawn closer to the

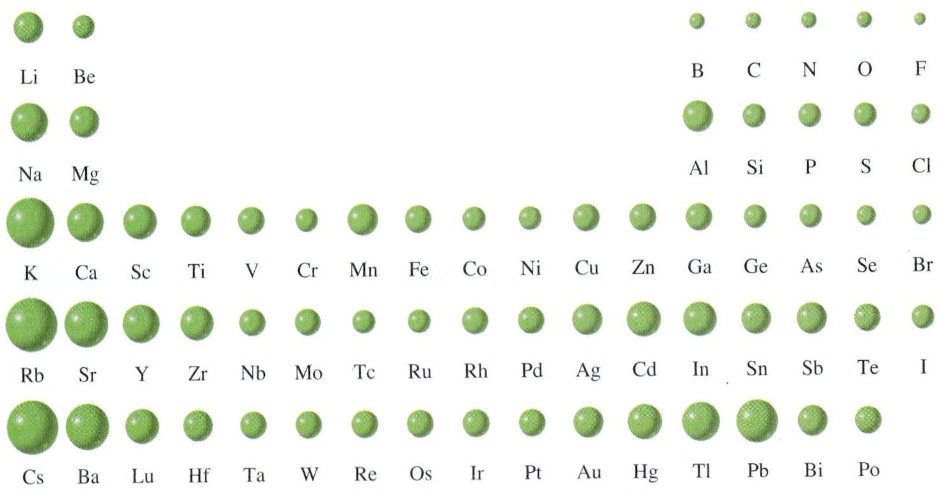

10 Learning Goal

Use the periodic table and its predictive power to estimate the relative sizes of atoms and ions, as well as relative magnitudes of ionization energy and electron affinity.

The radius of an atom is traditionally defined as one-half the distance between atoms in a covalent bond. The covalent bond is discussed in Section 3.1.

Figure 2.12

Variation in the size of atoms as a function of their position in the periodic table. Note particularly the decrease in size from left to right in the periodic table and the increase in size as we proceed down the table, although some exceptions do exist. (Lanthanide and actinide elements are not included here.)

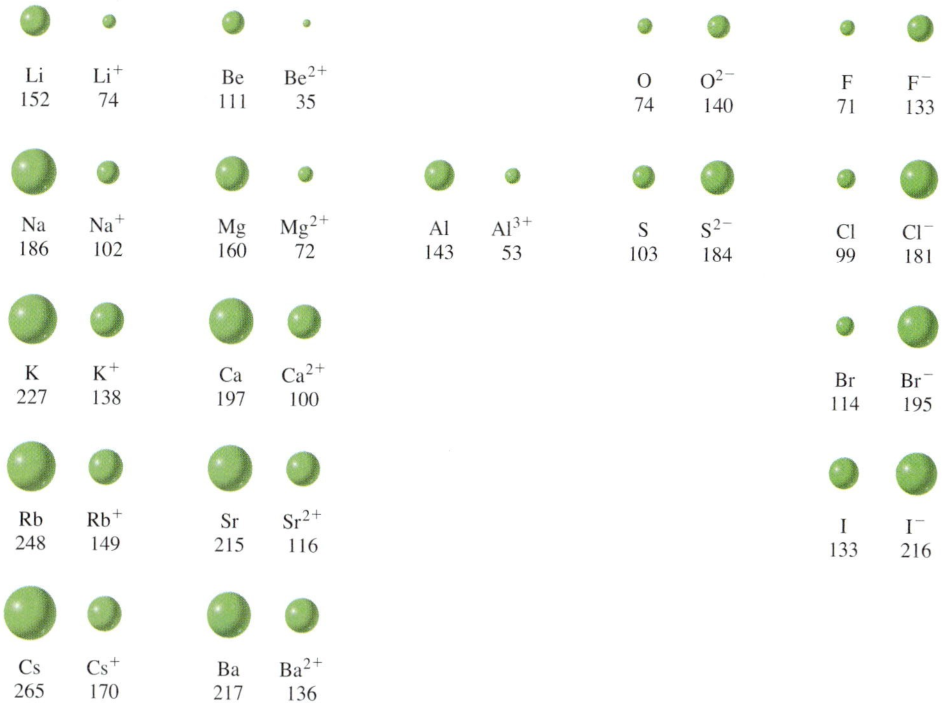

Li 152 · Li⁺ 74 · Be 111 · Be²⁺ 35 · O 74 · O²⁻ 140 · F 71 · F⁻ 133

Na 186 · Na⁺ 102 · Mg 160 · Mg²⁺ 72 · Al 143 · Al³⁺ 53 · S 103 · S²⁻ 184 · Cl 99 · Cl⁻ 181

K 227 · K⁺ 138 · Ca 197 · Ca²⁺ 100 · Br 114 · Br⁻ 195

Rb 248 · Rb⁺ 149 · Sr 215 · Sr²⁺ 116 · I 133 · I⁻ 216

Cs 265 · Cs⁺ 170 · Ba 217 · Ba²⁺ 136

Figure 2.13

Relative size of ions and their parent atoms. Atomic radii are provided in units of picometers.

nucleus. This results in contraction of the atomic radius and therefore a decrease in atomic size. This effect is apparent as we go *across* the periodic table within a period. Atomic size decreases from left to right in the periodic table. See how many exceptions you can find in Figure 2.12.

Animation
Atomic Radius

Ion Size

Positive ions (cations) are smaller than the parent atom. The cation has more protons than electrons (an increased nuclear charge). The excess nuclear charge pulls the remaining electrons closer to the nucleus. Also, cation formation often results in the loss of all outer shell electrons, resulting in a significant decrease in radius.

Negative ions (anions) are larger than the parent atom. The anion has more electrons than protons. Owing to the excess negative charge, the nuclear "pull" on each individual electron is reduced. The electrons are held less tightly, resulting in a larger anion radius in contrast to the neutral atom.

Ions with multiple positive charge (such as Cu^{2+}) are even *smaller* than their corresponding monopositive ion (Cu^+); ions with multiple negative charge (such as O^{2-}) are *larger* than their corresponding less negative ion.

Figure 2.13 depicts the relative sizes of several atoms and their corresponding ions.

Ionization Energy

The energy required to remove an electron from an isolated atom is the **ionization energy.** The process for sodium is represented as follows:

$$\text{ionization energy} + Na \longrightarrow Na^+ + e^-$$

The magnitude of the ionization energy should correlate with the strength of the attractive force between the nucleus and the outermost electron.

- Reading *down* a group, note that the ionization energy decreases because the atom's size is increasing. The outermost electron is progressively farther from the nuclear charge, hence easier to remove.

• Reading *across* a period, note that atomic size decreases because the outermost electrons are closer to the nucleus, more tightly held, and more difficult to remove. Therefore the ionization energy generally increases.

A correlation does exist between trends in atomic size and ionization energy. Atomic size generally *decreases* from the bottom to top of a group and from left to right in a period. Ionization energies generally *increase* in the same periodic way. Note also that ionization energies are highest for the noble gases (Figure 2.14a). A high value for ionization energy means that it is difficult to remove electrons from the atom, and this, in part, accounts for the extreme stability and nonreactivity of noble gases.

Electron Affinity

The energy released when a single electron is added to an isolated atom is the **electron affinity.** If we consider ionization energy in relation to positive ion formation (remember that the magnitude of the ionization energy tells us the ease of *removal* of an electron, hence the ease of forming positive ions), then electron affinity provides a measure of the ease of forming negative ions. A large electron affinity (energy released) indicates that the atom becomes more stable as it becomes a negative ion (through gaining an electron). Consider the gain of an electron by a bromine atom:

$$Br + e^- \longrightarrow Br^- + energy$$

Electron affinity

Periodic trends for electron affinity are as follows:

• Electron affinities generally decrease down a group.
• Electron affinities generally increase across a period.

Remember these trends are not absolute. Exceptions exist, as seen in the irregularities in Figure 2.14b.

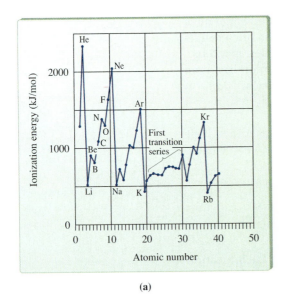

(a)

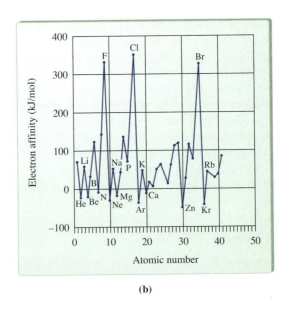

(b)

Figure 2.14

(a) The ionization energies of the first forty elements versus their atomic numbers. Note the very high values for elements located on the right in the periodic table and low values for those on the left. Some exceptions to the trends are evident. (b) The periodic variation of electron affinity. Note the very low values for the noble gases and the elements on the far left of the periodic table. These elements do not form negative ions. In contrast, F, Cl, and Br readily form negative ions.

QUESTION 2.19

Rank Be, N, and F in order of increasing

a. atomic size
b. ionization energy
c. electron affinity

QUESTION 2.20

Rank Cl, Br, I, and F in order of increasing

a. atomic size
b. ionization energy
c. electron affinity

SUMMARY

2.1 Composition of the Atom

The basic structural unit of an element is the *atom*, which is the smallest unit of an element that retains the chemical properties of that element.

The atom has two distinct regions. The *nucleus* is a small, dense, positively charged region in the center of the atom composed of positively charged *protons* and uncharged *neutrons*. Surrounding the nucleus is a diffuse region of negative charge occupied by *electrons*, the source of the negative charge. Electrons are very low in mass in comparison to protons and neutrons.

The *atomic number* (Z) is equal to the number of protons in the atom. The *mass number* (A) is equal to the sum of the protons and neutrons (the mass of the electrons is insignificant).

Isotopes are atoms of the same element that have different masses because they have different numbers of neutrons (different mass numbers). The chemical behavior of isotopes is identical to that of any other isotope of the same element.

Ions are electrically charged particles that result from a gain or loss of one or more electrons by the parent atom. *Anions*, negative ions, are formed by a gain of one or more electrons by the parent atom. *Cations*, positive ions, are formed by a loss of one or more electrons from the parent atom.

2.2 Development of Atomic Theory

The first experimentally based theory of atomic structure was proposed by John Dalton. Although Dalton pictured atoms as indivisible, the experiments of William Crookes, Eugene Goldstein, and J. J. Thomson indicated that the atom is composed of charged particles: protons and electrons. The third fundamental atomic particle is the neutron. An experiment conducted by Hans Geiger led Ernest Rutherford to propose that the majority of the mass and positive charge of the atom is located in a small, dense region, the *nucleus*, with small, negatively charged electrons occupying a much larger, diffuse space outside of the nucleus.

2.3 The Periodic Law and the Periodic Table

The *periodic law* is an organized "map" of the elements that relates their structure to their chemical and physical properties. It states that the elements, when arranged according to their atomic numbers, show a distinct periodicity (regular variation) of their properties. The periodic table is the result of the periodic law.

The modern periodic table exists in several forms. The most important variation is in group numbering. The tables in this text use the two most commonly accepted numbering systems.

A horizontal row of elements in the periodic table is referred to as a *period*. The periodic table consists of seven periods. The lanthanide series is part of period 6; the actinide series is part of period 7.

The columns of elements in the periodic table are called *groups* or *families*. The elements of a particular family share many similarities in physical and chemical properties because of the similarities in electronic structure. Some of the most important groups are named; for example, *alkali metals* (IA or 1), *alkaline earth metals* (IIA or 2), *halogens* (VIIA or 17), and *noble gases* (VIII or 18).

Group A elements are called *representative elements*; Group B elements are *transition elements*. A bold zigzag line runs from top to bottom of the table, beginning to the left of boron (B) and ending between polonium (Po) and astatine (At). This line acts as the boundary between *metals* to the left and *nonmetals* to the right. Elements straddling the boundary, *metalloids*, have properties intermediate between those of metals and nonmetals.

2.4 Electron Arrangement and the Periodic Table

The outermost electrons in an atom are *valence electrons*. For representative elements, the number of valence electrons in an atom corresponds to the group or family number (old numbering system using Roman numerals). Metals tend to have fewer valence electrons than nonmetals.

The *electron configuration* of the elements is predictable, using the aufbau principle. Knowing the electron configuration, we can identify valence electrons and begin to predict the kinds of reactions that the elements will undergo.

Elements in the last family, noble gases, have either two valence electrons (helium) or eight valence electrons (neon, argon, krypton, xenon, and radon). Their most important properties are their extreme stability and lack of reactivity. A full valence level is responsible for this unique stability.

2.5 The Octet Rule

The *octet rule* tells us that in chemical reactions, elements will gain, lose, or share the minimum number of electrons necessary to achieve the electron configuration of the nearest noble gas.

Metallic elements tend to form cations. The ion is *isoelectronic* with its nearest noble gas neighbor and has a stable octet of electrons in its outermost energy level. Nonmetallic elements tend to gain electrons to become isoelectronic with the nearest noble gas element, forming anions.

2.6 Trends in the Periodic Table

Atomic size decreases from left to right and from bottom to top in the periodic table. *Cations* are smaller than the parent atom. *Anions* are larger than the parent atom. Ions with multiple positive charge are even smaller than their corresponding monopositive ion; ions with multiple negative charge are larger than their corresponding less negative ion.

The energy required to remove an electron from the atom is the *ionization energy*. Down a group, the ionization energy generally decreases. Across a period, the ionization energy generally increases.

The energy released when a single electron is added to a neutral atom in the gaseous state is known as the *electron affinity*. Electron affinities generally decrease proceeding down a group and increase proceeding across a period.

KEY TERMS

alkali metal (2.3)
alkaline earth metal (2.3)
anion (2.1)
atom (2.1)
atomic mass (2.1)
atomic number (2.1)
atomic orbital (2.4)
cathode rays (2.2)
cation (2.1)
electron (2.1)
electron affinity (2.6)
electron configuration (2.4)
energy levels (2.4)
group (2.3)
halogen (2.3)
ion (2.1)
ionization energy (2.6)
isoelectronic (2.5)
isotope (2.1)
mass number (2.1)
metal (2.3)
metalloid (2.3)
neutron (2.1)
noble gas (2.3)
nonmetal (2.3)
nucleus (2.1)
octet rule (2.5)
period (2.3)
periodic law (2.3)
proton (2.1)
quantization (2.4)
representative element (2.3)
sublevel (2.4)
transition element (2.3)
valence electron (2.4)

QUESTIONS AND PROBLEMS

Composition of the Atom

Foundations

2.21 Calculate the number of protons, neutrons, and electrons in
   a. $^{16}_{8}O$
   b. $^{31}_{15}P$

2.22 Calculate the number of protons, neutrons, and electrons in
   a. $^{136}_{56}Ba$
   b. $^{209}_{84}Po$

2.23 State the mass and charge of the
   a. electron
   b. proton
   c. neutron

2.24 Calculate the number of protons, neutrons, and electrons in
   a. $^{37}_{17}Cl$
   b. $^{23}_{11}Na$
   c. $^{84}_{36}Kr$

2.25 a. What is an ion?
   b. What process results in the formation of a cation?
   c. What process results in the formation of an anion?

2.26 a. What are isotopes?
   b. What is the major difference among isotopes of an element?
   c. What is the major similarity among isotopes of an element?

Applications

2.27 How many protons are in the nucleus of the isotope Rn-220?

2.28 How many neutrons are in the nucleus of the isotope Rn-220?

2.29 Selenium-80 is a naturally occurring isotope. It is found in over-the-counter supplements.
   a. How many protons are found in one atom of selenium-80?
   b. How many neutrons are found in one atom of selenium-80?

2.30 Iodine-131 is an isotope used in thyroid therapy.
   a. How many protons are found in one atom of iodine-131?
   b. How many neutrons are found in one atom of iodine-131?

2.31 Write symbols for each isotope:
   a. Each atom contains 1 proton and 0 neutrons.
   b. Each atom contains 6 protons and 8 neutrons.

2.32 Write symbols for each isotope:
   a. Each atom contains 1 proton and 2 neutrons.
   b. Each atom contains 92 protons and 146 neutrons.

2.33 Fill in the blanks:

Symbol	No. of Protons	No. of Neutrons	No. of Electrons	Charge
Example:				
$^{40}_{20}$Ca	20	20	20	0
$^{23}_{11}$Na	11	___	11	0
$^{32}_{16}$S^{2-}	16	16	___	2−
___	8	8	8	0
$^{24}_{12}$Mg^{2+}	___	12	___	2+
___	19	20	18	___

2.34 Fill in the blanks:

Atomic Symbol	No. of Protons	No. of Neutrons	No. of Electrons	Charge
Example:				
$^{27}_{13}$Al	13	14	13	0
$^{39}_{19}$K	19	___	19	0
$^{31}_{15}$P^{3-}	15	16	___	___
___	29	34	27	2+
$^{55}_{26}$Fe^{2+}	___	29	___	2+
___	8	8	10	___

2.35 Fill in the blanks:
a. An isotope of an element differs in mass because the atom has a different number of _____.
b. The atomic number gives the number of _____ in the nucleus.
c. The mass number of an atom is due to the number of _____ and _____ in the nucleus.
d. A charged atom is called a(n) _____.
e. Electrons surround the _____ and have a _____ charge.

2.36 Label each of the following statements as true or false:
a. An atom with an atomic number of 7 and a mass of 14 is identical to an atom with an atomic number of 6 and a mass of 14.
b. Neutral atoms have the same number of electrons as protons.
c. The mass of an atom is due to the sum of the number of protons, neutrons, and electrons.

Development of Atomic Theory

Foundations

2.37 What are the major postulates of Dalton's atomic theory?
2.38 Which points of Dalton's theory are no longer current?
2.39 Note the major accomplishment of each of the following:
a. Chadwick
b. Goldstein
c. Crookes
2.40 Note the major accomplishment of each of the following:
a. Geiger
b. Thomson
c. Rutherford

Applications

2.41 List at least three properties of the electron.
2.42 Describe the process that occurs when electrical energy is applied to a sample of hydrogen gas.
2.43 What is a cathode ray? Which subatomic particle is detected?

2.44 Pictured is a cathode ray tube. Show the path that an electron would follow in the tube.

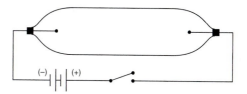

The Periodic Law and the Periodic Table

Foundations

2.45 Provide the name of the element represented by each of the following symbols:
a. Na
b. K
c. Mg
d. B
2.46 Provide the name of the element represented by each of the following symbols:
a. Ca
b. Cu
c. Co
d. Si
2.47 Which group of the periodic table is known as alkali metals? List them.
2.48 Which group of the periodic table is known as alkaline earth metals? List them.
2.49 Which group of the periodic table is known as halogens? List them.
2.50 Which group of the periodic table is known as noble gases? List them.

Applications

2.51 Label each of the following statements as true or false:
a. Elements of the same group have similar properties.
b. Atomic size decreases from left to right across a period.
2.52 Label each of the following statements as true or false:
a. Ionization energy increases from top to bottom within a group (See Figure 2.14.)
b. Representative metals are located on the left in the periodic table.
2.53 For each of the elements Na, Ni, Al, P, Cl, and Ar, provide the following information:
a. Which are metals?
b. Which are representative metals?
c. Which tend to form positive ions?
d. Which are inert or noble gases?
2.54 For each of the elements Ca, K, Cu, Zn, Br, and Kr, provide the following information:
a. Which are metals?
b. Which are representative metals?
c. Which tend to form positive ions?
d. Which are inert or noble gases?

Electron Arrangement and the Periodic Table

Foundations

2.55 How many valence electrons are found in an atom of each of the following elements?
a. H
b. Na
c. B
d. F
e. Ne
f. He

2.56 How many valence electrons are found in an atom of each of the following elements?
  a. Mg
  b. K
  c. C
  d. Br
  e. Ar
  f. Xe

2.57 Distinguish between a principal energy level and a sublevel.

2.58 Distinguish between a sublevel and an orbital.

2.59 Sketch a diagram and describe our current model of an *s* orbital.

2.60 How is a 2*s* orbital different from a 1*s* orbital?

2.61 How many *p* orbitals can exist in a given principal energy level?

2.62 Sketch diagrams of a set of *p* orbitals. How does a p_x orbital differ from a p_y orbital? From a p_z orbital?

2.63 How does a 3*p* orbital differ from a 2*p* orbital?

2.64 What is the maximum number of electrons that an orbital can hold?

Applications

2.65 What is the maximum number of electrons in each of the following energy levels?
  a. $n = 1$
  b. $n = 2$
  c. $n = 3$

2.66 a. What is the maximum number of *s* electrons that can exist in any one principal energy level?
  b. How many *p* electrons?
  c. How many *d* electrons?
  d. How many *f* electrons?

2.67 In which orbital is the highest energy electron located in each of the following elements?
  a. Al
  b. Na
  c. Sc
  d. Ca
  e. Fe
  f. Cl

2.68 Using only the periodic table or list of elements, write the electron configuration of each of the following atoms:
  a. B
  b. S
  c. Ar
  d. V
  e. Cd
  f. Te

2.69 Which of the following electron configurations are not possible? Why?
  a. $1s^2 1p^2$
  b. $1s^2 2s^2 2p^2$
  c. $2s^2, 2s^2, 2p^6, 2d^1$
  d. $1s^2, 2s^3$

2.70 For each incorrect electron configuration in Question 2.69, assume that the number of electrons is correct, identify the element, and write the correct electron configuration.

The Octet Rule

Foundations

2.71 Give the most probable ion formed from each of the following elements:
  a. Li
  b. O
  c. Ca
  d. Br
  e. S
  f. Al

2.72 Using only the periodic table or list of elements, write the electron configuration of each of the following ions:
  a. I^-
  b. Ba^{2+}
  c. Se^{2-}
  d. Al^{3+}

2.73 Which of the following pairs of atoms and/or ions are isoelectronic with one another?
  a. O^{2-}, Ne
  b. S^{2-}, Cl^-

2.74 Which of the following pairs of atoms and/or ions are isoelectronic with one another?
  a. F^-, Cl^-
  b. K^+, Ar

Applications

2.75 Which atom or ion in each of the following groups would you expect to find in nature?
  a. Na, Na^+, Na^-
  b. S^{2-}, S^-, S^+
  c. Cl, Cl^-, Cl^+

2.76 Which atom or ion in each of the following groups would you expect to find in nature?
  a. K, K^+, K^-
  b. O^{2-}, O, O^{2+}
  c. Br, Br^-, Br^+

2.77 Write the electron configuration of each of the following biologically important ions:
  a. Ca^{2+}
  b. Mg^{2+}

2.78 Write the electron configuration of each of the following biologically important ions:
  a. K^+
  b. Cl^-

Trends in the Periodic Table

Foundations

2.79 Arrange each of the following lists of elements in order of increasing atomic size:
  a. N, O, F
  b. Li, K, Cs
  c. Cl, Br, I

2.80 Arrange each of the following lists of elements in order of increasing atomic size:
  a. Al, Si, P, Cl, S
  b. In, Ga, Al, B, Tl
  c. Sr, Ca, Ba, Mg, Be
  d. P, N, Sb, Bi, As

2.81 Arrange each of the following lists of elements in order of increasing ionization energy:
  a. N, O, F
  b. Li, K, Cs
  c. Cl, Br, I

2.82 Arrange each of the following lists of elements in order of decreasing electron affinity:
  a. Na, Li, K
  b. Br, F, Cl
  c. S, O, Se

Applications

2.83 Explain why a positive ion is always smaller than its parent atom.

2.84 Explain why a negative ion is always larger than its parent atom.

2.85 Explain why a fluoride ion is commonly found in nature but a fluorine atom is not.

2.86 Explain why a sodium ion is commonly found in nature but a sodium atom is not.

FOR FURTHER UNDERSTANDING

1. A natural sample of chromium, taken from the ground, will contain four isotopes: Cr-50, Cr-52, Cr-53, and Cr-54. Predict which isotope is in greatest abundance. Explain your reasoning.
2. Crookes's cathode ray tube experiment inadvertently supplied the basic science for a number of modern high-tech devices. List a few of these devices and describe how they involve one or more aspects of this historic experiment.
3. Name five elements that you came in contact with today. Were they in combined form or did they exist in the form of atoms? Were they present in pure form or in mixtures? If mixtures, were they heterogeneous or homogeneous? Locate each in the periodic table by providing the group and period designation, for example: Group IIA (2), period 3.
4. The periodic table is incomplete. It is possible that new elements will be discovered from experiments using high-energy particle accelerators. Predict as many properties as you can that might characterize the element that would have an atomic number of 118. Can you suggest an appropriate name for this element?
5. The element titanium is now being used as a structural material for bone and socket replacement (shoulders, knees). Predict properties that you would expect for such applications; go to the library or Internet and look up the properties of titanium and evaluate your answer.
6. Imagine that you have undertaken a voyage to an alternate universe. Using your chemical skills, you find a collection of elements quite different than those found here on earth. After measuring their properties and assigning symbols for each, you wish to organize them as Mendeleev did for our elements. Design a periodic table using the information you have gathered:

Symbol	Mass (amu)	Reactivity	Electrical Conductivity
A	2.0	High	High
B	4.0	High	High
C	6.0	Moderate	Trace
D	8.0	Low	0
E	10.0	Low	0
F	12.0	High	High
G	14.0	High	High
H	16.0	Moderate	Trace
I	18.0	Low	0
J	20.0	None	0
K	22.0	High	High
L	24.0	High	High

Predict the reactivity and conductivity of an element with a mass of 30.0 amu. Which element in our universe does this element most closely resemble?

3

LEARNING GOALS

1 Classify compounds as having ionic, covalent, or polar covalent bonds.

2 Write the formula of a compound when provided with the name of the compound.

3 Name common inorganic compounds using standard conventions and recognize the common names of frequently used substances.

4 Predict differences in physical state, melting and boiling points, solid-state structure, and solution chemistry that result from differences in bonding.

5 Draw Lewis structures for covalent compounds.

6 Describe the relationship between bond order, bond energy, and bond length.

7 Predict the geometry of molecules using the octet rule and Lewis structures.

8 Understand the role that molecular geometry plays in determining the solubility and melting and boiling points of compounds.

9 Use the principles of VSEPR theory and molecular geometry to predict relative melting points, boiling points, and solubilities of compounds.

Structure and Properties of Ionic and Covalent Compounds

*A*ll of us, at one time or another, have wondered at the magnificent sight of thousands of migrating birds, flying in formation, heading south for the winter and returning each spring.

Less visible, but no less impressive, are the schools of fish that travel thousands of miles, returning to the same location year after year. Almost instantly, when faced with some external stimulus such as a predator, they snap into a formation that rivals an army drill team for precision.

The questions of how these life-forms know when and where they are going and how they establish their formations have perplexed scientists for many years. The explanations so far are really just hypotheses.

Some clues to the mystery may be hidden in very tiny particles of magnetite, Fe_3O_4. Magnetite contains iron that is naturally magnetic, and collections of these particles behave like a compass needle; they line up in formation aligned with the earth's magnetic field.

Magnetotactic bacteria contain magnetite in the form of magnetosomes, small particles of Fe_3O_4. Fe_3O_4 is a compound whose atoms are joined by chemical bonds. Electrons in the iron atoms have an electron configuration that results in single electrons (not pairs of electrons) occupying orbitals. These unpaired electrons impart magnetic properties to the compound.

The normal habitat of magnetotactic bacteria is either freshwater or the ocean; the bacteria orient themselves to the earth's magnetic field and swim to the nearest pole (north or south). This causes them to swim into regions of nutrient-rich sediment.

Could the directional device, the simple F_3O_4 unit, also be responsible for direction finding in higher organisms in much the same way that an explorer uses a compass? Perhaps so! Recent studies have shown evidence of magnetosomes in the brains of birds, tuna, green turtles, and dolphins.

Most remarkably, at least one study has shown evidence that magnetite is present in the human brain.

These preliminary studies offer hope of unraveling some of the myth and mystery of guidance and communication in living systems. The answers may involve a very basic compound like those we will study in this chapter.

Behavior and communication in the animal kingdom is not well understood. What environmental factors, other than magnetotactic bacteria, may affect animal behavior?

3.1 Chemical Bonding

When two or more atoms form a chemical compound, the atoms are held together in a characteristic arrangement by attractive forces. The **chemical bond** is the force of attraction between any two atoms in a compound. The attraction is the force that overcomes the repulsion of the positively charged nuclei of the two atoms.

Interactions involving valence electrons are responsible for the chemical bond. We shall focus our attention on these electrons and the electron arrangement of atoms both before and after bond formation.

Lewis Symbols

The **Lewis symbol,** or Lewis structure, developed by G. N. Lewis early in the twentieth century, is a convenient way of representing atoms singly or in combination. Its principal advantage is that *only* valence electrons (those that may participate in bonding) are shown. Lewis symbolism is based on the octet rule that was described in Chapter 2.

Recall that the number of valence electrons can be determined from the position of the element in the periodic table (see Figure 2.7).

To draw Lewis structures, we first write the chemical symbol of the atom; this symbol represents the nucleus and all of the lower energy nonvalence electrons. The valence electrons are indicated by dots arranged around the atomic symbol. For example,

H ·	He :
Hydrogen	Helium
Li ·	Be :
Lithium	Beryllium
· B ·	· C ·
Boron	Carbon
· N ·	· O ·
Nitrogen	Oxygen
: F ·	: Ne :
Fluorine	Neon

The two valence electrons in helium occupy one orbital, the 1s orbital. The two valence electrons in beryllium also occupy one orbital, the 2s orbital. Consequently these valence electrons are paired in the Lewis structure. Refer to the electron configurations in Chapter 2.

Note particularly that the number of dots corresponds to the number of valence electrons in the outermost shell of the atoms of the element. Each of the four "sides" of the chemical symbol represent an atomic orbital capable of holding one or two valence electrons. Because each atomic orbital can hold no more than two electrons, we can show a maximum of two dots on each side of the element's symbol. Using the same logic employed in writing electron configurations in Chapter 2, we place one dot on each side then sequentially add a second dot, filling each side in turn. This process is limited by the total number of available valence electrons. Each unpaired dot (representing an unpaired electron) is available to form a chemical bond with another element, producing a compound. Figure 3.1 depicts the Lewis dot structures for the representative elements.

Principal Types of Chemical Bonds: Ionic and Covalent

Two principal classes of chemical bonds exist: ionic and covalent. Both involve valence electrons.

Ionic bonding involves a transfer of one or more electrons from one atom to another, leading to the formation of an ionic bond. **Covalent bonding** involves a sharing of electrons resulting in the covalent bond.

Before discussing each type, we should recognize that the distinction between ionic and covalent bonding is not always clear-cut. Some compounds are clearly

Animation

Ionic and Covalent Bonding

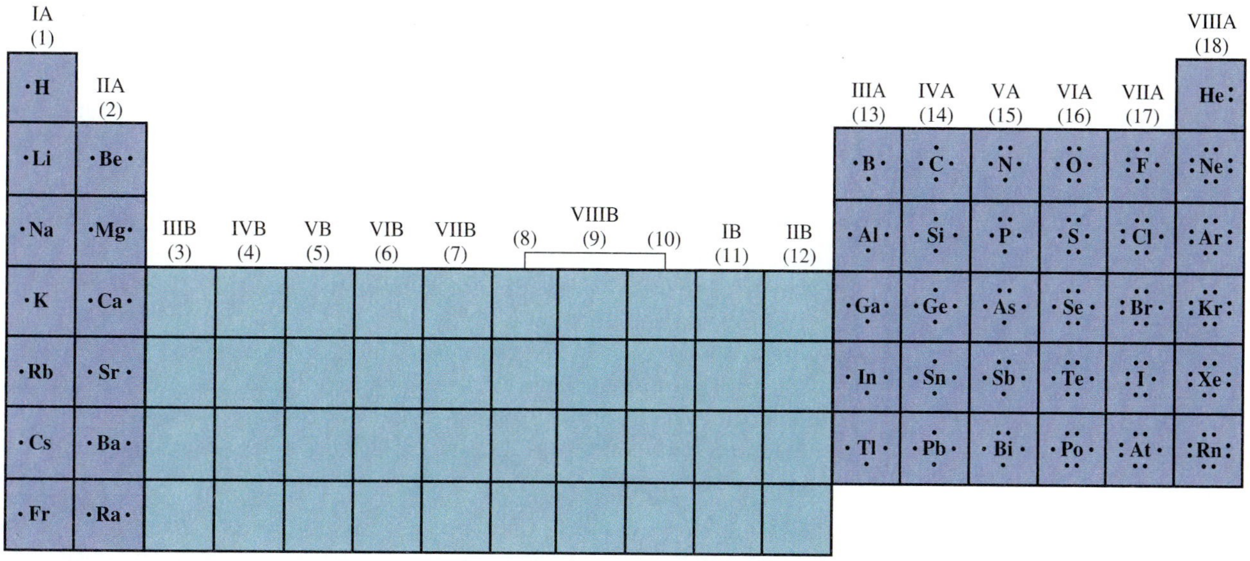

Figure 3.1

Lewis dot symbols for representative elements. Each unpaired electron is a potential bond.

ionic, and some are clearly covalent, but many others possess both ionic and covalent characteristics.

Ionic Bonding

Representative elements form ions that obey the octet rule. Ions of opposite charge attract each other and this attraction is the essence of the ionic bond. Consider the reaction of a sodium atom and a chlorine atom to produce sodium chloride:

$$Na + Cl \longrightarrow NaCl$$

Recall that the sodium atom has

- low ionization energy (it readily loses an electron) and
- low electron affinity (it does not want more electrons).

If sodium loses its valence electron, it will become isoelectronic (same number of electrons) with neon, a very stable noble gas atom. This tells us that the sodium atom would be a good electron donor, forming the sodium ion:

$$Na \cdot \longrightarrow Na^+ + e^-$$

Recall that the chlorine atom has

- high ionization energy (it will not easily give up an electron) and
- high electron affinity (it readily accepts another electron).

Chlorine will gain one more electron. By doing so, it will complete an octet (eight outermost electrons) and be isoelectronic with argon, a stable noble gas. Therefore, chlorine behaves as a willing electron acceptor, forming a chloride ion:

$$: \overset{..}{\underset{..}{Cl}} \cdot + e^- \longrightarrow [\, : \overset{..}{\underset{..}{Cl}} : \,]^-$$

The electron released by sodium (*electron donor*) is the electron received by chlorine (*electron acceptor*):

$$Na \cdot \longrightarrow Na^+ + e^-$$
$$e^- + \cdot \overset{..}{\underset{..}{Cl}} : \longrightarrow [\, : \overset{..}{\underset{..}{Cl}} : \,]^-$$

1 Learning Goal

Classify compounds as having ionic, covalent, or polar covalent bonds.

Refer to Section 2.6 for a discussion of ionization energy and electron affinity.

The resulting ions of opposite charge, Na^+ and Cl^-, are attracted to each other (opposite charges attract) and held together by this *electrostatic force* as an **ion pair:** Na^+Cl^-.

This electrostatic force, the attraction of opposite charges, is quite strong and holds the ions together. It is the ionic bond.

The essential features of ionic bonding are the following:

- Atoms of elements with low ionization energy and low electron affinity tend to form positive ions.
- Atoms of elements with high ionization energy and high electron affinity tend to form negative ions.
- Ion formation takes place by an electron transfer process.
- The positive and negative ions are held together by the electrostatic force between ions of opposite charge in an ionic bond.
- Reactions between representative metals and nonmetals (elements far to the left and right, respectively, in the periodic table) tend to result in ionic bonds.

Although ionic compounds are sometimes referred to as ion pairs, in the solid state, these ion pairs do not actually exist as individual units. The positive ions exert attractive forces on several negative ions, and the negative ions are attracted to several positive centers. Positive and negative ions arrange themselves in a regular three-dimensional repeating array to produce a stable arrangement known as a **crystal lattice.** The lattice structure for sodium chloride is shown in Figure 3.2.

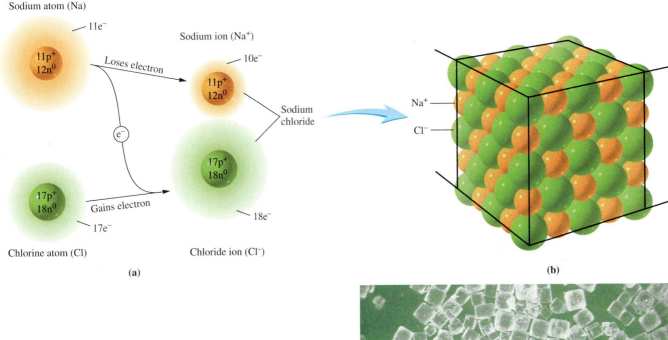

Figure 3.2

The arrangement of ions in a crystal of NaCl (sodium chloride, table salt). (a) A sodium atom loses one electron to become a smaller sodium ion, and a chlorine atom gains that electron, becoming a larger chloride ion. (b) Attraction of Na^+ and Cl^- forms NaCl ion pairs that aggregate in a three-dimensional crystal lattice structure. (c) A microscopic view of NaCl crystals shows their cubic geometry. Each tiny crystal contains billions of sodium and chloride ions.

Covalent Bonding

The octet rule is not just for ionic compounds. Covalently bonded compounds share electrons to complete the octet of electrons for each of the atoms participating in the bond. Consider the bond formed between two hydrogen atoms, producing the diatomic form of hydrogen: H_2. Individual hydrogen atoms are not stable, and two hydrogen atoms readily combine to produce diatomic hydrogen:

$$H + H \longrightarrow H_2$$

If a hydrogen atom were to gain a second electron, it would be isoelectronic with the stable electron configuration of helium. However, because two identical hydrogen atoms have an equal tendency to gain or lose electrons, an electron transfer from one atom to the other is unlikely to occur under normal conditions. Each atom may attain a noble gas structure only by *sharing* its electron with the other, as shown with Lewis symbols:

$$H\cdot + \cdot H \longrightarrow H : H$$

When electrons are shared rather than transferred, the *shared electron pair* is referred to as a *covalent bond* (Figure 3.3). Compounds characterized by covalent bonding are called *covalent compounds*. Covalent bonds tend to form between atoms with similar tendencies to gain or lose electrons. The most obvious examples are the diatomic molecules H_2, N_2, O_2, F_2, Cl_2, Br_2, and I_2. Bonding in these molecules is *totally covalent* because there can be no net tendency for electron transfer between identical atoms. The formation of F_2, for example, may be represented as

$$:\overset{\cdot}{\underset{\cdot\cdot}{F}}\cdot \quad \cdot\overset{\cdot}{\underset{\cdot\cdot}{F}}: \longrightarrow :\overset{\cdot\cdot}{\underset{\cdot\cdot}{F}}:\overset{\cdot\cdot}{\underset{\cdot\cdot}{F}}:$$

As in H_2, a single covalent bond is formed. The bonding electron pair is said to be *localized*, or largely confined to the region between the two fluorine nuclei.

Two atoms do not have to be identical to form a covalent bond. Consider compounds such as the following:

$H:\overset{\cdot\cdot}{\underset{\cdot\cdot}{F}}:$	$H:\overset{\cdot\cdot}{O}:H$	$H:\overset{\displaystyle H}{\underset{\displaystyle H}{C}}:H$	$H:\overset{\cdot\cdot}{\underset{}{\overset{\displaystyle H}{N}}}:H$
Hydrogen fluoride	**Water**	**Methane**	**Ammonia**
7e⁻ from F	6e⁻ from O	4e⁻ from C	5e⁻ from N
1e⁻ from H	2e⁻ from 2H	4e⁻ from 4H	3e⁻ from 3H
8e⁻ for F	8e⁻ for O	8e⁻ for C	8e⁻ for N
2e⁻ for H	2e⁻ for H	2e⁻ for H	2e⁻ for H

In each of these cases, bond formation satisfies the octet rule. A total of eight electrons surround each atom other than hydrogen. Hydrogen has only two electrons (corresponding to the electronic structure of helium).

QUESTION 3.1

Using Lewis symbols, write an equation predicting the product of the reaction of

a. $Li + Br$ b. $Mg + Cl$

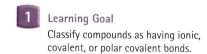

Animation
Covalent Bond Sharing Electrons

A diatomic compound is one that is composed of two atoms joined by a covalent bond.

Fourteen valence electrons are arranged so that each fluorine atom is surrounded by eight electrons. The octet rule is satisfied for each fluorine atom.

Hydrogen atoms approach at high velocity

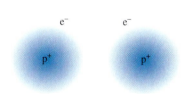
Hydrogen nuclei begin to attract each other's electrons.

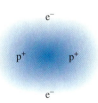
Hydrogen atoms form the hydrogen molecule; atoms are held together by the shared electrons, the covalent bond.

Figure 3.3

Covalent bonding in hydrogen.

QUESTION 3.2

Using Lewis symbols, write an equation predicting the product of the reaction of

a. Na + O   b. Na + S

Polar Covalent Bonding and Electronegativity

The Polar Covalent Bond

Covalent bonding is the sharing of an electron pair by two atoms. However, just as we may observe in our day-to-day activities, sharing is not always equal. In a molecule like H_2 (or N_2, or any other diatomic molecule composed of only one element), the electrons, on average, spend the same amount of time in the vicinity of each atom; the electrons have no preference because both atoms are identical.

Now consider a diatomic molecule composed of two different elements; HF is a common example. It has been experimentally shown that the electrons in the H—F bond are not equally shared; the electrons spend more time in the vicinity of the fluorine atom. This unequal sharing can be described in various ways:

Partial electron transfer: This describes the bond as having both covalent and ionic properties.

Unequal electron density: The density of electrons around F is greater than the density of electrons around H.

Polar covalent bond is the preferred term for a bond made up of unequally shared electron pairs. One end of the bond (in this case, the F atom) is more electron-rich (higher electron density), hence, more negative. The other end of the bond (in this case, the H atom) is less electron-rich (lower electron density), hence, more positive. These two ends, one somewhat positive (δ^+) and the other somewhat negative (δ^-) may be described as electronic poles, hence the term polar covalent bonds.

The water molecule is perhaps the best-known example of a molecule that exhibits polar covalent bonding (Figure 3.4). In Section 3.4, we will see that the water molecule itself is polar and this fact is the basis for many of water's unique properties.

Once again, we can use the predictive power of the periodic table to help us determine whether a particular bond is polar or nonpolar covalent. We already know that elements that tend to form negative ions (by gaining electrons) are found to the right of the table, whereas positive ion formers (that may lose electrons) are located on the left side of the table. Elements whose atoms strongly attract electrons are described as electronegative elements. Linus Pauling, a chemist noted for his theories on chemical bonding, developed a scale of relative *electronegativities* that correlates reasonably well with the positions of the elements in the periodic table.

Electronegativity

Electronegativity (E_n) is a measure of the ability of an atom to attract electrons in a chemical bond. Elements with high electronegativity have a greater ability to attract electrons than elements with low electronegativity. Pauling developed a method to assign values of electronegativity to many of the elements in the periodic table. These values range from a low of 0.7 to a high of 4.0, 4.0 being the most electronegative element.

Figure 3.5 shows that the most electronegative elements (excluding the nonreactive noble gas elements) are located in the upper right corner of the periodic table, whereas the least electronegative elements are found in the lower left corner

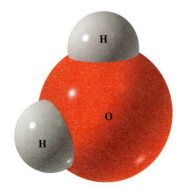

(a)

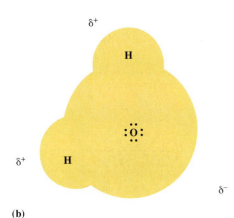

(b)

Figure 3.4

Polar covalent bonding in water. Oxygen is electron-rich (δ^-) and hydrogen is electron-deficient (δ^+) due to unequal electron sharing. Water has two polar covalent bonds.

IA (1)	IIA (2)	IIIB (3)	IVB (4)	VB (5)	VIB (6)	VIIB (7)	VIIIB (8)	(9)	(10)	IB (11)	IIB (12)	IIIA (13)	IVA (14)	VA (15)	VIA (16)	VIIA (17)
H 2.1																
Li 1.0	Be 1.5											B 2.0	C 2.5	N 3.0	O 3.5	F 4.0
Na 0.9	Mg 1.2											Al 1.5	Si 1.8	P 2.1	S 2.5	Cl 3.0
K 0.8	Ca 1.0	Sc 1.3	Ti 1.5	V 1.6	Cr 1.6	Mn 1.5	Fe 1.8	Co 1.9	Ni 1.9	Cu 1.9	Zn 1.6	Ga 1.6	Ge 1.8	As 2.0	Se 2.4	Br 2.8
Rb 0.8	Sr 1.0	Y 1.2	Zr 1.4	Nb 1.6	Mo 1.8	Tc 1.9	Ru 2.2	Rh 2.2	Pd 2.2	Ag 1.9	Cd 1.7	In 1.7	Sn 1.8	Sb 1.9	Te 2.1	I 2.5
Cs 0.7	Ba 0.9	La* 1.1	Hf 1.3	Ta 1.5	W 1.7	Re 1.9	Os 2.2	Ir 2.2	Pt 2.2	Au 2.4	Hg 1.9	Tl 1.8	Pb 1.9	Bi 1.9	Po 2.0	At 2.2
Fr 0.7	Ra 0.9	Ac† 1.1														

Color legend:
- Less than 1.0
- Between 1.0–3.0
- Greater than or equal to 3.0

*Lanthanides: 1.1 – 1.3
†Actinides: 1.1 – 1.5

Figure 3.5

Electronegativities of the elements.

of the table. In general, electronegativity values increase as we proceed left to right and bottom to top of the table. Like other periodic trends, numerous exceptions occur.

If we picture the covalent bond as a competition for electrons between two positive centers, it is the difference in electronegativity, ΔE_n, that determines the extent of polarity. Consider: H_2 or H—H

$$\Delta E_n = \left[\begin{array}{c}\text{Electronegativity} \\ \text{of hydrogen}\end{array}\right] - \left[\begin{array}{c}\text{Electronegativity} \\ \text{of hydrogen}\end{array}\right]$$

$$\Delta E_n = 2.1 - 2.1 = 0$$

The bond in H_2 is nonpolar covalent. Bonds between *identical* atoms are *always* nonpolar covalent. Also, Cl_2 or Cl—Cl

$$\Delta E_n = \left[\begin{array}{c}\text{Electronegativity} \\ \text{of chlorine}\end{array}\right] - \left[\begin{array}{c}\text{Electronegativity} \\ \text{of chlorine}\end{array}\right]$$

$$\Delta E_n = 3.0 - 3.0 = 0$$

The bond in Cl_2 is nonpolar covalent. Now consider HCl or H—Cl

$$\Delta E_n = \left[\begin{array}{c}\text{Electronegativity} \\ \text{of chlorine}\end{array}\right] - \left[\begin{array}{c}\text{Electronegativity} \\ \text{of hydrogen}\end{array}\right]$$

$$\Delta E_n = 3.0 - 2.1 = 0.9$$

The bond in HCl is polar covalent.

An electronegativity difference of 1.9 is generally accepted as the boundary between polar covalent and ionic bonds. Although, strictly speaking, any electronegativity difference, no matter how small, produces a polar bond, the degree of polarity for bonds with electronegativity differences less than 0.5 is minimal. Consequently, we shall classify these bonds as nonpolar.

Linus Pauling is the only person to receive two Nobel Prizes in very unrelated fields: the chemistry award in 1954 and eight years later, the Nobel Peace Prize. His career is a model of interdisciplinary science, with important contributions ranging from chemical physics to molecular biology.

By convention, the electronegativity difference is calculated by subtracting the less electronegative element's value from the value for the more electronegative element. In this way, negative numbers are avoided.

Animation
Electronegativity

QUESTION 3.3

Is the bond formed between carbon and nitrogen polar covalent? Why?

QUESTION 3.4

Is the bond formed between silicon and chlorine polar covalent? Why?

3.2 Naming Compounds and Writing Formulas of Compounds

Nomenclature is the assignment of a correct and unambiguous name to each and every chemical compound. Assignment of a name to a structure or deducing the structure from a name is a necessary first step in any discussion of these compounds.

Ionic Compounds

The **formula** is the representation of the fundamental compound using chemical symbols and numerical subscripts. It is the "shorthand" symbol for a compound— for example,

$$NaCl \quad \text{and} \quad MgBr_2$$

The formula identifies the number and type of the various atoms that make up the compound unit. The number of like atoms in the unit is shown by the use of a subscript. The presence of only one atom is understood when no subscript is present.

The formula $NaCl$ indicates that each ion pair consists of one sodium cation (Na^+) and one chloride anion (Cl^-). Similarly, the formula $MgBr_2$ indicates that one magnesium ion and two bromide ions combine to form the compound.

In Chapter 2, we learned that positive ions are formed from elements that

- are located at the left of the periodic table,
- are referred to as *metals*, and
- have low ionization energies, low electron affinities, and hence easily *lose* electrons.

Elements that form negative ions, on the other hand,

- are located at the right of the periodic table (but exclude the noble gases),
- are referred to as *nonmetals*, and
- have high ionization energies, high electron affinities, and hence easily *gain* electrons.

In short, metals and nonmetals usually react to produce ionic compounds resulting from the transfer of one or more electrons from the metal to the nonmetal.

Writing Formulas of Ionic Compounds from the Identities of the Component Ions

2 Learning Goal

Write the formula of a compound when provided with the name of the compound.

The *formula* of an ionic compound is the smallest whole-number ratio of ions in the substance. It is important to be able to write the formula of an ionic compound when provided with the identities of the ions that make up the compound. The charge of each ion can usually be determined from the group (family) of the periodic table in which the parent element is found. The cations and anions must combine so that the resulting formula unit has a net charge of zero.

Consider the following examples.

EXAMPLE 3.1 Predicting the Formula of an Ionic Compound

Predict the formula of the ionic compound formed from the reaction of sodium and oxygen atoms.

SOLUTION

Sodium is in group IA (or 1); it has *one* valence electron. Loss of this electron produces Na^+. Oxygen is in group VIA (or 16); it has *six* valence electrons. A gain of two electrons (to create a stable octet) produces O^{2-}. Two positive charges are necessary to counterbalance two negative charges on the oxygen anion. Because each sodium ion carries a 1+ charge, two sodium ions are needed for each O^{2-}. The subscript 2 is used to indicate that the formula unit contains two sodium ions. Thus the formula of the compound is Na_2O.

2 Learning Goal

Write the formula of a compound when provided with the name of the compound.

EXAMPLE 3.2 Predicting the Formula of an Ionic Compound

Predict the formula of the compound formed by the reaction of aluminum and oxygen atoms.

SOLUTION

Aluminum is in group IIIA (or 13) of the periodic table; we predict that it has three valence electrons. Loss of these electrons produces Al^{3+}. Oxygen is in group VIA (or 16) of the periodic table and has six valence electrons. A gain of two electrons (to create a stable octet) produces O^{2-}. How can we combine Al^{3+} and O^{2-} to yield a unit of zero charge? It is necessary that *both* the cation and anion be multiplied by factors that will result in a zero net charge:

$$2 \times (+3) = +6 \quad \text{and} \quad 3 \times (-2) = -6$$
$$2 \times Al^{3+} = +6 \quad \text{and} \quad 3 \times O^{2-} = -6$$

Hence the formula is Al_2O_3.

QUESTION 3.5

Predict the formulas of the compounds formed from the combination of ions of the following elements:

a. lithium and bromine
b. calcium and bromine
c. calcium and nitrogen

QUESTION 3.6

Predict the formulas of the compounds formed from the combination of ions of the following elements:

a. potassium and chlorine
b. magnesium and bromine
c. magnesium and nitrogen

TABLE 3.1 Common Monatomic Cations and Anions

Cation	Name	Anion	Name
H^-	Hydrogen ion	H^-	Hydride ion
Li^+	Lithium ion	F^-	Fluoride ion
Na^+	Sodium ion	Cl^-	Chloride ion
K^+	Potassium ion	Br^-	Bromide ion
Cs^+	Cesium ion	I^-	Iodide ion
Be^{2+}	Beryllium ion	O^{2-}	Oxide ion
Mg^{2+}	Magnesium ion	S^{2-}	Sulfide ion
Ca^{2+}	Calcium ion	N^{3-}	Nitride ion
Ba^{2+}	Barium ion	P^{3-}	Phosphide ion
Al^{3+}	Aluminum ion		
Ag^+	Silver ion		

Note: The ions important in biological systems are highlighted in magenta.

3 Learning Goal

Name common inorganic compounds using standard conventions and recognize the common names of frequently used substances.

Writing Names of Ionic Compounds from the Formula of the Compound

Nomenclature, the way in which compounds are named, is based on their formulas. The name of the cation appears first, followed by the name of the anion. **Monatomic ions** are ions consisting of a single atom. Common monatomic ions are listed in Table 3.1. The ions that are particularly important in biological systems are highlighted in magenta.

When naming compounds containing only monatomic ions, the positive ion has the name of the element; the negative ion is named by using the *stem* of the name of the element joined to the suffix *-ide*. Some examples follow.

Formula	Cation	and	Anion Stem	+	ide	=	Compound Name
NaCl	Sodium		Chlor	+	ide		Sodium chloride
Na_2O	Sodium		Ox	+	ide		Sodium oxide
Li_2S	Lithium		Sulf	+	ide		Lithium sulfide
$AlBr_3$	Aluminum		Brom	+	ide		Aluminum bromide
CaO	Calcium		Ox	+	ide		Calcium oxide

If the cation and anion exist in only one common charged form, there is no ambiguity between formula and name. Sodium chloride *must be* NaCl, and lithium sulfide *must be* Li_2S, so that the sum of positive and negative charges is zero. With many elements, such as the transition metals, several ions of different charge may exist. Fe^{2+}, Fe^{3+} and Cu^+, Cu^{2+} are two common examples. Clearly, an ambiguity exists if we use the name iron for both Fe^{2+} and Fe^{3+} or copper for both Cu^+ and Cu^{2+}. Two systems have been developed to avoid this problem: the *Stock system* (systematic name) and the *common nomenclature system.*

In the Stock system, a Roman numeral placed immediately after the name of the ion indicates the magnitude of the cation's charge. In the older common nomenclature system, the suffix *-ous* indicates the lower ionic charge, and the suffix *-ic* indicates the higher ionic charge. Consider the examples in Table 3.2.

Systematic names are easier and less ambiguous than common names. Whenever possible, we will use this system of nomenclature. The older, common names (-ous, -ic) are less specific; furthermore, they often use the Latin names of the elements (for example, iron compounds use *ferr-*, from *ferrum*, the Latin word for iron).

TABLE 3.2 Systematic (Stock) and Common Names for Iron and Copper Ions

For systematic name:

Formula	+Ion Charge	Cation Name	Compound Name
$FeCl_2$	2+	Iron(II)	Iron(II) chloride
$FeCl_3$	3+	Iron(III)	Iron(III) chloride
Cu_2O	1+	Copper(I)	Copper(I) oxide
CuO	2+	Copper(II)	Copper(II) oxide

For common nomenclature:

Formula	+Ion Charge	Cation Name	Common –ous/ic Name
$FeCl_2$	2+	Ferrous	Ferrous chloride
$FeCl_3$	3+	Ferric	Ferric chloride
Cu_2O	1+	Cuprous	Cuprous oxide
CuO	2+	Cupric	Cupric oxide

Polyatomic ions, such as the hydroxide ion, OH^-, are composed of two or more atoms bonded together. These ions, although bonded to other ions with ionic bonds, are themselves held together by covalent bonds. See Table 3.3.

TABLE 3.3 Common Polyatomic Cations and Anions

Ion	Name
NH_4^+	Ammonium
NO_2^-	Nitrite
NO_3^-	Nitrate
SO_3^{2-}	Sulfite
SO_4^{2-}	Sulfate
HSO_4^-	Hydrogen sulfate
OH^-	Hydroxide
CN^-	Cyanide
PO_4^{3-}	Phosphate
HPO_4^{2-}	Hydrogen phosphate
$H_2PO_4^-$	Dihydrogen phosphate
CO_3^{2-}	Carbonate
HCO_3^-	Bicarbonate
ClO^-	Hypochlorite
ClO_2^-	Chlorite
ClO_3^-	Chlorate
ClO_4^-	Perchlorate
CH_3COO^- (or $C_2H_3O_2^-$)	Acetate
MnO_4^-	Permanganate
$Cr_2O_7^{2-}$	Dichromate
CrO_4^{2-}	Chromate
O_2^{2-}	Peroxide

Note: The most commonly encountered ions are highlighted in magenta.

The polyatomic ion has an *overall* positive or negative charge. Some common polyatomic ions are listed in Table 3.3. The formulas, charges, and names of these polyatomic ions, especially those highlighted in magenta, should be memorized.

The following examples are formulas of several compounds containing polyatomic ions.

Formula	Cation	Anion	Name
NH_4Cl	NH_4^+	Cl^-	Ammonium chloride
$Ca(OH)_2$	Ca^{2+}	OH^-	Calcium hydroxide
Na_2SO_4	Na^+	SO_4^{2-}	Sodium sulfate
$NaHCO_3$	Na^+	HCO_3^-	Sodium bicarbonate

> Sodium bicarbonate may also be named sodium hydrogen carbonate, a preferred and less ambiguous name. Likewise, Na_2HPO_4 is named sodium hydrogen phosphate, and other ionic compounds are named similarly.

QUESTION 3.7

Name each of the following compounds:

a. KCN
b. MgS
c. $Mg(CH_3COO)_2$

QUESTION 3.8

Name each of the following compounds:

a. Li_2CO_3
b. $FeBr_2$
c. $CuSO_4$

Writing Formulas of Ionic Compounds from the Name of the Compound

It is also important to be able to write the correct formula when given the compound name. To do this, we must be able to predict the charge of monatomic ions and remember the charge and formula of polyatomic ions. Equally important, the relative number of positive and negative ions in the unit must result in a net (compound) charge of zero. Compounds are electrically neutral. Two examples follow.

2 Learning Goal

Write the formula of a compound when provided with the name of the compound.

EXAMPLE 3.3   Writing a Formula When Given the Name of the Compound

Write the formula of sodium sulfate.

SOLUTION

Step 1. The sodium ion is Na^+, a group I (or 1) element. The sulfate ion is SO_4^{2-} (from Table 3.3).

Step 2. Two positive charges, two sodium ions, are needed to cancel the charge on one sulfate ion (two negative charges).

Hence the formula is Na_2SO_4.

EXAMPLE 3.4   **Writing a Formula When Given the Name of the Compound**

2   **Learning Goal**
Write the formula of a compound when provided with the name of the compound.

Write the formula of ammonium sulfide.

SOLUTION

Step 1. The ammonium ion is NH_4^+ (from Table 3.3). The sulfide ion is S^{2-} (from its position on the periodic table).

Step 2. Two positive charges are necessary to cancel the charge on one sulfide ion (two negative charges).

Hence the formula is $(NH_4)_2S$.
    Note that parentheses must be used whenever a subscript accompanies a polyatomic ion.

QUESTION 3.9

Write the formula for each of the following compounds:

a. calcium carbonate
b. sodium bicarbonate
c. copper(I) sulfate

QUESTION 3.10

Write the formula for each of the following compounds:

a. sodium phosphate
b. potassium bromide
c. iron(II) nitrate

Covalent Compounds

Naming Covalent Compounds

Most covalent compounds are formed by the reaction of nonmetals. **Molecules** are compounds characterized by covalent bonding. We saw earlier that ionic compounds are not composed of single units but are part of a massive three-dimensional crystal structure in the solid state. Covalent compounds exist as discrete molecules in the solid, liquid, and gas states. This is a distinctive feature of covalently bonded substances.

    The conventions for naming covalent compounds follow:

3   **Learning Goal**
Name common inorganic compounds using standard conventions and recognize the common names of frequently used substances.

1. The names of the elements are written in the order in which they appear in the formula.
2. A prefix (Table 3.4) indicating the number of each kind of atom found in the unit is placed before the name of the element.
3. If only one atom of a particular kind is present in the molecule, the prefix mono- is usually omitted from the first element.
4. The stem of the name of the last element is used with the suffix -ide.
5. The final vowel in a prefix is often dropped before a vowel in the stem name.

By convention, the prefix *mono-* is often omitted from the second element as well (dinitrogen oxide, not dinitrogen monoxide). In other cases, common usage retains the prefix (carbon monoxide, not carbon oxide).

TABLE 3.4 Prefixes Used to Denote Numbers of Atoms in a Compound

Prefix	Number of Atoms
Mono-	1
Di-	2
Tri-	3
Tetra-	4
Penta-	5
Hexa-	6
Hepta-	7
Octa-	8
Nona-	9
Deca-	10

3 Learning Goal

Name common inorganic compounds using standard conventions and recognize the common names of frequently used substances.

EXAMPLE 3.5 Naming a Covalent Compound

Name the covalent compound N_2O_4.

SOLUTION

Step 1. two nitrogen atoms   four oxygen atoms

Step 2. di-   tetra-

Step 3. dinitrogen   tetr(a)oxide

The name is dinitrogen tetroxide.

The following are examples of other covalent compounds.

Formula	Name
N_2O	Dinitrogen monoxide
NO_2	Nitrogen dioxide
SiO_2	Silicon dioxide
CO_2	Carbon dioxide
CO	Carbon monoxide

QUESTION 3.11

Name each of the following compounds:

a. B_2O_3 c. ICl
b. NO d. PCl_3

QUESTION 3.12

Name each of the following compounds:

a. H_2S c. PCl_5
b. CS_2 d. P_2O_5

Writing Formulas of Covalent Compounds

Many compounds are so familiar that their *common names* are generally used. For example, H_2O is water, NH_3 is ammonia, C_2H_5OH (ethanol) is ethyl alcohol, and $C_6H_{12}O_6$ is glucose. It is useful to be able to correlate both systematic and common names with the corresponding molecular formula and vice versa.

When common names are used, formulas of covalent compounds can be written *only* from memory. You *must* remember that water is H_2O, ammonia is NH_3, and so forth. This is the major disadvantage of common names. Because of their widespread use, however, they cannot be avoided and must be memorized.

Compounds named by using Greek prefixes are easily converted to formulas. Consider the following examples.

EXAMPLE 3.6 Writing the Formula of a Covalent Compound

Write the formula of nitrogen monoxide.

SOLUTION

Nitrogen has no prefix; one is understood. Oxide has the prefix *mono*—one oxygen. Hence the formula is NO.

---

2 Learning Goal

Write the formula of a compound when provided with the name of the compound.

EXAMPLE 3.7 Writing the Formula of a Covalent Compound

Write the formula of dinitrogen tetroxide.

SOLUTION

Nitrogen has the prefix *di*—two nitrogen atoms. Oxygen has the prefix *tetr(a)*—four oxygen atoms. Hence the formula is N_2O_4.

QUESTION 3.13

Write the formula of each of the following compounds:

a. diphosphorus pentoxide
b. silicon dioxide

QUESTION 3.14

Write the formula of each of the following compounds:

a. nitrogen trifluoride
b. carbon monoxide

3.3 Properties of Ionic and Covalent Compounds

The differences in ionic and covalent bonding result in markedly different properties for ionic and covalent compounds. Because covalent molecules are distinct units, they have less tendency to form extended structures in the solid state. Ionic compounds, with ions joined by electrostatic attraction, do not have definable

4 Learning Goal

Predict differences in physical state, melting and boiling points, solid-state structure, and solution chemistry that result from differences in bonding.

LIFESTYLE Connection | Origin of the Elements

The current, most widely held theory of the origin of the universe is the "big bang" theory. An explosion of very dense matter was followed by expansion into space of the fragments resulting from this explosion. This is one of the scenarios that have been created by scientists fascinated by the origins of matter, the stars and planets, and life as we know it today.

The first fragments, or particles, were protons and neutrons moving with tremendous velocity and possessing large amounts of energy. Collisions involving these high-energy protons and neutrons formed deuterium atoms (^2H), which are isotopes of hydrogen. As the universe expanded and cooled, tritium (^3H), another hydrogen isotope, formed as a result of collisions of neutrons with deuterium atoms. Subsequent capture of a proton produced helium (He). Scientists theorize that a universe that was principally composed of hydrogen and helium persisted for perhaps 100,000 years until the temperature decreased sufficiently to allow the formation of a simple molecule, hydrogen, two atoms of hydrogen bonded together (H_2).

Many millions of years later, the effect of gravity caused these small units to coalesce, first into clouds and eventually into stars, with temperatures of millions of degrees. In this setting, these small collections of protons and neutrons combined to form larger atoms such as carbon (C) and oxygen (O), then sodium (Na), neon (Ne), magnesium (Mg), silicon (Si), and so forth. Subsequent explosions of stars provided the conditions that formed many larger atoms. These fragments, gathered together by the force of gravity, are the most probable origin of the planets in our own solar system.

The reactions that formed the elements as we know them today were a result of a series of *fusion reactions*, the joining of nuclei to produce larger atoms at very high temperatures (millions of degrees Celsius). These fusion reactions are similar to processes that are currently being studied as a possible alternative source of nuclear power. We shall study such nuclear processes in more detail in Chapter 9.

Nuclear reactions of this type do not naturally occur on the earth today. The temperature is simply too low. As a result, we have, for the most part, a collection of stable elements existing as chemical compounds, atoms joined together by chemical bonds while retaining their identity even in the combined state. Silicon exists all around us as sand and soil in a combined form, silicon dioxide; most metals exist as part of a chemical compound, such as iron ore. We are learning more about the structure and properties of these compounds in this chapter.

FOR FURTHER UNDERSTANDING

How does tritium differ from "normal" hydrogen?

Would you expect to find similar atoms on other planets?

---

units but form crystal lattices composed of enormous numbers of positive and negative ions in extended three-dimensional networks. The effects of this basic structural difference are summarized in this section.

Physical State

All ionic compounds (for example, NaCl, KCl, and $NaNO_3$) are solids at room temperature; covalent compounds may be solids (sugar), liquids (H_2O, ethanol), or gases (carbon monoxide, carbon dioxide). The three-dimensional crystal structures that are characteristic of ionic compounds hold them in rigid, solid arrangements, whereas molecules of covalent compounds may be fixed, as in a solid, or more mobile, a characteristic of liquids and gases.

Melting and Boiling Points

The **melting point** is the temperature at which a solid is converted to a liquid, and the **boiling point** is the temperature at which a liquid is converted to a gas at a specified pressure. Considerable energy is required to break apart an ionic crystal lattice with uncountable numbers of ionic interactions and convert the ionic substance to a liquid or a gas. As a result, the melting and boiling temperatures for ionic compounds are generally higher than those of covalent compounds, whose molecules interact less strongly in the solid state. A typical ionic compound, sodium chloride, has a melting point of 801°C; methane, a covalent compound, melts at $-182°C$. Exceptions to this general rule do exist; diamond, a covalent solid with an extremely high melting point, is a well-known example.

Structure of Compounds in the Solid State

Ionic solids are *crystalline,* characterized by regular structures, whereas covalent solids may either be crystalline or have no regular structure. In the latter case, they are said to be *amorphous.*

Solutions of Ionic and Covalent Compounds

In Chapter 1, we saw that mixtures are either heterogeneous or homogeneous. A homogeneous mixture is a solution. Many ionic solids dissolve in solvents, such as water. An ionic solid, if soluble, will form positive and negative ions in solution by **dissociation.**

Because ions in water are capable of carrying (conducting) a current of electricity, we refer to these compounds as **electrolytes,** and the solution is termed an **electrolytic solution.** Covalent solids dissolved in solution usually retain their neutral (molecular) character and are **nonelectrolytes.** The solution is not an electrical conductor.

The role of the solvent in the dissolution of solids is discussed in Section 3.5.

3.4 Drawing Lewis Structures of Molecules

A Strategy for Drawing Lewis Structures of Molecules

In Section 3.1, we used Lewis structures of individual atoms to help us understand the bonding process. To begin to explain the relationship between molecular structure and molecular properties, we will first need a set of guidelines to help us write Lewis structures for more complex molecules.

5 **Learning Goal**
Draw Lewis structures for covalent compounds.

1. *Use chemical symbols for the various elements to write the skeletal structure of the compound.* To accomplish this, place the bonded atoms next to one another. This is relatively easy for simple compounds; however, as the number of atoms in the compound increases, the possible number of arrangements increases dramatically. We may be told the pattern of arrangement of the atoms in advance; if not, we can make an intelligent guess and see if a reasonable Lewis structure can be constructed. Three considerations are very important here:
   - the least electronegative atom will be placed in the central position (the central atom),
   - hydrogen and fluorine (and the other halogens) often occupy terminal positions,
   - carbon often forms chains of carbon–carbon covalent bonds.
2. *Determine the number of valence electrons associated with each atom; combine them to determine the total number of valence electrons in the compound.* However, if we are representing polyatomic cations or anions, we must account for the charge on the ion. Specifically,
   - for polyatomic cations, subtract one electron for each unit of positive charge. This accounts for the fact that the positive charge arises from electron loss.
   - for polyatomic anions, add one electron for each unit of negative charge. This accounts for excess negative charge resulting from electron gain.
3. *Connect the central atom to each of the surrounding atoms using electron pairs.* Then complete the octets of all of the atoms bonded to the central atom. Recall that hydrogen needs only two electrons to complete its valence shell. Electrons not involved in bonding must be represented as lone pairs (see p. 90, NH_3) and the total number of electrons in the structure must equal the number of valence electrons computed in our second step.

The skeletal structure indicates only the relative positions of atoms in the molecule or ion. Bonding information results from the Lewis structure.

The central atom is often the element farthest to the left and/or lowest in the periodic table.
The central atom is often the element in the compound for which there is only one atom.
Hydrogen is *never* the central atom.

A MEDICAL Connection

Blood Pressure and the Sodium Ion/Potassium Ion Ratio

When you have a physical exam, the physician measures your blood pressure. This indicates the pressure of blood against the walls of the blood vessels each time the heart pumps. A blood pressure reading is always characterized by two numbers. With every heartbeat, there is an increase in pressure; this is the systolic blood pressure. When the heart relaxes between contractions, the pressure drops; this is the diastolic pressure. Thus the blood pressure is expressed as two values—for instance, 117/72—measured in millimeters of mercury. Hypertension is simply defined as high blood pressure. To the body, it means that the heart must work too hard to pump blood, and this can lead to heart failure or heart disease.

Heart disease accounts for 50% of all deaths in the United States. Epidemiological studies correlate the following major risk factors with heart disease: heredity, sex, race, age, diabetes, cigarette smoking, high blood cholesterol, and hypertension. Obviously, we can do little about our age, sex, and genetic heritage, but we can stop smoking, limit dietary cholesterol, and maintain normal blood pressure.

The number of Americans with hypertension is alarmingly high: 60 million adults and children. More than 10 million of these individuals take medication to control blood pressure, at a cost of nearly $2.5 billion each year. In many cases, blood pressure can be controlled without medication by increasing physical activity, losing weight, decreasing consumption of alcohol, and limiting intake of sodium.

It has been estimated that the average American ingests 7.5–10 g of salt (NaCl) each day. Because NaCl is about 40% (by mass) sodium ions, this amounts to 3–4 g of sodium daily. Until 1989, the Food and Nutrition Board of the National Academy of Sciences National Research Council's defined *e*stimated *s*afe and *a*dequate *d*aily *d*ietary *i*ntake (ESADDI) of sodium ion was 1.1–3.3 g. Clearly, Americans exceed this recommendation.

Recently, studies have shown that excess sodium is not the sole consideration in the control of blood pressure. More important is the sodium ion/potassium ion (Na^+/K^+) ratio. That ratio should be about 0.6; in other words, our diet should contain about 67% more potassium than sodium. Does the typical American diet fall within this limit? Definitely not! Young American males (25–30 years old) consume a diet with a $Na^+/K^+ = 1.07$, and the diet of females of the same age range has a $Na^+/K^+ = 1.04$. It is little wonder that so many Americans suffer from hypertension.

How can we restrict sodium in the diet, while increasing the potassium? A variety of foods are low in sodium and high in potassium. These include fresh fruits and vegetables and fruit juices, a variety of cereals, unsalted nuts, and cooked dried beans (legumes). The majority of the sodium that we ingest comes from commercially prepared foods. The consumer must read the nutritional information printed on cans and packages to determine whether the sodium levels are within acceptable limits.

FOR FURTHER UNDERSTANDING

Find several commercial food products on the shelves of your local grocery store; read the labels and calculate the sodium ion–potassium ion ratios.

Describe each product that you have chosen in terms of its suitability for inclusion in the diet of a person with moderately elevated blood pressure.

---

4. *If the octet rule is not satisfied for the central atom, move one or more electron pairs from the surrounding atoms.* Use these electrons to create double or triple bonds (see p. 89, CO_2) until all atoms have an octet.
5. *After you are satisfied with the Lewis structure that you have constructed, perform a final electron count.* This allows you to verify that the total number of electrons and the number around each atom are correct.

Now, let us see how these guidelines are applied in the examples that follow.

5 Learning Goal

Draw Lewis structures for covalent compounds.

EXAMPLE 3.8   Drawing Lewis Structures of Covalent Compounds

Draw the Lewis structure of carbon dioxide, CO_2.

SOLUTION

Draw a skeletal structure of the molecule, arranging the atoms in their most probable order.

Continued—

For CO_2, two possibilities exist:

C—O—O   and   O—C—O

Referring to Figure 3.5, we find that the electronegativity of oxygen is 3.5, whereas that of carbon is 2.5. Our strategy dictates that the least electronegative atom, in this case carbon, is the central atom. Hence the skeletal structure O—C—O may be presumed correct.

Next, we want to determine the number of valence electrons on each atom and add them to arrive at the total for the compound.

For CO_2,

$$1 \text{ C atom} \times 4 \text{ valence electrons} = 4 \text{ e}^-$$

$$2 \text{ O atoms} \times 6 \text{ valence electrons} = 12 \text{ e}^-$$

$$16 \text{ e}^- \text{ total}$$

Now, use electron pairs to connect the central atom, C, to each oxygen with a single bond.

$$\text{O} : \text{C} : \text{O}$$

Distribute the electrons around the atoms (in pairs if possible) to satisfy the octet rule, eight electrons around each element.

$$: \overset{..}{\text{O}} : \text{C} : \overset{..}{\text{O}} :$$

This structure satisfies the octet rule for each oxygen atom, but not the carbon atom (only four electrons surround the carbon).

However, when this structure is modified by moving two electrons from each oxygen atom to a position between C and O, each oxygen and carbon atom is surrounded by eight electrons. The octet rule is satisfied, and the structure below is the most probable Lewis structure for CO_2.

$$\overset{..}{\underset{..}{\text{O}}} :: \text{C} :: \overset{..}{\underset{..}{\text{O}}}$$

In this structure, four electrons (two electron pairs) are located between C and each O, and these electrons are shared in covalent bonds. Because a **single bond** is composed of two electrons (one electron pair) and because four electrons "bond" the carbon atom to each oxygen atom in this structure, there must be two bonds between each oxygen atom and the carbon atom, a **double bond:**

The notation for a single bond : is equivalent to—(one pair of electrons).

The notation for a double bond : : is equivalent to═(two pairs of electrons).

We may write CO_2 as shown above or, replacing dots with dashes to indicate bonding electron pairs,

$$\overset{..}{\underset{..}{\text{O}}} = \text{C} = \overset{..}{\underset{..}{\text{O}}}$$

As a final step, let us do some "electron accounting." There are eight electron pairs, and they correspond to sixteen valence electrons (8 pair $\times$ 2e$^-$/pair). Furthermore, there are eight electrons around each atom and the octet rule is satisfied. Therefore

$$\overset{..}{\underset{..}{\text{O}}} = \text{C} = \overset{..}{\underset{..}{\text{O}}}$$

is a satisfactory way to depict the structure of CO_2.

5 Learning Goal

Draw Lewis structures for covalent compounds.

EXAMPLE 3.9 | Drawing Lewis Structures of Covalent Compounds

Draw the Lewis structure of ammonia, NH_3.

SOLUTION

When trying to implement the first step in our strategy we may be tempted to make H our central atom because it is less electronegative than N. But, remember the margin note in this section:

> "Hydrogen is *never* the central atom."

Hence,

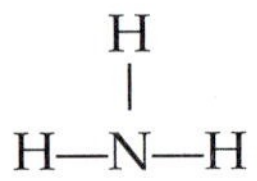

$$\begin{array}{c} H \\ | \\ H\!-\!N\!-\!H \end{array}$$

is our skeletal structure.

Applying our strategy to determine the total valence electrons for the molecule, we find that there are five valence electrons in nitrogen and one in each of the three hydrogens, for a total of eight valence electrons.

Applying our strategy for distribution of valence electrons results in the following Lewis diagram:

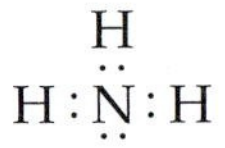

$$\begin{array}{c} H \\ H : \overset{..}{N} : H \end{array}$$

This satisfies the octet rule for nitrogen (eight electrons around N) and hydrogen (two electrons around each H) and is an acceptable structure for ammonia. Ammonia may also be written:

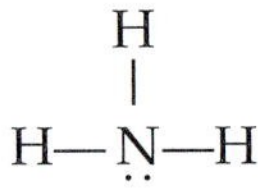

$$\begin{array}{c} H \\ | \\ H\!-\!\underset{..}{N}\!-\!H \end{array}$$

Note the pair of nonbonding electrons on the nitrogen atom. These are often called a **lone pair,** or *unshared* pair, of electrons. As we will see later in this section, lone pair electrons have a profound effect on molecular geometry. The geometry, in turn, affects the reactivity of the molecule.

QUESTION 3.15

Draw a Lewis structure for each of the following covalent compounds:

a. H_2O (water)
b. CH_4 (methane)

QUESTION 3.16

Draw a Lewis structure for each of the following covalent compounds:

a. C_2H_6 (ethane)
b. N_2 (nitrogen gas)

Multiple Bonds and Bond Energies

6 Learning Goal

Describe the relationship between bond order, bond energy, and bond length.

Hydrogen, oxygen, and nitrogen are present in the atmosphere as diatomic gases, H_2, O_2, and N_2. All are covalent molecules. Their stability and reactivity, however, are quite different. Hydrogen is an explosive material, sometimes used as a fuel.

Oxygen, although more stable than hydrogen, reacts with fuels in combustion. The explosion of the space shuttle *Challenger* resulted from the reaction of massive amounts of hydrogen and oxygen. Nitrogen, on the other hand, is extremely non-reactive. Because nitrogen makes up about 80% of the atmosphere, it dilutes the oxygen, which accounts for only about 20% of the atmosphere.

Breathing pure oxygen for long periods, although necessary in some medical situations, causes the breakdown of nasal and lung tissue over time. Oxygen diluted with nonreactive nitrogen is an ideal mixture for humans and animals to breathe.

Differences in reactivity among these three gases may be explained, in part, by the bonds holding the atoms together. The Lewis structure for H_2 (two valence electrons) is

$$H:H \quad \text{or} \quad H{-}H$$

For oxygen (twelve valence electrons, six on each atom), the only Lewis structure that satisfies the octet rule is

$$\ddot{O}:\,:\ddot{O} \quad \text{or} \quad \ddot{O}{=}\ddot{O}$$

The Lewis structure of N_2 (ten total valence electrons) must be

$$:N:\,:\,:N: \quad \text{or} \quad :N{\equiv}N:$$

Therefore,

N_2 has a *triple bond* (six bonding electrons).

O_2 has a *double bond* (four bonding electrons).

H_2 has a *single bond* (two bonding electrons).

A **triple bond,** in which three pairs of electrons are shared by two atoms, is very stable. More energy is required to break a triple bond than a double bond, and a double bond is stronger than a single bond. Stability is related to the bond energy. The **bond energy** is the amount of energy, in units of kilocalories or kilo-joules, required to break a bond holding two atoms together. Bond energy is there-fore a *measure* of stability. The values of bond energies decrease in the order *triple bond > double bond > single bond.*

The bond length is related to the presence or absence of multiple bonding. The distance of separation of two nuclei is greatest for a single bond, less for a double bond, and still less for a triple bond. The *bond length* decreases in the order *single bond > double bond > triple bond.*

> The term *bond order* is sometimes used to distinguish among single, double, and triple bonds. A bond order of 1 corresponds to a single bond, 2 corresponds to a double bond, and 3 corresponds to a triple bond.

QUESTION 3.17

Contrast a single and double bond with regard to

a. distance of separation of the bonded nuclei
b. strength of the bond

How are these two properties related?

QUESTION 3.18

Two nitrogen atoms in a nitrogen molecule are held together more strongly than the two chlorine atoms in a chlorine molecule. Explain this fact by comparing their respective Lewis structures.

Lewis Structures and Exceptions to the Octet Rule

The octet rule is remarkable in its ability to realistically model bonding and structure in covalent compounds. But, like any model, it does not adequately describe all systems. Beryllium, boron, and aluminum, in particular, tend to form compounds in which they are surrounded by fewer than eight electrons. This situation is termed an *incomplete octet*. Other molecules, such as nitric oxide:

$$\ddot{N}=\ddot{O}$$

are termed *odd electron* molecules. Note that it is impossible to pair all electrons to achieve an octet simply because the compound contains an odd number of valence electrons. Elements in the third period and beyond may involve *d* orbitals and form an *expanded octet,* with ten or even twelve electrons surrounding the central atom. Examples 3.10 and 3.11 illustrate common exceptions to the octet rule.

EXAMPLE 3.10 | **Drawing Lewis Structures of Covalently Bonded Compounds That Are Exceptions to the Octet Rule**

Draw the Lewis structure of beryllium hydride, BeH_2.

SOLUTION

A reasonable skeletal structure of BeH_2 is

$$H—Be—H$$

The total number of valence electrons in BeH_2 is

$$1 \text{ beryllium atom} \times 2 \text{ valence } e^-/\text{atom} = 2 \text{ } e^-$$
$$\underline{2 \text{ hydrogen atoms} \times 1 \text{ valence } e^-/\text{atom} = 2 \text{ } e^-}$$
$$4 \text{ } e^- \text{ total}$$

The resulting Lewis structure must be

$$H:Be:H \quad \text{ or } \quad H—Be—H$$

It is apparent that there is no way to satisfy the octet rule for Be in this compound. Consequently, BeH_2 is an exception to the octet rule. It contains an incomplete octet.

---

5 | **Learning Goal**
Draw Lewis structures for covalent compounds.

EXAMPLE 3.11 | **Drawing Lewis Structures of Covalently Bonded Compounds That Are Exceptions to the Octet Rule**

Draw the Lewis structure of phosphorus pentafluoride.

SOLUTION

A reasonable skeletal structure of PF_5 is

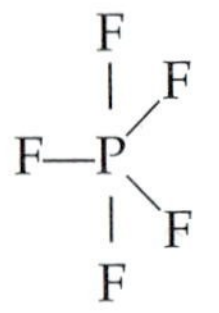

Phosphorus is a third-period element; it may have an expanded octet.

Continued—

The total number of valence electrons is:

$$1 \text{ phosphorus atom} \times 5 \text{ valence } e^-/\text{atom} = 5 \ e^-$$

$$5 \text{ fluorine atoms} \times 7 \text{ valence } e^-/\text{atom} = 35 \ e^-$$

$$40 \ e^- \text{ total}$$

Distributing the electrons around each F in the skeletal structure results in the Lewis structure:

$$\ddot{\underset{..}{:}}\ddot{F}: \\ | \quad \ddot{F}: \\ :\ddot{F}-P \\ | \quad \ddot{F}: \\ :\ddot{F}:$$

PF_5 is an example of a compound with an expanded octet.

Lewis Structures and Molecular Geometry: VSEPR Theory

The shape of a molecule plays a large part in determining its properties and re-activity. We may predict the shapes of various molecules by inspecting their Lewis structures for the orientation of their electron pairs. The covalent bond, for instance, in which bonding electrons are localized between the nuclear centers of the atoms, is *directional*; the bond has a specific orientation in space between the bonded atoms. Electrostatic forces in ionic bonds, in contrast, are *nondirectional*; they have no specific orientation in space. The specific orientation of electron pairs in covalent molecules imparts a characteristic shape to the molecules. Consider the following series of molecules whose Lewis structures are shown.

BeH_2 $\qquad$ $H:Be:H$

BF_3 $\qquad$ $:\ddot{F}: \\ :\ddot{F}:B:\ddot{F}:$

CH_4 $\qquad$ $H \\ :\ddot{} \\ H:C:H \\ H$

NH_3 $\qquad$ $H:\ddot{N}:H \\ H$

H_2O $\qquad$ $H:\ddot{O}:H$

The electron pairs around the central atom of the molecule arrange themselves to minimize electronic repulsion. This means that the electron pairs arrange themselves so that they can be as far as possible from each other. We may use this fact to predict molecular shape. This approach is termed the **valence shell electron pair repulsion (VSEPR) theory.**

Let's see how the VSEPR theory can be used to describe the bonding and structure of each of the preceding molecules.

BeH₂

As we saw in Example 3.10, beryllium hydride has two shared electron pairs around the beryllium atom. These electron pairs have minimum repulsion if they are located as far apart as possible while still bonding the hydrogens to the central

7 Learning Goal

Predict the geometry of molecules using the octet rule and Lewis structures.

Animation

Valence Shell Electron Pair Repulsion Theory

Animation

VSEPR and Molecular Geometry

Only four electrons surround the beryllium atom in BeH₂. Consequently, BeH₂ is a stable exception to the octet rule.

Figure 3.6

Bonding and geometry in beryllium hydride, BeH_2. (a) Linear geometry in BeH_2. (b) Ball-and-stick model of linear BeH_2.

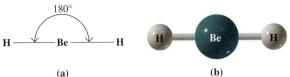

atom. This condition is met if the electron pairs are located on opposite sides of the molecule, resulting in a **linear structure,** 180° apart:

$$H:Be:H \quad \text{or} \quad H—Be—H$$

The *bond angle,* the angle between H—Be and Be—H bonds, formed by the two bonding pairs is 180° (Figure 3.6).

BF₃

> BF_3 has only six electrons around the central atom, B. It is one of a number of stable compunds that are exceptions to the octet rule.

Boron trifluoride has three shared electron pairs around the central atom. Placing the electron pairs in a plane, forming a triangle, minimizes the electron pair repulsion in this molecule, as depicted in Figure 3.7 and the following sketches:

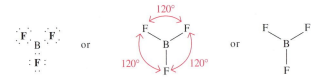

Such a structure is *trigonal planar,* and each F—B—F bond angle is 120°. We also find that compounds with central atoms in the same group of the periodic table have similar geometry. Aluminum, in the same group as boron, produces compounds such as AlH_3, which is also trigonal planar.

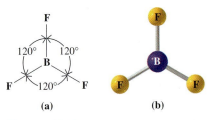

(a) (b)

Figure 3.7

Bonding and geometry in boron trifluoride, BF_3. (a) Trigonal planar geometry in BF_3. (b) Ball-and-stick model of trigonal planar BF_3.

CH₄

Methane has four shared pairs of electrons. Here, minimum electron repulsion is achieved by arranging the electrons at the corners of a tetrahedron (Figure 3.8). Each H—C—H bond angle is 109.5°. Methane has a three-dimensional **tetrahedral structure.** Silicon, in the same group as carbon, forms compounds such as $SiCl_4$ and SiH_4 that also have tetrahedral structures.

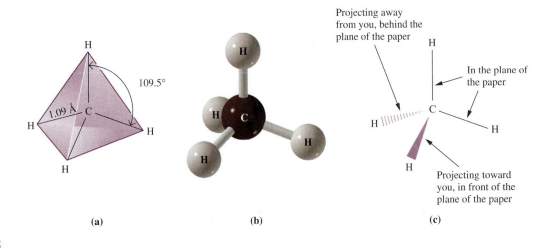

(a) (b) (c)

Figure 3.8

Representations of the three-dimensional structure of methane, CH_4. (a) Tetrahedral methane structure. (b) Ball-and-stick model of tetrahedral methane. (c) Three-dimensional representation of structure (b).

NH_3

Ammonia also has four electron pairs about the central atom. In contrast to methane, in which all four pairs are bonding, ammonia has three pairs of bonding electrons and one nonbonding lone pair of electrons. We might expect CH_4 and NH_3 to have similar but not identical electron pair arrangements. The lone pair in ammonia is more negative than the bonding pairs; some of the negative charge on the bonding pairs is offset by the hydrogen atoms with their positive nuclei. Thus the arrangement of electron pairs in ammonia is distorted.

The hydrogen atoms in ammonia are pushed closer together than in methane (Figure 3.9). The bond angle is 107° because lone pair–bond pair repulsions are greater than bond pair–bond pair repulsions. The structure or shape is termed *trigonal pyramidal*, and the molecule is termed a **trigonal pyramidal molecule.**

> CH_4, NH_3, and H_2O all have eight electrons around their central atoms; all obey the octet rule.

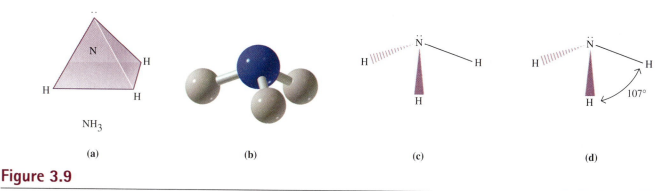

NH₃

(a)   (b)   (c)   (d)

Figure 3.9

The structure of the ammonia molecule. (a) Pyramidal ammonia structure. (b) Ball-and-stick model of pyramidal ammonia. (c) A three-dimensional sketch. (d) The H—N—H bond angle in ammonia.

H_2O

Water also has four electron pairs around the central atom; two pairs are bonding, and two pairs are nonbonding. These four electron pairs are approximately tetrahedral to each other; however, because of the difference between bonding and nonbonding electrons, noted earlier, the tetrahedral relationship is only approximate.

The **angular** (or *bent*) **structure** has a bond angle of 104.5°, which is 5° smaller than the tetrahedral angle, because of the repulsive effects of the lone pairs of electrons (as shown in Figure 3.10).

> Molecules with five and six electron pairs also exist. They may have structures that are *trigonal bipyramidal* (forming a six-sided figure) or *octahedral* (forming an eight-sided figure).

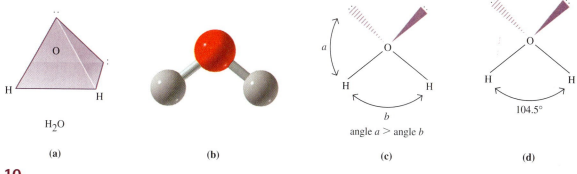

H₂O

(a)   (b)   (c)   (d)

Figure 3.10

The structure of the water molecule. (a) Angular water structure. (b) Ball-and-stick model of angular water. (c) A three-dimensional sketch. (d) The H—O—H bond angle in water.

TABLE 3.5 Molecular Structure: The Geometry of a Molecule Is Affected by the Number of Nonbonded Electron Pairs Around the Central Atom and the Number of Bonded Atoms

Bonded Atoms	Nonbonding Electron Pairs	Bond Angle	Molecular Structure	Example	Structure
2	0	180°	Linear	CO_2	
3	0	120°	Trigonal planar	SO_3	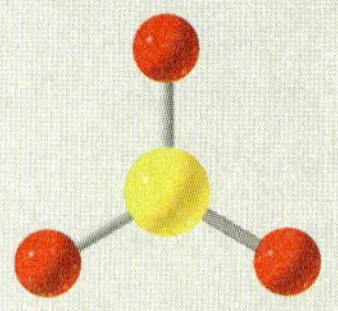
2	1	<120°	Angular	SO_2	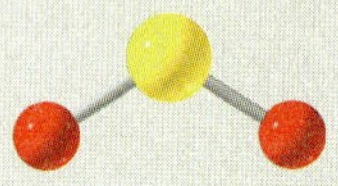
4	0	~109°	Tetrahedral	CH_4	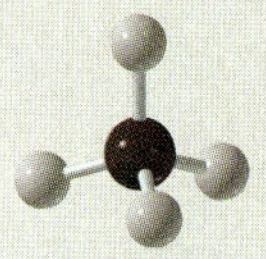
3	1	~107°	Trigonal pyramidal	NH_3	
2	2	~104.5°	Angular	H_2O	

The characteristics of linear, trigonal planar, angular, and tetrahedral structures are summarized in Table 3.5.

Periodic Structural Relationships

The molecules considered above contain the central atoms Be (Group IIA), B (Group IIIA), C (Group IVA), N (Group VA), and O (Group VIA). We may expect that a number of other compounds, containing the same central atom, will have structures with similar geometries. This is an approximation, not always true, but still useful in expanding our ability to write reasonable, geometrically accurate structures for a large number of compounds.

The periodic similarity of group members is also useful in predictions involving bonding. Consider Group VI, oxygen, sulfur, and selenium (Se). Each has six valence electrons. Each needs two more electrons to complete its octet. Each should react with hydrogen, forming H_2O, H_2S, and H_2Se.

If we recall that H_2O is an angular molecule with the following Lewis structure,

$$H : \overset{..}{\underset{..}{O}} :$$
$$H$$

it follows that H_2S and H_2Se would also be angular molecules with similar Lewis structures, or

$$H : \overset{..}{\underset{..}{S}} : \quad \text{and} \quad H : \overset{..}{\underset{..}{Se}} :$$
$$H \qquad\qquad\qquad H$$

This logic applies equally well to the other representative elements.

More Complex Molecules

A molecule such as dimethyl ether, CH_3—O—CH_3, has two different central atoms: oxygen and carbon. We could picture the parts of the molecule containing the CH_3 group (commonly referred to as the *methyl group*) as exhibiting tetrahedral geometry (analogous to methane):

The structure and properties of ethers are described in section 12.3.

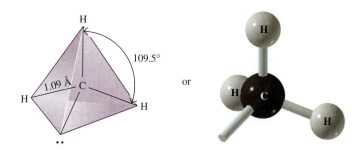

The part of the molecule connecting these two methyl groups (the oxygen) would have a bond angle similar to that of water (in which oxygen is also the central atom), approximately 104°, as seen in Figure 3.11. This is a reasonable way to represent the molecule dimethyl ether.

The structure and properties of amines are described in section 13.3.

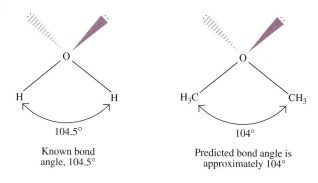

Known bond
angle, 104.5°

Predicted bond angle is
approximately 104°

Figure 3.11

A comparison of the bonding in water and dimethyl ether.

Trimethylamine, $(CH_3)_3N$, is a member of the amine family. As in the case of ether, two different central atoms are present. Carbon and nitrogen determine the geometry of amines. In this case, the methyl group should assume the tetrahedral geometry of methane, and the nitrogen atom should have the methyl groups in a pyramidal arrangement, similar to the hydrogen atoms in ammonia, as seen in Figure 3.12. This creates a pyramidal geometry around nitrogen. H—N—H bond angles in ammonia are 107°; experimental information shows a very similar C—N—C bond angle in trimethylamine.

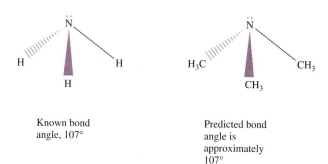

Known bond
angle, 107°

Predicted bond
angle is
approximately
107°

Figure 3.12

A comparison of the bonding in ammonia and trimethylamine.

QUESTION 3.19

Sketch the geometry of each of the following molecules (basing your structure on the Lewis electron dot representation of the molecule):

a. PH_3
b. SiH_4

QUESTION 3.20

Sketch the geometry of each of the following molecules (basing your structure on the Lewis electron dot representation of the molecule):

a. C_2H_4
b. C_2H_2

Animation

Ionic, Covalent, and Polar Covalent Compounds

It is essential to represent the molecule in its correct geometric form, using the Lewis and VSEPR theories, to understand its physical and chemical behavior. In the rest of this section, we use these models to predict molecular behavior.

Lewis Structures and Polarity

A molecule is *polar* if its centers of positive and negative charges do not coincide. Molecules whose positive and negative charges are separated when the molecules are placed in an electric field align themselves with the field. The molecule behaves as a *dipole* (having two "poles" or ends, one pole is more negative and the other pole is more positive) and is said to be polar.

Nonpolar molecules will not align with the electric field because their positive and negative centers are the same; no dipole exists. These molecules are nonpolar.

The hydrogen molecule is the simplest nonpolar molecule:

$$H:H \quad \text{or} \quad H-H$$

Both electrons, on average, are located at the center of the molecule and positively charged nuclei are on either side. The center of both positive and negative charge is at the center of the molecule; therefore the bond is nonpolar.

We may arrive at the same conclusion by considering the equality of electron sharing between the atoms being bonded. Electron sharing is related to the concept of electronegativity introduced in Section 3.1.

The atoms that comprise H_2 are identical; their electronegativity (electron attracting power) is the same. Thus the electrons remain at the center of the molecule, and the molecule is nonpolar.

Similarly, O_2, N_2, Cl_2, and F_2 are nonpolar molecules with nonpolar bonds. Arguments analogous to those made for hydrogen explain these observations as well.

Let's next consider hydrogen fluoride, HF. Fluorine is more electronegative than hydrogen. This indicates that the electrons are more strongly attracted to a fluorine atom than to a hydrogen atom. This results in a bond and molecule that are polar. The symbol

Less electronegative part of bond ⟶ ⟵ More electronegative part of bond

placed below a bond indicates the direction of polarity. The more negative end of the bond is near the head of the arrow, and the less negative end of the bond is next to the tail of the arrow. Symbols using the Greek letter *delta* may also be used to designate polarity. In this system, the more negative end of the bond is

Remember: Electronegativity deals with atoms in molecules, whereas electron affinity and ionization energy deal with isolated atoms.

designated δ^- (partial negative), and the less negative end is designated δ^+ (partial positive). The symbols are applied to the hydrogen fluoride molecule as follows:

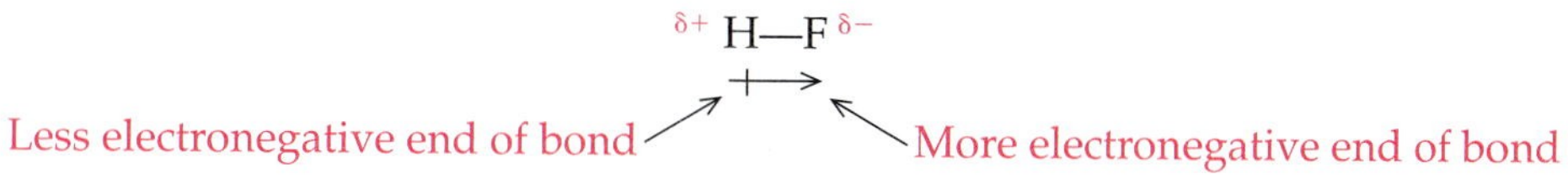

$$\delta^+ \text{ H—F } \delta^-$$

Less electronegative end of bond   More electronegative end of bond

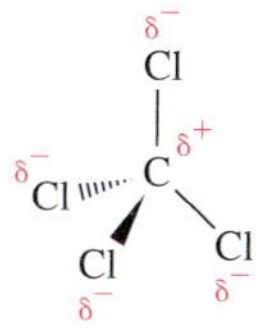

> The word *partial* implies less than a unit charge. Thus δ^+ and δ^- do not imply overall charge on a unit (such as the + or − sign on an ion); they are meant to show only the relative distribution of charge within a unit. The HF molecule shown is *neutral* but has an unequal charge distribution *within* the molecule.

HF is a **polar covalent molecule** characterized by **polar covalent bonding.** This implies that the electrons are shared unequally.

A molecule containing all nonpolar bonds must also be nonpolar. In contrast, a molecule containing polar bonds may be either polar or nonpolar depending on the relative arrangement of the bonds and any lone pairs of electrons.

Let's now examine the bonding in carbon tetrachloride. All four bonds of CCl_4 are polar because of the electronegativity difference between C and Cl. However, because of the symmetrical arrangement of the four C—Cl bonds, their polarities cancel, and the molecule is nonpolar covalent:

Now look at H_2O. Because of its angular (bent) structure, the polar bonds do not cancel, and the molecule is polar covalent:

Animation
Molecular Geometry and Polarity

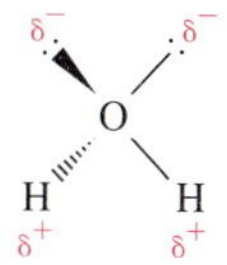

The electron density is shifted away from the hydrogens toward oxygen in the water molecule. In carbon tetrachloride, equal electron "pull" in all directions results in a nonpolar covalent molecule.

QUESTION 3.21

Predict which of the following bonds are polar, and, if polar, in which direction the electrons are pulled:

a. O—S
b. C≡N
c. Cl—Cl
d. I—Cl

QUESTION 3.22

Predict which of the following bonds are polar, and, if polar, in which direction the electrons are pulled:

a. Si—Cl
b. S—Cl
c. H—C
d. C—C

QUESTION 3.23

Predict whether each of the following molecules is polar:

    a. BCl_3
    b. NH_3
    c. HCl
    d. $SiCl_4$

QUESTION 3.24

Predict whether each of the following molecules is polar:

    a. CO_2
    b. SCl_2
    c. $BrCl$
    d. CS_2

3.5 Properties Based on Electronic Structure and Molecular Geometry

8 Learning Goal
Understand the role that molecular geometry plays in determining the solubility and melting and boiling points of compounds.

9 Learning Goal
Use the principles of VSEPR theory and molecular geometry to predict relative melting points, boiling points, and solubilities of compounds.

Intramolecular forces are attractive forces *within* molecules. They are the chemical bonds that determine the shape and polarity of individual molecules. **Intermolecular forces,** on the other hand, are forces *between* molecules.

It is important to distinguish between these two kinds of forces. *Intermolecular* forces determine such properties as the solubility of one substance in another and the freezing and boiling points of liquids. But, at the same time, we must realize that these forces are a direct consequence of the *intramolecular* forces in the individual units, the molecules.

In the following section we will see some of the consequences of bonding that are directly attributable to differences in intermolecular forces (solubility, boiling and melting points). In Section 7.5, we will investigate, in some detail, the nature of the intermolecular forces in water.

Solubility

The solute is the substance that is present in lesser quantity, and the solvent is the substance that is present in the greater amount (see Section 7.1).

Solubility is defined as the maximum amount of solute that dissolves in a given amount of solvent at a specified temperature. Polar molecules are most soluble in polar solvents, whereas nonpolar molecules are most soluble in nonpolar solvents. This is the rule of *"like dissolves like."* Substances of similar polarity are mutually soluble, and large differences in polarity lead to insolubility.

Case I: Ammonia and Water

The interaction of water and ammonia is an example of a particularly strong intermolecular force, the hydrogen bond; this phenomenon is discussed in Chapter 6.

Ammonia is soluble in water because both ammonia and water are polar molecules:

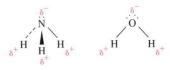

Dissolution of ammonia in water is a consequence of the intermolecular forces among the ammonia and water molecules. The δ^- end (a nitrogen) of the ammonia

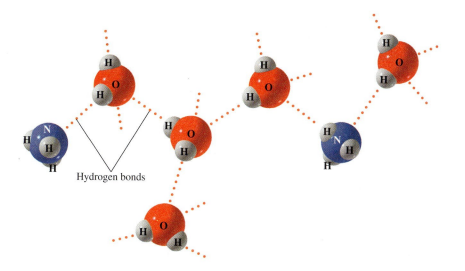

Figure 3.13

The interaction of polar covalent water molecules (the solvent) with polar covalent solute molecules such as ammonia, results in the formation of a solution.

molecule is attracted to the δ^+ end (a hydrogen) of the water molecule; at the same time the δ^+ end (a hydrogen) of the ammonia molecule is attracted to the δ^- end (an oxygen) of the water molecule. These attractive forces thus "pull" ammonia into water (and water into ammonia), and the ammonia molecules are randomly distributed throughout the solvent, forming a homogeneous solution (Figure 3.13).

Case II: Oil and Water

Oil and water do not mix; oil is a nonpolar substance composed primarily of molecules containing carbon and hydrogen. Water molecules, on the other hand, are quite polar. The potential solvent, water molecules, has partially charged ends, whereas the molecules of oil do not. As a result, water molecules exert their attractive forces on other water molecules, not on the molecules of oil; the oil remains insoluble, and because it is less dense than water, the oil simply floats on the surface of the water. This is illustrated in Figure 3.14.

Boiling Points of Liquids and Melting Points of Solids

Boiling a liquid requires energy. The energy is used to overcome the intermolecular attractive forces in the liquid, driving the molecules into the less associated gas phase. The amount of energy required is related to the boiling temperature. This, in turn, depends on the strength of the intermolecular attractive forces in the liquid, which parallel the polarity. This is not the only determinant of boiling point. Molecular mass is also an important consideration. The larger the mass of the molecule, the more difficult it becomes to convert the collection of molecules to a gas.

A similar argument can be made for the melting points of solids. The ease of conversion of a solid to a liquid also depends on the magnitude of the attractive forces in the solid. The situation actually becomes very complex for ionic solids because of the complexity of the crystal lattice.

As a general rule, polar compounds have strong attractive (intermolecular) forces, and their boiling and melting points tend to be higher than those of nonpolar substances of similar molecular mass.

Melting and boiling points of a variety of substances are included in Table 3.6.

Figure 3.14

The interaction of polar water molecules and nonpolar oil molecules. The familiar salad dressing, oil and vinegar, forms two layers. The oil does not dissolve in vinegar, an aqueous solution of acetic acid.

See Section 6.3, the solid state.

TABLE 3.6 Melting and Boiling Points of Selected Compounds in Relation to Their Bonding Type			
Formula (Name)	Bonding Type	M.P. (°C)	B.P. (°C)
N_2 (nitrogen)	Nonpolar covalent	−210	−196
O_2 (oxygen)	Nonpolar covalent	−219	−183
NH_3 (ammonia)	Polar covalent	−78	−33
H_2O (water)	Polar covalent	0	100
NaCl (sodium chloride)	Ionic	801	1413
KBr (potassium bromide)	Ionic	730	1435

QUESTION 3.25

Predict which compound in each of the following groups should have the higher melting and boiling points (Hint: Write the Lewis dot structure and determine whether the molecule is polar or nonpolar.):

a. H_2O and C_2H_4

b. CO and CH_4

c. NH_3 and N_2

d. Cl_2 and ICl

QUESTION 3.26

Predict which compound in each of the following groups should have the higher melting and boiling points (Hint: Write the Lewis dot structure and determine whether the molecule is polar or nonpolar.):

a. C_2H_6 and CH_4

b. CO and NO

c. F_2 and Br_2

d. $CHCl_3$ and CCl_4

SUMMARY

3.1 Chemical Bonding

When two atoms are joined to make a chemical compound, the force of attraction between the two atoms is the *chemical bond. Ionic bonding* is characterized by electron transfer before bond formation, forming an *ion pair*. In *covalent bonding*, electrons are shared between atoms in the bonding process. *Polar covalent bonding*, like covalent bonding, is based on the concept of electron sharing; however, the sharing is unequal and is based on the *electronegativity* difference between joined atoms. The *Lewis symbol*, showing only valence electrons, is a convenient way of representing atoms singly or in combination.

3.2 Naming Compounds and Writing Formulas of Compounds

The "shorthand" symbol for a compound is its *formula*. The formula identifies the number and type of atoms in the compound.

An ion that consists of only a single atom is said to be *monatomic. Polyatomic ions*, such as the hydroxide ion, OH^-, are composed of two or more atoms bonded together.

Names of ionic compounds are derived from the names of their ions. The name of the cation appears first, followed by the name of the anion. In the Stock system for naming an ion (the systematic name), a Roman numeral indicates the charge of the cation. In the older common nomenclature system, the suffix *-ous* indicates the lower of the ionic charges, and the suffix *-ic* indicates the higher ionic charge.

Most covalent compounds are formed by the reaction of nonmetals. Covalent compounds exist as *molecules*.

The convention used for naming covalent compounds is as follows:

• The names of the elements are written in the order in which they appear in the formula.

• A prefix indicating the number of each kind of atom found in the unit is placed before the name of the element.

- The stem of the name of the last element is used with the suffix -ide.

Many compounds are so familiar that their common names are used. It is useful to be able to correlate both systematic and common names with the corresponding molecular formula.

3.3 Properties of Ionic and Covalent Compounds

Covalently bonded molecules are discrete units, and they have less tendency to form extended structures in the solid state. Ionic compounds, with ions joined by electrostatic attraction, do not have definable units but form *crystal lattices* composed of positive and negative ions in extended three-dimensional networks.

The *melting point* is the temperature at which a solid is converted to a liquid; the *boiling point* is the temperature at which a liquid is converted to a gas. Melting and boiling temperatures for ionic compounds are generally higher than those of covalent compounds.

Ionic solids are crystalline, whereas covalent solids may be either crystalline or amorphous.

Many ionic solids dissolve in water, *dissociating* into positive and negative ions (an *electrolytic solution*). Because these ions can carry (conduct) a current of electricity, they are called *electrolytes*. Covalent solids in solution usually retain their neutral character and are *nonelectrolytes*.

3.4 Drawing Lewis Structures of Molecules

The procedure for drawing Lewis structures of molecules involves writing a skeletal structure of the molecule, arranging the atoms in their most probable order, determining the number of valence electrons on each atom, and combining them to get the total for the compound. The electrons are then distributed around the atoms (in pairs if possible) to satisfy the octet rule. At this point, electron pairs may be moved, creating double or triple bonds, to satisfy the octet rule for all atoms and produce the final structure.

The stability of a covalent compound is related to *bond energy*. The magnitude of the bond energy decreases in the order *triple bond > double bond > single bond*. The bond length decreases in the order *single bond > double bond > triple bond*.

The valence shell electron pair repulsion theory (VSEPR) states that electron pairs around the central atom of a molecule arrange themselves to minimize electronic repulsion; the electrons orient themselves as far as possible from each other. Two electron pairs around the central atom lead to a *linear* arrangement of the attached atoms; three indicate a *trigonal planar* arrangement, and four result in a *tetrahedral* geometry. Both *lone pair* and bonding pair electrons must be taken into account when predicting structure. Molecules

with fewer than four and as many as five or six electron pairs around the central atom also exist. They are exceptions to the octet rule.

A molecule is polar if its centers of positive and negative charges do not coincide. A *polar covalent* molecule has at least one *polar covalent bond*. An understanding of the concept of *electronegativity*, the relative electron-attracting power of atoms in molecules, helps us to assess the polarity of a bond.

A molecule containing all nonpolar bonds must be nonpolar. A molecule containing polar bonds may be either polar or nonpolar, depending on the relative position of the bonds.

3.5 Properties Based on Electronic Structure and Molecular Geometry

Attractions between molecules are called *intermolecular forces*. *Intramolecular forces*, on the other hand, are the attractive forces within molecules. The intermolecular forces determine such properties as the solubility of one substance in another and the freezing and boiling points of liquids.

Solubility is the maximum amount of solute that dissolves in a given amount of solvent at a specified temperature. Polar molecules are most soluble in polar solvents; nonpolar molecules are most soluble in nonpolar solvents. This is the rule of "like dissolves like."

As a general rule, polar compounds have strong intermolecular forces, and their boiling and melting points tend to be higher than nonpolar compounds of similar molecular mass.

KEY TERMS

angular structure (3.4)	lone pair (3.4)
boiling point (3.3)	melting point (3.3)
bond energy (3.4)	molecule (3.2)
chemical bond (3.1)	monatomic ion (3.2)
covalent bonding (3.1)	nomenclature (3.2)
crystal lattice (3.1)	nonelectrolyte (3.3)
dissociation (3.3)	polar covalent bonding (3.4)
double bond (3.4)	polar covalent molecule (3.4)
electrolyte (3.3)	polyatomic ion (3.2)
electrolytic solution (3.3)	single bond (3.4)
electronegativity (3.1)	solubility (3.5)
formula (3.2)	tetrahedral structure (3.4)
intermolecular force (3.5)	trigonal pyramidal
intramolecular force (3.5)	molecule (3.4)
ionic bonding (3.1)	triple bond (3.4)
ion pair (3.1)	valence shell electron pair
Lewis symbol (3.1)	repulsion (VSEPR)
linear structure (3.4)	theory (3.4)

QUESTIONS AND PROBLEMS

Chemical Bonding

Foundations

3.27 Classify each of the following compounds as ionic or covalent:
a. $MgCl_2$
b. CO_2
c. H_2S
d. NO_2

3.28 Classify each of the following compounds as ionic or covalent:
a. NaCl
b. CO
c. ICl
d. H_2

3.29 Classify each of the following compounds as ionic or covalent:
a. SiO_2
b. SO_2
c. SO_3
d. $CaCl_2$

3.30 Classify each of the following compounds as ionic or covalent:
a. NF_3
b. NaF
c. CsF
d. $SiCl_4$

Applications

3.31 Using Lewis symbols, write an equation predicting the product of the reaction of
a. S + H
b. P + H

3.32 Using Lewis symbols, write an equation predicting the product of the reaction of
a. Si + H
b. Ca + F

3.33 Explain, using Lewis symbols and the octet rule, why helium is so nonreactive.

3.34 Explain, using Lewis symbols and the octet rule, why neon is so nonreactive.

Naming Compounds and Writing Formulas of Compounds

Foundations

3.35 Name each of the following ions:
a. Cu^{2+}
b. Fe^{2+}
c. Fe^{3+}

3.36 Name each of the following ions:
a. S^{2-}
b. Cl^-
c. CO_3^{2-}

3.37 Write the symbol for each of the following monatomic ions:
a. the potassium ion
b. the bromide ion

3.38 Write the symbol for each of the following monatomic ions:
a. the calcium ion
b. the chromium(VI) ion

3.39 Write the formula for each of the following complex ions:
a. the sulfate ion
b. the nitrate ion

3.40 Write the formula for each of the following complex ions:
a. the phosphate ion
b. the bicarbonate ion

3.41 Write the correct formula for each of the following:
a. sodium chloride
b. magnesium bromide

3.42 Write the correct formula for each of the following:
a. copper(II) oxide
b. iron(III) oxide

3.43 Write the correct formula for each of the following:
a. silver cyanide
b. ammonium chloride

3.44 Write the correct formula for each of the following:
a. magnesium carbonate
b. magnesium bicarbonate

3.45 Name each of the following compounds:
a. $MgCl_2$
b. $AlCl_3$

3.46 Name each of the following compounds:
a. Na_2O
b. $Fe(OH)_3$

3.47 Name each of the following covalent compounds:
a. NO_2
b. SO_3

3.48 Name each of the following covalent compounds:
a. N_2O_4
b. CCl_4

Applications

3.49 Predict the formula of a compound formed from
a. aluminum and oxygen
b. lithium and sulfur

3.50 Predict the formula of a compound formed from
a. boron and hydrogen
b. magnesium and phosphorus

3.51 Predict the formula of a compound formed from
a. carbon and oxygen
b. sulfur and hydrogen

3.52 Predict the formula of a compound formed from
a. calcium and oxygen
b. silicon and hydrogen

3.53 Name each of the following:
a. NaClO
b. $NaClO_2$

3.54 Name each of the following:
a. $NaClO_3$
b. $NaClO_4$

Properties of Ionic and Covalent Compounds

Foundations

3.55 Contrast ionic and covalent compounds with respect to their solid-state structure.

3.56 Contrast ionic and covalent compounds with respect to their behavior in solution.

3.57 Contrast ionic and covalent compounds with respect to their relative boiling points.

3.58 Contrast ionic and covalent compounds with respect to their relative melting points.

Applications

3.59 Would KCl be expected to be a solid at room temperature? Why?

3.60 Would CCl_4 be expected to be a solid at room temperature? Why?

3.61 Would H_2O or CCl_4 be expected to have a higher boiling point? Why?

3.62 Would H_2O or CCl_4 be expected to have a higher melting point? Why?

Drawing Lewis Structures of Molecules

Foundations

3.63 Draw the appropriate Lewis structure for each of the following atoms:
  a. H
  b. He
  c. C
  d. N

3.64 Draw the appropriate Lewis structure for each of the following atoms:
  a. Be
  b. B
  c. F
  d. S

3.65 Draw the appropriate Lewis structure for each of the following ions:
  a. Li^+
  b. Mg^{2+}
  c. Cl^-
  d. P^{3-}

3.66 Draw the appropriate Lewis structure for each of the following ions:
  a. Be^{2+}
  b. Al^{3+}
  c. O^{2-}
  d. S^{2-}

Applications

3.67 Give the Lewis structure for each of the following compounds:
  a. NCl_3
  b. CH_3OH
  c. CS_2

3.68 Give the Lewis structure for each of the following compounds:
  a. HNO_3
  b. CCl_4
  c. PBr_3

3.69 Using the VSEPR theory, predict the geometry, polarity, and water solubility of each compound in Question 3.67.

3.70 Using the VSEPR theory, predict the geometry, polarity, and water solubility of each compound in Question 3.68.

3.71 Ethanol (ethyl alcohol or grain alcohol) has a molecular formula of C_2H_5OH. Represent the structure of ethanol using the Lewis electron dot approach.

3.72 Formaldehyde, H_2CO, in water solution has been used as a preservative for biological specimens. Represent the Lewis structure of formaldehyde.

3.73 Acetone, C_3H_6O, is a common solvent. It is found in such diverse materials as nail polish remover and industrial solvents. Draw its Lewis structure if its skeletal structure is

$$\begin{array}{c} O \\ | \\ C-C-C \end{array}$$

3.74 Ethylamine is an example of an important class of organic compounds. The molecular formula of ethylamine is $CH_3CH_2NH_2$. Draw its Lewis structure.

3.75 Predict whether the bond formed between each of the following pairs of atoms would be ionic, nonpolar, or polar covalent:
  a. S and O
  b. Si and P
  c. Na and Cl
  d. Na and O
  e. Ca and Br

3.76 Predict whether the bond formed between each of the following pairs of atoms would be ionic, nonpolar, or polar covalent:
  a. Cl and Cl
  b. H and H
  c. C and H
  d. Li and F
  e. O and O

3.77 Draw an appropriate covalent Lewis structure formed by the simplest combination of atoms in Problem 3.75 for each molecule that involves a nonpolar or polar covalent bond.

3.78 Draw an appropriate covalent Lewis structure formed by the simplest combination of atoms in Problem 3.76 for each molecule that involves a nonpolar or polar covalent bond.

Properties Based on Electronic Structure and Molecular Geometry

3.79 What is the relationship between the polarity of a bond and the polarity of the molecule?

3.80 What effect does polarity have on the solubility of a compound in water?

3.81 What effect does polarity have on the melting point of a pure compound?

3.82 What effect does polarity have on the boiling point of a pure compound?

3.83 Would you expect KCl to dissolve in water?

3.84 Would you expect ethylamine (Question 3.74) to dissolve in water?

FOR FURTHER UNDERSTANDING

1. Predict differences in our global environment that may have arisen if the freezing point and boiling point of water were 20°C higher than they are.
2. Would you expect the compound $C_2S_2H_4$ to exist? Why or why not?
3. Write a Lewis structure for the ammonium ion. Explain why its charge must be +1.
4. Which of the following compounds would be predicted to have the higher boiling point? Explain your reasoning.

Ethanol   Ethane

5. Why does the octet rule not work well for compounds of lanthanide and actinide elements? Suggest a number other than eight that may be more suitable.

4

LEARNING GOALS

1 Know the relationship between the mole and Avogadro's number and the usefulness of these quantities.

2 Perform calculations using Avogadro's number and the mole.

3 Write chemical formulas for common inorganic substances.

4 Calculate the formula weight and molar mass of a compound.

5 Know the major function served by the chemical equation, the basis for chemical calculations.

6 Classify chemical reactions by type: combination, decomposition, or replacement.

7 Recognize the various classes of chemical reactions: precipitation, reactions with oxygen, acid-base, and oxidation-reduction.

8 Balance chemical equations given the identity of products and reactants.

9 Calculate the number of moles or grams of product resulting from a given number of moles or grams of reactants or the number of moles or grams of reactant needed to produce a certain number of moles or grams of product.

10 Calculate theoretical and percent yield.

Calculations and the Chemical Equation

The calculation of chemical quantities based on chemical equations, termed *stoichiometry*, is the application of logic and arithmetic to chemical systems to answer questions such as the following:

A pharmaceutical company wishes to manufacture 1000 kg of a product next year. How much of each of the starting materials must be ordered? If the starting materials cost $20/g, how much money must be budgeted for chemicals in order to complete the project?

We often need to predict the quantity of a product produced from the reaction of a given amount of material. This calculation is not only possible but can be done with relative ease. It is equally possible to calculate how much of a material would be necessary to produce a desired amount of product.

What is required is a recipe: a procedure to follow. The basis for our recipe is the chemical equation. A properly written chemical equation provides all of the necessary information for the chemical calculation. The critical piece of information is the combining ratio of elements or compounds that must occur to produce a certain amount of product or products.

In this chapter, we define the mole, the fundamental unit of measure of chemical arithmetic, learn to write and balance chemical equations, and use these tools to perform calculations of chemical quantities.

Two solutions are mixed, forming a solid, termed a *precipitate*. Propose a strategy for separating the precipitate from the solution.

4.1 The Mole Concept and Atoms

Atoms are exceedingly small, yet their masses have been experimentally determined for each of the elements. The unit of measurement for these determinations is the **atomic mass unit,** abbreviated amu:

$$1 \text{ amu} = 1.661 \times 10^{-24} \text{ g}$$

The Mole and Avogadro's Number

1 Learning Goal
Know the relationship between the mole and Avogadro's number and the usefulness of these quantities.

The exact value of the atomic mass unit is defined in relation to a standard, just as the units of the metric system represent defined quantities. The carbon-12 isotope has been chosen and is assigned a mass of exactly 12 atomic mass units. Hence this standard reference point defines an atomic mass unit as exactly one-twelfth the mass of a carbon-12 atom.

The term atomic weight is not correct but is a fixture in common usage. Just remember that atomic weight is really "average atomic mass."

The periodic table provides atomic weights in atomic mass units. These atomic weights are average values, based on the contribution of all naturally occurring isotopes of the particular element. For example, the average mass of a carbon atom is 12.01 amu and

$$\frac{12.01 \cancel{\text{amu}} \text{ C}}{\text{C atom}} \times \frac{1.661 \times 10^{-24} \text{ g C}}{1 \cancel{\text{amu}} \text{ C}} = 1.995 \times \frac{10^{-23} \text{ g C}}{\text{C atom}}$$

The average mass of a helium atom is 4.003 amu and

$$\frac{4.003 \cancel{\text{amu}} \text{ He}}{\text{He atom}} \times \frac{1.661 \times 10^{-24} \text{ g He}}{1 \cancel{\text{amu}} \text{ He}} = 6.649 \times \frac{10^{-24} \text{ g He}}{\text{He atom}}$$

In everyday work, chemists use much larger quantities of matter (typically, grams or kilograms). A more practical unit for defining a "collection" of atoms is the **mole:**

$$1 \text{ mol of atoms} = 6.022 \times 10^{23} \text{ atoms of an element}$$

This number is **Avogadro's number.** Amedeo Avogadro, a nineteenth-century scientist, conducted a series of experiments that provided the basis for the mole concept. This quantity is based on the number of carbon-12 atoms in one mole of carbon-12.

The practice of defining a unit for a quantity of small objects is common; a *dozen* eggs, a *ream* of paper, and a *gross* of pencils are well-known examples. Similarly, a mole is 6.022×10^{23} individual units of anything. We could, if we desired, speak of a mole of eggs or a mole of pencils. However, in chemistry, we use the mole to represent a specific quantity of atoms, ions, or molecules.

The mole (mol) and the atomic mass unit (amu) are related. The atomic mass of an element corresponds to the average mass of a single atom in amu *and* the mass of a mole of atoms in grams.

The mass of 1 mol of atoms, in grams, is defined as the **molar mass.** Consider this relationship for sodium in Example 4.1.

2 Learning Goal
Perform calculations using Avogadro's number and the mole.

EXAMPLE 4.1 Relating Avogadro's Number to Molar Mass

Calculate the mass, in grams, of Avogadro's number of sodium atoms.

Continued—

SOLUTION

The periodic table indicates that the average mass of one sodium atom is 22.99 amu. This may be represented as:

$$\frac{22.99 \text{ amu Na}}{1 \text{ atom Na}}$$

To answer the question, we must calculate the molar mass in units of g/mol. We need two conversion factors, one to convert amu → grams and another to convert atoms → mol.

As previously noted, 1 amu is 1.661×10^{-24} g, and 6.022×10^{23} atoms of sodium is Avogadro's number. Similarly, these relationships may be formatted as

$$1.661 \times 10^{-24} \frac{\text{g Na}}{\text{amu}} \text{ and } 6.022 \times 10^{23} \frac{\text{atoms Na}}{\text{mol Na}}$$

Representing this information as a series of conversion factors, using the factor-label method,

$$22.99 \frac{\text{amu Na}}{\text{atom Na}} \times 1.661 \times 10^{-24} \frac{\text{g Na}}{\text{amu Na}} \times 6.022 \times 10^{23} \frac{\text{atoms Na}}{\text{mol Na}} = 22.99 \frac{\text{g Na}}{\text{mol Na}}$$

The average mass of one *atom* of sodium, in units of amu, is *numerically identical* to the mass of *Avogadro's number of atoms*, expressed in units of grams. Hence the molar mass of sodium is 22.99 g Na/mol.

Helpful Hint: Section 1.3 discusses the use of conversion factors.

The sodium example is not unique. The relationship holds for every element in the periodic table.

Because Avogadro's number of particles (atoms) is 1 mol, it follows that the average mass of one atom of hydrogen is 1.008 amu and the mass of 1 mol of hydrogen atoms is 1.008 g, or the average mass of one atom of carbon is 12.01 amu and the mass of 1 mol of carbon atoms is 12.01 g.

The difference in the mass of a mole of two different elements can be quite striking (Figure 4.1). For example, a mole of hydrogen atoms is 1.008 g, and a mole of lead atoms is 207.19 g.

Figure 4.1

The comparison of approximately one mole each of silver (as Morgan and Peace dollars), gold (as Canadian Maple Leaf coins), and copper (as pennies) shows the considerable difference in mass (as well as economic value) of equivalent moles of different substances.

QUESTION 4.1

Calculate the mass, in grams, of Avogadro's number of aluminum atoms.

QUESTION 4.2

Calculate the mass, in grams, of Avogadro's number of mercury atoms.

Calculating Atoms, Moles, and Mass

Performing calculations based on a chemical equation requires a facility for relating the number of atoms of an element to a corresponding number of moles of that

2 Learning Goal

Perform calculations using Avogadro's number and the mole.

element and ultimately to their mass in grams. Such calculations involve the use of conversion factors. The use of conversion factors was first described in Chapter 1. Some examples follow.

2 Learning Goal

Perform calculations using Avogadro's number and the mole.

EXAMPLE 4.2   Converting Moles to Atoms

How many iron atoms are present in 3.0 mol of iron metal?

SOLUTION

The calculation is based on choosing the appropriate conversion factor. The relationship

$$\frac{6.022 \times 10^{23} \text{ atoms Fe}}{1 \text{ mol Fe}}$$

follows directly from

$$1 \text{ mol Fe} = 6.022 \times 10^{23} \text{ atoms Fe}$$

Using this conversion factor,

$$\text{number of atoms of Fe} = 3.0 \text{ mol Fe} \times \frac{6.022 \times 10^{23} \text{ atoms Fe}}{1 \text{ mol Fe}}$$

$$= 18 \times 10^{23} \text{ atoms of Fe, or}$$

$$= 1.8 \times 10^{24} \text{ atoms of Fe}$$

EXAMPLE 4.3   Converting Atoms to Moles

Calculate the number of moles of sulfur represented by 1.81×10^{24} atoms of sulfur.

SOLUTION

$$1.81 \times 10^{24} \text{ atoms S} \times \frac{1 \text{ mol S}}{6.022 \times 10^{23} \text{ atoms S}} = 3.01 \text{ mol S}$$

Note that this conversion factor is the inverse of that used in Example 4.2. Remember, the conversion factor must cancel units that should not appear in the final answer.

EXAMPLE 4.4   Converting Moles of a Substance to Mass in Grams

What is the mass, in grams, of 3.01 mol of sulfur?

SOLUTION

We know from the periodic table that 1 mol of sulfur has a mass of 32.06 g. Setting up a suitable conversion factor between grams and moles results in

$$3.01 \text{ mol S} \times \frac{32.06 \text{ g S}}{1 \text{ mol S}} = 96.50 \text{ g S}$$

EXAMPLE 4.5   Converting Kilograms to Moles

2   Learning Goal

Perform calculations using Avogadro's number and the mole.

Calculate the number of moles of sulfur in 1.00 kg of sulfur.

SOLUTION

$$1.00 \ \cancel{kg \ S} \times \frac{10^3 \ \cancel{g \ S}}{1 \ \cancel{kg \ S}} \times \frac{1 \ mol \ S}{32.06 \ \cancel{g \ S}} = 31.2 \ mol \ S$$

EXAMPLE 4.6   Converting Grams to Number of Atoms

Calculate the number of atoms of sulfur in 1.00 g of sulfur.

SOLUTION

It is generally useful to map out a pattern for the required conversion. We are given the number of grams and need the number of atoms that correspond to that mass.

Begin by "tracing a path" to the answer:

$$\boxed{\text{grams sulfur}} \xrightarrow{\substack{\text{Step} \\ 1}} \boxed{\text{moles sulfur}} \xrightarrow{\substack{\text{Step} \\ 2}} \boxed{\text{atoms sulfur}}$$

Two transformations, or conversions, are required:

Step 1. Convert grams to moles.

Step 2. Convert moles to atoms.

To perform step 1, we could consider either

$$\frac{1 \ mol \ S}{32.06 \ g \ S} \quad or \quad \frac{32.06 \ g \ S}{1 \ mol \ S}$$

If we want grams to cancel, $\dfrac{1 \ mol \ S}{32.06 \ g \ S}$ is the correct choice, resulting in

$$\cancel{g \ S} \times \frac{1 \ mol \ S}{32.06 \ \cancel{g \ S}} = value \ in \ mol \ S$$

To perform step 2, the conversion of moles to atoms, the moles of S must cancel; therefore,

$$\cancel{mol \ S} \times \frac{6.022 \times 10^{23} \ atoms \ S}{1 \ \cancel{mol \ S}} = number \ of \ atoms \ S$$

which are the units desired in the solution.

Steps 1. and 2. are combined in a single calculation to produce a final result:

$$1.00 \ \cancel{g \ S} \times \frac{1 \ \cancel{mol \ S}}{32.06 \ \cancel{g \ S}} \times \frac{6.022 \times 10^{23} \ atoms \ S}{1 \ \cancel{mol \ S}} = 1.88 \times 10^{22} \ atoms \ S$$

The preceding examples demonstrate the use of a sequence of conversion factors to proceed from the information *provided* in the problem to the information *requested* by the problem.

Figure 4.2

Interconversion between numbers of moles, particles, and grams. The mole concept is central to chemical calculations involving measured quantities of matter.

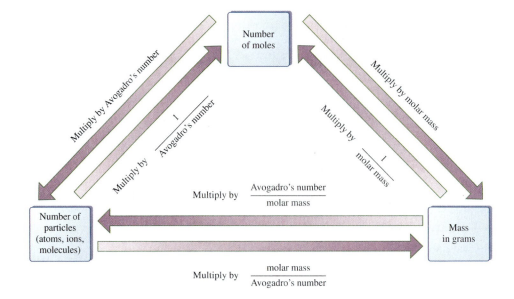

QUESTION 4.3

a. How many oxygen atoms are present in 2.50 mol of oxygen atoms?
b. How many oxygen atoms are present in 2.50 mol of oxygen molecules?

QUESTION 4.4

How many moles of sodium are represented by 9.03×10^{23} atoms of sodium?

QUESTION 4.5

What is the mass, in grams, of 3.50 mol of the element helium?

QUESTION 4.6

How many oxygen atoms are present in 40.0 g of oxygen molecules?

The conversion between the three principal measures of quantity of matter—the number of grams (mass), the number of moles, and the number of individual particles (atoms, ions, or molecules)—is essential to the art of problem solving in chemistry. Their interrelationship is depicted in Figure 4.2.

4.2 The Chemical Formula, Formula Weight, and Molar Mass

The Chemical Formula

3 **Learning Goal**

Write chemical formulas for common inorganic substances.

Compounds are pure substances. They are composed of two or more elements that are chemically combined. A **chemical formula** is a combination of symbols of the various elements that make up the compound. It serves as a convenient way to represent a compound. The chemical formula is based on the formula unit. The **formula unit** is the smallest collection of atoms that provides two important pieces of information:

• the identity of the atoms present in the compound and
• the relative numbers of each type of atom.

Let's look at the following formulas:

- *Hydrogen gas*, H_2. This indicates that two atoms of hydrogen are chemically bonded forming diatomic hydrogen, hence the subscript 2.
- *Water*, H_2O. Water is composed of molecules that contain two atoms of hydrogen (subscript 2) and one atom of oxygen (lack of a subscript means *one* atom).
- *Sodium chloride*, NaCl. One atom of sodium and one atom of chlorine combine to make sodium chloride.
- *Calcium hydroxide*, $Ca(OH)_2$. Calcium hydroxide contains one atom of calcium and two atoms each of oxygen and hydrogen. The subscript outside the parentheses applies to *all* atoms inside the parentheses.
- *Ammonium sulfate*, $(NH_4)_2SO_4$. Ammonium sulfate contains two ammonium ions (NH_4^+) and one sulfate ion (SO_4^{2-}). Each ammonium ion contains one nitrogen and four hydrogen atoms. The formula shows that ammonium sulfate contains two nitrogen atoms, eight hydrogen atoms, one sulfur atom, and four oxygen atoms.
- *Copper(II) sulfate pentahydrate*, $CuSO_4 \cdot 5H_2O$. This is an example of a compound that has water in its structure. Compounds containing one or more water molecules as an integral part of their structure are termed **hydrates.** Copper sulfate pentahydrate has five units of water (or ten H atoms and five O atoms) in addition to one copper atom, one sulfur atom, and four oxygen atoms for a total atomic composition of

1 copper atom
1 sulfur atom
9 oxygen atoms
10 hydrogen atoms

Note that the symbol for water is preceded by a dot, indicating that, although the water is a formula unit capable of standing alone, in this case, it is part of a larger structure. Copper sulfate also exists as a structure free of water, $CuSO_4$. This form is described as anhydrous (no water) copper sulfate. The physical and chemical properties of a hydrate often differ markedly from the anhydrous form (Figure 4.3).

Formula Weight and Molar Mass

Just as the atomic weight of an element is the average atomic mass for one atom of the naturally occurring element, expressed in atomic mass units, the **formula weight** of a compound is the sum of the atomic weights of all atoms in the compound, as represented by its formula. To calculate the formula weight of a compound we *must* know the correct formula. The formula weight is expressed in atomic mass units.

When working in the laboratory, we do not deal with individual molecules; instead, we use units of moles or grams. Eighteen grams of water (less than one ounce) contain approximately Avogadro's number of molecules (6.022×10^{23} molecules). Defining our working units as moles and grams makes good chemical sense.

We concluded earlier that the atomic mass of an element in amu from the periodic table corresponds to the mass of a mole of atoms of that element in units of grams/mol. It follows that **molar mass,** the mass of a mole of compound, is numerically equal to the formula weight in atomic mass units.

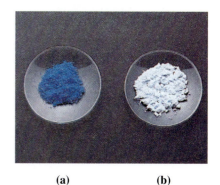

(a)   **(b)**

Figure 4.3

The marked difference in color of (a) hydrated and (b) anhydrous copper sulfate is clear evidence that they are different compounds.

It is possible to determine the correct molecular formula of a compound from experimental data. You can find further information online at www.mhhe.com/denniston in "Composition and Formulas of Compounds."

4   Learning Goal

Calculate the formula weight and molar mass of a compound.

EXAMPLE 4.7   Calculating Formula Weight and Molar Mass

Calculate the formula weight and molar mass of water, H_2O.

Continued—

EXAMPLE 4.7 —*Continued*

SOLUTION

Each water molecule contains two hydrogen atoms and one oxygen atom. The formula weight is

$$2 \text{ atoms of hydrogen} \times 1.008 \text{ amu/atom} = 2.016 \text{ amu}$$

$$\underline{1 \text{ atom of oxygen} \quad \times 16.00 \text{ amu/atom} = 16.00 \text{ amu}}$$

$$18.02 \text{ amu}$$

The average mass of a single molecule of H_2O is 18.02 amu and is the formula weight. Therefore the mass of a mole of H_2O is 18.02 g or 18.02 g/mol.

Helpful Hint: Adding 2.016 and 16.00 shows a result of 18.016 on your calculator. Proper use of significant figures (Chapter 1) dictates rounding that result to 18.02.

4 Learning Goal
Calculate the formula weight and molar mass of a compound.

EXAMPLE 4.8 Calculating Formula Weight and Molar Mass

Calculate the formula weight and molar mass of sodium sulfate.

SOLUTION

The sodium ion is Na^+, and the sulfate ion is SO_4^{2-}. Two sodium ions must be present to neutralize the negative charges on the sulfate ion. The formula is Na_2SO_4. Sodium sulfate contains two sodium atoms, one sulfur atom, and four oxygen atoms. The formula weight is

$$2 \text{ atoms of sodium} \times 22.99 \text{ amu/atom} = 45.98 \text{ amu}$$

$$1 \text{ atom of sulfur} \quad \times 32.06 \text{ amu/atom} = 32.06 \text{ amu}$$

$$\underline{4 \text{ atoms of oxygen} \times 16.00 \text{ amu/atom} = 64.00 \text{ amu}}$$

$$142.04 \text{ amu}$$

The average mass of a single unit of Na_2SO_4 is 142.04 amu and is the formula weight. Therefore the mass of a mole of Na_2SO_4 is 142.04 g, or 142.04 g/mol.

4 Learning Goal
Calculate the formula weight and molar mass of a compound.

EXAMPLE 4.9 Calculating Formula Weight and Molar Mass

Calculate the formula weight and molar mass of calcium phosphate.

SOLUTION

The calcium ion is Ca^{2+}, and the phosphate ion is PO_4^{3-}. To form a neutral unit, three Ca^{2+} must combine with two PO_4^{3-}; [3 × (+2)] calcium ion charges are balanced by [2 × (−3)], the phosphate ion charge. Thus, for calcium phosphate, $Ca_3(PO_4)_2$, the subscript 2 for phosphate dictates that there are two phosphorus atoms and eight oxygen atoms (2 × 4) in the formula unit. Therefore,

$$3 \text{ atoms of Ca} \times 40.08 \text{ amu/atom} = 120.24 \text{ amu}$$

$$2 \text{ atoms of P} \quad \times 30.97 \text{ amu/atom} = \;\; 61.94 \text{ amu}$$

$$\underline{8 \text{ atoms of O} \quad \times 16.00 \text{ amu/atom} = 128.00 \text{ amu}}$$

$$310.18 \text{ amu}$$

Continued—

The formula weight of calcium phosphate is 310.18 amu, and the molar mass is 310.18 g/mol.

QUESTION 4.7

Calculate the formula weight and molar mass of each of the following compounds:

a. NH_3 (ammonia)
b. $C_6H_{12}O_6$ (a sugar, glucose)
c. $CoCl_2 \cdot 6H_2O$ (cobalt chloride hexahydrate)

QUESTION 4.8

Calculate the formula weight and molar mass of each of the following compounds:

a. $C_2F_2Cl_4$ (a Freon gas)
b. C_3H_7OH (isopropyl alcohol, rubbing alcohol)
c. CH_3Br (bromomethane, a pesticide)

4.3 Chemical Equations and the Information They Convey

A Recipe for Chemical Change

The **chemical equation** is the shorthand notation for a chemical reaction. It describes all of the substances that react and all the products that form. **Reactants,** or starting materials, are all substances that undergo change in a chemical reaction; **products** are substances produced by a chemical reaction.

The chemical equation also describes the physical state of the reactants and products as solid, liquid, or gas. It tells us whether the reaction occurs and identifies the solvent and experimental conditions employed, such as heat, light, or electrical energy added to the system.

Most important, the relative number of moles of reactants and products appears in the equation. According to the **law of conservation of mass,** matter cannot be either gained or lost in a chemical reaction. The total mass of the products must be equal to the total mass of the reactants. In other words, the law of conservation of mass tells us that we must have a balanced chemical equation.

5 **Learning Goal**
Know the major function served by the chemical equation, the basis for chemical calculations.

Features of a Chemical Equation

Consider the decomposition of calcium carbonate:

$$CaCO_3(s) \xrightarrow{\Delta} CaO(s) + CO_2(g)$$

Calcium carbonate   Calcium oxide   Carbon dioxide

The factors involved in writing equations of this type are described as follows:

This equation reads: One mole of solid calcium carbonate decomposes upon heating to produce one mole of solid calcium oxide and one mole of gaseous carbon dioxide.

- *The identity of products and reactants must be specified using chemical symbols.* In some cases, it is possible to predict the products of a reaction. More often, the reactants and products must be verified by chemical analysis. (Generally, you will be given information regarding the identity of the reactants and products.)
- *Reactants are written to the left of the reaction arrow (→), and products are written to the right.* The direction in which the arrow points indicates the

direction in which the reaction proceeds. In the decomposition of calcium carbonate, the reactant on the left ($CaCO_3$) is converted to products on the right ($CaO + CO_2$) during the course of the reaction.

- **The physical states of reactants and products may be shown in parentheses.** For example:
  - $Cl_2(g)$ means that chlorine is in the gaseous state.
  - $Mg(s)$ indicates that magnesium is a solid.
  - $Br_2(l)$ indicates that bromine is present as a liquid.
  - $NH_3(aq)$ tells us that ammonia is present as an aqueous solution (dissolved in water).
- **The symbol Δ over the reaction arrow means that energy is necessary for the reaction to occur.** Often, this and other special conditions are noted above or below the reaction arrow. For example, "light" means that a light source provides energy necessary for the reaction. Such reactions are termed photochemical reactions.
- **The equation must be balanced.** All of the atoms of every reactant must also appear in the products, although in different compounds. We will treat this topic in detail later in this chapter.

According to the factors outlined, the equation for the decomposition of calcium carbonate may now be written as

$$CaCO_3(s) \xrightarrow{\Delta} CaO(s) + CO_2(g)$$

The Experimental Basis of a Chemical Equation

The chemical equation must represent a chemical change: One or more substances are changed into new substances, with different chemical and physical properties. Evidence for the reaction may be based on observations such as

See discussion of acid-base reactions in Chapter 8.

See A Medical Connection: Hot and Cold Packs in Chapter 5.

- the release of carbon dioxide gas when an acid is added to a carbonate,
- the formation of a solid (or precipitate) when solutions of iron ions and hydroxide ions are mixed,
- the production of heat when using hot packs for treatment of injury, and
- the change in color of a solution upon addition of a second substance.

Many reactions are not so obvious. Sophisticated instruments are now available to the chemist. These instruments allow the detection of subtle changes in chemical systems that would otherwise go unnoticed. Such instruments may measure

- heat or light absorbed or emitted as the result of a reaction,
- changes in the way the sample behaves in an electric or magnetic field before and after a reaction, and
- changes in electrical properties before and after a reaction.

Whether we use our senses or a million dollar computerized instrument, the "bottom line" is the same: We are measuring a change in one or more chemical or physical properties to understand the changes in a chemical system.

Disease can be described as a chemical system (actually a biochemical system) gone awry. Here, too, the underlying changes may not be obvious. Just as technology has helped chemists see subtle chemical changes in the laboratory, medical diagnosis has been revolutionized in our lifetimes using very similar technology. Some of these techniques are described in the Medical Connection: Magnetic Resonance Imaging, in Chapter 9.

Writing Chemical Reactions

6 Learning Goal
Classify chemical reactions by type: combination, decomposition, or replacement.

Chemical reactions, whether they involve the formation of precipitate, reaction with oxygen, acids and bases, or oxidation-reduction, generally follow one of a few simple patterns: combination, decomposition, and single- or double-replacement. Recognizing the underlying pattern will improve your ability to write and understand chemical reactions.

Combination Reactions

Combination reactions involve the joining of two or more elements or compounds, producing a product of different composition. The general form of a combination reaction is

$$A + B \longrightarrow AB$$

in which A and B represent reactant elements or compounds and AB is the product.
   Examples include

- combination of a metal and a nonmetal to form a salt,

$$Ca(s) + Cl_2(g) \longrightarrow CaCl_2(s)$$

- combination of hydrogen and chlorine molecules to produce hydrogen chloride,

$$H_2(g) + Cl_2(g) \longrightarrow 2HCl(g)$$

- formation of water from hydrogen and oxygen molecules,

$$2H_2(g) + O_2(g) \longrightarrow 2H_2O(g)$$

- reaction of magnesium oxide and carbon dioxide to produce magnesium carbonate,

$$MgO(s) + CO_2(g) \longrightarrow MgCO_3(s)$$

Decomposition Reactions

Decomposition reactions produce two or more products from a single reactant. The general form of these reactions is the reverse of a combination reaction:

$$AB \longrightarrow A + B$$

Some examples are

- the heating of calcium carbonate to produce calcium oxide and carbon dioxide,

$$CaCO_3(s) \longrightarrow CaO(s) + CO_2(g)$$

- the removal of water from a hydrated material (a *hydrate* is a substance that has water molecules incorporated in its structure),

Hydrated compounds are described on page 113.

$$CuSo_4 \cdot 5H_2O(s) \longrightarrow CuSO_4(s) + 5H_2O(g)$$

Replacement Reactions

Replacement reactions include both *single-replacement* and *double-replacement*. In a **single-replacement reaction,** one atom replaces another in the compound, producing a new compound:

$$A + BC \longrightarrow AC + B$$

Examples include

- the replacement of copper by zinc in copper sulfate,

$$Zn(s) + CuSO_4(aq) \longrightarrow ZnSO_4(aq) + Cu(s)$$

- the replacement of aluminum by sodium in aluminum nitrate,

$$3Na(s) + Al(NO_3)_3(aq) \longrightarrow 3NaNO_3(aq) + Al(s)$$

A **double-replacement reaction,** on the other hand, involves *two compounds* undergoing a "change of partners." Two compounds react by exchanging atoms to produce two new compounds:

$$AB + CD \longrightarrow AD + CB$$

Examples include

- the reaction of an acid (hydrochloric acid) and a base (sodium hydroxide) to produce water and salt, sodium chloride,

$$HCl(aq) + NaOH(aq) \longrightarrow H_2O(l) + NaCl(aq)$$

- the formation of solid barium sulfate from barium chloride and potassium sulfate,

$$BaCl_2(aq) + K_2SO_4(aq) \longrightarrow BaSO_4(s) + 2KCl(aq)$$

QUESTION 4.9

Classify each of the following reactions as decomposition (D), combination (C), single-replacement (SR), or double-replacement (DR):

a. $HNO_3(aq) + KOH(aq) \longrightarrow KNO_3(aq) + H_2O(l)$

b. $Al(s) + 3NiNO_3(aq) \longrightarrow Al(NO_3)_3(aq) + 3Ni(s)$

c. $KCN(aq) + HCl(aq) \longrightarrow HCN(aq) + KCl(aq)$

d. $MgCO_3(s) \longrightarrow MgO(s) + CO_2(g)$

QUESTION 4.10

Classify each of the following reactions as decomposition (D), combination (C), single-replacement (SR), or double-replacement (DR):

a. $2Al(OH)_3(s) \xrightarrow{\Delta} Al_2O_3(s) + 3H_2O(g)$

b. $Fe_2S_3(s) \xrightarrow{\Delta} 2Fe(s) + 3S(s)$

c. $Na_2CO_3(aq) + BaCl_2(aq) \xrightarrow{\Delta} BaCO_3(s) + 2NaCl(aq)$

d. $C(s) + O_2(g) \xrightarrow{\Delta} CO_2(g)$

TABLE 4.1 Solubilities of Some Common Ionic Compounds

Solubility Predictions

Sodium, potassium, and ammonium compounds are generally *soluble.*

Nitrates and acetates are generally *soluble.*

Chlorides, bromides, and iodides (halides) are generally *soluble.* However, halide compounds containing lead(II), silver(I), and mercury(I) are *insoluble.*

Carbonates and phosphates are generally *insoluble.* Sodium, potassium, and ammonium carbonates and phosphates are, however, *soluble.*

Hydroxides and sulfides are generally *insoluble.* Sodium, potassium, calcium, and ammonium compounds are, however, *soluble.*

Types of Chemical Reactions

Precipitation Reactions

Precipitation reactions include any chemical change in solution that results in one or more insoluble product(s). In aqueous solution reactions, the product is insoluble in water.

An understanding of precipitation reactions is useful in many ways. They may explain natural phenomena, such as the formation of stalagmites and stalactites in caves; they are simply precipitates in rocklike form. Kidney stones may result from the precipitation of calcium oxalate (CaC_2O_4). The routine act of preparing a solution requires that none of the solutes will react to form a precipitate.

How do you know whether a precipitate will form? Readily available solubility tables, such as Table 4.1, make prediction rather easy.

The following example illustrates the process.

EXAMPLE 4.10 **Predicting Whether Precipitation Will Occur**

Will a precipitate form if two solutions of the soluble salts NaCl and $AgNO_3$ are mixed?

SOLUTION

If two soluble salts react to form a precipitate, they will probably "exchange partners":

$$NaCl(aq) + AgNO_3(aq) \longrightarrow AgCl(?) + NaNO_3(?)$$

Next, refer to Table 4.1 to determine the solubility of AgCl and $NaNO_3$. We predict that $NaNO_3$ is soluble and AgCl is not:

$$NaCl(aq) + AgNO_3(aq) \longrightarrow AgCl(s) + NaNO_3(aq)$$

The fact that the solid AgCl is predicted classifies this double-replacement reaction as a precipitation reaction.

Helpful Hints: (aq) indicates a soluble species; (s) indicates an insoluble species.

QUESTION 4.11

Predict whether the following reactants, when mixed in aqueous solution, undergo a precipitation reaction. Write a balanced equation for each precipitation reaction.

a. potassium chloride and silver nitrate
b. potassium acetate and silver nitrate

QUESTION 4.12

Predict whether the following reactants, when mixed in aqueous solution, undergo a precipitation reaction. Write a balanced equation for each precipitation reaction.

a. sodium hydroxide and ammonium chloride
b. sodium hydroxide and iron(II) chloride

Acid–Base Reactions

Another approach to the classification of chemical reactions is based on the gain or loss of hydrogen ions. **Acid–base reactions** involve the transfer of a *hydrogen ion,* H^+, from one reactant (the acid) to another (the base).

Animation
Types of Reactions

Animation
Predicting Precipitation Reactions

7 **Learning Goal**
Recognize the various classes of chemical reactions: precipitation, reactions with oxygen, acid-base, and oxidation-reduction.

Precipitation reactions may be written as net ionic equations. You can find further information online at www.mhhe.com/denniston in "Writing Net Ionic Equations."

Animation
Precipitation of Barium Sulfate

Animation
Precipitation of Lead Iodide

See discussion of acid-base reactions in Chapter 8.

A common example of an acid-base reaction involves hydrochloric acid and sodium hydroxide:

$$HCl(aq) + NaOH(aq) \longrightarrow H_2O(l) + Na^+(aq) + Cl^-(aq)$$

| Acid | Base | Water | Salt |

A hydrogen ion is transferred from the acid to the base, producing water and a salt in solution.

Acid-base reactions may also be written as net ionic equations. You can find further information online at www.mhhe.com/denniston in "Writing Net Ionic Equations."

Reactions with Oxygen

Many substances react with oxygen. These reactions generally release energy. The combustion of gasoline is used for transportation. Fossil fuel combustion is used to heat homes and provide energy for industry. Reactions involving oxygen provide energy for all sorts of biochemical processes.

When organic (carbon-containing) compounds react with the oxygen in air (burning), carbon dioxide is usually produced. If the compound contains hydrogen, water is the other product.

Energetics of reactions is discussed in Chapter 5.

The reaction between oxygen and methane, CH_4, the major component of natural gas, is

$$CH_4(g) + 2O_2(g) \longrightarrow CO_2(g) + 2H_2O(g)$$

CO_2 and H_2O are waste products, and CO_2 may contribute to the greenhouse effect and global warming. The really important, unseen product is heat energy. That is why we use this reaction in our furnaces!

Inorganic substances also react with oxygen and produce heat, but these reactions usually proceed more slowly. *Corrosion* (rusting iron) is a familiar example:

See An Environmental Perspective: The Greenhouse Effect and Global Warming, Chapter 6.

$$4Fe(s) + 3O_2(g) \longrightarrow 2Fe_2O_3(s)$$
Rust

Some reactions of metals with oxygen are very rapid. A dramatic example is the reaction of magnesium with oxygen (see Figure 4.4):

$$2Mg(s) + O_2(g) \longrightarrow 2MgO(s)$$

Oxidation and Reduction

Oxidation is defined as a loss of electrons, loss of hydrogen atoms, or gain of oxygen atoms. *Sodium metal*, for example, is oxidized to *sodium ion*, losing one electron when it reacts with a nonmetal such as chlorine:

$$Na \longrightarrow Na^+ + e^-$$

Reduction is defined as a gain of electrons, gain of hydrogen atoms, or loss of oxygen atoms. A *chlorine atom* is reduced to a *chloride ion* by gaining one electron when it reacts with a metal such as sodium:

$$Cl + e^- \rightarrow Cl^-$$

Oxidation and reduction are complementary processes. The *oxidation half-reaction* produces an electron that is the reactant for the *reduction half-reaction*. The combination of two half-reactions, one oxidation and one reduction, produces the complete reaction:

Oxidation half-reaction:	$Na \longrightarrow Na^+ + e^-$
Reduction half-reaction:	$Cl + e^- \longrightarrow Cl^-$
Complete reaction:	$Na + Cl \longrightarrow Na^+ + Cl^-$

Figure 4.4

The reaction of magnesium metal and oxygen (in air) is a graphic example of a very rapid reaction at high temperature.

Oxidation-reduction reactions are often termed *redox reactions*.

Half-reactions, one oxidation and one reduction, are exactly that: one-half of a complete reaction. The two half-reactions combine to produce the complete reaction. Note that the electrons cancel: in the electron transfer process, no free electrons remain.

In the preceding reaction, sodium metal is the **reducing agent.** It releases electrons for the reduction of chlorine. Chlorine is the **oxidizing agent.** It accepts electrons from sodium, which is oxidized.

All reactions with oxygen, including the combustion of methane and the rusting of iron, described on p.120, are, in fact, oxidation-reduction reactions. In the process of rusting, the iron *loses* electrons; it is oxidized. Actually, the oxygen atoms *take* the electrons away from the iron atoms; the oxygen atoms are reduced.

The characteristics of oxidizing and reducing agents may be summarized as follows:

The reducing agent becomes oxidized and the oxidizing agent becomes reduced.

Oxidizing Agent
- Is reduced
- Gains electrons
- Causes oxidation

Reducing Agent
- Is oxidized
- Loses electrons
- Causes reduction

QUESTION 4.13

Write the oxidation half-reaction, the reduction half-reaction, and the complete reaction for the formation of calcium sulfide from the elements Ca and S.

QUESTION 4.14

Write the oxidation half-reaction, the reduction half-reaction, and the complete reaction for the formation of calcium iodide from calcium metal and I_2. Remember, the electron gain *must* equal the electron loss.

Applications of Oxidation and Reduction

Oxidation-reduction processes are important in many areas as diverse as industrial manufacturing and biochemical processes.

- **Corrosion.** The deterioration of metals caused by an oxidation-reduction process is termed **corrosion.** Metal atoms are converted to metal ions; the structure, hence the properties, changes dramatically, and usually for the worse (Figure 4.5).

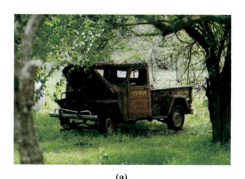
(a)

(b)

(c)

Figure 4.5

The rust (an oxide of iron) that diminishes structural strength and ruins the appearance of (a) automobiles, (b) bridges, and (c) other iron-based objects is a common example of an oxidation-reduction reaction.

- **Combustion of Fossil Fuels.** Burning fossil fuel is an extremely exothermic process. Energy is released to heat our homes, offices, and classrooms. The simplest fossil fuel is methane, CH_4, and its oxidation reaction is written:

$$CH_4(g) + 2O_2(g) \longrightarrow CO_2(g) + 2H_2O(g)$$

Methane is a hydrocarbon. The complete oxidation of any hydrocarbon (including those in gasoline, heating oil, liquid propane, and so forth) produces carbon dioxide and water. The energy released by these reactions is of paramount importance. The water and carbon dioxide are viewed as waste products, and the carbon dioxide contributes to the greenhouse effect (see An Environmental Perspective: The Greenhouse Effect and Global Warming on p. 181).

- **Bleaching.** Bleaching agents are most often oxidizing agents. Sodium hypochlorite (NaOCl) is the active ingredient in a variety of laundry products. It is an effective oxidizing agent. Products containing NaOCl are advertised for their stain-removing capabilities.

Stains are a result of colored compounds adhering to surfaces. Oxidation of these compounds produces products that are not colored or compounds that are subsequently easily removed from the surface, thus removing the stain.

- **Respiration.** There are many examples of biological oxidation-reduction reactions. For example, the electron-transport chain of aerobic respiration involves the reversible oxidation and reduction of iron atoms in cytochrome c:

$$\text{cytochrome } c \ (Fe^{3+}) + e^- \longrightarrow \text{cytochrome } c \ (Fe^{2+})$$

See Chapter 18 for the details of these energy-harvesting cellular oxidation-reduction reactions.

- **Metabolism.** When ethanol is metabolized in the liver, it is oxidized to acetaldehyde (the molecule partially responsible for hangovers). Continued oxidation of acetaldehyde produces acetic acid, which is eventually oxidized to CO_2 and H_2O. These reactions are catalyzed by liver enzymes.
- **Batteries:** The battery in your cellphone or portable music device is a cleverly designed device that enables electron transfer from the reducing agent to the oxidizing agent through an external circuit. The resulting electron flow (electrical current) powers your device.

To learn more about interesting applications involving oxidation-reduction reactions, see *A Medical Connection: Oxidizing Agents for Chemical Control of Microbes, A Medical Connection: Electrochemical Reactions in the Statue of Liberty and in Dental Fillings,* and *A Medical Connection: Turning the Human Body into a Battery* at www.mhhe.com/denniston.

4.4 Balancing Chemical Equations

8 Learning Goal

Balance chemical equations given the identity of products and reactants.

A chemical equation shows the *molar quantity* of reactants needed to produce a certain *molar quantity* of products.

The relative number of moles of each product and reactant is indicated by placing a whole-number *coefficient* before the formula of each substance in the chemical equation. A coefficient of 2 (for example, 2NaCl) indicates that 2 mol of sodium chloride are involved in the reaction. Also, $3NH_3$ signifies 3 mol of ammonia; it means that 3 mol of nitrogen atoms and 3×3, or 9, mol of hydrogen atoms are involved in the reaction. The coefficient 1 is understood, not written. Therefore H_2SO_4 would be interpreted as 1 mol of sulfuric acid, or 2 mol of hydrogen atoms, 1 mol of sulfur atoms, and 4 mol of oxygen atoms.

The equation

$$CaCO_3(s) \xrightarrow{\Delta} CaO(s) + CO_2(g)$$

is balanced as written. On the reactant side, we have

1 mol Ca

1 mol C

3 mol O

On the product side, there are

1 mol Ca

1 mol C

3 mol O

Therefore the law of conservation of mass is obeyed. The balanced equation reflects this fact.

Now consider the reaction of aqueous hydrogen chloride with solid calcium metal in aqueous solution:

$$HCl(aq) + Ca(s) \longrightarrow CaCl_2(aq) + H_2(g)$$

The equation, as written, is not balanced.

Reactants	**Products**
1 mol H atoms	2 mol H atoms
1 mol Cl atoms	2 mol Cl atoms
1 mol Ca atoms	1 mol Ca atoms

We need 2 mol of both H and Cl on the left, or reactant, side. An *incorrect* way of balancing the equation is as follows:

$$H_2Cl_2(aq) + Ca(s) \longrightarrow CaCl_2(aq) + H_2(g)$$

NOT a correct equation

The equation satisfies the law of conservation of mass; however, we have altered one of the reacting species. Hydrogen chloride is HCl, not H_2Cl_2. We must remember that *we cannot alter any chemical substance in the process of balancing the equation.* We can *only* introduce coefficients into the equation. Changing subscripts changes the identity of the chemicals involved, and that is not permitted. The equation must represent the reaction accurately. The correct equation is

$$2HCl(aq) + Ca(s) \longrightarrow CaCl_2(aq) + H_2(g)$$

Correct equation

This process is illustrated in Figure 4.6.

Many equations are balanced by trial and error. After the identities of the products and reactants, the physical state, and the reaction conditions are known, the following steps provide a method for correctly balancing a chemical equation:

Step 1. Count the number of moles of atoms of each element on both product and reactant side.

Step 2. Determine which elements are not balanced.

Step 3. Balance one element at a time using coefficients.

Step 4. After you believe that you have successfully balanced the equation, check, as in Step 1, to be certain that mass conservation has been achieved.

Let us apply these steps to the reaction of calcium with aqueous hydrogen chloride:

$$HCl(aq) + Ca(s) \longrightarrow CaCl_2(aq) + H_2(g)$$

The coefficients indicate *relative* numbers of moles: 10 mol of $CaCO_3$ produce 10 mol of CaO; 0.5 mol of $CaCO_3$ produce 0.5 mol of CaO; and so forth.

Animation
Conservation of Mass

Coefficients placed in front of the formula indicate the relative numbers of moles of compound (represented by the formula) that are involved in the reaction. Subscripts placed to the lower right of the atomic symbol indicate the relative number of atoms in the compound.

Water (H_2O) and hydrogen peroxide (H_2O_2) illustrate the effect a subscript can have. The two compounds show marked differences in physical and chemical properties.

Figure 4.6

Balancing the equation HCl + Ca → CaCl$_2$ + H$_2$. (a) Neither product is the correct chemical species. (b) The reactant, HCl, is incorrectly represented as H$_2$Cl$_2$. (c) This equation is correct; all species are correct, and the law of conservation of mass is obeyed.

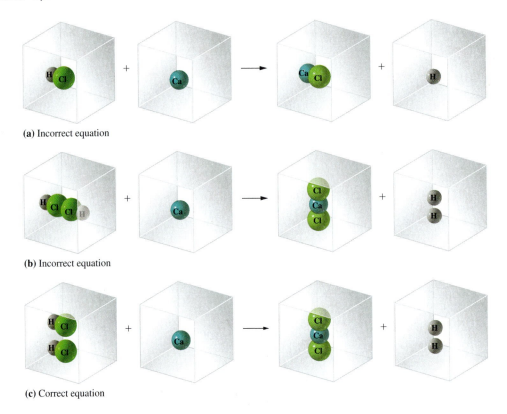

(a) Incorrect equation

(b) Incorrect equation

(c) Correct equation

Step 1.

Reactants	Products
1 mol H atoms	2 mol H atoms
1 mol Cl atoms	2 mol Cl atoms
1 mol Ca atoms	1 mol Ca atoms

Step 2. The numbers of moles of H and Cl are not balanced.

Step 3. Insertion of a 2 before HCl on the reactant side should balance the equation:

$$2HCl(aq) + Ca(s) \longrightarrow CaCl_2(aq) + H_2(g)$$

Step 4. Check for mass balance:

Reactants	Products
2 mol H atoms	2 mol H atoms
2 mol Cl atoms	2 mol Cl atoms
1 mol Ca atoms	1 mol Ca atoms

Hence the equation is balanced.

8 Learning Goal

Balance chemical equations given the identity of products and reactants.

EXAMPLE 4.11 Balancing Equations

Balance the following equation: Hydrogen gas and oxygen gas react explosively to produce gaseous water.

SOLUTION

Recall that hydrogen and oxygen are diatomic gases; therefore,

$$H_2(g) + O_2(g) \longrightarrow H_2O(g)$$

Continued—

Note that the moles of hydrogen atoms are balanced but that the moles of oxygen atoms are not; therefore we must first balance the moles of oxygen atoms:

$$H_2(g) + O_2(g) \longrightarrow 2H_2O(g)$$

Balancing moles of oxygen atoms creates an imbalance in the number of moles of hydrogen atoms, so

$$2H_2(g) + O_2(g) \longrightarrow 2H_2O(g)$$

The equation is balanced, with 4 mol of hydrogen atoms and 2 mol of oxygen atoms on each side of the reaction arrow.

EXAMPLE 4.12 Balancing Equations

8 **Learning Goal**
Balance chemical equations given the identity of products and reactants.

Balance the following equation: Propane gas, C_3H_8, a fuel, reacts with oxygen gas to produce carbon dioxide and water vapor. The reaction is

$$C_3H_8(g) + O_2(g) \rightarrow CO_2(g) + H_2O(g)$$

SOLUTION

First, balance the carbon atoms; there are 3 mol of carbon atoms on the left and only 1 mol of carbon atoms on the right. We need $3CO_2$ on the right side of the equation:

$$C_3H_8(g) + O_2(g) \longrightarrow 3CO_2(g) + H_2O(g)$$

Next, balance the hydrogen atoms; there are 2 mol of hydrogen atoms on the right and 8 mol of hydrogen atoms on the left. We need $4H_2O$ on the right:

$$C_3H_8(g) + O_2(g) \longrightarrow 3CO_2(g) + 4H_2O(g)$$

There are now 10 mol of oxygen atoms on the right and 2 mol of oxygen atoms on the left. To balance, we must have $5O_2$ on the left side of the equation:

$$C_3H_8(g) + 5O_2(g) \longrightarrow 3CO_2(g) + 4H_2O(g)$$

Remember: In every case, be sure to check the final equation for mass balance.

EXAMPLE 4.13 Balancing Equations

Balance the following equation: Butane gas, C_4H_{10}, a fuel used in pocket lighters, reacts with oxygen gas to produce carbon dioxide and water vapor. The reaction is

$$C_4H_{10}(g) + O_2(g) \longrightarrow CO_2(g) + H_2O(g)$$

SOLUTION

First, balance the carbon atoms; there are 4 mol of carbon atoms on the left and only 1 mol of carbon atoms on the right:

$$C_4H_{10}(g) + O_2(g) \longrightarrow 4CO_2(g) + H_2O(g)$$

— Continued —

EXAMPLE **4.13** —*Continued*

Next, balance hydrogen atoms; there are 10 mol of hydrogen atoms on the left and only 2 mol of hydrogen atoms on the right:

$$C_4H_{10}(g) + O_2(g) \longrightarrow 4CO_2(g) + 5H_2O(g)$$

There are now 13 mol of oxygen atoms on the right and only 2 mol of oxygen atoms on the left. Therefore, a coefficient of 6.5 is necessary for O_2.

$$C_4H_{10}(g) + 6.5O_2(g) \longrightarrow 4CO_2(g) + 5H_2O(g)$$

Fractional or decimal coefficients are often needed and used. However, the preferred form requires all integral coefficients. Multiplying each term in the equation by a suitable integer (2, in this case) satisfies this requirement. Hence,

$$2C_4H_{10}(g) + 13O_2(g) \longrightarrow 8CO_2(g) + 10H_2O(g)$$

The equation is balanced, with 8 mol of carbon atoms, 20 mol of hydrogen atoms, and 26 mol of oxygen atoms on each side of the reaction arrow.

Helpful Hint: When balancing equations, we find that it is often most efficient to begin by balancing the atoms in the most complicated formulas.

8 **Learning Goal**

Balance chemical equations given the identity of products and reactants.

EXAMPLE **4.14** **Balancing Equations**

Balance the following equation: Aqueous ammonium sulfate reacts with aqueous lead nitrate to produce aqueous ammonium nitrate and solid lead sulfate. The reaction is

$$(NH_4)_2SO_4(aq) + Pb(NO_3)_2(aq) \longrightarrow NH_4NO_3(aq) + PbSO_4(s)$$

SOLUTION

In this case, the polyatomic ions remain as intact units. Therefore, we can balance them as we would balance molecules rather than as atoms.

There are two ammonium ions on the left and only one ammonium ion on the right. Hence,

$$(NH_4)_2SO_4(aq) + Pb(NO_3)_2(aq) \longrightarrow 2NH_4NO_3(aq) + PbSO_4(s)$$

No further steps are necessary. The equation is now balanced. There are two ammonium ions, two nitrate ions, one lead ion, and one sulfate ion on each side of the reaction arrow.

QUESTION 4.15

Balance each of the following chemical equations:

a. $Fe(s) + O_2(g) \longrightarrow Fe_2O_3(s)$
b. $C_6H_6(l) + O_2(g) \longrightarrow CO_2(g) + H_2O(g)$

QUESTION 4.16

Balance each of the following chemical equations:

a. $S_2Cl_2(s) + NH_3(g) \longrightarrow N_4S_4(s) + NH_4Cl(s) + S_8(s)$
b. $C_2H_5OH(l) + O_2(g) \longrightarrow CO_2(g) + H_2O(g)$

A MEDICAL Connection | Carbon Monoxide Poisoning: A Case of Combining Ratios

A fuel, such as methane, CH_4, burned in an excess of oxygen produces carbon dioxide and water:

$$CH_4(g) + 2O_2(g) \longrightarrow CO_2(g) + 2H_2O(g)$$

The same combustion in the presence of insufficient oxygen produces carbon monoxide and water:

$$2CH_4(g) + 3O_2(g) \longrightarrow 2CO(g) + 4H_2O(g)$$

The combustion of methane, repeated over and over in millions of gas furnaces, is responsible for heating many of our homes in the winter. The furnace is designed to operate under conditions that favor the first reaction and minimize the second; excess oxygen is available from the surrounding atmosphere. Furthermore, the vast majority of exhaust gases (containing principally CO, CO_2, H_2O, and unburned fuel) are removed from the home through the chimney. However, if the chimney becomes obstructed or the burner malfunctions, carbon monoxide levels within the home can rapidly reach hazardous levels.

Why is exposure to carbon monoxide hazardous? Hemoglobin, an iron-containing compound, binds with O_2 and transports it throughout the body. Carbon monoxide also combines with hemoglobin, thereby blocking oxygen transport. The binding affinity of hemoglobin for carbon monoxide is about two hundred times as great as for O_2. Therefore, to maintain O_2 binding and transport capability, our exposure to carbon monoxide must be minimal. Proper ventilation and a suitable oxygen-to-fuel ratio are essential for any combustion process in the home, automobile, or workplace. In recent years, carbon monoxide sensors have been developed. These sensors sound an alarm when toxic levels of CO are reached. These warning devices have helped to create a safer indoor environment.

The example we have chosen is an illustration of what is termed the *law of multiple proportions*. This law states that identical reactants may produce different products, depending on their combining ratio. The experimental conditions (in this case, the quantity of available oxygen) determine the preferred path of the chemical reaction. In Section 4.5, we will learn how to use a properly balanced equation, representing the chemical change occurring, to calculate quantities of reactants consumed or products produced.

FOR FURTHER UNDERSTANDING

Why may new, more strict insulation standards for homes and businesses inadvertently increase the risk of carbon monoxide poisoning?

Explain the link between smoking and carbon monoxide that has motivated many states and municipalities to ban smoking in restaurants, offices, and other indoor spaces.

4.5 Calculations Using a Chemical Equation

General Principles

The calculation of quantities of products and reactants based on a balanced chemical equation is termed *stoichiometry*, and is important in many fields. The synthesis of drugs and other complex molecules on a large scale is conducted on the basis of balanced equations. This minimizes the waste of expensive chemical compounds used in these reactions. Similarly, the ratio of fuel and air in a home furnace or automobile must be adjusted carefully, according to their combining ratio, to maximize energy conversion, minimize fuel consumption, and minimize pollution.

In carrying out chemical calculations, we apply the following guidelines:

1. The chemical formulas of all reactants and products must be known.
2. The basis for the calculations is a balanced equation because the conservation of mass must be obeyed. If the equation is not properly balanced, the calculation is meaningless.
3. The calculations are performed in terms of moles. The coefficients in the balanced equation represent the relative number of moles of products and reactants.

We have seen that the number of moles of products and reactants often differs in a balanced equation. For example,

$$C(s) + O_2(g) \longrightarrow CO_2(g)$$

9 Learning Goal

Calculate the number of moles or grams of product resulting from a given number of moles or grams of reactants or the number of moles or grams of reactant needed to produce a certain number of moles or grams of product.

Animation
Stoichiometry

is a balanced equation. Two moles of reactants combine to produce one mole of product:

$$1 \text{ mol C} + 1 \text{ mol O}_2 \longrightarrow 1 \text{ mol CO}_2$$

However, 1 mol of C *atoms* and 2 mol of O *atoms* produce 1 mol of C *atoms* and 2 mol of O *atoms*. In other words, the number of moles of reactants and products may differ, but the number of moles of atoms cannot. The formation of CO_2 from C and O_2 may be described as follows:

$$C(s) + O_2(g) \longrightarrow CO_2(g)$$

$$1 \text{ mol C} + 1 \text{ mol O}_2 \longrightarrow 1 \text{ mol CO}_2$$

$$12.0 \text{ g C} + 32.0 \text{ g O}_2 \longrightarrow 44.0 \text{ g CO}_2$$

The mole is the basis of our calculations. However, moles are generally measured in grams (or kilograms). A facility for interconversion of moles and grams is fundamental to chemical arithmetic (see Figure 4.2). These calculations, discussed earlier in this chapter, are reviewed in Example 4.15.

Use of Conversion Factors

Conversion Between Moles and Grams

Conversion from moles to grams, and vice versa, requires only the formula weight of the compound of interest. Consider the following examples.

9 | **Learning Goal**

Calculate the number of moles or grams of product resulting from a given number of moles or grams of reactant needed to produce a certain number of moles or grams of product.

EXAMPLE 4.15 | **Converting Between Moles and Grams**

a. Convert 1.00 mol of oxygen gas, O_2, to grams.

SOLUTION

Use the following path:

| moles of oxygen | $\longrightarrow$ | grams of oxygen |

The molar mass of oxygen (O_2) is

$$\frac{32.0 \text{ g O}_2}{1 \text{ mol O}_2}$$

Therefore

$$1.00 \text{ mol O}_2 \times \frac{32.0 \text{ g O}_2}{1 \text{ mol O}_2} = 32.0 \text{ g O}_2$$

b. How many grams of carbon dioxide are contained in 10.0 mol of carbon dioxide?

SOLUTION

Use the following path:

| moles of carbon dioxide | $\longrightarrow$ | grams of carbon dioxide |

Continued—

The formula weight of CO_2 is

$$\frac{44.0 \text{ g } CO_2}{1 \text{ mol } CO_2}$$

and

$$10.0 \text{ mol } CO_2 \times \frac{44.0 \text{ g } CO_2}{1 \text{ mol } CO_2} = 4.40 \times 10^2 \text{ g } CO_2$$

c. How many moles of sodium are contained in 1 lb (454 g) of sodium metal?

SOLUTION

Use the following path:

| grams of sodium | $\longrightarrow$ | moles of sodium |

The number of moles of sodium atoms is

$$454 \text{ g Na} \times \frac{1 \text{ mol Na}}{22.99 \text{ g Na}} = 19.7 \text{ mol Na}$$

Helpful Hint: Note that each factor can be inverted producing a second possible factor. Only one will allow the appropriate unit cancellation.

QUESTION 4.17

Perform each of the following conversions:

a. 5.00 mol of water to grams of water
b. 25.0 g of LiCl to moles of LiCl

QUESTION 4.18

Perform each of the following conversions:

a. 1.00×10^{-5} mol of $C_6H_{12}O_6$ to micrograms of $C_6H_{12}O_6$
b. 35.0 g of $MgCl_2$ to moles of $MgCl_2$

Conversion of Moles of Reactants to Moles of Products

In Example 4.12, we balanced the equation for the reaction of propane and oxygen as follows:

$$C_3H_8(g) + 5O_2(g) \longrightarrow 3CO_2(g) + 4H_2O(g)$$

In this reaction, 1 mol of C_3H_8 corresponds to, or results in

5 mol of O_2 being consumed and

3 mol of CO_2 being formed and

4 mol of H_2O being formed.

This information may be written in the form of a conversion factor or ratio:

1 mol C_3H_8/5 mol O_2

Translated: One mole of C_3H_8 reacts with five moles of O_2.

1 mol C_3H_8/3 mol CO_2

Translated: One mole of C_3H_8 produces three moles of CO_2.

$$1 \text{ mol } C_3H_8 / 4 \text{ mol } H_2O$$

Translated: One mole of C_3H_8 produces four moles of H_2O.

Conversion factors, based on a chemical equation, permit us to perform a variety of calculations.

Let us look at a few examples, based on the combustion of propane and the equation that we balanced in Example 4.12.

EXAMPLE 4.16   Calculating Reacting Quantities

Calculate the number of grams of O_2 that will react with 1.00 mol of C_3H_8.

SOLUTION

Two conversion factors are necessary to solve this problem:

1. conversion from moles of C_3H_8 to moles of O_2 and
2. conversion of moles of O_2 to grams of O_2.

Therefore, our path is

$$\boxed{\text{moles } C_3H_8} \longrightarrow \boxed{\text{moles } O_2} \longrightarrow \boxed{\text{grams } O_2}$$

and

$$1.00 \text{ mol } C_3H_8 \times \frac{5 \text{ mol } O_2}{1 \text{ mol } C_3H_8} \times \frac{32.0 \text{ g } O_2}{1 \text{ mol } O_2} = 1.60 \times 10^2 \text{ g } O_2$$

---

9   Learning Goal

Calculate the number of moles or grams of product resulting from a given number of moles or grams of reactant or the number of moles or grams of reactant needed to produce a certain number of moles or grams of product.

EXAMPLE 4.17   Calculating Grams of Product from Moles of Reactant

Calculate the number of grams of CO_2 produced from the combustion of 1.00 mol of C_3H_8.

SOLUTION

Employ logic similar to that used in Example 4.16 and use the following path:

$$\boxed{\text{moles } C_3H_8} \longrightarrow \boxed{\text{moles } CO_2} \longrightarrow \boxed{\text{grams } CO_2}$$

Then

$$1.00 \text{ mol } C_3H_8 \times \frac{3 \text{ mol } CO_2}{1 \text{ mol } C_3H_8} \times \frac{44.0 \text{ g } CO_2}{1 \text{ mol } CO_2} = 132 \text{ g } CO_2$$

EXAMPLE 4.18   Relating Masses of Reactants and Products

Calculate the number of grams of C_3H_8 required to produce 36.0 g of H_2O.

Continued—

SOLUTION

It is necessary to convert

1. grams of H_2O to moles of H_2O,
2. moles of H_2O to moles of C_3H_8, and
3. moles of C_3H_8 to grams of C_3H_8.

Use the following path:

$$\boxed{\begin{array}{c}\text{grams}\\H_2O\end{array}} \longrightarrow \boxed{\begin{array}{c}\text{moles}\\H_2O\end{array}} \longrightarrow \boxed{\begin{array}{c}\text{moles}\\C_3H_8\end{array}} \longrightarrow \boxed{\begin{array}{c}\text{grams}\\C_3H_8\end{array}}$$

Then

$$36.0 \text{ g } H_2O \times \frac{1 \text{ mol } H_2O}{18.0 \text{ g } H_2O} \times \frac{1 \text{ mol } C_3H_8}{4 \text{ mol } H_2O} \times \frac{44.0 \text{ g } C_3H_8}{1 \text{ mol } C_3H_8} = 22.0 \text{ g } C_3H_8$$

QUESTION 4.19

The balanced equation for the combustion of ethanol (ethyl alcohol) is

$$C_2H_5OH(l) + 3O_2(g) \longrightarrow 2CO_2(g) + 3H_2O(g)$$

a. How many moles of O_2 will react with 1 mol of ethanol?
b. How many grams of O_2 will react with 1 mol of ethanol?

QUESTION 4.20

How many grams of CO_2 will be produced by the combustion of 1 mol of ethanol? (See Question 4.19.)

Let's consider an example that requires us to write and balance a chemical equation, use conversion factors, and calculate the amount of a reactant consumed in the chemical reaction.

EXAMPLE 4.19 Calculating a Quantity of Reactant

9 Learning Goal

Calculate the number of moles or grams of product resulting from a given number of moles or grams of reactants or the number of moles or grams of reactant needed to produce a certain number of moles or grams of product.

Calcium hydroxide may be used to neutralize (completely react with) aqueous hydrochloric acid. Calculate the number of grams of hydrochloric acid that would be neutralized by 0.500 mol of solid calcium hydroxide.

SOLUTION

The formula for calcium hydroxide is $Ca(OH)_2$ and that for hydrochloric acid is HCl. The unbalanced equation produces calcium chloride and water as products:

$$Ca(OH)_2(s) + HCl(aq) \longrightarrow CaCl_2(aq) + H_2O(l)$$

First, balance the equation:

$$Ca(OH)_2(s) + 2HCl(aq) \longrightarrow CaCl_2(aq) + 2H_2O(l)$$

Next, determine the necessary conversion:

1. moles of $Ca(OH)_2$ to moles of HCl and
2. moles of HCl to grams of HCl.

Continued—

Figure 4.7

An illustration of the law of conservation of mass. In this example, 1 mol of calcium hydroxide and 2 mol of hydrogen chloride react to produce 3 mol of product (2 mol of water and 1 mol of calcium chloride). The total mass, in grams, of reactant(s) consumed is equal to the total mass, in grams, of product(s) formed. *Note:* In reality, HCl does not exist as discrete molecules in water. The HCl separates to form H^+ and Cl^-. Ionization in water will be discussed with the chemistry of acids and bases in Chapter 8.

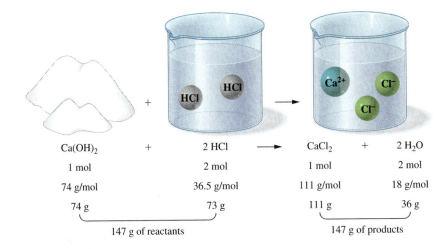

$Ca(OH)_2$	+	$2\,HCl$	$\longrightarrow$	$CaCl_2$	+	$2\,H_2O$
1 mol		2 mol		1 mol		2 mol
74 g/mol		36.5 g/mol		111 g/mol		18 g/mol
74 g		73 g		111 g		36 g

147 g of reactants   147 g of products

EXAMPLE 4.19 —*Continued*

Use the following path:

$$\boxed{\text{moles } Ca(OH)_2} \longrightarrow \boxed{\text{moles } HCl} \longrightarrow \boxed{\text{grams } HCl}$$

$$0.500 \text{ mol } Ca(OH)_2 \times \frac{2 \text{ mol HCl}}{1 \text{ mol } Ca(OH)_2} \times \frac{36.5 \text{ g HCl}}{1 \text{ mol HCl}} = 36.5 \text{ g HCl}$$

This reaction is illustrated in Figure 4.7.

Helpful Hints:
1. The reaction between an acid and a base produces a salt and water (Chapter 8).
2. Remember to balance the chemical equation; the proper coefficients are essential parts of the subsequent calculations.

---

9 **Learning Goal**

Calculate the number of moles or grams of product resulting from a given number of moles or grams of reactants or the number of moles or grams of reactant needed to produce a certain number of moles or grams of product.

EXAMPLE 4.20 Calculating Reactant Quantities

What mass of sodium hydroxide, NaOH, would be required to produce 8.00 g of the antacid milk of magnesia, $Mg(OH)_2$, by the reaction of $MgCl_2$ with NaOH?

SOLUTION

$$MgCl_2(aq) + 2NaOH(aq) \longrightarrow Mg(OH)_2(s) + 2NaCl(aq)$$

The equation tells us that 2 mol of NaOH form 1 mol of $Mg(OH)_2$. If we calculate the number of moles of $Mg(OH)_2$ in 8.00 g of $Mg(OH)_2$, we can determine the number of moles of NaOH necessary and then the mass of NaOH required:

$$\boxed{\text{mass } Mg(OH)_2} \longrightarrow \boxed{\text{moles } Mg(OH)_2} \longrightarrow \boxed{\text{moles } NaOH} \longrightarrow \boxed{\text{mass } NaOH}$$

$$58.3 \text{ g } Mg(OH)_2 = 1 \text{ mol } Mg(OH)_2$$

Therefore,

$$8.00 \text{ g } Mg(OH)_2 \times \frac{1 \text{ mol } Mg(OH)_2}{58.3 \text{ g } Mg(OH)_2} = 0.137 \text{ mol } Mg(OH)_2$$

Continued—

Two moles of NaOH react to give one mole of $Mg(OH)_2$. Therefore,

$$0.137 \text{ mol } Mg(OH)_2 \times \frac{2 \text{ mol NaOH}}{1 \text{ mol } Mg(OH)_2} = 0.274 \text{ mol NaOH}$$

40.0 g of NaOH = 1 mol of NaOH. Therefore,

$$0.274 \text{ mol NaOH} \times \frac{40.0 \text{ g NaOH}}{1 \text{ mol NaOH}} = 11.0 \text{ g NaOH}$$

The calculation may be done in a single step:

$$8.00 \text{ g } Mg(OH)_2 \times \frac{1 \text{ mol } Mg(OH)_2}{58.3 \text{ g } Mg(OH)_2} \times \frac{2 \text{ mol NaOH}}{1 \text{ mol } Mg(OH)_2} \times \frac{40.0 \text{ g NaOH}}{1 \text{ mol NaOH}} = 11.0 \text{ g NaOH}$$

Note once again that we have followed a logical and predictable path to the solution:

grams $Mg(OH)_2$ $\longrightarrow$ moles $Mg(OH)_2$ $\longrightarrow$ moles NaOH $\longrightarrow$ grams NaOH

Helpful Hint: Mass is a laboratory unit, whereas moles is a calculation unit. The laboratory balance is calibrated in units of mass (grams). Although moles are essential for calculation, often the starting point and objective are in mass units. As a result, our path is often grams → moles → grams.

A general problem-solving strategy is summarized in Figure 4.8. By systematically applying this strategy, you will be able to solve virtually any problem requiring calculations based on a chemical equation.

QUESTION 4.21

Metallic iron reacts with O_2 gas to produce iron(III) oxide.

a. Write and balance the equation.
b. Calculate the number of grams of iron needed to produce 5.00 g of product.

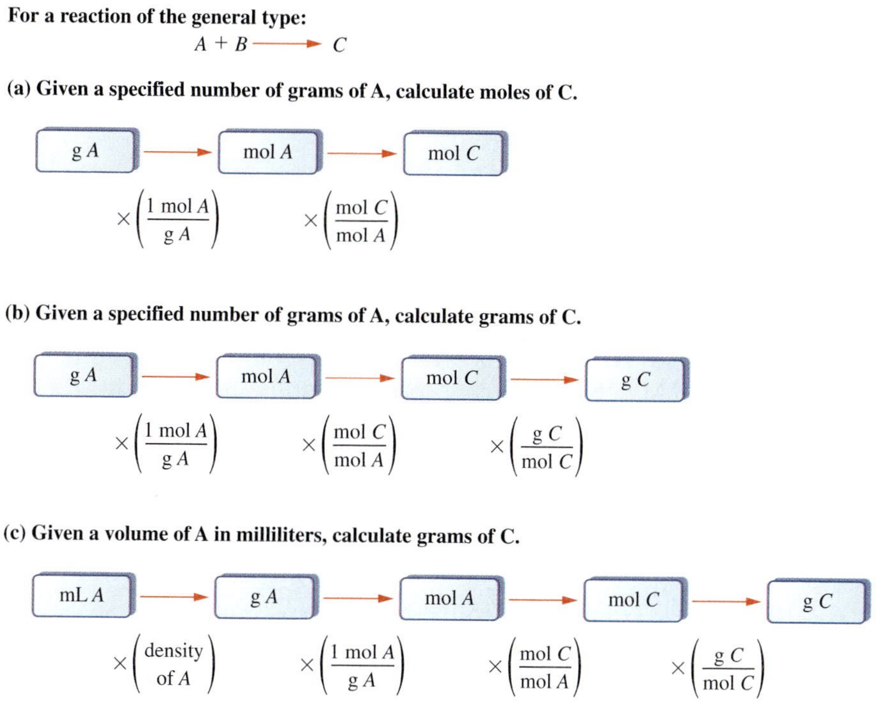

For a reaction of the general type:

$$A + B \longrightarrow C$$

(a) Given a specified number of grams of A, calculate moles of C.

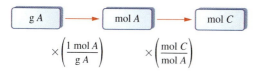

$$\boxed{g\ A} \longrightarrow \boxed{mol\ A} \longrightarrow \boxed{mol\ C}$$
$$\times \left(\frac{1 \text{ mol } A}{g\ A} \right) \qquad \times \left(\frac{mol\ C}{mol\ A} \right)$$

(b) Given a specified number of grams of A, calculate grams of C.

$$\boxed{g\ A} \longrightarrow \boxed{mol\ A} \longrightarrow \boxed{mol\ C} \longrightarrow \boxed{g\ C}$$
$$\times \left(\frac{1 \text{ mol } A}{g\ A} \right) \qquad \times \left(\frac{mol\ C}{mol\ A} \right) \qquad \times \left(\frac{g\ C}{mol\ C} \right)$$

(c) Given a volume of A in milliliters, calculate grams of C.

$$\boxed{mL\ A} \longrightarrow \boxed{g\ A} \longrightarrow \boxed{mol\ A} \longrightarrow \boxed{mol\ C} \longrightarrow \boxed{g\ C}$$
$$\times \left(\frac{\text{density}}{\text{of } A} \right) \quad \times \left(\frac{1 \text{ mol } A}{g\ A} \right) \quad \times \left(\frac{mol\ C}{mol\ A} \right) \quad \times \left(\frac{g\ C}{mol\ C} \right)$$

Figure 4.8

A general problem-solving strategy, using molar quantities.

A LIFESTYLE Connection | The Chemistry of Automobile Air Bags

Each year, thousands of individuals are killed or seriously injured in automobile accidents. Perhaps most serious is the front-end collision. The car decelerates or stops virtually on impact; the momentum of the passengers, however, does not stop, and the driver and passengers are thrown forward toward the dashboard and the front window. Suddenly, passive parts of the automobile, such as control knobs, the rearview mirror, the steering wheel, the dashboard, and the windshield, become lethal weapons.

Automobile engineers have been aware of these problems for a long time. They have made a series of design improvements to lessen the potential problems associated with front-end impact. Smooth switches rather than knobs, recessed hardware, and padded dashboards are examples. These changes, coupled with the use of lap and shoulder belts, which help to immobilize occupants of the car, have decreased the frequency and severity of the impact and lowered the death rate for this type of accident.

An almost ideal protection would be a soft, fluffy pillow, providing a cushion against impact. Such a device, an air bag inflated only on impact, is now standard equipment for the protection of the driver and front-seat passenger.

How does it work? Ideally, it inflates only when severe front-end impact occurs; it inflates very rapidly (in approximately 40 milliseconds), then deflates to provide a steady deceleration, cushioning the occupants from impact. A remarkably simple chemical reaction makes this a reality.

When solid sodium azide (NaN_3) is detonated by mechanical energy produced by an electric current, it decomposes to form solid sodium and nitrogen gas:

$$2NaN_3(s) \longrightarrow 2Na(s) + 3N_2(g)$$

The nitrogen gas inflates the air bag, cushioning the driver and front-seat passenger.

The solid sodium azide has a high density (characteristic of solids) and thus occupies a small volume. It can easily be stored in the center of a steering wheel or in the dashboard. The rate of detonation is very rapid. In milliseconds, it produces three moles of N_2 gas for every two moles of NaN_3. The N_2 gas occupies a relatively large volume because its density is low. This is a general property of gases.

Figuring out how much sodium azide is needed to produce enough nitrogen to properly inflate the bag is an example of a practical application of the chemical arithmetic that we are learning in this chapter.

FOR FURTHER UNDERSTANDING

Why is nitrogen gas preferred for this application?

If this reaction occurs in 5 milliseconds, how many seconds does this represent?

QUESTION 4.22

Barium carbonate decomposes to barium oxide and carbon dioxide upon heating.

a. Write and balance the equation.
b. Calculate the number of grams of carbon dioxide produced by heating 50.0 g of barium carbonate.

Theoretical and Percent Yield

10 Learning Goal

Calculate theoretical and percent yield.

The **theoretical yield** is the *maximum* amount of product that can be produced (in an ideal world). In the "real" world, it is difficult to produce the amount calculated as the theoretical yield. This is true for a variety of reasons. Some experimental error is unavoidable. Moreover, many reactions simply are not complete; some amount of reactant remains at the end of the reaction. We will study these processes, termed *equilibrium reactions* in Chapter 5.

A **percent yield,** the ratio of the actual and theoretical yields multiplied by 100%, is often used to show the relationship between predicted and experimental quantities. Thus

$$\% \text{ yield} = \frac{\text{actual yield}}{\text{theoretical yield}} \times 100\%$$

A MEDICAL Connection

Pharmaceutical Chemistry: The Practical Significance of Percent Yield

In recent years, the major pharmaceutical industries have introduced a wide variety of new drugs targeted to cure or alleviate the symptoms of a host of diseases that afflict humanity.

The vast majority of these drugs are synthetic; they are made in a laboratory or by an industrial process. These substances are complex molecules that are patiently designed and constructed from relatively simple molecules in a series of chemical reactions. A series of ten to twenty "steps," or sequential reactions, is not unusual to put together a final product that has the proper structure, geometry, and reactivity to treat a particular disease.

Although a great deal of research occurs to ensure that each of these steps in the overall process is efficient (having a large percent yield), the overall process is still very inefficient (low percent yield). This inefficiency, and the research needed to minimize it, at least in part determines the cost and availability of both prescription and over-the-counter preparations.

Consider a hypothetical five-step sequential synthesis. If each step has a percent yield of 80% our initial impression might be that this synthesis is quite efficient. However, on closer inspection, we find that quite the contrary is true.

The overall yield of the five-step reaction is the product of the decimal fraction of the percent yield of each of the sequential reactions. So, if the decimal fraction corresponding to 80% is 0.80:

$$0.80 \times 0.80 \times 0.80 \times 0.80 \times 0.80 = 0.33$$

Converting the decimal fraction to percentage:

$$0.33 \times 100\% = 33\% \text{ yield}$$

Many reactions are considerably less than 80% efficient, especially those that are used to prepare large molecules with complex arrangements of atoms. Imagine a more realistic scenario in which one step is only 20% efficient (a 20% yield) and the other four steps are 50%, 60%, 70%, and 80% efficient. Repeating the calculation with these numbers (after conversion to a decimal fraction):

$$0.20 \times 0.50 \times 0.60 \times 0.70 \times 0.80 = 0.0336$$

Converting the decimal fraction to a percentage:

$$0.0336 \times 100\% = 3.36\% \text{ yield}$$

a very inefficient process.

If we apply this logic to a fifteen- or twenty-step synthesis we gain some appreciation of the difficulty of producing modern pharmaceutical products. Add to this the challenge of predicting the most appropriate molecular structure that will have the desired biological effect and be relatively free of side effects. All these considerations give new meaning to the term *wonder drug* that has been attached to some of the more successful synthetic products.

We will study some of the elementary steps essential to the synthesis of a wide range of pharmaceutical compounds in later chapters, beginning with Chapter 10.

FOR FURTHER UNDERSTANDING

Explain the possible connection of this perspective to escalating costs of pharmaceutical products.

Can you describe other situations, not necessarily in the field of chemistry, where multiple-step processes contribute to inefficiency?

In Example 4.17, the theoretical yield of CO_2 is 132 g. For this reaction, let's assume that a chemist actually obtained 125 g CO_2. This is the actual yield and would normally be provided as part of the data in the problem.

Calculate the percent yield as follows:

$$\% \text{ yield} = \frac{\text{actual yield}}{\text{theoretical yield}} \times 100\%$$

$$= \frac{125 \text{ g } CO_2 \text{ actual}}{132 \text{ g } CO_2 \text{ theoretical}} \times 100\% = 94.7\%$$

10 Learning Goal
Calculate theoretical and percent yield.

EXAMPLE 4.21 Calculation of Percent Yield

Assume that the theoretical yield of iron in the process

$$2Al(s) + Fe_2O_3(s) \longrightarrow Al_2O_3(l) + 2Fe(l)$$

was 30.0 g.

If the actual yield of iron were 25.0 g in the process, calculate the percent yield.

SOLUTION

$$\% \text{ yield} = \frac{\text{actual yield}}{\text{theoretical yield}} \times 100\%$$

$$= \frac{25.0 \text{ g}}{30.0 \text{ g}} \times 100\%$$

$$= 83.3\%$$

QUESTION 4.23

Given the reaction represented by the balanced equation

$$Sn(s) + 2HF(aq) \longrightarrow SnF_2(s) + H_2(g)$$

a. Calculate the number of grams of SnF_2 produced by mixing 100.0 g Sn with excess HF.
b. If only 5.00 g SnF_2 were produced, calculate the % yield.

QUESTION 4.24

Given the reaction represented by the balanced equation

$$CH_4(g) + 3Cl_2(g) \longrightarrow 3HCl(g) + CHCl_3(g)$$

a. Calculate the number of grams of $CHCl_3$ produced by mixing 105 g Cl_2 with excess CH_4.
b. If 10.0 g $CHCl_3$ were produced, calculate the % yield.

SUMMARY

4.1 The Mole Concept and Atoms

Atoms are exceedingly small, yet their masses have been experimentally determined for each of the elements. The unit of measurement for these determinations is the *atomic mass unit*, abbreviated amu:

$$1 \text{ amu} = 1.661 \times 10^{-24} \text{ g}$$

The periodic table provides atomic masses in atomic mass units.

A more practical unit for defining a "collection" of atoms is the *mole:*

1 mol of atoms = 6.022×10^{23} atoms of an element

This number is referred to as *Avogadro's number.*

The mole and the atomic mass unit are related. The atomic mass of a given element corresponds to the average mass of a single atom in atomic mass units and the mass of a mole of atoms in grams. The mass of one mole of atoms is termed the *molar mass* of the element. One mole of atoms of any element contains the same number, Avogadro's number, of atoms.

4.2 The Chemical Formula, Formula Weight, and Molar Mass

Compounds are pure substances that are composed of two or more elements that are chemically combined. They are represented by their *chemical formula,* a combination of symbols of the various elements that make up the compounds. The chemical formula is based on the *formula unit.* This is the smallest collection of atoms that provides the identity of the atoms present in the compound and the relative numbers of each type of atom.

Just as a mole of atoms is based on atomic mass, a mole of a compound is based on the formula mass or *formula weight.* The formula weight is calculated by addition of the masses of all the atoms or ions of which the unit is composed. To calculate the formula weight, the formula unit must be known. The formula weight of one mole of a compound is its *molar mass* in units of g/mol.

4.3 Chemical Equations and the Information They Convey

A *chemical equation* is the shorthand notation for a chemical reaction. It describes all of the substances that react to produce the product(s). *Reactants,* or starting materials, are all substances that undergo change in a chemical reaction; *products* are substances produced by a chemical reaction.

According to the *law of conservation of mass,* matter can neither be gained nor lost in a chemical reaction. The law of conservation of mass states that we must have a balanced chemical equation.

Features of a suitable equation include the following:

- The identity of products and reactants must be specified.
- Reactants are written to the left of the reaction arrow ($\rightarrow$) and products to the right.
- The physical states of reactants and products are shown in parentheses.
- The symbol Δ over the reaction arrow means that energy is necessary for the reaction to occur.
- The equation must be balanced.

Chemical reactions involve the *combination* of reactants to produce products, the *decomposition* of reactant(s) into products, or the *replacement* of one or more elements in a compound to yield products. Replacement reactions are subclassified as either *single-* or *double-*replacement.

Reactions that produce products with similar characteristics are often classified as a single group. The formation of an insoluble solid, a *precipitate,* is very common. Such reactions are precipitation reactions.

Chemical reactions that have a common reactant may be grouped together. Reactions involving oxygen, *combustion reactions,* are such a class.

Another approach to the classification of chemical reactions is based on the transfer of hydrogen ions (+ charge)

or electrons (− charge). *Acid-base reactions* involve the transfer of a hydrogen ion, H^+, from one reactant to another. Another important reaction type, *oxidation-reduction,* takes place because of the transfer of negative charge, one or more electrons, from one reactant to another.

4.4 Balancing Chemical Equations

A chemical equation enables us to determine the quantity of reactants needed to produce a certain molar quantity of products. The chemical equation expresses these quantities in terms of moles.

The relative number of moles of each product and reactant is indicated by placing a whole-number coefficient before the formula of each substance in the chemical equation.

Many equations are balanced by trial and error. If the identities of the products and reactants, the physical state, and the reaction conditions are known, the following steps provide a method for correctly balancing a chemical equation:

- Count the number of atoms of each element on both product and reactant sides.
- Determine which atoms are not balanced.
- Balance one element at a time using coefficients.
- After you believe that you have successfully balanced the equation, check to be certain that mass conservation has been achieved.

4.5 Calculations Using a Chemical Equation

Calculations involving chemical quantities are based on the following requirements:

- The basis for the calculations is a balanced equation.
- The calculations are performed in terms of moles.
- The conservation of mass must be obeyed.

The mole is the basis for calculations. However, masses are generally measured in grams (or kilograms). Therefore, you must be able to interconvert moles and grams to perform chemical arithmetic.

KEY TERMS

acid-base reaction (4.3)	law of conservation of
atomic mass unit (4.1)	mass (4.3)
Avogadro's number (4.1)	molar mass (4.1, 4.2)
chemical equation (4.3)	mole (4.1)
chemical formula (4.2)	oxidation (4.3)
combination reaction (4.3)	oxidizing agent (4.3)
corrosion (4.3)	percent yield (4.5)
decomposition reaction (4.3)	product (4.3)
double-replacement	reactant (4.3)
reaction (4.3)	reducing agent (4.3)
formula unit (4.2)	reduction (4.3)
formula weight (4.2)	single-replacement reaction (4.3)
hydrate (4.2)	theoretical yield (4.5)

QUESTIONS AND PROBLEMS

The Mole Concept and Atoms

Foundations

4.25 We purchase eggs by the dozen. Name several other familiar packaging units.

4.26 One dozen eggs is a convenient consumer unit. Explain why the mole is a convenient chemist's unit.

4.27 What is the average molar mass of
   a. Si
   b. Ag

4.28 What is the average molar mass of
   a. S
   b. Na

4.29 What is the mass of Avogadro's number of argon atoms?

4.30 What is the mass of Avogadro's number of iron atoms?

Applications

4.31 How many grams are contained in 2.00 mol of neon atoms?

4.32 How many grams are contained in 3.00 mol of carbon atoms?

4.33 What is the mass in grams of 1.00 mol of helium atoms?

4.34 What is the mass in grams of 1.00 mol of nitrogen atoms?

4.35 Calculate the number of moles corresponding to
   a. 20.0 g He
   b. 0.040 kg Na
   c. 3.0 g Cl_2

4.36 Calculate the number of moles corresponding to
   a. 0.10 g Ca
   b. 4.00 g Fe
   c. 2.00 kg N_2

The Chemical Formula, Formula Weight, and Molar Mass

Foundations

4.37 Distinguish between the terms *molecule* and *ion pair*.

4.38 Distinguish between the terms *formula weight* and *molecular weight*.

4.39 Calculate the molar mass, in grams per mole, of each of the following formula units:
   a. NaCl
   b. Na_2SO_4
   c. $Fe_3(PO_4)_2$

4.40 Calculate the molar mass, in grams per mole, of each of the following formula units:
   a. S_8
   b. $(NH_4)_2SO_4$
   c. CO_2

4.41 Calculate the molar mass, in grams per mole, of oxygen gas, O_2.

4.42 Calculate the molar mass, in grams per mole, of ozone, O_3.

4.43 Calculate the molar mass of $CuSO_4 \cdot 5H_2O$.

4.44 Calculate the molar mass of $CaCl_2 \cdot 2H_2O$.

Applications

4.45 Calculate the number of moles corresponding to
   a. 15.0 g NaCl
   b. 15.0 g Na_2SO_4

4.46 Calculate the number of moles corresponding to
   a. 15.0 g NH_3
   b. 16.0 g O_2

4.47 Calculate the mass in grams corresponding to
   a. 1.000 mol H_2O
   b. 2.000 mol NaCl

4.48 Calculate the mass in grams corresponding to
   a. 0.400 mol NH_3
   b. 0.800 mol $BaCO_3$

4.49 Calculate the mass in grams corresponding to
   a. 10.0 mol He
   b. 1.00×10^2 mol H_2

4.50 Calculate the mass in grams corresponding to
   a. 2.00 mol CH_4
   b. 0.400 mol $Ca(NO_3)_2$

4.51 How many moles are in 50.0 g of each of the following substances?
   a. KBr
   b. $MgSO_4$

4.52 How many moles are in 50.0 g of each of the following substances?
   a. Br_2
   b. NH_4Cl

4.53 How many moles are in 50.0 g of each of the following substances?
   a. CS_2
   b. $Al_2(CO_3)_3$

4.54 How many moles are in 50.0 g of each of the following substances?
   a. $Sr(OH)_2$
   b. $LiNO_3$

Chemical Equations and the Information They Convey

4.55 What is the meaning of the subscript in a chemical formula?

4.56 What is the meaning of the coefficient in a chemical equation?

4.57 Give an example of
   a. a decomposition reaction
   b. a single-replacement reaction

4.58 Give an example of
   a. a combination reaction
   b. a double-replacement reaction

4.59 Give an example of a precipitate-forming reaction.

4.60 Give an example of a reaction in which oxygen is a reactant.

4.61 What is the meaning of Δ over the reaction arrow?

4.62 What is the meaning of (*s*), (*l*), or (*g*) immediately following the symbol for a chemical substance?

Balancing Chemical Equations

Foundations

4.63 When you are balancing an equation, why must the subscripts in the chemical formula remain unchanged?

4.64 Describe the process of checking to ensure that an equation is properly balanced.

4.65 On which side of the reaction arrow are reactants found?

4.66 On which side of the reaction arrow are products found?

Applications

4.67 Balance each of the following equations:
   a. $C_2H_6(g) + O_2(g) \longrightarrow CO_2(g) + H_2O(g)$
   b. $K_2O(s) + P_4O_{10}(s) \longrightarrow K_3PO_4(s)$
   c. $MgBr_2(aq) + H_2SO_4(aq) \longrightarrow HBr(g) + MgSO_4(aq)$

4.68 Balance each of the following equations:
   a. $C_6H_{12}O_6(s) + O_2(g) \longrightarrow CO_2(g) + H_2O(g)$
   b. $H_2O(l) + P_4O_{10}(s) \longrightarrow H_3PO_4(aq)$
   c. $PCl_5(g) + H_2O(l) \longrightarrow HCl(aq) + H_3PO_4(aq)$

4.69 Complete, then balance, each of the following equations:
   a. $Ca(s) + F_2(g) \longrightarrow$
   b. $Mg(s) + O_2(g) \longrightarrow$
   c. $H_2(g) + N_2(g) \longrightarrow$

4.70 Complete, then balance, each of the following equations:
   a. $Li(s) + O_2(g) \longrightarrow$
   b. $Ca(s) + N_2(g) \longrightarrow$
   c. $Al(s) + S(s) \longrightarrow$

4.71 Balance each of the following equations:
  a. $C_4H_{10}(g) + O_2(g) \longrightarrow H_2O(g) + CO_2(g)$
  b. $Au_2S_3(s) + H_2(g) \longrightarrow Au(s) + H_2S(g)$
  c. $Al(OH)_3(s) + HCl(aq) \longrightarrow AlCl_3(aq) + H_2O(l)$
  d. $(NH_4)_2Cr_2O_7(s) \longrightarrow Cr_2O_3(s) + N_2(g) + H_2O(g)$
  e. $C_2H_5OH(l) + O_2(g) \longrightarrow CO_2(g) + H_2O(g)$

4.72 Balance each of the following equations:
  a. $Fe_2O_3(s) + CO(g) \longrightarrow Fe_3O_4(s) + CO_2(g)$
  b. $C_6H_6(l) + O_2(g) \longrightarrow CO_2(g) + H_2O(g)$
  c. $I_4O_9(s) + I_2O_6(s) \longrightarrow I_2(s) + O_2(g)$
  d. $KClO_3(s) \rightarrow KCl(s) + O_2(g)$
  e. $C_6H_{12}O_6(s) \rightarrow C_2H_6O(l) + CO_2(g)$

4.73 Write a balanced equation for each of the following reactions:
  a. Ammonia is formed by the reaction of nitrogen and hydrogen.
  b. Hydrochloric acid reacts with sodium hydroxide to produce water and sodium chloride.
  c. Nitric acid reacts with calcium hydroxide to produce water and calcium nitrate.
  d. Butane (C_4H_{10}) reacts with oxygen to produce water and carbon dioxide.

4.74 Write a balanced equation for each of the following reactions:
  a. Glucose, a sugar, $C_6H_{12}O_6$, is oxidized in the body to produce water and carbon dioxide.
  b. Sodium carbonate, upon heating, produces sodium oxide and carbon dioxide.
  c. Sulfur, present as an impurity in coal, is burned in oxygen to produce sulfur dioxide.
  d. Hydrofluoric acid (HF) reacts with glass (SiO_2) in the process of etching to produce silicon tetrafluoride and water.

Calculations Using a Chemical Equation

4.75 How many grams of boron oxide, B_2O_3, can be produced from 20.0 g diborane (B_2H_6)?

$$B_2H_6(l) + 3O_2(g) \longrightarrow B_2O_3(s) + 3H_2O(l)$$

4.76 How many grams of Al_2O_3 can be produced from 15.0 g Al?

$$4Al(s) + 3O_2(g) \longrightarrow 2Al_2O_3(s)$$

4.77 For the reaction

$$N_2(g) + H_2(g) \longrightarrow NH_3(g)$$

  a. Balance the equation.
  b. How many moles of H_2 would react with 1 mol of N_2?
  c. How many moles of product would form from 1 mol of N_2?
  d. If 14.0 g of N_2 were initially present, calculate the number of moles of H_2 required to react with all of the N_2.
  e. For conditions outlined in part (d), how many grams of product would form?

4.78 Aspirin (acetylsalicylic acid) may be formed from salicylic acid and acetic acid as follows:

$$C_7H_6O_3(aq) + CH_3COOH(aq) \longrightarrow C_9H_8O_4(s) + H_2O(l)$$

Salicylic   Acetic   Aspirin
acid   acid

  a. Is this equation balanced? If not, complete the balancing.
  b. How many moles of aspirin may be produced from 1.00×10^2 mol salicylic acid?
  c. How many grams of aspirin may be produced from 1.00×10^2 mol salicylic acid?

  d. How many grams of acetic acid would be required to react completely with 1.00×10^2 mol salicylic acid?

4.79 The proteins in our bodies are composed of molecules called amino acids. One amino acid is methionine; its molecular formula is $C_5H_{11}NO_2S$. Calculate
  a. the formula weight of methionine
  b. the number of oxygen atoms in a mole of this compound
  c. the mass of oxygen in a mole of the compound
  d. the mass of oxygen in 50.0 g of the compound

4.80 Triglycerides (Chapter 15) are used in biochemical systems to store energy; they can be formed from glycerol and fatty acids. The molecular formula of glycerol is $C_3H_8O_3$. Calculate
  a. the formula weight of glycerol
  b. the number of oxygen atoms in a mole of this compound
  c. the mass of oxygen in a mole of the compound
  d. the mass of oxygen in 50.0 g of the compound

4.81 The burning of acetylene (C_2H_2) in oxygen is the reaction in an oxyacetylene torch. How much oxygen is needed to burn 20.0 kg of acetylene? The unbalanced equation is

$$C_2H_2(g) + O_2(g) \longrightarrow CO_2(g) + H_2O(g)$$

4.82 The reaction of calcium hydride with water can be used to prepare hydrogen gas:

$$CaH_2(s) + 2H_2O(l) \longrightarrow Ca(OH)_2(aq) + 2H_2(g)$$

How many moles of hydrogen gas are produced in the reaction of 1.00×10^2 g calcium hydride with water?

4.83 A rocket can be powered by the reaction between dinitrogen tetroxide and hydrazine:

$$N_2O_4(l) + 2N_2H_4(l) \longrightarrow 3N_2(g) + 4H_2O(g)$$

An engineer designed the rocket to hold 1.00 kg N_2O_4 and excess N_2H_4. How much N_2 would be produced according to the engineer's design?

4.84 A 4.00-g sample of Fe_3O_4 reacts with O_2 to produce Fe_2O_3:

$$4Fe_3O_4(s) + O_2(g) \longrightarrow 6Fe_2O_3(s)$$

Determine the number of grams of Fe_2O_3 produced.

4.85 If the % yield of nitrogen gas in Problem 4.83 is 75.0%, what is the actual yield of nitrogen?

4.86 If the % yield of Fe_2O_3 in Problem 4.84 is 90.0%, what is the actual yield of Fe_2O_3?

FOR FURTHER UNDERSTANDING

1. Which of the following has fewer moles of carbon: 100 g of $CaCO_3$ or 0.5 mol of CCl_4?
2. Which of the following has fewer moles of carbon: 6.02×10^{22} molecules of C_2H_6 or 88 g of CO_2?
3. How many molecules are found in each of the following?
   a. 1.0 lb of sucrose, $C_{12}H_{22}O_{11}$ (table sugar)
   b. 1.57 kg of N_2O (anesthetic)
4. How many molecules are found in each of the following?
   a. 4×10^5 tons of SO_2 (produced by the 1980 eruption of the Mount St. Helens volcano)
   b. 25.0 lb of SiO_2 (major constituent of sand)

5

LEARNING GOALS

1 Correlate the terms *endothermic* and *exothermic* with heat flow between a *system* and its *surroundings*.

2 State the meaning of the terms *enthalpy* and *entropy* and know their implications.

3 Describe experiments that yield thermochemical information and calculate fuel values based on experimental data.

4 Describe the concept of reaction rate and the role of kinetics in chemical change.

5 Describe the importance of *activation energy* and the *activated complex* in determining reaction rate.

6 Predict the way reactant structure, concentration, temperature, and catalysis affect the rate of a chemical reaction.

7 Recognize and describe equilibrium situations.

8 Use LeChatelier's principle to predict changes in equilibrium position.

Energy, Rate, and Equilibrium

In Chapter 4, we calculated quantities of matter involved in chemical change, assuming that all of the reacting material was consumed and that only products of the reaction remain at the end of the reaction. Often this is not true. Furthermore, not all chemical reactions take place at the same speed; some occur almost instantaneously (explosions), whereas others may proceed for many years (corrosion).

Two concepts play important roles in determining the extent and speed of a chemical reaction: thermodynamics, which deals with energy changes in chemical reactions, and kinetics, which describes the rate or speed of a chemical reaction.

Although both thermodynamics and kinetics involve energy, they are two separate considerations. A reaction may be thermodynamically favored but very slow; conversely, a reaction may be very fast because it is kinetically favorable yet produce very little (or no) product because it is thermodynamically unfavorable.

In this chapter, we investigate the fundamentals of thermodynamics and kinetics, with an emphasis on the critical role that energy changes play in chemical reactions. We consider physical change and chemical change, including the conversions that take place among the states of matter (solid, liquid, and gas). We use these concepts to explain the behavior of reactions that do not go to completion, equilibrium reactions. We demonstrate how equilibrium composition can be altered using LeChatelier's principle.

Modern windmills are less polluting generators of electricity. What advantages and disadvantages are associated with this technology?

5.1 Energy, Work, and Heat

Energy: The Basics

Energy, the ability to do work, may be categorized as either **kinetic energy,** the energy of motion, or **potential energy,** the energy of position. Kinetic energy may be considered energy in process; potential energy is stored energy. All energy is either kinetic or potential.

Another useful way of classifying energy is by form. The principal forms of energy include light, heat, electrical, mechanical, and chemical energy. All of these forms of energy share the following set of characteristics:

- In chemical reactions, energy cannot be created or destroyed.
- Energy may be converted from one form to another.
- Conversion of energy from one form to another always occurs with less than 100% efficiency. Energy is not lost (remember, energy cannot be destroyed) but rather, is not useful. We buy gasoline to move our car from place to place; however, much of the energy stored in the gasoline is released as heat.
- All chemical reactions involve either a "gain" or a "loss" of energy.

Energy absorbed or liberated in chemical reactions is usually in the form of heat energy. Heat energy may be represented in units of *calories* or *joules;* their relationship is

$$1 \text{ calorie (cal)} = 4.18 \text{ joules (J)}$$

One calorie is defined as the amount of heat energy required to increase the temperature of 1 gram of water 1°C.

Heat energy measurement is a quantitative measure of heat content. It is an extensive property, dependent upon the quantity of material. Temperature, as we have mentioned, is an intensive property, independent of quantity.

Not all substances have the same capacity for holding heat; 1 gram of iron and 1 gram of water do *not* contain the same amount of heat energy even if they are at the same temperature. One gram of iron will absorb and store 0.108 calorie of heat energy when the temperature is raised 1°C. In contrast, 1 gram of water will absorb almost ten times as much energy, 1.00 calorie, when the temperature is increased an equivalent amount.

> The *kilocalorie* (kcal) is the familiar nutritional calorie. It is also known as the large Calorie; note that in this term the *C* is uppercase to distinguish it from the calorie used in scientific calculations, termed the *small* calorie. The large calorie is 1000 small calories. Refer to Section 5.2 and A Lifestyle Connection: Food Calories for more information.

> Water in the environment (lakes, oceans, and streams) has a powerful effect on the climate because of its ability to store large quantities of energy. In summer, water stores heat energy, moderating temperatures of the surrounding area. In winter, some of this stored energy is released to the air as the water temperature falls; this prevents the surroundings from experiencing extreme changes in temperature.

Thermodynamics

Thermodynamics is the study of energy, work, and heat. It may be applied to chemical change, such as the calculation of the quantity of heat obtainable from the combustion of one gallon of fuel oil. Similarly, energy released or consumed in physical change, such as the boiling or freezing of water, may be determined.

There are three basic laws of thermodynamics, but only the first two will be of concern here. They help us to understand why some chemical reactions may occur spontaneously, whereas others do not.

QUESTION 5.1

Is temperature an intensive or extensive property? [Refer to Chapter 1 for a discussion of intensive and extensive properties.]

QUESTION 5.2

Is energy an intensive or extensive property? [Refer to Chapter 1 for a discussion of intensive and extensive properties.]

A light bulb illuminates a room. What energy conversion process is taking place?

The fire in a fireplace warms the room. What energy conversion process is taking place?

The Chemical Reaction and Energy

John Dalton believed that chemical change involved joining, separating, or rearranging atoms. Two centuries later, this statement stands as an accurate description of chemical reactions. However, we now know much more about the nonmaterial energy changes that are an essential part of every reaction.

Our current model of chemical reactions includes the following:

- molecules and atoms in a reaction mixture are in constant, random motion;
- these molecules and atoms frequently collide with each other;
- only some collisions, those with sufficient energy, will break bonds in molecules; and
- when reactant bonds are broken, new bonds may be formed and products result.

These ideas are basic to the **kinetic-molecular theory;** we will study this theory in more detail in Chapter 6.

It is worth noting that we cannot measure an absolute value for energy stored in a chemical system. We can measure only the *change* in energy (energy absorbed or released) as a chemical reaction occurs. Also, it is often both convenient and necessary to establish a boundary between the *system* and its *surroundings.*

The **system** contains the process under study. The **surroundings** encompass the rest of the universe. Energy is lost from the system to the surroundings or energy may be gained by the system at the expense of the surroundings. This energy change, most often in the form of heat, may be determined because the temperature of the system or surroundings will change, and this change can be measured. This process is illustrated in Figure 5.1.

Consider the combustion of methane in a Bunsen burner, the system. The temperature of the air surrounding the burner increases, indicating that some of the potential energy of the system has been converted to heat energy. The heat energy of the system (methane, oxygen, and the Bunsen burner) is being lost to the surroundings.

Now, an exact temperature measurement of the air before and after the reaction is difficult. However, if we could insulate a portion of the surroundings to isolate and trap the heat, we could calculate a useful quantity, the heat of the reaction.

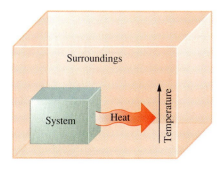

(a)

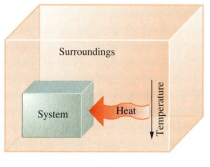

(b)

Figure 5.1

Illustration of heat flow in (a) exothermic and (b) endothermic reactions.

Experimental strategies for measuring temperature change and calculating heats of reactions, termed *calorimetry*, are discussed in Section 5.2.

Exothermic and Endothermic Reactions

1 Learning Goal
Correlate the terms *endothermic* and *exothermic* with heat flow between a *system* and its *surroundings*.

The first law of thermodynamics states that the energy of the universe is constant. It is the law of conservation of energy. The study of energy changes in chemical reactions is a very practical application of the first law. Consider, for example, the generalized reaction:

$$A{-}B + C{-}D \longrightarrow A{-}D + C{-}B$$

An **exothermic reaction** releases energy to the surroundings. The surroundings become warmer.

Each chemical bond is stored chemical energy (potential energy). For the reaction to take place, bond $A{-}B$ and bond $C{-}D$ must break; this process *always* requires energy. At the same time, bonds $A{-}D$ and $C{-}B$ must form; this process always releases energy.

If the energy required to break the $A{-}B$ and $C{-}D$ bonds is *less* than the energy given off when the $A{-}D$ and $C{-}B$ bonds form, the reaction will release the excess energy. The energy is a *product*, and the reaction is called an exothermic (*Gr. exo*, out, and *Gr. therm*, heat) reaction. This conversion of chemical energy to heat is represented in Figure 5.2a.

The combustion of methane, represented by a *thermochemical equation*, is an example of an exothermic reaction.

$$CH_4(g) + 2O_2(g) \longrightarrow CO_2(g) + 2H_2O(g) + \boxed{211 \text{ kcal}}$$

Exothermic reaction

In an exothermic reaction, heat is released *from the system* to the surroundings. In an endothermic reaction, heat is absorbed *by the system* from the surroundings.

This thermochemical equation reads: the combustion of one mole of methane releases 211 kcal of heat.

An **endothermic reaction** absorbs energy from the surroundings. The surroundings become colder.

If the energy required to break the $A{-}B$ and $C{-}D$ bonds is *greater* than the energy released when the $A{-}D$ and $C{-}B$ bonds form, the reaction will need an external supply of energy (perhaps from a Bunsen burner). Insufficient energy is available in the system to initiate the bond breaking. Such a reaction is called an endothermic (*Gr. endo*, to take on, and *Gr. therm*, heat) reaction, and energy is a *reactant*. The conversion of heat energy to chemical energy is represented in Figure 5.2b.

The decomposition of ammonia into nitrogen and hydrogen is one example of an endothermic reaction:

$$\boxed{22 \text{ kcal}} + 2NH_3(g) \longrightarrow N_2(g) + 3H_2(g)$$

Endothermic reaction

Figure 5.2

(a) An exothermic reaction. ΔE represents the energy released during the progress of the exothermic reaction: $A + B \longrightarrow C + D + \Delta E$. (b) An endothermic reaction. ΔE represents the energy absorbed during the progress of the endothermic reaction: $\Delta E + A + B \longrightarrow C + D$.

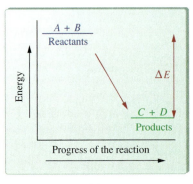

(a)

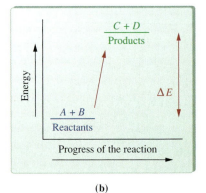

(b)

A MEDICAL Connection | Hot and Cold Packs

Hot packs provide "instant warmth" for hikers and skiers and are used in treatment of injuries such as pulled muscles. Cold packs are in common use today for the treatment of injuries and the reduction of swelling. These useful items are an excellent example of basic science producing a technologically useful product.

Both hot and cold packs depend on large energy changes taking place during a chemical reaction. Cold packs rely on an endothermic reaction, and hot packs generate heat energy from an exothermic reaction.

A cold pack is fabricated as two separate compartments within a single package. One compartment contains NH_4NO_3, and the other contains water. When the package is squeezed, the inner boundary between the two compartments ruptures, allowing the components to mix, and the following reaction occurs:

$$6.7 \text{ kcal/mol} + NH_4NO_3(s) \longrightarrow NH_4^+(aq) + NO_3^-(aq)$$

This reaction is endothermic; heat taken from the surroundings produces the cooling effect.

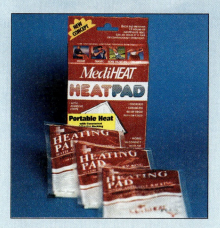

Heating pads.

The design of a hot pack is similar. Here, finely divided iron powder is mixed with oxygen. Production of iron oxide results in the evolution of heat:

$$4Fe + 3O_2 \longrightarrow 2Fe_2O_3 + 198 \text{ kcal/mol}$$

This reaction occurs via an oxidation-reduction mechanism (see Chapter 4). The iron atoms are oxidized, O_2 is reduced. Electrons are transferred from the iron atoms to O_2, and Fe_2O_3 forms exothermically. The rate of the reaction is slow; therefore, the heat is liberated gradually over a period of several hours.

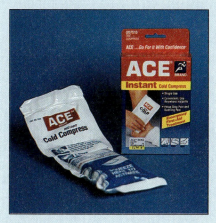

A cold pack.

FOR FURTHER UNDERSTANDING

What is the sign of ΔH for each equation in this story?

Would reactions with small rate constants be preferred for applications such as those described here? Why or why not?

This thermochemical equation reads: the decomposition of two moles of ammonia requires 22 kcal of heat.

The examples used here show the energy absorbed or released as heat energy. Depending on the reaction and the conditions under which the reaction is run, the energy may take the form of light energy or electrical energy. A firefly releases energy as a soft glow of light on a summer evening. An electrical current results from a chemical reaction in a battery, enabling your car to start.

Enthalpy

Enthalpy, or H, is the term used to represent heat. The *change in enthalpy* is the energy difference between the products and reactants of a chemical reaction and is symbolized as ΔH. By convention, energy released is represented with a negative

2 **Learning Goal**

State the meaning of the terms *enthalpy* and *entropy* and know their implications.

Animation

Heat Flow in Endothermic and Exothermic Reactions

In these discussions, we consider that the enthalpy change and energy change are identical. This is true for most common reactions carried out in a lab, with minimal volume change.

sign (indicating an exothermic reaction), and energy absorbed is shown with a positive sign (indicating an endothermic reaction).

For the combustion of methane, an exothermic process, energy is a *product* in the thermochemical equation, and

$$\Delta H = -211 \text{ kcal}$$

For the decomposition of ammonia, an endothermic process, energy is a *reactant* in the thermochemical equation, and

$$\Delta H = +22 \text{ kcal}$$

Spontaneous and Nonspontaneous Reactions

It seems that all exothermic reactions should be spontaneous. After all, an external supply of energy does not appear to be necessary; energy is a product of the reaction. It also seems that all endothermic reactions should be nonspontaneous: energy is a reactant that we must provide. However, these hypotheses are not supported by experimentation.

Experimental measurement has shown that most *but not all* exothermic reactions are spontaneous. Likewise, most *but not all*, endothermic reactions are not spontaneous. There must be some factor in addition to enthalpy that will help us to explain the less obvious cases of nonspontaneous exothermic reactions and spontaneous endothermic reactions. This other factor is entropy.

2 **Learning Goal**

State the meaning of the terms *enthalpy* and *entropy* and know their implications.

A *system* is a part of the universe upon which we wish to focus our attention. For example, it may be a beaker containing reactants and products.

Chapter 6 compares the physical properties of solids, liquids, and gases.

Animation

Entropy and Rubber Bands

Entropy

The first law of thermodynamics considers the enthalpy of chemical reactions. The second law states that the universe spontaneously tends toward increasing disorder or randomness.

Entropy is a measure of the randomness of a chemical system. The entropy of a substance is represented by the symbol S. A random, or disordered, system is characterized by *high entropy*; a well-organized system has *low entropy*.

What do we mean by disorder in chemical systems? Disorder is simply the absence of a regular repeating pattern. Disorder or randomness increases as we convert from the solid to the liquid to the gaseous state. As we have seen, solids often have an ordered crystalline structure, liquids have, at best, a loose arrangement, and gas particles are virtually random in their distribution. Therefore gases have high entropy, and crystalline solids have very low entropy. Figures 5.3 and 5.4 illustrate properties of entropy in systems.

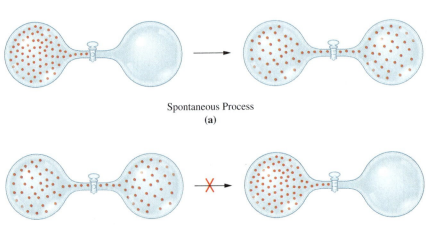

Spontaneous Process
(a)

Nonspontaneous Process
(b)

Figure 5.3

(a) Gas particles, trapped in the left chamber, spontaneously diffuse into the right chamber, initially under vacuum, when the valve is opened. (b) It is unimaginable that the gas particles will rearrange themselves and reverse the process to create a vacuum. This can only be accomplished using a pump, that is, by doing work on the system.

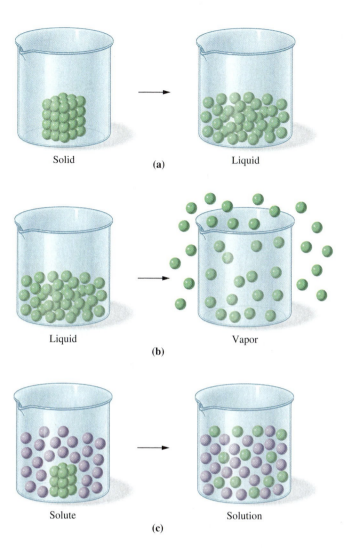

Figure 5.4

Processes such as (a) melting, (b) vaporization, and (c) dissolution increase entropy, or randomness, of particles.

Solid   (a)   Liquid

Liquid   (b)   Vapor

Solute   (c)   Solution

QUESTION 5.5

Are the following processes exothermic or endothermic?

a. Fuel oil is burned in a furnace.
b. $C_6H_{12}O_6(s) \longrightarrow 2C_2H_5OH(l) + 2CO_2(g)$, $\Delta H = -16$ kcal
c. $N_2O_5(g) + H_2O(l) \longrightarrow 2HNO_3(l) + 18.3$ kcal

QUESTION 5.6

Are the following processes exothermic or endothermic?

a. When solid NaOH is dissolved in water, the solution gets hotter.
b. $S(s) + O_2(g) \longrightarrow SO_2(g)$, $\Delta H = -71$ kcal
c. $N_2(g) + 2O_2(g) + 16.2$ kcal $\longrightarrow 2NO_2(g)$

The second law describes the entire universe or any isolated system within the universe. On a more personal level, we all fall victim to the law of increasing disorder. Chaos in our room or workplace is certainly not our intent! It happens almost effortlessly. However, reversal of this process requires work and energy. The same is true at the molecular level. The gradual deterioration of our cities' infrastructure (roads, bridges, water mains, and so forth) is an all-too-familiar example.

A LIFESTYLE Connection | Triboluminescence: Sparks in the Dark with Candy

Generations of children have inadvertently discovered the phenomenon of triboluminescence. Crushing a wintergreen candy (Lifesavers) with the teeth in a dark room (in front of a few friends or a mirror) or simply rubbing two pieces of candy together may produce the effect—transient sparks of light!

Triboluminescence is simply the production of light upon fracturing a solid. It is easily observed and straightforward to describe but difficult to explain. It is believed to result from charge separation produced by the disruption of a crystal lattice. The charge separation has a very short lifetime. When the charge distribution returns to equilibrium, energy is released, and that energy is the light that is observed.

Dr. Linda M. Sweeting and several other groups of scientists have tried to reproduce these events under controlled circumstances. Crystals similar to the sugars in wintergreen candy are prepared with a very high level of purity. Some theories attribute the light emission to impurities in a crystal rather than to the crystal itself. Devices have been constructed that will crush the crystal with a uniform and reproducible force. Light-measuring devices, spectrophotometers, accurately measure the various wavelengths of light and the intensity of the light at each wavelength.

Through the application of careful experimentation and measurement of light-emitting properties of a variety of related compounds, these scientists hope to develop a theory of light emission from fractured solids.

This is one more example of the scientific method improving our understanding of everyday occurrences.

Charles Schulz's "Peanuts" vision of triboluminescence.

Reprinted by permission of UFS, Inc.

FOR FURTHER UNDERSTANDING

Can you suggest an experiment that would support or refute the hypothesis that impurities in crystals are responsible for triboluminescence?

Would you expect that amorphous substances would exhibit triboluminescence? Why or why not?

Millions of dollars (translated into energy and work) are needed annually just to try to maintain the status quo.

The entropy of a reaction is measured as a difference, ΔS, between the entropies, S, of products and reactants.

The drive toward increased entropy, along with a tendency to achieve a lower potential energy, is responsible for spontaneous chemical reactions. Reactions that are exothermic and whose products are more disordered (higher in entropy) will occur spontaneously, whereas endothermic reactions producing products of lower entropy will not be spontaneous. If they are to take place at all, they will need some energy input.

The two situations described above are clear-cut and unambiguous. In any other situation the reaction may or may not be spontaneous. It depends on the relative size of the enthalpy and entropy values.

QUESTION 5.7

Which substance has the greatest entropy, $He(g)$ or $Na(s)$? Explain your reasoning.

QUESTION 5.8

Which substance has the greatest entropy, $H_2O(l)$ or $H_2O(g)$? Explain your reasoning.

EXAMPLE 5.1	Determining Whether a Process Is Exothermic or Endothermic

An ice cube is dropped into a glass of water at room temperature. The ice cube melts. Is the melting of the ice exothermic or endothermic?

SOLUTION

Consider the ice cube as the system and the water, the surroundings. For the cube to melt, it must gain energy and its energy source must be the water. The heat flow is from surroundings to system. The system gains energy (+energy); hence, the melting process (physical change) is endothermic.

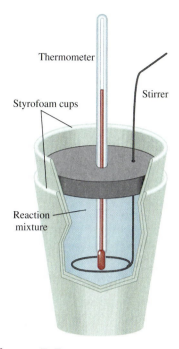

Thermometer

Stirrer

Styrofoam cups

Reaction mixture

QUESTION 5.9

Predict whether a reaction with positive ΔH and negative ΔS will be spontaneous or nonspontaneous. Explain your reasoning.

QUESTION 5.10

Predict whether a reaction with negative ΔH and positive ΔS will be spontaneous or nonspontaneous. Explain your reasoning.

Figure 5.5

A "coffee cup" calorimeter used for the measurement of heat change in chemical reactions. The concentric Styrofoam cups insulate the system from its surroundings. Heat released by the chemical reaction enters the solution, raising its temperature, which is measured by using a thermometer.

5.2 Experimental Determination of Energy Change in Reactions

Calorimetry is the measurement of heat energy changes in a chemical reaction. This technique involves the measurement of the change in the temperature of a quantity of water or solution that is in contact with the reaction of interest and isolated from the surroundings. A device used for these measurements is a *calorimeter*, which measures heat changes in calories.

Something as simple as a pair of Styrofoam coffee cups can be used to construct a calorimeter, and it produces surprisingly accurate results. It is a good insulator, and, when filled with solution, it can be used to measure temperature changes as the result of a chemical reaction in that solution (Figure 5.5). The change in the temperature of the solution, caused by the reaction, can be used to calculate the gain or loss of heat energy for the reaction.

In an exothermic reaction, heat released by the reaction is absorbed by the surrounding solution. In an endothermic reaction, the reactants absorb heat from the solution.

The **specific heat** of a substance is defined as the number of calories of heat needed to raise the temperature of 1 g of the substance 1 degree Celsius. Knowing the specific heat of the water or the aqueous solution along with the total number of grams of solution and the temperature increase (measured as the difference between the final and initial temperatures of the solution), enables the experimenter to calculate the heat released during the reaction.

The solution behaves as a "trap" or "sink" for energy released in the exothermic process. The temperature increase indicates a gain in heat energy. An endothermic reaction, on the other hand, takes heat energy away from the solution, lowering its temperature.

3 **Learning Goal**

Describe experiments that yield thermochemical information and calculate fuel values based on experimental data.

A LIFESTYLE Connection | Food Calories

The body gets its energy through the processes known collectively as metabolism, which will be discussed in detail in subsequent chapters on biochemistry and nutrition. The primary energy sources for the body are carbohydrates, fats, and proteins, which we obtain from the foods we eat. The amount of energy available from a given foodstuff is related to the Calories (C) available in the food. Calories are a measure of the energy and heat content that can be derived from food. One (food) Calorie (symbolized by C) equals 1000 (metric) calories (symbolized by c):

$$1 \text{ Calorie} = 1000 \text{ calories} = 1 \text{ kilocalorie}$$

The energy available in food can be measured by totally burning the food; in other words, we are using the food as a fuel. The energy given off in the form of heat is directly related to the amount of chemical energy, energy stored in chemical bonds, that is available in the food and that the food could provide to the body through the various metabolic pathways.

The classes of food molecules are not equally energy-rich. For instance, when oxidized via metabolic pathways, carbohydrates and proteins provide the cell with 4 Calories per gram, whereas fats generate approximately 9 Calories per gram.

In addition, as with all processes, not all available energy can be efficiently extracted from food; a certain percentage is always lost. The average person requires between 2000 and 3000 Calories per day to maintain normal body functions such as the regulation of body temperature, muscle movement, and so on. If a person takes in more Calories than the body uses, the Calorie-containing substances will be stored as fat, and the person will gain weight. Conversely, if a person uses more Calories than are ingested, the individual will lose weight.

Excess Calories are stored in the form of fat, the form that provides the greatest amount of energy per gram. Too many Calories lead to too much fat. Similarly, a lack of Calories (in the form of food) forces the body to raid its storehouse, the fat. Weight is lost in this process as the fat is consumed. Unfortunately, it always seems easier to add fat to the storehouse than to remove it.

The "rule of thumb" is that 3500 Calories are equivalent to approximately 1 pound of body weight. You have to take in 3500 Calories more than you use to gain a pound, and you have to expend 3500 Calories more than you normally use to lose a pound. If you eat as little as 100 Calories a day above your body's needs, you could gain about 10–11 pounds per year:

$$\frac{100 \text{ C}}{\text{day}} \times \frac{365 \text{ day}}{1 \text{ year}} \times \frac{1 \text{ lb}}{3500 \text{ C}} = \frac{10.4 \text{ lb}}{\text{year}}$$

A frequently recommended procedure for increasing the rate of weight loss involves a combination of dieting (taking in fewer Calories) and exercise. The numbers of Calories used in several activities are

Activity	Energy Output (Calories/minute-pound of body weight)
Running	0.110
Swimming	0.045
Jogging	0.070
Bicycling	0.050
Tennis	0.045
Walking	0.028
Golfing	0.036
Driving a car	0.015
Standing or sitting	0.011
Sleeping	0.008

FOR FURTHER UNDERSTANDING

Sarah runs 1 hour each day, and Nancy swims 2 hours each day. Assuming that Sarah and Nancy are the same weight, which girl burns more calories in 1 week?

Would you expect a runner to burn more calories in summer or winter? Why?

The quantity of heat absorbed or released by the reaction (Q) is the product of the mass of solution in the calorimeter (m_s), the specific heat of the solution (SH_s), and the change in temperature (ΔT_s) of the solution as the reaction proceeds from the initial to final state.

The heat is calculated by using the following equation:

$$Q = m_s \times \Delta T_s \times SH_s$$

The subscript s refers to measurements involving a solution.

with units

$$\text{calories} = \text{gram} \times {}^\circ C \times \frac{\text{calories}}{\text{gram-}^\circ C}$$

The details of the experimental approach are illustrated in Example 5.2.

EXAMPLE 5.2   Calculating Energy Involved in Calorimeter Reactions

3   Learning Goal
Describe experiments that yield thermochemical information and calculate fuel values based on experimental data.

If 0.050 mol of hydrochloric acid (HCl) is mixed with 0.050 mol of sodium hydroxide (NaOH) in a "coffee cup" calorimeter, the temperature of 1.00×10^2 g of the resulting solution increases from 25.0°C to 31.5°C. If the specific heat of the solution is 1.00 cal/g solution °C, calculate the quantity of energy involved in the reaction. Also, is the reaction endothermic or exothermic?

SOLUTION

The change in temperature is

$$\Delta T_s = T_{s\ final} - T_{s\ initial}$$

$$= 31.5°C - 25.0°C = 6.5°C$$

$$Q = m_s \times \Delta T_s \times SH_s$$

$$Q = 1.00 \times 10^2 \text{ g solution} \times 6.5°C \times \frac{1.00 \text{ cal}}{\text{g solution °C}}$$

$$= 6.5 \times 10^2 \text{ cal}$$

6.5×10^2 cal (or 0.65 kcal) of heat energy were released by this acid-base reaction to the surroundings, the solution; the reaction is exothermic.

EXAMPLE 5.3   Calculating Energy Involved in Calorimeter Reactions

If 0.10 mol of ammonium chloride (NH_4Cl) is dissolved in water producing 1.00×10^2 g solution, the water temperature decreases from 25.0°C to 18.0°C. If the specific heat of the resulting solution is 1.00 cal/g-°C, calculate the quantity of energy involved in the process. Also, is the dissolution of ammonium chloride endothermic or exothermic?

SOLUTION

The change in temperature is

$$\Delta T = T_{s\ final} - T_{s\ initial}$$

$$= 18.0°C - 25.0°C = -7.0°C$$

$$Q = m_s \times \Delta T_s \times SH_s$$

$$Q = 1.00 \times 10^2 \text{ g solution} \times (-7.0°C) \times \frac{1.00 \text{ cal}}{\text{g solution °C}}$$

$$= -7.0 \times 10^2 \text{ cal}$$

7.0×10^2 cal (or 0.70 kcal) of heat energy were absorbed by the dissolution process because the solution lost (– sign) 7.0×10^2 cal of heat energy to the system. The reaction is endothermic.

QUESTION 5.11

Refer to Example 5.2 and calculate the temperature change that would have been observed if 50.0 g solution were in the calorimeter instead of 1.00×10^2 g solution.

QUESTION 5.12

Refer to Example 5.2 and calculate the temperature change that would have been observed if 1.00×10^2 g of another liquid, producing a solution with a specific heat capacity of 0.800 cal/g-°C, was substituted for the water in the calorimeter.

QUESTION 5.13

Convert the energy released in Example 5.2 to joules.

QUESTION 5.14

Convert the energy absorbed in Example 5.3 to joules.

Animation
Calories in a Sandwich

Note: Refer to A Lifestyle Connection: Food Calories, p. 150.

3 **Learning Goal**
Describe experiments that yield thermochemical information and calculate fuel values based on experimental data.

Animation
Bomb Calorimeter

Many chemical reactions that produce heat are combustion reactions. In our bodies, many food substances (carbohydrates, proteins, and fats, Chapters 14, 15, and 16) are oxidized to release energy. **Fuel value** is the amount of energy per gram of food.

The fuel value of food is an important concept in nutrition science. The fuel value is generally reported in units of *nutritional Calories*. One **nutritional Calorie** is equivalent to one kilocalorie (1000 calories). It is also known as the *large Calorie* (uppercase C).

Energy necessary for our daily activity and bodily function comes largely from the "combustion" of carbohydrates. Chemical energy from foods that is not used to maintain normal body temperature or in muscular activity is stored in the bonds of chemical compounds known collectively as fat. Thus, "high-calorie" foods are implicated in obesity.

A special type of calorimeter, a *bomb calorimeter*, is useful for the measurement of the fuel value (Calories) of foods. Such a device is illustrated in Figure 5.6. Its design is similar, in principle, to that of the "coffee cup" calorimeter discussed earlier. It incorporates the insulation from the surroundings, solution pool, reaction chamber, and thermometer. Oxygen gas is added as one of the reactants, and an electrical igniter is inserted to initiate the reaction. However, it is not open to the atmosphere. In the sealed container, the reaction continues until the sample is completely oxidized. All of the heat energy released during the reaction is captured in the water.

Figure 5.6

A bomb calorimeter that may be used to measure heat released upon combustion of a sample. This device is commonly used to determine the fuel value of foods. The bomb calorimeter is similar to the "coffee cup" calorimeter. However, note the electrical component necessary to initiate the combustion reaction.

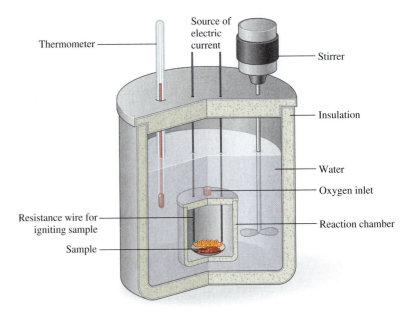

A LIFESTYLE Connection | The Cost of Energy? More Than You Imagine

When we purchase gasoline for our automobiles or oil for the furnace, we are certainly buying matter. That matter is only a storage device; we are really purchasing the energy stored in the chemical bonds. Combustion, burning in oxygen, releases the stored potential energy in a form suited to its function: mechanical energy to power a vehicle or heat energy to warm a home.

Energy release is a consequence of change. In fuel combustion, this change results in the production of waste products that may be detrimental to our environment. This necessitates the expenditure of time, money, and *more* energy to clean up our surroundings.

If we are paying a considerable price for our energy supply, it would be nice to believe that we are at least getting full value for our expenditure. Even that is not the case. Removal of energy from molecules also extracts a price. For example, a properly tuned automobile engine is perhaps 30% efficient. That means that less than one-third of the available energy actually moves the car. The other two-thirds is released into the atmosphere as wasted energy, mostly heat energy. The law of conservation of energy tells us that the energy is not destroyed, but it is certainly not available to us in a useful form.

Can we build a 100% efficient energy transfer system? Is there such a thing as cost-free energy? No, on both counts. It is theoretically impossible, and the laws of thermodynamics, which we discuss in this chapter, tell us why this is so.

FOR FURTHER UNDERSTANDING

If energy is conserved, why should we worry about running out of energy?

In what ways could solar energy be an improvement over fossil fuels?

EXAMPLE 5.4 Calculating the Fuel Value of Foods

One gram of glucose (a common sugar or carbohydrate) was burned in a bomb calorimeter. The temperature of 1.00×10^3 g H_2O was raised from 25.0°C to 28.8°C ($\Delta T_w = 3.8$°C). Calculate the fuel value of glucose.

3 Learning Goal

Describe experiments that yield thermochemical information and calculate fuel values based on experimental data.

SOLUTION

Recall that the fuel value is the number of nutritional Calories liberated by the combustion of 1 g of material and 1 g of material was burned in the calorimeter. Then

$$\text{Fuel value} = Q = m_w \times \Delta T_w \times SH_w$$

Water is the surroundings in the calorimeter; it has a specific heat capacity equal to 1.00 cal/g H_2O °C.

$$\text{Fuel value} = Q = \text{g } H_2O \times °C \times \frac{1.00 \text{ cal}}{\text{g } H_2O \text{ °C}}$$

$$= 1.00 \times 10^3 \text{ g } H_2O \times 3.8°C \times \frac{1.00 \text{ cal}}{\text{g } H_2O \text{ °C}}$$

$$= 3.8 \times 10^3 \text{ cal}$$

and

$$3.8 \times 10^3 \text{ cal} \times \frac{1 \text{ nutritional Calorie}}{10^3 \text{ cal}} = 3.8 \text{ C (nutritional Calories, or kcal)}$$

The fuel value of glucose is 3.8 kcal/g.

The subscript w refers to measurements of water in the bomb calorimeter.

QUESTION 5.15

A 1.0-g sample of a candy bar (which contains lots of sugar!) was burned in a bomb calorimeter. A 3.0°C temperature increase was observed for 1.00×10^3 g of water. The entire candy bar weighed 2.5 ounces. Calculate the fuel value (in nutritional Calories) of the sample and the total caloric content of the candy bar.

QUESTION 5.16

If the fuel value of 1.00 g of a certain carbohydrate (sugar) is 3.00 nutritional Calories, how many grams of water must be present in the calorimeter to record a 5.00°C change in temperature?

4 Learning Goal

Describe the concept of reaction rate and the role of kinetics in chemical and physical change.

Animation

Kinetics

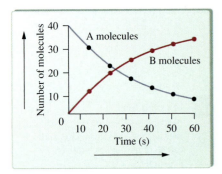

Figure 5.7

For a hypothetical reaction $A \longrightarrow B$, the number of A molecules (reactant molecules) decreases over time and the number of B molecules (product molecules) increases over time.

5.3 Kinetics

Thermodynamics help us to decide whether a chemical reaction is spontaneous. Knowing that a reaction can occur spontaneously tells us nothing about the time that it may take.

Chemical **kinetics** is the study of the **rate** (or speed) **of chemical reactions.** Kinetics also gives an indication of the *mechanism* of a reaction, a step-by-step description of how reactants become products. Kinetic information may be represented as the *disappearance* of reactants or *appearance* of products over time. A typical graph of number of molecules versus time is shown in Figure 5.7.

Information about the rate at which various chemical processes occur is useful. For example, what is the "shelf life" of processed foods? When will slow changes in composition make food unappealing or even unsafe? Many drugs lose their potency with time because the active ingredient decomposes into other substances. The rate of hardening of dental filling material (via a chemical reaction) influences the dentist's technique. Our very lives depend on the efficient transport of oxygen to each of our cells and the rapid use of the oxygen for energy-harvesting reactions.

The diagram in Figure 5.8 is a useful way of depicting the kinetics of a reaction at the molecular level.

Often a color change, over time, can be measured. Such changes are useful in assessing the rate of a chemical reaction (Figure 5.9).

Let's see what actually happens when two chemical compounds react and which experimental conditions affect the rate of a reaction.

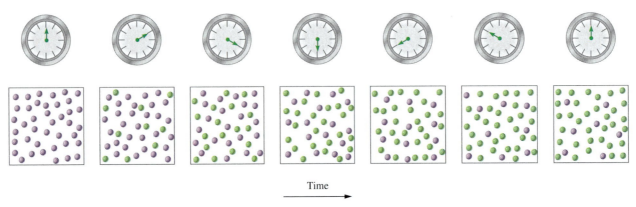

Time

Figure 5.8

An alternate way of representing the information contained in Figure 5.7.

Figure 5.9

The conversion of reddish brown Br_2 in solution to colorless Br^- over time.

Time

The Chemical Reaction

Consider the exothermic reaction that we discussed in Section 5.1:

$$CH_4(g) + 2O_2(g) \longrightarrow CO_2(g) + 2H_2O(l) + \text{211 kcal}$$

For the reaction to proceed, C—H and O—O bonds must be broken, and C—O and H—O bonds must be formed. Sufficient energy must be available to cause the bonds to break if the reaction is to take place. This energy is provided by the collision of molecules. If sufficient energy is available at the temperature of the reaction, one or more bonds will break, and the atoms will recombine in a lower energy arrangement, in this case, as carbon dioxide and water. A collision producing product molecules is termed an *effective collision*. Only effective collisions lead to chemical reaction.

Animation
Rates of Chemical Reactions

Activation Energy and the Activated Complex

The minimum amount of energy required to initiate a chemical reaction is called the **activation energy** for the reaction.

We can picture the chemical reaction in terms of the changes in potential energy that occur during the reaction. Figure 5.10a graphically shows these changes for an exothermic reaction. Important characteristics of this graph include the following:

- The reaction proceeds from reactants to products through an extremely unstable state that we call the **activated complex.** The activated complex cannot be isolated from the reaction mixture but may be thought of as a

Animation
Activation Energy

5 **Learning Goal**
Describe the importance of *activation energy* and the *activated complex* in determining reaction rate.

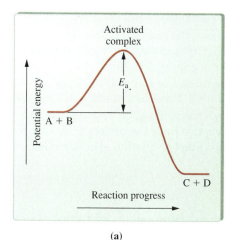

(a)

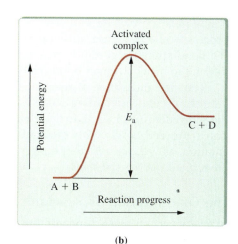

(b)

Figure 5.10

(a) The change in potential energy as a function of reaction time for an exothermic chemical reaction. Note particularly the energy barrier associated with the formation of the activated complex. This energy barrier (E_a) is the activation energy. (b) The change in potential energy as a function of reaction time for an endothermic chemical reaction. In contrast to the exothermic reaction in (a), the energy of the products is greater than the energy of the reactants.

short-lived group of atoms structured so that it quickly and easily breaks apart into the products of the reaction.

- Formation of the activated complex requires energy. The difference between the energy of reactants and that of the activated complex is the activation energy. This energy must be provided by the collision of the reacting molecules or atoms at the temperature of the reaction.
- Because this is an exothermic reaction, the overall energy change must be a *net* release of energy. The *net* release of energy is the difference in energy between products and reactants.

For an endothermic reaction, such as the decomposition of water,

$$\text{energy} + 2H_2O(l) \longrightarrow 2H_2(g) + O_2(g)$$

the change of potential energy with reaction time is shown in Figure 5.10b.

The reaction takes place slowly because of the large activation energy required for the conversion of water into the elements hydrogen and oxygen.

> **This reaction will take place when an electrical current is passed through water. The process is called *electrolysis*.**

QUESTION 5.17

The act of striking a match illustrates the role of activation energy in a chemical reaction. Explain.

QUESTION 5.18

Distinguish between the terms *net energy* and *activation energy*.

Factors That Affect Reaction Rate

> **6 Learning Goal**
> Predict the way reactant structure, concentration, temperature, and catalysis affect the rate of a chemical reaction.

Five major factors influence reaction rate:

- the structure of the reacting species,
- the concentration of reactants,
- the temperature of the reactants,
- the physical state of the reactants, and
- the presence of a catalyst.

Structure of the Reacting Species

Reactions among ions in solution are usually very rapid. Ionic compounds in solution are dissociated; consequently, their bonds are already broken, and the activation energy for their reaction should be very low. On the other hand, reactions involving covalently bonded compounds may proceed more slowly. Covalent bonds must be broken and new bonds formed. The activation energy for this process would be significantly higher than that for the reaction of free ions.

Bond strengths certainly play a role in determining reaction rates. It is more difficult to break stronger bonds so the activation energy is higher.

The size and shape of reactant molecules influence the rate of the reaction. Large molecules, containing bulky groups of atoms, may block the reactive part of the molecule from interacting with another reactive substance, causing the reaction to proceed slowly. Only molecular collisions that have the correct collision orientation, as well as sufficient energy, lead to product formation. These collisions are termed *effective collisions*.

The Concentration of Reactants

> *Concentration is introduced in Section 7.1, and units and calculations are discussed in Sections 7.2 and 7.3.*

In a very general sense, **concentration** is a measure of the quantity of a substance of interest (perhaps a reactant or product) contained in a specified volume of a mixture.

The rate of a chemical reaction is often a complex function of the concentration of one or more of the reacting substances. The rate will generally *increase* as concentration *increases* simply because a higher concentration means more reactant molecules in a given volume and therefore a greater number of collisions per unit time. If we assume that other variables are held constant, a larger number of collisions leads to a larger number of effective collisions. For example, the rate at which a fire burns depends on the concentration of oxygen in the atmosphere surrounding the fire, as well as on the concentration of the fuel (perhaps methane or propane). A common fire-fighting strategy is the use of fire extinguishers filled with carbon dioxide. The carbon dioxide *dilutes* the oxygen to a level where combustion can no longer be sustained.

The Temperature of Reactants

The rate of a reaction *increases* as the temperature increases because the average kinetic energy of the reacting particles is directly proportional to the Kelvin temperature. Increasing the speed of particles increases the likelihood of collision, and the higher kinetic energy means that a higher percentage of these collisions will result in product formation (effective collisions). A 10°C rise in temperature often doubles the reaction rate.

The Physical State of Reactants

The rate of a reaction depends on the physical state of the reactants: solid, liquid, or gas. For a reaction to occur, the reactants must collide frequently and have sufficient energy to react. In the solid state, the atoms, ions, or molecules are restricted in their motion. In the gaseous and liquid states, the particles have both free motion and proximity to each other. There are many more collisions per unit time. Consequently, there are more effective collisions as well. Hence reactions tend to be fastest in the liquid and gaseous states and slowest in the solid state.

These factors will also be considered in our discussion of the states of matter (Chapter 6).

The Presence of a Catalyst

A **catalyst** is a substance that *increases* the reaction rate. If added to a reaction mixture, the catalytic substance undergoes no net change, nor does it alter the outcome of the reaction. However, the catalyst interacts with the reactants to create an alternative pathway for production of products. This alternative path has a lower activation energy. This makes it easier for the reaction to take place and thus increases the rate. This effect is illustrated in Figure 5.11.

Catalysis is important industrially; it may often make the difference between profit and loss in the sale of a product. For example, catalysis is useful in converting double bonds to single bonds. An important application of this principle involves the

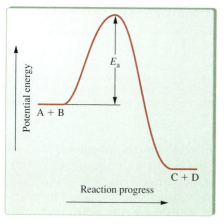

(a) Uncatalyzed reaction

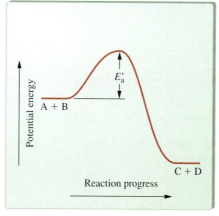

(b) Catalyzed reaction

Figure 5.11

The effect of a catalyst on the magnitude of the activation energy of a chemical reaction: (a) uncatalyzed reaction, (b) catalyzed reaction. Note that the presence of a catalyst decreases the activation energy ($E'_a < E_a$), thus increasing the rate of the reaction.

Figure 5.12

The synthesis of ammonia, an important industrial product, is facilitated by a solid-phase catalyst (the Haber process). H_2 and N_2 bind to the surface, their bonds are weakened, dissociation and reformation as ammonia occur, and the newly formed ammonia molecules leave the surface. This process is repeated over and over, with no change in the catalyst.

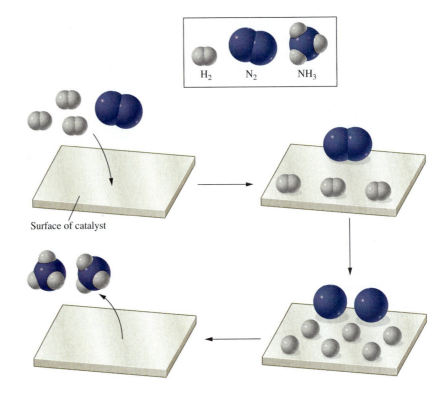

Surface of catalyst

Section 16.10 describes enzyme catalysis.

process of hydrogenation. Hydrogenation converts one or more of the carbon–carbon double bonds of unsaturated fats (e.g., corn oil, olive oil) to single bonds characteristic of saturated fats (such as margarine). The use of a metal catalyst, such as nickel, in contact with the reaction mixture dramatically increases the rate of the reaction.

Thousands of essential biochemical reactions in our bodies are controlled and speeded up by biological catalysts called *enzymes.*

A molecular level view of the action of a solid catalyst widely used in industrial synthesis of ammonia is presented in Figure 5.12.

QUESTION 5.19

Would you imagine that a substance might act as a poison if it interfered with the function of an enzyme? Why?

QUESTION 5.20

Bacterial growth decreases markedly in a refrigerator. Why?

5.4 Equilibrium

Rate and Reversibility of Reactions

 Learning Goal
Recognize and describe equilibrium situations.

We have assumed that most chemical and physical changes considered thus far proceed to completion. A complete reaction is one in which all reactants have been converted to products. However, many important chemical reactions do not go to completion. As a result, after no further obvious change is taking place, measurable quantities of reactants and products remain. Reactions of this type (incomplete reactions) are called **equilibrium reactions.**

Animation
Equilibrium

Examples of physical and chemical equilibria abound in nature. Many environmental systems depend on fragile equilibria. The amount of oxygen dissolved in a certain volume of lake water (the oxygen concentration) is governed by the principles of equilibrium. The lives of plants and animals within this system are critically related to the levels of dissolved oxygen in the water.

The very form and function of the earth is a consequence of a variety of complex equilibria. Stalactite and stalagmite formations in caves are made up of solid calcium carbonate ($CaCO_3$). They owe their existence to an equilibrium process described by the following equation:

$$Ca^{2+}(aq) + 2HCO_3^-(aq) \rightleftharpoons CaCO_3(s) + CO_2(aq) + H_2O(l)$$

Chemical Equilibrium

The Reaction of N_2 and H_2 Illustrates a Dynamic Equilibrium

When we mix nitrogen gas (N_2) and hydrogen gas (H_2) at an elevated temperature (perhaps 500°C), some of the molecules will collide with sufficient energy to break N—N and H—H bonds. Rearrangement of the atoms will produce the product (NH_3):

$$N_2(g) + 3H_2(g) \rightleftharpoons 2NH_3(g)$$

Beginning with a mixture of hydrogen and nitrogen, the rate of the reaction is initially rapid, because the reactant concentration is high; as the reaction proceeds, the concentration of reactants decreases. At the same time, the concentration of the product, ammonia, is increasing. At equilibrium, the *rate of depletion* of hydrogen and nitrogen *is equal to* the *rate of depletion* of ammonia. In other words, *the rates of the forward and reverse reactions are equal.*

The concentration of the various species is fixed at equilibrium because product is being *consumed and formed at the same rate.* In other words, the reaction continues indefinitely (dynamic), but the concentration of products and reactants is fixed (equilibrium). This is a **dynamic equilibrium.** The composition of this reaction mixture as a function of time is depicted in Figure 5.13.

Dynamic equilibrium can be particularly dangerous for living cells because it represents a situation in which nothing is getting done. There is no gain. Let's consider an exothermic reaction designed to produce a net gain of energy for the cell. In a dynamic equilibrium, the rate of the forward (energy-releasing) reaction is equal to the rate of the backward (energy-requiring) reaction. Thus there is no net gain of energy to fuel cellular activity, and the cell will die.

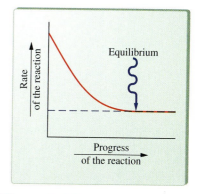

Figure 5.13

The change of the rate of reaction as a function of time. The rate of reaction, initially rapid, decreases as the concentration of reactant decreases and approaches a limiting value at equilibrium.

Animation
Dynamic Equilibrium

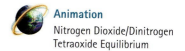
Animation
Nitrogen Dioxide/Dinitrogen Tetraoxide Equilibrium

QUESTION 5.21

Construct an example of a dynamic equilibrium using a subway car at rush hour.

QUESTION 5.22

A certain change in reaction conditions for a process increased the rate of the forward reaction much more than that of the reverse reaction. Did the amount of product increase, decrease, or remain the same? Why?

QUESTION 5.23

How could one determine when a reaction has reached equilibrium?

QUESTION 5.24

Does the attainment of equilibrium imply that no further change is taking place in the system?

LeChatelier's Principle

In the nineteenth century, the French chemist, LeChatelier, discovered that changes in equilibrium depend on the amount of "stress" applied to a system. The stress may take the form of an increase or decrease in the temperature of the system at equilibrium or perhaps a change in the amount of reactant or product present in a fixed volume (the concentration of reactant or product).

LeChatelier's principle states that if a stress is placed on a system at equilibrium, the system will respond by altering the equilibrium composition to minimize the stress.

Consider the equilibrium discussed earlier:

$$N_2(g) + 3H_2(g) \rightleftharpoons 2NH_3(g)$$

If the reactants and products are present in a fixed volume (such as 1L) and more NH_3 (the *product*) is introduced into the container, the system will be stressed—the equilibrium will be disturbed. The system will try to alleviate the stress (as we all do) by *removing* as much of the added material as possible. How can it accomplish this? By converting some NH_3 to H_2 and N_2. The equilibrium shifts to the left, and the dynamic equilibrium is soon reestablished.

Adding extra H_2 or N_2 would apply the stress to the other side of the equilibrium. To minimize the stress, the system would "use up" some of the excess H_2 or N_2 to make product, NH_3. The equilibrium would shift to the right.

In summary,

$$N_2(g) + 3H_2(g) \rightleftharpoons 2NH_3(g)$$

Product introduced: Equilibrium shifted
⟵

Reactant introduced: Equilibrium shifted
⟶

What would happen if some of the ammonia molecules were *removed* from the system? The loss of ammonia represents a stress on the system; to relieve that stress, the ammonia would be replenished by the reaction of hydrogen and nitrogen. The equilibrium would shift to the right.

Effect of Concentration

Addition of extra product or reactant to a fixed reaction volume is just another way of saying that we have increased the concentration of product or reactant. Removal of material from a fixed volume decreases the concentration. Therefore, changing the concentration of one or more components of a reaction mixture is a way to alter the equilibrium composition of an equilibrium mixture (Figure 5.14).

Figure 5.14

The effect of concentration on equilibrium position of the reaction:

$FeSCN^{2+}(aq) \rightleftharpoons Fe^{3+}(aq) + SCN^-(aq)$
(red)   (yellow)   (colorless)

Solution (a) represents this reaction at equilibrium; addition of SCN^- shifts the equilibrium to the left (b) intensifying the red color. Removal of SCN^- shifts the equilibrium to the right (c) shown by the disappearance of the red color.

(a)   (b)   (c)

Let's look at some additional experimental variables that may change an equilibrium composition.

Effect of Heat

The change in equilibrium composition caused by the addition or removal of heat from an equilibrium mixture can be explained by treating heat as a product or reactant. The reaction of nitrogen and hydrogen is an exothermic reaction:

$$N_2(g) + 3H_2(g) \rightleftharpoons 2NH_3(g) + 22\text{ kcal}$$

Adding heat to the reaction is similar to increasing the amount of product. The equilibrium will shift to the left, increasing the amounts of N_2 and H_2 and decreasing the amount of NH_3. If the reaction takes place in a fixed volume, the concentrations of N_2 and H_2 increase and the NH_3 concentration decreases.

Removal of heat produces the reverse effect. More ammonia is produced from N_2 and H_2, and the concentrations of these reactants must decrease.

In the case of an endothermic reaction such as

$$39\text{ kcal} + 2N_2(g) + O_2(g) \rightleftharpoons 2N_2O(g)$$

addition of heat is analogous to the addition of reactant, and the equilibrium shifts to the right. Removal of heat would shift the reaction to the left, favoring the formation of reactants.

The dramatic effect of heat on the position of equilibrium is shown in Figure 5.15.

Effect of Pressure

Only gases are affected significantly by changes in pressure because gases are free to expand and compress in accordance with Boyle's law. However, liquids and solids are not compressible, so their volumes are unaffected by pressure.

Expansion and compression of gases and Boyle's law are discussed in Section 6.1.

Therefore, pressure changes will alter equilibrium composition only in reactions that involve a gas or variety of gases as products and/or reactants. Again, consider the ammonia example,

$$N_2(g) + 3H_2(g) \rightleftharpoons 2NH_3(g)$$

One mole of N_2 and three moles of H_2 (total of four moles of reactants) convert to two moles of NH_3 (two moles of product). An increase in pressure favors a decrease in volume and formation of product. This decrease in volume is made possible by a shift to the right in equilibrium composition. Two moles of ammonia require less volume than four moles of reactant.

The industrial process for preparing ammonia, the Haber process, uses pressures of several hundred atmospheres to increase the yield.

Figure 5.15

The effect of heat on equilibrium position. For the reaction:

$$CoCl_4^{2-}(aq) + 6H_2O(l) \rightleftharpoons$$
(blue)

$$Co(H_2O)_6^{2+}(aq) + 4Cl^-(aq)$$
(pink)

heating the solution favors the blue $CoCl_4^{2-}$ species; cooling favors the pink $Co(H_2O)_6^{2+}$ species.

A decrease in pressure allows the volume to expand. The equilibrium composition shifts to the left and ammonia decomposes to form more nitrogen and hydrogen.

In contrast, the decomposition of hydrogen iodide,

$$2HI(g) \rightleftharpoons H_2(g) + I_2(g)$$

is unaffected by pressure. The number of moles of gaseous product and reactant are identical. No volume advantage is gained by a shift in equilibrium composition.

In summary,

- Pressure affects the equilibrium composition only of reactions that involve at least one gaseous substance.
- Additionally, the relative number of moles of gaseous products and reactants must differ.
- The equilibrium composition will shift to increase the number of moles of gas when the pressure decreases; it will shift to decrease the number of moles of gas when the pressure increases.

Effect of a Catalyst

A catalyst has no effect on equilibrium composition. A catalyst increases the rates of both forward and reverse reactions to the same extent. The equilibrium composition *and* equilibrium concentration do not change when a catalyst is used, but the equilibrium composition is achieved in a shorter time. The role of a solid-phase catalyst in the synthesis of ammonia is shown in Figure 5.12.

8 | Learning Goal
Use LeChatelier's principle to predict changes in equilibrium position.

EXAMPLE 5.5   Predicting Changes in Equilibrium Composition

Earlier in this section, we considered the geologically important reaction that occurs in rock and soil:

$$Ca^{2+}(aq) + 2HCO_3^-(aq) \rightleftharpoons CaCO_3(s) + CO_2(aq) + H_2O(l)$$

Predict the effect on the equilibrium composition for each of the following changes.

a. The $[Ca^{2+}]$ is increased.
b. The $[HCO_3^-]$ is decreased.
c. A catalyst is added.

SOLUTION

a. The concentration of reactant increases; the equilibrium shifts to the right, and more products are formed.
b. The concentration of reactant decreases; the equilibrium shifts to the left, and more reactants are formed.
c. A catalyst has no effect on the equilibrium composition.

QUESTION 5.25

For the hypothetical equilibrium reaction

$$A(g) + B(g) \rightleftharpoons C(g) + D(g)$$

predict whether the amount of *A* in a 5.0-L container would increase, decrease, or remain the same for each of the following changes.

a. Addition of excess *B*
b. Addition of excess *C*
c. Removal of some *D*
d. Addition of a catalyst

QUESTION 5.26

For the hypothetical equilibrium reaction

$$A(g) + B(g) \rightleftharpoons C(g) + D(g)$$

predict whether the amount of *A* in a 5.0-L container would increase, decrease, or remain the same for each of the following changes.

a. Removal of some *B*
b. Removal of some *C*
c. Addition of excess *D*
d. Removal of a catalyst

SUMMARY

5.1 Energy, Work, and Heat

Thermodynamics is the study of energy, work, and heat. Thermodynamics can be applied to the study of chemical reactions because we can determine the quantity of heat flow (by measuring the temperature change) between the *system* and the *surroundings*. *Exothermic reactions* release energy and products that are lower in energy than the reactants. *Endothermic reactions* require energy input. Heat energy is represented as *enthalpy, H*. The energy gain or loss is the change in enthalpy, ΔH, and is one factor that is useful in predicting whether a reaction is spontaneous or nonspontaneous.

Entropy, S, is a measure of the randomness of a system. A random, or disordered system has high entropy; a well-ordered system has low entropy. The change in entropy in a chemical reaction, ΔS, is also a factor in predicting reaction spontaneity.

5.2 Experimental Determination of Energy Change in Reactions

A *calorimeter* measures heat changes (in calories or joules) in chemical reactions.

The *specific heat* of a substance is the number of calories of heat needed to raise the temperature of 1 g of the substance 1 degree Celsius.

The amount of energy per gram of food is referred to as its *fuel value*. Fuel values are commonly reported in units of *nutritional Calories* (1 nutritional Calorie = 1 kcal). A bomb calorimeter is useful for measuring the fuel value of foods.

5.3 Kinetics

Chemical *kinetics* is the study of the *rate* or speed of a chemical reaction. Energy for reactions is provided by molecular collisions. If this energy is sufficient, bonds may break, and atoms may recombine in a different arrangement, producing product. A collision producing one or more product molecules is termed an effective collision.

The minimum amount of energy needed for a reaction is the *activation energy*. The reaction proceeds from reactants to products through an intermediate state, the *activated complex*.

Experimental conditions influencing the reaction rate include the structure of the reacting species, the concentration of reactants, the temperature of reactants, the physical state of reactants, and the presence or absence of a catalyst.

A *catalyst* increases the rate of a reaction. The catalytic substance undergoes no net change in the reaction, nor does it alter the outcome of the reaction.

5.4 Equilibrium

Many chemical reactions do not completely convert reactants to products. A mixture of products and reactants exists, and its composition will remain constant until the experimental conditions are changed. This mixture is in a state of *chemical equilibrium*. The reaction continues indefinitely (dynamic), but the concentrations of products and reactants are fixed (equilibrium) because the rates of the forward and reverse reactions are equal. This is a *dynamic equilibrium*.

LeChatelier's principle states that if a stress is placed on an equilibrium system, the system will respond by altering the equilibrium to minimize the stress.

KEY TERMS

activated complex (5.3)
activation energy (5.3)
calorimetry (5.2)
catalyst (5.3)
concentration (5.3)
dynamic equilibrium (5.4)
endothermic reaction (5.1)
enthalpy (5.1)
entropy (5.1)
equilibrium reaction (5.4)

exothermic reaction (5.1)
fuel value (5.2)
kinetics (5.3)
LeChatelier's principle (5.4)
nutritional Calorie (5.2)
rate of chemical reaction (5.3)
specific heat (5.2)
surroundings (5.1)
system (5.1)
thermodynamics (5.1)

QUESTIONS AND PROBLEMS

Energy, Work, and Heat

Fundamentals

5.27 What is the energy unit most commonly employed in chemistry?

5.28 Which energy unit is commonly employed in nutrition science?

5.29 Describe what is meant by an exothermic reaction.

5.30 Describe what is meant by an endothermic reaction.

5.31 Describe how a calorimeter is used to distinguish between exothermic and endothermic reactions.

5.32 Construct a diagram of a coffee-cup calorimeter.

5.33 Why does a calorimeter have a "double-walled" container?

5.34 Explain why the fuel value of foods is an important factor in nutrition science.

5.35 State the first law of thermodynamics.

5.36 State the second law of thermodynamics.

5.37 Explain what is meant by the term *enthalpy*.

5.38 Explain what is meant by the term *entropy*.

Applications

5.39 5.00 g of octane are burned in a bomb calorimeter containing 2.00×10^2 g H_2O. How much energy, in calories, is released if the water temperature increases $6.00°C$?

5.40 0.0500 mol of a nutrient substance is burned in a bomb calorimeter containing 2.00×10^2 g H_2O. If the formula weight of this nutrient substance is 114 g/mol, what is the fuel value (in nutritional Calories) if the temperature of the water increased by $5.70°C$.

5.41 Calculate the energy released, in joules, in Question 5.39 (recall conversion factors, Chapter 1).

5.42 Calculate the fuel value, in kilojoules, in Question 5.40 (recall conversion factors, Chapter 1).

5.43 Predict whether each of the following processes increases or decreases entropy, and explain your reasoning.
  a. melting of a solid metal
  b. boiling of water

5.44 Predict whether each of the following processes increases or decreases entropy, and explain your reasoning.
  a. burning a log in a fireplace
  b. condensation of water vapor on a cold surface

5.45 Isopropyl alcohol, commonly known as rubbing alcohol, feels cool when applied to the skin. Explain why.

5.46 Energy is required to break chemical bonds during a reaction. When is energy released?

Kinetics

Fundamentals

5.47 Provide an example of a reaction that is extremely slow, taking days, weeks, or years to complete.

5.48 Provide an example of a reaction that is extremely fast, perhaps quicker than the eye can perceive.

5.49 Define the term *activated complex* and explain its significance in a chemical reaction.

5.50 Define and explain the term *activation energy* as it applies to chemical reactions.

Applications

5.51 Describe the general characteristics of a catalyst.

5.52 Select one enzyme from a later chapter in this book and describe its biochemical importance.

5.53 Sketch a potential energy diagram similar to Figure 5.11 for a reaction that shows the effect of a catalyst on an exothermic reaction.

5.54 Sketch a potential energy diagram similar to Figure 5.11 for a reaction that shows the effect of a catalyst on an endothermic reaction.

5.55 Give at least two examples from the life sciences in which the rate of a reaction is critically important.

5.56 Give at least two examples from everyday life in which the rate of a reaction is an important consideration.

5.57 Describe how an increase in the concentration of reactants increases the rate of a reaction.

5.58 Describe how an increase in the temperature of reactants increases the rate of a reaction.

5.59 Describe how a catalyst speeds up a chemical reaction.

5.60 Explain how a catalyst can be involved in a chemical reaction without being consumed in the process.

Equilibrium

Fundamentals

5.61 Explain LeChatelier's principle.

5.62 How can LeChatelier's principle help us to increase yields of chemical reactions?

5.63 Describe the meaning of the term *dynamic equilibrium*.

5.64 What is the relationship between the forward and reverse rates of a reaction at equilibrium?

Applications

5.65 For the reaction

$$CH_4(g) + Cl_2(g) \rightleftharpoons CH_3Cl(g) + HCl(g) + 26.4 \text{ kcal}$$

predict the effect on the equilibrium (will it shift to the left or to the right, or will there be no change?) for each of the following changes.
  a. The temperature is increased.
  b. The pressure is increased by decreasing the volume of the container.
  c. A catalyst is added.

5.66 For the reaction

$$47 \text{ kcal} + 2SO_3(g) \rightleftharpoons 2SO_2(g) + O_2(g)$$

predict the effect on the equilibrium (will it shift to the left or to the right, or will there be no change?) for each of the following changes.
  a. The temperature is increased.
  b. The pressure is increased by decreasing the volume of the container.
  c. A catalyst is added.

5.67 Label each of the following statements as true or false and explain why.

    a. A slow reaction is an incomplete reaction.

    b. The rates of forward and reverse reactions are never the same.

5.68 Label each of the following statements as true or false and explain why.

    a. A reaction is at equilibrium when no reactants remain.

    b. A reaction at equilibrium is undergoing continual change.

5.69 Use LeChatelier's principle to predict whether the amount of PCl_3 in a 1.00-L container is increased, is decreased, or remains the same for the equilibrium

$$PCl_3(g) + Cl_2(g) \rightleftharpoons PCl_5(g) + \text{heat}$$

when each of the following changes is made.

    a. PCl_5 is added.

    b. Cl_2 is added.

    c. PCl_5 is removed.

    d. The temperature is decreased.

    e. A catalyst is added.

5.70 Use LeChatelier's principle to predict the effects, if any, of each of the following changes on the equilibrium system, described below, in a closed container.

$$C(s) + 2H_2(g) \rightleftharpoons CH_4(g) + 18 \text{ kcal}$$

    a. adding more H_2.

    b. removing CH_4.

    c. increasing the temperature.

    d. adding a catalyst.

5.71 Will an increase in pressure increase, decrease, or have no effect on the concentration of $H_2(g)$ in the reaction:

$$C(s) + H_2O(g) \rightleftharpoons CO(g) + H_2(g)$$

5.72 Will an increase in pressure increase, decrease, or have no effect on the concentration of $NO(g)$ in the reaction:

$$N_2(g) + O_2(g) \rightleftharpoons 2NO(g)$$

5.73 True or false: The equilibrium will shift to the right when a catalyst is added to the mixture described in Question 5.71. Explain your reasoning.

5.74 True or false: The equilibrium for an endothermic reaction will shift to the right when the reaction mixture is heated. Explain your reasoning.

5.75 A bottle of carbonated beverage slowly goes "flat" (loses CO_2) after it is opened. Explain, using LeChatelier's principle.

5.76 Carbonated beverages quickly go flat (lose CO_2) when heated. Explain, using LeChatelier's principle.

FOR FURTHER UNDERSTANDING

1. Can the following statement ever be true? "Heating a reaction mixture increases the rate of a certain reaction but decreases the yield of product from the reaction." Explain why or why not.

2. Molecules must collide for a reaction to take place. Sketch a model of the orientation and interaction of HI and Cl that is most favorable for the reaction:

$$HI(g) + Cl(g) \longrightarrow HCl(g) + I(g)$$

3. Human behavior often follows LeChatelier's principle. Provide one example and explain in terms of LeChatelier's principle.

4. A clever device found in some homes is a figurine that is blue on dry, sunny days and pink on damp, rainy days. These figurines are coated with substances containing chemical species that undergo the following equilibrium reaction:

$$Co(H_2O)_6^{2+}(aq) + 4Cl^-(aq) \rightleftharpoons CoCl_4^{2-}(aq) + 6H_2O(l)$$

    a. Which substance is blue?

    b. Which substance is pink?

    c. How is LeChatelier's principle applied here?

5. You have spent the entire morning in a 20°C classroom. As you ride the elevator to the cafeteria, six persons enter the elevator after being outside on a subfreezing day. You suddenly feel chilled. Explain the heat flow situation in the elevator in thermodynamic terms.

6

OUTLINE

LEARNING GOALS

1 Describe the major points of the kinetic molecular theory of gases.

2 Explain the relationship between the kinetic molecular theory and the physical properties of macroscopic quantities of gases.

3 Describe the behavior of gases expressed by the gas laws: Boyle's law, Charles's law, combined gas law, Avogadro's law, the ideal gas law, and Dalton's law.

4 Use gas law equations to calculate conditions and changes in conditions of gases.

5 Describe properties of the liquid state in terms of the properties of the individual molecules that comprise the liquid.

6 Describe the processes of melting, boiling, evaporation, and condensation.

7 Describe the dipolar attractions known collectively as van der Waals forces.

8 Describe hydrogen bonding and its relationship to boiling and melting temperatures.

9 Relate the properties of the various classes of solids (ionic, covalent, molecular, and metallic) to the structure of these solids.

States of Matter
Gases, Liquids, and Solids

Heating the air in these balloons is sufficient to cause them to float majestically in the air. In what ways does a hot air balloon differ from the airship described on this page?

One of the largest and most luxurious airships of the 1930s, the *Hindenburg*, completed thirty-six transatlantic flights within a year after its construction. It was the flagship of a new era of air travel. But, on May 6, 1937, while making a landing approach near Lakehurst, New Jersey, the hydrogen-filled airship exploded and burst into flames. In this tragedy, thirty-seven of the ninety-six passengers were killed and many others were injured (see Figure 6.1).

We may never know the exact cause. Many believe that the massive ship (it was more than 800 feet long) struck an overhead power line. Others speculate that lightning ignited the hydrogen and some believe that sabotage may have been involved.

In retrospect, such an accident was inevitable. Hydrogen gas is very reactive, it combines with oxygen readily and rapidly, and this reaction liberates a large amount of energy. An explosion is the result of rapid, energy-releasing reactions.

Why was hydrogen chosen? Hydrogen is the lightest element. One mole of hydrogen has a mass of 2 grams. Hydrogen can be easily prepared in pure form, an essential requirement; more than seven million cubic feet of hydrogen were needed for each airship. Hydrogen has a low density; hence it provides great lift. The lifting power of a gas is based on the difference in density of the gas and the surrounding air (air is composed of gases with much greater molar masses; N_2 is 28 g and O_2 is 32 g). Engineers believed that the hydrogen would be safe when enclosed by the hull of the airship.

Today, airships are filled with helium (its molar mass is 4 g) and are used principally for advertising and television. A Goodyear blimp can be seen hovering over almost every significant outdoor sporting event.

Can a substance such as O_2 or N_2 also exist as a liquid, or even as a solid? We will see that a reduction in temperature or an increase in pressure can force atoms or molecules closer together, allowing them to behave as liquids or solids. Dry ice, for example, is solid carbon dioxide. Changes in state are described as physical changes. When a substance undergoes a change in state, many of its physical properties change. For example, when ice forms from liquid water, changes occur in density and hardness, but it is still water. Table 6.1 summarizes the important differences in physical properties among gases, liquids, and solids.

Figure 6.1

The Hindenburg, filled with hydrogen gas, explodes in Lakehurst, New Jersey, in 1937.

TABLE 6.1 A Comparison of Physical Properties of Gases, Liquids, and Solids

	Gas	Liquid	Solid
Volume and Shape	Expands to fill the volume of its container; consequently, it takes the shape of the container	Has a fixed volume at a given mass and temperature; volume principally dependent on its mass and secondarily on temperature; it assumes the shape of its container	Has a fixed volume; volume principally dependent on its mass and secondarily on temperature; it has a definite shape
Density	Low (typically ~10^{-3} g/mL)	High (typically ~1 g/mL)	High (typically 1–10 g/mL)
Compressibility	High	Very low	Virtually incompressible
Particle Motion	Virtually free	Molecules or atoms "slide" past each other	Vibrate about a fixed position
Intermolecular Distance	Very large	Molecules or atoms are close to each other	Molecules, ions, or atoms are close to each other

6.1 The Gaseous State

Ideal Gas Concept

An **ideal gas** is simply a model of the way that particles (molecules or atoms) behave at the microscopic level. The behavior of the individual particles can be inferred from the macroscopic behavior of samples of real gases. We can easily measure the temperature, volume, pressure, and quantity (mass) of real gases. Similarly, when we systematically change one of these properties, we can determine the effect on each of the others. For example, putting more molecules in a balloon (the act of blowing up a balloon) causes its volume to increase in a predictable way. Careful measurements show a direct proportionality between the quantity of molecules and the volume of the balloon, an observation made by Amadeo Avogadro more than 200 years ago.

We owe a great deal of credit to the efforts of scientists Boyle, Charles, Avogadro, Dalton, and Gay-Lussac, whose careful work elucidated the relationships among gas properties. Their efforts are summarized in the ideal gas law and are the subject of the first section of this chapter.

Description of a Gas

The most important gas laws (Boyle's law, Charles's law, Avogadro's law, Dalton's law, and the ideal gas law) involve relationships between pressure (P), volume (V), temperature (T), and number of moles (n) of gas. For any gas, knowing the values of these four variables, along with the identity of the gas, will characterize the gas.

Pressure

Gas pressure is a result of the force exerted by the collision of particles with the walls of the container. **Pressure** is force per unit area. The pressure of a gas may be measured with a **barometer,** invented by Evangelista Torricelli in the mid-1600s. The most common type of barometer is the mercury barometer depicted in Figure 6.2. A tube, sealed at one end, is filled with mercury and inverted in a dish of mercury. The pressure of the atmosphere pushing down on the mercury surface in the dish supports the column of mercury. The height of the column is proportional to the atmospheric pressure. The tube can be calibrated to give a numerical reading in millimeters, centimeters, or inches of mercury. A commonly used unit of measurement is the atmosphere (atm). One standard atmosphere (1 atm) of pressure is equivalent to a height of mercury that is equal to

760 mm Hg (millimeters of mercury)

76.0 cm Hg (centimeters of mercury)

1 mm of Hg is also = 1 torr, in honor of Torricelli.

The English system equivalent is a pressure of 14.7 lb/in.² (pounds per square inch) or 29.9 in. Hg (inches of mercury). A recommended, yet less frequently used, systematic unit is the pascal (or kilopascal), named in honor of Blaise Pascal, a seventeenth-century French mathematician and scientist:

$$1 \text{ atm} = 1.01 \times 10^5 \text{ Pa (pascal)} = 101 \text{ kPa (kilopascal)}$$

Atmospheric pressure is due to the cumulative force of the air molecules (N_2 and O_2, for the most part) that are attracted to the earth's surface by gravity.

Volume

A gas will expand to occupy any available space. It follows that the volume of any gas is determined by the volume of its container. A gas, even one under high pressure, is mostly empty space. The volume occupied by the gas should not be confused with the volume of the gas particles (atoms or molecules); the particle volume is assumed to be zero for most gas calculations.

The container volume, hence, the volume of the gas can be measured in a variety of ways. For large containers, with a regular geometric shape, the dimensions can be measured and the volume calculated. For small containers, a few liters or less, the container can be filled with water, and the volume of water measured, perhaps with a graduated cylinder.

Temperature

The temperature of a gas is related to the kinetic energy of the gas particles (see p. 170) and is most often measured with a thermometer. In practice, any of the common scales for temperature (Celsius, Fahrenheit, or Kelvin) are used. However, only the Kelvin scale can be used in gas calculations; gas particle energies are directly related to Kelvin, not Celsius or Fahrenheit temperatures.

Number of Moles of Gas

The quantity of a gas, not to be confused with its volume, is measured as a mass, perhaps grams, and then converted to a number of moles of gas. Moles

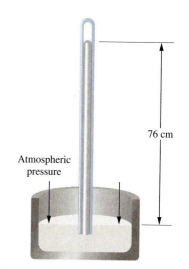

Figure 6.2

A mercury barometer of the type invented by Torricelli. The height of the column of mercury is a function of the magnitude of the surrounding atmospheric pressure. The mercury in the tube is supported by atmospheric pressure.

See the relationship among grams, moles, and Avogadro's number in Chapter 4, p. 108.

and Avogadro's number of gas particles are related, and it is the number of gas particles and their total energy that determine such properties as the pressure that the gas exerts on the walls of its container, the temperature of the gas, and the volume that the gas occupies.

QUESTION 6.1

Express each of the following in units of atmospheres:

a. 725 mm Hg
b. 29.0 cm Hg
c. 555 torr

QUESTION 6.2

Express each of the following in units of atmospheres:

a. 10.0 torr
b. 61.0 cm Hg
c. 275 mm Hg

Kinetic Molecular Theory of Gases

1 Learning Goal
Describe the major points of the kinetic molecular theory of gases.

The kinetic molecular theory of gases provides a reasonable explanation of the behavior of gases that we study in this chapter. The macroscopic properties result from the action of the individual molecules comprising the gas.

The **kinetic molecular theory** can be summarized as follows:

1. Gases are made up of small atoms or molecules that are in constant, random motion.
2. The distance of separation among these atoms or molecules is very large in comparison to the size of the individual atoms or molecules. In other words, a gas is mostly empty space.
3. All of the atoms and molecules behave independently. No attractive or repulsive forces exist between atoms or molecules in a gas.
4. Atoms and molecules collide with each other and with the walls of the container without *losing* energy. The energy is *transferred* from one atom or molecule to another.
5. The average kinetic energy of the atoms or molecules increases or decreases in proportion to the absolute temperature.

Kinetic energy (K.E.) is equal to $1/2\ mv^2$, in which m = mass and v = velocity. Thus increased velocity at higher temperature correlates with an increase in kinetic energy.

Properties of Gases and the Kinetic Molecular Theory

2 Learning Goal
Explain the relationship between the kinetic molecular theory and the physical properties of macroscopic quantities of gases.

We know that gases are easily *compressible*. The reason is that a gas is mostly empty space, providing space for the particles to be pushed closer together.

Gases will *expand* to fill any available volume because they move freely with sufficient energy to overcome their attractive forces.

Gases have a *low density*. Density is defined as mass per volume. Because gases are mostly empty space, they have a low mass per volume.

Gases readily *diffuse* through each other simply because they are in continuous motion and paths are readily available because of the large space between adjacent atoms or molecules. Light molecules diffuse rapidly; heavier molecules diffuse more slowly.

Gases exert *pressure* on their containers. Pressure is a force per unit area resulting from collisions of gas particles with the walls of their container.

Gases behave most *ideally at low pressures and high temperatures*. At low pressures, the average distance of separation among atoms or molecules is greatest, minimizing interactive forces. At high temperatures, the atoms and molecules are in rapid motion and are able to overcome interactive forces more easily.

Boyle's Law: Relationship of Pressure and Volume

The Irish scientist, Robert Boyle, found that the volume of a gas varies *inversely* with the pressure exerted by the gas if the number of moles and temperature of gas are held constant. This relationship is known as **Boyle's law.**

Mathematically, the *product* of pressure (P) and volume (V) is a constant:

$$PV = k_1$$

This relationship is illustrated in Figure 6.3.

Boyle's law is often used to calculate the volume resulting from a pressure change or vice versa. We consider

$$P_iV_i = k_1$$

the *initial* condition and

$$P_fV_f = k_1$$

the final condition. Because PV, initial or final, is constant and is equal to k_1,

$$P_iV_i = P_fV_f$$

Consider a gas occupying a volume of 10.0 L at 1.00 atm of pressure. The product, $PV = (10.0\ L)(1.00\ atm)$, is a constant, k_1. Doubling the pressure, to 2.0 atm, decreases the volume to 5.0 L:

$$(2.0\ atm)(V_x) = (10.0\ L)(1.00\ atm)$$

$$V_x = 5.0\ L$$

Tripling the pressure decreases the volume by a factor of 3:

$$(3.0\ atm)(V_x) = (10.0\ L)(1.00\ atm)$$

$$V_x = 3.3\ L$$

Animation
Boyle's Law

3 **Learning Goal**
Describe the behavior of gases expressed by the gas laws: Boyle's law, Charles's law, combined gas law, Avogadro's law, the ideal gas law, and Dalton's law.

You can find further information on solving gas law problems online at www.mhhe.com/denniston in "A Review of Mathematics."

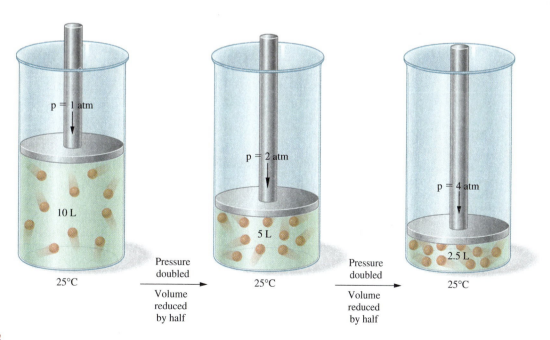

Figure 6.3

An illustration of Boyle's law. Note the inverse relationship of pressure and volume.

4 Learning Goal
Use gas law equations to calculate conditions and changes in conditions of gases.

EXAMPLE 6.1 Calculating a Final Pressure

A sample of oxygen, at 25°C, occupies a volume of 5.00×10^2 mL at 1.50 atm pressure. What pressure must be applied to compress the gas to a volume of 1.50×10^2 mL, with no temperature change?

SOLUTION

Boyle's law applies directly, because there is no change in temperature or number of moles (no gas enters or leaves). Begin by identifying each term in the Boyle's law expression:

$$P_i = 1.50 \text{ atm}$$

$$V_i = 5.00 \times 10^2 \text{ mL}$$

$$V_f = 1.50 \times 10^2 \text{ mL}$$

$$P_i V_i = P_f V_f$$

and solve

$$P_f = \frac{P_i V_i}{V_f}$$

$$= \frac{(1.50 \text{ atm})(5.00 \times 10^2 \text{ mL})}{1.50 \times 10^2 \text{ mL}}$$

$$= 5.00 \text{ atm}$$

Helpful Hints:

1. Go to www.mhhe.com/denniston for a review of the mathematics used here.
2. The calculation can be done with any volume units. It is important only that the units be the *same* on both sides of the equation.

QUESTION 6.3

Complete the following table:

	Initial Pressure (atm)	Final Pressure (atm)	Initial Volume (L)	Final Volume (L)
a.	X	5.0	1.0	7.5
b.	5.0	X	1.0	0.20

QUESTION 6.4

Complete the following table:

	Initial Pressure (atm)	Final Pressure (atm)	Initial Volume (L)	Final Volume (L)
a.	1.0	0.50	X	0.30
b.	1.0	2.0	0.75	X

3 Learning Goal
Describe the behavior of gases expressed by the gas laws: Boyle's law, Charles's law, combined gas law, Avogadro's law, the ideal gas law, and Dalton's law.

Charles's Law: Relationship of Temperature and Volume

Jacques Charles, a French scientist, studied the relationship between gas volume and temperature. This relationship, **Charles's law,** states that the volume of a gas varies *directly* with the absolute temperature (K) if the pressure and number of moles of gas are constant.

Mathematically, the *ratio* of volume (*V*) and temperature (*T*) is a constant:

$$\frac{V}{T} = k_2$$

In a way analogous to Boyle's law, we may establish a set of initial conditions,

$$\frac{V_i}{T_i} = k_2$$

and final conditions,

$$\frac{V_f}{T_f} = k_2$$

Because k_2 is a constant, we may equate them, resulting in

$$\frac{V_i}{T_i} = \frac{V_f}{T_f}$$

and use this expression to solve some practical problems.

Consider a gas occupying a volume of 10.0 L at 273 K. The ratio *V/T* is a constant, k_2. Doubling the temperature, to 546 K, increases the volume to 20.0 L as shown here:

$$\frac{10.0 \text{ L}}{273 \text{ K}} = \frac{V_f}{546 \text{ K}}$$

$$V_f = 20.0 \text{ L}$$

Tripling the temperature, to 819 K, increases the volume by a factor of 3:

$$\frac{10.0 \text{ L}}{273 \text{ K}} = \frac{V_f}{819 \text{ K}}$$

$$V_f = 30.0 \text{ L}$$

These relationships are illustrated in Figure 6.4.

Temperature is a measure of the energy of molecular motion. The Kelvin scale is *absolute*, that is, directly proportional to molecular motion. Celsius and Fahrenheit are simply numerical scales based on the melting and boiling points of water. It is for this reason that Kelvin is used for energy-dependent relationships such as the gas laws.

Animation
Charles's Law

You can find further information on solving gas law problems online at www.mhhe.com/denniston in "A Review of Mathematics."

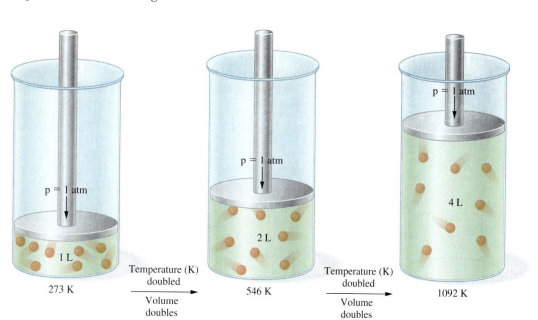

Figure 6.4

An illustration of Charles's law. Note the direct relationship between volume and temperature.

4 Learning Goal
Use gas law equations to calculate conditions and changes in conditions of gases.

You can find further information on solving gas law problems online at www.mhhe.com/denniston in "A Review of Mathematics."

EXAMPLE 6.2 Calculating a Final Volume

A balloon filled with helium has a volume of 4.0×10^3 L at 25°C. What volume will the balloon occupy at 50°C if the pressure surrounding the balloon remains constant?

SOLUTION

Remember, the temperature must be converted to Kelvin before Charles's law is applied:

$$T_i = 25°C + 273 = 298 \text{ K}$$

$$T_f = 50°C + 273 = 323 \text{ K}$$

$$V_i = 4.0 \times 10^3 \text{ L}$$

$$V_f = ?$$

Using

$$\frac{V_i}{T_i} = \frac{V_f}{T_f}$$

and substituting our data, we get

$$V_f = \frac{(V_i)(T_f)}{T_i} = \frac{(4.0 \times 10^3 \text{ L})(323 \text{ K})}{298 \text{ K}} = 4.3 \times 10^3 \text{ L}$$

QUESTION 6.5

A sample of nitrogen gas has a volume of 3.00 L at 25°C. What volume will it occupy at each of the following temperatures if the pressure and number of moles are constant?

a. 100°C
b. 150°F
c. 273 K

QUESTION 6.6

A sample of nitrogen gas has a volume of 3.00 L at 25°C. What volume will it occupy at each of the following temperatures if the pressure and number of moles are constant?

a. 546 K
b. 0°C
c. 373 K

Figure 6.5

Charles's law predicts that the volume of air in the balloon will increase when heated. We assume that the volume of the balloon is fixed; consequently, some air will be pushed out. The air remaining in the balloon is less dense (same volume, less mass) and the balloon will rise. When the heater is turned off the air cools, the density increases, and the balloon returns to earth.

The behavior of a hot-air balloon is a commonplace consequence of Charles's law. The balloon rises because air expands when heated (Figure 6.5). The volume of the balloon is fixed because the balloon is made of an inelastic material; as a result, when the air expands some of the air must be forced out. Hence the density of the remaining air is less (less mass contained in the same volume), and the balloon rises. Turning down the heat reverses the process, and the balloon descends.

Combined Gas Law

Boyle's law describes the inverse proportional relationship between volume and pressure; Charles's law shows the direct proportional relationship between volume and temperature. Often, a sample of gas (a fixed number of moles of gas) undergoes change involving volume, pressure, and temperature simultaneously. It would be useful to have one equation that describes such processes.

The **combined gas law** is such an equation. It is derived from Boyle's law and Charles's law and takes the form:

$$\frac{P_i V_i}{T_i} = \frac{P_f V_f}{T_f}$$

Let's look at two examples that use this expression.

Animation
Interactive Gas Laws

3 Learning Goal
Describe the behavior of gases expressed by the gas laws: Boyle's law, Charles's law, combined gas law, Avogadro's law, the ideal gas law, and Dalton's law.

EXAMPLE 6.3 Using the Combined Gas Law

Calculate the volume of N_2 that results when 0.100 L of the gas is heated from 300. K to 350. K at 1.00 atm.

SOLUTION

Summarize the data:

$P_i = 1.00$ atm $P_f = 1.00$ atm

$V_i = 0.100$ L $V_f = ?$ L

$T_i = 300.$ K $T_f = 350.$ K

$$\frac{P_i V_i}{T_i} = \frac{P_f V_f}{T_f}$$

that can be rearranged as

$$P_f V_f T_i = P_i V_i T_f$$

and

$$V_f = \frac{P_i V_i T_f}{P_f T_i}$$

Because $P_i = P_f$

$$V_f = \frac{V_i T_f}{T_i}$$

Substituting gives

$$V_f = \frac{(0.100 \text{ L})(350.\ \cancel{K})}{300.\ \cancel{K}}$$

$$= 0.117 \text{ L}$$

Helpful Hints:
1. Go to www.mhhe.com/denniston for a review of the mathematics used here.
2. In this case, because the pressure is constant, the combined gas law reduces to Charles's law.

4 Learning Goal
Use gas law equations to calculate conditions and changes in conditions of gases.

4 Learning Goal

Use gas law equations to calculate conditions and changes in conditions of gases.

EXAMPLE 6.4 Using the Combined Gas Law

A sample of helium gas has a volume of 1.27 L at 149 K and 5.00 atm. When the gas is compressed to 0.320 L at 50.0 atm, the temperature increases markedly. What is the final temperature?

SOLUTION

Summarize the data:

$$P_i = 5.00 \text{ atm} \qquad P_f = 50.0 \text{ atm}$$

$$V_i = 1.27 \text{ L} \qquad V_f = 0.320 \text{ L}$$

$$T_i = 149 \text{ K} \qquad T_f = ? \text{ K}$$

The combined gas law expression is

$$\frac{P_i V_i}{T_i} = \frac{P_f V_f}{T_f}$$

which we rearrange as

$$P_f V_f T_i = P_i V_i T_f$$

and

$$T_f = \frac{P_f V_f T_i}{P_i V_i}$$

Substituting yields

$$T_f = \frac{(50.0 \text{ atm})(0.320 \text{ L})(149 \text{ K})}{(5.00 \text{ atm})(1.27 \text{ L})}$$

$$= 375 \text{ K}$$

Helpful Hint: Go to www.mhhe.com/denniston for a review of the mathematics used here.

QUESTION 6.7

Hydrogen sulfide, H_2S, has the characteristic odor of rotten eggs. If a sample of H_2S gas at 760. torr and 25.0°C in a 2.00-L container is allowed to expand into a 10.0-L container at 25.0°C, what is the pressure in the 10.0-L container?

QUESTION 6.8

Cyclopropane, C_3H_6, is used as a general anesthetic. If a sample of cyclopropane stored in a 2.00-L container at 10.0 atm and 25.0°C is transferred to a 5.00-L container at 5.00 atm, what is the resulting temperature?

Avogadro's Law: Relationship of Number of Moles and Volume

3 Learning Goal

Describe the behavior of gases expressed by the gas laws: Boyle's law, Charles's law, combined gas law, Avogadro's law, the ideal gas law, and Dalton's law.

The relationship between the volume and number of moles of a gas at constant temperature and pressure is known as **Avogadro's law.** It states that equal volumes of any ideal gas contain the same number of moles if measured under the same conditions of temperature and pressure.

Mathematically, the *ratio* of volume (V) to number of moles (n) is a constant:

$$\frac{V}{n} = k_3$$

Consider 1 mol of gas occupying a volume of 10.0 L; using logic similar to the application of Boyle's and Charles's laws, 2 mol of the gas would occupy 20.0 L, 3 mol would occupy 30.0 L, and so forth. As we have done with the previous laws, we can formulate a useful expression relating initial and final conditions:

$$\frac{V_i}{n_i} = \frac{V_f}{n_f}$$

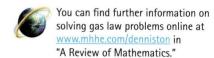

You can find further information on solving gas law problems online at www.mhhe.com/denniston in "A Review of Mathematics."

EXAMPLE 6.5 Using Avogadro's Law

If 5.50 mol of CO occupy 20.6 L, how many liters will 16.5 mol of CO occupy at the same temperature and pressure?

4 Learning Goal

Use gas law equations to calculate conditions and changes in conditions of gases.

SOLUTION

The quantities moles and volume are related through Avogadro's law. Summarizing the data:

$V_i = 20.6$ L $V_f = ?$ L

$n_i = 5.50$ mol $n_f = 16.5$ mol

Using the mathematical expression for Avogadro's law:

$$\frac{V_i}{n_i} = \frac{V_f}{n_f}$$

and rearranging as

$$V_f = \frac{V_i n_f}{n_i}$$

then substituting yields

$$V_f = \frac{(20.6 \text{ L})(16.5 \text{ mol})}{(5.50 \text{ mol})}$$

$$= 61.8 \text{ L of CO}$$

QUESTION 6.9

1.00 mole of hydrogen gas occupies 22.4 L. How many moles of hydrogen are needed to fill a 100.0 L container at the same pressure and temperature?

QUESTION 6.10

How many moles of hydrogen are needed to triple the volume occupied by 0.25 mol of hydrogen, assuming no changes in pressure or temperature?

Molar Volume of a Gas

The volume occupied by *1 mol* of any gas is referred to as its **molar volume.** At **standard temperature and pressure (STP),** the molar volume of any gas is 22.4 L.

STP conditions are defined as follows:

$$T = 273 \text{ K (or } 0°\text{C)}$$

$$P = 1 \text{ atm}$$

Thus, 1 mol of N_2, O_2, H_2, or He all occupy the *same volume, 22.4 L*, at STP.

Gas Densities

It is also possible to compute the density of various gases at STP. If we recall that density is the mass/unit volume,

$$d = \frac{m}{V}$$

and that 1 mol of helium weighs 4.00 g,

$$d_{\text{He}} = \frac{4.00 \text{ g}}{22.4 \text{ L}} = 0.178 \text{ g/L at STP}$$

or, because 1 mol of nitrogen weighs 28.0 g, then

$$d_{N_2} = \frac{28.0 \text{ g}}{22.4 \text{ L}} = 1.25 \text{ g/L at STP}$$

Heating a gas, such as air, will decrease its density and have a lifting effect as well.

The large difference in gas densities of helium and nitrogen (which makes up about 80% of the air) accounts for the lifting power of helium. A balloon filled with helium will rise through a predominantly nitrogen atmosphere because its gas density is less than 15% of the density of the surrounding atmosphere:

$$\frac{0.178 \text{ g/L}}{1.25 \text{ g/L}} \times 100\% = 14.2\%$$

The Ideal Gas Law

Animation

Ideal Gas Law

3 Learning Goal

Describe the behavior of gases expressed by the gas laws: Boyle's law, Charles's law, combined gas law, Avogadro's law, the ideal gas law, and Dalton's law.

Boyle's law (relating volume and pressure), Charles's law (relating volume and temperature), and Avogadro's law (relating volume to the number of moles) may be combined into a single expression relating all four terms. This expression is the **ideal gas law:**

$$PV = nRT$$

in which R, based on k_1, k_2, and k_3 (Boyle's, Charles's, and Avogadro's law constants), is a constant and is referred to as the *ideal gas constant:*

$$R = 0.0821 \text{ L-atm K}^{-1}\text{mol}^{-1}$$

Mathematically, 0.0821 L-atm/K mol is identical to 0.0821 L-atm K^{-1} mol^{-1}.

if the units

P in atmospheres

V in liters

n in number of moles

T in Kelvin

are used.

Consider some examples of the application of the ideal gas equation.

4 Learning Goal

Use gas law equations to calculate conditions and changes in conditions of gases.

EXAMPLE 6.6 Calculating a Molar Volume

Demonstrate that the molar volume of oxygen gas at STP is 22.4 L.

Continued—

SOLUTION

$$PV = nRT$$

$$V = \frac{nRT}{P}$$

At standard temperature and pressure,

$$T = 273 \text{ K}$$

$$P = 1.00 \text{ atm}$$

The other constants are

$$n = 1.00 \text{ mol}$$

$$R = 0.0821 \text{ L-atm K}^{-1} \text{mol}^{-1}$$

Then

$$V = \frac{(1.00 \text{ mol})(0.0821 \text{ L-atm K}^{-1} \text{mol}^{-1})(273 \text{ K})}{(1.00 \text{ atm})}$$

$$= 22.4 \text{ L}$$

EXAMPLE 6.7   Calculating the Number of Moles of a Gas

Calculate the number of moles of helium in a 1.00-L balloon at 27°C and 1.00 atm of pressure.

4 Learning Goal

Use gas law equations to calculate conditions and changes in conditions of gases.

SOLUTION

$$PV = nRT$$

$$n = \frac{PV}{RT}$$

If

$$P = 1.00 \text{ atm}$$

$$V = 1.00 \text{ L}$$

$$T = 27°C + 273 = 300. \text{ K}$$

$$R = 0.0821 \text{ L-atm K}^{-1} \text{mol}^{-1}$$

then

$$n = \frac{(1.00 \text{ atm})(1.00 \text{ L})}{(0.0821 \text{ L-atm K}^{-1} \text{mol}^{-1})(300. \text{ K})}$$

$$n = 0.0406 \text{ mol or } 4.06 \times 10^{-2} \text{ mol}$$

EXAMPLE 6.8   Converting Mass to Volume

Oxygen used in hospitals and laboratories is often obtained from cylinders containing liquefied oxygen. If a cylinder contains 1.00×10^2 kg of liquid oxygen, how many liters of oxygen can be produced at 1.00 atm of pressure at room temperature (20.0°C)?

4 Learning Goal

Use gas law equations to calculate conditions and changes in conditions of gases.

Continued—

EXAMPLE 6.8 —Continued

SOLUTION

$$PV = nRT$$

$$V = \frac{nRT}{P}$$

Using conversion factors, we obtain

$$n_{O_2} = 1.00 \times 10^2 \text{ kg O}_2 \times \frac{10^3 \text{ g O}_2}{1 \text{ kg O}_2} \times \frac{1 \text{ mol O}_2}{32.0 \text{ g O}_2}$$

Then

$$n = 3.13 \times 10^3 \text{ mol O}_2$$

and

$$T = 20.0°C + 273 = 293 \text{ K}$$

$$P = 1.00 \text{ atm}$$

then

$$V = \frac{(3.13 \times 10^3 \text{ mol})(0.0821 \text{ L-atm K}^{-1} \text{ mol}^{-1})(293 \text{ K})}{1.00 \text{ atm}}$$

$$= 7.53 \times 10^4 \text{ L}$$

QUESTION 6.11

What volume is occupied by 10.0 g N_2 at 30.0°C and a pressure of 750 torr?

QUESTION 6.12

A 20.0-L gas cylinder contains 4.80 g H_2 at 25°C. What is the pressure of this gas?

QUESTION 6.13

How many moles of N_2 gas will occupy a 5.00-L container at standard temperature and pressure?

QUESTION 6.14

At what temperature will 2.00 mol He fill a 2.00-L container at standard pressure?

Dalton's Law of Partial Pressures

3 Learning Goal

Describe the behavior of gases expressed by the gas laws: Boyle's law, Charles's law, combined gas law, Avogadro's law, the ideal gas law, and Dalton's law.

Our discussion of gases so far has presumed that we are working with a single pure gas. A *mixture* of gases exerts a pressure that is the *sum* of the pressures that each gas would exert if it were present alone under the same conditions. This is known as **Dalton's law** of partial pressures.

Stated another way, the total pressure of a mixture of gases is the sum of the **partial pressures.**

$$P_i = p_1 + p_2 + p_3 + \cdots$$

AN ENVIRONMENTAL Perspective | The Greenhouse Effect and Global Warming

A greenhouse is a bright, warm, and humid environment for growing plants, vegetables, and flowers even during the cold winter months. It functions as a closed system in which the concentration of water vapor is elevated and visible light streams through the windows; this creates an ideal climate for plant growth.

Some of the visible light is absorbed by plants and soil in the greenhouse and radiated as infrared radiation. This radiated energy is blocked by the glass or absorbed by water vapor and carbon dioxide (CO_2). This trapped energy warms the greenhouse and is a form of solar heating: light energy is converted to heat energy.

On a global scale, the same process takes place. Although more than half of the sunlight that strikes the earth's surface is reflected back into space, the fraction of light that is absorbed produces sufficient heat to sustain life. How does this happen? Greenhouse gases, such as CO_2, trap energy radiated from the earth's surface and store it in the atmosphere. This moderates our climate. The earth's surface would be much colder and more inhospitable if the atmosphere was not able to capture some reasonable amount of solar energy.

Can we have too much of a good thing? It appears so. Since 1900, the atmospheric concentration of CO_2 has increased from 296 parts per million (ppm) to more than 350 ppm (approximately a 17% increase). The energy demands of technological and population growth have caused massive increases in the combustion of organic matter and carbon-based fuels (coal, oil, and natural gas), adding over 50 billion tons of CO_2 to that already present in the atmosphere. Photosynthesis naturally removes CO_2 from the atmosphere. However, the removal of forestland to create living space and cropland decreases the amount of vegetation available to consume atmospheric CO_2 through photosynthesis. The rapid destruction of the Amazon rain forest is just the latest of many examples.

If our greenhouse model is correct, an increase in CO_2 levels should produce global warming, perhaps changing our climate in unforeseen and undesirable ways.

FOR FURTHER UNDERSTANDING

What steps might be taken to decrease levels of CO_2 in the atmosphere over time?

In what ways might our climate and our lives change as a consequence of significant global warming?

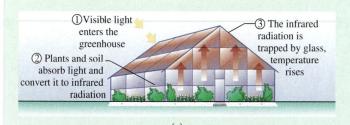

(a)

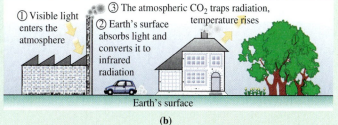

(b)

(a) A greenhouse traps solar radiation as heat. (b) Our atmosphere also acts as a solar collector. Carbon dioxide, like the windows of a greenhouse, allows the visible light to enter and traps the heat.

in which P_t = total pressure and $p_1, p_2, p_3, \ldots$, are the partial pressures of the component gases. For example, the total pressure of our atmosphere is equal to the sum of the pressures of N_2 and O_2 (the principal components of air):

$$P_{air} = p_{N_2} + p_{O_2}$$

Other gases, such as argon (Ar), carbon dioxide (CO_2), carbon monoxide (CO), and methane (CH_4) are present in the atmosphere at very low partial pressures. However, their presence may result in dramatic consequences; one such gas is carbon dioxide. Classified as a "greenhouse gas," it exerts a significant effect on our climate. Its role is described in An Environmental Perspective: The Greenhouse Effect and Global Warming.

> The ideal gas law applies to mixtures of gases as well as to pure gases.

Ideal Gases Versus Real Gases

To this point, we have assumed, in both theory and calculations, that all gases behave as ideal gases. However, in reality there is no such thing as an ideal gas. As we noted at the beginning of this section, the ideal gas is a model (a very useful one) that describes the behavior of individual atoms and molecules; this behavior translates to the collective properties of measurable quantities of these atoms and molecules. Limitations of the model arise from the fact that interactive forces, even between the widely spaced particles of gas, are not totally absent in any sample of gas.

Attractive forces are present in gases composed of polar molecules. Nonuniform charge distribution on polar molecules creates positive and negative regions, resulting in electrostatic attraction and deviation from ideality.

See Sections 3.5 and 6.2 for a discussion of interactions of polar molecules.

Calculations involving polar gases such as HF, NO, and SO_2 based on ideal gas equations (which presume no such interactions) are approximations. However, at low pressures, such approximations certainly provide useful information. Nonpolar molecules, on the other hand, are only weakly attracted to each other and behave much more ideally in the gas phase.

6.2 The Liquid State

Molecules in the liquid state are close to one another. Attractive forces are large enough to keep the molecules together in contrast to gases, whose cohesive forces are so low that a gas expands to fill any volume. However, these attractive forces in a liquid are not large enough to restrict movement, as in solids.

5 Learning Goal

Describe properties of the liquid state in terms of the properties of the individual molecules that comprise the liquid.

QUESTION 6.15

Compare the strength of intermolecular forces in liquids with those in gases.

QUESTION 6.16

Compare the strength of intermolecular forces in liquids with those in solids.

Let's look at the various properties of liquids in more detail.

Compressibility

Liquids are practically incompressible. The molecules are so close to one another that even the application of many atmospheres of pressure does not significantly decrease the volume. This makes liquids ideal for the transmission of force, as in the brake lines of an automobile. The force applied by the driver's foot on the brake pedal does not compress the brake fluid in the lines; rather, it transmits the force directly to the brake pads, and the friction between the brake pads and rotors (that are attached to the wheel) stops the car.

Viscosity

The **viscosity** of a liquid is a measure of its resistance to flow. Viscosity is a function of both the attractive forces between molecules and molecular geometry.

Molecules with complex structures, which do not "slide" smoothly past each other, and polar molecules, tend to have higher viscosity than less structurally

A MEDICAL Connection | Blood Gases and Respiration

Respiration must deliver oxygen to cells and the waste product, carbon dioxide, to the lungs to be exhaled. Dalton's law of partial pressures helps to explain how this process occurs.

Gases (such as O_2 and CO_2) move from a region of higher partial pressure to one of lower partial pressure to establish an equilibrium. At the interface of the lung, the membrane barrier between the blood and the surrounding atmosphere, the following situation exists: The atmospheric O_2 partial pressure is high, and the atmospheric CO_2 partial pressure is low. The reverse is true on the other side of the membrane (blood). Thus CO_2 is efficiently removed from the blood, and O_2 is efficiently moved into the bloodstream.

At the other end of the line, capillaries are distributed in close proximity to the cells that need to expel CO_2 and gain O_2. The partial pressure of CO_2 is high in these cells, and the partial pressure of O_2 is low, having been used up by the energy-harvesting reaction, the oxidation of glucose:

$$C_6H_{12}O_6 + 6O_2 \longrightarrow 6CO_2 + 6H_2O + energy$$

The O_2 diffuses into the cells (from a region of high to low partial pressure), and the CO_2 diffuses from the cells to the blood (again from a region of high to low partial pressure).

The net result is a continuous process proceeding according to Dalton's law. With each breath we take, oxygen is distributed to the cells and used to generate energy, and the waste product, CO_2, is expelled by the lungs.

FOR FURTHER UNDERSTANDING

Carbon dioxide and carbon monoxide can be toxic, but for different reasons. Use the Internet to research this topic and:

Explain why carbon dioxide is toxic.

Explain why carbon monoxide is toxic.

---

complex, less polar liquids. Glycerol, which is used in a variety of skin treatments, has the structural formula:

$$
\begin{array}{c}
\text{H} \\
| \\
\text{H}-\text{C}-\text{O}-\text{H} \\
| \\
\text{H}-\text{C}-\text{O}-\text{H} \\
| \\
\text{H}-\text{C}-\text{O}-\text{H} \\
| \\
\text{H}
\end{array}
$$

It is quite viscous, owing to its polar nature and its significant intermolecular attractive forces. This is certainly desirable in a skin treatment because its viscosity keeps it on the area being treated. Gasoline, on the other hand, is much less viscous and readily flows through the gas lines of your auto; it is composed of nonpolar molecules.

Viscosity generally decreases with increasing temperature. The increased kinetic energy at higher temperatures overcomes some of the intermolecular attractive forces. The temperature effect is an important consideration in the design of products that must remain fluid at low temperatures, such as motor oils and transmission fluids found in automobiles.

Surface Tension

The **surface tension** of a liquid is a measure of the attractive forces exerted among molecules at the surface of a liquid. Only the surface molecules are not totally surrounded by other liquid molecules (the top of the molecule faces the atmosphere). These surface molecules are surrounded and attracted by fewer liquid molecules than those below and to each side. Hence, the net attractive forces on surface

Animation

Surface Tension of Water

molecules pull them downward into the body of the liquid. As a result, the surface molecules behave as a "skin" that covers the interior.

This increased surface force is responsible for the spherical shape of drops of liquid. Drops of water "beading" on a polished surface, such as a waxed automobile, illustrate this effect.

Because surface tension is related to the attractive forces exerted among molecules, surface tension generally decreases with an increase in temperature or a decrease in the polarity of molecules that make up the liquid.

Substances known as **surfactants** can be added to a liquid to decrease surface tension. Common surfactants include soaps and detergents that reduce water's surface tension; this promotes the interaction of water with grease and dirt, making it easier to remove them.

Vapor Pressure of a Liquid

Evaporation, condensation, and the meaning of the term *boiling point* are all related to the concept of liquid vapor pressure. Consider the following example. A liquid, such as water, is placed in a sealed container. After a time, the contents of the container are analyzed. Both liquid water and water vapor are found at room temperature, when we might expect water to be found only as a liquid. In this closed system, some of the liquid water was converted to a gas:

$$\text{energy} + H_2O(l) \longrightarrow H_2O(g)$$

How did this happen? The temperature is too low for conversion of a liquid to a gas by boiling. According to the kinetic theory, liquid molecules are in continuous motion, with their *average* kinetic energy directly proportional to the Kelvin temperature. The word *average* is the key. Although the average kinetic energy is too low to allow "average" molecules to escape from the liquid phase to the gas phase, there exists a range of molecules with different energies, some low and some high, that make up the "average" (Figure 6.6). Thus, some of these high-energy molecules possess sufficient energy to escape from the bulk liquid.

At the same time, a fraction of these gaseous molecules lose energy (perhaps by collision with the walls of the container) and return to the liquid state:

$$H_2O(g) \longrightarrow H_2O(l) + \text{energy}$$

The process of conversion of liquid to gas, at a temperature too low to boil, is **evaporation.** The reverse process, conversion of the gas to the liquid state, is **condensation.** After some time, the rates of evaporation and condensation become *equal,* and this sets up a dynamic equilibrium between liquid and vapor states. The **vapor pressure of a liquid** is defined as the pressure exerted by the vapor *at equilibrium.*

$$H_2O(g) \rightleftharpoons H_2O(l)$$

The equilibrium process of evaporation and condensation of water is depicted in Figure 6.7.

The boiling point of a liquid is defined as the temperature at which the vapor pressure of the liquid becomes equal to the atmospheric pressure. The "normal" atmospheric pressure is 760 torr, or 1 atm, and the **normal boiling point** is the temperature at which the vapor pressure of the liquid is equal to 1 atm.

It follows from the definition that the boiling point of a liquid is not constant. It depends on the atmospheric pressure. At high altitudes, where the atmospheric pressure is low, the boiling point of a liquid, such as water, is lower than the normal boiling point (for water, 100°C). High atmospheric pressure increases the boiling point.

Apart from its dependence on the surrounding atmospheric pressure, the boiling point depends on the nature of the attractive forces between the liquid mole-

Animation
Vapor Pressure

6 Learning Goal
Describe the processes of melting, boiling, evaporation, and condensation.

The process of evaporation of perspiration from the skin produces a cooling effect, because heat is stored in the evaporating molecules.

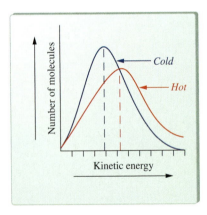

Figure 6.6

The temperature dependence of liquid vapor pressure is illustrated. The average molecular kinetic energy increases with temperature. Note that the average values are indicated by dashed lines. The small number of high-energy molecules may evaporate.

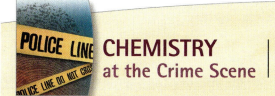

CHEMISTRY at the Crime Scene | Explosives at the Airport

The images flash across our television screens: a "bomb sniffing" dog being led through an airport or train station, pausing to sniff packages or passengers, looking for anything of a suspicious nature. Or, perhaps, we see a long line of people waiting to pass through a scanning device surrounded by what appears to be hundreds of thousands of dollars worth of electronic gadgetry.

At one level, we certainly know what is happening. These steps are taken to increase the likelihood that our trip, as well as everyone else's, will be as safe and worry-free as possible. From a scientific standpoint, we may wonder how these steps actually detect explosive materials. What do the dog and some electronic devices have in common? How can a dog sniff a solid or a liquid? Surely everyone knows that the nose can only sense gases, and explosive devices are solids or liquids, or a combination of the two.

One potential strategy is based on the concept of vapor pressure, which you have just studied. We now know that liquids, such as water, have a measurable vapor pressure at room temperature. In fact, most liquids and many solids have vapor pressures large enough to allow detection of the molecules in the gas phase. The challenge is finding devices that are sufficiently sensitive and selective, enabling them to detect low concentrations of molecules characteristic of explosives, without becoming confused by thousands of other compounds routinely present in the air.

Each explosive device has its own "signature," a unique mix of chemicals used in its manufacture and assembly. If only one, or perhaps a few, of these compounds has a measurable vapor pressure, it may be detected with a sensitive measuring device.

Dogs are renowned for their keen sense of smell, and some breeds are better than others. Dogs can be trained to signal the presence of certain scents by barking or exhibiting unusual agitation. A qualified handler can recognize these cues and alert appropriate authorities.

Scientific instruments are designed to mimic the scenario described here. A device, the mass spectrometer, can detect very low concentrations of molecules in the air. Additionally, it can distinguish certain "target" molecules, because each different compound has its own unique molar mass. Detection of molecules of interest generates an electrical signal, and an alarm is sounded.

Compounds with high vapor pressures are most easily detected. Active areas of forensic research involve designing a new generation of instruments that are even more sensitive and selective than those currently available. Decreased cost and increased portability and reliability will enable many sites, not currently being monitored, to have the same level of protection as major transit facilities.

FOR FURTHER UNDERSTANDING

Would you expect nonpolar or polar molecules of similar mass to be more easily detected? Why?

Why must an explosives detection device be highly selective?

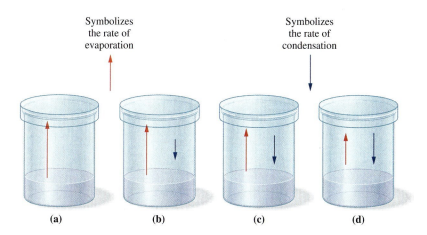

Symbolizes the rate of evaporation

Symbolizes the rate of condensation

(a) (b) (c) (d)

Figure 6.7

Liquid water in equilibrium with water vapor. (a) Initiation: process of evaporation exclusively.
(b, c) After a time, both evaporation and condensation occur, but evaporation predominates.
(d) Dynamic equilibrium established. Rates of evaporation and condensation are equal.

cules. Polar liquids, such as water, with large intermolecular attractive forces have *higher* boiling points than nonpolar liquids, such as gasoline, which exhibit weak attractive forces.

QUESTION 6.17

Distinguish between the terms *evaporation* and *condensation.*

QUESTION 6.18

Distinguish between the terms *evaporation* and *boiling.*

---

van der Waals Forces

7 Learning Goal
Describe the dipolar attractions known collectively as van der Waals forces.

Physical properties of liquids, such as those discussed in the previous section, can be explained in terms of their intermolecular forces. We have seen (see Section 3.5) that attractive forces between polar molecules, **dipole–dipole interactions,** significantly decrease vapor pressure and increase the boiling point. However, nonpolar substances can exist as liquids as well; many are liquids and even solids at room temperature. What is the nature of the attractive forces in these nonpolar compounds?

In 1930, Fritz London demonstrated that he could account for a weak attractive force between any two molecules, whether polar or nonpolar. He postulated that the electron distribution in molecules is not fixed; electrons are in continuous motion, relative to the nucleus. So, for a short time, a nonpolar molecule could experience an *instantaneous dipole,* a short-lived polarity caused by a temporary dislocation of the electron cloud. These temporary dipoles could interact with other temporary dipoles, just as permanent dipoles interact in polar molecules. We now call these intermolecular forces **London forces.**

London forces and dipole–dipole interactions are collectively known as **van der Waals forces.** London forces exist among polar and nonpolar molecules because electrons are in constant motion in all molecules. Dipole–dipole attractions occur only among polar molecules. In addition to van der Waals forces, a special type of dipole–dipole force, the *hydrogen bond,* has a very significant effect on molecular properties, particularly in biological systems.

Hydrogen Bonding

8 Learning Goal
Describe hydrogen bonding and its relationship to boiling and melting temperatures.

Recall that the most electronegative elements are in the upper right corner of the periodic table, and these elements exert strong electron attraction in molecules as described in Chapter 3.

Intramolecular hydrogen bonding between polar regions helps keep proteins folded in their proper three-dimensional structure. See Chapter 16.

Typical forces in polar liquids, discussed above, are only about 1–2% as strong as ionic and covalent bonds. However, certain liquids have boiling points that are much higher than we would predict from these dipolar interactions alone. This indicates the presence of some strong intermolecular force. This attractive force is due to **hydrogen bonding.** Molecules in which a hydrogen atom is bonded to a small, highly electronegative atom such as nitrogen, oxygen, or fluorine exhibit this effect. The presence of a highly electronegative atom bonded to a hydrogen atom creates a large dipole:

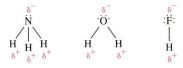

This arrangement of atoms produces a very polar bond, often resulting in a polar molecule with strong intermolecular attractive forces. Although the hydrogen

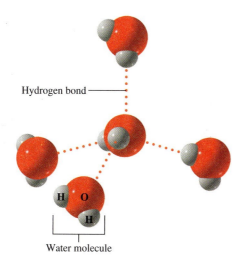

Hydrogen bond

H O

H

Water molecule

Figure 6.8

Hydrogen bonding in water. Note that the central water molecule is hydrogen bonded to four other water molecules. The attractive force between the hydrogen (δ^+) part of one water molecule and the oxygen (δ^-) part of another water molecule constitutes the hydrogen bond.

bond is weaker than bonds formed *within* molecules (covalent and polar covalent *intra*molecular forces), it is the strongest attractive force *between* molecules (intermolecular force).

Consider the boiling points of four small molecules:

CH_4	NH_3	H_2O	HF
−161°C	−33°C	+100°C	+19.5°C

Clearly, ammonia, water, and hydrogen fluoride boil at significantly higher temperatures than methane. The N—H, O—H, and F—H bonds are far more polar than the C—H bond, owing to the high electronegativity of N, O, and F. The complex network of attractive forces among water molecules (depicted in Figure 6.8) accounts for water's unusually high boiling point.

Hydrogen bonding has an extremely important influence on the behavior of many biological systems. Molecules such as proteins and DNA require extensive hydrogen bonding to maintain their structures and hence functions. DNA (deoxyribonucleic acid, Chapter 17) is a giant among molecules with intertwined chains of atoms held together by millions of hydrogen bonds.

6.3 The Solid State

The close packing of the particles of a solid results from attractive forces that are strong enough to restrict motion. This occurs because the kinetic energy of the particles is insufficient to overcome the attractive forces among particles. The particles are "locked" together in a defined and highly organized fashion. This results in fixed shape and volume, although, at the atomic level, vibrational motion is observed.

Properties of Solids

Solids are virtually incompressible, owing to the small distance between particles. Most will convert to liquids at a higher temperature, when the increased heat energy overcomes some of the attractive forces within the solid. The temperature at which a solid is converted to the liquid phase is its **melting point.** The melting point depends on the strength of the attractive forces in the solid, hence its structure. As we might expect, polar solids have higher melting points than nonpolar solids of the same molecular weight.

9 Learning Goal

Relate the properties of the various classes of solids (ionic, covalent, molecular, and metallic) to the structure of these solids.

Figure 6.9

Crystalline solids.

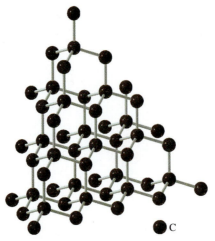

(a) The crystal structure of diamond

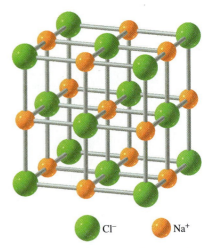

Cl⁻   Na⁺

(b) The crystal structure of sodium chloride

(c) The crystal structure of methane, a frozen molecular solid. Only one methane molecule is shown in detail.

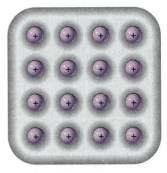

(d) The crystal structure of a metallic solid. The gray area represents mobile electrons around fixed metal cations.

A solid may be a **crystalline solid,** having a regular repeating structure, or an **amorphous solid,** having no organized structure. Diamond and sodium chloride (Figure 6.9) are examples of crystalline substances; glass, plastic, and concrete are examples of amorphous solids.

QUESTION 6.19

Explain why solids are essentially incompressible.

QUESTION 6.20

Distinguish between amorphous and crystalline solids.

Types of Crystalline Solids

Crystalline solids may exist in one of four general groups:

1. *Ionic solids.* The units that comprise an **ionic solid** are positive and negative ions. Electrostatic forces hold the crystal together. They generally have high melting points and are hard and brittle. A common example of an ionic solid is sodium chloride.
2. *Covalent solids.* The units that comprise a **covalent solid** are atoms held together by covalent bonds. They have very high melting points (1200°C to

A LIFESTYLE Connection | An Extraordinary Molecule

Think for a moment. What is the only common molecule that exists in all three physical states of matter (solid, liquid, and gas) under natural conditions on earth? This molecule is absolutely essential for life; in fact, life probably arose in this substance. It is the most abundant molecule in the cells of living organisms (70–95%) and covers 75% of the earth's surface. Without it, cells quickly die, and without it the earth would not be a fit environment in which to live. By now you have guessed that we are talking about the water molecule. It is so abundant on earth that we take this deceptively simple molecule for granted.

What are some of the properties of water that cause it to be essential to life as we know it? Water has the ability to stabilize temperatures on the earth and in the body. This ability is due in part to the energy changes that occur when water changes physical state; but ultimately, this ability is due to the polar nature of the water molecule.

Life can exist only within a fairly narrow range of temperatures. Above or below that range, the chemical reactions necessary for life, and thus life itself, will cease. Water can moderate temperature fluctuation and maintain the range necessary for life, and one property that allows it to do so is its unusually high specific heat, 1 cal/g °C. This means that water can absorb or lose more heat energy than many other substances without a significant temperature change. This is because in the liquid state, every water molecule is hydrogen bonded to other water molecules. Because a temperature increase is really just a measure of increased (more rapid) molecular movement, we must get the water molecules moving more rapidly, independent of one another, to register a temperature increase. Before we can achieve this independent, increased activity, the hydrogen bonds between molecules must be broken. Much of the heat energy that water absorbs is involved in breaking hydrogen bonds and is *not* used to increase molecular movement. Thus a great deal of heat is needed to raise the temperature of water even a little bit.

Water also has a very high heat of vaporization. It takes 540 calories to change 1 g of liquid water at 100°C to a gas and even more, 603 cal/g, when the water is at 37°C, human body temperature. That is about twice the heat of vaporization of alcohol. As water molecules evaporate, the surface of the liquid cools because only the highest-energy (or "hottest") molecules leave as a gas. Only the "hottest" molecules have enough energy to break the hydrogen bonds that bind them to other water molecules. Indeed, evaporation of water molecules from the surfaces of lakes and oceans helps to maintain stable temperatures in those bodies of water. Similarly, evaporation of perspiration from body surfaces helps to prevent overheating on a hot day or during strenuous exercise.

Even the process of freezing helps stabilize and moderate temperatures. This is especially true in the fall. Water releases

Surface of a body of water.

heat when hydrogen bonds are formed. This is an example of an exothermic process. Thus, when water freezes, solidifying into ice, additional hydrogen bonds are formed, and heat is released into the environment. As a result, the temperature change between summer and winter is more gradual, allowing organisms to adjust to the change.

One last feature that we take for granted is the fact that when we put ice in our iced tea on a hot summer day, the ice floats. This means that the solid state of water is actually *less* dense than the liquid state! In fact, it is about 10% less dense, having an open lattice structure with each molecule of water bonded to the maximum of four other water molecules. What would happen if ice did sink? All bodies of water, including the mighty oceans would eventually freeze solid, killing all aquatic and marine plant and animal life. Even in the heat of summer, only a few inches of ice at the surface would thaw. Instead, the ice forms at the surface and provides a layer of insulation that prevents the water below from freezing.

As we continue our study of chemistry, we will refer again and again to this amazing molecule.

FOR FURTHER UNDERSTANDING

Why is the high heat of vaporization of water important to our bodies?

Why is it cooler at the ocean shore than in the desert during summer?

Figure 6.10

The structure of ice, a molecular solid. Hydrogen bonding among water molecules produces a regular open structure that is less dense than liquid water.

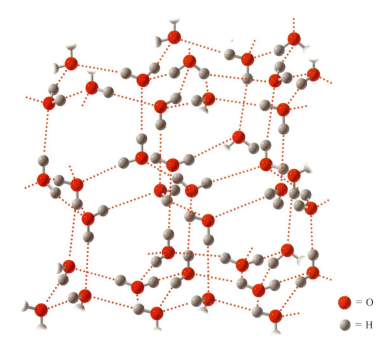

$\bullet$ = O

$\bullet$ = H

Intermolecular forces are also discussed in Sections 3.5 and 6.2.

2000°C or more is not unusual) and are extremely hard. They are insoluble in most solvents. Diamond is a covalent solid composed of covalently bonded carbon atoms. Diamonds are used for industrial cutting because they are so hard and as gemstones because of their crystalline beauty.

3. *Molecular solids.* The units that make up a **molecular solid,** molecules, are held together by intermolecular attractive forces (London forces, dipole–dipole interactions, and hydrogen bonding). Molecular solids are usually soft and have low melting points. They are frequently volatile and are poor electrical conductors. Common examples are ice (solid water; Figure 6.10) and frozen methane (Figure 6.9, c).

4. *Metallic solids.* The units that comprise a **metallic solid** are metal atoms held together by metallic bonds. **Metallic bonds** are formed by the overlap of orbitals of metal atoms, resulting in regions of high electron density surrounding the positive metal nuclei. Electrons in these regions are extremely mobile. They are able to move freely from atom to atom through pathways that are, in reality, overlapping atomic orbitals. This results in the high *conductivity* (ability to carry electrical current) exhibited by many metallic solids. Silver and copper are common examples of metallic solids. Metals are easily shaped and are used for a variety of purposes. Most of them are practical applications such as hardware, cookware, and surgical and dental tools. Others are purely for enjoyment and decoration, such as silver and gold jewelry.

SUMMARY

6.1 The Gaseous State

The *kinetic molecular theory* describes an *ideal gas* in which gas particles exhibit no interactive or repulsive forces and the volumes of the individual gas particles are assumed to be negligible.

Boyle's law states that the volume of a gas varies inversely with the pressure exerted by the gas if the number of moles and temperature of gas are held constant ($PV = k_1$).

Charles's law states that the volume of a gas varies directly with the absolute temperature (K) if the pressure and number of moles of gas are constant ($V/T = k_2$).

Avogadro's law states that equal volumes of any gas contain the same number of moles if measured at constant temperature and pressure ($V/n = k_3$).

The volume occupied by 1 mol of any gas is its *molar volume.* At *standard temperature and pressure* (STP), the molar volume of any ideal gas is 22.4 L. STP conditions are defined as 273 K (or 0°C) and 1 atm pressure.

Boyle's law, Charles's law, and Avogadro's law may be combined into a single expression relating all four terms, the *ideal gas law: PV = nRT.* R is the ideal gas constant (0.0821 L-atm K^{-1} mol^{-1}) if the units P (atmospheres), V (liters), n (number of moles), and T (Kelvin) are used.

The *combined gas law* provides a convenient expression for gas law calculations involving the most common variables: pressure, volume, and temperature.

Dalton's law of partial pressures states that a mixture of gases exerts a pressure that is the sum of the pressures that each gas would exert if it were present alone under similar conditions ($P_t = p_1 + p_2 + p_3 + \cdots$).

6.2 The Liquid State

Liquids are practically incompressible because of the closeness of the molecules. The *viscosity* of a liquid is a measure of its resistance to flow. Viscosity generally decreases with increasing temperature. The *surface tension* of a liquid is a measure of the attractive forces at the surface of a liquid. *Surfactants* decrease surface tension.

The conversion of liquid to vapor at a temperature below the boiling point of the liquid is *evaporation*. Conversion of the gas to the liquid state is *condensation*. The *vapor pressure of the liquid* is defined as the pressure exerted by the vapor at equilibrium at a specified temperature. The *normal boiling point* of a liquid is the temperature at which the vapor pressure of the liquid is equal to 1 atm.

Molecules in which a hydrogen atom is bonded to a small, highly electronegative atom such as nitrogen, oxygen, or fluorine exhibit *hydrogen bonding*. Hydrogen bonding in liquids is responsible for lower than expected vapor pressures and higher than expected boiling points. The presence of *van der Waals forces* and hydrogen bonds significantly affects the boiling points of liquids as well as the melting points of solids.

6.3 The Solid State

Solids have fixed shapes and volumes. They are *incompressible*, owing to the closeness of the particles. Solids may be *crystalline*, having a regular, repeating structure, or *amorphous*, having no organized structure.

Crystalline solids may exist as *ionic solids, covalent solids, molecular solids,* or *metallic solids*. Electrons in metallic solids are extremely mobile, resulting in the high *conductivity* (ability to carry electrical current) exhibited by many metallic solids.

KEY TERMS

amorphous solid (6.3)
Avogadro's law (6.1)

barometer (6.1)
Boyle's law (6.1)

Charles's law (6.1)
combined gas law (6.1)
condensation (6.2)
covalent solid (6.3)
crystalline solid (6.3)
Dalton's law (6.1)
dipole–dipole interactions (6.2)
evaporation (6.2)
hydrogen bonding (6.2)
ideal gas (6.1)
ideal gas law (6.1)
ionic solid (6.3)
kinetic molecular theory (6.1)
London forces (6.2)
melting point (6.3)

metallic bond (6.3)
metallic solid (6.3)
molar volume (6.1)
molecular solid (6.3)
normal boiling point (6.2)
partial pressure (6.1)
pressure (6.1)
standard temperature and
   pressure (STP) (6.1)
surface tension (6.2)
surfactant (6.2)
van der Waals forces (6.2)
vapor pressure of a
   liquid (6.2)
viscosity (6.2)

QUESTIONS AND PROBLEMS

Kinetic Molecular Theory

Foundations

6.21 Compare and contrast the gas, liquid, and solid states with regard to the average distance of particle separation.

6.22 Compare and contrast the gas, liquid, and solid states with regard to the nature of the interactions among the particles.

6.23 Describe the molecular/atomic basis of gas pressure.

6.24 Describe the measurement of gas pressure.

Applications

6.25 Why are gases easily compressible?

6.26 Why are gas densities much lower than those of liquids or solids?

6.27 Do gases exhibit more ideal behavior at low or high pressures? Why?

6.28 Do gases exhibit more ideal behavior at low or high temperatures? Why?

6.29 Use the kinetic molecular theory to explain why dissimilar gases mix more rapidly at high temperatures than at low temperatures.

6.30 Use the kinetic molecular theory to explain why aerosol cans carry instructions warning against heating or disposing of the container in a fire.

6.31 Predict and explain any observed changes taking place when an inflated balloon is cooled (perhaps refrigerated).

6.32 Predict and explain any observed changes taking place when an inflated balloon is heated (perhaps microwaved).

Gas Laws

Foundations

6.33 The pressure on a fixed mass of a gas is tripled at constant temperature. Will the volume increase, decrease, or remain the same?

6.34 By what factor will the volume of the gas in Question 6.33 change?

Applications

6.35 Explain why the Kelvin scale is used for gas law calculations.

6.36 The temperature on a summer day may be 90°F. Convert this value to Kelvins.

6.37 Will the volume of gas increase, decrease, or remain the same if the temperature is increased and the pressure is decreased? Explain.

6.38 Will the volume of gas increase, decrease, or remain the same if the temperature is decreased and the pressure is increased? Explain.

A sample of helium gas was placed in a cylinder and the volume of the gas was measured as the pressure was slowly increased. The results of this experiment are shown graphically.

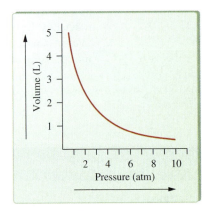

Boyle's Law

Questions 6.39–6.42 are based on this experiment.

6.39 At what pressure does the gas occupy a volume of 5 L?

6.40 What is the volume of the gas at a pressure of 5 atm?

6.41 Calculate the Boyle's law constant at a volume of 2 L.

6.42 Calculate the Boyle's law constant at a pressure of 2 atm.

6.43 Calculate the pressure, in atmospheres, required to compress a sample of helium gas from 20.9 L (at 1.00 atm) to 4.00 L.

6.44 A balloon filled with helium gas at 1.00 atm occupies 15.6 L. What volume would the balloon occupy in the upper atmosphere, at a pressure of 0.150 atm?

Applications

6.45 The temperature of a gas is raised from 25°C to 50°C. Will the volume double if the mass and pressure do not change? Why or why not?

6.46 Verify your answer to Question 6.45 by calculating the temperature needed to double the volume of the gas.

6.47 Determine the change in volume that takes place when a 2.00-L sample of $N_2(g)$ is heated from 250°C to 500°C.

6.48 Determine the change in volume that takes place when a 2.00-L sample of $N_2(g)$ is heated from 250 K to 500 K.

Use the combined gas law,

$$\frac{P_i V_i}{T_i} = \frac{P_f V_f}{T_f}$$

to answer Questions 6.49 and 6.50.

6.49 Rearrange the combined gas law expression to solve for the final volume.

6.50 Rearrange the combined gas law expression to solve for the final temperature.

6.51 If 2.25 L of a gas at 16°C and 1.00 atm are compressed at a pressure of 125 atm at 20°C, calculate the new volume of the gas.

6.52 A balloon filled with helium gas occupies 2.50 L at 25°C and 1.00 atm. When released, it rises to an altitude where the temperature is 20°C and the pressure is only 0.800 atm. Calculate the new volume of the balloon.

6.53 If 5.00 g helium gas is added to a 1.00 L balloon containing 1.00 g of helium gas, what is the new volume of the balloon? Assume no change in temperature or pressure.

6.54 How many grams of helium must be added to a balloon containing 8.00 g helium gas to double its volume? Assume no change in temperature or pressure.

Molar Volume and the Ideal Gas Law

Foundations

6.55 Will 1.00 mol of a gas always occupy 22.4 L?

6.56 H_2O and CH_4 are gases at 150°C. Which exhibits more ideal behavior? Why?

6.57 What are the units and numerical value of standard temperature?

6.58 What are the units and numerical value of standard pressure?

Applications

6.59 A sample of nitrogen gas, stored in a 4.0-L container at 32°C, exerts a pressure of 5.0 atm. Calculate the number of moles of nitrogen gas in the container.

6.60 Seven moles of carbon monoxide are stored in a 30.0-L container at 65°C. What is the pressure of the carbon monoxide in the container?

6.61 Calculate the volume of 44.0 g of carbon monoxide at STP.

6.62 Calculate the volume of 44.0 g of carbon dioxide at STP.

6.63 Calculate the number of moles of a gas in a 7.55-L container at 45°C, if the gas exerts a pressure of 725 mm Hg.

6.64 Calculate the pressure exerted by 1.00 mol of gas, contained in a 7.55-L cylinder at 45°C.

6.65 A sample of argon (Ar) gas occupies 65.0 mL at 22°C and 750 torr. What is the volume of this Ar gas sample at STP?

6.66 A sample of O_2 gas occupies 257 mL at 20°C and 1.20 atm. What is the volume of this O_2 gas sample at STP?

6.67 Calculate the molar volume of Ar gas at STP.

6.68 Calculate the molar volume of O_2 gas at STP.

Dalton's Law

Foundations

6.69 State Dalton's law in words.

6.70 State Dalton's law in equation form.

Applications

6.71 A gas mixture has three components: N_2, F_2, and He. Their partial pressures are 0.40 atm, 0.16 atm, and 0.18 atm, respectively. What is the pressure of the gas mixture?

6.72 A gas mixture has a total pressure of 0.56 atm and consists of He and Ne. If the partial pressure of the He in the mixture is 0.27 atm, what is the partial pressure of the Ne in the mixture?

The Liquid State

Foundations

6.73 What is the relationship between the temperature of a liquid and the vapor pressure of that liquid?

6.74 What is the relationship between the strength of the attractive forces in a liquid and its vapor pressure?

6.75 Describe the process at the molecular level that accounts for the property of viscosity.

6.76 Describe the process at the molecular level that accounts for the property of surface tension.

Applications

Questions 6.77–6.80 are based on the following:

| methane | chloromethane | methanol |

6.77 Which of these molecules exhibit London forces? Why?
6.78 Which of these molecules exhibit dipole–dipole forces? Why?
6.79 Which of these molecules exhibit hydrogen bonding? Why?
6.80 Which of these molecules would you expect to have the highest boiling point? Why?

The Solid State

6.81 Describe one property that is characteristic of
  a. ionic solids
  b. covalent solids
6.82 Describe one property that is characteristic of
  a. molecular solids
  b. metallic solids
6.83 Predict whether beryllium or carbon would be a better conductor of electricity in the solid state. Why?
6.84 Why is diamond used as an industrial cutting tool?
6.85 Mercury and chromium are toxic substances. Which element is more likely to be an air pollutant? Why?

6.86 Why is the melting point of silicon much higher than that of argon, even though argon has a greater molar mass?

FOR FURTHER UNDERSTANDING

1. An elodea plant, commonly found in tropical fish aquaria, is found to produce 5.0×10^{22} molecules of oxygen per hour. What volume of oxygen (STP) would be produced in an 8-hour period?
2. A chemist measures the volume of 1.00 mol of helium gas at STP and obtains a value of 22.4 L. After changing the temperature to 137 K, the experimental value was found to be 11.05 L. Verify the chemist's results using the ideal gas law and explain any apparent discrepancies.
3. A chemist measures the volumes of 1.00 mol of H_2 and 1.00 mol of CO and finds that they differ by 0.10 L. Which gas produced the larger volume? Do the results contradict the ideal gas law? Why or why not?
4. A 100.0-g sample of water was decomposed using an electric current (electrolysis) producing hydrogen gas and oxygen gas. Write the balanced equation for the process and calculate the volume of each gas produced (STP). Explain any relationship you may observe between the volumes obtained and the balanced equation for the process.
5. An autoclave is used to sterilize surgical equipment. It is far more effective than steam produced from boiling water in the open atmosphere because it generates steam at a pressure of 2 atm. Explain why an autoclave is such an efficient sterilization device.

7

LEARNING GOALS

1 Distinguish among the terms *solution*, *solute*, and *solvent*.

2 Describe various kinds of solutions, and give examples of each.

3 Describe the relationship between solubility and equilibrium.

4 Calculate solution concentration in units of weight/volume percent, weight/weight percent, parts per thousand, and parts per million.

5 Calculate solution concentration using molarity.

6 Perform dilution calculations.

7 Describe and explain concentration-dependent solution properties.

8 Describe the ways in which the chemical and physical properties of water make it a truly unique solvent.

9 Explain the role of electrolytes in blood and their relationship to the process of dialysis.

Solutions

A significant deterrent to achieving optimal athletic performance in strenuous competitive sports, especially in warm weather, is dehydration—the loss of body fluids through perspiration. Perspiration is an aqueous solution and the loss of ions dissolved in water (the solvent) has as great a negative impact on bodily function as the loss of the water itself.

This fact was not lost on researchers at the University of Florida in the 1960s. The University of Florida football team (the Florida 'Gators) plays much of its schedule in a very warm climate. This, coupled with the intense physical effort required and the heavy padding that must be worn to prevent injury, makes dehydration of the athletes a real concern.

The remedy, at that time, was to drink large volumes of water while consuming salt tablets to replace the lost ions and high sugar foods (often oranges) to provide energy. How much of each was pure guesswork. Often the cure was worse than the original problem, and cramping resulted.

The team doctor, along with medical researchers at the university, had an idea. Would it be possible to mix all three components in some proportion to produce a solution that achieved the desired effect without the unwanted side effect? The result of this research was a solution of ionic compounds and sugar dissolved in water that mimicked the composition of the perspiration being lost, with flavoring added to make it palatable. The identity of the ions, sodium, potassium, and chloride was certainly important, as were the concentrations of the ions and producing a solution that achieved the desired effect without inducing cramps.

The football team began using the solution and was victorious in the 1966 Orange Bowl, offering some proof that Gatorade had a positive effect on the team's performance. Gatorade was the first "sports drink" (obviously named for the Florida 'Gators.) Currently, several competing brands are available in the marketplace, differing slightly in composition, concentration, and flavor. They are widely used by college and professional teams, as well as the "weekend athlete."

In this chapter, we will learn more about solutions, their composition, and their concentration. We will see why some substances are soluble in water while others are not. We will see why the concentration of ions, such as sodium and potassium, is critical to the function and the integrity of the cells in our bodies.

Carbonated beverages are a commonplace example of a solution of a gas dissolved in a liquid solvent. Based on your everyday experience, can you predict whether the solubility of carbon dioxide in water (or cola) would increase or decrease as the temperature of the solution increases?

7.1 Properties of Solutions

1 Learning Goal
Distinguish among the terms *solution*, *solute*, and *solvent*.

A **solution** is a homogeneous (or uniform) mixture of two or more substances. A solution is composed of one or more *solutes*, dissolved in a *solvent*. The **solute** is a compound of a solution that is present in lesser quantity than the solvent. The **solvent** is the solution component present in the largest quantity. For example, when sugar (the solute) is added to water (the solvent), the sugar dissolves in the water to produce a solution. In those instances in which the solvent is water, we refer to the homogeneous mixture as an **aqueous solution,** from the Latin *aqua*, meaning water.

The dissolution of a solid in a liquid is perhaps the most common example of solution formation. However, it is also possible to form solutions in gases and solids as well as in liquids. For example,

- Air is a gaseous mixture, but it is also a solution; oxygen and a number of trace gases are dissolved in the gaseous solvent, nitrogen.
- Alloys, such as brass and silver and the gold used to make jewelry, are also homogeneous mixtures of two or more kinds of metal atoms in the solid state.

Although solid and gaseous solutions are important in many applications, our emphasis will be on *liquid solutions* because so many important chemical reactions take place in liquid solutions.

General Properties of Liquid Solutions

2 Learning Goal
Describe various kinds of solutions, and give examples of each.

Section 3.3 discusses properties of compounds.

Animation
Dissolution of Compounds

Animation
Strong, Weak, and Nonelectrolytes

Particles in electrolyte solutions are ions, making the solution an electrical conductor.

Particles in nonelectrolyte solution are individual molecules. No ions are formed in the dissolution process.

Recall that matter in solution, as in gases, is in continuous, random motion (Section 6.1).

Section 3.5 relates properties and molecular geometry.

Liquid solutions are clear and transparent with no visible particles of solute. They may be colored or colorless, depending on the properties of the solute and solvent. Note that the terms *clear* and *colorless* do not mean the same thing; a clear solution has only one state of matter that can be detected; *colorless* simply means the absence of color.

Recall that solutions of **electrolytes** are formed from solutes that are soluble ionic compounds. These compounds dissociate in solution to produce ions that behave as charge carriers. Solutions of electrolytes are good conductors of electricity, for example, sodium chloride dissolving in water:

$$NaCl(s) \xrightarrow{H_2O} Na^+(aq) + Cl^-(aq)$$

Solid sodium chloride   Dissolved sodium chloride

In contrast, solutions of **nonelectrolytes** are formed from nondissociating *molecular* solutes (nonelectrolytes), and these solutions are nonconducting, for example, dissolving sugar in water:

$$C_6H_{12}O_6(s) \xrightarrow{H_2O} C_6H_{12}O_6(aq)$$

Solid glucose   Dissolved glucose

A *true solution* is a homogeneous mixture with uniform properties throughout. In a true solution, the solute cannot be isolated from the solution by filtration. The particle size of the solute is about the same as that of the solvent, and solvent and solute pass directly through the filter paper. Furthermore, solute particles will not "settle out" after a time. All of the molecules of solute and solvent are intimately mixed. The continuous particle motion in solution maintains the homogeneous, random distribution of solute and solvent particles.

Volumes of solute and solvent are not additive; 1 L of alcohol mixed with 1 L of water does not result in exactly 2 L of solution. The volume of pure liquid is determined by the way in which the individual molecules "fit together." When two or more kinds of molecules are mixed, the interactions become more complex. Solvent interacts with solvent, solute interacts with solvent, and solute may interact with other solute. This will be important to remember when we solve concentration problems later.

Figure 7.1

The Tyndall effect. The container on the left contains a colloidal suspension, which scatters the light. This scattered light is visible as a haze. The container on the right contains a true solution; no scattered light is observed.

Solutions and Colloids

How can you recognize a solution? A beaker containing a clear liquid may be a pure substance, a true solution, or a colloid. Only chemical analysis, determining the identity of all substances in the liquid, can distinguish between a pure substance and a solution. A pure substance has *one* component; pure water is an example. A true solution will contain more than one substance, with the tiny particles homogeneously intermingled.

A **colloidal suspension** also consists of solute particles distributed throughout a solvent. However, the distribution is not completely homogeneous, owing to the size of the colloidal particles. Particles with diameters of 1×10^{-9} m (1 nm) to 2×10^{-7} m (200 nm) are colloids. Particles smaller than 1 nm are solution particles; those larger than 200 nm are precipitates (solid in contact with solvent).

See Section 4.3 for more information on precipitates.

To the naked eye, a colloidal suspension and a true solution appear identical; neither solute nor colloid can be seen by the naked eye. However, a simple experiment, using only a bright light source, can readily make the distinction based upon differences in their interaction with light. Colloid particles are large enough to scatter light; solute particles are not. When a beam of light passes through a colloidal suspension, the large particles scatter light, and the liquid appears hazy. We see this effect in sunlight passing through fog. Fog is a colloidal suspension of tiny particles of liquid water dispersed throughout a gas, air. The haze is light scattered by droplets of water. You may have noticed that your automobile headlights are not very helpful in foggy weather. Visibility becomes worse rather than better because light scattering increases.

The light-scattering ability of colloidal suspensions is termed the *Tyndall effect*. True solutions, with very tiny particles, do not scatter light—no haze is observed—and true solutions are easily distinguished from colloidal suspensions by observing their light-scattering properties (Figure 7.1).

A **suspension** is a heterogeneous mixture that contains particles much larger than a colloidal suspension; over time, these particles may settle, forming a second phase. A suspension is not a true solution, nor is it a precipitate.

QUESTION 7.1

Describe how you would distinguish experimentally between a pure substance and a true solution.

QUESTION 7.2

Describe how you would distinguish experimentally between a true solution and a colloidal suspension.

Section 3.5 describes molecular interactions in detail.

The term *qualitative* implies identity, and the term *quantitative* relates to quantity.

Degree of Solubility

In our discussion of the relationship of polarity and solubility, the rule *"like dissolves like"* was described as the fundamental condition for solubility. Polar solutes are soluble in polar solvents, and nonpolar solutes are soluble in nonpolar solvents. Thus, knowing a little bit about the structure of the molecule enables us to predict qualitatively the solubility of the compound.

The *degree* of **solubility,** *how much* solute can dissolve in a given volume of solvent, is a quantitative measure of solubility. It is difficult to predict the solubility of each and every compound. However, general solubility trends are based on the following considerations:

- *The magnitude of difference between polarity of solute and solvent.* The greater the difference, the less soluble the solute.
- *Temperature.* An increase in temperature usually, but not always, increases solubility. Often, the effect is dramatic. For example, an increase in temperature from 0°C to 100°C increases the water solubility of KCl from 28 g/100 mL to 58 g/100 mL.
- *Pressure.* Pressure has little effect on the solubility of solids and liquids in liquids. However, the solubility of a gas in liquid is directly proportional to the applied pressure. Carbonated beverages, for example, are made by dissolving carbon dioxide in the beverage under high pressure (hence the term *carbonated*).

When a solution contains all the solute that can be dissolved at a particular temperature, it is a **saturated solution.** When solubility values are given—for example, 13.3 g of potassium nitrate in 100 mL of water at 24°C—they refer to the concentration of a saturated solution.

As we have already noted, *increasing* the temperature generally increases the amount of solute a given solution may hold. Conversely, *cooling* a saturated solution often results in a decrease in the amount of solute in solution. The excess solute falls to the bottom of the container as a **precipitate** (a solid in contact with the solution). Occasionally, on cooling, the excess solute may remain in solution for a time. Such a solution is described as a **supersaturated solution.** This type of solution is inherently unstable. With time, excess solute will precipitate, and the solution will revert to a saturated solution, which is stable.

Solubility and Equilibrium

3 Learning Goal

Describe the relationship between solubility and equilibrium.

When an excess of solute is added to a solvent, it begins to dissolve and continues until it establishes a *dynamic equilibrium* between dissolved and undissolved solute.

Initially, the rate of dissolution is large. After a time, the rate of the reverse process, precipitation, increases. The rates of dissolution and precipitation eventually become equal, and there is no further change in the composition of the solution. There is, however, a continual exchange of solute particles between solid and liquid phases because particles are in constant motion. The solution is saturated. The most precise definition of a saturated solution is a solution that is in equilibrium with undissolved solute.

The concept of equilibrium was introduced in Section 5.4

Solubility of Gases: Henry's Law

The concept of partial pressure is a consequence of Dalton's law, discussed in Section 6.1

When a liquid and a gas are allowed to come to equilibrium, the amount of gas dissolved in the liquid reaches some maximum level. This quantity can be predicted from a very simple relationship. **Henry's law** states that the number of moles of a gas dissolved in a liquid at a given temperature is proportional to the partial pressure of the gas. In other words, the gas solubility is directly proportional to the pressure of that gas in the atmosphere that is in contact with the liquid.

Carbonated beverages are bottled at high pressures of carbon dioxide. When the cap is removed, the fizzing results from the fact that the partial pressure of carbon dioxide in the atmosphere is much less than that used in the bottling process. As a result, the equilibrium quickly shifts to one of lower gas solubility.

Gases are most soluble at low temperatures, and gas solubility decreases markedly at higher temperatures. This explains many common observations. For example, a chilled container of carbonated beverage that is opened quickly goes flat as it warms to room temperature. As the beverage warms up, the solubility of the carbon dioxide decreases.

Henry's law helps to explain the process of respiration. Respiration depends on a rapid and efficient exchange of oxygen and carbon dioxide between the atmosphere and blood. This transfer occurs through the lungs. The process, oxygen entering the blood and carbon dioxide released to the atmosphere, is accomplished in air sacs called *alveoli,* which are surrounded by an extensive capillary system. Equilibrium is quickly established between alveolar air and capillary blood. The temperature of the blood is effectively constant. Therefore, the equilibrium concentration of both oxygen and carbon dioxide are determined by the partial pressures of the gases (Henry's law). The oxygen is transported to cells, a variety of reactions takes place, and the waste product of respiration, carbon dioxide, is brought back to the lungs to be expelled into the atmosphere.

See A Medical Connection: Blood Gases and Respiration, Chapter 6.

QUESTION 7.3

Explain why, over time, a bottle of soft drink goes "flat" after it is opened.

QUESTION 7.4

Would the soft drink in Question 7.3 go "flat" faster if the bottle warmed to room temperature? Why?

7.2 Concentration Based on Mass

Solution **concentration** is defined as the amount of solute dissolved in a given amount of solution. The concentration of a solution has a profound effect on the properties of a solution, both *physical* (melting and boiling points) and *chemical* (solution reactivity). Solution concentration may be expressed in many different units. Here we consider concentration units based on percentage.

4 Learning Goal
Calculate solution concentration in units of weight/volume percent, weight/weight percent, parts per thousand, and parts per million.

Weight/Volume Percent

The concentration of a solution is defined as the amount of solute dissolved in a specified amount of solution,

$$\text{concentration} = \frac{\text{amount of solute}}{\text{amount of solution}}$$

If we define the amount of solute as the *mass* of solute (in grams) and the amount of solution in *volume* units (milliliters), concentration is expressed as the ratio

$$\text{concentration} = \frac{\text{grams of solute}}{\text{milliliters of solution}}$$

This concentration can then be expressed as a percentage by multiplying the ratio by the factor 100%. This results in

$$\% \text{ concentration} = \frac{\text{grams of solute}}{\text{milliliters of solution}} \times 100\%$$

A LIFESTYLE Connection | Scuba Diving: Nitrogen and the Bends

A deep-water diver's worst fear is interruption of the oxygen supply through equipment malfunction, forcing his or her rapid rise to the surface in search of air. If a diver must ascend too rapidly, he or she may suffer a condition known as "the bends."

Key to understanding this problem is recognition of the tremendous increase in pressure that divers withstand as they descend, because of the weight of the water above them. At the surface, the pressure is approximately 1 atm. At a depth of 200 feet, the pressure is approximately six times as great.

At these pressures, the solubility of nitrogen in the blood increases dramatically. Oxygen solubility increases as well, although its effect is less serious (O_2 is 20% of air, N_2 is 80%). As the diver quickly rises, the pressure decreases rapidly, and the nitrogen "boils" out of the blood, stopping blood flow and impairing nerve transmission. The joints of the body lock in a bent position, hence the name of the condition: the bends.

To minimize the problem, scuba tanks are often filled with mixtures of helium and oxygen rather than nitrogen and oxygen. Helium has a much lower solubility in blood and, like nitrogen, is inert.

FOR FURTHER UNDERSTANDING

Why are divers who slowly rise to the surface less likely to be adversely affected?

What design features would be essential in deep-water manned exploration vessels?

The percent concentration expressed in this way is called **weight/volume percent,** or **% (W/V).** Thus

$$\% \left(\frac{W}{V}\right) = \frac{\text{grams of solute}}{\text{milliliters of solution}} \times 100\%$$

Consider the following examples.

EXAMPLE 7.1 Calculating Weight/Volume Percent

4 Learning Goal
Calculate solution concentration in units of weight/volume percent, weight/weight percent, parts per thousand, and parts per million.

Calculate the percent composition, or % (W/V), of 3.00×10^2 mL of solution containing 15.0 g of glucose.

SOLUTION

There are 15.0 g of glucose, the solute, and 3.00×10^2 mL of total solution. Therefore, substituting in our expression for weight/volume percent:

*Continued—*

$$\% \left(\frac{W}{V} \right) = \frac{15.0 \text{ g glucose}}{3.00 \times 10^2 \text{ mL solution}} \times 100\%$$

$$= 5.00\% \left(\frac{W}{V} \right) \text{glucose}$$

EXAMPLE 7.2 Calculating the Weight of Solute from a Weight/Volume Percent

Calculate the number of grams of NaCl in 5.00×10^2 mL of a 10.0% solution.

SOLUTION

Begin by substituting the data from the problem:

$$10.0\% \left(\frac{W}{V} \right) = \frac{X \text{ g NaCl}}{5.00 \times 10^2 \text{ mL solution}} \times 100\%$$

Cross-multiplying to simplify,

$$X \text{ g NaCl} \times 100\% = \left(10.0\% \frac{W}{V} \right) (5.00 \times 10^2 \text{ mL solution})$$

$$X \text{ g NaCl} \times 100\% = \left(10.0\% \frac{g}{mL} \right) (5.00 \times 10^2 \text{ mL solution})$$

Dividing both sides by 100% to isolate grams NaCl on the left side of the equation,

$$X = 50.0 \text{ g NaCl}$$

4 Learning Goal

Calculate solution concentration in units of weight/volume percent, weight/weight percent, parts per thousand, and parts per million.

If the units of mass are other than grams, or if the solution volume is in units other than milliliters, the proper conversion factor must be used to arrive at the units used in the equation.

Section 1.3 discusses units and unit conversion.

QUESTION 7.5

Calculate the % (W/V) of 0.0600 L of solution containing 10.0 g NaCl.

QUESTION 7.6

Calculate the volume (in milliliters) of a 25.0% (W/V) solution containing 10.0 g NaCl.

QUESTION 7.7

Calculate the % (W/V) of 0.200 L of solution containing 15.0 g KCl.

QUESTION 7.8

Calculate the mass (in grams) of sodium hydroxide required to make 2.00 L of a 1.00% (W/V) solution.

QUESTION 7.9

Calculate the % (W/V) of oxygen gas if 20.0 g of oxygen gas are diluted with 80.0 g of nitrogen gas in a 78.0-L container at standard temperature and pressure.

QUESTION 7.10

Calculate the % (W/V) of argon gas if 50.0 g of argon gas are diluted with 80.0 g of helium gas in a 476-L container at standard temperature and pressure.

Weight/Weight Percent

The **weight/weight percent,** or % **(W/W),** is most useful for mixtures of solids, whose weights (masses) are easily obtained. The expression used to calculate weight/weight percentage is analogous in form to % (W/V):

$$\% \left(\frac{W}{W} \right) = \frac{\text{grams solute}}{\text{grams solution}} \times 100\%$$

EXAMPLE 7.3 Calculating Weight/Weight Percent

Calculate the % (W/W) of platinum in a gold ring that contains 14.00 g gold and 4.500 g platinum.

SOLUTION

Using our definition of weight/weight percent

$$\% \left(\frac{W}{W} \right) = \frac{\text{grams solute}}{\text{grams solution}} \times 100\%$$

Substituting,

$$= \frac{4.500 \text{ g platinum}}{4.500 \text{ g platinum} + 14.00 \text{ g gold}} \times 100\%$$

$$= \frac{4.500 \text{ g}}{18.50 \text{ g}} \times 100\%$$

$$= 24.32\% \text{ platinum}$$

QUESTION 7.11

Calculate the % (W/W) of oxygen gas in Question 7.9.

QUESTION 7.12

Calculate the % (W/W) of argon gas in Question 7.10.

Parts Per Thousand (ppt) and Parts Per Million (ppm)

The calculation of concentration in parts per thousand or parts per million is based on the same logic as weight/weight percent. Percentage is actually the number of parts of solute in 100 parts of solution. For example, a 5.00% (W/W) is made up of 5.00 g solute in 100 g solution.

$$5.00\% \text{ (W/W)} = \frac{5.00 \text{ g solute}}{100 \text{ g solution}} \times 100\%$$

It follows that a 5.00 parts per thousand (ppt) solution is made up of 5.00 g solute in 1000 g solution.

$$5.00 \text{ ppt} = \frac{5.00 \text{ g solute}}{1000 \text{ g solution}} \times 10^3 \text{ ppt}$$

Using similar logic, a 5.00 parts per million solution (ppm) is made up of 5.00 g solute in 1,000,000 g solution.

$$5.00 \text{ ppm} = \frac{5.00 \text{ g solute}}{1,000,000 \text{ g solution}} \times 10^6 \text{ ppm}$$

The general expressions are

$$\text{ppt} = \frac{\text{grams solute}}{\text{grams solution}} \times 10^3 \text{ ppt}$$

and

$$\text{ppm} = \frac{\text{grams solute}}{\text{grams solution}} \times 10^6 \text{ ppm}$$

Ppt and ppm are most often used for expressing the concentrations of very dilute solutions. They are widely used in environmental chemistry to express the concentration of hazardous substances, such as heavy metals (such as Pb^{2+}, Hg^{2+}, Cd^{2+}, among others) and pesticides in lakes, streams, and drinking water.

EXAMPLE 7.4 Calculating ppt and ppm

A 1.00 g sample of stream water was found to contain 1.0×10^{-6} g lead. Calculate the concentration of lead in the stream water in units of % (W/W), ppt, and ppm. Which is the most suitable unit?

SOLUTION

weight percent: $\% \text{ (W/W)} = \dfrac{\text{grams solute}}{\text{grams solution}} \times 100\%$

$\% \text{ (W/W)} = \dfrac{1.0 \times 10^{-6} \text{ g Pb}}{1.0 \text{ g solution}} \times 100\%$

$\% \text{ (W/W)} = 1.0 \times 10^{-4}\%$

parts per thousand: $\text{ppt} = \dfrac{\text{grams solute}}{\text{grams solution}} \times 10^3 \text{ ppt}$

$\text{ppt} = \dfrac{1.0 \times 10^{-6} \text{ g Pb}}{1.0 \text{ g solution}} \times 10^3 \text{ ppt}$

$\text{ppt} = 1.0 \times 10^{-3} \text{ ppt}$

parts per million $\text{ppm} = \dfrac{\text{grams solute}}{\text{grams solution}} \times 10^6 \text{ ppm}$

$\text{ppm} = \dfrac{1.0 \times 10^{-6} \text{ g Pb}}{1.0 \text{ g solution}} \times 10^6 \text{ ppm}$

$\text{ppm} = 1.0 \text{ ppm}$

Parts per million is the most reasonable unit.

4 **Learning Goal**
Calculate solution concentration in units of weight/volume percent, weight/weight percent, parts per thousand, and parts per million.

QUESTION **7.13**

Calculate the ppt and ppm of oxygen gas in Question 7.9.

QUESTION **7.14**

Calculate the ppt and ppm of argon gas in Question 7.10.

7.3 Concentration Based on Moles

In our discussion of the chemical arithmetic of reactions in Chapter 4, we saw that the chemical equation represents the relative number of *moles* of reactants producing products. When chemical reactions occur in solution, it is most useful to represent their concentrations on a *molar* basis.

Molarity

5 Learning Goal

Calculate solution concentration using molarity.

The most common mole-based concentration unit is molarity. **Molarity,** symbolized M, is defined as the number of moles of solute per liter of solution, or

$$M = \frac{\text{moles solute}}{\text{L solution}}$$

EXAMPLE 7.5 Calculating Molarity from Moles

Calculate the molarity of 2.0 L of solution containing 5.0 mol NaOH.

SOLUTION

Using our expression for molarity

$$M = \frac{\text{moles solute}}{\text{L solution}}$$

Substituting,

$$M_{\text{NaOH}} = \frac{5.0 \text{ mol solute}}{2.0 \text{ L solution}}$$

$$= 2.5 \ M$$

Section 1.3 discussed units and unit conversion.

Remember the need for conversion factors to convert from mass to number of moles. Consider the following example.

EXAMPLE 7.6 Calculating Molarity from Mass

5 Learning Goal

Calculate solution concentration using molarity.

If 5.00 g glucose are dissolved in 1.00×10^2 mL of solution, calculate the molarity, M, of the glucose solution.

Continued—

SOLUTION

To use our expression for molarity, it is necessary to convert from units of grams of glucose to moles of glucose. The molar mass of glucose is 1.80×10^2 g/mol. Therefore

$$5.00 \text{ g} \times \frac{1 \text{ mol}}{1.80 \times 10^2 \text{ g}} = 2.78 \times 10^{-2} \text{ mol glucose}$$

and we must convert mL to L:

$$1.00 \times 10^2 \text{ mL} \times \frac{1 \text{ L}}{10^3 \text{ mL}} = 1.00 \times 10^{-1} \text{ L}$$

Substituting these quantities:

$$M_{glucose} = \frac{2.78 \times 10^{-2} \text{ mol}}{1.00 \times 10^{-1} \text{ L}}$$

$$= 2.78 \times 10^{-1} \ M$$

EXAMPLE 7.7   Calculating Volume from Molarity

Calculate the volume of a 0.750 M sulfuric acid (H_2SO_4) solution containing 0.120 mol of solute.

5 Learning Goal
Calculate solution concentration using molarity.

SOLUTION

Substituting in our basic expression for molarity, we obtain

$$0.750 \ M \ H_2SO_4 = \frac{0.120 \text{ mol } H_2SO_4}{X \text{ L}}$$

$$X \text{ L} = 0.160 \text{ L}$$

QUESTION 7.15

Calculate the number of moles of solute in 5.00×10^2 mL of 0.250 M HCl.

QUESTION 7.16

Calculate the number of grams of silver nitrate required to prepare 2.00 L of 0.500 M $AgNO_3$.

Dilution

Laboratory reagents are often purchased as concentrated solutions (for example, 12 M HCl or 6 M NaOH) for reasons of safety, economy, and space limitations. We must often *dilute* such a solution to a larger volume to prepare a less

6 Learning Goal
Perform dilution calculations.

concentrated solution for the experiment at hand. The approach to such a calculation is as follows.

We define

$$M_1 = \text{molarity of solution } \textit{before} \text{ dilution}$$

$$M_2 = \text{molarity of solution } \textit{after} \text{ dilution}$$

$$V_1 = \text{volume of solution } \textit{before} \text{ dilution}$$

$$V_2 = \text{volume of solution } \textit{after} \text{ dilution}$$

and

$$M = \frac{\text{moles solute}}{\text{L solution}}$$

This equation can be rearranged:

$$\text{moles solute} = (M)\,(\text{L solution})$$

The number of moles of solute *before* and *after* dilution is unchanged, because dilution involves only addition of extra solvent:

$$\text{moles}_1 \text{ solute} = \text{moles}_2 \text{ solute}$$

Initial	Final
condition	condition

or

$$(M_1)(\text{L}_1 \text{ solution}) = (M_2)(\text{L}_2 \text{ solution})$$

$$(M_1)(V_1) = (M_2)(V_2)$$

Knowing any three of these terms enables us to calculate the fourth.

6 Learning Goal

Perform dilution calculations.

EXAMPLE 7.8 **Calculating Molarity After Dilution**

Calculate the molarity of a solution made by diluting 0.050 L of 0.10 M HCl solution to a volume of 1.0 L.

SOLUTION

Summarize the information provided in the problem:

$$M_1 = 0.10\ M$$

$$M_2 = X\ M$$

$$V_1 = 0.050\ \text{L}$$

$$V_2 = 1.0\ \text{L}$$

Then, using the dilution expression:

$$(M_1)\,(V_1) = (M_2)\,(V_2)$$

solve for M_2, the final solution concentration:

$$M_2 = \frac{(M_1)\,(V_1)}{V_2}$$

Substituting,

$$X\,M = \frac{(0.10\ M)\,(0.050\ \text{L})}{(1.0\ \text{L})}$$

$$= 0.0050\ M \text{ or } 5.0 \times 10^{-3}\ M \text{ HCl}$$

EXAMPLE 7.9 Calculating a Dilution Volume

Calculate the volume, in liters, of water that must be added to dilute 20.0 mL of 12.0 M HCl to 0.100 M HCl.

6 **Learning Goal**
Perform dilution calculations.

SOLUTION

Summarize the information provided in the problem:

$$M_1 = 12.0\ M$$

$$M_2 = 0.100\ M$$

$$V_1 = 20.0\ \text{mL (0.0200 L)}$$

$$V_2 = V_{final}$$

Then, using the dilution expression:

$$(M_1)(V_1) = (M_2)(V_2)$$

solve for V_2, the final volume:

$$V_2 = \frac{(M_1)(V_1)}{(M_2)}$$

Substituting,

$$V_{final} = \frac{(12.0\ M)(0.0200\ \text{L})}{0.100\ M}$$

$$= 2.40\ \text{L solution}$$

Note that this is the *total final volume*. The amount of water added equals this volume *minus* the original solution volume, or

$$2.40\ \text{L} - 0.0200\ \text{L} = 2.38\ \text{L water}$$

QUESTION 7.17

How would you prepare 1.0×10^2 mL of 2.0 M HCl, starting with concentrated (12.0 M) HCl?

QUESTION 7.18

What volume of 0.200 M sugar solution can be prepared from 50.0 mL of 0.400 M solution?

The dilution equation is valid with any concentration units, such as % (W/V) *as well as* molarity, which was used in Examples 7.8 and 7.9. However, you must use the same units for both initial *and* final concentration values. Only in this way can you cancel units properly.

7.4 Concentration-Dependent Solution Properties

Colligative properties are solution properties that depend on the *concentration of solute particles*, rather than the *identity of the solute*.

There are four colligative properties of solutions:

7 **Learning Goal**
Describe and explain concentration-dependent solution properties.

1. vapor pressure lowering
2. freezing point depression

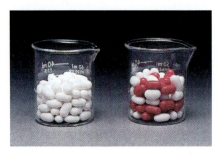

Figure 7.2

An illustration of Raoult's law: lowering of vapor pressure by addition of solute molecules. White units represent solvent molecules, and red units are solute molecules. Solute molecules present a barrier to the escape of solvent molecules, thus decreasing the vapor pressure.

Recall that the concept of liquid vapor pressure was discussed in Section 6.2.

Section 5.4 discusses equilibrium.

The symbols k_f and k_b represent proportionality constants.

3. boiling point elevation
4. osmotic pressure

Each of these properties has widespread practical application. We look at each in some detail in the following sections.

Vapor Pressure Lowering

Raoult's law states that, when a nonvolatile solute is added to a solvent, the vapor pressure of the solvent decreases in proportion to the concentration of the solute.

Raoult's law may be explained in molecular terms by using the following logic: Vapor pressure of a solution results from the escape of solvent molecules from the liquid to the gas phase, thus increasing the partial pressure of the gas phase solvent molecules until the equilibrium vapor pressure is reached. Solute molecules hinder the escape of solvent molecules, thus lowering the equilibrium vapor pressure (Figure 7.2).

Perhaps the most important consequence of Raoult's law is the effect of the solute on the freezing and boiling points of a solution. When a nonvolatile solute is added to a solvent, the freezing point of the resulting solution decreases (a lower temperature is required to convert the liquid to a solid). The boiling point of the solution is found to increase (it requires a higher temperature to form the gaseous state).

Freezing Point Depression and Boiling Point Elevation

The freezing point depression may be explained by examining the equilibrium between solid and liquid states. At the freezing point, ice is in equilibrium with liquid water:

$$H_2O\ (l) \underset{(r)}{\overset{(f)}{\rightleftharpoons}} H_2O\ (s)$$

The solute molecules interfere with the rate at which liquid water molecules associate to form the solid state, decreasing the rate of the forward reaction. For a true equilibrium, the rate of the forward (f) and reverse (r) processes must be equal. Lowering the temperature eventually slows the rate of the reverse (r) process sufficiently to match the rate of the forward reaction. At the lower temperature, equilibrium is established, and the solution freezes.

Boiling point elevation can be explained by considering the definition of boiling point, that is, the temperature at which the vapor pressure of a liquid equals atmospheric pressure. Raoult's law states that the vapor pressure of a solution is decreased by the presence of a solute. Therefore, a higher temperature is necessary to raise the vapor pressure to atmospheric pressure, hence the boiling point elevation.

The extent of the freezing point depression (ΔT_f) is proportional to the solute concentration over a limited range of concentration:

$$\Delta T_f = k_f \times (\text{solute concentration})$$

The boiling point elevation (ΔT_b) is also proportional to the solute concentration:

$$\Delta T_b = k_b \times (\text{solute concentration})$$

If the value of the proportionality factor (k_f or k_b) is known for the solvent of interest, the magnitude of the freezing point depression or boiling point elevation can be calculated for a solution of known concentration.

Solute concentration must be in *mole*-based units. The number of particles (molecules or ions) is critical here, not the mass of solute. One *heavy* molecule will have exactly the same effect on the freezing or boiling point as one *light* molecule. A mole-based unit will correctly represent the number of *particles* in solution because it is related directly to Avogadro's number.

Practical applications that take advantage of freezing point depression of solutions by solutes include the following:

- Salt is spread on roads to melt ice in winter. The salt lowers the freezing point of the water, so it exists in the liquid phase below its normal freezing point, 0°C or 32°F.
- Solutes such as ethylene glycol, "antifreeze," are added to car radiators in the winter to prevent freezing by lowering the freezing point of the coolant.

We have referred to the concentration of *particles* in our discussion of colligative properties. Why did we stress this term? The reason is that there is a very important difference between electrolytes and nonelectrolytes. That difference is the way in which they behave when they dissolve. For example, if we dissolve 1 mol of glucose ($C_6H_{12}O_6$) in 1 L of water,

$$1\ C_6H_{12}O_6(s) \xrightarrow{\text{H}_2\text{O}} 1\ C_6H_{12}O_6(aq)$$

1 mol (Avogadro's number, 6.022×10^{23} particles) of glucose is present in solution. *Glucose is a covalently bonded nonelectrolyte.* Dissolving 1 mol of sodium chloride in 1 L of water,

$$1\ NaCl(s) \xrightarrow{\text{H}_2\text{O}} 1\ Na^+(aq) + 1\ Cl^-(aq)$$

produces 2 mol of particles (1 mol of sodium ions and 1 mol of chloride ions). *Sodium chloride is an ionic electrolyte.*

$$1\ \text{mol glucose} \longrightarrow 1\ \text{mol of particles in solution}$$

$$1\ \text{mol sodium chloride} \longrightarrow 2\ \text{mol of particles in solution}$$

It follows that 1 mol of sodium chloride will decrease the vapor pressure, increase the boiling point, or depress the freezing point of 1 L of water *twice as much* as 1 mol of glucose in the same quantity of water.

QUESTION 7.19

Comparing pure water and a 10% (W/V) glucose solution, which has the higher freezing point?

QUESTION 7.20

Comparing pure water and a 10% (W/V) glucose solution, which has the higher boiling point?

Osmotic Pressure

Osmosis

The cell membrane mediates the interaction of the cell with its environment and is responsible for the controlled passage of material into and out of the cell. One of the principle means of transport is termed *diffusion.* **Diffusion** is the net movement of solute or solvent molecules from an area of high concentration to an area of low concentration. This region where the concentration decreases over a distance is termed the **concentration gradient.** Because of the structure of the cell membrane, only small molecules are able to diffuse freely across this barrier. Large molecules and highly charged ions are restricted by that barrier. In other words, the cell membrane is behaving in a selective fashion. Such membranes are termed **selectively permeable membranes.**

Because a cell membrane is selectively permeable, it is not always possible for solutes to pass through it in response to a concentration gradient. In such cases, the

Figure 7.3

Osmosis across a membrane. The solvent, water, diffuses from an area of lower solute concentration (side A) to an area of higher solute concentration (side B).

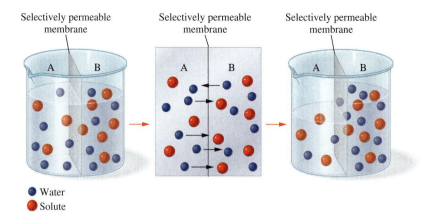

Water

Solute

Animation

Osmosis

solvent diffuses through the membrane. Such membranes, permeable to solvent but not to solute, are specifically called **semipermeable membranes. Osmosis** is the diffusion of a solvent (water in biological systems) through a semipermeable membrane in response to a water concentration gradient.

Suppose that we place a 0.5 M glucose solution in a dialysis bag that is composed of a membrane with pores that allow the passage of water molecules but not glucose molecules. Consider what will happen when we place this bag into a beaker of pure water. We have created a gradient in which there is a higher concentration of glucose inside the bag than outside, but the glucose cannot diffuse through the bag to achieve equal concentration on both sides of the membrane.

Now let's think about this situation in another way. We have a higher concentration of water molecules outside the bag (where there is only pure water) than inside the bag (where some of the water molecules are occupied in dipole-dipole interactions with solute particles and are consequently unable to move freely in the system). Because water can diffuse through the membrane, a net diffusion of water will occur through the membrane into the bag. This is the process of osmosis (Figure 7.3).

As you have probably already guessed, this system can never reach equilibrium (equal concentrations inside and outside the bag). Regardless of how much water diffuses into the bag, diluting the glucose solution, the concentration of glucose will always be higher inside the bag (and the accompanying free water concentration will always be lower).

What happens when the bag has taken in as much water as it can, when it has expanded as much as possible? Now the walls of the bag exert a force that will stop the *net* flow of water into the bag. **Osmotic pressure** is the pressure that must be exerted to stop the flow of water across a selectively permeable membrane by osmosis. Stated more precisely, the osmotic pressure of a solution is the net pressure with which water enters it by osmosis from a pure water compartment when the two compartments are separated by a semipermeable membrane.

Osmotic concentration or **osmolarity** is the term used to describe the osmotic strength of a solution. It depends only on the ratio of the number of solute particles to the number of solvent particles. Thus the chemical nature and size of the solute are not important, only the concentration, expressed in molarity. For instance, a 2 M solution of glucose (a sugar of molar mass 180) has the same osmolarity as a 2 M solution of albumin (a protein of molar mass 60,000).

Blood plasma has an osmolarity equivalent to a 0.30 M glucose solution or a 0.15 M NaCl solution. This latter is true because NaCl in solution dissociates into Na^+ and Cl^- and thus contributes twice the number of solute particles as a molecule that does not ionize. If red blood cells, which have an osmolarity equal to blood plasma, are placed in a 0.30 M glucose solution, no net osmosis will occur because the osmolarity and water concentration inside the red blood cell are equal to those of the 0.30 M glucose solution. The solutions inside and outside the red

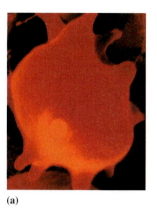

(a)   (b)   (c)

Figure 7.4

Scanning electron micrographs of red blood cells exposed to (a) hypertonic, (b) isotonic, and (c) hypotonic solutions.

blood cell are said to be **isotonic** (*iso* means "same," and *tonic* means "strength") **solutions.** Because the osmolarity is the same inside and outside, the red blood cell will remain the same size (Figure 7.4b).

What happens if we now place the red blood cells into a **hypotonic solution,** in other words, a solution having a lower osmolarity than the cytoplasm of the cell? In this situation there will be a net movement of water into the cell as water diffuses down its concentration gradient. The membrane of the red blood cell does not have the strength to exert a sufficient pressure to stop this flow of water, and the cell will swell and burst (Figure 7.4c). Alternatively, if we place the red blood cells into a **hypertonic solution** (one with a greater osmolarity than the cell), water will pass out of the cells, and they will shrink dramatically in size (Figure 7.4a).

These principles have important applications in the delivery of intravenous (IV) solutions into an individual. Normally, any fluids infused intravenously must have the correct osmolarity; they must be isotonic with the blood cells and the blood plasma. Such infusions are frequently either 5.5% dextrose (glucose) or "normal saline." The first solution is composed of 5.5 g of glucose per 100 mL of solution (0.30 M), and the latter of 9.0 g of NaCl per 100 mL of solution (0.15 M). In either case, they have the same osmotic pressure and osmolarity as the plasma and blood cells and can therefore be safely administered without upsetting the osmotic balance between the blood and the blood cells.

Practical examples of osmosis abound, including the following:

- A sailor, lost at sea in a lifeboat, dies of dehydration while surrounded by water. Because of its high salt concentration, seawater dehydrates the cells of the body as a result of the large osmotic pressure difference between itself and intracellular fluids.
- A cucumber, soaked in brine, shrivels into a pickle. The water in the cucumber is drawn into the brine (salt) solution because of a difference in osmotic pressure (Figure 7.5).
- A Medical Connection: Oral Rehydration Therapy describes one of the most lethal and pervasive examples of cellular fluid imbalance.

QUESTION 7.21

Describe the direction of water flow in the cucumber/pickle story (above).

QUESTION 7.22

What is the semipermeable membrane in Question 7.21?

QUESTION 7.23

Describe the direction of water flow in the sailor story (above).

(a)

(b)

Figure 7.5

A cucumber (a) in an acidic salt solution undergoes considerable shrinkage on its way to becoming a pickle (b) because of osmosis.

A MEDICAL Connection | Oral Rehydration Therapy

Diarrhea kills millions of children before they reach the age of five years. This is particularly true in third world countries where sanitation, water supplies, and medical care are poor. In the case of diarrhea, death results from fluid loss, electrolyte imbalance, and hypovolemic shock (multiple organ failure due to insufficient perfusion). Cholera is one of the best-understood bacterial diarrheas. The organism *Vibrio cholera*, seen in the micrograph below, survives passage through the stomach and reproduces in the intestine, where it produces a toxin called choleragen. The toxin causes excessive excretion of Na^+, Cl^-, and HCO_3^- from epithelial cells lining the intestine. The increased ion concentration (hypertonic solution) outside the cell results in movement of massive quantities of water into the intestinal lumen. This causes the severe, abundant, clear vomit and diarrhea that can result in the loss of 10–15 L of fluid per day. Over the four- to six-day progress of the disease, a patient may lose from one to two times his or her body mass!

The need for fluid replacement is obvious. Oral rehydration is preferred to intravenous administration of fluids and electrolytes because it is noninvasive. In many third world countries, it is the only therapy available in remote areas. The rehydration formula includes 50–80 g/L rice (or other starch), 3.5 g/L sodium chloride, 2.5 g/L sodium bicarbonate, and 1.5 g/L potassium chloride. Oral rehydration takes advantage of the cotransport of Na^+ and glucose across the cells lining the intestine. Thus, the channel protein brings glucose into the cells, and Na^+ is carried along. Movement of these materials into the cells will help alleviate the osmotic imbalance, reduce the diarrhea, and correct the fluid and electrolyte imbalance.

The disease runs its course in less than a week. In fact, antibiotics are not used to combat cholera. The only effective therapy is oral rehydration, which reduces mortality to less than 1%. A much better option is prevention. In the photo below, a woman is shown filtering water through sari cloth. This simple practice has been shown to reduce the incidence of cholera significantly.

A woman is shown filtering water through sari cloth.

FOR FURTHER UNDERSTANDING

Explain dehydration in terms of osmosis.

Explain why even severely dehydrated individuals continue to experience further fluid loss.

Vibrio cholera.

QUESTION 7.24

What is the semipermeable membrane in Question 7.23?

7.5 Water as a Solvent

Water is by far the most abundant substance on earth. It is an excellent solvent for most inorganic substances. It is often referred to as the "universal solvent" and is the principal biological solvent. Approximately 60% of the adult human body is water, and maintenance of this level is essential for survival. These characteristics are a direct consequence of the molecular structure of water.

As we saw in Chapter 3, water is a bent molecule with a 104.5° bond angle. This angular structure, resulting from the effect of the two lone pairs of electrons around the oxygen atom, is responsible for the polar nature of water. The polarity, in turn, gives water its unique properties.

Because water molecules are polar, water is an excellent solvent for other polar substances ("like dissolves like"). Because much of the matter on earth is polar, hence at least somewhat water-soluble, water has been described as the universal solvent. It is readily accessible and easily purified. It is nontoxic and quite nonreactive. The high boiling point of water, 100°C, compared with molecules of similar size such as N_2 (b.p. = −196°C), is also explained by water's polar character. Strong dipole–dipole interactions between a δ^+ hydrogen of one molecule and δ^- oxygen of a second, referred to as hydrogen bonding, create an interactive molecular network in the liquid phase (see Figure 6.8). The strength of these interactions requires more energy (higher temperature) to boil water. The higher than expected boiling point enhances water's value as a solvent; often, reactions are carried out at higher temperatures to increase their rates. Other solvents, with lower boiling points, would simply boil away, and the reaction would stop.

This idea is easily extended to our own chemistry—because 60% of our bodies is water, we should appreciate the polarity of water on a hot day. As a biological solvent in the human body, water is involved in the transport of ions, nutrients, and waste into and out of cells. Water is also the solvent for biochemical reactions in cells and the digestive tract. Water is a reactant or product in some biochemical processes.

8 Learning Goal
Describe the ways in which the chemical and physical properties of water make it a truly unique solvent.

Animation
The Properties of Water

See A Lifestyle Connection: An Extraordinary Molecule, in Chapter 6.

Refer to Chapter 3 for a more complete description of the bonding, structure, and polarity of water.

Recall the discussion of intermolecular forces in Chapters 3 and 6.

EXAMPLE 7.10 Predicting Structure from Observable Properties

Sucrose is a common sugar and we know that it is used as a sweetener when dissolved in many beverages. What does this allow us to predict about the structure of sucrose?

SOLUTION

Sucrose is used as a sweetener in teas, coffee, and a host of soft drinks. The solvent in all of these beverages is water, a polar molecule. The rule "like dissolves like" implies that sucrose must also be a polar molecule. Without even knowing the formula or structure of sucrose, we can infer this important information from a simple experiment—dissolving sugar in our morning cup of coffee.

8 Learning Goal
Describe the ways in which the chemical and physical properties of water make it a truly unique solvent.

See Figure 14.11 for the structure of sucrose.

QUESTION 7.25

Predict whether carbon monoxide or carbon dioxide would be more soluble in water. Explain your answer. (Hint: Refer to Section 6.2, the discussion of interactions in the liquid state.)

QUESTION 7.26

Predict whether ammonia or methane would be more soluble in water. Explain your answer. (Hint: Refer to Section 6.2, the discussion of interactions in the liquid state.)

7.6 Electrolytes in Body Fluids

The concentrations of cations, anions, and other substances in biological fluids are critical to health. Consequently, the osmolarity of body fluids is carefully regulated by the kidney.

Representation of Concentration of Ions in Solution

The concentration of ions in solution may be represented in a variety of ways. The most common include moles per liter (molarity) and equivalents per liter.

When discussing solutions of ionic compounds, molarity emphasizes the number of individual particles (molecules or ions). For example, a 1 molar solution of NaCl contains Avogadro's number, 6.022×10^{23}, of Na^+ per liter and Avogadro's number of Cl^- per liter. In contrast, equivalents per liter emphasize charge; one equivalent of Na^+ contains Avogadro's number of positive charge.

We defined 1 mol as the number of grams of an atom, molecule, or ion corresponding to Avogadro's number of particles. One **equivalent** of an ion is the number of grams of the ion corresponding to Avogadro's number of electrical charges. Some examples follow:

$$1 \text{ mol } Na^+ = 1 \text{ equivalent } Na^+ \qquad (\text{one } Na^+ = 1 \text{ unit of charge/ion})$$
$$1 \text{ mol } Cl^- = 1 \text{ equivalent } Cl^- \qquad (\text{one } Cl^- = 1 \text{ unit of charge/ion})$$
$$1 \text{ mol } Ca^{2+} = 2 \text{ equivalents } Ca^{2+} \qquad (\text{one } Ca^{2+} = 2 \text{ units of charge/ion})$$
$$1 \text{ mol } CO_3^{2-} = 2 \text{ equivalents } CO_3^{2-} \qquad (\text{one } CO_3^{2-} = 2 \text{ units of charge/ion})$$
$$1 \text{ mol } PO_4^{3-} = 3 \text{ equivalents } PO_4^{3-} \qquad (\text{one } PO_4^{3-} = 3 \text{ units of charge/ion})$$

Moles per liter can be changed to equivalents per liter (or the reverse) by using conversion factors.

Milliequivalents (meq) or milliequivalents/liter (meq/L) are often used when describing small amounts or low concentrations of ions. These units are routinely used when describing ions in blood, urine, and blood plasma.

EXAMPLE 7.11 Calculating Ion Concentration

Calculate the number of equivalents per liter (eq/L) of phosphate ion, PO_4^{3-}, in a solution that is 5.0×10^{-3} M phosphate.

SOLUTION

It is necessary to use two conversion factors:

$$\text{mol } PO_4^{3-} \longrightarrow \text{mol charge}$$

and

$$\text{mol charge} \longrightarrow \text{eq } PO_4^{3-}$$

Arranging these factors in sequence yields:

$$\frac{5.0 \times 10^{-3} \text{ mol } PO_4^{3-}}{1 \text{ L}} \times \frac{3 \text{ mol charge}}{1 \text{ mol } PO_4^{3-}} \times \frac{1 \text{ eq } PO_4^{3-}}{1 \text{ mol charge}} = \frac{1.5 \times 10^{-2} \text{ eq } PO_4^{3-}}{\text{L}}$$

EXAMPLE 7.12   Calculating Electrolyte Concentrations

A typical concentration of calcium ion in blood plasma is 4 meq/L. Represent this concentration in moles/L.

9 **Learning Goal**

Explain the role of electrolytes in blood and their relationship to the process of dialysis.

SOLUTION

The calcium ion has a 2+ charge (recall that calcium is in Group IIA of the periodic table hence, a 2+ charge on the calcium ion).

We will need three conversion factors:

$$\text{meq (milliequivalents)} \longrightarrow \text{eq (equivalents)}$$

$$\text{eq (equivalents)} \longrightarrow \text{moles of charge}$$

$$\text{moles of charge} \longrightarrow \text{moles of calcium ion}$$

Using dimensional analysis, as in Example 7.11,

$$\frac{4 \text{ meq } Ca^{2+}}{1 \text{ L}} \times \frac{1 \text{ eq } Ca^{2+}}{10^3 \text{ meq } Ca^{2+}} \times \frac{1 \text{ mol charge}}{1 \text{ eq } Ca^{2+}} \times \frac{1 \text{ mol } Ca^{2+}}{2 \text{ mol charge}} = \frac{2 \times 10^{-3} \text{ mol } Ca^{2+}}{\text{L}}$$

QUESTION 7.27

Sodium chloride [0.9% (W/V)] is a solution administered intravenously to replace fluid loss. It is frequently used to avoid dehydration. The sodium ion concentration is 15.4 meq/L. Calculate the sodium ion concentration in moles/L.

QUESTION 7.28

A potassium chloride solution that also contains 5% (W/V) dextrose is administered intravenously to treat some forms of malnutrition. The potassium ion concentration in this solution is 40 meq/L. Calculate the potassium ion concentration in moles/L.

Blood: An Ionic Solution and More

The two most important cations in body fluids are Na^+ and K^+. Sodium ion is the most abundant cation in the blood and intercellular fluids, whereas potassium ion is the most abundant intracellular cation. In blood and intercellular fluid, the Na^+ concentration is 135 milliequivalents/L and the K^+ concentration is 3.5–5.0 meq/L. Inside the cell, the situation is reversed. The K^+ concentration is 125 meq/L and the Na^+ concentration is 10 meq/L.

If osmosis and simple diffusion were the only mechanisms for transporting water and ions across cell membranes, these concentration differences would not occur. One positive ion would be just as good as any other. However, the situation is more complex than this. Large protein molecules embedded in cell membranes actively pump sodium ions to the outside of the cell and potassium ions into the cell. This is termed *active transport* because cellular energy must be expended to transport those ions. Proper cell function in the regulation of muscles and the nervous system depends on the sodium ion/potassium ion ratio inside and outside the cell.

If the Na^+ concentration in the blood becomes too low, urine output decreases, the mouth feels dry, the skin becomes flushed, and a fever may develop. The blood level of Na^+ may be elevated when large amounts of water are lost. Diabetes, certain high-protein diets, and diarrhea may cause an elevated blood Na^+ level. In extreme cases, elevated Na^+ levels may cause confusion, stupor, or coma.

Concentrations of K^+ in the blood may rise to dangerously high levels following any injury that causes large numbers of cells to rupture, releasing their

A MEDICAL Connection | Hemodialysis

As we have seen in this section, blood is the medium for the exchange of both nutrients and waste products. The membranes of the kidneys remove waste materials such as urea and uric acid and excess salts and large quantities of water. This process of waste removal is termed **dialysis,** a process similar in function to osmosis (Section 7.4). Semipermeable membranes in the kidneys, dialyzing membranes, allow small molecules (principally water and urea) and ions in solution to pass through and ultimately collect in the bladder. From there, they can be eliminated from the body.

Unfortunately, a variety of diseases can cause partial or complete kidney failure. Should the kidneys fail to perform their primary function, dialysis of waste products, urea and other waste products rapidly increase in concentration in the blood. This can become a life-threatening situation in a very short time.

The most effective short-term treatment of kidney failure is the use of a machine, an artificial kidney, that mimics the function of the kidney. The artificial kidney removes waste from the blood using the process of hemodialysis (blood dialysis). The blood is pumped through a long semipermeable membrane, the dialysis membrane. The dialysis process is similar to osmosis. However, in addition to water molecules, larger molecules (including the waste products in the blood) and ions can pass across the membrane from the blood into a dialyzing fluid. The dialyzing fluid is isotonic with normal blood; it also is similar in its concentration of all other essential blood components. The waste materials move across the dialysis membrane (from a higher to a lower concentration, as in diffusion). A successful dialysis procedure selectively removes the waste from the body without upsetting the critical electrolyte balance in the blood.

Hemodialysis, although lifesaving, is not by any means a pleasant experience. The patient's water intake must be severely limited to minimize the number of times each week that treatment must be used. Many dialysis patients require two or three treatments per week and each session may require one-half (or more) day of hospitalization, especially when the patient suffers from complicating conditions such as diabetes.

Improvements in technology, as well as the growth and sophistication of our health care delivery systems over the past several years, have made dialysis treatment much more patient friendly. Dialysis centers, specializing in the treatment of kidney patients, are now found in most major population centers. Smaller, more automated dialysis units are available for home use, under the supervision of a nursing practitioner. With the remarkable progress in kidney transplant success, dialysis is becoming, more and more, a temporary solution, sustaining life until a suitable kidney donor match can be found.

FOR FURTHER UNDERSTANDING

In what way is dialysis similar to osmosis?

In what way does dialysis differ from osmosis?

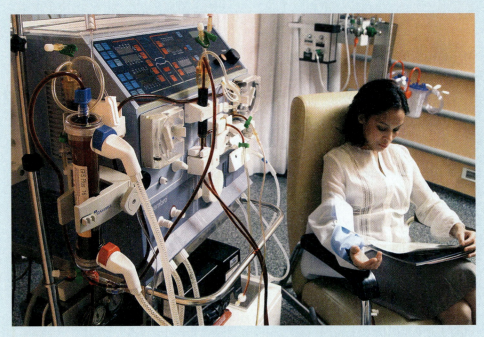

Dialysis patient.

intracellular K^+. This may lead to death by heart failure. Similarly, very low levels of K^+ in the blood may also cause death from heart failure. This may occur following prolonged exercise that results in excessive sweating. When this happens, both body fluids and electrolytes must be replaced. Salt tablets containing both NaCl and KCl taken with water and drinks such as Gatorade effectively provide water and electrolytes and prevent serious symptoms.

The cationic charge in blood is neutralized by two major anions, Cl^- and HCO_3^-. The chloride ion plays a role in acid-base balance, maintenance of osmotic pressure within an acceptable range, and oxygen transport by hemoglobin. The bicarbonate anion is the form in which most waste CO_2 is carried in the blood.

A variety of proteins is also found in the blood. Because of their larger size, they exist in colloidal suspension. These proteins include blood clotting factors, immunoglobulins (antibodies) that help us fight infection, and albumins that act as carriers of nonpolar, hydrophobic substances (fatty acids and steroid hormones) that cannot dissolve in water.

Additionally, blood is the medium for the exchange of nutrients and waste products. Nutrients, such as the polar sugar, glucose, enter the blood from the intestine or the liver. Because glucose molecules are polar, they dissolve in body fluids and are circulated to tissues throughout the body. Nonpolar nutrients are transported with the help of carrier proteins. Similarly, nitrogen-containing waste products, such as urea, are passed from cells to the blood. They are continuously and efficiently removed from the blood by the kidneys.

In cases of loss of kidney function, mechanical devices—dialysis machines— mimic the action of the kidney. The process of blood dialysis—hemodialysis—is discussed in A Medical Connection: Hemodialysis on page 216.

SUMMARY

7.1 Properties of Solutions

A majority of chemical reactions, and virtually all important organic and biochemical reactions, take place not as a combination of two or more pure substances, but rather as reactants dissolved in solution, *solution reactions.*

A *solution* is a homogeneous (or uniform) mixture of two or more substances. A solution is composed of one or more *solutes*, dissolved in a *solvent*. When the solvent is water, the solution is called an *aqueous solution.*

Liquid solutions are clear and transparent with no visible particles of solute. They may be colored or colorless, depending on the properties of the solute and solvent.

In solutions of *electrolytes,* the solutes are ionic compounds that dissociate in solution to produce ions. They are good conductors of electricity. Solutions of *nonelectrolytes* are formed from nondissociating molecular solutes (nonelectrolytes), and their solutions are nonconducting.

The rule "like dissolves like" is the fundamental condition for solubility. Polar solutes are soluble in polar solvents, and nonpolar solutes are soluble in nonpolar solvents.

The degree of solubility depends on the difference between the polarity of solute and solvent, the temperature, and the pressure. Pressure considerations are significant only for solutions of gases.

When a solution contains all the solute that can be dissolved at a particular temperature, it is *saturated.* Excess solute falls to the bottom of the container as a *precipitate.* Occasionally, on cooling, the excess solute may remain in solution for a time before precipitation. Such a solution is a *supersaturated solution.* When excess solute, the precipitate, contacts solvent, the dissolution process reaches a state of dynamic equilibrium. *Colloidal suspensions* have particle sizes between those of true solutions and precipitates. A *suspension* is a heterogeneous mixture that contains particles much larger than a colloidal suspension. Over time, these particles may settle, forming a second phase.

Henry's law describes the solubility of gases in liquids. At a given temperature, the solubility of a gas is proportional to the partial pressure of the gas.

7.2 Concentration Based on Mass

The amount of solute dissolved in a given amount of solution is the solution *concentration.* The more widely used percentage-based concentration units are *weight/volume percent* and *weight/weight percent.* Parts per thousand (ppt) and parts per million (ppm) are used with very dilute solutions.

7.3 Concentration Based on Moles

Molarity, symbolized M, is defined as the number of moles of solute per liter of solution.

Dilution is often used to prepare less concentrated solutions. The expression for this calculation is $(M_1)(V_1) = (M_2)(V_2)$. Knowing any three of these terms enables one to calculate the fourth. The concentration of solute may be represented as moles per liter (molarity) or any other suitable concentration units. However, both concentrations must be in the same units when using the dilution equation.

7.4 Concentration–Dependent Solution Properties

Solution properties that depend on the concentration of solute particles, rather than on the identity of the solute, are *colligative properties.*

There are four colligative properties of solutions; all depend on the concentration of particles in solution.

1. *Vapor pressure lowering. Raoult's law* states that when a solute is added to a solvent, the vapor pressure of the solvent decreases in proportion to the concentration of the solute.
2. and 3. Freezing point depression and boiling point elevation. When a nonvolatile solid is added to a solvent, the freezing point of the resulting solution decreases, and the boiling point increases. The magnitudes of both the freezing point depression (ΔT_f) and the boiling point elevation (ΔT_b) are proportional to the solute concentration over a limited range of concentrations.
4. *Osmosis and osmotic pressure. Osmosis* is the movement of solvent from a dilute solution to a more concentrated solution through a *semipermeable membrane.* The pressure that must be applied to the more concentrated solution to stop this flow is the *osmotic pressure.*

In biological systems, if the concentration of the fluid surrounding red blood cells is higher than that inside the cell (a *hypertonic* solution), water flows from the cell, causing it to collapse. Too low a concentration of this fluid relative to the solution within the cell (a *hypotonic* solution) will cause cell rupture.

Two solutions are *isotonic* if they have identical osmotic pressures. In an isotonic solution, the osmotic pressure differential across the cell is zero, and no cell disruption occurs.

7.5 Water as a Solvent

The role of water in the solution process deserves special attention. It is often referred to as the "universal solvent" because of the large number of ionic and polar covalent compounds that are at least partially soluble in water. It is the principal biological solvent. These characteristics are a direct consequence of the molecular geometry and structure of water and its ability to undergo hydrogen bonding.

7.6 Electrolytes in Body Fluids

The concentrations of cations, anions, and other substances in biological fluids are critical to health. As a result, the osmolarity of body fluids is carefully regulated by the kidney using the process of *dialysis.*

KEY TERMS

aqueous solution (7.1)
colligative property (7.4)
colloidal suspension (7.1)
concentration (7.2)
concentration gradient (7.4)
dialysis (7.6)
diffusion (7.4)
electrolyte (7.1)
equivalent (7.4)
Henry's law (7.1)
hypertonic solution (7.4)
hypotonic solution (7.4)
isotonic solution (7.4)
molarity (7.3)
nonelectrolyte (7.1)
osmolarity (7.4)
osmosis (7.4)

osmotic pressure (7.4)
precipitate (7.1)
Raoult's law (7.4)
saturated solution (7.1)
selectively permeable
   membrane (7.4)
semipermeable membrane (7.4)
solubility (7.1)
solute (7.1)
solution (7.1)
solvent (7.1)
supersaturated solution (7.1)
suspension (7.1)
weight/volume percent
   (% [W/V]) (7.2)
weight/weight percent
   (% [W/W]) (7.2)

QUESTIONS AND PROBLEMS

Concentration Based on Mass

Fundamentals

7.29 Calculate the composition of each of the following solutions in weight/volume %:
   a. 20.0 g NaCl in 1.00 L solution
   b. 33.0 g sugar, $C_6H_{12}O_6$, in 5.00×10^2 mL solution

7.30 Calculate the composition of each of the following solutions in weight/volume %:
   a. 0.700 g KCl per 1.00 mL
   b. 1.00 mol $MgCl_2$ in 2.50×10^2 mL solution

7.31 Calculate the composition of each of the following solutions in weight/weight %:
   a. 21.0 g NaCl in 1.00×10^2 g solution
   b. 21.0 g NaCl in 5.00×10^2 mL solution ($d = 1.12$ g/mL)

7.32 Calculate the composition of each of the following solutions in weight/weight %:
   a. 1.00 g KCl in 1.00×10^2 g solution
   b. 50.0 g KCl in 5.00×10^2 mL solution ($d = 1.14$ g/mL)

Applications

7.33 How many grams of solute are needed to prepare each of the following solutions?
   a. 2.50×10^2 g of 0.900% (W/W) NaCl
   b. 2.50×10^2 g of 1.25% (W/W) $NaC_2H_3O_2$ (sodium acetate)

7.34 How many grams of solute are needed to prepare each of the following solutions?
   a. 2.50×10^2 g of 5.00% (W/W) NH_4Cl (ammonium chloride)
   b. 2.50×10^2 g of 3.50% (W/W) Na_2CO_3

7.35 A solution was prepared by dissolving 14.6 g of KNO_3 in sufficient water to produce 75.0 mL of solution. What is the weight/volume % of this solution?

7.36 A solution was prepared by dissolving 12.4 g of $NaNO_3$ in sufficient water to produce 95.0 mL of solution. What is the weight/volume % of this solution?

7.37 How many grams of sugar would you use to prepare 100 mL of a 1.00 weight/volume % solution?

7.38 How many mL of 4.0 weight/volume % $Mg(NO_3)_2$ solution would contain 1.2 g of magnesium nitrate?

7.39 A solution contains 1.0 mg of Cu^{2+} per 0.50 kg solution. Calculate the concentration in ppt.

7.40 A solution contains 1.0 mg of Cu^{2+} per 0.50 kg solution. Calculate the concentration in ppm.

Concentration Based on Moles

7.41 Calculate the molarity of each solution in Problem 7.29.

7.42 Calculate the molarity of each solution in Problem 7.30.

7.43 Calculate the number of grams of solute that would be needed to make each of the following solutions:
 a. 2.50×10^2 mL of 0.100 M NaCl
 b. 2.50×10^2 mL of 0.200 M $C_6H_{12}O_6$ (glucose)

7.44 Calculate the number of grams of solute that would be needed to make each of the following solutions:
 a. 2.50×10^2 mL of 0.100 M NaBr
 b. 2.50×10^2 mL of 0.200 M KOH

7.45 Calculate the molarity of a sucrose (table sugar, $C_{12}H_{22}O_{11}$) solution that contains 50.0 g of sucrose per liter.

7.46 A saturated silver chloride solution contains 1.58×10^{-4} g of silver chloride per 1.00×10^2 mL of solution. What is the molarity of this solution?

7.47 It is desired to prepare 0.500 L of a 0.100 M solution of NaCl from a 1.00 M stock solution. How many milliliters of the stock solution must be taken for the dilution?

7.48 50.0 mL of a 0.250 M sucrose solution was diluted to 5.00×10^2 mL. What is the molar concentration of the resulting solution?

7.49 A 50.0-mL portion of a stock solution was diluted to 500.0 mL. If the resulting solution was 2.00 M, what was the molarity of the original stock solution?

7.50 A 6.00-mL portion of an 8.00 M stock solution is to be diluted to 0.400 M. What will be the final volume after dilution?

Concentration–Dependent Solution Properties

Foundations

7.51 What is meant by the term *colligative property*?

7.52 Name and describe four colligative solution properties.

7.53 Explain, in terms of solution properties, why salt is used to melt ice in the winter.

7.54 Explain, in terms of solution properties, why a wilted plant regains its "health" when watered.

Applications

7.55 In what way do colligative properties and chemical properties differ?

7.56 Look up the meaning of the term "colligative." Why is it an appropriate title for these properties?

7.57 State Raoult's law.

7.58 What is the major importance of Raoult's law?

7.59 Why does one mole of $CaCl_2$ lower the freezing point of water more than one mole of NaCl?

7.60 Using salt to try to melt ice on a day when the temperature is $-20°C$ will be unsuccessful. Why?

Answer questions 7.61–7.66 by comparing two solutions: 0.50 M sodium chloride (an ionic compound) and 0.50 M sucrose (a covalent compound).

7.61 Which solution has the higher melting point?

7.62 Which solution has the higher boiling point?

7.63 Which solution has the higher vapor pressure?

7.64 Each solution is separated from water by a semipermeable membrane. Which solution has the higher osmotic pressure?

7.65 Two solutions, A and B, are separated by a semipermeable membrane. For each case, predict whether there will be a net flow of water in one direction and, if so, which direction.
 a. A is pure water and B is 5% glucose.
 b. A is 0.10 M glucose and B is 0.10 M KCl.

7.66 Two solutions, A and B, are separated by a semipermeable membrane. For each case, predict whether there will be a net flow of water in one direction and, if so, which direction.
 a. A is 0.10 M NaCl and B is 0.10 M KCl.
 b. A is 0.10 M NaCl and B is 0.20 M glucose.

7.67 Label each solution as isotonic, hypotonic, or hypertonic in comparison to 0.9% (W/W) NaCl (0.15 M NaCl).
 a. 0.15 M $CaCl_2$ b. 0.35 M glucose

7.68 Label each solution as isotonic, hypotonic, or hypertonic in comparison to 0.9% (W/W) NaCl (0.15 M NaCl).
 a. 0.15 M glucose   b. 3% NaCl

Water as a Solvent

7.69 What properties make water such a useful solvent?

7.70 Sketch the "interactive network" of water molecules in the liquid state.

7.71 Solutions of ammonia in water are sold as window cleaners. Why do these solutions have a long "shelf life"?

7.72 Why does water's abnormally high boiling point help to make it a desirable solvent?

7.73 Sketch the interaction of a water molecule with a sodium ion.

7.74 Sketch the interaction of a water molecule with a chloride ion.

7.75 What type of solute dissolves readily in water?

7.76 What type of solute dissolves readily in gasoline?

Electrolytes in Body Fluids

7.77 Explain why a dialysis solution must have a low sodium ion concentration if it is designed to remove excess sodium ion from the blood.

7.78 Explain why a dialysis solution must have an elevated potassium ion concentration when loss of potassium ion from the blood is a concern.

7.79 Describe conditions that can lead to elevated concentrations of sodium in the blood.

7.80 Describe conditions that can lead to dangerously low concentrations of potassium in the blood.

FOR FURTHER UNDERSTANDING

1. Which of the following compounds would cause the greater freezing point depression, per mole, in H_2O: $C_6H_{12}O_6$ (glucose) or NaCl?

2. Which of the following compounds would cause the greater boiling point elevation, per mole, in H_2O: $MgCl_2$ or $HOCH_2CH_2OH$ (ethylene glycol, antifreeze)? (Hint: $HOCH_2CH_2OH$ is covalent.)

3. Analytical chemists often take advantage of differences in solubility to separate ions. For example, adding Cl^- to a solution of Cu^{2+} and Ag^+ causes AgCl to precipitate; Cu^{2+} remains in solution. Filtering the solution results in a separation. Design a scheme to separate the cations Ca^{2+} and Pb^{2+}.

4. Using the strategy outlined in the above problem, design a scheme to separate the anions S^{2-} and CO_3^{2-}.

5. Design an experiment that would enable you to measure the degree of solubility of a salt such as KI in water.

6. How could you experimentally distinguish between a saturated solution and a supersaturated solution?

7. Blood is essentially an aqueous solution, but it must transport a variety of nonpolar substances (hormones, for example). Colloidal proteins, termed *albumins,* facilitate this transport. Must these albumins be polar or nonpolar? Why?

8

OUTLINE

LEARNING GOALS

1. Identify acids and bases and acid-base reactions.

2. Describe the role of the solvent in acid-base reactions.

3. Write equations describing acid-base dissociation and label the conjugate acid-base pairs.

4. Calculate pH from concentration data.

5. Calculate hydronium and/or hydroxide ion concentration from pH data.

6. Provide examples of the importance of pH in chemical and biochemical systems.

7. Describe the meaning and utility of neutralization reactions.

8. Describe the applications of buffers to chemical and biochemical systems, particularly blood chemistry.

Acids and Bases

An acid-base reaction involves the transfer of one or more positively charged units, protons or hydrogen ions.

Acids and bases include some of the most important compounds in nature. Historically, it was recognized that certain compounds, acids, had a sour taste, were able to dissolve some metals, and caused vegetable dyes to change color. Bases have long been recognized by their bitter taste, slippery feel, and corrosive nature. Bases react strongly with acids and cause many metal ions in solution to form solid precipitates.

Acid-base chemistry is an integral part of aqueous solution chemistry (Chapter 7). Pure water has acid-base properties; it is the most common solvent for acid-base reactions. An increase in the acidity of natural water systems from various sources of pollution can have a profound effect on plant and animal life in a body of water. The ability of fish to successfully reproduce, various species of water plants to grow, and coral reefs to maintain their structure all critically depend on the acidity of the water.

Digestion of proteins is aided by stomach acid (hydrochloric acid) and many biochemical processes such as enzyme catalysis depend on the proper level of acidity. A wide variety of chemical reactions depend critically on the acid-base composition of the solution. This is especially true of the biochemical reactions occurring in the cells of our bodies. For this reason, the level of acidity must be very carefully regulated. This is done with substances called buffers.

The level of acidity or basicity of a solution is measured and reported using a logarithm-based scale, the pH scale. This scale represents hydrogen ion concentration data as a number ranging from 0 (very acidic) through 7 (neutral) to 14 (very basic). Since it is easily measured, it serves as a useful indicator of water quality.

In this chapter, we will learn the fundamentals of acid-base chemistry, see how pH is measured and calculated, study acid-base reactions and buffers, and look at some interesting applications of acid-base chemistry.

Two solutions containing food coloring and differing only in pH, are shown. Food coloring is added to many products to improve "eye appeal." After reading this chapter, explain why buffering compounds are often used in food products.

8.1 Acids and Bases

1 Learning Goal
Identify acids and bases and acid-base reactions.

The properties of acids and bases are related to their chemical structure. All acids have common characteristics that enable them to increase the hydrogen ion concentration in water. All bases lower the hydrogen ion concentration in water.

Two theories, one developed from the other, help us to understand the unique chemistry of acids and bases.

Arrhenius Theory of Acids and Bases

One of the earliest definitions of acids and bases is the **Arrhenius theory.** According to this theory, an **acid,** dissolved in water, dissociates to form *hydrogen ions or protons* (H^+), and a **base,** dissolved in water, dissociates to form *hydroxide ions* (OH^-). For example, hydrochloric acid dissociates in solution according to the reaction

$$HCl(aq) \longrightarrow H^+(aq) + Cl^-(aq)$$

Sodium hydroxide, a base, produces hydroxide ions in solution:

$$NaOH(aq) \longrightarrow Na^+(aq) + OH^-(aq)$$

The Arrhenius theory satisfactorily explains the behavior of many acids and bases. However, a substance such as ammonia, NH_3, has basic properties but cannot be an Arrhenius base, because it contains no OH^-. The *Brønsted-Lowry theory* explains this mystery and gives us a broader view of acid-base theory by considering the central role of the solvent in the dissociation process.

Brønsted–Lowry Theory of Acids and Bases

The **Brønsted-Lowry theory** defines an **acid** as a proton (H^+) donor and a **base** as a proton acceptor.

Hydrochloric acid in solution *donates* a proton to the solvent, water, thus behaving as a Brønsted-Lowry acid:

$$HCl(aq) + H_2O(l) \longrightarrow H_3O^+(aq) + Cl^-(aq)$$

H_3O^+ is referred to as the hydrated proton or **hydronium ion.**

The basic properties of ammonia are clearly accounted for by the Brønsted-Lowry theory. Ammonia *accepts* a proton from the solvent water, producing OH^-. An equilibrium mixture of NH_3, H_2O, NH_4^+, and OH^- results.

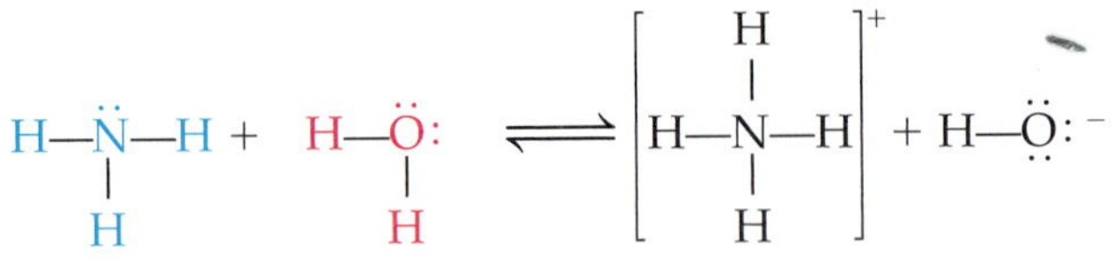

$$NH_3(aq) + H-OH(l) \rightleftharpoons NH_4^+(aq) + OH^-(aq)$$

For aqueous solutions, the Brønsted-Lowry theory adequately describes the behavior of acids and bases. We shall limit our discussion of acid-base chemistry to aqueous solutions and use the Brønsted-Lowry definition described here.

Acid-Base Properties of Water

2 Learning Goal
Describe the role of the solvent in acid-base reactions.

The role that the solvent, water, plays in acid-base reactions is noteworthy. In the example above, the water molecule accepts a proton from the HCl molecule. The water is behaving as a proton acceptor, a base.

However, when water is a solvent for ammonia (NH_3), a base, the water molecule donates a proton to the ammonia molecule. The water, in this situation, is acting as a proton donor, an acid.

Water, owing to the fact that it possesses *both* acid and base properties, is termed **amphiprotic.** Water is the most commonly used solvent for acids and bases. Solute–solvent interactions between water and acids or bases promote both the solubility and the dissociation of acids and bases.

Acid and Base Strength

The terms *acid or base strength* and *acid or base concentration* are easily confused. *Strength* is a measure of the *degree of dissociation* of an acid or base in solution, independent of its concentration. The degree of dissociation is the fraction of acid or base molecules that produces ions in solution. Concentration, as we have learned, refers to the amount of solute (in this case, the amount of acid or base) per quantity of solution.

The strength of acids and bases in water depends on the extent to which they react with the solvent, water. Acids and bases are classified as *strong* when the reaction with water is virtually 100% complete and as *weak* when the reaction with water is much less than 100% complete.

Important strong acids include

Hydrochloric acid	$HCl(aq) + H_2O(l) \longrightarrow H_3O^+(aq) + Cl^-(aq)$
Nitric acid	$HNO_3(aq) + H_2O(l) \longrightarrow H_3O^+(aq) + NO_3^-(aq)$
Sulfuric acid	$H_2SO_4(aq) + H_2O(l) \longrightarrow H_3O^+(aq) + HSO_4^-(aq)$

Note that the equation for the dissociation of each of these acids is written with a single arrow. This indicates that the reaction has little or no tendency to proceed in the reverse direction to establish equilibrium. Virtually all of the acid molecules are dissociated to form ions.

All common strong bases are *metal hydroxides.* Strong bases completely dissociate in aqueous solution to produce hydroxide ions and metal cations. Of the common metal hydroxides, only NaOH and KOH are soluble in water and are readily usable strong bases:

Sodium hydroxide	$NaOH(aq) \longrightarrow Na^+(aq) + OH^-(aq)$
Potassium hydroxide	$KOH(aq) \longrightarrow K^+(aq) + OH^-(aq)$

Weak acids and weak bases dissolve in water principally in the molecular form. Only a small percentage of the molecules dissociates to form the hydronium or hydroxide ion.

Two important weak acids are

Acetic acid	$CH_3COOH(aq) + H_2O(l) \rightleftharpoons H_3O^+(aq) + CH_3COO^-(aq)$
Carbonic acid	$H_2CO_3(aq) + H_2O(l) \rightleftharpoons H_3O^+(aq) + HCO_3^-(aq)$

We have already mentioned the most common weak base, ammonia. Many organic compounds function as weak bases. Several examples of weak bases follow:

Pyridine	$C_5H_5N(aq) + H_2O(l) \rightleftharpoons C_5H_5NH^+(aq) + OH^-(aq)$
Aniline	$C_6H_5NH_2(aq) + H_2O(l) \rightleftharpoons C_6H_5NH_3^+(aq) + OH^-(aq)$
Methylamine	$CH_3NH_2(aq) + H_2O(l) \rightleftharpoons CH_3NH_3^+(aq) + OH^-(aq)$

The fundamental chemical difference between strong and weak acids or bases is their equilibrium ion concentration. A strong acid, such as HCl, does not, in aqueous solution, exist to any measurable degree in equilibrium with its ions, H_3O^+ and Cl^-. On the other hand, a weak acid, such as acetic acid, establishes a dynamic equilibrium with its ions, H_3O^+ and CH_3COO^-.

Concentration of solutions is discussed in Section 7.3.

The concentration of an acid or base does affect the degree of dissociation. However, the major factor in determining the degree of dissociation is the strength of the acid or base.

Reversibility of reactions is discussed in Section 5.4.

Animation
Ionization of a Strong Base and a Weak Base

Animation
Dissociation of Acetic Acid

The double arrow implies an equilibrium between dissociated and undissociated species.

Many organic compounds have acid or base properties. The chemistry of organic acids and bases will be discussed in Chapter 13 (Carboxylic Acids, Esters, Amines, and Amides).

3 Learning Goal

Write equations describing acid-base dissociation and label the conjugate acid-base pairs.

Conjugate Acids and Bases

The Brønsted-Lowry theory contributed several fundamental ideas that broadened our understanding of solution chemistry. First of all, an acid-base reaction is a charge-transfer process. Second, the transfer process usually involves the solvent. Water may accept or donate a proton. Last, and perhaps most important, the acid-base reaction is seen as a reversible process. This leads to the possibility of a reversible, dynamic equilibrium (see Section 5.4).

Consequently, any acid-base reaction can be represented by the general equation

$$HA + B \rightleftharpoons BH^+ + A^-$$
$$\text{(acid)} \quad \text{(base)}$$

In the forward reaction, the acid (HA) donates a proton (H^+) to the base (B) leading to the formation of BH^+ and A^-. However, in the reverse reaction, the BH^+ behaves as an acid; it donates its "extra" proton to A^-. A^- is therefore a base in its own right because it accepts the proton.

These *product* acids and bases are termed *conjugate acids and bases*.

A **conjugate acid** is the species formed when a base accepts a proton.

A **conjugate base** is the species formed when an acid donates a proton.

The acid and base on the opposite sides of the equation are collectively termed a **conjugate acid-base pair.** In the above equation:

BH^+ is the conjugate acid of the base B.

A^- is the conjugate base of the acid HA.

B and BH^+ constitute a conjugate acid-base pair.

HA and A^- constitute a conjugate acid-base pair.

Rewriting our model equation:

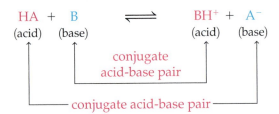

Although we show the forward and reverse arrows to indicate the reversibility of the reaction, seldom are the two processes "equal but opposite." One reaction, either forward or reverse, is usually favored. Consider the reaction of hydrochloric acid in water:

Forward reaction: significant

$$HCl(aq) + H_2O(l) \rightleftharpoons H_3O^+(aq) + Cl^-(aq)$$
$$\text{(acid)} \quad \text{(base)} \quad \quad \text{(acid)} \quad \text{(base)}$$

Reverse reaction: not significant

HCl is a much better proton donor than H_3O^+. Consequently, the forward reaction predominates, the reverse reaction is inconsequential, and hydrochloric acid is termed a *strong acid*. The dissociation of hydrochloric acid is so favorable that we describe it as 100% dissociated and use only a single forward arrow to represent its behavior in water:

$$HCl(aq) + H_2O(l) \longrightarrow H_3O^+(aq) + Cl^-(aq)$$

The degree of dissociation, or strength, of acids and bases has a profound influence on their aqueous chemistry. For example, vinegar (a 5% [W/V] solution of acetic acid in water) is a consumable product; aqueous hydrochloric acid in water is not. Why? Acetic acid is a *weak* acid and, as a result, a dilute solution does no damage to the mouth and esophagus. The following section looks at the strength of acids and bases in solution in more detail.

QUESTION 8.1

Write an equation for the reaction of each of the following with water:

a. HF (a weak acid)
b. NH_3 (a weak base)

QUESTION 8.2

Write an equation for the reaction of each of the following with water:

a. H_2S (a weak acid)
b. CH_3NH_2 (a weak base)

QUESTION 8.3

Select the conjugate acid-base pairs for each reaction in Question 8.1.

QUESTION 8.4

Select the conjugate acid-base pairs for each reaction in Question 8.2.

The relative strength of an acid or base is determined by the ease with which it donates or accepts a proton. Acids with the greatest proton-donating capability (strongest acids) have the weakest conjugate bases. Good proton acceptors (strong bases) have weak conjugate acids. This relationship is clearly indicated in Figure 8.1. This figure can be used to help us compare and predict relative acid-base strength.

EXAMPLE 8.1   Predicting Relative Acid–Base Strengths

a. Write the conjugate acid of HS^-.

SOLUTION

The conjugate acid may be constructed by adding a proton (H^+) to the base structure, consequently, H_2S.

b. Using Figure 8.2, identify the stronger base, HS^- or F^-.

SOLUTION

HS^- is the stronger base because it is located farther down the right-hand column.

c. Using Figure 8.2, identify the stronger acid, H_2S or HF.

SOLUTION

HF is the stronger acid because its conjugate base is weaker *and* because it is located farther up the left-hand column.

3   **Learning Goal**

Write equations describing acid-base dissociation and label the conjugate acid-base pairs.

Figure 8.1

Conjugate acid-base pairs. Strong acids have weak conjugate bases; strong bases have weak conjugate acids. Note that, in every case, the conjugate base has one fewer H^+ than the corresponding conjugate acid.

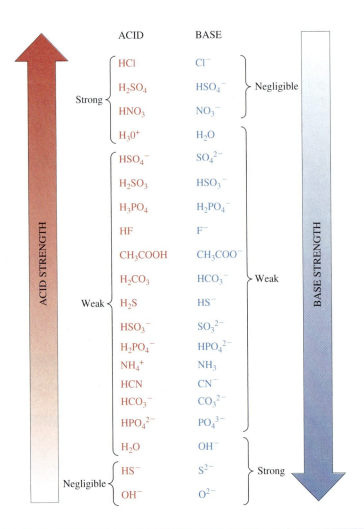

Identify the stronger acid in each pair.

a. H_2O or NH_4^+
b. H_2SO_4 or H_2SO_3

QUESTION 8.6

Identify the stronger base in each pair.

a. CO_3^{2-} or PO_4^{3-}
b. HCO_3^- or HPO_4^{2-}

Solutions of acids and bases used in the laboratory must be handled with care. Acids burn because of their exothermic reaction with water present on and in the skin. Bases react with proteins, which are principal components of the skin and eyes.

Such solutions are more hazardous if they are strong or concentrated. A strong acid or base produces more H_3O^+ or OH^- than the corresponding weak acid or base. More concentrated acids or bases contain more H_3O^+ or OH^- than do less concentrated solutions of the same strength.

The Dissociation of Water

Aqueous solutions of acids and bases are electrolytes. The dissociation of an acid or base produces ions that can conduct an electrical current. As a result of the differences in the degree of dissociation, *strong acids and bases are strong electrolytes; weak acids and bases are weak electrolytes.* The conductivity of these solutions is principally dependent on the solute and not on the solvent (water).

Although pure water is virtually 100% molecular, a small number of water molecules do ionize. This process occurs by the transfer of a proton from one water molecule to another, producing a hydronium ion and a hydroxide ion:

$$H_2O(l) + H_2O(l) \rightleftharpoons H_3O^+(aq) + OH^-(aq)$$

This process is the **autoionization,** or self-ionization, of water. Water is therefore a *very* weak electrolyte and a very poor conductor of electricity. Water has *both* acid and base properties; dissociation produces both the hydronium and hydroxide ion.

Pure water at room temperature has a hydronium ion concentration of 1.0×10^{-7} M. One hydroxide ion is produced for each hydronium ion. Therefore, the hydroxide ion concentration is also 1.0×10^{-7} M. Molar equilibrium concentration is conveniently indicated by brackets around the species whose concentration is represented:

$$[H_3O^+] = 1.0 \times 10^{-7} M$$

$$[OH^-] = 1.0 \times 10^{-7} M$$

The product of hydronium and hydroxide ion concentration in pure water is referred to as the **ion product for water,** symbolized by K_w.

$$K_w = \text{ion product} = [H_3O^+][OH^-]$$
$$= [1.0 \times 10^{-7}][1.0 \times 10^{-7}]$$
$$= 1.0 \times 10^{-14}$$

The ion product is constant because its value does not depend on the nature or concentration of the solute, as long as the temperature does not change. The ion product is a temperature-dependent quantity.

The nature and concentration of the solutes added to water do alter the relative concentrations of H_3O^+ and OH^- present, but the product, $[H_3O^+][OH^-]$, always equals 1.0×10^{-14} at 25°C. This relationship is the basis for a scale that is useful in the measurement of the level of acidity or basicity of solutions. This scale, the pH scale, is discussed next.

8.2 pH: A Measurement Scale for Acids and Bases

A Definition of pH

The **pH scale** gauges the hydronium ion concentration and reflects the degree of acidity or basicity of a solution. The pH scale is somewhat analogous to the temperature scale used for assignment of relative levels of hot or cold. The temperature scale was developed to allow us to indicate how cold or how hot an object is. The pH scale specifies how acidic or how basic a solution is. The pH scale has values that range from 0 (very acidic) to 14 (very basic). A pH of 7, the middle of the scale, is neutral, neither acidic nor basic.

To help us to develop a concept of pH, let's consider the following:

- Addition of an acid (proton donor) to water *increases* the $[H_3O^+]$ and decreases the $[OH^-]$.
- Addition of a base (proton acceptor) to water *decreases* the $[H_3O^+]$ by increasing the $[OH^-]$.
- $[H_3O^+] = [OH^-]$ when *equal* amounts of acid and base are present.
- In all three cases, $[H_3O^+][OH^-] = 1.0 \times 10^{-14} =$ the ion product for water at 25°C.

3 **Learning Goal**

Write equations describing acid-base dissociation and label the conjugate acid-base pairs.

pH values greater than 14 and less than zero are possible, but largely meaningless, due to ion association characteristics of very concentrated solutions.

Figure 8.2

The measurement of pH. (a) A strip of test paper impregnated with indicator (a material that changes color as the acidity of the surroundings changes) is put in contact with the solution of interest. The resulting color is matched with a standard color chart (colors shown as a function of pH) to obtain the approximate pH. (b) A pH meter uses a sensor (a pH electrode) that develops an electrical potential that is proportional to the pH of the solution.

(a)

(b)

Measuring pH

The pH of a solution can be calculated if the concentration of either H_3O^+ or OH^- is known. Alternatively, measurement of pH allows the calculation of H_3O^+ or OH^- concentration. The pH of aqueous solutions may be approximated by using indicating paper (pH paper) that develops a color related to the solution pH. Alternatively, a pH meter can give us a much more exact pH measurement. A sensor measures an electrical property of a solution that is proportional to pH (Figure 8.2).

Calculating pH

One of our objectives in this chapter is to calculate the pH of a solution when the hydronium or hydroxide ion concentration is known, and to calculate $[H_3O^+]$ or $[OH^-]$ from the pH.

The pH of a solution is defined as the negative logarithm of the molar concentration of the hydronium ion:

$$pH = -\log [H_3O^+]$$

4 Learning Goal
Calculate pH from concentration data.

EXAMPLE 8.2 Calculating pH from Acid Molarity

Calculate the pH of a $1.0 \times 10^{-3}\ M$ solution of HCl.

SOLUTION

HCl is a strong acid. If 1 mol HCl dissolves and dissociates in 1 L of aqueous solution, it produces 1 mol H_3O^+ (a 1 M solution of H_3O^+). Therefore a $1.0 \times 10^{-3}\ M$ HCl solution has $[H_3O^+] = 1.0 \times 10^{-3}\ M$, and

$$pH = -\log [H_3O^+]$$
$$= -\log [1.0 \times 10^{-3}]$$
$$= -[-3.00] = 3.00$$

5 Learning Goal
Calculate hydronium and/or hydroxide ion concentration from pH data.

EXAMPLE 8.3 Calculating $[H_3O^+]$ from pH

Calculate the $[H_3O^+]$ of a solution of hydrochloric acid with pH = 4.00.

Continued—

SOLUTION

We use the pH expression:

$$pH = -\log [H_3O^+]$$

$$4.00 = -\log [H_3O^+]$$

Multiplying both sides of the equation by -1, we get

$$-4.00 = \log [H_3O^+]$$

Taking the antilogarithm of both sides (the reverse of a logarithm),

$$\text{antilog} -4.00 = [H_3O^+]$$

The antilog is the exponent of 10; therefore

$$1.0 \times 10^{-4} \, M = [H_3O^+]$$

EXAMPLE 8.4   **Calculating the pH of a Base**

Calculate the pH of a $1.0 \times 10^{-5} \, M$ solution of NaOH.

4 Learning Goal
Calculate pH from concentration data.

SOLUTION

NaOH is a strong base. If 1 mol NaOH dissolves and dissociates in 1 L of aqueous solution, it produces 1 mol OH^- (a 1 M solution of OH^-). Therefore a $1.0 \times 10^{-5} \, M$ NaOH solution has $[OH^-] = 1.0 \times 10^{-5} \, M$. To calculate pH, we need $[H_3O^+]$. Recall that

$$K_w = [H_3O^+][OH^-] = 1.0 \times 10^{-14}$$

Solving this equation for $[H_3O^+]$,

$$[H_3O^+] = \frac{1.0 \times 10^{-14}}{[OH^-]}$$

Substituting the information provided in the problem,

$$= \frac{1.0 \times 10^{-14}}{1.0 \times 10^{-5}}$$

$$= 1.0 \times 10^{-9} \, M$$

The solution is now similar to that in Example 8.2:

$$pH = -\log [H_3O^+]$$

$$= -\log [1.0 \times 10^{-9}]$$

$$= 9.00$$

EXAMPLE 8.5   **Calculating Both Hydronium and Hydroxide Ion Concentrations from pH**

Calculate the $[H_3O^+]$ and $[OH^-]$ of a sodium hydroxide solution with a pH = 10.00.

5 Learning Goal
Calculate hydronium and/or hydroxide ion concentration from pH data.

Continued—

EXAMPLE 8.5 —*Continued*

SOLUTION

First, calculate $[H_3O^+]$:

$$pH = -\log [H_3O^+]$$
$$10.00 = -\log [H_3O^+]$$
$$-10.00 = \log [H_3O^+]$$
$$antilog -10 = [H_3O^+]$$
$$1.0 \times 10^{-10} M = [H_3O^+]$$

To calculate the $[OH^-]$, we need to solve for $[OH^-]$ by using the following expression:

$$K_w = [H_3O^+][OH^-] = 1.0 \times 10^{-14}$$
$$[OH^-] = \frac{1.0 \times 10^{-14}}{[H_3O^+]}$$

Substituting the $[H_3O^+]$ from the first part,

$$[OH^-] = \frac{1.0 \times 10^{-14}}{[1.0 \times 10^{-10}]}$$
$$= 1.0 \times 10^{-4} M$$

Often, the pH or $[H_3O^+]$ will not be a whole number (pH = 1.5, pH = 5.3, $[H_3O^+] = 1.5 \times 10^{-3}$ and so forth). With the advent of inexpensive and versatile calculators, calculations with noninteger numbers pose no great problems. Consider Examples 8.6 and 8.7.

4 Learning Goal
Calculate pH from concentration data.

EXAMPLE 8.6 Calculating pH with Noninteger Numbers

Calculate the pH of a sample of lake water that has a $[H_3O^+] = 6.5 \times 10^{-5} M$.

SOLUTION

$$pH = -\log [H_3O^+]$$
$$= -\log [6.5 \times 10^{-5}]$$
$$= 4.19$$

The pH, 4.19, is low enough to suspect acid rain. (See An Environmental Perspective: Acid Rain in this chapter.)

5 Learning Goal
Calculate hydronium and/or hydroxide ion concentration from pH data.

EXAMPLE 8.7 Calculating $[H_3O^+]$ from pH

The measured pH of a sample of lake water is 6.40. Calculate $[H_3O^+]$.

SOLUTION

An alternative mathematical form of

$$pH = -\log [H_3O^+]$$

Continued—

is the expression

$$[H_3O^+] = 10^{-pH}$$

which we will use when we must solve for $[H_3O^+]$.

$$[H_3O^+] = 10^{-6.40}$$

Performing the calculation on your calculator results in 3.98×10^{-7} or $4.0 \times 10^{-7}\ M = [H_3O^+]$.

Examples 8.2–8.7 illustrate the most frequently used pH calculations. It is important to remember that in the case of a base, you must convert the $[OH^-]$ to $[H_3O^+]$, using the expression for the ion product for the solvent, water. It is also useful to remember the following points:

- The pH of a 1 M solution of any strong acid is 0.
- The pH of a 1 M solution of any strong base is 14.
- Each tenfold change in concentration changes the pH by one unit. A tenfold change in concentration is equivalent to moving the decimal point one place.
- A *decrease* in acid concentration *increases* the pH.
- A *decrease* in base concentration *decreases* the pH.

Figure 8.3 provides a convenient overview of solution pH.

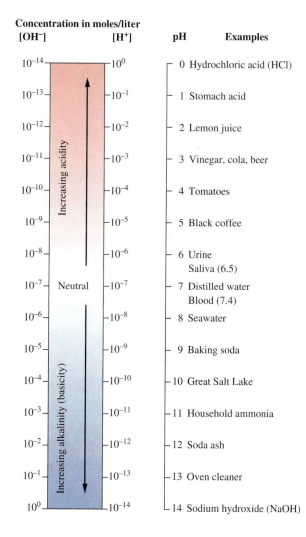

Concentration in moles/liter

[OH⁻]	[H⁺]	pH	Examples
10^{-14}	10^{0}	0	Hydrochloric acid (HCl)
10^{-13}	10^{-1}	1	Stomach acid
10^{-12}	10^{-2}	2	Lemon juice
10^{-11}	10^{-3}	3	Vinegar, cola, beer
10^{-10}	10^{-4}	4	Tomatoes
10^{-9}	10^{-5}	5	Black coffee
10^{-8}	10^{-6}	6	Urine / Saliva (6.5)
10^{-7}	10^{-7}	7	Distilled water / Blood (7.4)
10^{-6}	10^{-8}	8	Seawater
10^{-5}	10^{-9}	9	Baking soda
10^{-4}	10^{-10}	10	Great Salt Lake
10^{-3}	10^{-11}	11	Household ammonia
10^{-2}	10^{-12}	12	Soda ash
10^{-1}	10^{-13}	13	Oven cleaner
10^{0}	10^{-14}	14	Sodium hydroxide (NaOH)

Increasing acidity

Neutral

Increasing alkalinity (basicity)

Figure 8.3

The pH scale. A pH of 7 is neutral ($[H_3O^+] = [OH^-]$). Values less than 7 are acidic (H_3O^+ predominates) and values greater than 7 are basic (OH^- predominates).

A MEDICAL Connection | Drug Delivery

When a doctor prescribes medicine to treat a disease or relieve its symptoms, the medication may be administered in a variety of ways. Drugs may be taken orally, injected into a muscle or a vein, or absorbed through the skin. Specific instructions are often provided to regulate the particular combination of drugs that can or cannot be taken. The diet, both before and during the drug therapy, may be of special importance.

To appreciate why drugs are administered in a specific way, it is necessary to understand a few basic facts about medications and how they interact with the body.

Drugs function by undergoing one or more chemical reactions in the body. Few compounds react in only one way, to produce a limited set of products, even in the simple environment of a beaker or flask. Imagine the number of possible reactions that a drug can undergo in a complex chemical factory like the human body. In many cases, a drug can react in a variety of ways other than its intended path. These alternative paths are side reactions, sometimes producing *side effects* such as nausea, vomiting, insomnia, or drowsiness. Side effects may be unpleasant and may actually interfere with the primary function of the drug.

The development of safe, effective medication, with minimal side effects, is a slow and painstaking process and determining the best drug delivery system is a critical step. For example, a drug that undergoes an unwanted side reaction in an acidic solution would not be very effective if administered orally. The acidic digestive fluids in the stomach could prevent the drug from even reaching the intended organ, let alone retaining its potency. The drug could be administered through a vein into the blood; blood is not acidic, in contrast to digestive fluids. In this way, the drug may be delivered intact to the intended site in the body, where it is free to undergo its primary reaction.

Drug delivery has become a science in its own right. Pharmacology, the study of drugs and their uses in the treatment of disease, has a goal of creating drugs that are highly selective. In other words, they will undergo only one reaction, the intended reaction. Encapsulation of drugs, enclosing them within larger molecules or collections of molecules, may protect them from unwanted reactions as they are transported to their intended site.

FOR FURTHER UNDERSTANDING

Do a little research on the Internet and determine: What is the major undesirable side effect of aspirin?

What is a recently discovered, desirable side effect of aspirin?

QUESTION 8.7

Calculate the $[OH^-]$ of the solution in Example 8.2.

QUESTION 8.8

Calculate the $[OH^-]$ of the solution in Example 8.3.

QUESTION 8.9

Calculate the pH corresponding to a solution of sodium hydroxide with $[OH^-]$ of $1.0 \times 10^{-2}\ M$.

QUESTION 8.10

Calculate the pH corresponding to a solution of sodium hydroxide with $[OH^-]$ of $1.0 \times 10^{-6}\ M$.

QUESTION 8.11

Calculate the $[H_3O^+]$ corresponding to pH $= 8.50$.

QUESTION 8.12

Calculate the $[H_3O^+]$ corresponding to pH = 4.50.

The Importance of pH and pH Control

Solution pH and pH control play a major role in many facets of our lives. Consider a few examples:

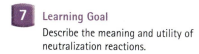
6 Learning Goal

Provide examples of the importance of pH and pH control in chemical and biochemical systems.

- *Agriculture:* Crops grow best in a soil of proper pH. Proper fertilization involves the maintenance of a suitable pH.
- *Physiology:* If the pH of our blood were to shift by one unit, we would die. Many biochemical reactions in living organisms are extremely pH dependent.
- *Industry:* From manufacture of processed foods to the manufacture of automobiles, industrial processes often require rigorous pH control.
- *Municipal services:* Purification of drinking water and treatment of sewage must be carried out at their optimum pH.
- *Acid rain:* Nitric acid and sulfuric acid, resulting largely from the reaction of components of vehicle emissions and electric power generation (nitrogen and sulfur oxides) with water, are carried down by precipitation and enter aquatic systems (lakes and streams), lowering the pH of the water. A less than optimum pH poses serious problems for native fish populations.

See An Environmental Perspective: Acid Rain in this chapter.

The list could continue on for many pages. However, in summary, any change that takes place in aqueous solution generally has at least some pH dependence.

8.3 Reactions Between Acids and Bases

Neutralization

The reaction of an acid with a base to produce a salt and water is referred to as **neutralization.** In the strictest sense, neutralization requires equal numbers of moles of H_3O^+ and OH^- to produce a neutral solution (no excess acid or base).

Consider the reaction of a solution of hydrochloric acid and sodium hydroxide:

$$HCl(aq) + NaOH(aq) \longrightarrow NaCl(aq) + H_2O(l)$$
Acid   Base   Salt   Water

Our objective is to make the balanced equation represent the process actually occurring. We recognize that HCl, NaOH, and NaCl are dissociated in solution:

$$H^+(aq) + Cl^-(aq) + Na^+(aq) + OH^-(aq) \longrightarrow Na^+(aq) + Cl^-(aq) + H_2O(l)$$

We further know that Na^+ and Cl^- are unchanged in the reaction; they are termed *spectator ions.* If we write only those components that actually change, ignoring the spectator ions, we produce a *net, balanced ionic equation:*

$$H^+(aq) + OH^-(aq) \longrightarrow H_2O(l)$$

If we realize that the H^+ occurs in aqueous solution as the hydronium ion, H_3O^+, the most correct form of the net, balanced ionic equation is

$$H_3O^+(aq) + OH^-(aq) \longrightarrow 2H_2O(l)$$

The equation for any strong acid/strong base neutralization reaction is the same as this equation.

7 Learning Goal

Describe the meaning and utility of neutralization reactions.

Equation balancing is discussed in Chapter 4.

You can find further information about net ionic equations online at www.mhhe.com/denniston in "Writing Net Ionic Equations."

Animation

The Neutralization Reaction of NaOH and HCl

A neutralization reaction may be used to determine the concentration of an unknown acid or base solution. The technique of **titration** involves the addition of measured amounts of a **standard solution** (one whose concentration is known with certainty) to neutralize the second, unknown solution. From the volumes of the two solutions and the concentration of the standard solution, the concentration of the unknown solution may be determined. Consider the following application.

7 Learning Goal

Describe the meaning and utility of neutralization reactions.

EXAMPLE 8.8 **Determining the Concentration of a Solution of Hydrochloric Acid**

Step 1. A known volume (perhaps 25.00 mL) of the unknown acid is measured into a flask using a pipet.

Step 2. An **indicator,** a substance that changes color as the solution reaches a certain pH (Figures 8.4 and 8.5), is added to the unknown solution. We

Continued—

Figure 8.4

The color of the petals of the hydrangea is formed by molecules that behave as acid-base indicators. The color is influenced by the pH of the soil in which the hydrangea is grown. The plant (a) was grown in soil with higher pH (more basic) than plant (b).

(a) (b)

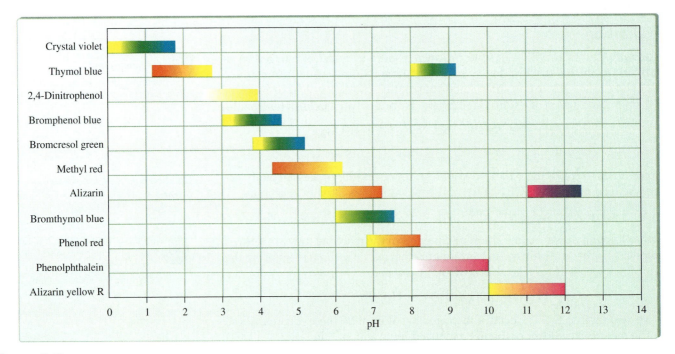

Figure 8.5

The relationship between pH and color of a variety of compounds, some of which are commonly used as acid-base indicators. Many indicators are naturally occurring substances.

must know, from prior experience, the expected pH at the equivalence point (see step 4, below). For this titration, phenolphthalein or phenol red would be a logical choice because the equivalence-point pH is known to be seven.

Step 3. A solution of sodium hydroxide (perhaps 0.1000 *M*) is carefully added to the unknown solution using a **buret** (Figure 8.6), which is a long glass tube calibrated in milliliters. A stopcock at the bottom of the buret regulates the amount of liquid dispensed. The standard solution is added until the indicator changes color.

Step 4. At this point, the **equivalence point,** the number of moles of hydroxide ion added is equal to the number of moles of hydronium ion present in the unknown acid.

Step 5. The volume dispensed by the buret (perhaps 35.00 mL) is measured and used in the calculation of the unknown acid concentration.

Step 6. The calculation is as follows:

Pertinent information for this titration includes:

Volume of the unknown acid solution, 25.00 mL

Volume of sodium hydroxide solution added, 35.00 mL

Concentration of the sodium hydroxide solution, 0.1000 *M*

Furthermore, from the balanced equation, we know that HCl and NaOH react in a 1:1 combining ratio.

Using a strategy involving conversion factors

$$35.00 \text{ mL NaOH} \times \frac{1 \text{ L NaOH}}{10^3 \text{ mL NaOH}} \times \frac{0.1000 \text{ mol NaOH}}{\text{L NaOH}} = 3.500 \times 10^{-3} \text{ mol NaOH}$$

Knowing that HCl and NaOH undergo a 1:1 reaction,

$$3.500 \times 10^{-3} \text{ mol NaOH} \times \frac{1 \text{ mol HCl}}{1 \text{ mol NaOH}} = 3.500 \times 10^{-3} \text{ mol HCl}$$

3.500×10^{-3} mol HCl are contained in 25.00 mL of HCl solution. Thus,

$$\frac{3.500 \times 10^{-3} \text{ mol HCl}}{25.00 \text{ mL HCl soln}} \times \frac{10^3 \text{ mL HCl soln}}{1 \text{ L HCl soln}} = 1.400 \times 10^{-1} \text{ mol HCl/L HCl soln}$$

$$= 0.1400 \text{ } M$$

The titration of an acid with a base is depicted in Figure 8.7.

An alternate problem-solving strategy produces the same result:

$$(M_{acid})(V_{acid}) = (M_{base})(V_{base})$$

and

$$M_{acid} = M_{base}\left(\frac{V_{base}}{V_{acid}}\right)$$

$$M_{acid} = (0.1000 \text{ } M)\left(\frac{35.00 \text{ mL}}{25.00 \text{ mL}}\right)$$

$$M_{acid} = 0.1400 \text{ } M$$

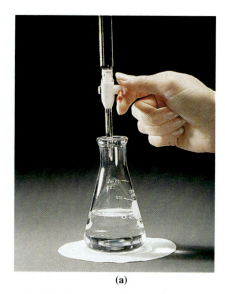

(a)

(b)

Figure 8.6

An acid–base titration. (a) An exact volume of a standard solution (in this example, a base) is added to a solution of unknown concentration (in this example, an acid). (b) From the volume (read from the buret) and concentration of the standard solution, coupled with the mass or volume of the unknown, the concentration of the unknown may be calculated.

Animation
Titration of HCl with NaOH

AN ENVIRONMENTAL
Perspective | Acid Rain

Acid rain is a global environmental problem that has raised public awareness of the chemicals polluting the air through the activities of our industrial society. Normal rain has a pH of about 5.6 as a result of the chemical reaction between carbon dioxide gas and water in the atmosphere. The following equation shows this reaction:

$$CO_2(g) \quad + \quad H_2O(l) \quad \rightleftharpoons \quad H_2CO_3(aq)$$
$$\text{Carbon dioxide} \quad \text{Water} \quad \text{Carbonic acid}$$

Acid rain refers to conditions that are much more acidic than this. In upstate New York, the rain has as much as twenty-five times the acidity of normal rainfall. One rainstorm, recorded in West Virginia, produced rainfall that measured 1.5 on the pH scale. This is approximately the pH of stomach acid or about ten thousand times more acidic than "normal rain" (remember that the pH scale is logarithmic; a 1 pH unit decrease represents a tenfold increase in hydronium ion concentration).

Acid rain is destroying life in streams and lakes. More than half the highland lakes in the western Adirondack Mountains have no native game fish. In addition to these 300 lakes, 140 lakes in Ontario have suffered a similar fate. It is estimated that 48,000 other lakes in Ontario and countless others in the northeastern and central United States are threatened. Our forests are endangered as well. The acid rain decreases soil pH, which in turn alters the solubility of minerals needed by plants. Studies have shown that about 40% of the red spruce and maple trees in New England have died. Increased acidity of rainfall appears to be the major culprit.

What is the cause of this acid rain? The combustion of fossil fuels (gas, oil, and coal) by power plants produces oxides of sulfur and nitrogen. Nitrogen oxides, in excess of normal levels, arise mainly from conversion of atmospheric nitrogen to nitrogen oxides in the engines of gasoline and diesel powered vehicles. Sulfur oxides result from the oxidation of sulfur

pH Values for a Variety of Substances Compared with the pH of Acid Rain

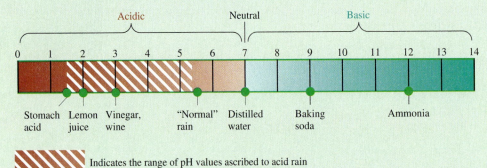

QUESTION 8.13

Calculate the molar concentration of a sodium hydroxide solution if 40.00 mL of this solution were required to neutralize 20.00 mL of a 0.2000 M solution of hydrochloric acid.

QUESTION 8.14

Calculate the molar concentration of a sodium hydroxide solution if 36.00 mL of this solution were required to neutralize 25.00 mL of a 0.2000 M solution of hydrochloric acid.

Polyprotic Substances

Not all acid-base reactions occur in a 1:1 combining ratio (as hydrochloric acid and sodium hydroxide in the previous example). Acid-base reactions with other than 1:1 combining ratios occur between what are termed *polyprotic substances*.

Damage caused by acid rain.

Nitrogen dioxide (which causes the brown color of smog) then reacts with water to form nitric acid:

$$3NO_2(g) + H_2O(l) \longrightarrow 2HNO_3(aq) + NO(g)$$

Similar chemistry is seen with sulfur oxides. Coal may contain as much as 3% sulfur. When the coal is burned, the sulfur also burns. This produces choking, acrid sulfur dioxide gas:

$$S(s) + O_2(g) \longrightarrow SO_2(g)$$

By itself, sulfur dioxide can cause serious respiratory problems for people with asthma or other lung diseases, but matters are worsened by the reaction of SO_2 with atmospheric oxygen:

$$2SO_2(g) + O_2(g) \longrightarrow 2SO_3(g)$$

Sulfur trioxide will react with water in the atmosphere:

$$SO_3(g) + H_2O(l) \longrightarrow H_2SO_4(aq)$$

The product, sulfuric acid, is even more irritating to the respiratory tract. When the acid rain created by the reactions shown above falls to earth, the impact is significant.

It is easy to balance these chemical equations, but decades could be required to balance the ecological systems that we have disrupted by our massive consumption of fossil fuels. A sudden decrease of even 25% in the use of fossil fuels would lead to worldwide financial chaos. Development of alternative fuel sources, such as solar energy and safe nuclear power, will help to reduce our dependence on fossil fuels and help us to balance the global equation.

FOR FURTHER UNDERSTANDING

Criticize this statement: "Passing and enforcing strong legislation against sulfur and nitrogen oxide emission will solve the problem of acid rain in the United States."

Research the literature to determine the percentage of electricity that is produced from coal in your state of residence.

in fossil fuels. The sulfur atoms were originally a part of the amino acids and proteins of plants and animals that became, over the millennia, our fuel. These react with water, as does the CO_2 in normal rain, but the products are strong acids: sulfuric and nitric acids. Let's look at the equations for these processes.

In the atmosphere, nitric oxide (NO) can react with oxygen to produce nitrogen dioxide as shown:

$$2NO(g) + O_2(g) \longrightarrow 2NO_2(g)$$

Nitric oxide   Oxygen   Nitrogen dioxide

Polyprotic substances donate (as acids) or accept (as bases) more than one proton per formula unit.

Reactions of Polyprotic Substances

HCl dissociates to produce one H^+ ion for each HCl. For this reason, it is termed a *monoprotic acid*. Its reaction with sodium hydroxide is

$$HCl(aq) + NaOH(aq) \longrightarrow H_2O(l) + Na^+(aq) + Cl^-(aq)$$

Sulfuric acid, in contrast, is a *diprotic acid*. Each unit of H_2SO_4 produces two H^+ ions (the prefix *di-* indicating two). Its reaction with sodium hydroxide is

$$H_2SO_4(aq) + 2NaOH(aq) \longrightarrow 2H_2O(l) + 2Na^+(aq) + SO_4^{2-}(aq)$$

Phosphoric acid is a *triprotic acid*. Each unit of H_3PO_4 produces three H^+ ions. Its reaction with sodium hydroxide is

$$H_3PO_4(aq) + 3NaOH(aq) \longrightarrow 3H_2O(l) + 3Na^+(aq) + PO_4^{3-}(aq)$$

Figure 8.7

Commercial products that claim improved function owing to their ability to control pH.

Animation

Interactive Properties of Buffers

We ignore Na^+ in the description of the buffer. Na^+ does not actively participate in the reaction.

The acetate ion is the conjugate base of acetic acid.

Dissociation of Polyprotic Substances

Sulfuric acid, and other diprotic acids, dissociate in two steps:

Step 1. $H_2SO_4(aq) + H_2O(l) \longrightarrow H_3O^+(aq) + HSO_4^-(aq)$

Step 2. $HSO_4^-(aq) + H_2O(l) \rightleftharpoons H_3O^+(aq) + SO_4^{2-}(aq)$

Notice that H_2SO_4 behaves as a strong acid (Step 1) and HSO_4^- behaves as a weak acid, indicated by a double arrow (Step 2).

Phosphoric acid dissociates in three steps, all forms behaving as weak acids.

Step 1. $H_3PO_4(aq) + H_2O(l) \rightleftharpoons H_3O^+(aq) + H_2PO_4^-(aq)$

Step 2. $H_2PO_4^-(aq) + H_2O(l) \rightleftharpoons H_3O^+(aq) + HPO_4^{2-}(aq)$

Step 3. $HPO_4^{2-}(aq) + H_2O(l) \rightleftharpoons H_3O^+(aq) + PO_4^{3-}(aq)$

Bases exhibit this property as well.
NaOH produces one OH^- ion per formula unit:

$$NaOH(aq) \longrightarrow Na^+(aq) + OH^-(aq)$$

$Ba(OH)_2$, barium hydroxide, produces two OH^- ions per formula unit:

$$Ba(OH)_2(aq) \longrightarrow Ba^{2+}(aq) + 2OH^-(aq)$$

8.4   Acid–Base Buffers

A **buffer solution** contains components that enable the solution to resist large changes in pH when either acids or bases are added. Buffer solutions may be prepared in the laboratory to maintain optimum conditions for a chemical reaction. Buffers are routinely used in commercial products to maintain optimum conditions for product behavior (Figure 8.7).

Buffer solutions also occur naturally. Blood, for example, is a complex natural buffer solution maintaining a pH of approximately 7.4, optimum for oxygen transport. The major buffering agent in blood is the mixture of carbonic acid (H_2CO_3) and bicarbonate ions (HCO_3^-).

The Buffer Process

The basis of buffer action is the establishment of an equilibrium between either a weak acid and its conjugate base or a weak base and its conjugate acid. Let's consider the case of a weak acid and its salt.

A common buffer solution may be prepared from acetic acid (CH_3COOH) and sodium acetate (CH_3COONa). Sodium acetate is a salt that is the source of the conjugate base CH_3COO^-. An *equilibrium* is established in solution between the weak acid and the conjugate base.

$$\underset{\substack{\text{Acetic acid}\\\text{(weak acid)}}}{CH_3COOH(aq)} + \underset{\text{Water}}{H_2O(l)} \rightleftharpoons \underset{\text{Hydronium ion}}{H_3O^+(aq)} + \underset{\substack{\text{Acetate ion}\\\text{(conjugate base)}}}{CH_3COO^-(aq)}$$

A buffer solution functions in accordance with LeChatelier's principle, which states that an equilibrium system, when stressed, will shift its equilibrium to relieve that stress. This principle is illustrated by the following examples.

Addition of Base (OH⁻) to a Buffer Solution

Addition of a basic substance to a buffer solution causes the following changes.

• OH^- from the base reacts with H_3O^+ producing water.

• Molecular acetic acid *dissociates* to replace the H_3O^+ consumed by the base, maintaining the pH close to the initial level.

This is an example of LeChatelier's principle because the loss of H_3O^+ (the *stress*) is compensated for by the dissociation of acetic acid to produce more H_3O^+.

Addition of Acid (H_3O^+) to a Buffer Solution

Addition of an acidic solution to a buffer results in the following changes.

• H_3O^+ from the acid increases the overall $[H_3O^+]$.
• The system reacts to this stress, in accordance with LeChatelier's principle, to form more molecular acetic acid; the acetate ion combines with H_3O^+. Thus, the H_3O^+ concentration and therefore, the pH, remain close to the initial level.

These effects may be summarized as follows:

$$CH_3COOH(aq) + H_2O(l) \rightleftharpoons H_3O^+(aq) + CH_3COO^-(aq)$$

OH^- added, equilibrium shifts to the right

$\longrightarrow$

H_3O^+ added, equilibrium shifts to the left

$\longleftarrow$

Buffer Capacity

Buffer capacity is a measure of the ability of a solution to resist large changes in pH when a strong acid or strong base is added. More specifically, buffer capacity is described as the amount of strong acid or strong base that a buffer can neutralize without significantly changing its pH. Buffering capacity against base is a function of the concentration of the weak acid (in this case, CH_3COOH). Buffering capacity against acid is dependent on the concentration of the anion of the salt, the conjugate base (CH_3COO^- in this example). Buffer solutions are often designed to have identical buffer capacity for both acids and bases. This is achieved when, in the above example, $[CH_3COO^-]/[CH_3COOH] = 1$. As an added bonus, making the $[CH_3COO^-]$ and $[CH_3COOH]$ as large as is practical ensures a high buffer capacity for both added acid and added base.

Animation
Effect of Addition of a Strong Acid and a Strong Base on a Buffer

QUESTION 8.15

Explain how the molar concentration of H_2CO_3 in the blood would change if the partial pressure of CO_2 in the lungs were to increase. (Refer to A Medical Connection: Control of Blood pH on page 240.)

QUESTION 8.16

Explain how the molar concentration of H_2CO_3 in the blood would change if the partial pressure of CO_2 in the lungs were to decrease. (Refer to A Medical Connection: Control of Blood pH on page 240.)

QUESTION 8.17

Explain how the molar concentration of hydronium ion in the blood would change under each of the conditions described in Questions 8.15 and 8.16.

QUESTION 8.18

Explain how the pH of blood would change under each of the conditions described in Questions 8.15 and 8.16.

A MEDICAL Connection | Control of Blood pH

A pH of 7.4 is maintained in blood partly by a carbonic acid–bicarbonate buffer system based on the following equilibrium:

$$H_2CO_3(aq) + H_2O(l) \rightleftharpoons H_3O^+(aq) + HCO_3^-(aq)$$

Carbonic acid   Bicarbonate ion
(weak acid)   (salt)

The regulation process based on LeChatelier's principle is similar to the acetic acid–sodium acetate buffer, which we have already discussed.

Red blood cells transport O_2, bound to hemoglobin, to the cells of body tissue. The metabolic waste product, CO_2, is picked up by the blood and delivered to the lungs.

The CO_2 in the blood also participates in the carbonic acid–bicarbonate buffer equilibrium. Carbon dioxide reacts with water in the blood to form carbonic acid:

$$CO_2(aq) + H_2O(l) \rightleftharpoons H_2CO_3(aq)$$

As a result, the buffer equilibrium becomes more complex:

$$CO_2(aq) + 2H_2O(l) \rightleftharpoons H_2CO_3(aq) + H_2O(l) \rightleftharpoons H_3O^+(aq) + HCO_3^-(aq)$$

Through this sequence of relationships, the concentration of CO_2 in the blood affects the blood pH.

Higher than normal CO_2 concentrations shift the above equilibrium to the right (LeChatelier's principle), increasing $[H_3O^+]$ and lowering the pH. The blood becomes too acidic, leading to numerous medical problems. A situation of high blood CO_2 levels and low pH is termed *acidosis*. Respiratory acidosis results from various diseases (emphysema, pneumonia) that restrict the breathing process, causing the buildup of waste CO_2 in the blood.

Lower than normal CO_2 levels, on the other hand, shift the equilibrium to the left, decreasing $[H_3O^+]$ and making the pH more basic. This condition is termed *alkalosis* (from "alkali," implying basic). Hyperventilation, or rapid breathing, is a common cause of respiratory alkalosis.

FOR FURTHER UNDERSTANDING

Carbon monoxide and carbon dioxide are both toxic in high concentrations, but for different reasons. Do literature (or Internet) research and describe how breathing carbon monoxide is harmful. Contrast this with carbon dioxide's toxic effects (described above).

SUMMARY

8.1 Acids and Bases

One of the earliest definitions of acids and bases is the *Arrhenius theory*. According to this theory, an *acid* dissociates to form hydrogen ions, H^+, and a *base* dissociates to form hydroxide ions, OH^-. The *Brønsted-Lowry theory* defines an acid as a proton (H^+) donor and a base as a proton acceptor.

Water, the solvent in many acid-base reactions, is *amphiprotic*. It has both acid and base properties.

The strength of acids and bases in water depends on their degree of dissociation, the extent to which they react with the solvent, water. Acids and bases are strong when the reaction with water is virtually 100% complete and weak when the reaction with water is much less than 100% complete.

Weak acids and weak bases dissolve in water principally in the molecular form. Only a small percentage of the molecules dissociate to form the *hydronium* ion or *hydroxide* ion.

Aqueous solutions of acids and bases are electrolytes. The dissociation of the acid or base produces ions, which conduct an electrical current. Strong acids and bases are strong electrolytes. Weak acids and bases are weak electrolytes.

Although pure water is virtually 100% molecular, a small number of water molecules do ionize. This process occurs by the transfer of a proton from one water molecule to another, producing a hydronium ion and a hydroxide ion. This process is the *autoionization*, or self-ionization, of water.

Pure water at room temperature has a hydronium ion concentration of 1.0×10^{-7} M. One hydroxide ion is produced for each hydronium ion. Therefore, the hydroxide ion concentration is also 1.0×10^{-7} M. The product of hydronium and hydroxide ion concentration (1.0×10^{-14}) is the *ion product for water*, K_w.

8.2 pH: A Measurement Scale for Acids and Bases

The *pH scale* correlates the hydronium ion concentration with a number, the pH, that serves as a useful indicator of the degree of acidity or basicity of a solution. The pH of a solution is defined as the negative logarithm of the molar concentration of the hydronium ion ($pH = -\log [H_3O^+]$).

8.3 Reactions Between Acids and Bases

The reaction of an acid with a base to produce a salt and water is referred to as *neutralization*. Neutralization requires equal numbers of moles of H_3O^+ and OH^- to produce a neutral solution (no excess acid or base). A neutralization reaction may be used to determine the concentration of an unknown acid or base solution. The technique of *titration* involves the addition of measured amounts of a *standard solution* (one whose concentration is known) from a *buret* to neutralize the second, unknown solution. The *equivalence point* is signaled by an *indicator*.

8.4 Acid–Base Buffers

A *buffer solution* contains components that enable the solution to resist large changes in pH when acids or bases are added. The basis of buffer action is an equilibrium between either a weak acid and its salt or a weak base and its salt.

A buffer solution follows LeChatelier's principle, which states that an equilibrium system, when stressed, will shift its equilibrium to alleviate that stress.

Buffering against base is a function of the concentration of the weak acid for an acidic buffer. Buffering against acid is dependent on the concentration of the anion of the salt.

KEY TERMS

acid (8.1)
amphiprotic (8.1)
Arrhenius theory (8.1)
autoionization (8.1)
base (8.1)
Brønsted-Lowry theory (8.1)
buffer capacity (8.4)
buffer solution (8.4)
buret (8.3)
conjugate acid (8.1)
conjugate acid-base pair (8.1)

conjugate base (8.1)
equivalence point (8.3)
hydronium ion (8.1)
indicator (8.3)
ion product for water (8.1)
neutralization (8.3)
pH scale (8.2)
polyprotic substance (8.3)
standard solution (8.3)
titration (8.3)

QUESTIONS AND PROBLEMS

Acids and Bases

Foundations

8.19 **a.** Define an acid according to the Arrhenius theory.
b. Define an acid according to the Brønsted-Lowry theory.

8.20 **a.** Define a base according to the Arrhenius theory.
b. Define a base according to the Brønsted-Lowry theory.

8.21 What are the essential differences between the Arrhenius and Brønsted-Lowry theories?

8.22 Why is ammonia described as a Brønsted-Lowry base and not an Arrhenius base?

Applications

8.23 Write an equation for the reaction of each of the following with water:
a. HNO_2
b. HCN

8.24 Write an equation for the reaction of each of the following with water:
a. HNO_3
b. $HCOOH$

8.25 Select the conjugate acid-base pairs for each reaction in Question 8.23.

8.26 Select the conjugate acid-base pairs for each reaction in Question 8.24.

8.27 Label each of the following as a strong or weak acid (consult Figure 8.1, if necessary):
a. H_2SO_3
b. H_2CO_3
c. H_3PO_4

8.28 Label each of the following as a strong or weak base (consult Figure 8.2, if necessary):
a. KOH
b. CN^-
c. SO_4^{2-}

8.29 Identify the conjugate acid-base pairs in each of the following chemical equations:
a. $NH_4^+ (aq) + CN^- (aq) \rightleftharpoons NH_3 (aq) + HCN(aq)$
b. $CO_3^{2-} (aq) + HCl (aq) \rightleftharpoons HCO_3^- (aq) + Cl^- (aq)$

8.30 Identify the conjugate acid-base pairs in each of the following chemical equations:
a. $HCOOH (aq) + NH_3 (aq) \rightleftharpoons HCOO^- (aq) + NH_4^+ (aq)$
b. $HCl (aq) + OH^- (aq) \rightleftharpoons H_2O (l) + Cl^- (aq)$

8.31 Distinguish between the terms acid-base *strength* and acid-base *concentration*.

8.32 Of the diagrams shown here, which one represents:
a. a concentrated strong acid
b. a dilute strong acid
c. a concentrated weak acid
d. a dilute weak acid

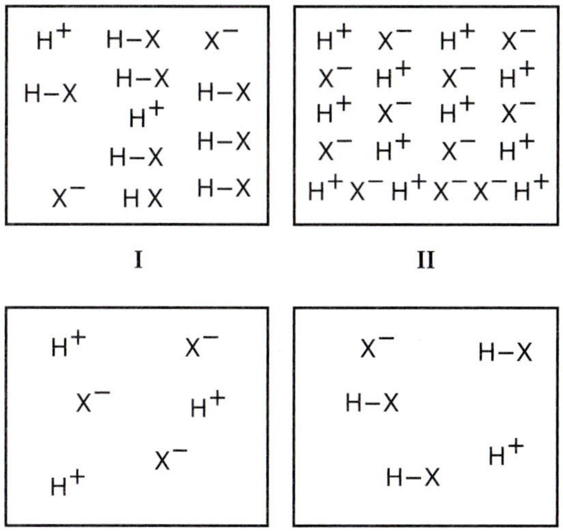

8.33 Classify each of the following as a Brønsted acid, Brønsted base, or both:
a. H_3O^+
b. OH^-
c. H_2O

8.34 Classify each of the following as a Brønsted acid, Brønsted base, or both:
   a. NH_4^+
   b. NH_3

8.35 Classify each of the following as a Brønsted acid, Brønsted base, or both:
   a. H_2CO_3
   b. HCO_3^-
   c. CO_3^{2-}

8.36 Classify each of the following as a Brønsted acid, Brønsted base, or both:
   a. H_2SO_4
   b. HSO_4^-
   c. SO_4^{2-}

8.37 Write the formula of the conjugate acid of CN^-.

8.38 Write the formula of the conjugate acid of Br^-.

8.39 Write the formula of the conjugate base of HI.

8.40 Write the formula of the conjugate base of HCOOH.

pH of Acid and Base Solutions

8.41 Calculate the $[H_3O^+]$ of an aqueous solution that is
   a. $1.0 \times 10^{-7} M$ in OH^-
   b. $1.0 \times 10^{-3} M$ in OH^-

8.42 Calculate the $[H_3O^+]$ of an aqueous solution that is
   a. $1.0 \times 10^{-9} M$ in OH^-
   b. $1.0 \times 10^{-5} M$ in OH^-

8.43 Label each solution in Problem 8.41 as acidic, basic, or neutral.

8.44 Label each solution in Problem 8.42 as acidic, basic, or neutral.

8.45 Calculate the pH of a solution that has
   a. $[H_3O^+] = 1.0 \times 10^{-7}$
   b. $[OH^-] = 1.0 \times 10^{-9}$

8.46 Calculate the pH of a solution that has
   a. $[H_3O^+] = 1.0 \times 10^{-10}$
   b. $[OH^-] = 1.0 \times 10^{-5}$

8.47 Calculate *both* $[H_3O^+]$ and $[OH^-]$ for a solution that is
   a. pH = 1.00
   b. pH = 9.00

8.48 Calculate *both* $[H_3O^+]$ and $[OH^-]$ for a solution that is
   a. pH = 5.00
   b. pH = 7.20

8.49 Calculate *both* $[H_3O^+]$ and $[OH^-]$ for a solution that is
   a. pH = 1.30
   b. pH = 9.70

8.50 Calculate *both* $[H_3O^+]$ and $[OH^-]$ for a solution that is
   a. pH = 5.50
   b. pH = 7.00

8.51 What is a neutralization reaction?

8.52 Describe the purpose of a titration.

8.53 The pH of urine may vary between 4.5 and 8.2. Determine the H_3O^+ concentration and OH^- concentration if the measured pH is
   a. 6.00
   b. 5.20
   c. 7.80

8.54 The hydronium ion concentration in blood of three different patients was

Patient	$[H_3O^+]$
A	5.0×10^{-8}
B	3.1×10^{-8}
C	3.2×10^{-8}

What is the pH of each patient's blood? If the normal range is 7.30–7.50, which, if any, of these patients have an abnormal blood pH?

8.55 Determine how many times more acidic a solution is at
   a. pH 2 relative to pH 4
   b. pH 7 relative to pH 11
   c. pH 2 relative to pH 12

8.56 Determine how many times more basic a solution is at
   a. pH 6 relative to pH 4
   b. pH 10 relative to pH 9
   c. pH 11 relative to pH 6

8.57 What is the H_3O^+ concentration of a solution with a pH of
   a. 5.0
   b. 12.0
   c. 5.5

8.58 What is the OH^- concentration of each solution in Question 8.57?

8.59 Calculate the pH of a solution with a H_3O^+ concentration of
   a. $1.0 \times 10^{-6} M$
   b. $1.0 \times 10^{-8} M$
   c. $5.6 \times 10^{-4} M$

8.60 What is the OH^- concentration of each solution in Question 8.59?

8.61 Calculate the pH of a solution that has $[H_3O^+] = 7.5 \times 10^{-4} M$.

8.62 Calculate the pH of a solution that has $[H_3O^+] = 6.6 \times 10^{-5} M$.

8.63 Calculate the pH of a solution that has $[OH^-] = 5.5 \times 10^{-4} M$.

8.64 Calculate the pH of a solution that has $[OH^-] = 6.7 \times 10^{-9} M$.

Buffer Solutions

Foundations

8.65 Which of the following are capable of forming a buffer solution?
   a. NH_3 and NH_4Cl
   b. HNO_3 and KNO_3

8.66 Which of the following are capable of forming a buffer solution?
   a. HBr and $MgCl_2$
   b. H_2CO_3 and $NaHCO_3$

8.67 Define
   a. buffer solution
   b. acidosis (refer to A Medical Connection: Control of Blood pH on page 240)

8.68 Define
   a. alkalosis (refer to A Medical Connection: Control of Blood pH on page 240)
   b. standard solution

Applications

8.69 For the equilibrium situation involving acetic acid,

$$CH_3COOH(aq) + H_2O(l) \rightleftharpoons CH_3COO^-(aq) + H_3O^+(aq)$$

explain the equilibrium shift occurring for the following changes:
   a. A strong acid is added to the solution.
   b. The solution is diluted with water.

8.70 For the equilibrium situation involving acetic acid,

$$CH_3COOH(aq) + H_2O(l) \rightleftharpoons CH_3COO^-(aq) + H_3O^+(aq)$$

explain the equilibrium shift occurring for the following changes:
   a. A strong base is added to the solution.
   b. More acetic acid is added to the solution.

FOR FURTHER UNDERSTANDING

1. Acid rain is a threat to our environment because it can increase the concentration of toxic metal ions, such as Cd^{2+} and Cr^{3+}, in rivers and streams. If cadmium and chromium are present in sediment as $Cd(OH)_2$ and $Cr(OH)_3$, write reactions that demonstrate the effect of acid rain. Use the library or Internet to

find the properties of cadmium and chromium responsible for their environmental impact.

2. Aluminum carbonate is more soluble in acidic solution, forming aluminum cations. Write a reaction (or series of reactions) that explains this observation.

3. Carbon dioxide reacts with the hydroxide ion to produce the bicarbonate anion. Write the Lewis dot structures for each reactant and product. Label each as a Brønsted acid or base. Explain the reaction using the Brønsted theory. Why would the Arrhenius theory provide an inadequate description of this reaction?

4. Maalox is an antacid composed of $Mg(OH)_2$ and $Al(OH)_3$. Explain the origin of the trade name Maalox. Write chemical reactions that demonstrate the antacid activity of Maalox.

5. Acid rain has been described as a regional problem, whereas the greenhouse effect is a global problem. Do you agree with this statement? Why or why not?

9

OUTLINE

LEARNING GOALS

1　Enumerate the characteristics of alpha, beta, and gamma radiation.

2　Write balanced equations for nuclear processes.

3　Calculate the amount of radioactive substance remaining after a specified period of time has elapsed.

4　Describe how nuclear energy can generate electricity: fission, fusion, and the breeder reactor.

5　Cite examples of the use of radioactive isotopes in medicine.

6　Describe the use of ionizing radiation in cancer therapy.

7　Discuss the preparation and use of radioisotopes in diagnostic imaging studies.

8　Explain the difference between natural and artificial radioactivity.

9　Describe the characteristics of radioactive materials that relate to radiation exposure and safety.

10　Be familiar with common techniques for the detection of radioactivity.

11　Know the common units of radiation intensity: the curie, roentgen, rad, and rem.

The Nucleus and Radioactivity

Our earlier discussion of the atom and atomic structure revealed a nucleus containing protons and neutrons surrounded by electrons. Until now, we have treated the nucleus as simply a region of positive charge in the center of the atom. The focus of our interest has been the electrons and their arrangement around the nucleus.

However, the behavior of nuclei may have as great an effect on our everyday lives as any of the thousands of synthetic compounds developed over the past several decades. Examples of nuclear technology range from everyday items (smoke detectors) to sophisticated instruments for medical diagnosis and treatment and electrical power generation (nuclear power plants).

Beginning in 1896 with Becquerel's discovery of radiation emitted from uranium ore, the technology arising from this and related findings has produced both risks and benefits. Although early discoveries of radioactivity and its properties expanded our fundamental knowledge and brought fame to the investigators, it was not accomplished without a price. Several early investigators died prematurely of cancer and other diseases caused by the radiation they studied.

Even today, the existence of nuclear energy and its associated technology is a mixed blessing. On one side, the horrors of Nagasaki and Hiroshima, the fear of nuclear proliferation, and potential contamination of populated areas resulting from the peaceful application of nuclear energy are critical problems facing society. Conversely, hundreds of thousands of lives have been saved because of the early detection of disease such as cancer by diagnosis based on the interaction of radiation and the body and the cure of cancer using techniques such as cobalt-60 treatment. Furthermore, nuclear energy is an alternative energy source, providing an opportunity for us to compensate for the depletion of oil reserves.

Archaeology is one area of study that has progressed due to the availability of radiocarbon dating, a topic that you will encounter in this chapter. Can you name other applications of this useful technology?

9.1 Natural Radioactivity

Animation
Radioactive Decay

Be careful not to confuse the mass number (a simple count of the neutrons and protons) with the atomic mass, which includes the contribution of electrons and is a true *mass* figure.

Isotopes are introduced in Section 2.1.

Radioactivity is the process by which some atoms emit energy and particles. The energy and particles are termed *radiation*. Nuclear radiation occurs as a result of an alteration in nuclear composition or structure. This process occurs in a nucleus that is unstable and hence radioactive. Radioactivity is a nuclear event: *matter and energy released during this process come from the nucleus.*

We shall designate the nucleus using *nuclear symbols,* analogous to the *atomic symbols* that were introduced in Section 2.1. The nuclear symbols consist of the *symbol* of the element, the *atomic number* (the number of protons in the nucleus), and the *mass number,* which is defined as the sum of neutrons and protons in the nucleus.

With the use of nuclear symbols, the fluorine nucleus is represented as

$$\text{Mass number} \longrightarrow {}^{19}_{9}\text{F} \longleftarrow \text{Symbol of the element}$$
$$\text{Atomic number} \longrightarrow$$
$$\text{(or nuclear charge)}$$

This symbol is equivalent to writing *fluorine-19.* This alternative representation is frequently used to denote specific isotopes of elements.

Not all nuclei are unstable. Only unstable nuclei undergo change and produce radioactivity, the process of radioactive decay. Recall that different atoms of the same element having different masses exist as *isotopes.* One isotope of an element may be radioactive, whereas others of the same element may be quite stable. It is important to distinguish between the terms *isotope* and *nuclide.* The term *isotope* refers to any atoms that have the same atomic number but different mass number. The term **nuclide** refers to any atom characterized by an atomic number and a mass number.

Many elements in the periodic table occur in nature as mixtures of isotopes. Two common examples include carbon (Figure 9.1),

$$^{12}_{6}\text{C} \qquad\qquad ^{13}_{6}\text{C} \qquad\qquad ^{14}_{6}\text{C}$$

Carbon-12   Carbon-13   Carbon-14

and hydrogen,

$$^{1}_{1}\text{H} \qquad\qquad ^{2}_{1}\text{H} \qquad\qquad ^{3}_{1}\text{H}$$

Hydrogen-1   Hydrogen-2   Hydrogen-3

Protium   Deuterium   Tritium
              (symbol D)   (symbol T)

Protium is a stable isotope and makes up more than 99.9% of naturally occurring hydrogen. Deuterium (D) can be isolated from hydrogen; it can form compounds such as "heavy water," D_2O. Heavy water is a potential source of deuterium for fusion processes. Tritium (T) is rare and unstable, hence radioactive.

Figure 9.1

Three isotopes of carbon. Each nucleus contains the same number of protons. Only the number of neutrons is different; hence, each isotope has a different mass.

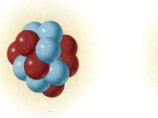

Carbon-12 has
6 protons and
6 neutrons

Carbon-13 has
6 protons and
7 neutrons

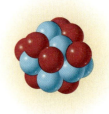

Carbon-14 has
6 protons and
8 neutrons

In writing the symbols for a nuclear process, it is essential to indicate the particular isotope involved. This is why the mass number and atomic number are used. These values tell us the number of neutrons in the species, hence the isotope's identity.

Three types of natural radiation emitted by unstable nuclei are *alpha particles*, *beta particles*, and *gamma rays*.

Alpha Particles

An **alpha particle** (α) contains two protons and two neutrons. An alpha particle is identical to the nucleus of the helium atom (He) or a *helium ion* (He^{2+}), which also contains two protons (atomic number = 2) and two neutrons (mass number − atomic number = 2). Having no electrons to counterbalance the nuclear charge, the alpha particle may be symbolized as

$$_2^4He^{2+} \quad \text{or} \quad _2^4H \quad \text{or} \quad \alpha$$

Alpha particles have a relatively large mass compared to other nuclear particles. Consequently, alpha particles emitted by radioisotopes are relatively slow-moving particles (approximately 10% of the speed of light), and they are stopped by barriers as thin as a few pages of this book.

Beta Particles and Positrons

The **beta particle** (β), in contrast, is a fast-moving electron traveling at approximately 90% of the speed of light as it leaves the nucleus. It is formed in the nucleus by the conversion of a neutron into a proton. The beta particle is represented as

$$_{-1}^0e \quad \text{or} \quad _{-1}^0\beta \quad \text{or} \quad \beta$$

The subscript −1 is written in the same position as the atomic number and, like the atomic number (number of protons), indicates the charge of the particle.

Beta particles are smaller and faster than alpha particles. They are more penetrating and are stopped only by more dense materials such as wood, metal, or several layers of clothing.

A positron has the same mass as a beta particle but carries a positive charge and is symbolized as $_{+1}^0e$. Positrons are produced by the conversion of a proton to a neutron (the proton, in effect, loses its positive charge).

Gamma Rays

Gamma rays (γ) are the most energetic form of radiation; gamma rays are a form of light energy and result from nuclear processes. In contrast, alpha radiation and beta radiation are matter. Because gamma radiation has no protons, neutrons, or electrons, the symbol for a gamma ray is simply

$$\gamma$$

Gamma radiation is highly energetic and is the most penetrating form of nuclear radiation. Barriers of lead, concrete, or, more often, a combination of the two are required for protection from this type of radiation.

Properties of Alpha, Beta, and Gamma Radiation

Important properties of alpha, beta, and gamma radiation are summarized in Table 9.1.

Alpha, beta, and gamma radiation are collectively termed *ionizing radiation*. **Ionizing radiation** produces a trail of ions throughout the material that it penetrates. The ionization process changes the chemical composition of the material. When the material is living tissue, radiation-induced illness may result (Section 9.6).

Alpha, beta, and gamma radiation have widespread use in the field of medicine.

Animation
Types of Radioactive Decay

Learning Goal
1 Enumerate the characteristics of alpha, beta, and gamma radiation.

Other radiation particles, such as neutrinos and deuterons, will not be discussed here.

Animation
Alpha, Beta, and Gamma Emission

Learning Goal
1 Enumerate the characteristics of alpha, beta, and gamma radiation.

A LIFESTYLE Connection | An Extraordinary Woman in Science

The path to a successful career in science, or any other field for that matter, is seldom smooth or straight. That was certainly true for Madame Marie Sklodowska Curie. Her lifelong ambition was to raise a family and do something interesting for a career. This was a lofty goal for a nineteenth-century woman.

The political climate in Poland, coupled with the prevailing attitudes toward women and careers, especially careers in science, certainly did not make it any easier for Mme. Curie. To support herself and her sister, she toiled at menial jobs until moving to Paris to resume her studies.

It was in Paris that she met her future husband and fellow researcher, Pierre Curie. Working with crude equipment in a laboratory that was primitive, even by the standards of the time, she and Pierre made a most revolutionary discovery only two years after Henri Becquerel discovered radioactivity. Radioactivity, the emission of energy from certain substances, was released from *inside* the atom and was independent of the molecular form of the substance. The absolute proof of this assertion came only after the Curies processed over one *ton* of a material (pitchblende) to isolate less than a gram of pure radium. The difficult conditions under which this feat was accomplished are perhaps best stated by Sharon Bertsch McGrayne in her book *Nobel Prize Women in Science* (Birch Lane Press, New York, p. 23):

> The only space large enough at the school was an abandoned dissection shed. The shack was stifling hot in summer and freez-ing cold in winter. It had no ventilation system for removing poisonous fumes, and its roof leaked. A chemist accustomed to Germany's modern laboratories called it "a cross between a stable and a potato cellar and, if I had not seen the work table with the chemical apparatus, I would have thought it a practical joke." This ramshackle shed became the symbol of the Marie Curie legend.

The pale green glow emanating from the radium was beautiful to behold. Mme. Curie would go to the shed in the middle of the night to bask in the light of her accomplishment. She did not realize that this wonderful accomplishment would, in time, be responsible for her death. Ironically, the field of medicine has been a major beneficiary of advances in nuclear and radio-chemistry, despite the toxic properties of those same radioactive materials.

Mme. Curie received not one, but two Nobel Prizes, one in physics and one in chemistry. She was the first woman in France to earn the rank of professor.

FOR FURTHER UNDERSTANDING

What is the "pale green glow" that Mme. Curie observed?

Compare and contrast opportunities for women in science today vs. the late nineteenth century.

The penetrating power of alpha radiation is very low. Damage to internal organs from this form of radiation is negligible except when an alpha particle emitter is actually ingested. Beta particles have much higher velocities than alpha particles; still, they have limited penetrating power. They cause skin and eye damage and, to a lesser extent, damage to internal organs. Shielding is required when working with beta emitters. Pregnant women must take special precautions. The great penetrating power and high energy of gamma radiation make it particularly difficult to shield. Hence, it can damage internal organs.

Anyone working with any type of radiation must take precautions. Radiation safety is required, monitored, and enforced in the United States under provisions of the Occupational Safety and Health Act (OSHA).

Animation
Alpha, Beta, and Gamma Rays

TABLE 9.1 A Summary of the Major Properties of Alpha, Beta, and Gamma Radiation

Name and Symbol	Identity	Charge	Mass (amu)	Velocity	Penetration
Alpha (α)	Helium nucleus	+2	4.0026	5–10% of the speed of light	Low
Beta (β)	Electron	−1	0.000549	Up to 90% of the speed of light	Medium
Gamma (γ)	Radiant energy	0	0	Speed of light	High

What part of the atom is the source of α and β radiation?

What part of the atom is the source of γ radiation?

9.2 Writing a Balanced Nuclear Equation

Nuclear equations represent nuclear change in much the same way as chemical equations represent chemical change.

A **nuclear equation** can be used to represent the process of radioactive decay. In radioactive decay, a *nuclide* breaks down, producing a *new nuclide, smaller particles, and/or energy.* The concept of mass balance, required when writing chemical equations, is also essential for nuclear equations. When writing a balanced equation, remember that

- the total mass on each side of the reaction arrow must be identical, and
- the sum of the atomic numbers on each side of the reaction arrow must be identical.

2 Learning Goal
Write balanced equations for nuclear processes.

Alpha Decay

Consider the decay of one isotope of uranium, $^{234}_{92}U$, into thorium and an alpha particle. Because an alpha particle is lost in this process, this decay is called *alpha decay.*
Examine the balanced equation for this nuclear reaction:

$$^{238}_{92}U \longrightarrow {}^{234}_{90}Th + {}^{4}_{2}He$$

Uranium-238   Thorium-234   Helium-4

The sum of the mass numbers on the right ($234 + 4 = 238$) is equal to the mass number on the left. The atomic numbers on the right ($90 + 2 = 92$) are equal to the atomic number on the left.

Beta Decay

Beta decay is illustrated by the decay of one of the less abundant nitrogen isotopes, $^{16}_{7}N$. Upon decomposition, nitrogen-16 produces oxygen-16 and a beta particle. Conceptually, a neutron = proton + electron. In beta decay, one neutron in nitrogen-16 is converted to a proton and the electron, the beta particle, is released. The reaction is represented as

$$^{16}_{7}N \longrightarrow {}^{16}_{8}O + {}^{0}_{-1}e$$

or

$$^{16}_{7}N \longrightarrow {}^{16}_{8}O + \beta$$

Note that the mass number of the beta particle is zero, because the electron includes no protons or neutrons. Sixteen nuclear particles are accounted for on both sides of the reaction arrow. Note also that the product nuclide has the same mass number as the parent nuclide but the atomic number has *increased* by one unit.

The atomic number on the left ($+7$) is counterbalanced by [$8 + (-1)$] or ($+7$) on the right. Therefore the equation is correctly balanced.

Positron Emission

The decay of carbon-11 to a stable isotope, boron-11, is one example of positron emission.

$$^{11}_{6}\text{C} \longrightarrow {}^{11}_{5}\text{B} + {}^{0}_{1}\text{e}$$

or

$$^{11}_{6}\text{C} \longrightarrow {}^{11}_{5}\text{B} + {}^{0}_{1}\beta$$

A **positron** has the same mass as an electron, or beta particle, but opposite (+) charge. In contrast to beta emission, the product nuclide has the same mass number as the parent nuclide, but the atomic number has *decreased* by one unit.

The atomic number on the left (+6) is counterbalanced by [5 + (+1)] or (+6) on the right. Therefore, the equation is correctly balanced.

Gamma Production

If *gamma radiation* were the only product of nuclear decay, there would be no measurable change in the mass or identity of the radioactive nuclei because the gamma emitter has simply gone to a lower energy state. An example of an isotope that decays in this way is technetium-99m. It is described as a **metastable isotope,** meaning that it is unstable and increases its stability through gamma decay without change in the mass or charge of the isotope. The letter *m* is used to denote a metastable isotope. The decay equation for $^{99m}_{43}\text{Tc}$ is

$$^{99m}_{43}\text{Tc} \longrightarrow {}^{99}_{43}\text{Tc} + \gamma$$

More often, gamma radiation is produced along with other products. For example, iodine-131 decays as follows:

$$^{131}_{53}\text{I} \longrightarrow {}^{131}_{54}\text{Xe} + {}^{0}_{-1}\beta + \gamma$$

| Iodine-131 | Xenon-131 | Beta particle | Gamma ray |

This reaction may also be represented as

$$^{131}_{53}\text{I} \longrightarrow {}^{131}_{54}\text{Xe} + {}^{0}_{-1}\text{e} + \gamma$$

An isotope of xenon, a beta particle, and gamma radiation are produced.

Predicting Products of Nuclear Decay

It is possible to use a nuclear equation to predict one of the products of a nuclear reaction if the others are known. Consider the following example, in which we represent the unknown product as ?:

$$^{40}_{19}\text{K} \longrightarrow ? + {}^{0}_{-1}\text{e}$$

Step 1. The mass number of this isotope of potassium is 40. Therefore the sum of the mass number of the products must also be 40, and ? must have a mass number of 40.

Step 2. Likewise, the atomic number on the left is 19, and the sum of the unknown atomic number plus the charge of the beta particle (−1) must equal 19.

Step 3. The unknown atomic number must be 20, because [20 + (−1) = 19]. The unknown is

$$^{40}_{20}?$$

If we consult the periodic table, the element that has atomic number 20 is calcium; therefore ? = $^{40}_{20}\text{Ca}$.

2 Learning Goal

Write balanced equations for nuclear processes.

EXAMPLE 9.1 Predicting the Products of Radioactive Decay

Determine the identity of the unknown product of the alpha decay of curium-245:

$$^{245}_{96}\text{Cm} \longrightarrow {}^{4}_{2}\text{He} + ?$$

SOLUTION

Step 1. The mass number of the curium isotope is 245. Therefore the sum of the mass numbers of the products must also be 245, and ? must have a mass number of 241.

Step 2. Likewise, the atomic number on the left is 96, and the sum of the unknown atomic number plus the atomic number of the alpha particle (2) must equal 96.

Step 3. The unknown atomic number must be 94, because [94 + 2 = 96]. The unknown is

$$^{241}_{94}?$$

Referring to the periodic table, we find that the element that has atomic number 94 is plutonium; therefore ? = $^{241}_{94}\text{Pu}$.

QUESTION 9.3

Complete each of the following nuclear equations:

a. $^{85}_{36}\text{Kr} \longrightarrow ? + {}^{0}_{-1}\text{e}$
b. $? \longrightarrow {}^{4}_{2}\text{He} + \; + {}^{222}_{86}\text{Rn}$

QUESTION 9.4

Complete each of the following nuclear equations:

a. $^{239}_{92}\text{U} \longrightarrow ? + {}^{0}_{-1}\text{e}$
b. $^{11}_{5}\text{B} \longrightarrow {}^{7}_{3}\text{Li} + ?$

9.3 Properties of Radioisotopes

Why are some isotopes radioactive, but others are not? Do all radioactive isotopes decay at the same rate? Are all radioactive materials equally hazardous? We address these and other questions in this section.

Nuclear Structure and Stability

A measure of nuclear stability is the **binding energy** of the nucleus. The binding energy of the nucleus is the energy required to break up a nucleus into its component protons and neutrons. This must be very large because identically charged protons in the nucleus exert extreme repulsive forces on one another. These forces must be overcome if the nucleus is to be stable. When a nuclide decays, some energy is released because the products are more stable than the parent nuclide. This released energy is the source of the high-energy radiation emitted and the basis for all nuclear technology.

Why are some isotopes more stable than others? The answer to this question is not completely clear. Evidence obtained so far points to several important factors that describe stable nuclei:

- Nuclear stability correlates with the ratio of neutrons to protons in the isotope. For example, for light atoms, a neutron:proton ratio of 1 characterizes a stable atom.
- Nuclei with large numbers of protons (84 or more) tend to be unstable.
- Naturally occurring isotopes containing 2, 8, 20, 50, 82, or 126 protons or neutrons are stable. These *magic numbers* seem to indicate the presence of energy levels in the nucleus, analogous to electronic energy levels in the atom.
- Isotopes with even numbers of protons or neutrons are generally more stable than those with odd numbers of protons or neutrons.
- All isotopes (except hydrogen-1) with more protons than neutrons are unstable. However, the reverse is not true.

Half-Life

The **half-life** ($t_{1/2}$) is the time required for one-half of a given quantity of a substance to undergo change. Not all radioactive isotopes decay at the same rate. The rate of nuclear decay is generally represented in terms of the half-life of the isotope. Each isotope has its own characteristic half-life that may be as short as a few millionths of a second or as long as billions of years. Half-lives of some naturally occurring and synthetic isotopes are given in Table 9.2.

The stability of an isotope is indicated by the isotope's half-life. Isotopes with short half-lives decay rapidly; they are very unstable. This is not meant to imply that substances with long half-lives are less hazardous. Often, just the reverse is true.

Imagine that we begin with 100 mg of a radioactive isotope that has a half-life of 24 hours. After one half-life, or 24 hours, 1/2 of 100 mg will have decayed to other products, and 50 mg remain. After two half-lives (48 hours), 1/2 of the remaining material has decayed, leaving 25 mg, and so forth:

$$100 \text{ mg} \xrightarrow[\text{Half-life} \atop (24 \text{ h})]{\text{One}} 50 \text{ mg} \xrightarrow[\text{Half-life} \atop (48 \text{ h total})]{\text{A Second}} 25 \text{ mg} \longrightarrow \text{etc.}$$

3 Learning Goal

Calculate the amount of radioactive substance remaining after a specified period of time has elapsed.

Animation

Radioactive Half-Life

Refer to the discussion of radiation exposure and safety in Sections 9.6 and 9.7.

TABLE 9.2 Half-Lives of Selected Radioisotopes

Name	Symbol	Half-Life
Carbon-14	$^{14}_{6}C$	5730 years
Cobalt-60	$^{60}_{27}Co$	5.3 years
Hydrogen-3	$^{3}_{1}H$	12.3 years
Iodine-131	$^{131}_{53}I$	8.1 days
Iron-59	$^{59}_{26}Fe$	45 days
Molybdenum-99	$^{99}_{42}Mo$	67 hours
Sodium-24	$^{24}_{11}Na$	15 hours
Strontium-90	$^{90}_{38}Sr$	28 years
Technetium-99m	$^{99m}_{43}Tc$	6 hours
Uranium-235	$^{235}_{92}U$	710 million years

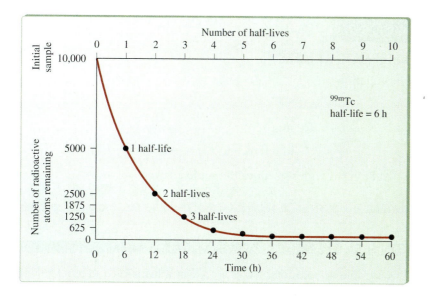

Figure 9.2

The decay curve for the medically useful radioisotope technetium-99m. Note that the number of radioactive atoms remaining—hence the radioactivity—approaches zero.

Decay of a radioisotope that has a reasonably short $t_{1/2}$ is experimentally determined by following its activity as a function of time. Graphing the results produces a radioactive decay curve as shown in Figure 9.2.

The mass of any radioactive substance remaining after a period may be calculated with knowledge of the initial mass and the half-life of the isotope, following the scheme just outlined. The general equation for this process is:

$$m_f = m_i(0.5)^n$$

where m_f = final or remaining mass
        m_i = initial mass
          n = number of half-lives

EXAMPLE 9.2   Predicting the Extent of Radioactive Decay

A 50.0-mg supply of iodine-131, used in hospitals in the treatment of hyperthyroidism, was stored for 32.4 days. If the half-life of iodine-131 is 8.1 days, how many milligrams remain?

SOLUTION

First calculate n, the number of half-lives elapsed using the half-life as a conversion factor:

$$n = 32.4 \text{ days} \times \frac{1 \text{ half-life}}{8.1 \text{ days}} = 4.0 \text{ half-lives}$$

Then calculate the amount remaining:

first   second   third   fourth

50.0 mg $\longrightarrow$ 25.0 mg $\longrightarrow$ 12.5 mg $\longrightarrow$ 6.25 mg $\longrightarrow$ 3.13 mg

half-life   half-life   half-life   half-life

Hence, 3.13 mg of iodine-131 remain after 32.4 days.

Continued—

3   **Learning Goal**

Calculate the amount of radioactive substance remaining after a specified period of time has elapsed.

EXAMPLE 9.2 —*Continued*

AN ALTERNATE STRATEGY

Use the equation

$$m_f = m_i(0.5)^n$$

$$m_f = 50.0 \text{ mg}(0.5)^4$$

$$m_f = 3.13 \text{ mg of iodine-131 remain after 32.4 days.}$$

Note that both strategies produce the same answer.

QUESTION 9.5

A 100.0-ng sample of sodium-24 was stored in a lead-lined cabinet for 2.5 days. How much sodium-24 remained? See Table 9.2 for the half-life of sodium-24.

QUESTION 9.6

If 10 ng of technetium-99m are administered to a patient, how much will remain one day later, assuming that no technetium has been eliminated by any other process? See Table 9.2 for the half-life of technetium-99m.

9.4 Nuclear Power

Energy Production

4 Learning Goal

Describe how nuclear energy can generate electricity: fission, fusion, and the breeder reactor.

Einstein predicted that a small amount of nuclear mass corresponds to a very large amount of energy that is released when the nucleus breaks apart. Einstein's equation is

$$E = mc^2$$

in which

$$E = \text{energy}$$

$$m = \text{mass}$$

$$c = \text{speed of light}$$

This kinetic energy, when rapidly released, is the basis for the greatest instruments of destruction developed by humankind, nuclear bombs. However, when heat energy is released in a controlled fashion, as in a nuclear power plant, the heat energy converts liquid water into steam. The steam, in turn, drives an electrical generator, producing electricity.

Nuclear Fission

Animation

Nuclear Fission

Fission (splitting) occurs when a heavy nuclear particle is split into smaller nuclei by a smaller nuclear particle (such as a neutron). This splitting process is accompanied by the release of large amounts of energy.

A nuclear power plant uses a fissionable material (capable of undergoing fission), such as uranium-235, as fuel. The energy released by the fission process in the nuclear core heats water in an adjoining chamber, producing steam. The high pressure of the steam drives a turbine, which converts this heat energy into

AN ENVIRONMENTAL
Perspective

Nuclear Waste Disposal

Nuclear waste arises from a variety of sources. A major source is the spent fuel from nuclear power plants. Medical laboratories generate significant amounts of low-level waste from tracers and therapy. Even household items with limited lifetimes, such as certain types of smoke detectors, use a tiny amount of radioactive material.

Virtually everyone is aware, through television and newspapers, of the problems of solid waste (nonnuclear) disposal that our society faces. For the most part, this material will degrade in some reasonable amount of time. Still, we are disposing of trash and garbage at a rate that far exceeds nature's ability to recycle it.

Now imagine the problem with nuclear waste. We cannot alter the rate at which it decays. This is defined by the half-life. We can't heat it, stir it, or add a catalyst to speed up the process as we can with chemical reactions. Furthermore, the half-lives of many nuclear waste products are very long: plutonium, for example, has a half-life in excess of 24,000 years. Ten half-lives represents the approximate time required for the radioactivity of a substance to reach background levels. So we are talking about a *very* long storage time.

Where on earth can something so very hazardous be contained and stored with reasonable assurance that it will lie undisturbed for a quarter of a million years? Perhaps this is a rhetorical question. Scientists, engineers, and politicians have debated this question for almost fifty years. As yet, no permanent disposal site has been agreed upon. Most agree that the best solution is burial in a stable rock formation, but there is no firm agreement on the location. Fear of earthquakes, which may release large quantities of radioactive materials into the underground water system, is the most serious consideration. Such a disaster could render large sections of the country unfit for habitation.

Many argue for the continuation of temporary storage sites with the hope that the progress of science and technology will, in the years ahead, provide a safer and more satisfactory long-term solution.

A photograph of the earth, taken from the moon, clearly illustrates the limits of resources and the limits to waste disposal.

The nuclear waste problem, important for its own sake, also affects the development of future societal uses of nuclear chemistry. Before we can fully enjoy its benefits, we must learn to use and dispose of it safely.

FOR FURTHER UNDERSTANDING

Summarize the major arguments supporting expanded use of nuclear power for electrical energy.

Enumerate the characteristics of an "ideal" solution to the nuclear waste problem.

---

electricity using an electric power generator. The energy transformation may be summarized as follows:

$$\text{nuclear energy} \longrightarrow \text{heat energy} \longrightarrow \text{mechanical energy} \longrightarrow \text{electrical energy}$$

| Nuclear reactor | Steam | Turbine | Electricity |

The fission reaction, once initiated, is self-perpetuating. For example, neutrons are used to initiate the reaction:

$$_{0}^{1}\text{n} + _{92}^{235}\text{U} \longrightarrow _{92}^{236}\text{U} \longrightarrow _{36}^{92}\text{Kr} + _{56}^{141}\text{Ba} + 3_{0}^{1}\text{n} + \text{energy}$$

| Fuel | Unstable | Products of reaction |

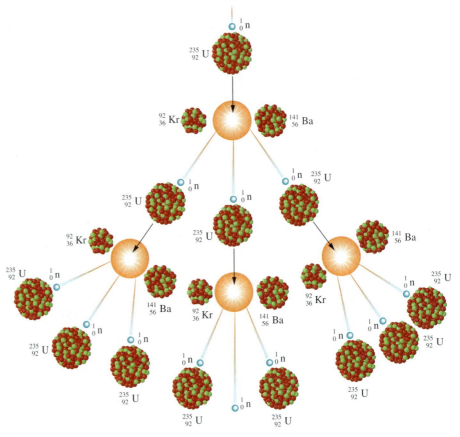

Figure 9.3

The fission of uranium-235 producing a chain reaction. Note that the number of available neutrons, which "trigger" the decomposition of the fissionable nuclei to release energy, increases at each step in the "chain." In this way, the reaction builds in intensity. Control rods stabilize (or limit) the extent of the chain reaction to a safe level.

Note that three neutrons are released as product for each single reacting neutron. Each of the three neutrons produced is available to initiate another fission process. Nine neutrons are released from this process. These, in turn, react with other nuclei. The fission process continues and intensifies, producing very large amounts of energy (Figure 9.3). This process of intensification is referred to as a **chain reaction.**

To maintain control over the process and to prevent dangerous overheating, rods fabricated from cadmium or boron are inserted into the core. These rods, which are controlled by the reactor's main operating system, absorb free neutrons as needed, thereby moderating the reaction.

A nuclear fission reactor may be represented as a series of energy transfer zones, as depicted in Figure 9.4. A view of the core of a fission reactor is shown in Figure 9.5.

Animation
Nuclear Chain Reaction

Nuclear Fusion

Fusion (meaning *to join together*) results from the combination of two small nuclei to form a larger nucleus with a concurrent release of large amounts of energy. The best example of a fusion reactor is the sun. Continuous fusion processes furnish our solar system with light and heat.

An example of a fusion reaction is the combination of two isotopes of hydrogen, deuterium ($^{2}_{1}H$) and tritium ($^{3}_{1}H$), to produce helium, a neutron, and energy:

$$^{2}_{1}H + ^{3}_{1}H \longrightarrow ^{4}_{2}He + ^{1}_{0}n + \text{energy}$$

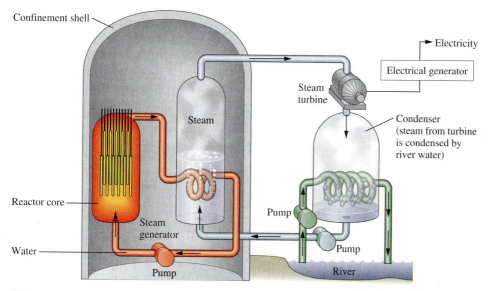

Figure 9.4

A representation of the "energy zones" of a nuclear reactor. Heat produced by the reactor core is carried by water in a second zone to a boiler. Water in the boiler (third zone) is converted to steam, which drives a turbine to convert heat energy to electrical energy. The isolation of these zones from each other allows heat energy transfer without actual physical mixing. This minimizes the transport of radioactive material into the environment.

Although fusion is capable of producing tremendous amounts of energy, no commercially successful fusion plant exists in the United States. Safety concerns relating to problems of containment of the reaction, resulting directly from the technological problems associated with containing high temperatures (millions of degrees) and pressures required to sustain a fusion process, have slowed the development of fusion reactors.

Breeder Reactors

A **breeder reactor** is a variation of a fission reactor that literally manufactures its own fuel. A perceived shortage of fissionable isotopes makes the breeder an attractive alternative to conventional fission reactors. A breeder reactor uses $^{238}_{92}U$, which is abundant but nonfissionable. In a series of steps, the uranium-238 is converted to plutonium-239, which *is* fissionable and undergoes a fission chain reaction, producing energy. The attractiveness of a reactor that makes its own fuel from abundant starting materials is offset by the high cost of the system, potential environmental damage, and fear of plutonium proliferation. Plutonium can be readily used to manufacture nuclear bombs. Currently, only France and Japan operate breeder reactors for electrical power generation.

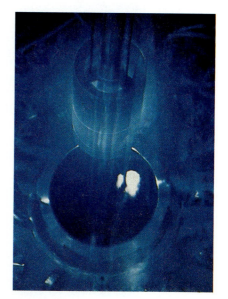

Figure 9.5

The core of a nuclear reactor located at Oak Ridge National Laboratories in Tennessee.

9.5 Medical Applications of Radioactivity

The use of radiation in the treatment of various forms of cancer, as well as the newer area of **nuclear medicine,** the use of radioisotopes in diagnosis, has become widespread in the past quarter century. Let's look at the properties of radiation that make it an indispensable tool in modern medical care.

 Animation
Operation of a Nuclear Power Plant

5 **Learning Goal**
Cite examples of the use of radioactive isotopes in medicine.

A LIFESTYLE Connection | Radiocarbon Dating

Natural radioactivity is useful in establishing the approximate age of objects of archaeological, anthropological, or historical interest. Radiocarbon dating is the estimation of the age of objects through measurement of isotopic ratios of carbon.

Radiocarbon dating is based on the measurement of the relative amounts (or ratio) of $^{14}_{6}C$ and $^{12}_{6}C$ present in an object. $^{14}_{6}C$ is formed in the upper atmosphere by the bombardment of $^{14}_{7}N$ by high-speed neutrons (cosmic radiation):

$$^{14}_{7}N + ^{1}_{0}n \longrightarrow ^{14}_{6}C + ^{1}_{1}H$$

The carbon-14, along with the more abundant carbon-12, is converted into living plant material by the process of photosynthesis. Carbon proceeds up the food chain as the plants are consumed by animals, including humans. When a plant or animal dies, the uptake of both carbon-14 and carbon-12 ceases. However, the amount of carbon-14 slowly decreases because carbon-14

is radioactive ($t_{1/2}$ = 5730 years). Carbon-14 decay produces nitrogen:

$$^{14}_{6}C \longrightarrow ^{14}_{7}N + ^{0}_{-1}e$$

When an artifact is found and studied, the relative amounts of carbon-14 and carbon-12 are determined. By using suitable equations involving the $t_{1/2}$ of carbon-14, it is possible to approximate the age of the artifact.

This technique has been widely used to increase our knowledge about the history of the earth, to establish the age of objects, and even to detect art forgeries. Early paintings were made with pigments fabricated from vegetable dyes (plant material that, while alive, metabolized carbon).

The carbon-14 dating technique is limited to objects that are less than fifty thousand years old, or approximately nine half-lives, which is a practical upper limit. Older objects that have geological or archaeological significance may be dated using naturally occurring isotopes having much longer half-lives.

Examples of useful dating isotopes are listed below:

Isotopes Useful in Radioactive Dating

Isotope	Half-Life (years)	Upper Limit (years)	Dating Applications
Carbon-14	5730	5×10^4	Charcoal, organic material, artwork
Tritium (^3_1H)	12.3	1×10^2	Aged wines, artwork
Potassium-40	1.3×10^9	Age of earth (4×10^9)	Rocks, planetary material
Rhenium-187	4.3×10^{10}	Age of earth (4×10^9)	Meteorites
Uranium-238	4.5×10^9	Age of earth (4×10^9)	Rocks, earth's crust

Radiocarbon dating was used in the authentication study of the Shroud of Turin. It is a minimally destructive technique and is valuable in estimating the age of historical artifacts.

FOR FURTHER UNDERSTANDING

Describe the process used to determine the age of the wooden coffin of King Tut.

What property of carbon enables us to assess the age of a painting?

Cancer Therapy Using Radiation

6 Learning Goal

Describe the use of ionizing radiation in cancer therapy.

When high-energy radiation, such as gamma radiation, passes through a cell, it may collide with one of the molecules in the cell and cause it to lose one or more electrons, causing a series of events that result in the production of ion pairs. For this reason, such radiation is termed *ionizing radiation* (Section 9.1).

Ions produced in this way are highly energetic. Consequently, they may damage biological molecules and cause changes in cellular biochemical processes.

Interaction of ionizing radiation with intracellular water produces free electrons and other particles that can damage DNA. This may result in diminished or altered cell function or, in extreme cases, the death of the cell.

An organ that is cancerous is composed of both healthy cells and malignant cells. Tumor cells are more susceptible to the effects of gamma radiation than normal cells because they are undergoing cell division more frequently. Therefore, exposure of the tumor area to carefully targeted and controlled dosages of high-energy gamma radiation from cobalt-60 (a high-energy gamma ray source) will kill a higher percentage of abnormal cells than normal cells. If the dosage is administered correctly, a sufficient number of malignant cells will die, destroying the tumor, and enough normal cells will survive to maintain the function of the affected organ.

Gamma radiation can cure cancer. Paradoxically, the exposure of healthy cells to gamma radiation can actually cause cancer. For this reason, radiation therapy for cancer is a treatment that requires unusual care and sophistication.

Nuclear Medicine

The diagnosis of a host of biochemical irregularities or diseases of the human body has been made routine through the use of radioactive tracers. Medical **tracers** are small amounts of radioactive substances used as probes to study internal organs. Medical techniques involving tracers are **nuclear imaging** procedures.

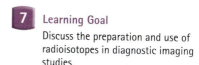

Learning Goal

Discuss the preparation and use of radioisotopes in diagnostic imaging studies.

A small amount of the tracer, an isotope of an element that is known to be attracted to the organ of interest, is administered to the patient. For a variety of reasons, such as ease of administration of the isotope to the patient and targeting the organ of interest, the isotope is often a part of a larger molecule or ion. Because the isotope is radioactive, its path may be followed by using suitable detection devices. A "picture" of the organ is obtained, often far more detailed than is possible with conventional X-rays. Such techniques are noninvasive; surgery is not required to investigate the condition of the internal organ, eliminating the risk of an operation.

Animation

Nuclear Medical Techniques

The radioactive isotope of an element chosen for tracer studies has chemical behavior similar to any other isotope of the same element. For example, iodine-127, the most abundant nonradioactive isotope of iodine, is used by the body in the synthesis of thyroid hormones and tends to concentrate in the thyroid gland. Both radioactive iodine-131 and iodine-127 behave in the same way, making it possible to use iodine-131 to study the thyroid. The rate of uptake of the radioactive isotope gives valuable information regarding underactivity or overactivity (hypoactive or hyperactive thyroid).

Isotopes with short half-lives are preferred for tracer studies. These isotopes emit their radiation in a more concentrated burst (short half-life materials have greater activity), facilitating their detection. If the radioactive decay is easily detected, the method is more sensitive and thus capable of providing more information. Furthermore, an isotope with a short half-life decays to background more rapidly. This is a mechanism for removal of the radioactivity from the body. If the radioactive element is also rapidly metabolized and excreted, this is obviously beneficial as well.

The following examples illustrate the use of imaging procedures for diagnosis of disease.

- *Bone disease and injury.* The most widely used isotope for bone studies is technetium-99m, which is incorporated into a variety of ions and molecules that direct the isotope to the tissue being investigated. Technetium compounds containing phosphate are preferentially adsorbed on the surface of bone. New bone formation (common to virtually all bone injuries) increases the incorporation of the technetium compound. As a result, an enhanced image appears at the site of the injury. Bone tumors behave in a similar fashion.

TABLE 9.3 Isotopes Commonly Used in Nuclear Medicine

Area of Body	Isotope	Use
Blood	Red blood cells tagged with chromium-51	Determine blood volume in body
Bone	*Technetium-99m, barium-131	Allow early detection of the extent of bone tumors and active sites of rheumatoid arthritis
Brain	*Technetium-99m	Detect and locate brain tumors and stroke
Coronary artery	Thallium-201	Determine the presence and location of obstructions in coronary arteries
Heart	*Technetium-99m	Determine cardiac output, size, and shape
Kidney	*Technetium-99m	Determine renal function and location of cysts; a common follow-up procedure for kidney transplant patients
Liver-spleen	*Technetium-99m	Determine size and shape of liver and spleen; location of tumors
Lung	Xenon-133	Determine whether lung fills properly; locate region of reduced ventilation and tumors
Thyroid	Iodine-131	Determine rate of iodine uptake by thyroid

*The destination of this isotope is determined by the identity of the compound in which it is incorporated.

- *Cardiovascular diseases.* Thallium-201 is used in the diagnosis of coronary artery disease. The isotope is administered intravenously and delivered to the heart muscle in proportion to the blood flow. Areas of restricted flow are observed as having lower levels of radioactivity, indicating some type of blockage.
- *Pulmonary disease.* Xenon is one of the noble gases. Radioactive xenon-133 may be inhaled by the patient. The radioactive isotope will be transported from the lungs and distributed through the circulatory system. Monitoring the distribution, as well as the reverse process, the removal of the isotope from the body (exhalation), can provide evidence of obstructive pulmonary disease, such as cancer or emphysema.

Examples of useful isotopes and the organ(s) in which they tend to concentrate are summarized in Table 9.3.

For many years, imaging with radioactive tracers was used exclusively for diagnosis. Recent applications have expanded to other areas of medicine. Imaging is now used extensively to guide surgery, assist in planning radiation therapy, and support the technique of angioplasty.

QUESTION 9.7

Technetium-99m is used in diagnostic imaging studies involving the brain. What fraction of the radioisotope remains after 12 hours have elapsed? See Table 9.2 for the half-life of technetium-99m.

QUESTION 9.8

Barium-131 is a radioisotope used to study bone formation. A patient ingested barium-131. How much time will elapse until only one-fourth of the barium-131 remains, assuming that none of the isotope is eliminated from the body through normal processes? The half-life of barium-131 is 11.6 minutes.

A MEDICAL Connection | Magnetic Resonance Imaging

The Nobel Prize in physics was awarded to Otto Stern in 1943 and to Isidor Rabi in 1944. They discovered that certain atomic nuclei have a property known as spin, analogous to the spin associated with electrons which was previously known. The spin of electrons is responsible for the magnetic properties of atoms. Spinning nuclei behave as tiny magnets, producing magnetic fields as well.

One very important aspect of this phenomenon is the fact that the atoms in close proximity to the spinning nuclei (its chemical environment) exert an effect on nuclear spin. In effect, measurable differences in spin are indicators of their surroundings. This relationship has been exhaustively studied for one atom in particular, hydrogen, and magnetic resonance techniques have become useful tools for the study of molecules containing hydrogen.

Human organs and tissue are made up of compounds containing hydrogen atoms. In the 1970s and 1980s, the experimental technique was extended beyond tiny laboratory samples of pure compounds to the most complex sample possible—the human body. The result of these experiments is termed *magnetic resonance imaging (MRI).*

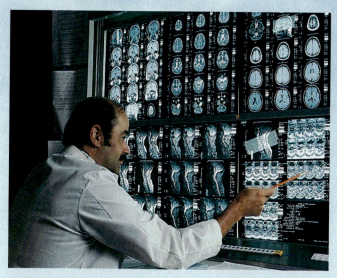

Dr. Paul Barnett of the Greater Baltimore Medical Center studies images obtained using MRI.

MRI is noninvasive to the body, requires no use of radioactive substances, and is quick, safe, and painless. A person is placed in a cavity surrounded by a magnetic field, and an image (based on the extent of radio frequency energy absorption) is generated, stored, and sorted in a computer. Differences between normal and malignant tissue, atherosclerotic thickening of an aortal wall, and a host of other problems may be seen clearly in the final image. For other applications of MRI, see *A Medical Connection: Seeing a Thought* online at www.mhhe.com/denniston.

Advances in MRI technology have provided medical practitioners with a powerful tool in diagnostic medicine. This is but one more example of basic science leading to technological advancement.

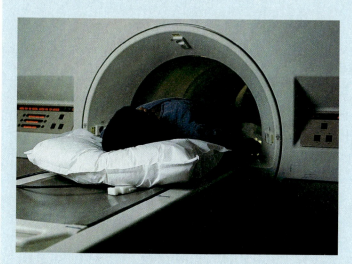

A patient entering an MRI scanner.

FOR FURTHER UNDERSTANDING

Why is hydrogen a useful atom to study in biological systems?

Why would MRI provide minimal information about bone tissue?

Making Isotopes for Medical Applications

In early experiments with radioactivity, the radioactive isotopes were naturally occurring. For this reason, the radioactivity produced by these unstable isotopes is described as **natural radioactivity.** If, on the other hand, a normally stable, non-radioactive nucleus is made radioactive, the resulting radioactivity is termed **artificial radioactivity.** The stable nucleus is made unstable by the introduction of "extra" protons, neutrons, or both.

7 Learning Goal

Discuss the preparation and use of radioisotopes in diagnostic imaging studies.

8 Learning Goal

Explain the difference between natural and artificial radioactivity.

Figure 9.6

A portion of a linear accelerator located at Brookhaven National Laboratory in New York. Particles can be accelerated at velocities close to the speed of light and accurately strike small "target" nuclei. At such facilities, rare isotopes can be synthesized and their properties studied.

The process of forming radioactive substances is often accomplished in the core of a **nuclear reactor,** in which an abundance of small nuclear particles, particularly neutrons, is available. Alternatively, extremely high-velocity charged particles (such as alpha and beta particles) may be produced in **particle accelerators,** such as a cyclotron. Accelerators are extremely large and use magnetic and electric fields to "push and pull" charged particles toward their target at very high speeds. A portion of the accelerator at the Brookhaven National Laboratory is shown in Figure 9.6.

Many isotopes that are useful in medicine are produced by particle bombardment. A few examples include the following:

- Gold-198, used as a tracer in the liver, is prepared by neutron bombardment.

$$^{197}_{79}\text{Au} + ^{1}_{0}\text{n} \longrightarrow ^{198}_{79}\text{Au}$$

- Gallium-67, used in the diagnosis of Hodgkin's disease, is prepared by proton bombardment.

$$^{66}_{30}\text{Zn} + ^{1}_{1}\text{p} \longrightarrow ^{67}_{31}\text{Ga}$$

Some medically useful isotopes, with short half-lives, must be prepared near the site of the clinical test. Preparation and shipment from a reactor site would waste time and result in an isotopic solution that had already undergone significant decay, resulting in diminished activity.

A common example is technetium-99m. It has a half-life of only six hours. It is prepared in a small generator, often housed in a hospital's radiology laboratory (Figure 9.7). The generator contains radioactive molybdate ion (MoO_4^{2-}). Molybdenum-99 is more stable than technetium-99m; it has a half-life of 67 hours.

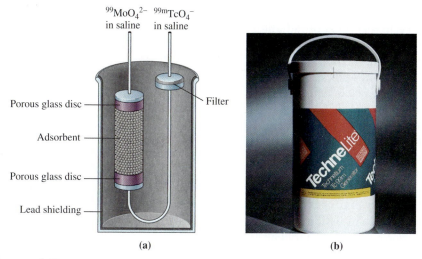

(a)   (b)

Figure 9.7

Preparation of technetium-99m. (a) A diagram depicting the conversion of $^{99}MoO_4^{2-}$ to $^{99m}TcO_4^-$ through radioactive decay. The radioactive pertechnetate ion is periodically removed from the generator in saline solution and used in tracer studies. (b) A photograph of a commercially available technetium-99m generator suitable for use in a hospital laboratory.

The molybdenum in molybdate ion decays according to the following nuclear equation:

$$^{99}_{42}Mo \longrightarrow {}^{99m}_{43}Tc + {}^{0}_{-1}e$$

Chemically, radioactive molybdate MoO_4^{2-} converts to radioactive pertechnetate ion (TcO_4^-). The radioactive TcO_4^- is removed from the generator when needed. It is administered to the patient as an aqueous salt solution that has an osmotic pressure identical to that of human blood.

9.6   Biological Effects of Radiation

It is necessary to use suitable precautions in working with radioactive substances. The chosen protocol is based on an understanding of the effects of radiation, dosage levels and "tolerable levels," the way in which radiation is detected and measured, and the basic precepts of radiation safety.

Radiation Exposure and Safety

In working with radioactive materials, the following factors must be considered.

The Magnitude of the Half-Life

In considering safety, isotopes with short half-lives have, at the same time, one major disadvantage and one major advantage.

On one hand, short-half-life radioisotopes produce a larger amount of radioactivity per unit time than a long-half-life substance. For example, consider equal amounts of hypothetical isotopes that produce alpha particles. One has a half-life of ten days; the other has a half-life of one hundred days. After one half-life, each substance will produce exactly the same number of alpha particles. However, the first substance generates the alpha particles in only one-tenth of the time,

9 Learning Goal

Describe the characteristics of radioactive materials that relate to radiation exposure and safety.

Higher levels of exposure in a short time produce clearer images.

hence it emits ten times as much radiation per unit time. Equal exposure times will result in a higher level of radiation exposure for substances with short half-lives and lower levels for substances with long half-lives.

On the other hand, materials with short half-lives (weeks, days, or less) may be safer to work with, especially if an accident occurs. Over time (depending on the magnitude of the half-life), radioactive isotopes will decay to **background radiation** levels. This is the level of radiation attributable to our surroundings on a day-to-day basis.

Virtually all matter is composed of both radioactive and nonradioactive isotopes. Small amounts of radioactive material in the air, water, soil, and so forth make up a part of the background levels. Cosmic rays from outer space continually bombard us with radiation, contributing to the total background. Owing to the inevitability of background radiation, there can be no situation on earth where we observe zero radiation levels.

An isotope with a short half-life, for example 5.0 min, may decay to background in as few as ten half-lives

$$10 \text{ half-lives} \times \frac{5.0 \text{ min}}{1 \text{ half-life}} = 50 \text{ min}$$

See An Environmental Perspective: Nuclear Waste Disposal on page 255.

A spill of such material could be treated by waiting ten half-lives, perhaps by going to lunch. When you return to the laboratory, the material that was spilled will be no more radioactive than the floor itself. An accident with plutonium-239, which has a half-life of 24,000 years, would be quite a different matter! After fifty minutes, virtually all of the plutonium-239 would still remain. Long-half-life isotopes, by-products of nuclear technology, pose the greatest problems for safe disposal. Finding a site that will remain undisturbed "forever" is quite a formidable task.

QUESTION 9.9

Describe the advantage of using isotopes with short half-lives for tracer applications in a medical laboratory.

QUESTION 9.10

Can you think of any disadvantage associated with the use of isotopes described in Question 9.9? Explain.

Shielding

Alpha and beta particles are relatively low in penetrating power and require low-level **shielding.** A lab coat and gloves are generally sufficient protection from this low-penetration radiation. On the other hand, shielding made of lead, concrete, or both is required for gamma rays (and X-rays, which are also high-energy radiation). Extensive manipulation of gamma emitters is often accomplished in laboratory and industrial settings by using robotic control: computer-controlled mechanical devices that can be programmed to perform virtually all manipulations normally carried out by humans.

Distance from the Radioactive Source

Radiation intensity varies *inversely* with the *square* of the distance from the source. Doubling the distance from the source *decreases* the intensity by a factor of four (2^2). Again, the use of robot manipulators is advantageous, allowing a greater distance between the operator and the radioactive source.

AN ENVIRONMENTAL Perspective | Radon and Indoor Air Pollution

Marie and Pierre Curie first discovered that air in contact with radium compounds became radioactive. Later experiments by Ernest Rutherford and others isolated the radioactive substance from the air. This substance was an isotope of the noble gas radon (Rn).

We now know that radium (Ra) produces radon by spontaneous decay:

$$^{226}_{88}\text{Ra} \longrightarrow {}^{4}_{2}\text{He} + {}^{222}_{86}\text{Rn}$$

Radium in trace quantities is found in the soil and rock and is unequally distributed in the soil. The decay product, radon, is emitted from the soil to the surrounding atmosphere. Radon is also found in higher concentrations when uranium is found in the soil. This is not surprising because radium is formed as a part of the stepwise decay of uranium.

If someone constructs a building over soil or rock that has a high radium content (or uses stone with a high radium content to build the foundation!), radon gas can percolate through the basement and accumulate in the house. Couple this with the need to build more energy-efficient, well-insulated dwellings, and the radon levels in buildings in some regions of the country can become quite high.

Radon itself is radioactive; however, its radiation is not the major problem. Because it is a gas and chemically inert, it is rapidly exhaled after breathing. However, radon decays to polonium:

$$^{222}_{86}\text{Rn} \longrightarrow {}^{4}_{2}\text{He} + {}^{218}_{84}\text{Po}$$

This polonium isotope is radioactive and is a nonvolatile heavy metal that can attach itself to bronchial or lung tissue, emitting hazardous radiation and producing other isotopes that are also radioactive.

In the United States, homes are now being tested and monitored for radon. In many states, proof of acceptable levels of radon is a condition of sale of the property. Studies continue to attempt to find reasonable solutions to the radon problem. Current recommendations include sealing cracks and openings in basements, increasing ventilation, and evaluating sites before construction of buildings. Debate continues within the scientific community regarding a safe and attainable indoor air quality standard for radon.

FOR FURTHER UNDERSTANDING

Why is indoor radon more hazardous than outdoor radon?

Polonium-218 has a very long half-life. Explain why this constitutes a potential health problem.

Time of Exposure

The effects of radiation are cumulative. Generally, potential damage is directly proportional to the time of exposure. Workers exposed to moderately high levels of radiation on the job may be limited in the time that they can perform that task. For example, workers involved in the cleanup of the Three Mile Island nuclear plant, incapacitated in 1979, observed strict limits on the amount of time that they could be involved in the cleanup activities.

Types of Radiation Emitted

Alpha and beta emitters are generally less hazardous than gamma emitters, owing to differences in energy and penetrating power that require less shielding. However, ingestion or inhalation of an alpha emitter or beta emitter can, over time, cause serious tissue damage; the radioactive substance is in direct contact with sensitive tissue. An Environmental Perspective: Radon and Indoor Air Pollution (above) expands on this problem.

Waste Disposal

Virtually all applications of nuclear chemistry create radioactive waste and, along with it, the problems of safe handling and disposal. Most disposal sites, at present, are considered temporary, until a long-term safe solution can be found.

Figure 9.8

Photograph of the construction of one-million-gallon capacity storage tanks for radioactive waste. Located in Hanford, Washington, they are now covered with 6–8 ft of earth.

10 Learning Goal

Be familiar with common techniques for the detection of radioactivity.

CT represents *Computer-aided Tomography:* the computer reconstructs a series of measured images of tissue density (tomography). Small differences in tissue density may indicate the presence of a tumor.

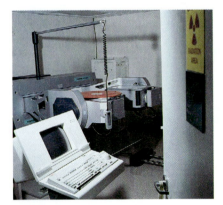

Figure 9.9

An imaging laboratory at the Greater Baltimore Medical Center.

Figure 9.8 conveys a sense of the enormity of the problem. Also, An Environmental Perspective: Nuclear Waste Disposal on page 255, examines this problem in more detail.

9.7 Measurement of Radiation

The changes that take place when radiation interacts with matter (such as photographic film) provide the basis of operation for various radiation detection devices.

The principal detection methods involve the use of either photographic film to create an image of the location of the radioactive substance or a counter that allows the measurement of intensity of radiation emitted from some source by converting the radiation energy to an electrical signal.

Nuclear Imaging

This approach is often used in nuclear medicine. An isotope is administered to a patient, perhaps iodine-131, which is used to study the thyroid gland, and the isotope begins to concentrate in the organ of interest. Nuclear images (photographs) of that region of the body are taken at periodic intervals using a special type of film. The emission of radiation from the radioactive substance creates the image, in much the same way as light causes the formation of images on conventional film in a camera. Upon development of the series of photographs, a record of the organ's uptake of the isotope over time enables the radiologist to assess the condition of the organ.

Computer Imaging

The coupling of rapid developments in the technology of television and computers, resulting in the marriage of these two devices, has brought about a versatile alternative to photographic imaging.

A specialized television camera, sensitive to emitted radiation from a radioactive substance administered to a patient, develops a continuous and instantaneous record of the voyage of the isotope throughout the body. The signal, transmitted to the computer, is stored, sorted, and portrayed on a monitor. Advantages include increased sensitivity, allowing a lower dose of the isotope, speed through elimination of the developing step, and versatility of application, limited perhaps only by the creativity of the medical practitioners.

A particular type of computer imaging, useful in diagnostic medicine, is the CT scanner. The CT scanner measures the interaction of X-rays with biological tissue, gathering huge amounts of data and processing the data to produce detailed information, all in a relatively short time. Such a device may be less hazardous than conventional X-ray techniques because it generates more useful information per unit of radiation. It often produces a superior image. A photograph of a CT scanner is shown in Figure 9.9, and an image of a damaged spinal bone, taken by a CT scanner, is shown in Figure 9.10.

The Geiger Counter

A Geiger counter is an instrument that detects ionizing radiation. Ions, produced by radiation passing through a tube filled with an ionizable gas, can conduct an electrical current between two electrodes. This current flow can be measured and is proportional to the level of radiation (Figure 9.11). Such devices, which were routinely used in laboratory and industrial monitoring, have been largely replaced by more sophisticated devices, often used in conjunction with a computer.

Film Badges

A common sight in any hospital or medical laboratory or any laboratory that routinely uses radioisotopes is the film badge worn by all staff members exposed in any way to low-level radioactivity.

A film badge is merely a piece of photographic film that is sensitive to energies corresponding to radioactive emissions. It is shielded from light, which would interfere, and mounted in a clip-on plastic holder that can be worn throughout the workday. The badges are periodically collected and developed. The degree of darkening is proportional to the amount of radiation to which the worker has been exposed, just as a conventional camera produces images on film in proportion to the amount of light that it "sees."

Proper record keeping thus allows the laboratory using radioactive substances to maintain an ongoing history of each individual's exposure and, at the same time, promptly pinpoint any hazards that might otherwise go unnoticed.

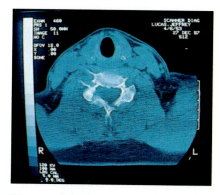

Figure 9.10

Damage observed in a spinal bone on a CT scan image.

Units of Radiation Measurement

The amount of radiation emitted by a source or received by an individual is reported in a variety of ways, using units that describe different aspects of radiation. The *curie* and the *roentgen* describe the intensity of the emitted radiation, whereas the *rad* and the *rem* describe the biological effects of radiation.

The **curie** is a measure of the amount of radioactivity in a radioactive source. The curie is independent of the nature of the radiation (alpha, beta, or gamma) and its effect on biological tissue. A curie is defined as the amount of radioactive material that produces 3.7×10^{10} atomic disintegrations per second.

The **roentgen** is a measure of very high energy ionizing radiation (X-ray and gamma ray) only. The roentgen is defined as the amount of radiation needed to produce 2×10^9 ion pairs when passing through one cm^3 of air at 0°C. The roentgen is a measure of radiation's interaction with air and gives no information about the effect on biological tissue.

The **rad,** or *radiation absorbed dosage,* provides more meaningful information than either of the previous units of measure. It takes into account the nature of the absorbing material. It is defined as the dosage of radiation able to transfer 2.4×10^{-3} cal of energy to one kg of matter.

The **rem,** or *roentgen equivalent for man,* describes the biological damage caused by the absorption of different kinds of radiation by the human body. The *rem* is obtained by multiplication of the *rad* by a factor called the *relative biological effect (RBE).* The RBE is a function of the type of radiation (alpha, beta, or gamma). Although a beta particle is more penetrating than an alpha particle, an alpha particle is approximately ten times more damaging to biological tissue. As a result, the RBE

11 **Learning Goal**

Know the common units of radiation intensity: the curie, roentgen, rad, and rem.

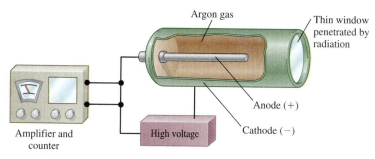

Figure 9.11

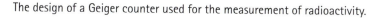

The design of a Geiger counter used for the measurement of radioactivity.

Figure 9.12

Relative yearly radiation dosages for individuals in the continental United States. Red, yellow, and green shading indicates higher levels of background radiation. Blue shading indicates regions of lower background exposure.

is ten for alpha particles and one for beta particles. Relative yearly radiation dosages received by Americans are shown in Figure 9.12.

The **lethal dose (LD$_{50}$)** of radiation is defined as the acute dosage of radiation that would be fatal for 50% of the exposed population within 30 days. An estimated lethal dose is 500 rems. Some biological effects, however, may be detectable at a level as low as 25 rem.

QUESTION 9.11

From a clinical standpoint, what advantages does expressing radiation in rems have over the use of other radiation units?

QUESTION 9.12

Is the roentgen unit used in the measurement of alpha particle radiation? Why or why not?

SUMMARY

9.1 Natural Radioactivity

Radioactivity is the process by which atoms emit energetic, ionizing particles or rays. These particles or rays are termed radiation. Nuclear radiation occurs because the nucleus is unstable, hence radioactive. Nuclear symbols consist of the elemental symbol, the atomic number, and the mass number.

Not all *nuclides* are unstable. Only unstable nuclides undergo change and produce radioactivity in the process of radioactive decay. Three types of natural radiation emitted by unstable nuclei are *alpha particles*, *beta particles*, and *gamma rays*. This radiation is collectively termed *ionizing radiation*.

9.2 Writing a Balanced Nuclear Equation

A *nuclear equation* represents a nuclear process such as radioactive decay. The total of the mass numbers on each side of the reaction arrow must be identical, and the sum of the atomic numbers of the reactants must equal the sum of the atomic numbers of the products. Nuclear equations can be used to predict products of nuclear reactions.

9.3 Properties of Radioisotopes

The *binding energy* of the nucleus is a measure of nuclear stability. When an isotope decays, energy is released. Nuclear stability correlates with the ratio of neutrons to protons in the isotope. Nuclei with large numbers of protons tend to be unstable, and isotopes containing 2, 8, 20, 50, 82, or 126 protons or neutrons (magic numbers) are stable. Also, isotopes with even numbers of protons or neutrons

are generally more stable than those with odd numbers of protons or neutrons.

The *half-life, $t_{1/2}$,* is the time required for one-half of a given quantity of a substance to undergo change. Each isotope has its own characteristic half-life. The degree of stability of an isotope is indicated by the isotope's half-life. Isotopes with short half-lives decay rapidly; they are very unstable.

9.4 Nuclear Power

Einstein predicted that a small amount of nuclear mass would convert to a very large amount of energy when the nucleus breaks apart. Fission reactors are used to generate electrical power. Technological problems with *fusion* and *breeder* reactors have prevented their commercialization in the United States.

9.5 Medical Applications of Radioactivity

The use of radiation in the treatment of various forms of cancer and in the newer area of *nuclear medicine* has become widespread in the past quarter century.

Ionizing radiation causes changes in cellular biochemical processes that may damage or kill the cell. A cancerous organ is composed of both healthy and malignant cells. Exposure of the tumor area to controlled dosages of high-energy gamma radiation from cobalt-60 will kill a higher percentage of abnormal cells than normal cells and is a valuable cancer therapy.

The diagnosis of a host of biochemical irregularities or diseases of the human body has been made routine through the use of radioactive tracers. *Tracers* are small amounts of radioactive substances used as probes to study internal organs. Because the isotope is radioactive, its path may be followed by using suitable detection devices. A "picture" of the organ is obtained, far more detailed than is possible with conventional X-rays.

The radioactivity produced by unstable isotopes is described as *natural radioactivity*. A normally stable, non-radioactive nucleus can be made radioactive, and this is termed *artificial radioactivity* (the process produces synthetic isotopes). Synthetic isotopes are often used in clinical situations. Isotopic synthesis may be carried out in the core of a *nuclear reactor* or in a *particle accelerator*. Short-lived isotopes, such as technetium-99m, are often produced directly at the site of clinical testing.

9.6 Biological Effects of Radiation

Safety considerations are based on the magnitude of the *half-life, shielding,* distance from the radioactive source, time of exposure, and type of radiation emitted. We are never entirely free of the effects of radioactivity. *Background radiation* is normal radiation attributable to our surroundings.

Virtually all applications of nuclear chemistry create radioactive waste and, along with it, the problems of safe handling and disposal. Most disposal sites are considered temporary, until a long-term safe solution can be found.

9.7 Measurement of Radiation

The changes that take place when radiation interacts with matter provide the basis for various radiation detection devices. Photographic imaging, computer imaging, the Geiger counter, and film badges represent the most frequently used devices for detecting and measuring radiation.

Commonly used radiation units include the *curie,* a measure of the amount of radioactivity in a radioactive source; the *roentgen,* a measure of high-energy radiation (X-ray and gamma ray); the *rad* (radiation absorbed dosage), which takes into account the nature of the absorbing material; and the *rem* (roentgen equivalent for man), which describes the biological damage caused by the absorption of different kinds of radiation by the human body. The *lethal dose* of radiation, LD_{50}, is defined as the dose that would be fatal for 50% of the exposed population within thirty days.

KEY TERMS

alpha particle (9.1)	curie (9.7)
artificial radioactivity (9.5)	fission (9.4)
background radiation (9.6)	fusion (9.4)
beta particle (9.1)	gamma ray (9.1)
binding energy (9.3)	half-life ($t_{1/2}$) (9.3)
breeder reactor (9.4)	ionizing radiation (9.1)
chain reaction (9.4)	lethal dose (LD_{50}) (9.7)

metastable isotope (9.2)	positron (9.2)
natural radioactivity (9.5)	rad (9.7)
nuclear equation (9.2)	radioactivity (9.1)
nuclear imaging (9.5)	rem (9.7)
nuclear medicine (9.5)	roentgen (9.7)
nuclear reactor (9.5)	shielding (9.6)
nuclide (9.1)	tracer (9.5)
particle accelerator (9.5)	

QUESTIONS AND PROBLEMS

Natural Radioactivity

Fundamentals

9.13 Describe the meaning of the term *natural radioactivity*.

9.14 What is background radiation?

9.15 What is the composition of an alpha particle?

9.16 What is alpha decay?

9.17 What is the composition of a beta particle?

9.18 What is the composition of a positron?

9.19 In what way do beta particles and positrons differ?

9.20 In what way are beta particles and positrons similar?

9.21 What are the major differences between alpha and beta particles?

9.22 What are the major differences between alpha particles and gamma radiation?

9.23 How do nuclear reactions and chemical reactions differ?

9.24 We can control the rate of chemical reactions. Can we control the rate of natural radiation?

Applications

9.25 Write the nuclear symbol for an alpha particle.

9.26 Write the nuclear symbol for a beta particle.

9.27 Write the nuclear symbol for uranium-235.

9.28 How many protons and neutrons are contained in the nucleus of uranium-235?

9.29 How many protons and neutrons are contained in each of the three naturally occurring isotopes of hydrogen?

9.30 How many protons and neutrons are contained in each of the three naturally occurring isotopes of carbon?

9.31 Write the nuclear symbol for nitrogen-15.

9.32 Write the nuclear symbol for carbon-14.

Writing a Balanced Nuclear Equation

Fundamentals

9.33 Write a nuclear reaction to represent cobalt-60 decaying to nickel-60 plus a beta particle plus a gamma ray.

9.34 Write a nuclear reaction to represent radium-226 decaying to radon-222 plus an alpha particle.

9.35 Complete the following nuclear reactions:
   a. $^{23}_{11}Na + ^{2}_{1}H \longrightarrow ? + ^{1}_{1}H$
   b. $^{238}_{92}U + ^{14}_{7}N \longrightarrow ? + 6^{1}_{0}n$
   c. $^{24}_{10}Ne \longrightarrow \beta + ?$

9.36 Complete the following nuclear reactions:
   a. $^{190}_{78}Pt \longrightarrow \alpha + ?$
   b. $? \longrightarrow ^{140}_{56}Ba + ^{0}_{1}e$
   c. $? \longrightarrow ^{214}_{90}Th + ^{4}_{2}He$

Applications

9.37 Element 107 was synthesized by bombarding bismuth-209 with chromium-54. Write the equation for this process if one product is a neutron.

9.38 Element 109 was synthesized by bombarding bismuth-209 with iron-58. Write the equation for this process if one product is a neutron.

9.39 Write a balanced nuclear equation for beta emission by magnesium-27.

9.40 Write a balanced nuclear equation for alpha decay of bismuth-212.

9.41 Write a balanced nuclear equation for positron emission by nitrogen-12.

9.42 Write a balanced nuclear equation for the formation of polonium-206 through alpha decay.

Properties of Radioisotopes

Fundamentals

9.43 What is the difference between natural radioactivity and artificial radioactivity?

9.44 Is the fission of uranium-235 an example of natural or artificial radioactivity?

9.45 Summarize the major characteristics of nuclei for which we predict a high degree of stability.

9.46 Explain why the binding energy of a nucleus is expected to be large.

Applications

9.47 If 3.2 mg of the radioisotope iodine-131 is administered to a patient, how much will remain in the body after 24 days, assuming that no iodine has been eliminated from the body by any other process? (See Table 9.2 for the half-life of iodine-131.)

9.48 A patient receives 9.0 ng of a radioisotope with a half-life of 12 hours. How much will remain in the body after 2.0 days, assuming that radioactive decay is the only path for removal of the isotope from the body?

9.49 A sample containing 1.00×10^2 mg of iron-59 is stored for 135 days. What mass of iron-59 will remain at the end of the storage period? (See Table 9.2 for the half-life of iron-59.)

9.50 An instrument for cancer treatment containing a cobalt-60 source was manufactured in 1978. In 1995, it was removed from service and, in error, was buried in a landfill with the source still in place. What percentage of its initial radioactivity will remain in the year 2010? (See Table 9.2 for the half-life of cobalt-60.)

9.51 The half-life of molybdenum-99 is 67 hr. A 200-μg quantity decays, over time, to 25 μg. How much time has elapsed?

9.52 The half-life of strontium-87 is 2.8 hr. What percentage of this isotope will remain after 8 hours and 24 minutes?

Nuclear Power

Fundamentals

9.53 Which type of nuclear process splits nuclei to release energy?

9.54 Which type of nuclear process combines small nuclei to release energy?

9.55 **a.** Describe the process of fission.
   b. How is this reaction useful as the basis for the production of electrical energy?

9.56 **a.** Describe the process of fusion.
   b. How could this process be used for the production of electrical energy?

Applications

9.57 Write a balanced nuclear equation for a fusion reaction.

9.58 What are the major disadvantages of a fission reactor for electrical energy production?

9.59 What is meant by the term *breeder reactor*?

9.60 What are the potential advantages and disadvantages of breeder reactors?

9.61 Describe what is meant by the term *chain reaction*.

9.62 Why are cadmium rods used in a fission reaction?

9.63 What is the greatest barrier to development of fusion reactors?

9.64 What type of nuclear reaction fuels our solar system?

Medical Applications of Radioactivity

9.65 The isotope indium-111 is used in medical laboratories as a label for blood platelets. To prepare indium-111, silver-108 is bombarded with an alpha particle, forming an intermediate isotope of indium. Write a nuclear equation for the process, and identify the intermediate isotope of indium.

9.66 Radioactive molybdenum-99 is used to produce the tracer isotope, technetium-99m. Write a nuclear equation for the formation of molybdenum-99 from stable molybdenum-98 bombarded with neutrons.

9.67 Describe an application of each of the following isotopes:
   a. technetium-99m
   b. xenon-133

9.68 Describe an application of each of the following isotopes:
   a. iodine-131
   b. thallium-201

9.69 Why is radiation therapy an effective treatment for certain types of cancer?

9.70 Describe how medically useful isotopes may be prepared.

9.71 What is the source of background radiation?

9.72 Why do high-altitude jet flights increase a person's exposure to background radiation?

Answer questions 9.73 through 9.80 based on the assumption that you are employed by a clinical laboratory that prepares radioactive isotopes for medical diagnostic tests. Consider α, β, positron, and γ emission.

9.73 What would be the effect on your level of radiation exposure if you increase your average distance from a radioactive source?

9.74 Would wearing gloves have any significant effect? Why?

9.75 Would limiting your time of exposure have a positive effect? Why?

9.76 Would wearing a lab apron lined with thin sheets of lead have a positive effect? Why?

9.77 Would the use of robotic manipulation of samples have an effect? Why?

9.78 Would the use of concrete rather than wood paneling help to protect workers in other parts of the clinic? Why?

9.79 Would the thickness of the concrete in question 9.78 be an important consideration? Why?

9.80 Suggest a protocol for radioactive waste disposal.

Measurement of Radiation

9.81 X-ray technicians often wear badges containing photographic film. How is this film used to indicate exposure to X-rays?

9.82 Why would a Geiger counter be preferred to film for assessing the immediate danger resulting from a spill of some solution containing a radioisotope?

9.83 What is meant by the term *relative biological effect*?

9.84 What is meant by the term *lethal dose* of radiation?

FOR FURTHER UNDERSTANDING

1. Isotopes used as radioactive tracers have chemical properties that are similar to those of a nonradioactive isotope of the same element. Explain why this is a critical consideration in their use.

2. A chemist proposes a research project to discover a catalyst that will speed up the decay of radioactive isotopes that are waste

products of a medical laboratory. Such a discovery would be a potential solution to the problem of nuclear waste disposal. Critique this proposal.

3. A controversial solution to the disposal of nuclear waste involves burial in sealed chambers far below the earth's surface. Describe potential pros and cons of this approach.

4. What type of radioactive decay is favored if the number of protons in the nucleus is much greater than the number of neutrons? Explain.

5. If the proton-to-neutron ratio in question 4 (above) were reversed, what radioactive decay process would be favored? Explain.

6. Radioactive isotopes are often used as "tracers" to follow an atom through a chemical reaction, and the following is an example. Acetic acid reacts with methyl alcohol by eliminating a molecule of water to form methyl acetate. Explain how you would use the radioactive isotope oxygen-18 to show whether the oxygen atom in the water product comes from the —OH of the acid or the —OH of the alcohol.

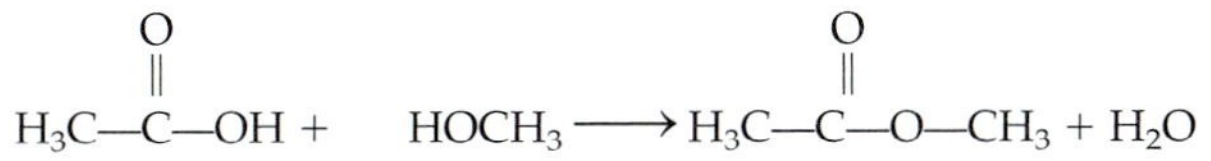

$$H_3C-\overset{\overset{\textstyle O}{\|}}{C}-OH + HOCH_3 \longrightarrow H_3C-\overset{\overset{\textstyle O}{\|}}{C}-O-CH_3 + H_2O$$

Acetic acid   Methyl alcohol   Methyl acetate

LEARNING GOALS

1 Compare and contrast organic and inorganic compounds.

2 Draw structures that represent each of the families of organic compounds.

3 Write the names and draw the structures of the common functional groups.

4 Write condensed and structural formulas for saturated hydrocarbons.

5 Describe the relationship between the structure and physical properties of saturated hydrocarbons.

6 Use the basic rules of the I.U.P.A.C. Nomenclature System to name alkanes and substituted alkanes.

7 Draw constitutional (structural) isomers of simple organic compounds.

8 Write the names and draw the structures of simple cycloalkanes.

9 Write equations for combustion reactions of alkanes.

10 Write equations for halogenation reactions of alkanes.

An Introduction to Organic Chemistry
The Saturated Hydrocarbons

Organic chemistry is the study of carbon-containing compounds. The term *organic* was coined in 1807 by the Swedish chemist Jöns Jakob Berzelius. At that time, it was thought that all organic compounds, such as fats, sugars, coal, and petroleum, were formed by living or once living organisms. All early attempts to synthesize these compounds in the laboratory failed, and it was thought that a vital force, available only in living cells, was needed for their formation.

This idea began to change in 1828 when a twenty-seven-year-old German physician, whose first love was chemistry, synthesized the organic molecule, urea, from inorganic starting materials. This man was Friedrich Wöhler, the "father of organic chemistry."

As a child, Wöhler didn't do particularly well in school because he spent so much time doing chemistry experiments at home. Eventually, he did earn his medical degree, but he decided to study chemistry in the laboratory of Berzelius rather than practice medicine.

After a year, he returned to Germany to teach and, as it turned out, to do the experiment that made him famous. The goal of the experiment was to prepare ammonium cyanate from a mixture of potassium cyanate and ammonium sulfate. He heated a solution of the two salts and crystallized the product. But the product didn't look like ammonium cyanate. It was a white crystalline material that looked exactly like urea! Urea is a waste product of protein breakdown in the body and is excreted in the urine. Wöhler recognized urea crystals because he had previously purified them from the urine of dogs and humans. Excited about his accidental discovery, he wrote to his teacher and friend Berzelius, "I can make urea without the necessity of a kidney, or even an animal, whether man or dog."

Fossil fuels, such as coal and oil, were formed through tens of millions of years from the remains of ancient plants and animals. So dependent is society on these fuels that we search constantly for new sources, and even fight over known reservoirs of this nonrenewable resource? If you were a Department of Energy analyst, what type of energy policy would you recommend for the United States?

$$NH_4^+ \; [N{=}C{=}O]^-$$

Ammonium cyanate
(inorganic salt)

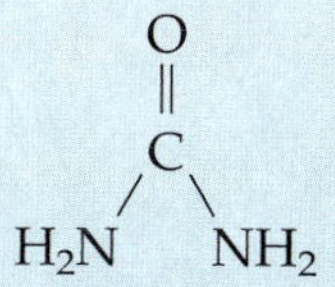

$$\overset{\displaystyle O}{\underset{\displaystyle H_2N \qquad NH_2}{\overset{\displaystyle \|}{C}}}$$

Urea
(organic compound)

Ironically, Wöhler, the first person to synthesize an organic compound from inorganic substances, devoted the rest of his career to inorganic chemistry. However, other chemists continued this work. As a result, the "vital force theory" was laid to rest, and modern organic chemistry was born.

—*Continued next page*

Continued—

In this and later chapters, we will study the amazing array of organic molecules (molecules made up of carbon, hydrogen, and a few other elements), many of which are essential to life. As we will see, all the structural and functional molecules of the cell, including the phospholipids that make up the cell membrane and the enzymes that speed up biological reactions, are organic molecules. Smaller organic molecules, such as the sugars, glucose and fructose, are used as fuel by our cells, whereas others, such as penicillin and aspirin, are useful in the treatment of disease. All these organic compounds, and many more, are the subject of the remaining chapters of this text.

10.1 The Chemistry of Carbon

The number of possible carbon-containing compounds is almost limitless. The importance of these organic compounds is reflected in the fact that over half of this book is devoted to the study of molecules made with this single element.

Why are there so many organic compounds? There are several reasons. First, carbon can form *stable, covalent* bonds with other carbon atoms. Just think about the three *allotropic forms* of elemental carbon shown in Figure 10.1.

Graphite is made of carbon-to-carbon bonds organized in planar units that slide over one another, making it an excellent lubricant. In diamond, carbon-to-carbon bonds form a large, three-dimensional network, producing an extremely hard substance used in jewelry and cutting tools. The most unusual is buckminsterfullerene or the *buckey ball*, which consists of sixty carbon atoms in the shape of a soccer ball.

A second reason for the vast number of organic compounds is that carbon atoms can form stable bonds with other elements. Several families of organic compounds (alcohols, aldehydes, ketones, esters, and ethers) contain oxygen atoms bonded to carbon. Others contain nitrogen, sulfur, or halogens. These elements confer a wide variety of new chemical and physical properties on organic compounds.

Third, carbon can form double or triple bonds with other carbon atoms to produce a variety of organic molecules with very different properties. Finally, the number of ways in which carbon and other atoms can be arranged is nearly limitless. In addition to linear chains of carbon atoms, ring structures and branched

Allotropes are forms of an element that have the same physical state but different properties.

In future chapters, we will discuss families of organic molecules containing oxygen atoms (alcohols, aldehydes, ketones, carboxylic acids, ethers, and esters), nitrogen atoms (amides and amines), sulfur atoms, and halogen atoms.

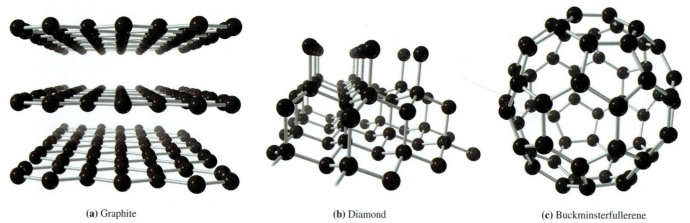

(a) Graphite   **(b)** Diamond   **(c)** Buckminsterfullerene

Figure 10.1

Three allotropic forms of elemental carbon.

chains are common. Two organic compounds may even have the same number and kinds of atoms but completely different structures and thus different properties. Such organic molecules are called *isomers*.

Important Differences Between Organic and Inorganic Compounds

The bonds between carbon and another atom are almost always *covalent bonds*, whereas the bonds in many inorganic compounds are *ionic bonds*. Ionic bonds result from the *transfer* of one or more electrons from one atom to another and the attraction between the positive and negative ions formed by the electron transfer. Ionic compounds often form three-dimensional crystals that may dissociate in water to form ions. These are called electrolytes. Covalent bonds are formed by *sharing* one or more pairs of electrons, producing individual units called molecules. Most covalent compounds are nonelectrolytes.

As a result of these differences, ionic substances usually have much higher melting and boiling points than covalent compounds. They are more likely to dissolve in water than in a less polar solvent, whereas organic compounds, which are typically nonpolar or only moderately polar, are less soluble, or insoluble in water. In Table 10.1, the physical properties of the organic compound butane are compared with those of sodium chloride, an inorganic compound of similar formula weight.

1 Learning Goal
Compare and contrast organic and inorganic compounds.

Polar covalent compounds, such as HCl, dissociate in water and, thus, are electrolytes. Carboxylic acids, the family of organic compounds that we will study in Chapter 13, are weak electrolytes when dissolved in water.

QUESTION 10.1

A student is presented with a sample of an unknown substance and asked to classify the substance as organic or inorganic. What tests should the student carry out?

QUESTION 10.2

What results would the student expect if the sample in Question 10.1 were an inorganic compound? What results would the student expect if it were an organic compound?

TABLE 10.1 Comparison of the Major Properties of a Typical Organic and an Inorganic Compound: Butane Versus Sodium Chloride		
Property	**Organic Compounds (e.g., Butane)**	**Inorganic Compounds (e.g., Sodium Chloride)**
Formula weight	58	58.5
Bonding	Covalent (C_4H_{10})	Ionic (Na^+ and Cl^- ions)
Physical state at room temperature and atmospheric pressure	Gas	Solid
Boiling point	Low ($-0.5°C$)	High ($1413°C$)
Melting point	Low ($-139°C$)	High ($801°C$)
Solubility in water	Insoluble	High (36 g/100 mL)
Solubility in organic solvents (e.g., hexane)	High	Insoluble
Flammability	Flammable	Nonflammable
Electrical conductivity	Nonconductor	Conducts electricity in solution and in molten liquid

AN ENVIRONMENTAL
Perspective

Frozen Methane: Treasure or Threat?

Methane is the simplest hydrocarbon, but it has some unusual behaviors. One of these is the ability to form a clathrate, which is an unusual type of matter in which molecules of one substance form a cage around molecules of another substance. For instance, water molecules can form a latticework around methane molecules to form frozen methane hydrate, possibly one of the biggest reservoirs of fossil fuel on earth.

Typically we wouldn't expect a nonpolar molecule, such as methane, to interact with a polar molecule, such as water. So, then, how is this structure formed? As we have studied earlier, water molecules interact with one another by strong hydrogen bonding. In the frozen state, these hydrogen-bonded water molecules form an open latticework. The nonpolar methane molecule is simply trapped inside one of the spaces within the lattice.

Vast regions of the ocean floor are covered by such ice fields of frozen methane. The U.S. Geological Survey estimates that the amount of methane hydrate in the United States is worth over two hundred times the conventional natural gas resources in this country!

But is it safe to harvest the methane from this ice? Caution will certainly be required. Methane is flammable, and, like carbon dioxide, it is a greenhouse gas. In fact, it is about twenty times more efficient at trapping heat than carbon dioxide. (Greenhouse gases are discussed in greater detail in An Environmental Perspective: The Greenhouse Effect and Global Warming in Chapter 6.) So the U.S. Department of Energy, which is working with industry to develop ways to harvest the methane, must figure out how to do that without releasing much into the atmosphere where it could intensify global warming.

It may be that a huge release of methane from these frozen reserves was responsible for a major global warming that occurred fifty-five million years ago and lasted for one hundred thousand years. NASA scientists using computer

Three-dimensional structure of methane hydrate.

simulations hypothesize that a shift of the continental plates may have released vast amounts of methane gas from the ocean floor. This methane raised the temperature of earth by about 13°F. The persistence of methane in the atmosphere warmed the earth enough to melt the ice in the oceans and at polar caps and completely change the global climate. If this theory turns out to be true, it highlights the importance of controlling the amount of methane, as well as carbon dioxide, that we release into the air. Certainly, harvesting the frozen methane of the oceans, if we chose to do it, must be done with great care.

FOR FURTHER UNDERSTANDING

News stories alternatively describe frozen methane as a "New Frontier" and "Armageddon." Explain these opposing views.

What are the ethical considerations involved in mining for frozen methane?

---

2 Learning Goal

Draw structures that represent each of the families of organic compounds.

Alkanes contain only carbon-to-carbon single bonds (C—C); alkenes have at least one carbon-to-carbon double bond (C=C); and alkynes have at least one carbon-to-carbon triple bond (C≡C).

Families of Organic Compounds

The most general classification of organic compounds divides them into hydrocarbons and substituted hydrocarbons. A **hydrocarbon** molecule contains only carbon and hydrogen. A **substituted hydrocarbon** is one in which one or more hydrogen atoms are replaced by another atom or group of atoms.

Hydrocarbons can be further subdivided into aliphatic and aromatic hydrocarbons (Figure 10.2). The four families of **aliphatic hydrocarbons** are alkanes, cycloalkanes, alkenes, and alkynes.

Alkanes are **saturated hydrocarbons** because they contain only carbon and hydrogen and have only carbon-to-hydrogen and carbon-to-carbon single bonds. Alkenes and alkynes are **unsaturated hydrocarbons** because they contain at least one carbon-to-carbon double or triple bond, respectively.

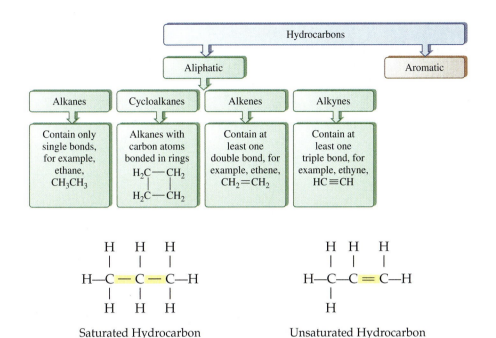

Figure 10.2

The family of hydrocarbons is divided into two major classes: aliphatic and aromatic. Aliphatic hydrocarbons are further subdivided into four major subclasses: alkanes, cycloalkanes, alkenes, and alkynes.

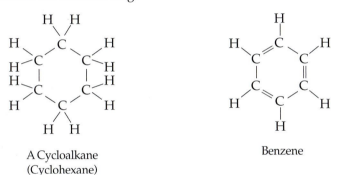

Saturated Hydrocarbon   Unsaturated Hydrocarbon

Recall that each of the lines in these structures represents a shared pair of electrons. See Chapter 3.

Some hydrocarbons are cyclic. Cycloalkanes consist of carbon atoms bonded to one another to produce a ring. **Aromatic hydrocarbons** contain a benzene ring or a derivative of the benzene ring.

A Cycloalkane
(Cyclohexane)

Benzene

We will learn a more accurate way to represent benzene in Section 11.5.

A substituted hydrocarbon is produced by replacing one or more hydrogen atoms with a functional group. A **functional group** is an atom or group of atoms arranged in a particular way that is primarily responsible for the chemical and physical properties of the molecule in which it is found. Although hydrocarbons have little biological activity, the addition of a functional group confers unique and interesting properties that give the molecule important biological or medical properties.

All compounds that have a particular functional group are members of the same family. We have just seen that alkenes are characterized by the presence of carbon-to-carbon double bonds. Similarly, all alcohols contain a hydroxyl group (—OH). The addition of the hydroxyl group to a hydrocarbon such as ethane causes the new molecule, ethanol, to be polar. As a result, ethanol (CH_3CH_2OH) is liquid at room temperature and highly soluble in water, whereas the alkane of similar molecular weight, propane ($CH_3CH_2CH_3$), is a gas at room temperature and is completely insoluble in water. Other common functional groups are shown in Table 10.2, along with an example of a molecule from each family.

The chemistry of organic and biological molecules is usually controlled by the functional group in the molecule. Just as members of the same family of the periodic table exhibit similar chemistry, organic molecules with the same functional group exhibit similar chemistry. Although it would be impossible to learn the chemistry of each organic molecule, it is relatively easy to learn the chemistry of

3 Learning Goal

Write the names and draw the structures of the common functional groups.

This is analogous to the classification of the elements within the periodic table. See Chapter 2.

TABLE 10.2 Common Functional Groups

Type of Compound	Structural Formula	Condensed Formula	Examples Structural Formula	I.U.P.A.C. Name	Common Name
Alcohol	R—O—H	ROH	CH_3CH_2—O—H	Ethanol	Ethyl alcohol
Aldehyde	R—C(=O)—H	RCHO	CH_3C(=O)—H	Ethanal	Acetaldehyde
Amide	R—C(=O)—N(H)—H	$RCONH_2$	CH_3C(=O)—N(H)—H	Ethanamide	Acetamide
Amine	R—N(H)—H	RNH_2	CH_3CH_2N(H)—H	Ethanamine	Ethyl amine
Carboxylic acid	R—C(=O)—O—H	RCOOH	CH_3C(=O)—O—H	Ethanoic acid	Acetic acid
Ester	R—C(=O)—O—R′	RCOOR′	CH_3C(=O)—OCH_3	Methyl ethanoate	Methyl acetate
Ether	R—O—R′	ROR′	CH_3OCH_3	Methoxymethane	Dimethyl ether
Halide	R—Cl (or —Br, —F, —I)	RCl	CH_3CH_2Cl	Chloroethane	Ethyl chloride
Ketone	R—C(=O)—R′	RCOR′	CH_3CCH_3	Propanone	Acetone
Alkene	R₂C=CR₂ (RHC=CHR)	RCHCHR	$CH_3CH{=}CH_2$	Propene	Propylene
Alkyne	R—C≡C—R	RCCR	$CH_3C{\equiv}CH$	Propyne	Methyl acetylene

the families of molecules containing each functional group. In this way, you can learn the chemistry of all members of a family of organic compounds, or biological molecules, just by learning the chemistry of its characteristic functional group or groups.

10.2 Alkanes

Alkanes are saturated hydrocarbons; that is, alkanes contain only carbon and hydrogen bonded together through carbon–hydrogen and carbon–carbon single bonds. C_nH_{2n+2} is the general formula for alkanes. In this formula, n is the number of carbon atoms in the molecule. For example, $n = 3$ would indicate 3 carbon atoms and $2(3) + 2$ or 8 hydrogen atoms and a formula of C_3H_8.

Structure and Physical Properties

4 Learning Goal

Write condensed and structural formulas for saturated hydrocarbons.

Four types of formulas, each providing different information, are used in organic chemistry: the molecular formula, the structural formula, the condensed formula, and the line formula.

The **molecular formula** tells the kind and number of each type of atom in a molecule but does not show the bonding pattern. Consider the molecular formulas for simple alkanes:

CH_4 C_2H_6 C_3H_8 C_4H_{10}

Methane   Ethane   Propane   Butane

For the first three compounds, there is only one possible arrangement of the atoms. However, for C_4H_{10} there are two possible arrangements. How do we know which is correct? The problem is solved by using the **structural formula,** which shows each atom and bond in a molecule. The following are the structural formulas for methane, ethane, propane, and the two isomers of butane:

Methane   Ethane   Propane   Butane   2-Methylpropane (isobutane)

> Recall that a covalent bond, representing a pair of shared electrons, can be drawn as a line between two atoms. For the structure to be correct, each carbon atom must show four pairs of shared electrons.

The advantage of a structural formula is that it shows the complete structure, but for large molecules it is time-consuming to draw and requires too much space. The compromise is the **condensed formula.** It shows all the atoms in a molecule and places them in a sequential order that indicates which atoms are bonded to which. The following are the condensed formulas for the preceding five compounds.

CH_4 CH_3CH_3 $CH_3CH_2CH_3$ $CH_3(CH_2)_2CH_3$ $(CH_3)_3CH$

Methane   Ethane   Propane   Butane   2-Methylpropane (isobutane)

The names and formulas of the first ten straight-chain alkanes are shown in Table 10.3.

The simplest representation of a molecule is the **line formula.** In the line formula, we assume that there is a carbon atom at any location where two or more lines intersect. We also assume that there is a carbon at the end of any line and that each

TABLE 10.3 Names and Formulas of the First Ten Straight-Chain Alkanes

Name	Molecular Formula	Condensed Formula	Melting Point, °C*	Boiling Point, °C*
Alkanes	C_nH_{2n+2}			
Methane	CH_4	CH_4	−182.5	−162.2
Ethane	C_2H_6	CH_3CH_3	−183.9	−88.6
Propane	C_3H_8	$CH_3CH_2CH_3$	−187.6	−42.1
Butane	C_4H_{10}	$CH_3CH_2CH_2CH_3$ or $CH_3(CH_2)_2CH_3$	−137.2	−0.5
Pentane	C_5H_{12}	$CH_3CH_2CH_2CH_2CH_3$ or $CH_3(CH_2)_3CH_3$	−129.8	36.1
Hexane	C_6H_{14}	$CH_3CH_2CH_2CH_2CH_2CH_3$ or $CH_3(CH_2)_4CH_3$	−95.2	68.8
Heptane	C_7H_{16}	$CH_3CH_2CH_2CH_2CH_2CH_2CH_3$ or $CH_3(CH_2)_5CH_3$	−90.6	98.4
Octane	C_8H_{18}	$CH_3CH_2CH_2CH_2CH_2CH_2CH_2CH_3$ or $CH_3(CH_2)_6CH_3$	−56.9	125.6
Nonane	C_9H_{20}	$CH_3CH_2CH_2CH_2CH_2CH_2CH_2CH_2CH_3$ or $CH_3(CH_2)_7CH_3$	−53.6	150.7
Decane	$C_{10}H_{22}$	$CH_3CH_2CH_2CH_2CH_2CH_2CH_2CH_2CH_2CH_3$ or $CH_3(CH_2)_8CH_3$	−29.8	174.0

*Melting and boiling points as reported in the National Institute of Standards and Technology Chemistry Webbook, which can be found at http://webbook.nist.gov/.

carbon in the structure is bonded to the correct number of hydrogen atoms. Compare the structural and line formulas for butane and 2-methylpropane, shown here:

Butane

2-Methylpropane

Each carbon atom forms four single covalent bonds, but each hydrogen atom has only a single covalent bond. Although a carbon atom may be involved in single, double, or triple bonds, it always shares four pairs of electrons. The Lewis dot structure of the simplest alkane, methane, shows the four shared pairs of electrons (Figure 10.3a). When carbon is involved in four single bonds, the *bond angle*, the angle between two atoms or substituents attached to carbon, is 109.5°, as predicted by the valence shell electron pair repulsion (VSEPR) theory. Thus, alkanes contain carbon atoms that have tetrahedral geometry.

Molecular geometry is described in Section 3.4.

A tetrahedron is a geometric solid having the structure shown in Figure 10.3b. There are many different ways to draw the tetrahedral carbon (Figures 10.3c–10.3e). In Figure 10.3c, solid lines, dashes, and wedges are used to represent the structure of methane. Dashes go back into the page away from you; wedges come out of the page toward you; and solid lines are in the plane of the page. The structure in Figure 10.3d is the same as that in Figure 10.3c; it just leaves a lot more to the imagination. Figure10.3e is a ball-and-stick model of the methane molecule. Three-dimensional drawings of two other simple alkanes are shown in Figure 10.4.

Figure 10.3

The tetrahedral carbon atom: (a) Lewis dot structure; (b) a tetrahedron; (c) the tetrahedral carbon drawn with dashes and wedges; (d) the stick drawing of the tetrahedral carbon atom; (e) ball-and-stick model of methane.

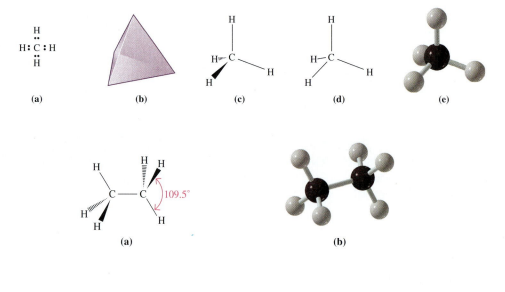

Figure 10.4

(a) Drawing and (b) ball-and-stick model of ethane. All of the carbon atoms have a tetrahedral arrangement, and all bond angles are approximately 109.5°.
(c) Drawing and (d) ball-and-stick model of a more complex alkane, butane.

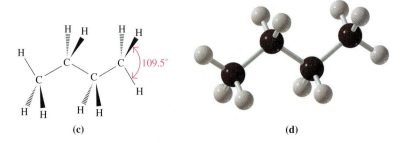

All hydrocarbons are nonpolar molecules. As a result they are not water-soluble but are soluble in nonpolar organic solvents. Furthermore, they have relatively low melting points and boiling points and are generally less dense than water. In general, the longer the hydrocarbon chain (greater the formula weight), the higher the melting and boiling points and the greater the density (see Table 10.3).

5 **Learning Goal**
Describe the relationship between the structure and physical properties of saturated hydrocarbons.

EXAMPLE 10.1 **Using Different Types of Formulas to Represent Organic Compounds**

Draw the structural and condensed formulas of the following line structure:

4 **Learning Goal**
Write condensed and structural formulas for saturated hydrocarbons.

SOLUTION

Remember that each intersection of lines represents a carbon atom and that each line ends in a carbon atom. This gives us the following carbon skeleton:

$$\underset{\displaystyle C-C-C-\overset{\displaystyle C}{|}-C-\overset{\displaystyle C}{|}-C}{}$$

By adding the correct number of hydrogen atoms to the carbon skeleton, we are able to complete the structural formula of this compound.

$$H-\overset{H}{\underset{H}{C}}-\overset{H}{\underset{H}{C}}-\overset{H}{\underset{H}{C}}-\overset{\overset{H}{|}}{\underset{H}{C}}(CH)-\overset{H}{\underset{H}{C}}-\overset{\overset{H}{|}}{\underset{H}{C}}(CH)-\overset{H}{\underset{H}{C}}-H$$

From the structural formula we can write the condensed formula as follows:

$$CH_3CH_2CH_2CH(CH_3)CH_2CH(CH_3)CH_3$$

Molecules having the same molecular formula but different arrangements of atoms are called structural or constitutional isomers. The carbon chains of these isomers may be straight (Table 10.3) or branched. Compare the following structural isomers of C_6H_{14}. Hexane has a straight chain or carbon backbone:

$$CH_3CH_2CH_2CH_2CH_2CH_3$$
Hexane

Structural or constitutional isomers are explored in detail later in this section.

All of the other isomers have one or more of the carbon atoms branching from the main carbon chain:

$\overset{\displaystyle CH_3}{|}$	$\overset{\displaystyle CH_3}{|}$	$\overset{\displaystyle CH_3}{|}$	$\overset{\displaystyle CH_3}{|}$
$CH_3CHCH_2CH_2CH_3$	$CH_3CH_2CHCH_2CH_3$	$CH_3CHCHCH_3$	$CH_3CCH_2CH_3$
		$\underset{\displaystyle CH_3}{|}$	$\underset{\displaystyle CH_3}{|}$
2-Methylpentane	3-Methylpentane	2,3-Dimethylbutane	2,2-Dimethylbutane

These branched-chain forms of the molecule have a much smaller surface area than the straight-chain. As a result, the van der Waals forces attracting the molecules to

van der Waals forces are discussed in Section 6.2.

TABLE 10.4 Melting and Boiling Points of Five Alkanes of Molecular Formula C_6H_{14}

Name	Condensed Formula	Boiling Point* °C	Melting Point* °C
Hexane	$CH_3CH_2CH_2CH_2CH_2CH_3$	68.8	−95.2
2-Methylpentane	$CH_3CH(CH_3)CH_2CH_2CH_3$	60.9	−153.2
3-Methylpentane	$CH_3CH_2CH(CH_3)CH_2CH_3$	63.3	−118
2,3-Dimethylbutane	$CH_3CH(CH_3)CH(CH_3)CH_3$	58.1	−130.2
2,2-Dimethylbutane	$CH_3C(CH_3)_2CH_2CH_3$	49.8	−100.2

*Melting and boiling points as reported in the National Institute of Standards and Technology Chemistry Webbook, which can be found at http://webbook.nist.gov/.

one another are less strong and these molecules have lower melting and boiling points than the straight-chain isomers. The melting and boiling points of the five structural isomers of C_6H_{14} are shown in Table 10.4.

QUESTION 10.3

Draw the structural formula for each of the molecules in Table 10.4.

QUESTION 10.4

Draw the line formula for each of the molecules in Table 10.4.

Alkyl Groups

Alkyl groups are alkanes with one fewer hydrogen atom. The name of the alkyl group is derived from the name of the alkane containing the same number of carbon atoms. The *-ane* ending of the alkane name is replaced by the *-yl* ending. Thus, $—CH_3$ is a methyl group and $—CH_2CH_3$ is an ethyl group. The dash in these two structures represents the point at which the alkyl group can bond to another atom. The first five continuous-chain alkyl groups are presented in Table 10.5.

Carbon atoms are classified according to the number of other carbon atoms to which they are attached. A **primary carbon (1°)** is directly bonded to one other carbon. A **secondary carbon (2°)** is bonded to two other carbon atoms, a **tertiary**

TABLE 10.5 Names and Formulas of the First Five Continuous-Chain Alkyl Groups

Alkyl Group Structure	Name
$—CH_3$	Methyl
$—CH_2CH_3$	Ethyl
$—CH_2CH_2CH_3$	Propyl
$—CH_2CH_2CH_2CH_3$	Butyl
$—CH_2CH_2CH_2CH_2CH_3$	Pentyl

TABLE 10.6 Structures and Names of Some Branched-Chain Alkyl Groups

Structure	Classification	Common Name	I.U.P.A.C. Name
$CH_3CH—$ $\vert$ CH_3	2°	Isopropyl*	1-Methylethyl
CH_3 $\vert$ $CH_3CHCH_2—$	1°	Isobutyl*	2-Methylpropyl
CH_3 $\vert$ $CH_3CH_2CH—$	2°	*sec*-Butyl†	1-Methylpropyl
CH_3 $\vert$ $CH_3C—$ $\vert$ CH_3	3°	*t*-Butyl or *tert*-Butyl‡	1,1-Dimethylethyl

*The prefix *iso*- (isomeric) is used when there are two methyl groups at the end of the alkyl group.
†The prefix *sec*- (secondary) indicates that there are two carbons bonded to the carbon that attaches the alkyl group to the parent molecule.
‡The prefix *t*- or *tert*- (tertiary) means that three carbons are attached to the carbon that attaches the alkyl group to the parent molecule.

carbon (3°) is bonded to three other carbon atoms, and a **quaternary carbon (4°)** to four.

Alkyl groups are classified according to the number of carbons attached to the carbon atom that joins the alkyl group to a molecule.

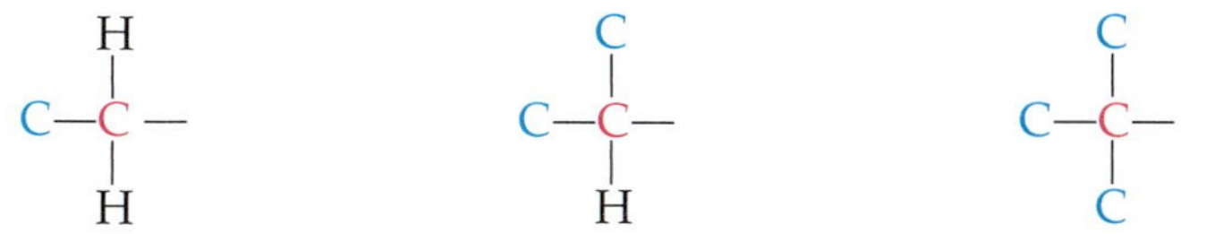

Primary alkyl group — Secondary alkyl group — Tertiary alkyl group

All of the continuous-chain alkyl groups are primary alkyl groups (see Table 10.5). Several branched-chain alkyl groups are shown in Table 10.6. Notice that the isopropyl and *sec*-butyl groups are secondary alkyl groups, the isobutyl group is a primary alkyl group, and the *t*-butyl (*tert*-butyl) is a tertiary alkyl group.

QUESTION 10.5

Classify each of the carbon atoms in the following structures as either primary, secondary, or tertiary.

a. 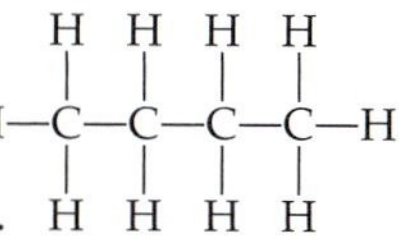

b. 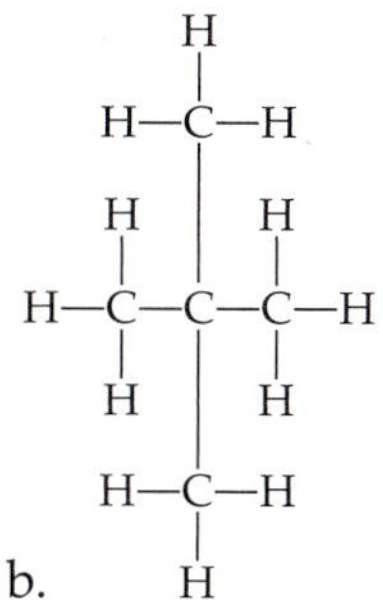

c. 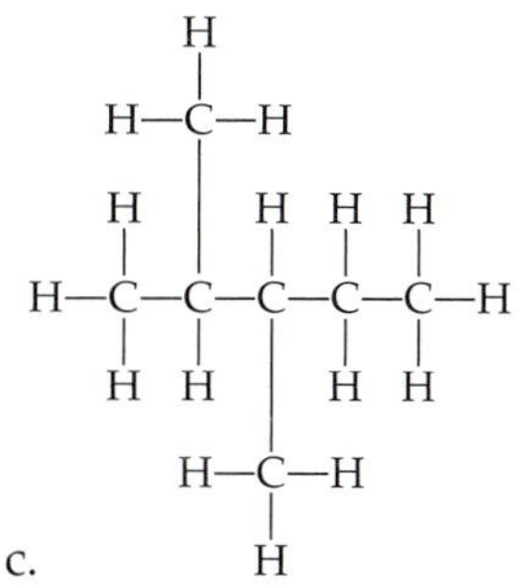

QUESTION 10.6

Classify each of the carbon atoms in the following structures as either primary, secondary, or tertiary.

   a. $CH_3CH_2C(CH_3)_2CH_2CH_3$
   b. $CH_3CH_2CH_2CH_2CH(CH_3)CH(CH_3)CH_3$

Nomenclature

6   **Learning Goal**

Use the basic rules of the I.U.P.A.C. Nomenclature System to name alkanes and substituted alkanes.

Historically, organic compounds were named by the chemist who discovered them. Often the names reflected the source of the compound. For instance, the antibiotic penicillin is named for the mold *Penicillium notatum*, which produces it. The pain reliever aspirin was made by adding an acetate group to a compound first purified from the bark of a willow tree and later from the meadowsweet plant (*Spirea ulmaria*). Thus, the name aspirin comes from *a* (acetate) and *spir* (genus of meadowsweet).

These names are easy for us to remember because we come into contact with these compounds often. However, as the number of compounds increased, organic chemists realized that historical names were not adequate because they revealed nothing about the structure of a compound. Thousands of such compounds and their common names had to be memorized! What was needed was a set of nomenclature (naming) rules that would produce a unique name for every organic compound. Furthermore, the name should be so descriptive that, by knowing the name, a student or scientist could write the structure.

Throughout this book, we will primarily use the I.U.P.A.C. Nomenclature System. When common names are used, they will be shown in parentheses beneath the I.U.P.A.C. name.

The International Union of Pure and Applied Chemistry (I.U.P.A.C.) is the organization responsible for establishing and maintaining a standard, universal system for naming organic compounds. The system of nomenclature developed by this group is called the **I.U.P.A.C. Nomenclature System.** The following rules are used for naming alkanes by the I.U.P.A.C. system.

1. Determine the name of the **parent compound,** the longest, continuous, carbon chain in the compound. Refer to Tables 10.3 and 10.7 to determine the parent name. Notice that these names are made up of a prefix related to the number of carbons in the chain and the suffix *-ane*, indicating that the molecule is an alkane (Table 10.7). Write down the name of the parent compound, leaving

It is important to learn the prefixes for the carbon chain lengths. We will use them in the nomenclature for all organic molecules.

TABLE 10.7 Carbon Chain Length and Prefixes Used in the I.U.P.A.C. Nomenclature System		
Carbon Chain Length	**Prefix**	**Alkane Name**
1	Meth-	Meth*ane*
2	Eth-	Eth*ane*
3	Prop-	Prop*ane*
4	But-	But*ane*
5	Pent-	Pent*ane*
6	Hex-	Hex*ane*
7	Hept-	Hept*ane*
8	Oct-	Oct*ane*
9	Non-	Non*ane*
10	Dec-	Dec*ane*

space before the name to identify the substituents. Parent chains are highlighted in yellow in the following examples:

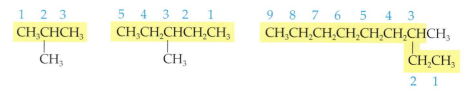

Parent name: Propane   Pentane   Nonane

2. Number the parent chain to give the lowest number to the carbon bonded to the first group encountered on the parent chain, regardless of the numbers that result for the other substituents.

3. Name and number each atom or group attached to the parent compound. The number tells you the position of the group on the main chain, and the name tells you which type of substituent is present at that position. For example, it may be one of the halogens [F-(fluoro), Cl-(chloro), Br-(bromo), and I-(iodo)] or an alkyl group (Tables 10.5 and 10.6). In the following examples, the parent chain is highlighted in yellow:

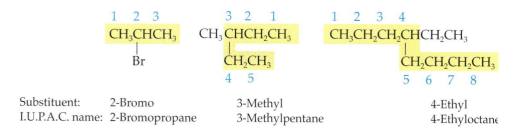

Substituent: 2-Bromo   3-Methyl   4-Ethyl
I.U.P.A.C. name: 2-Bromopropane   3-Methylpentane   4-Ethyloctane

4. If the same substituent occurs more than once in the compound, a separate position number is given for each, and the prefixes *di-, tri-, tetra-, penta-,* and so forth are used, as shown in the following examples:

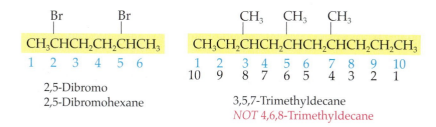

2,5-Dibromo
2,5-Dibromohexane

3,5,7-Trimethyldecane
NOT 4,6,8-Trimethyldecane

5. Place the names of the substituents in alphabetical order before the name of the parent compound, which you wrote down in Step 1. Numbers are separated by commas, and numbers are separated from names by hyphens. By convention, halogen substituents are placed before alkyl substituents.

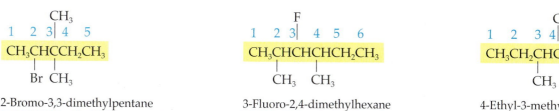

2-Bromo-3,3-dimethylpentane
NOT 3,3-Dimethyl-2-bromopentane

3-Fluoro-2,4-dimethylhexane
NOT 2,4-Dimethyl-3-fluorohexane

4-Ethyl-3-methyloctane
NOT 3-Methyl-4-ethyloctane

6 Learning Goal

Use the basic rules of the I.U.P.A.C. Nomenclature System to name alkanes and substituted alkanes.

EXAMPLE 10.2 Naming Substituted Alkanes Using the I.U.P.A.C. System

Name the following alkanes using I.U.P.A.C. nomenclature.

SOLUTION

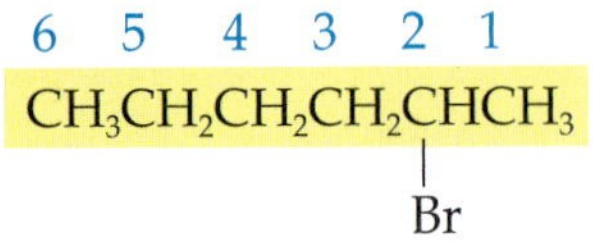

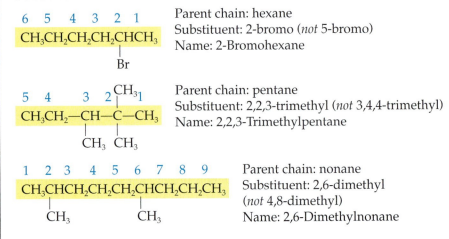

Parent chain: hexane
Substituent: 2-bromo (*not* 5-bromo)
Name: 2-Bromohexane

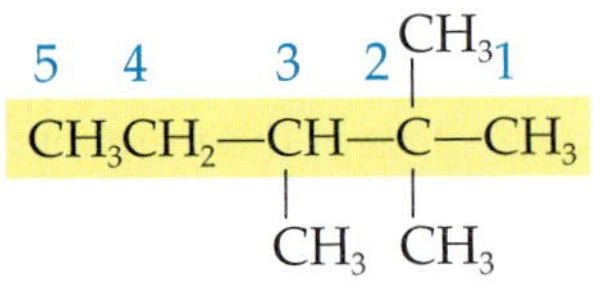

Parent chain: pentane
Substituent: 2,2,3-trimethyl (*not* 3,4,4-trimethyl)
Name: 2,2,3-Trimethylpentane

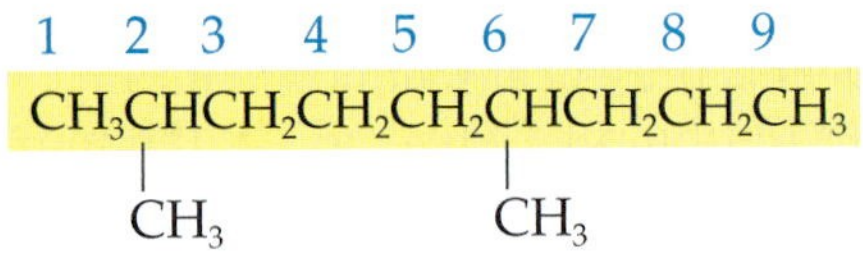

Parent chain: nonane
Substituent: 2,6-dimethyl (*not* 4,8-dimethyl)
Name: 2,6-Dimethylnonane

QUESTION 10.7

Name the following compounds, using the I.U.P.A.C. Nomenclature System:

a. 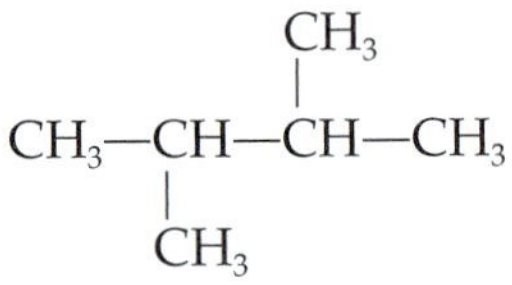

c. 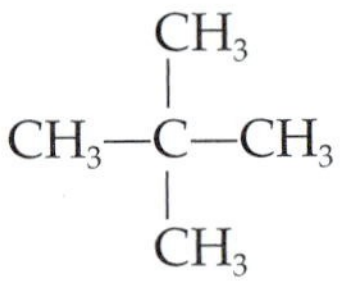

b. 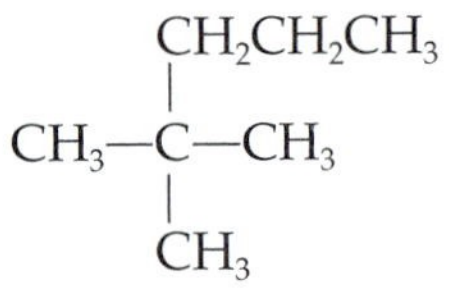

d. 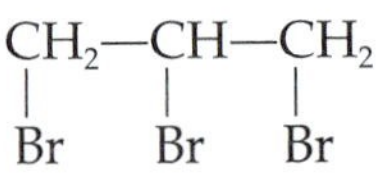

QUESTION 10.8

Name the following compounds, using the I.U.P.A.C. Nomenclature System:

a. 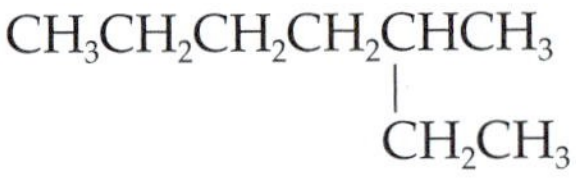

c. 

b. 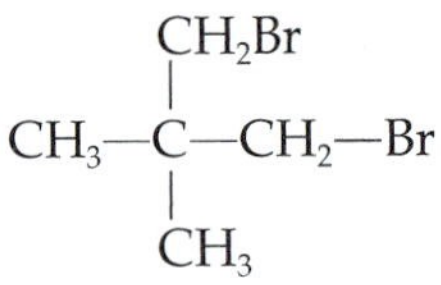

d. 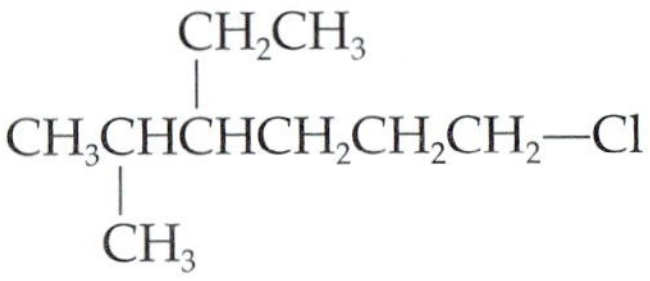

Having learned to name a compound using the I.U.P.A.C. system, we can easily write the structural formula of a compound, given its name. First, draw and number the parent carbon chain. Add the substituent groups to the correct carbon and finish the structure by adding the correct number of hydrogen atoms.

EXAMPLE 10.3 Drawing the Structure of a Compound Using the I.U.P.A.C. Name

Draw the structural formula for 1-bromo-4-methylhexane.

SOLUTION

Begin by drawing the six-carbon parent chain and indicating the four bonds for each carbon atom.

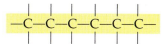

Next, number each carbon atom:

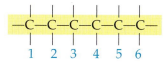

Now add the substituents. In this example a bromine atom is bonded to carbon-1, and a methyl group is bonded to carbon-4:

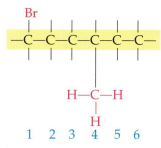

Finally, add the correct number of hydrogen atoms so that each carbon has four covalent bonds:

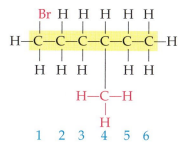

As a final check of your accuracy, use the I.U.P.A.C. system to name the compound that you have just drawn, and compare the name with that in the original problem.

The molecular formula and condensed formula can be written from the structural formula shown. The molecular formula is $C_7H_{15}Br$, and the condensed formula is $BrCH_2CH_2CH_2CH(CH_3)CH_2CH_3$.

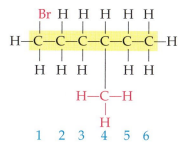

4 Learning Goal
Write condensed and structural formulas for saturated hydrocarbons.

6 Learning Goal
Use the basic rules of the I.U.P.A.C. Nomenclature System to name alkanes and substituted alkanes.

Constitutional or Structural Isomers

As we saw earlier, there are two arrangements of the atoms represented by the molecular formula C_4H_{10}: butane and 2-methylpropane. Molecules having the same molecular formula but a different arrangement of atoms are called **constitutional**

7 Learning Goal
Draw constitutional (structural) isomers of simple organic compounds.

isomers or **structural isomers.** These isomers are unique compounds because of their structural differences, and they have different physical and chemical properties. For instance, the data in Table 10.4 show that the branched-chain isomers of C_6H_{14} have lower boiling and melting points than the straight-chain isomer, hexane. These differences reflect the different shapes of the molecules.

7 **Learning Goal**

Draw constitutional (structural) isomers of simple organic compounds.

EXAMPLE 10.4 **Drawing Constitutional or Structural Isomers of Alkanes**

Write all the constitutional isomers having the molecular formula C_6H_{14}.

SOLUTION

1. Begin with the continuous six-carbon chain structure:

$$\overset{1}{C}H_3-\overset{2}{C}H_2-\overset{3}{C}H_2-\overset{4}{C}H_2-\overset{5}{C}H_2-\overset{6}{C}H_3$$

Isomer A

2. Now try five-carbon chain structures with a methyl group attached to one of the internal carbon atoms of the chain:

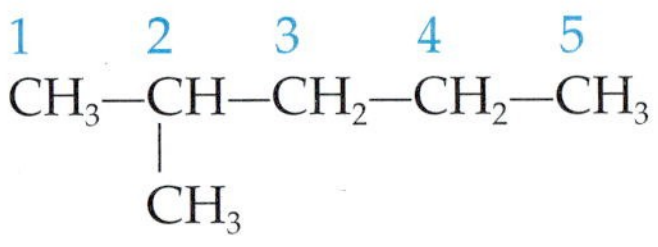

$$\overset{1}{C}H_3-\overset{2}{C}H-\overset{3}{C}H_2-\overset{4}{C}H_2-\overset{5}{C}H_3 \quad \text{and} \quad \overset{1}{C}H_3-\overset{2}{C}H_2-\overset{3}{C}H-\overset{4}{C}H_2-\overset{5}{C}H_3$$
$$\qquad\qquad | \qquad\qquad\qquad\qquad\qquad\qquad\qquad | $$
$$\qquad\quad CH_3 \qquad\qquad\qquad\qquad\qquad\qquad\qquad CH_3$$

Isomer B   Isomer C

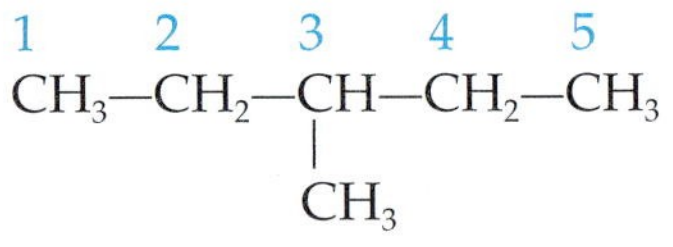

3. Next consider the possibilities for a four-carbon structure to which two methyl groups (—CH_3) may be attached:

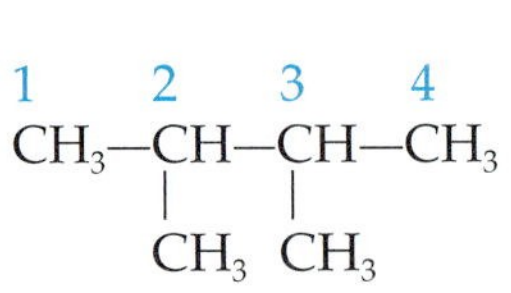

$$\qquad\qquad\qquad\qquad\qquad\qquad CH_3$$
$$\overset{1}{C}H_3-\overset{2}{C}H-\overset{3}{C}H-\overset{4}{C}H_3 \quad \text{and} \quad \overset{1}{C}H_3-\overset{2}{C}-\overset{3}{C}H_2-\overset{4}{C}H_3$$
$$\qquad\quad | \quad | \qquad\qquad\qquad\qquad\qquad | $$
$$\qquad CH_3 \; CH_3 \qquad\qquad\qquad\qquad CH_3$$

Isomer D   Isomer E

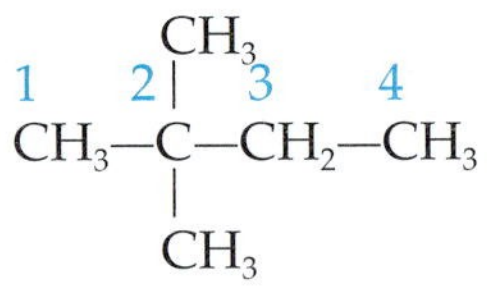

These are the five possible constitutional isomers of C_6H_{14}. At first, it may seem that other isomers are also possible. But careful comparison will show that they are duplicates of those already constructed. For example, rather than add two methyl groups, a single ethyl group (—CH_2CH_3) could be added to the four-carbon chain:

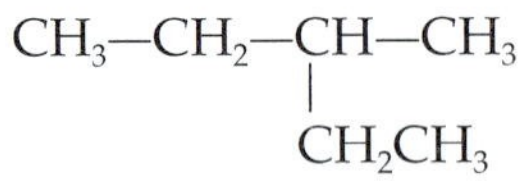

$$CH_3-CH_2-CH-CH_3$$
$$\qquad\qquad\qquad | $$
$$\qquad\qquad CH_2CH_3$$

But close examination will show that this is identical to isomer C. Perhaps we could add one ethyl group and one methyl group to a three-carbon parent chain, with the following result:

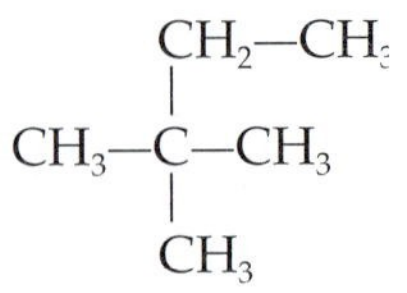

$$\qquad\qquad CH_2-CH_3$$
$$\qquad\qquad\quad | $$
$$CH_3-C-CH_3$$
$$\qquad\quad | $$
$$\qquad CH_3$$

Again we find that this structure is the same as one of the isomers we have already identified, isomer E.

Continued—

To check whether you have accidentally made duplicate isomers, name them using the I.U.P.A.C. system. All isomers must have different I.U.P.A.C. names. So if two names are identical, the structures are also identical. Use the I.U.P.A.C. system to name the isomers in this example, and prove to yourself that the last two structures are simply duplicates of two of the original five isomers.

QUESTION 10.9

Draw a complete structural formula for each of the straight-chain isomers of the following alkanes:

a. C_4H_9Br

b. $C_4H_8Br_2$

QUESTION 10.10

Name all of the isomers that you obtained in Question 10.9.

10.3 Cycloalkanes

Cycloalkanes are a family having C—C single bonds in a ring structure. They have the general molecular formula C_nH_{2n} and thus have two fewer hydrogen atoms than the corresponding alkane (C_nH_{2n+2}). The structures and names of some simple cycloalkanes are shown in Figure 10.5.

In the I.U.P.A.C. system, cycloalkanes are named by applying the following simple rules.

- Determine the name of the alkane with the same number of carbon atoms as there are within the ring and add the prefix *cyclo-*. For example, cyclopentane is the cycloalkane that has five carbon atoms.

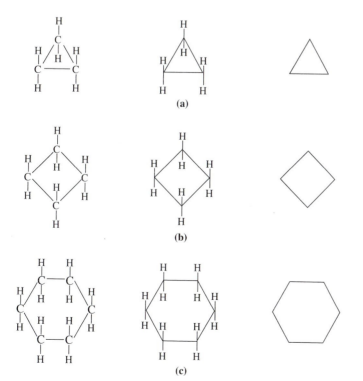

Figure 10.5

Cycloalkanes: (a) cyclopropane; (b) cyclobutane; (c) cyclohexane. All of the cycloalkanes are shown using structural formulas (left column), condensed structural formulas (center column), and line formulas (right column).

- If the cycloalkane is substituted, place the names of the groups in alphabetical order before the name of the cycloalkane. No number is needed if there is only one substituent.
- If more than one group is present, use numbers that result in the *lowest possible position numbers.*

8 Learning Goal
Write the names and draw the structures of simple cycloalkanes.

EXAMPLE 10.5 Naming a Substituted Cycloalkane Using the I.U.P.A.C. Nomenclature System

Name the following cycloalkanes using I.U.P.A.C. nomenclature.

SOLUTION

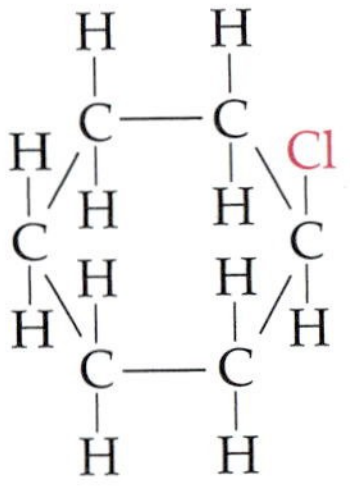

Parent chain: cyclohexane

Substituent: chloro (no number is required because there is only one substituent)

Name: Chlorocyclohexane

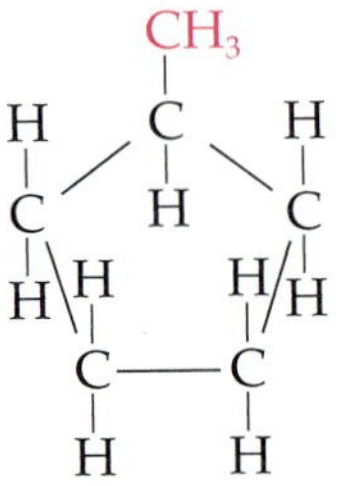

Parent chain: cyclopentane

Substituent: methyl (no number is required because there is only one substituent)

Name: Methylcyclopentane

These cycloalkanes could also be shown as line formulas, as shown below. Each line represents a carbon–carbon bond. A carbon atom and the correct number of hydrogen atoms are assumed to be at the point where the lines meet and at the end of a line.

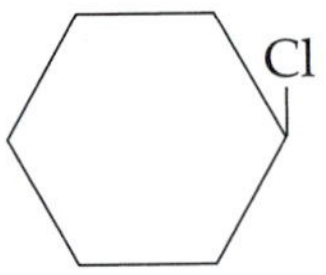

Chlorocyclohexane

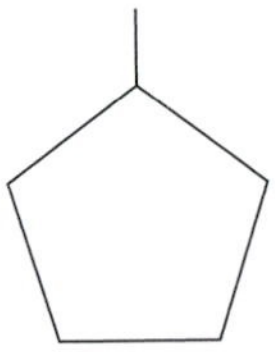

Methylcyclopentane

QUESTION 10.11

Name each of the following substituted cycloalkanes using the I.U.P.A.C. Nomenclature System:

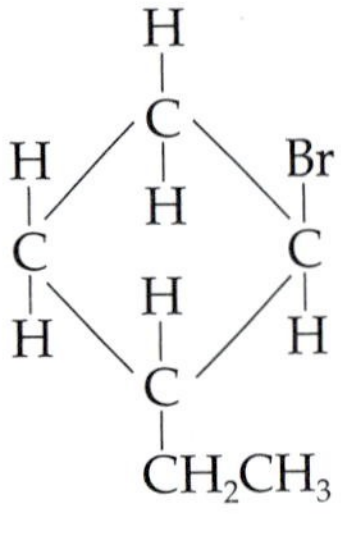

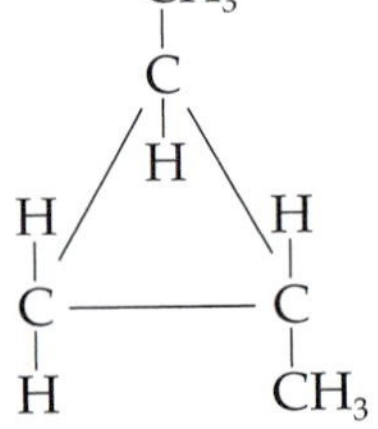

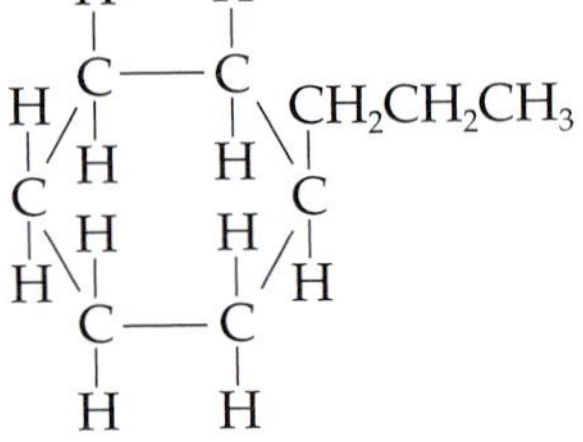

AN ENVIRONMENTAL Perspective

The Petroleum Industry and Gasoline Production

Petroleum consists primarily of alkanes and small amounts of alkenes and aromatic hydrocarbons. Substituted hydrocarbons, such as phenol, are also present in very small quantities. Although the composition of petroleum varies with the source of the petroleum (United States, Persian Gulf, etc.), the mixture of hydrocarbons can be separated into its component parts on the basis of differences in the boiling points of various hydrocarbons (distillation). Often several successive distillations of various fractions of the original mixture are required to completely purify the desired component. In the first distillation, the petroleum is separated into several fractions, each of which consists of a mix of hydrocarbons. Each fraction can be further purified by successive distillations. On an industrial scale, these distillations are carried out in columns that may be hundreds of feet high.

The gasoline fraction of petroleum, called straight-run gasoline, consists primarily of alkanes and cycloalkanes with six to twelve carbon atoms in the skeleton. This fraction has very poor fuel performance. Branched-chain alkanes are superior to straight-chain alkanes as fuels because they are more volatile and burn less rapidly and more efficiently in the cylinder. Alkenes and aromatic hydrocarbons are also good fuels. Methods have been developed to convert hydrocarbons of higher and lower molecular weights than gasoline to the appropriate molecular weight range and to convert straight-chain hydrocarbons into branched ones. *Catalytic cracking* fragments a large hydrocarbon into smaller ones. *Catalytic reforming* results in the rearrangement of a hydrocarbon into a more useful form.

The quality of a fuel is measured as its octane rating. Heptane is a very poor fuel and is given an octane rating of zero. 2,2,4-Trimethylpentane (commonly called isooctane) is an excellent fuel and is given an octane rating of one hundred. Gasoline octane ratings are experimentally determined by comparison with these two compounds in test engines.

One environmental consequence of our dependence on oil and oil by-products is the release of oil through accidental and even purposeful spills. To learn more about oil spills and cleanup

Mining the sea for hydrocarbons.

strategies, see *An Environmental Perspective: Oil-Eating Microbes* at www.mhhe.com/denniston.

FOR FURTHER UNDERSTANDING

Explain why the mixture of hydrocarbons in crude oil can be separated by distillation.

Draw the structures of heptane and 2,2,4-trimethylpentane (isooctane).

QUESTION 10.12

Name each of the following substituted cycloalkanes using the I.U.P.A.C. Nomenclature System:

10.4 Reactions of Alkanes and Cycloalkanes

Combustion

9 Learning Goal

Write equations for combustion reactions of alkanes.

Alkanes, cycloalkanes, and other hydrocarbons can be oxidized (by burning) in the presence of excess molecular oxygen. In this reaction, called **combustion,** they burn at high temperatures, producing carbon dioxide and water and releasing large amounts of energy as heat.

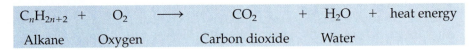

$$C_nH_{2n+2} \;+\; O_2 \;\longrightarrow\; CO_2 \;+\; H_2O \;+\; \text{heat energy}$$

Alkane   Oxygen   Carbon dioxide   Water

Combustion reactions are discussed in Section 4.3.

The following examples show a combustion reaction for a simple alkane and a simple cycloalkane:

$$CH_4 + 2O_2 \longrightarrow CO_2 + 2H_2O + \text{heat energy}$$

Methane

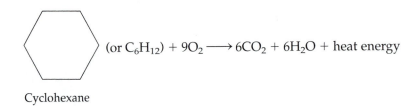

$$(\text{or } C_6H_{12}) + 9O_2 \longrightarrow 6CO_2 + 6H_2O + \text{heat energy}$$

Cyclohexane

EXAMPLE 10.6   Balancing Equations for the Combustion of Alkanes

9 Learning Goal

Write equations for combustion reactions of alkanes.

Balance the following equation for the combustion of hexane:

$$C_6H_{14} + O_2 \longrightarrow CO_2 + H_2O$$

SOLUTION

First, balance the carbon atoms; there are 6 mol of carbon atoms on the left and only 1 mol of carbon atoms on the right:

$$C_6H_{14} + O_2 \longrightarrow 6CO_2 + H_2O$$

Next, balance hydrogen atoms; there are 14 mol of hydrogen atoms on the left and only 2 mol of hydrogen atoms on the right:

$$C_6H_{14} + O_2 \longrightarrow 6CO_2 + 7H_2O$$

Now there are 19 mol of oxygen atoms on the right and only 2 mol of oxygen atoms on the left. Therefore, a coefficient of 9.5 is needed for O_2.

$$C_6H_{14} + 9.5O_2 \longrightarrow 6CO_2 + 7H_2O$$

Although decimal coefficients are sometimes used, it is preferable to have all integral coefficients. Multiplying each term in the equation by 2 will satisfy this requirement, giving us the following balanced equation:

$$2C_6H_{14} + 19O_2 \longrightarrow 12CO_2 + 14H_2O$$

The equation is now balanced with 12 mol of carbon atoms, 28 mol of hydrogen atoms and 38 mol of oxygen atoms on each side of the equation.

CHEMISTRY at the Crime Scene | Arson and Alkanes

In September of 2005, Thomas Sweatt was sentenced to life in prison for setting 45 fires in the Washington, D.C. area. Aside from millions of dollars in property damage, two people died as a result of these fires. Mr. Sweatt confessed to the fires, which terrorized the Washington metropolitan area over a two-year period, stating that he was "addicted to setting fires."

Authorities estimate that one-third of all fires are arson and the Federal Bureau of Investigation reports that arson is more common in the United States than anywhere else in the world. Although arsonists may set fires as terrorist acts, to defraud insurance companies, to gain revenge, or to cover up another crime, other arsonists are mentally ill pyromaniacs like Mr. Sweatt.

As soon as a fire is extinguished and the scene is secure, investigators immediately gather evidence to determine the cause of the fire. They study the pattern of the fire to determine the point of origin. This is critical, because this is where they must sample for the presence of accelerants, flammable substances that cause fires to burn hotter and spread more quickly. The most common accelerants are mixtures of hydrocarbons, including gasoline, kerosene, or diesel fuel.

Because all of these accelerants contain molecules that evaporate, they may be detected at the point of origin by trained technicians or "sniffer dogs." However, a much more advanced technology is also available; it is called *headspace gas chromatography*. Gas chromatography separates and identifies components of a sample based on differences in their boiling points. Each gas mixture produces its own unique "chemical fingerprint" or chromatogram. Crime scene technicians collect debris from the point of origin and seal it in an airtight vial. In the laboratory, they heat the vial so the hydrocarbons evaporate and are trapped in the headspace of the vial. These gases are then collected with a needle and syringe and injected into the gas chromatogram for analysis and identification.

To be absolutely certain that an accelerant has been used, the crime scene technicians also collect debris from control sites away from the point of origin. The reason for this is that pyrolysis, the decomposition or transformation of a compound

caused by heat, may produce products that simulate accelerants. If an accelerant is found at the point of origin and not among the other pyrolysis products, it can be concluded that arson was the cause of the blaze.

Of course, the next priority is to catch the arsonist. Crime scene technicians collect and analyze physical evidence, including fingerprints, footprints, and other artifacts, found at the crime scene. In the case of Mr. Sweatt, it was DNA fingerprint evidence from articles of clothing left at the crime scenes that led to his capture and conviction and ended his two-year arson spree.

FOR FURTHER UNDERSTANDING

Investigate the composition of gasoline, diesel fuel, and kerosene and explain how they can be distinguished from one another using gas chromatography.

Other accelerants that have been used by arsonists include nail polish remover (acetone), grain alcohol (ethanol), and rubbing alcohol (2-propanol or isopropyl alcohol.) What properties do these substances share that make them useful as accelerants?

The energy they release, along with their availability and relatively low cost, makes hydrocarbons very useful as fuels. In fact, combustion is essential to our very existence. It is the process by which we heat our homes, run our cars, and generate electricity. Although combustion of fossil fuels is vital to industry and society, it also represents a threat to the environment. The buildup of CO_2 may contribute to global warming and change the face of the earth in future generations.

See An Environmental Perspective: The Greenhouse Effect and Global Warming in Chapter 6.

Halogenation

10 Learning Goal
Write equations for halogenation reactions of alkanes.

Alkanes and cycloalkanes can also react with a halogen (usually chlorine or bromine) in a reaction called **halogenation.** Halogenation is a **substitution reaction,** that is, a reaction that results in the replacement of one group for another. In this reaction, a halogen atom is substituted for one of the hydrogen atoms in the alkane. The products of this reaction are an **alkyl halide** or *haloalkane* and a hydrogen halide.

The halogenation reaction is of great value because it converts unreactive alkanes into versatile starting materials for the synthesis of desired compounds. This is important in the pharmaceutical industry for the synthesis of some drugs. In addition, alkyl halides having two or more halogen atoms are useful solvents, refrigerants, insecticides, and herbicides.

Halogenation can occur only in the presence of heat and/or light, as indicated by the reaction conditions noted over the reaction arrows. The general equation for the halogenation of an alkane follows. The R in the general structure for the alkane may be either a hydrogen atom or an alkyl group.

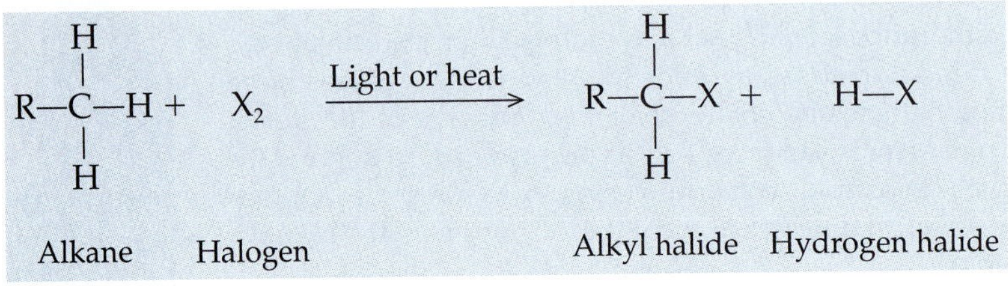

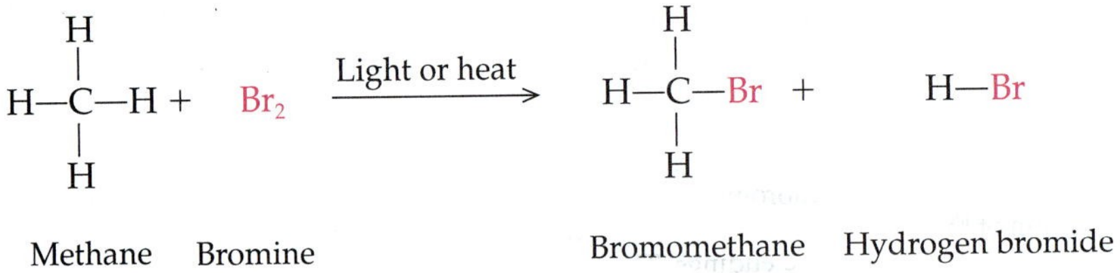

$$CH_3CH_3 + Cl_2 \xrightarrow{\text{Light}} CH_3CH_2{-}Cl + H{-}Cl$$

Ethane Chlorine Chloroethane Hydrogen chloride

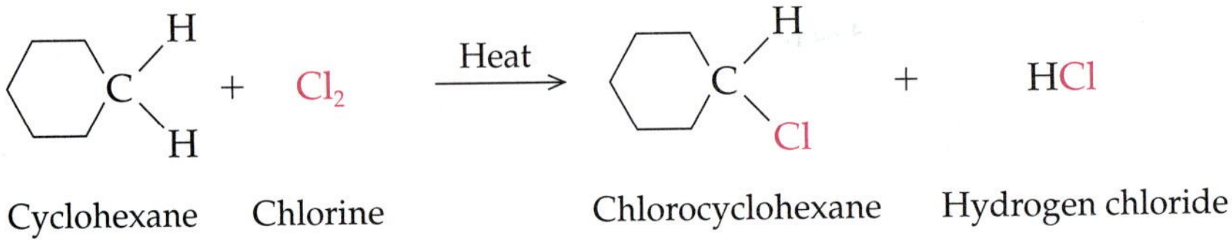

The alkyl halide may continue to react forming a mixture of products substituted at multiple sites or substituted multiple times at the same site.

If the halogenation reaction is allowed to continue, the alkyl halide formed may react with other halogen atoms. When this happens, a mixture of products may be formed. For instance, bromination of methane will produce bromomethane (CH_3Br), dibromomethane (CH_2Br_2), tribromomethane ($CHBr_3$), and tetrabromomethane (CBr_4).

In more complex alkanes, halogenation can occur to some extent at all positions to give a mixture of monosubstituted products. For example, bromination of propane produces a mixture of 1-bromopropane and 2-bromopropane.

QUESTION 10.13

Write a balanced equation for each of the following reactions. Show all possible products.

a. the complete combustion of cyclobutane
b. the monobromination of propane
c. the complete combustion of ethane
d. the monochlorination of butane

A MEDICAL Connection | Polyhalogenated Hydrocarbons Used as Anesthetics

Polyhalogenated hydrocarbons are hydrocarbons containing two or more halogen atoms. Some polyhalogenated compounds are notorious for the problems they have caused humankind. For instance, some insecticides such as DDT, chlordane, kepone, and lindane do not break down rapidly in the environment. As a result, these toxic compounds accumulate in biological tissue of a variety of animals, including humans, and may cause neurological damage, birth defects, or even death.

Other halogenated hydrocarbons are very useful in medicine. They were among the first anesthetics (pain relievers) used routinely in medical practice. These chemicals played a central role as the studies of medicine and dentistry advanced into modern times.

$$CH_3CH_2-Cl \qquad CH_3-Cl$$

Chloroethane (ethyl chloride)

Chloromethane (methyl chloride)

Chloroethane and chloromethane are local anesthetics. A local anesthetic deadens the feeling in a portion of the body. Applied topically (on the skin), chloroethane and chloromethane numb the area. Rapid evaporation of these anesthetics lowers the skin temperature, deadening the local nerve endings. They act rapidly, but the effect is brief, and feeling is restored quickly.

$$CHCl_3$$

Trichloromethane (chloroform)

In the past, chloroform was used as both a general and a local anesthetic. When administered through inhalation, it rapidly causes loss of consciousness. However, the effects of this powerful anesthetic are of short duration. Chloroform is no longer used because it was shown to be carcinogenic. Recently chloroform and other trihalomethanes have been found in relatively high concentrations in indoor swimming pools. To learn more about this issue, see *A Medical Connection: Chloroform in Your Swimming Pool?* online at www.mhhe.com/denniston.

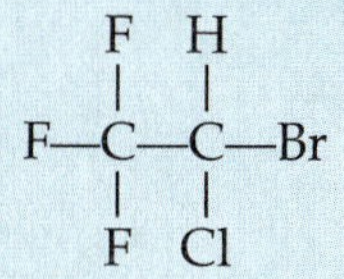

2-Bromo-2-chloro-1,1,1-trifluoroethane (Halothane)

Halothane is a general anesthetic that is administered by inhalation. It is considered a very safe anesthetic and is widely used.

FOR FURTHER UNDERSTANDING

In the first 24 hours following administration, 70% of the halothane is eliminated from the body in exhaled gases. Explain why halothane is so readily eliminated in exhaled gases.

Rapid evaporation of chloroethane from the skin surface causes cooling that causes local deadening of nerve endings. Explain why the skin surface cools dramatically as a result of evaporation of chloroethane.

QUESTION 10.14

Write a balanced equation for each of the following reactions. Show all possible products.

a. the complete combustion of decane
b. the monochlorination of cyclobutane
c. the monobromination of pentane
d. the complete combustion of hexane

QUESTION 10.15

Provide the I.U.P.A.C. names for the products of the reactions in Question 10.13b and 10.13d.

QUESTION 10.16

Provide the I.U.P.A.C. names for the products of the reactions in Question 10.14b and 10.14c.

SUMMARY OF REACTIONS

Reactions of Alkanes

Combustion:

$$C_nH_{2n+2} + O_2 \longrightarrow CO_2 + H_2O + \text{heat energy}$$

Alkane **Oxygen**   **Carbon** **Water**
                            dioxide

Halogenation:

$$
\begin{array}{ccc}
& \text{H} & & & \text{H} \\
& | & & & | \\
\text{R}-\text{C}-\text{H} & + & \text{X}_2 & \xrightarrow{\text{light or heat}} & \text{R}-\text{C}-\text{X} & + & \text{H}-\text{X} \\
& | & & & | \\
& \text{H} & & & \text{H}
\end{array}
$$

Alkane   **Halogen**   **Alkyl**   **Hydrogen**
                                      halide   **halide**

and numbering the carbon chain to provide the lowest possible number for all substituents. The substituent names and numbers are used as prefixes before the name of the parent compound.

Constitutional or *structural isomers* are molecules that have the same molecular formula but different structures. They have different physical and chemical properties because the atoms are bonded to one another in different patterns.

10.3 Cycloalkanes

Cycloalkanes are a family of organic molecules having C—C single bonds in a ring structure. They are named by adding the prefix *cyclo-* to the name of the alkane parent compound.

10.4 Reactions of Alkanes and Cycloalkanes

Alkanes can participate in *combustion* reactions. In complete combustion reactions, they are oxidized to produce carbon dioxide, water, and heat energy. They can also undergo *halogenation* reactions to produce *alkyl halides*.

SUMMARY

10.1 The Chemistry of Carbon

The modern science of organic chemistry began with Wöhler's synthesis of urea in 1828. At that time, people believed that it was impossible to synthesize an organic molecule outside a living system. We now define organic chemistry as the study of carbon-containing compounds. The differences between the ionic bond, which is characteristic of many inorganic substances, and the covalent bond in organic compounds are responsible for the great contrast in properties and reactivity between organic and inorganic compounds. All organic compounds are classified as either *hydrocarbons* or *substituted hydrocarbons*. In substituted hydrocarbons, a hydrogen atom is replaced by a functional group. A *functional group* is an atom or group of atoms arranged in a particular way that imparts specific chemical or physical properties to a molecule. The major families of organic molecules are defined by the specific functional groups that they contain.

KEY TERMS

aliphatic hydrocarbon (10.1)
alkane (10.2)
alkyl group (10.2)
alkyl halide (10.4)
aromatic hydrocarbon (10.1)
combustion (10.4)
condensed formula (10.2)
constitutional isomers (10.2)
cycloalkane (10.3)
functional group (10.1)
halogenation (10.4)
hydrocarbon (10.1)
I.U.P.A.C. Nomenclature
  System (10.2)

line formula (10.2)
molecular formula (10.2)
parent compound (10.2)
primary (1°) carbon (10.2)
quaternary (4°) carbon (10.2)
saturated hydrocarbon (10.1)
secondary (2°) carbon (10.2)
structural formula (10.2)
structural isomer (10.2)
substituted hydrocarbon (10.1)
substitution reaction (10.4)
tertiary (3°) carbon (10.2)
unsaturated hydrocarbon
  (10.1)

10.2 Alkanes

Alkanes are *saturated hydrocarbons*, that is, hydrocarbons that have only carbon and hydrogen atoms that are bonded together by carbon–carbon and carbon–hydrogen single bonds. They have the general molecular formula C_nH_{2n+2} and are nonpolar, water-insoluble compounds with low melting and boiling points. In the *I.U.P.A.C. Nomenclature System* the alkanes are named by determining the number of carbon atoms in the parent compound

QUESTIONS AND PROBLEMS

The Chemistry of Carbon

Foundations

10.17 Why is the number of organic compounds nearly limitless?
10.18 What are allotropes?
10.19 What are the three allotropic forms of carbon?
10.20 Describe the three allotropes of carbon.
10.21 Why do ionic substances generally have higher melting and boiling points than covalent substances?

10.22 Why are ionic substances more likely to be water-soluble?

Applications

10.23 Rank the following compounds from highest to lowest boiling points:
  a. H_2O CH_4 LiCl
  b. C_2H_6 C_3H_8 NaCl

10.24 Rank the following compounds from highest to lowest melting points:
  a. H_2O CH_4 KCl
  b. C_6H_{14} $C_{16}H_{34}$ NaCl

10.25 What would be the physical state of each of the compounds in Question 10.23 at room temperature?

10.26 Which of the compounds in Question 10.24 would be soluble in water?

10.27 Consider the differences between organic and inorganic compounds as you answer each of the following questions.
  a. Which compounds make good electrolytes?
  b. Which compounds exhibit ionic bonding?
  c. Which compounds have lower melting points?
  d. Which compounds are more likely to be soluble in water?
  e. Which compounds are flammable?

10.28 Describe the major differences between ionic and covalent bonds.

10.29 Give the structural formula for each of the following:

  a. 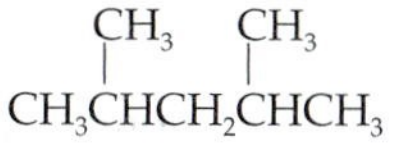 $CH_3CHCH_2CHCH_3$ (with CH_3, CH_3 substituents)

  b. 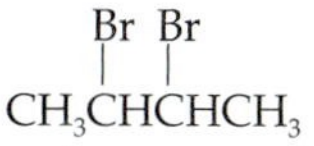 $CH_3CHCHCH_3$ (with Br, Br substituents)

10.30 Give the structural formula for each of the following:

  a. 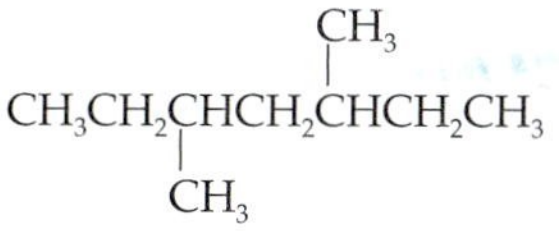 $CH_3CH_2CHCH_2CHCH_2CH_3$ (with CH_3, CH_3 substituents)

  b. 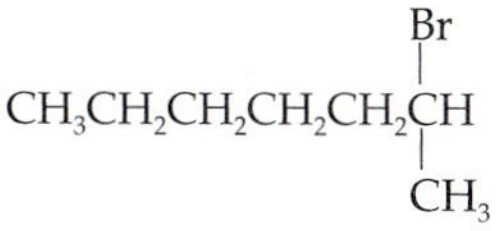 $CH_3CH_2CH_2CH_2CH_2CH$ (with Br, CH_3 substituents)

10.31 Condense each of the following structural formulas:

  a.

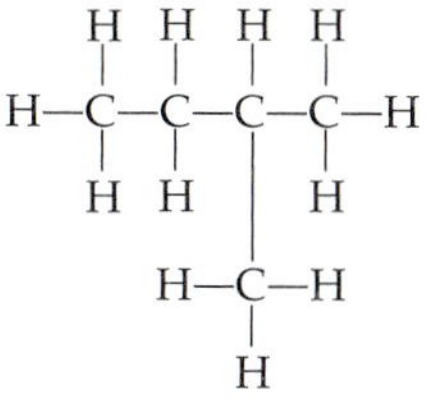

  b.

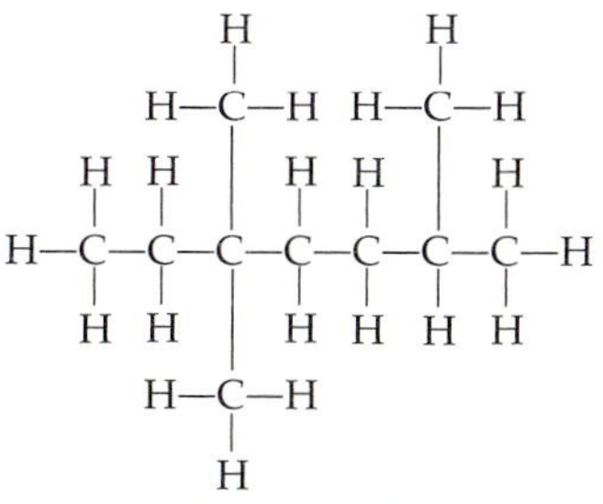

10.32 Condense each of the following structural formulas:

  a.

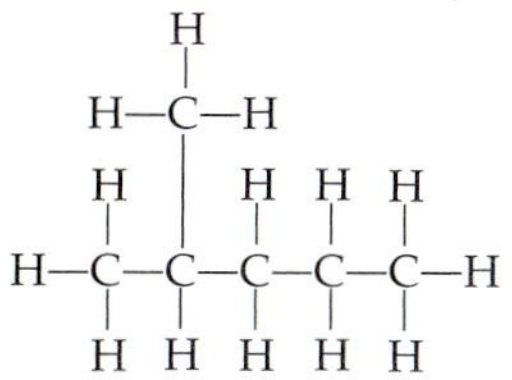

  b.

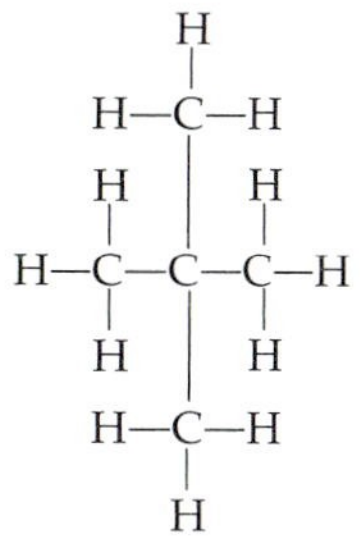

10.33 Convert the following structural formulas into line formulas:

  a.

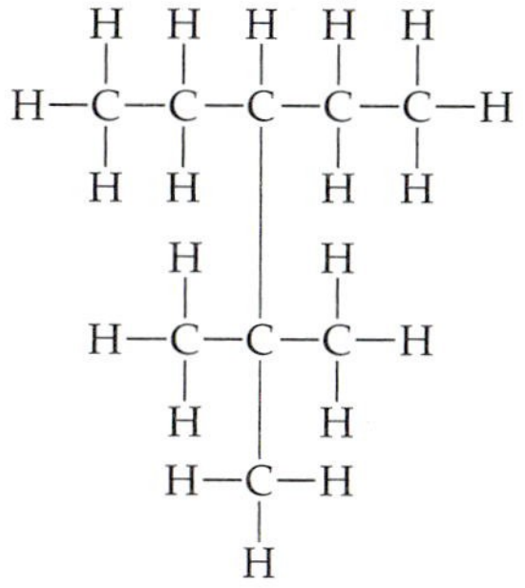

  b.

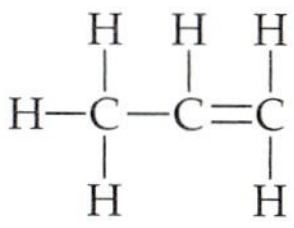

  c.

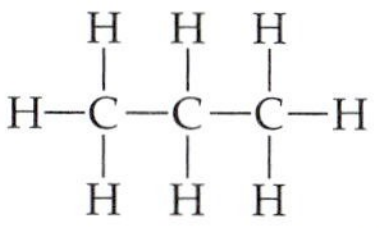

10.34 Convert the structural formulas in Question 10.33 into condensed structural formulas.

10.35 Convert the following structural formulas into line formulas:

  a.

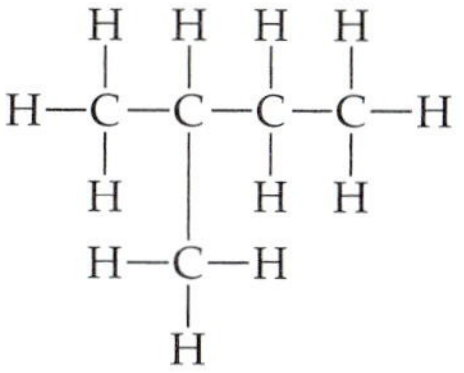

  b.

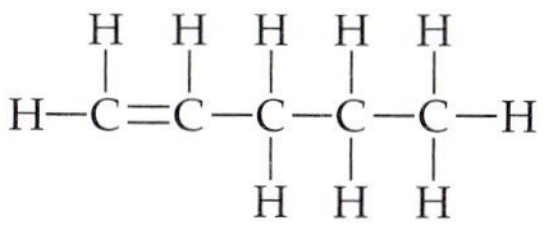

  c.

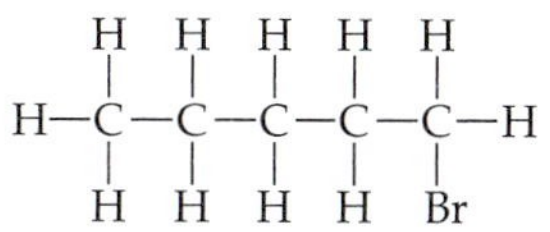

10.36 Convert the following line formulas into condensed structural formulas:

  a.

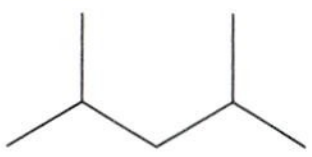

  b.

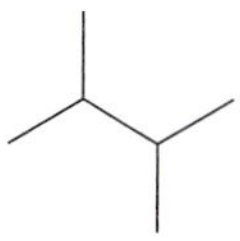

  c.

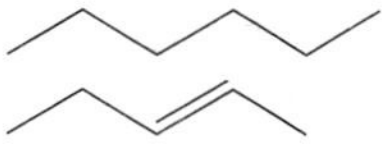

  d.

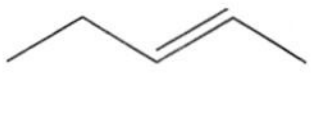

10.37 Which of the following structures are not possible? State your reasons.

a.
$$CH_3CHCH_2CH_3$$
with CH_3 above

b. $CH_3CHCH_2CH_3$ with CH_3 above

c. $CH_3CHCH_2CHCH_3$ with CH_3 and CH_3 above, and CH_3 below

d. $CH_3CH_2CH_2CH_2CH_3$

e. $CH_2CH_3CH_2CH_3$ with CH_3 below and CH_2CH_3 below

f. $CH_3CH_2CH_2CH_3$

10.38 Using the octet rule, explain why carbon forms four bonds in a stable compound.

10.39 Convert the following condensed structural formulas into structural formulas:

a. $CH_3CH(CH_3)CH(CH_3)CH_2CH_3$
b. $CH_3CH_2CH_2CH_2CH_3$
c. $CH_3CH_2CH(CH_2CH_3)CH_2CH_2CH_3$
d. $CH_3CH(CH_3)CH(CH_3)CH_2CH_2CH_2CH_3$

10.40 Convert the following condensed structural formulas into structural formulas:

a. $CH_3CH_2CH(CH_3)CH_2CH_2CH_2CH(CH_2CH_3)CH_3$
b. $CH_3C(CH_3)_2CH_2CH_3$
c. $CH_3CH(CH_3)CH(CH_3)CH(CH_3)CH_2CH_3$
d. $CH_3C(CH_3)_2CH(CH_2CH_3)CH_2CH_2CH_3$

10.41 Using structural formulas, draw a typical alcohol, aldehyde, ketone, carboxylic acid, and amine. (*Hint:* Refer to Table 10.2).

10.42 Name the functional group in each of the following molecules:

a. $CH_3CH_2CH_2—OH$
b. $CH_3CH_2CH_2—NH_2$
c. $CH_3CH_2CH_2—C=O$ with H below
d. $CH_3CH_2CH_2—C=O$ with OH below
e. $CH_3CH_2CH_2—C=O$ with OCH_2CH_3 below
f. $CH_3CH_2—O—CH_2CH_3$
g. $CH_3CH_2CH_2—I$

10.43 Give the general formula for each of the following:

a. An alkane
b. An alkyne
c. An alkene
d. A cycloalkane
e. A cycloalkene

10.44 Of the classes of compounds listed in Problem 10.43, which are saturated? Which are unsaturated?

10.45 What functional group characterizes each of the following families of organic compounds. (*Hint:* Refer to Table 10.2.)

a. A carboxylic acid
b. An amine
c. An alcohol

10.46 Folic acid is a vitamin required by the body for nucleic acid synthesis. The structure of folic acid is given below. Circle and identify as many functional groups as possible.

Folic acid

10.47 Aspirin is a pain reliever that has been used for over a hundred years. Today, small doses of aspirin are recommended to reduce the risk of heart attack. The structure of aspirin is given below. Circle and identify the functional groups in the molecule:

Aspirin
Acetylsalicylic acid

10.48 Vitamin C is one of the water-soluble vitamins. Although most animals can produce vitamin C, humans are not able to do so. We must get our supply of the vitamin from our diet. The structure of vitamin C is shown below. Circle and identify the functional groups in the molecule:

Vitamin C
Ascorbic acid

Alkanes

Foundations

10.49 Why are hydrocarbons not water soluble?

10.50 Describe the relationship between the length of hydrocarbon chains and the melting points of the compounds.

10.51 Rank the following compounds from highest to lowest boiling points:

a. heptane   butane   hexane   ethane
b. $CH_3CH_2CH_2CH_2CH_3$ $CH_3CH_2CH_3$
   $CH_3CH_2CH_2CH_2CH_2CH_2CH_2CH_2CH_3$

10.52 Rank the following compounds from highest to lowest melting points:

a. decane   propane   methane   ethane
b. $CH_3CH_2CH_2CH_2CH_3$ $CH_3(CH_2)_8CH_3$
   $CH_3(CH_2)_6CH_3$

10.53 What would be the physical state of each of the compounds in Problem 10.51 at room temperature?

10.54 What would be the physical state of each of the compounds in Problem 10.52 at room temperature?

Applications

10.55 Draw each of the following:

a. 2-Bromobutane
b. 2-Chloro-2-methylpropane
c. 2,2-Dimethylhexane

10.56 Draw each of the following:

a. Dichlorodiiodomethane
b. 1,4-Diethylcyclohexane
c. 2-Iodo-2,4,4-trimethylpentane

10.57 Draw each of the following compounds using structural formulas:
  a. 2,2-Dibromobutane
  b. 2-Iododecane
  c. 1,2-Dichloropentane
  d. 1-Bromo-2-methylpentane

10.58 Draw each of the following compounds using structural formulas:
  a. 1,1,1-Trichlorodecane
  b. 1,2-Dibromo-1,1,2-trifluoroethane
  c. 3,3,5-Trimethylheptane
  d. 1,3,5-Trifluoropentane

10.59 Name each of the following using the I.U.P.A.C. Nomenclature System:

  a. 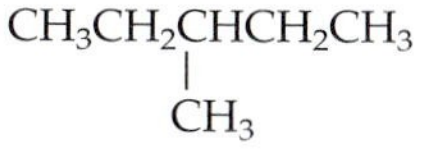 CH$_3$CH$_2$CHCH$_2$CH$_3$
          |
        CH$_3$

  b. 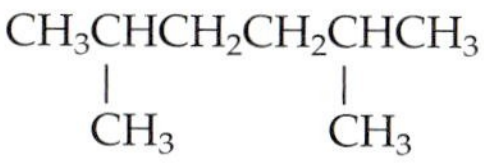 CH$_3$CHCH$_2$CH$_2$CHCH$_3$
       | |
      CH$_3$ CH$_3$

  c. 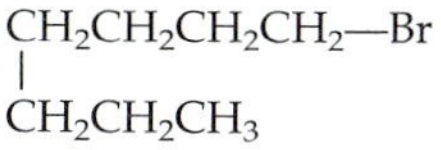 CH$_2$CH$_2$CH$_2$CH$_2$—Br
      |
    CH$_2$CH$_2$CH$_3$

  d.  Cl—CH$_2$CH$_2$CHCH$_3$
               |
           CH$_3$

10.60 Provide the I.U.P.A.C. name for each of the following compounds:

  a. 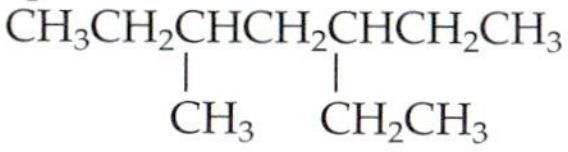 CH$_3$CH$_2$CHCH$_2$CHCH$_2$CH$_3$
          | |
       CH$_3$ CH$_2$CH$_3$

  b. 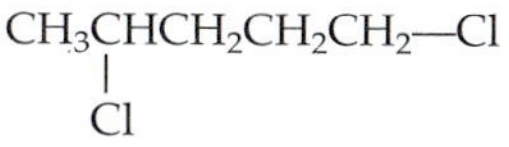 CH$_3$CHCH$_2$CH$_2$CH$_2$—Cl
      |
    Cl

        CH$_3$
        |
  c. 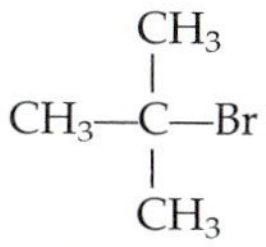 CH$_3$—C—Br
        |
        CH$_3$

10.61 Give the I.U.P.A.C. name for each of the following:

  a. 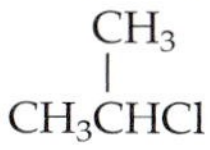   CH$_3$
         |
    CH$_3$CHCl

  b.    I
        |
    CH$_3$CHCH$_2$CH$_3$

  c. 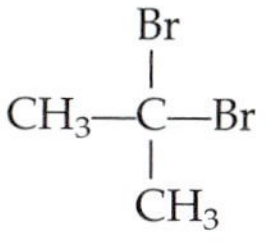   Br
        |
    CH$_3$—C—Br
        |
        CH$_3$

  d. 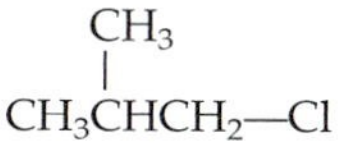   CH$_3$
        |
    CH$_3$CHCH$_2$—Cl

  e. 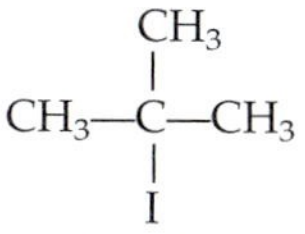   CH$_3$
        |
    CH$_3$—C—CH$_3$
        |
        I

10.62 Name the following using the I.U.P.A.C. Nomenclature System:

        Cl
        |
  a. 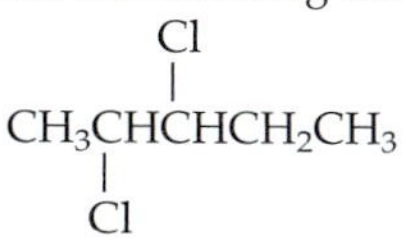 CH$_3$CHCHCH$_2$CH$_3$
         |
       Cl

          CH$_3$
          |
  b. 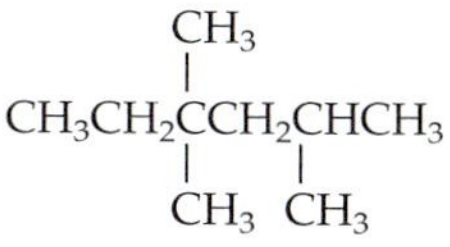 CH$_3$CH$_2$CCH$_2$CHCH$_3$
         | |
        CH$_3$ CH$_3$

        CH$_3$ CH$_3$
        | |
  c. 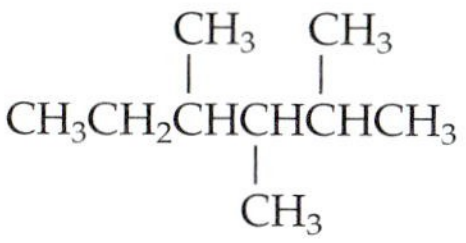 CH$_3$CH$_2$CHCHCHCH$_3$
              |
           CH$_3$

    Br
    |
  d. 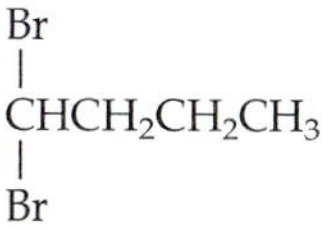 CHCH$_2$CH$_2$CH$_3$
    |
    Br

10.63 Which of the following pairs of compounds are identical? Which are constitutional isomers? Which are completely unrelated?

  a. 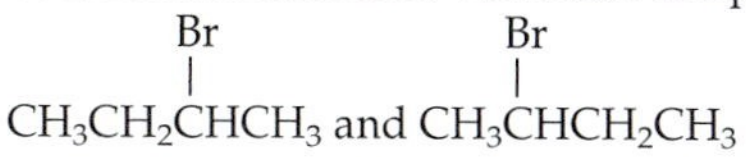   Br   Br
          | |
   CH$_3$CH$_2$CHCH$_3$ and CH$_3$CHCH$_2$CH$_3$

  b. 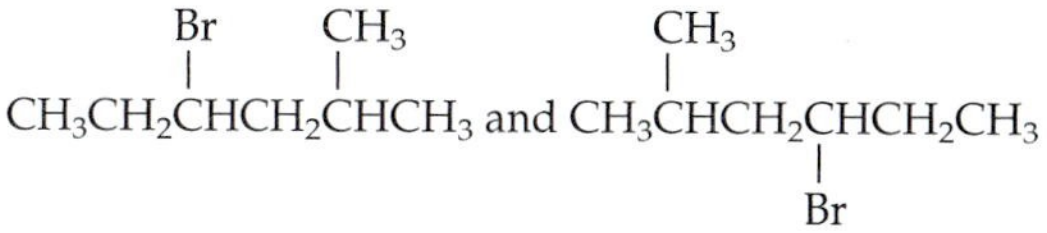   Br   CH$_3$ CH$_3$
        | | |
   CH$_3$CH$_2$CHCH$_2$CHCH$_3$ and CH$_3$CHCH$_2$CHCH$_2$CH$_3$
                              |
                              Br

  c. 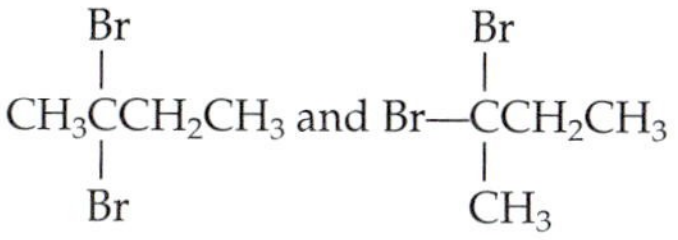   Br   Br
       | |
   CH$_3$CCH$_2$CH$_3$ and Br—CCH$_2$CH$_3$
       | |
       Br   CH$_3$

  d. 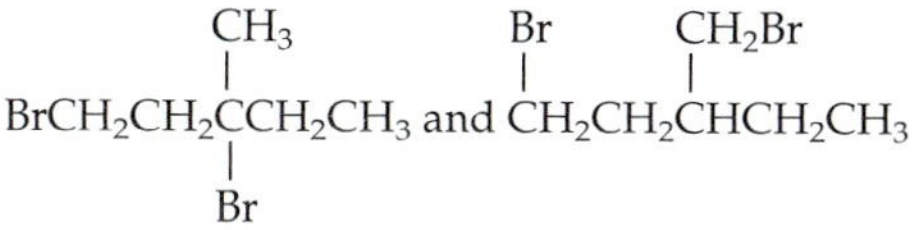   CH$_3$ Br   CH$_2$Br
           | | |
   BrCH$_2$CH$_2$CCH$_2$CH$_3$ and CH$_2$CH$_2$CHCH$_2$CH$_3$
            |
           Br

10.64 Which of the following pairs of molecules are identical compounds? Which are constitutional isomers?

  a. 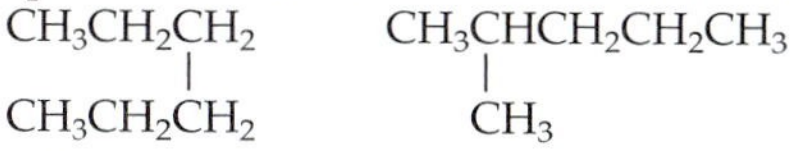 CH$_3$CH$_2$CH$_2$ CH$_3$CHCH$_2$CH$_2$CH$_3$
           | |
    CH$_3$CH$_2$CH$_2$ CH$_3$

  b. CH$_3$CH$_2$CH$_2$CH$_2$CH$_2$CH$_2$CH$_3$ CH$_3$CH$_2$CH$_2$CH$_2$CH$_2$
                                                 |
                                        CH$_3$CH$_2$

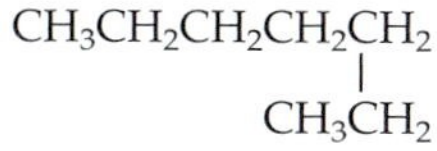

Cycloalkanes

Foundations

10.65 Describe the structure of a cycloalkane.

10.66 Describe the I.U.P.A.C. rules for naming cycloalkanes.

10.67 What is the general formula for a cycloalkane?

10.68 How does the general formula of a cycloalkane compare with that of an alkane?

Applications

10.69 Name each of the following cycloalkanes, using the I.U.P.A.C. system:

  a.

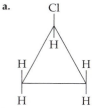

  b.

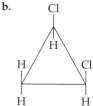

c.

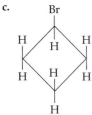

10.70 Name each of the following cycloalkanes using the I.U.P.A.C. system:

a.

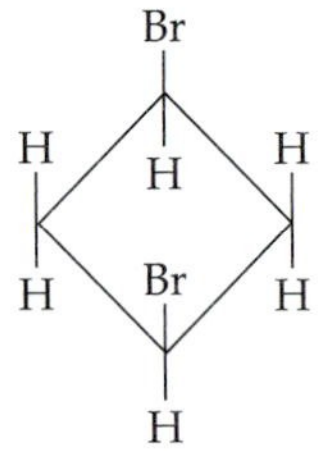

b.

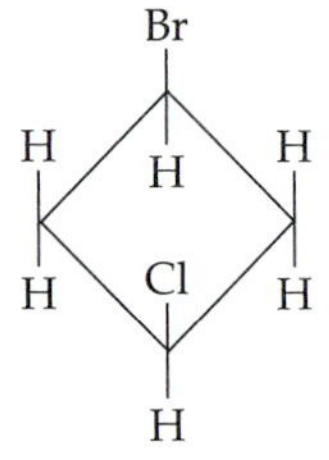

c.

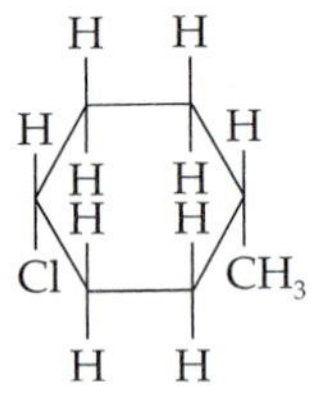

d.

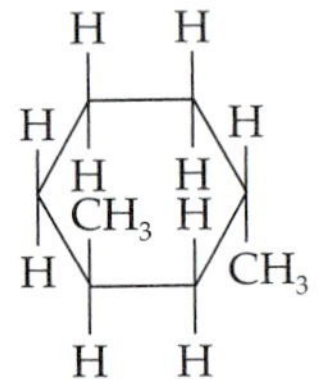

10.71 Draw the structure of each of the following cycloalkanes:
a. 1-Bromo-2-methylcyclobutane
b. Iodocyclopropane
c. 1-Bromo-3-chlorocyclopentane
d. 1,2-Dibromo-3-methycyclohexane

10.72 Which of the following names are correct and which are incorrect? If incorrect, write the correct name.
a. 1,4,5-Tetrabromocyclohexane
b. 1,3-Dimethylcyclobutane
c. 1,2-Dichlorocyclopentane
d. 3-Bromocyclopentane

Reactions of Alkanes and Cycloalkanes

Foundations

10.73 Write a balanced equation for the complete combustion of each of the following:
a. propane
b. heptane
c. nonane
d. decane

10.74 Write a balanced equation for the complete combustion of each of the following:
a. pentane
b. hexane

c. octane
d. ethane

Applications

10.75 Complete each of the following reactions by supplying the missing reactant or product as indicated by a question mark:

a. $2CH_3CH_2CH_2CH_3 + 13O_2 \xrightarrow{\text{Heat}} ?$ (Complete combustion)

b. $CH_3-\overset{\overset{\displaystyle CH_3}{|}}{\underset{\underset{\displaystyle CH_3}{|}}{C}}-H + Br_2 \xrightarrow{\text{Light}} ?$ (Give all possible monobrominated products)

c. ⬡ $+ \xrightarrow{?} Cl-$⬡ $+ HCl$

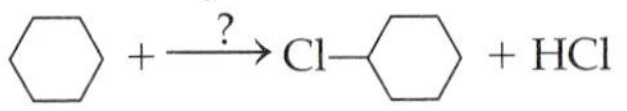

10.76 Give all the possible monochlorinated products for the following reaction:

$$CH_3\overset{\overset{\displaystyle CH_3}{|}}{CH}CH_2CH_3 + Cl_2 \xrightarrow{\text{Light}} ?$$

Name the products, using I.U.P.A.C. nomenclature.

10.77 Draw the constitutional isomers of molecular formula C_6H_{14} and name each using the I.U.P.A.C. system:
a. Which one gives two and only two monobromo derivatives when it reacts with Br_2 and light? Name the products, using the I.U.P.A.C. system.
b. Which give three and only three monobromo products? Name the products, using the I.U.P.A.C. system.
c. Which give four and only four monobromo products? Name the products, using the I.U.P.A.C. system.

10.78 a. Draw and name all of the isomeric products obtained from the monobromination of propane with Br_2/light. If halogenation were a completely random reaction and had an equal probability of occurring at any of the C—H bonds in a molecule, what percentage of each of these monobromo products would be expected?
b. Answer part (a) using 2-methylpropane as the starting material.

10.79 A mole of hydrocarbon formed eight moles of CO_2 and eight moles of H_2O upon combustion. Determine the molecular formula of the hydrocarbon and give the balanced combustion reaction.

10.80 Highly substituted alkyl fluorides, called perfluoroalkanes, are often used as artificial blood substitutes. These perfluoroalkanes have the ability to transport O_2 through the bloodstream as blood does. Some even have twice the O_2 transport capability and are used to treat gangrenous tissue. The structure of perfluorodecalin is shown below. How many moles of fluorine must be reacted with one mole of decalin to produce perfluorodecalin?

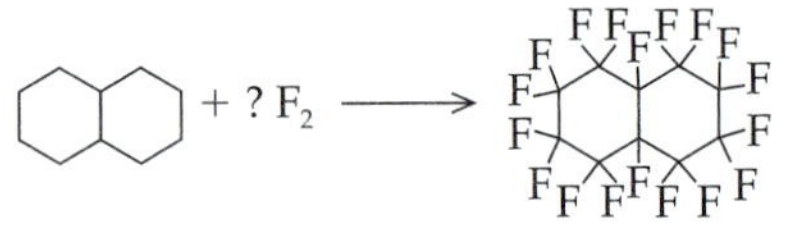

⬡⬡ $+ ? F_2 \longrightarrow$

Decalin   Perfluorodecalin

FOR FURTHER UNDERSTANDING

1. You are given two unlabeled bottles, each of which contains a colorless liquid. One contains hexane and the other contains water. What physical properties could you use to identify the two liquids? What chemical property could you use to identify them?

2. You are given two beakers, each of which contains a white crystalline solid. Both are soluble in water. How would you determine which of the two solids is an ionic compound and which is a covalent compound?

3. Chlorofluorocarbons (CFCs) are man-made compounds made up of carbon and the halogens, fluorine and chlorine. One of the most widely used is Freon-12 (CCl_2F_2). It was introduced as a refrigerant in the 1930s. This was an important advance because Freon-12 replaced ammonia and sulfur dioxide, two toxic chemicals that were previously used in refrigeration systems. Freon-12 was hailed as a perfect replacement because it has a boiling point of $-30°C$ and is almost completely inert. To what family of organic molecules do CFCs belong? Design a strategy for the synthesis of Freon-12.

4. Over time, CFC production increased dramatically as their uses increased. They were used as propellants in spray cans, as gases to expand plastic foam, and in many other applications. By 1985, production of CFCs reached 850,000 tons. Much of this leaked into the atmosphere and in that year the concentration of CFCs reached 0.6 parts per billion. Another observation was made by groups of concerned scientists: as the level of CFCs rose, the ozone level in the upper atmosphere declined. Does this correlation between CFC levels and ozone levels prove a relationship between these two phenomena? Explain your reasoning.

5. Although manufacture of CFCs was banned on December 31, 1995, the C—F and C—Cl bonds of CFCs are so strong that the molecules may remain in the atmosphere for 120 years. Within 5 years, they diffuse into the upper stratosphere where ultraviolet photons can break the C—Cl bonds. This process releases chlorine atoms, as shown here for Freon-12:

$$CCl_2F_2 + photon \longrightarrow CClF_2 + Cl$$

The chlorine atoms are extremely reactive because of their strong tendency to acquire a stable octet of electrons. The following reactions occur when a chlorine atom reacts with an ozone molecule (O_3). First, chlorine pulls an oxygen atom away from ozone:

$$Cl + O_3 \longrightarrow ClO + O_2$$

Then ClO, a highly reactive molecule, reacts with an oxygen atom:

$$ClO + O \longrightarrow Cl + O_2$$

Write an equation representing the overall reaction (sum of the two reactions). How would you describe the role of Cl in these reactions?

11

LEARNING GOALS

1 Describe the physical properties of alkenes and alkynes.

2 Draw the structures and write the I.U.P.A.C. names of simple alkenes and alkynes.

3 Write the names and draw the structures of simple geometric isomers of alkenes.

4 Write equations predicting the products of the simple addition reactions of alkenes and alkynes: hydrogenation, halogenation, and hydration.

5 Apply Markovnikov's rule to predict the major and minor products of hydration reactions of unsymmetrical alkenes.

6 Write equations representing the formation of addition polymers of alkenes.

7 Draw the structures and write the names of common aromatic hydrocarbons.

8 Describe heterocyclic aromatic compounds and list several biological molecules in which they are found.

The Unsaturated Hydrocarbons

Alkenes, Alkynes, and Aromatics

F or many years it was suspected that there existed a gas that stimulated fruit ripening and had other effects on plants. The ancient Chinese observed that their fruit ripened more quickly if incense was burned in the room. Early in this century, shippers realized that they could not store oranges and bananas on the same ships because some "emanation" given off by the oranges caused the bananas to ripen too early.

Puerto Rican pineapple growers and Philippine mango growers independently developed a traditional practice of building bonfires near their crops. They believed that the smoke caused the plants to bloom synchronously.

In the mid-nineteenth century, streetlights were fueled with natural gas. Occasionally the pipes leaked, releasing gas into the atmosphere. On some of these occasions, the leaves fell from all the shade trees in the region surrounding the gas leak.

What is the gas responsible for these diverse effects on plants? In 1934, R. Gane demonstrated that the simple alkene ethene (ethylene) was the "emanation" responsible for fruit ripening. More recently, it has been shown that ethene induces and synchronizes flowering in pineapples and mangos, induces senescence (aging) and loss of leaves in trees, and effects a wide variety of other responses in various plants.

We can be grateful to ethene for the fresh, unbruised fruits that we can purchase at the grocery store. These fruits are picked when they are not yet ripe, while they are still firm. They then can be shipped great distances and gassed with ethylene when they reach their destination. Under the influence of the ethene, the fruit ripens and is displayed in the store.

In this chapter we will study the unsaturated hydrocarbons. This group of organic compounds includes the alkenes, such as ethene, which all contain at least one carbon-carbon double bond and the alkynes, which all contain at least one carbon-carbon triple bond.

Ancient Chinese folk tales claimed that fruit ripened more quickly if incense was burned in the room. Today, fruit growers pick and ship unripe fruit to market and grocers then treat it with ethene gas. Explain the roles of science and technology in the journey from Chinese folklore to modern food handling procedures.

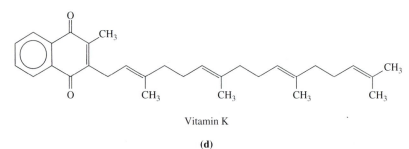

Palmitoleic acid

(a)

Vitamin A

(b)

Vitamin A

(c)

Vitamin K

(d)

Figure 11.1

(a) Structural formula of the sixteen-carbon monounsaturated fatty acid, palmitoleic acid.
(b) Condensed formula of vitamin A, which is required for vision. (c) Line formula of vitamin A. (d) Line formula of vitamin K, a lipid-soluble vitamin required for blood clotting. The six-member ring with the circle represents a benzene ring. See Figure 11.4 for other representations of the benzene ring.

You can find further information on lipid-soluble vitamins online at www.mhhe.com/denniston in "Lipid-Soluble Vitamins."

11.1 Alkenes and Alkynes: Structure and Physical Properties

1 Learning Goal

Describe the physical properties of alkenes and alkynes.

Unsaturated hydrocarbons are those that contain at least one carbon–carbon double or triple bond. They include alkenes, alkynes, and aromatic compounds. All alkenes have at least one carbon–carbon double bond; all alkynes have at least one carbon–carbon triple bond. Aromatic compounds are particularly stable cyclic compounds that contain a benzene ring.

Many important biological molecules are characterized by the presence of double bonds (Figure 11.1). For instance, we classify fatty acids as either monoun-saturated (having one double bond), polyunsaturated (having two or more double bonds), or saturated (having only single bonds). Vitamin A (retinol), a vitamin required for vision, contains a nine-carbon conjugated hydrocarbon chain. Vitamin K, a vitamin required for blood clotting, contains an aromatic ring.

Alkenes and **alkynes** are unsaturated hydrocarbons. The characteristic functional group of an alkene is the carbon–carbon double bond. The functional group that characterizes alkynes is the carbon–carbon triple bond. The following general formulas compare the structures of alkanes, alkenes, and alkynes.

General formulas:

	Alkane	Alkene	Alkyne
	C_nH_{2n+2}	C_nH_{2n}	C_nH_{2n-2}

Structural formulas:

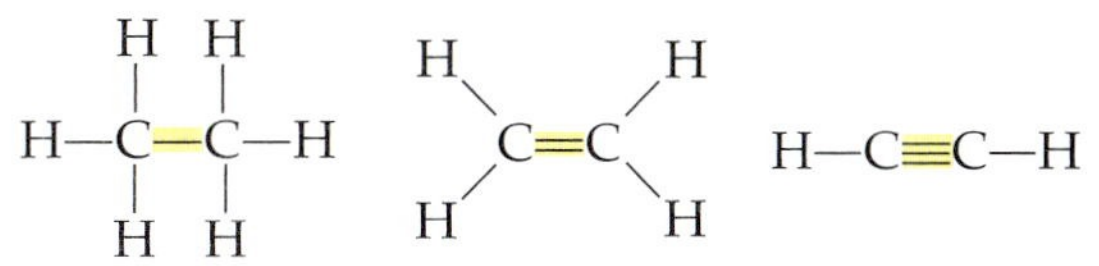

	Ethane	Ethene	Ethyne
	(ethane)	(ethylene)	(acetylene)

Molecular formulas: C_2H_6 C_2H_4 C_2H_2

Condensed formulas: CH_3CH_3 $H_2C=CH_2$ $HC\equiv CH$

These compounds have the same number of carbon atoms but differ in the number of hydrogen atoms. Alkenes contain two fewer hydrogens than the corresponding alkanes, and alkynes contain two fewer hydrogens than the corresponding alkenes.

In alkanes, the four bonds to the central carbon have tetrahedral geometry. When carbon is bonded by one double bond and two single bonds, as in ethene (an alkene), the molecule is *planar,* because all atoms lie in a single plane. Each bond angle is approximately 120°. When two carbon atoms are bonded by a triple bond, as in ethyne (an alkyne), each bond angle is 180°. Thus, the molecule is linear, and all atoms are positioned in a straight line.

The VSEPR theory and molecular geometry are covered in more detail in Section 3.4.

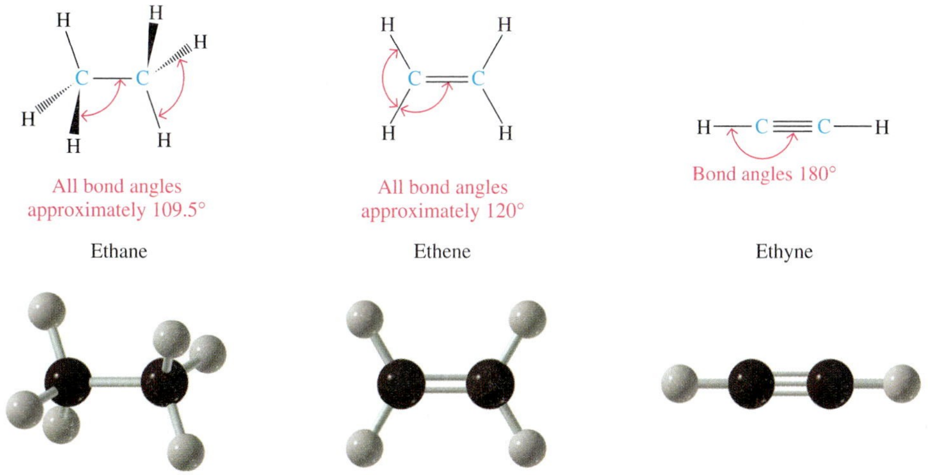

All bond angles approximately 109.5°

Ethane

All bond angles approximately 120°

Ethene

Bond angles 180°

Ethyne

Like alkanes, alkenes, alkynes, and aromatic compounds are nonpolar and have properties similar to alkanes of the same carbon chain length. Because they are nonpolar, the "like dissolves like" rule tells us that they are not soluble in water, but they are very soluble in nonpolar solvents such as other hydrocarbons.

11.2 Alkenes and Alkynes: Nomenclature

To determine the name of an alkene or alkyne using the I.U.P.A.C. Nomenclature System, use the following simple rules:

• Name the parent compound using the longest continuous carbon chain containing the double bond (alkenes) or triple bond (alkynes).

2 Learning Goal

Draw the structures and write the I.U.P.A.C. names of simple alkenes and alkynes.

• Replace the *-ane* ending of the alkane with the *-ene* ending for an alkene or the *-yne* ending for an alkyne. For example,

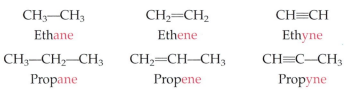

• Number the chain to give the lowest number for the first of the two carbons containing the double bond or triple bond. For example,

CH₃CH₂CH=CH₂ CH≡CCH₂CH₂CH₃

1-Butene
(*NOT* 3-Butene)

1-Pentyne
(*NOT* 4-Pentyne)

The position of the double bond, not the substituent, determines the numbering of the carbon chain.

• Determine the name and carbon number of each group bonded to the parent alkene or alkyne, and place the name and number in front of the name of the parent compound. Remember that with alkenes and alkynes, the double or triple bond takes precedence over a halogen or alkyl group, as shown in the following examples:

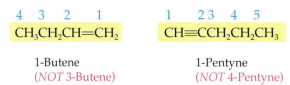

2-Chloro-2-butene

2-Bromo-3-hexyne

2,6-Dimethyl-3-octene

3-Chloro-4-methyl-3-hexene

Alkenes with many double bonds are often referred to as polyenes (*poly*—many; *enes*—double bonds).

• Alkenes having more than one double bond are called alkadienes (two double bonds) or alkatrienes (three double bonds), as seen in these examples:

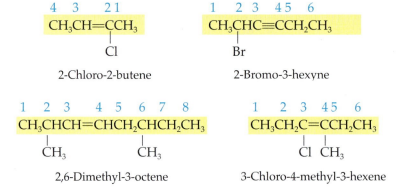

2,4-Hexadiene

1,4-Pentadiene

3-Methyl-1,4-cyclohexadiene

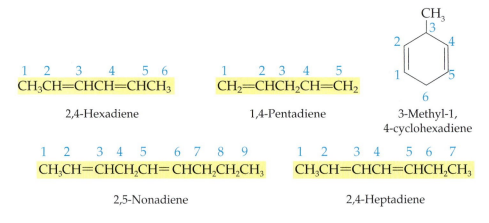

2,5-Nonadiene

2,4-Heptadiene

2 Learning Goal

Draw the structures and write the I.U.P.A.C. names of simple alkenes and alkynes.

EXAMPLE 11.1 Naming Alkenes and Alkynes Using I.U.P.A.C. Nomenclature

Name the following alkene and alkyne using I.U.P.A.C. nomenclature.

— *Continued*—

SOLUTION

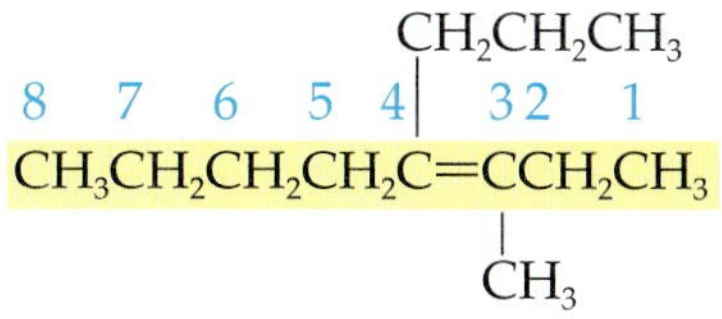

$$\underset{8\quad7\quad6\quad5\quad\overset{|}{4}\quad\;3\;2\quad1}{\underset{\displaystyle\underset{CH_3}{|}}{CH_3CH_2CH_2CH_2C{=}CCH_2CH_3}}\quad\overset{CH_2CH_2CH_3}{}$$

Longest chain containing the double bond: octene

Position of double bond: 3-octene
(*NOT* 5-octene)

Substituents: 3-methyl and 4-propyl

Name: 3-Methyl-4-propyl-3-octene

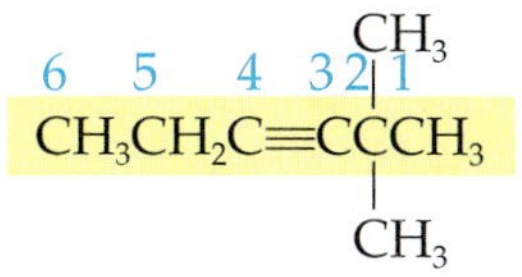

$$\underset{6\quad5\quad4\quad3\,2\,1}{\underset{\displaystyle\underset{CH_3}{|}}{CH_3CH_2C{\equiv}CCH_3}}\quad\overset{CH_3}{}$$

Longest chain containing the triple bond: hexyne

Position of triple bond: 3-hexyne (must be!)

Substituents: 2,2-dimethyl

Name: 2,2-Dimethyl-3-hexyne

EXAMPLE 11.2   Naming Cycloalkenes Using I.U.P.A.C. Nomenclature

Name the following cycloalkenes using I.U.P.A.C. nomenclature.

2   **Learning Goal**
Draw the structures and write the I.U.P.A.C. names of simple alkenes and alkynes.

SOLUTION

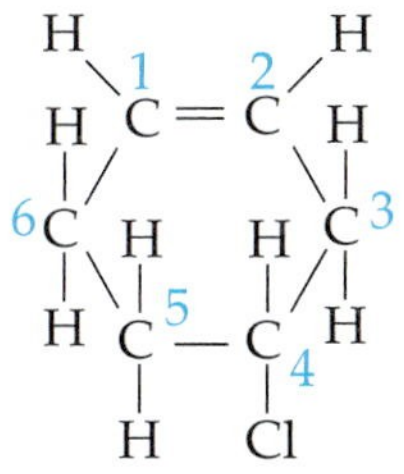

Parent chain: cyclohexene

Position of double bond: carbon-1 (carbons of the double bond are numbered 1 and 2)

Substituents: 4-chloro (*NOT* 5-chloro)

Name: 4-Chlorocyclohexene

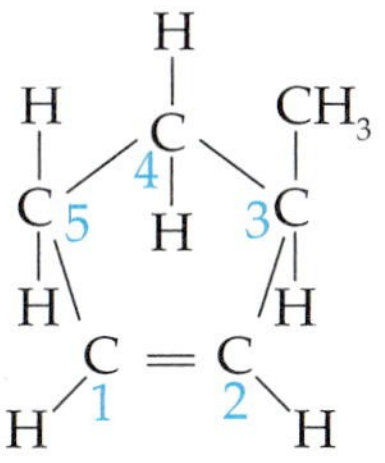

Parent chain: cyclopentene

Position of double bond: carbon-1

Substituent: 3-methyl (*NOT* 5-methyl)

Name: 3-Methylcyclopentene

QUESTION 11.1

Draw a complete structural formula for each of the following compounds:

a. 1-Bromo-3-hexyne
b. 2-Butyne
c. Dichloroethyne
d. 9-Iodo-1-nonyne

QUESTION 11.2

Name the following compounds using the I.U.P.A.C. Nomenclature System:

a. $CH_3C{\equiv}CCH_2CH_3$
b. $\underset{\quad\;\;\underset{Br}{|}\;\;\underset{Br}{|}}{CH_3CH_2CHCHCH_2C{\equiv}CH}$

A MEDICAL Connection | Killer Alkynes in Nature

There are many examples of alkynes that are beneficial to humans. Among these are *parasalamide*, a pain reliever, *paragyline*, an antihypertensive, and *17-ethynylestradiol*, a synthetic estrogen that is used as an oral contraceptive.

But in addition to these medically useful alkynes, there are in nature a number that are toxic. Some are extremely toxic to mammals, including humans; others are toxic to fungi, fish, or insects. All of these compounds are plant products that may help protect the plant from destruction by predators.

Capillin is produced by the oriental wormwood plant. Research has shown that a dilute solution of capillin inhibits the growth of certain fungi. Because fungal growth can damage or destroy a plant, the ability to make capillin may provide a survival advantage to the plants. Perhaps one day, it may be developed to combat fungal infections in humans.

Ichthyothereol is a fast-acting poison commonly found in plants referred to as fish-poison plants. Ichthyothereol is a very toxic polyacetylenic alcohol that inhibits energy production in the mitochondria. Latin American native tribes use these plants to coat the tips of the arrows used to catch fish. Although ichthyothereol is poisonous to the fish, fish caught by this method pose no risk to the people who eat them!

An extract of the leaves of English ivy reportedly has antibacterial, analgesic, and sedative effects. The compound thought to be responsible for these characteristics, as well as antifungal activity, is *falcarinol*. Falcarinol, isolated from a tree in Panama, also has been reported by the Molecular Targets Drug Discovery Program to have antitumor activity. Perhaps one day, this compound, or a derivative of it, will be useful in treating cancer in humans.

Cicutoxin has been described as the most lethal toxin native to North America. It is a neurotoxin that is produced by the water hemlock *(Cicuta maculata)*, which is in the same family of plants as parsley, celery, and carrots. Cicutoxin is present in all

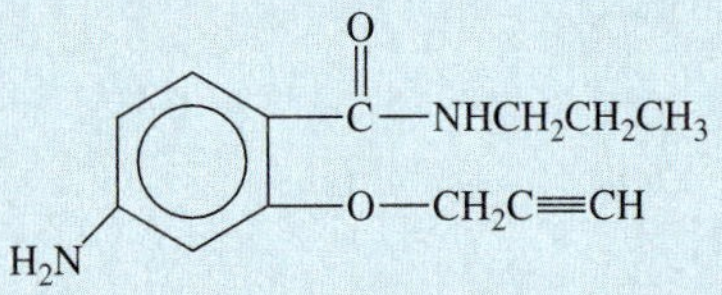

Parasalamide

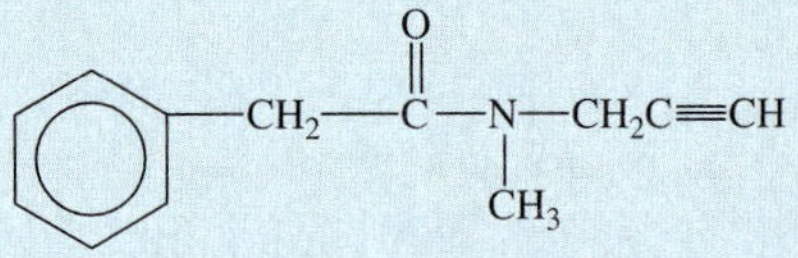

Paragyline

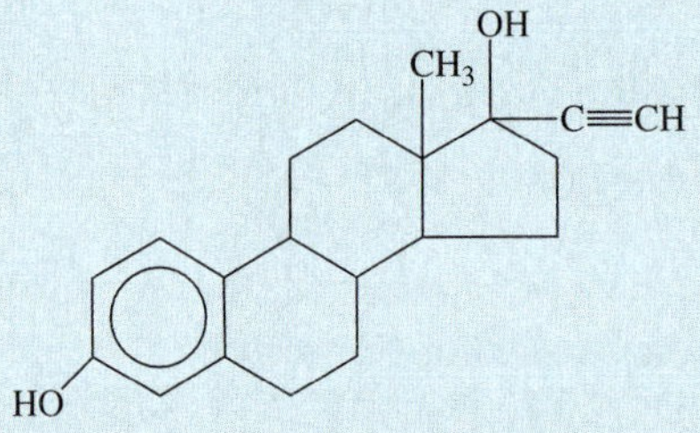

17-Ethynylestradiol

Alkynes used for medicinal purposes.

c. 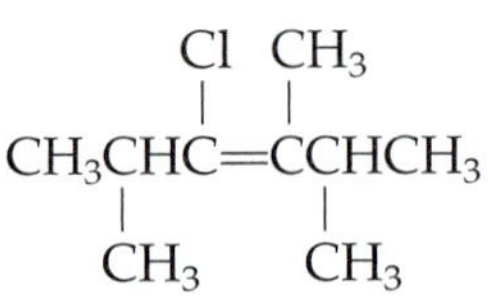 CH₃CHC=CCHCH₃

d. 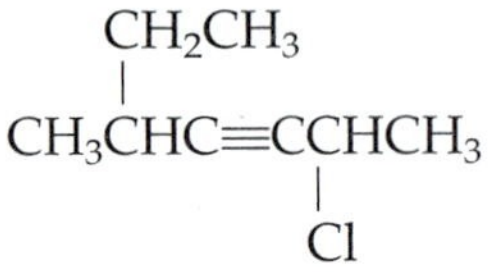 CH₃CHC≡CCHCH₃

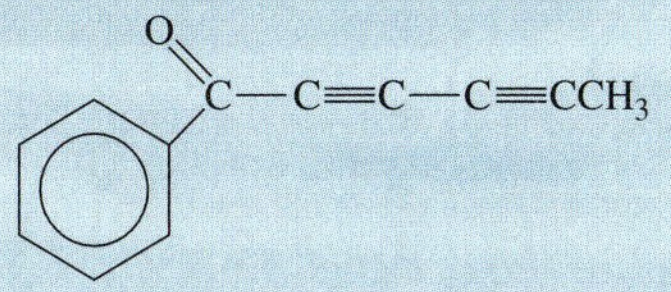

Capillin

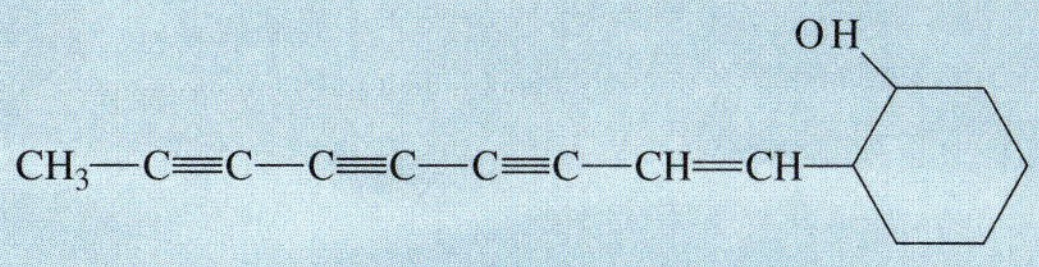

Ichthyothereol

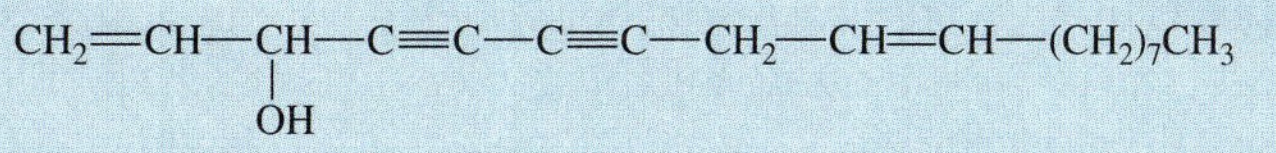

Falcarinol

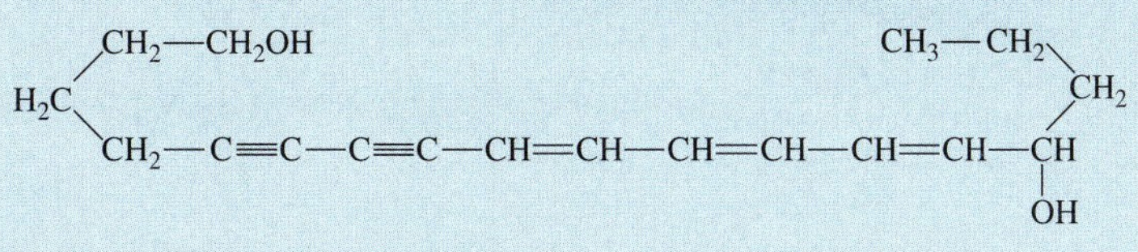

Cicutoxin

Alkynes that exhibit toxic activity.

cicutoxin poisoning include dilation of pupils, muscle twitching, rapid pulse and breathing, violent convulsions, coma, and death. Onset of symptoms is rapid and death may occur within two to three hours. No antidote exists for cicutoxin poisoning. The only treatment involves controlling convulsions and seizures to preserve normal heart and lung function. Fortunately, cicutoxin poisoning is a very rare occurrence. Occasionally, animals may graze on the plants in the spring, resulting in death within fifteen minutes. Humans seldom come into contact with the water hemlock. The most recent cases have involved individuals foraging for wild ginseng, or other wild roots, and mistaking the water hemlock root for an edible plant.

Cicuta maculata, or water hemlock, produces the most deadly toxin indigenous to North America.

parts of the plants but is most concentrated in the root. Eating a portion as small as 2–3 cm^2 can be fatal to adults. Cicutoxin acts directly on the nervous system. Signs and symptoms of

FOR FURTHER UNDERSTANDING

Circle and name the functional groups in parasalamide and paragyline.

The fungus *Tinea pedis* causes athlete's foot. Describe an experiment you might carry out to determine whether capillin might be effective against athlete's foot.

11.3 Geometric Isomers

The carbon–carbon double bond is rigid because of the shapes of the orbitals involved in its formation. As a result, rotation around the carbon–carbon double bond is restricted. As a consequence, these molecules form geometric or *cis-trans* isomers. In alkenes, **geometric isomers** occur when there are two different groups on each of the carbon atoms attached by the double bond. If two similar groups are on the same side of the double bond, the molecule is a *cis* isomer. If two similar groups are on opposite sides of the double bond, the molecule is a *trans* isomer.

 Learning Goal

Write the names and draw the structures of simple geometric isomers of alkenes.

Consider the two isomers of 1,2-dichloroethene:

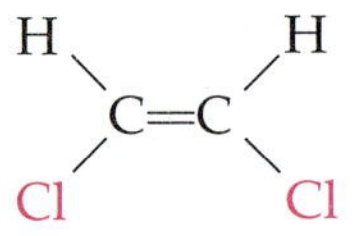

cis-1, 2-Dichloroethene 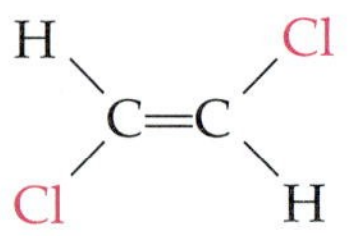   trans-1, 2-Dichloroethene

In these molecules, each carbon atom of the double bond is also bonded to two different atoms: a hydrogen atom and a chlorine atom. In the molecule on the left, both chlorine atoms are on the same side of the double bond; this is the *cis* isomer and the complete name for this molecule is *cis*-1,2-dichloroethene. In the molecule on the right, the chlorine atoms are on opposite sides of the double bond; this is the *trans* isomer and the complete name of this molecule is *trans*-1,2-dichloroethene.

If one of the two carbon atoms of the double bond has two identical substituents, there are no *cis-trans* isomers for that molecule. Consider the example of 1,1-dichloroethene:

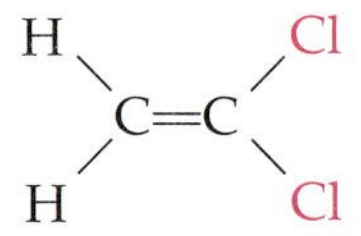

1, 1-Dichloroethene

EXAMPLE 11.3   Naming *cis* and *trans* Compounds

3 Learning Goal

Write the names and draw the structures of simple geometric isomers of alkenes.

Name the following geometric isomers.

SOLUTION

The longest chain of carbon atoms in each of the following molecules is highlighted in yellow. *The chain must also contain the carbon–carbon double bond.* The location of functional groups relative to the double bond is used in determining the appropriate prefix, *cis* or *trans*, to be used in naming each of the molecules.

Parent chain: heptene

Position of double bond: 3-

Substituents: 3,4-dichloro

Configuration: *trans*

Name: *trans*-3,4-Dichloro-3-heptene

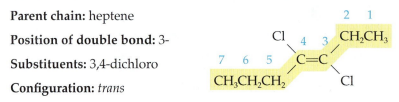

Parent chain: octene

Position of double bond: 3-

Substituents: 3,4-dimethyl

Configuration: *cis*

Name: *cis*-3,4-Dimethyl-3-octene

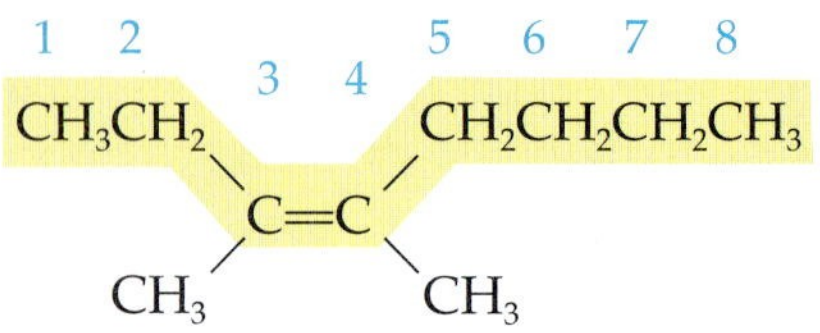

QUESTION 11.3

In each of the following pairs of molecules, identify the *cis* isomer and the *trans* isomer.

a.

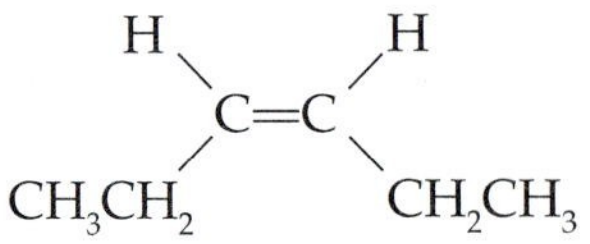

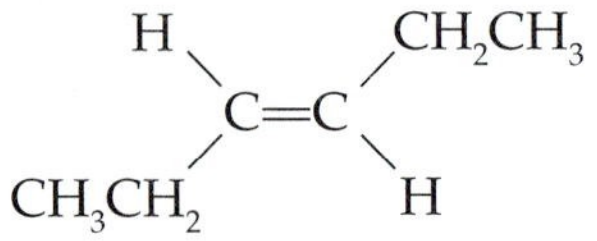

b.

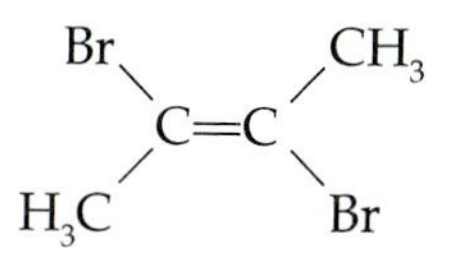

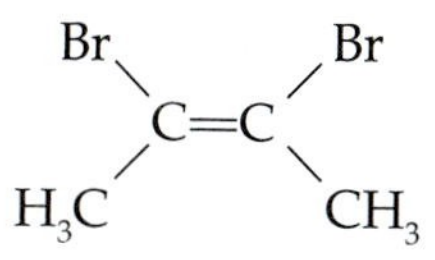

$$\begin{array}{c} Br \\ \diagdown \\ H_3C \diagup \end{array} C=C \begin{array}{c} CH_3 \\ \diagup \\ \diagdown Br \end{array} \qquad \begin{array}{c} Br \\ \diagdown \\ H_3C \diagup \end{array} C=C \begin{array}{c} Br \\ \diagup \\ \diagdown CH_3 \end{array}$$

QUESTION 11.4

Provide the complete I.U.P.A.C. name for each of the compounds in Question 11.3.

EXAMPLE 11.4   Identifying Geometric Isomers

Determine whether each of the following molecules can exist as *cis-trans* isomers: (a) 3-ethyl-3-hexene, and (b) 3-methyl-2-pentene.

3   **Learning Goal**

Write the names and draw the structures of simple geometric isomers of alkenes.

SOLUTION

a. Examine the structure of 3-ethyl-3-hexene:

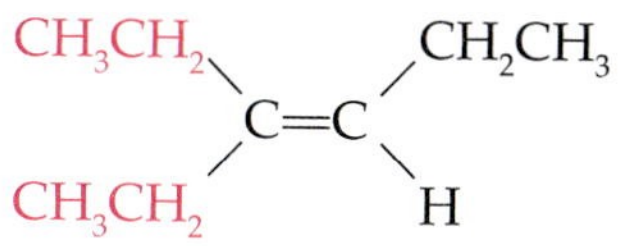

$$\begin{array}{c} CH_3CH_2 \\ \diagdown \\ CH_3CH_2 \diagup \end{array} C=C \begin{array}{c} CH_2CH_3 \\ \diagup \\ \diagdown H \end{array}$$

We see that one of the carbons of the carbon–carbon double bond is bonded to two ethyl groups. Because this carbon is bonded to two identical groups, there can be no *cis* or *trans* isomers of this compound.

b. Examination of the structure of 3-methyl-2-pentene reveals that both *cis* and *trans* isomers can be drawn.

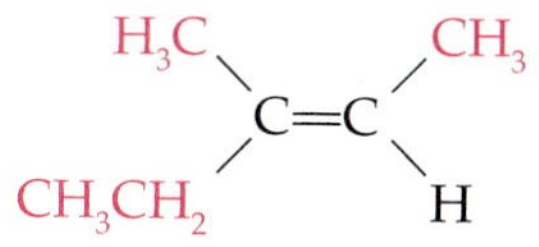

$$\begin{array}{c} H_3C \\ \diagdown \\ CH_3CH_2 \diagup \end{array} C=C \begin{array}{c} CH_3 \\ \diagup \\ \diagdown H \end{array} \qquad \begin{array}{c} H_3C \\ \diagdown \\ CH_3CH_2 \diagup \end{array} C=C \begin{array}{c} H \\ \diagup \\ \diagdown CH_3 \end{array}$$

cis-3-Methyl-2-pentene 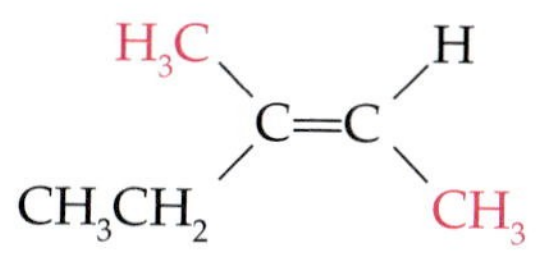   *trans*-3-Methyl-2-pentene

Each of the carbon atoms involved in the double bond is attached to two different groups. As a result, we can determine which is the *cis* isomer and which is the *trans* isomer based on the positions of the methyl groups relative to the double bond.

QUESTION 11.5

Which of the following molecules can exist as both *cis* and *trans* isomers? Explain your reasoning.

a.

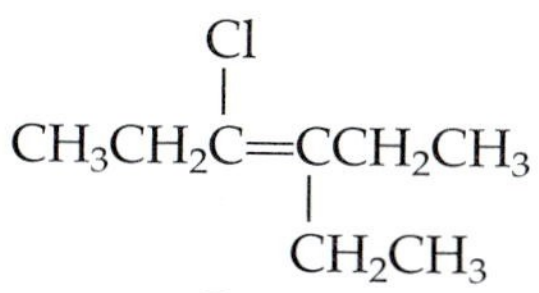

$$\begin{array}{c} Cl \\ | \\ CH_3CH_2C=CCH_2CH_3 \\ | \\ CH_2CH_3 \end{array}$$

b.

$$\begin{array}{c} Br \\ | \\ CH_3CH_2C=CBr \\ | \\ Br \end{array}$$

c.

$$\begin{array}{c} Cl \quad Cl \\ | \quad | \\ CH_3C=CCH_3 \end{array}$$

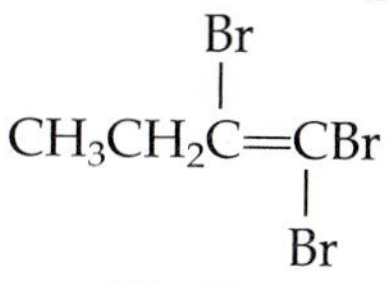

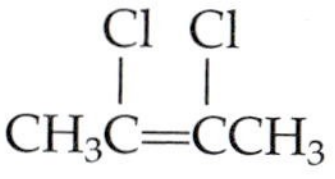

AN ENVIRONMENTAL Perspective | Alkenes in Nature

There are surprising numbers of polyenes, alkenes with several double bonds, found in nature. These molecules, which have wildly different properties and functions, are built from one or more five-carbon units called *isoprene*.

$$
\begin{array}{c}
CH_3 \\
| \\
CH_2{=}C{-}CH{=}CH_2
\end{array}
$$

Isoprene

The molecules that are produced are called *isoprenoids*, or *terpenes*. Terpenes include steroids; chlorophyll and carotenoid pigments that function in photosynthesis; and lipid-soluble vitamins A, D, E, and K (Figure 11.1).

Many other terpenes are plant products familiar to us because of their distinctive aromas. *Geraniol*, the familiar scent of geraniums, is a molecule made up of two isoprene units. Purified from plant sources, geraniol is the active ingredient in several natural insect repellents. These can be applied directly to the skin to provide four hours of protection against a variety of insects, including mosquitoes, ticks, and fire ants.

D-*Limonene* is the most abundant component of the oil extracted from the rind of citrus fruits. Because of its pleasing orange aroma, D-limonene is used as a flavor and fragrance

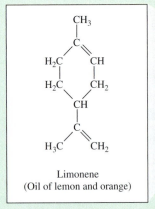

Geraniol
(Roses and geraniums)

Limonene
(Oil of lemon and orange)

Many plant products, familiar to us because of their distinctive aromas, are isoprenoids, which are alkenes having several double bonds.

additive in foods. However, the most rapidly expanding use of the compound is as a solvent. In this role, D-limonene can be used in place of more toxic solvents, such as mineral spirits, methyl ethyl ketone, acetone, toluene, and fluorinated and chlorinated organic solvents. It can also be formulated as a water-based cleaning product, such as Orange Glo, that can be used in place of more caustic cleaning solutions. There is a form of limonene that is a molecular mirror image of D-limonene. It is called L-limonene and has a pine or turpentine aroma.

The terpene *myrcene* is found in bayberry. It is used in perfumes and scented candles because it adds a refreshing, spicy

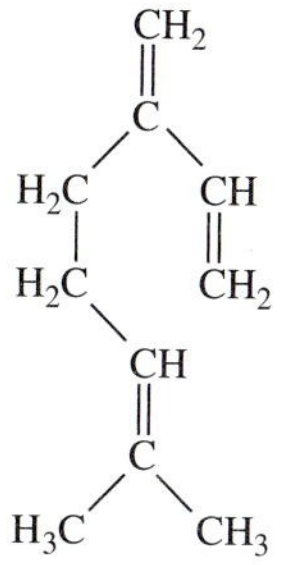

Myrcene
(Oil of bayberry)

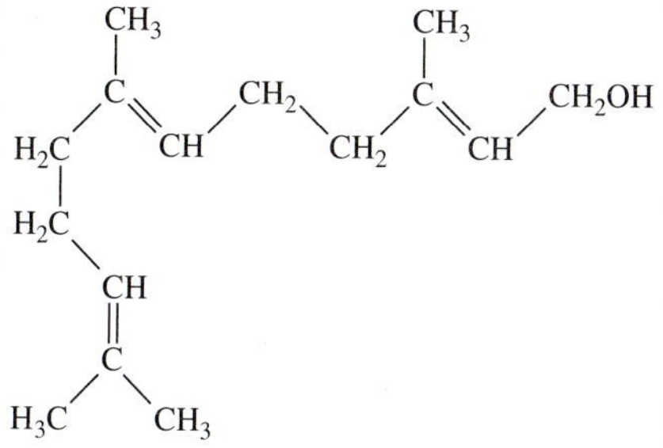

Farnesol
(Lily of the valley)

aroma to them. Trace amounts of myrcene may be used as a flavor component in root beer.

Farnesol is a terpene found in roses, orange blossom, wild cyclamen, and lily of the valley. Cosmetics companies began to use farnesol in skin care products in the early 1990s. It is claimed that farnesol smoothes wrinkles and increases skin elasticity. It is also thought to reduce skin aging by promoting regeneration of cells and activation of the synthesis of molecules, such as collagen, that are required for healthy skin.

FOR FURTHER UNDERSTANDING

Products are often advertised to have "natural" ingredients. What, if any, value is there in a natural product compared to a synthetic product.

In hopes of identifying new chemicals of use in medicine, ethnobotanists travel to remote regions of the earth to investigate new plants and the way people use those plants for food, shelter, and medical purposes. Investigate the history of the development of the pain reliever, aspirin, to learn about the accidental discovery of its biological activity.

QUESTION 11.6

Draw each of the *cis-trans* isomers in Question 11.5 and provide the complete names using the I.U.P.A.C. Nomenclature System.

QUESTION 11.7

Draw condensed formulas for each of the following compounds:

a. *cis*-3-Octene
b. *trans*-5-Chloro-2-hexene
c. *trans*-2,3-Dichloro-2-butene

QUESTION 11.8

Name each of the following compounds, using the I.U.P.A.C. system. Be sure to indicate *cis* or *trans* where applicable.

a.

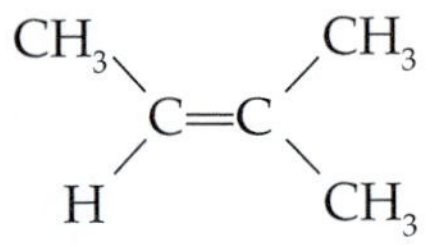

$$CH_3 \backslash \quad / CH_3$$
$$\quad C=C$$
$$H / \quad \backslash CH_3$$

b.

$$CH_3CH_2 \backslash \quad / CH_2CH_3$$
$$\quad C=C$$
$$CH_3 / \quad \backslash H$$

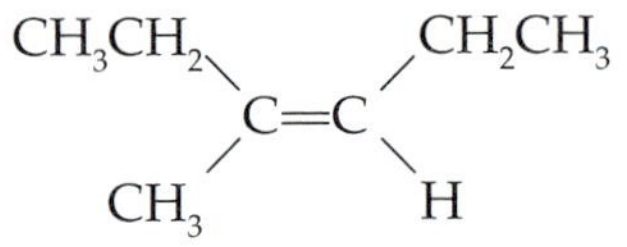

c.

$$CH_3 \backslash \quad / H$$
$$\quad C=C \quad CH_3$$
$$H / \quad \backslash CH_2\overset{|}{C}CH_3$$
$$\quad\quad\quad\quad\quad\overset{|}{CH_3}$$

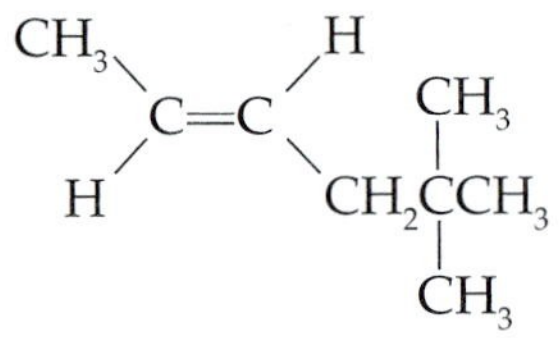

Cis-Trans Isomers in the Diet

The recent debate over the presence of *cis* and *trans* isomers of fatty acids in our diet points out the relevance of geometric isomers to our lives. Fatty acids are long-chain carboxylic acids found in vegetable oils (unsaturated fats) and animal fats (saturated fats). Oleic acid,

We will learn more about the properties of carboxylic acids in Section 13.1 and about the biological significance of fatty acids in Section 15.2.

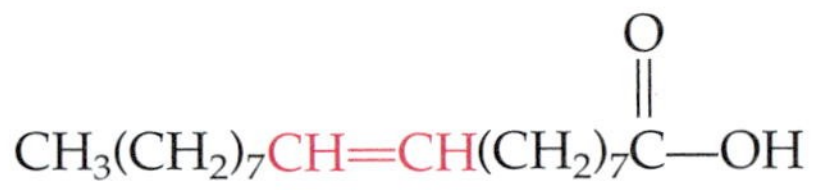

$$\overset{\displaystyle O}{\overset{\displaystyle \|}{CH_3(CH_2)_7CH{=}CH(CH_2)_7C{-}OH}}$$

is the naturally occurring fatty acid in olive oil. Its I.U.P.A.C. name, *cis*-9-octadecenoic acid, reveals that this is a *cis*-fatty acid.

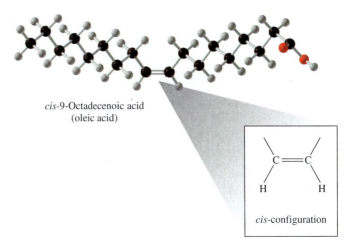

cis-9-Octadecenoic acid
(oleic acid)

cis-configuration

Although oleic acid is V-shaped as a result of the configuration of the double bond, its geometric isomer, *trans*-9-octadecenoic acid, is a rigid linear molecule.

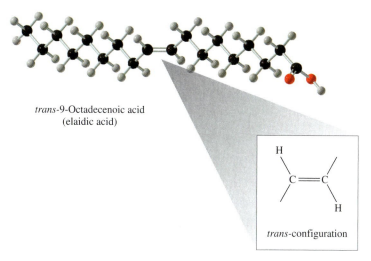

trans-9-Octadecenoic acid
(elaidic acid)

trans-configuration

The majority of *trans*-fatty acids found in the diet result from hydrogenation, which is a reaction used to convert oils into solid fats, such as margarine. It has recently been reported that *trans*-fatty acids in the diet elevate levels of "bad" or LDL cholesterol and lower the levels of "good" or HDL cholesterol, thereby increasing the risk of heart disease. Other studies suggest that *trans*-fatty acids may also increase the risk of type 2 diabetes.

cis- and trans-Fatty acids will be discussed in greater detail in Chapter 15. Hydrogenation is described in Section 11.4.

11.4 Reactions Involving Alkenes and Alkynes

Reactions of alkenes involve the carbon–carbon double bond. The key reaction of the double bond is the **addition reaction.** This involves the addition of two atoms or groups of atoms to a double bond. The major alkene addition reactions include addition of hydrogen (H_2), halogens (Cl_2 or Br_2), or water (HOH). A generalized addition reaction is shown here. The R in these structures represents any alkyl group.

4 **Learning Goal**
Write equations predicting the products of the simple addition reactions of alkenes and alkynes: hydrogenation, halogenation, and hydration.

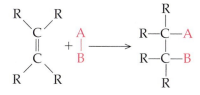

Note that the double bond is replaced by a single bond. The former double bond carbons receive a new single bond to a new atom, producing either an alkane or a substituted alkane.

Hydrogenation: Addition of H_2

Hydrogenation is the addition of a molecule of hydrogen (H_2) to a carbon–carbon double bond to give an alkane. In this reaction the double bond is broken, and two new C—H single bonds result. Platinum, palladium, or nickel is needed as a catalyst to speed up the reaction. Heat and/or pressure may also be required.

Recall that a catalyst itself undergoes no net change in the course of a chemical reaction (see Section 5.3).

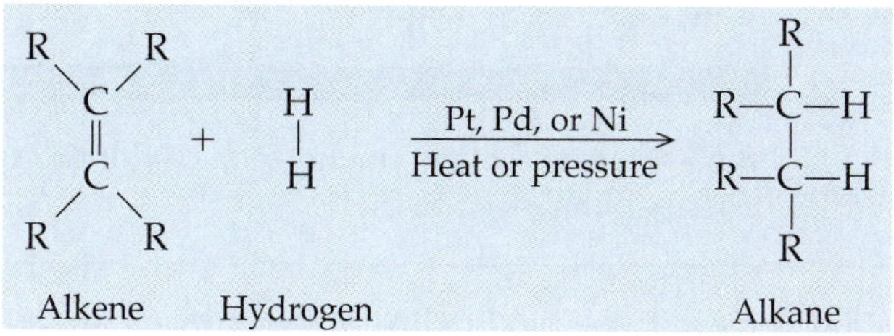

$$
\underset{\text{Alkene}}{\overset{R}{\underset{R}{\overset{|}{C}}}\underset{}{\overset{}{\underset{}{\|}}}\overset{R}{\underset{R}{\overset{|}{C}}}} + \underset{\text{Hydrogen}}{\overset{H}{\underset{H}{\overset{|}{}}}} \xrightarrow[\text{Heat or pressure}]{\text{Pt, Pd, or Ni}} \underset{\text{Alkane}}{\overset{R}{\underset{R}{\overset{|}{\underset{|}{\overset{R-C-H}{R-C-H}}}}}}
$$

Note that the alkene is gaining two hydrogens. Thus, hydrogenation is a reduction reaction (see Section 4.3).

EXAMPLE 11.5 Writing Equations for the Hydrogenation of Alkenes

Write a balanced equation showing the hydrogenation of *trans*-2-pentene.

SOLUTION

Begin by drawing the structure of *trans*-2-pentene and of diatomic hydrogen (H_2) and indicating the catalyst.

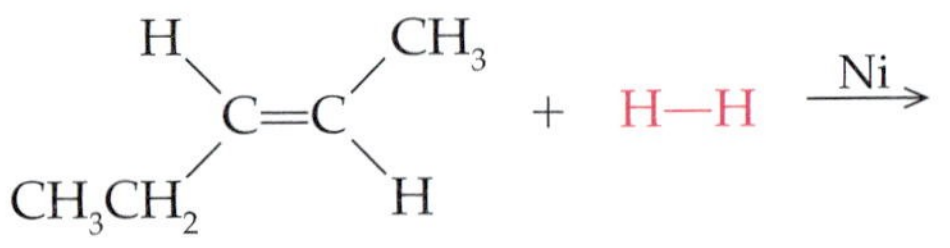

trans-2-Pentene   Hydrogen

Knowing that one hydrogen atom will form a covalent bond with each of the carbon atoms of the carbon–carbon double bond, we can write the product and complete the equation.

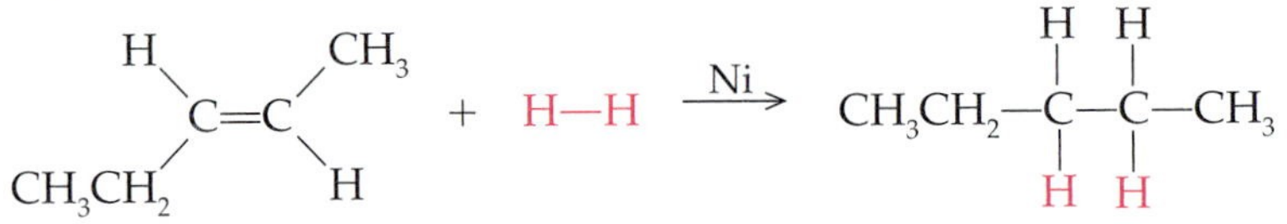

trans-2-Pentene   Hydrogen   Pentane

QUESTION 11.9

The *trans* isomer of 2-pentene was used in Example 11.5. Would the result be any different if the *cis* isomer had been used?

QUESTION 11.10

Write balanced equations for the hydrogenation of 1-butene and *cis*-2-butene.

The conditions for the hydrogenation of alkynes are very similar to those for the hydrogenation of alkenes. Two moles of hydrogen add to the triple bond of the alkyne to produce an alkane, as seen in the following equation:

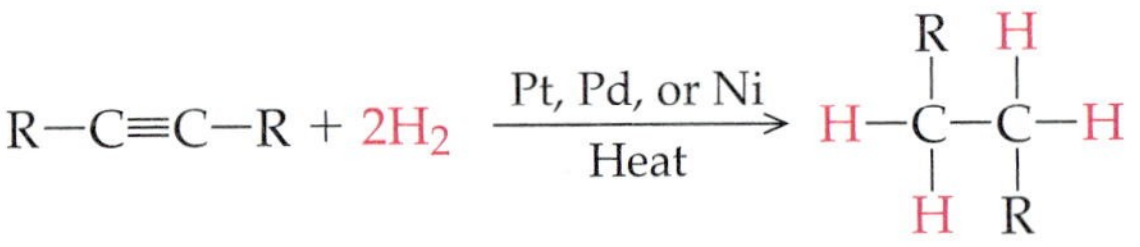

Alkyne   Hydrogen   Alkane
(2 mol)

QUESTION 11.11

Write balanced equations for the complete hydrogenation of each of the following alkynes:

a. $H_3CC{\equiv}CCH_3$ b. $H_3CC{\equiv}CCH_2CH_3$

QUESTION 11.12

Using the I.U.P.A.C. Nomenclature System, name each of the products and reactants in the reactions described in Question 11.11.

Saturated and unsaturated dietary fats are discussed in Section 15.2.

Hydrogenation is used in the food industry to produce margarine, which is a mixture of hydrogenated vegetable oils (Figure 11.2). Vegetable oils are unsaturated,

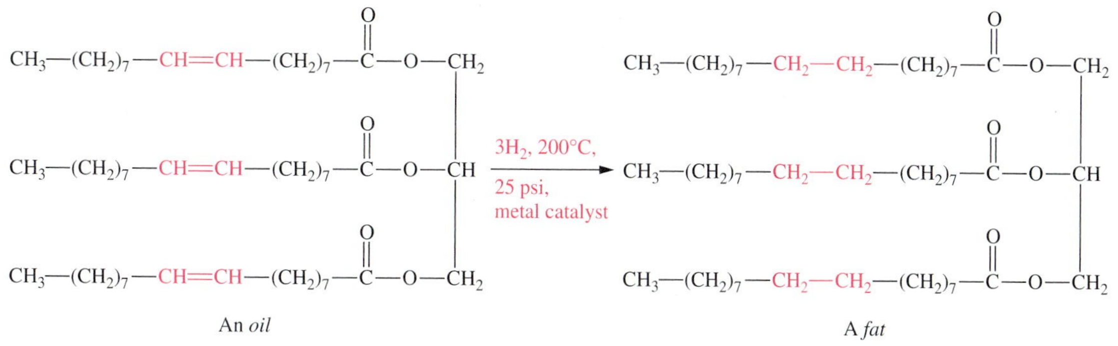

$$CH_3-(CH_2)_7-CH=CH-(CH_2)_7-\overset{\overset{O}{\|}}{C}-O-CH_2$$

$$CH_3-(CH_2)_7-CH=CH-(CH_2)_7-\overset{\overset{O}{\|}}{C}-O-CH$$

$$CH_3-(CH_2)_7-CH=CH-(CH_2)_7-\overset{\overset{O}{\|}}{C}-O-CH_2$$

An *oil*

3H₂, 200°C, 25 psi, metal catalyst

$$CH_3-(CH_2)_7-CH_2-CH_2-(CH_2)_7-\overset{\overset{O}{\|}}{C}-O-CH_2$$

$$CH_3-(CH_2)_7-CH_2-CH_2-(CH_2)_7-\overset{\overset{O}{\|}}{C}-O-CH$$

$$CH_3-(CH_2)_7-CH_2-CH_2-(CH_2)_7-\overset{\overset{O}{\|}}{C}-O-CH_2$$

A *fat*

Figure 11.2

Conversion of a typical oil to a fat involves hydrogenation. In this example, triolein (an oil) is converted to tristearin (a fat).

that is, they contain many double bonds and as a result have low melting points and are liquid at room temperature. The hydrogenation of these double bonds to single bonds increases the melting point of these oils and results in a fat, such as Crisco, that remains solid at room temperature. A similar process with highly refined corn oil and added milk solids produces corn oil margarine. As we saw in the previous section, such margarine may contain *trans*-fatty acids as a result of hydrogenation.

Halogenation: Addition of X₂

Chlorine (Cl_2) or bromine (Br_2) can be added to a double bond. This reaction, called **halogenation,** proceeds readily and does not require a catalyst:

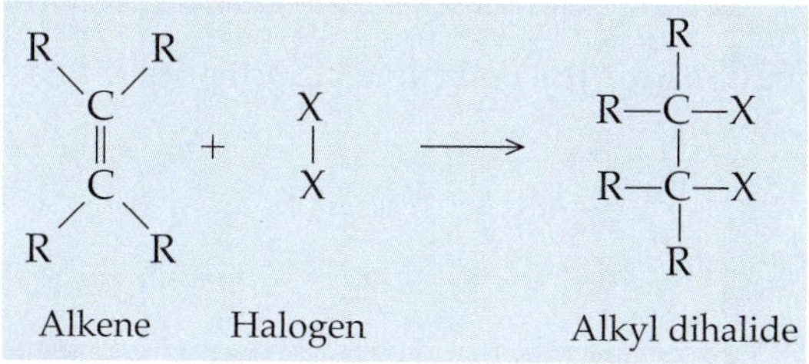

Alkene   Halogen   Alkyl dihalide

EXAMPLE 11.6   Writing Equations for the Halogenation of Alkenes

Write a balanced equation showing the bromination of *trans*-2-butene.

SOLUTION

Begin by drawing the structure of *trans*-2-butene and of diatomic bromine (Br_2).

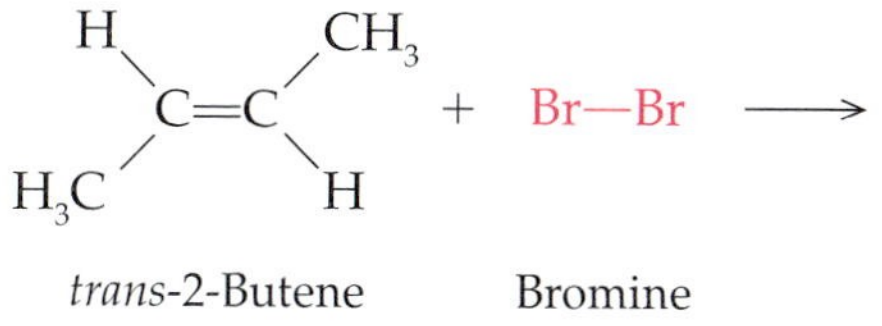

trans-2-Butene   Bromine

Knowing that one bromine atom will form a covalent bond with each of the carbon atoms of the carbon–carbon double bond, we can write the product and complete the equation.

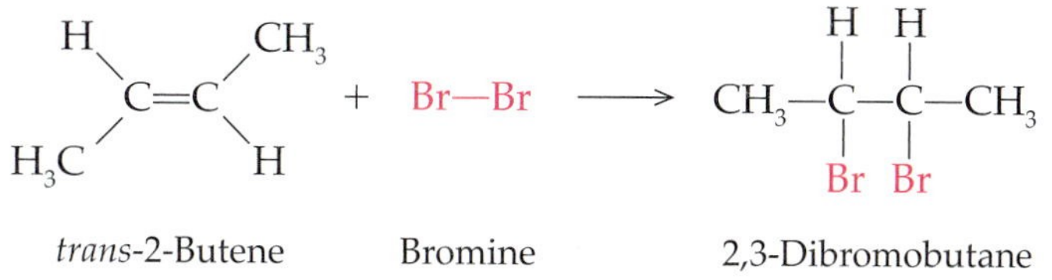

trans-2-Butene   Bromine   2,3-Dibromobutane

4 Learning Goal

Write equations predicting the products of the simple addition reactions of alkenes and alkynes: hydrogenation, halogenation, and hydration.

QUESTION 11.13

Write a balanced equation for the addition of bromine to each of the following alkenes. Draw and name the products and reactants for each reaction.

  a. $CH_3CH{=}CH_2$
  b. $CH_3CH{=}CHCH_3$

QUESTION 11.14

Using the I.U.P.A.C. Nomenclature System, name each of the products and reactants in the reactions described in Question 11.13.

Alkynes also react with the halogens bromine or chlorine. Two moles of halogen add to the triple bond to produce a tetrahaloalkane:

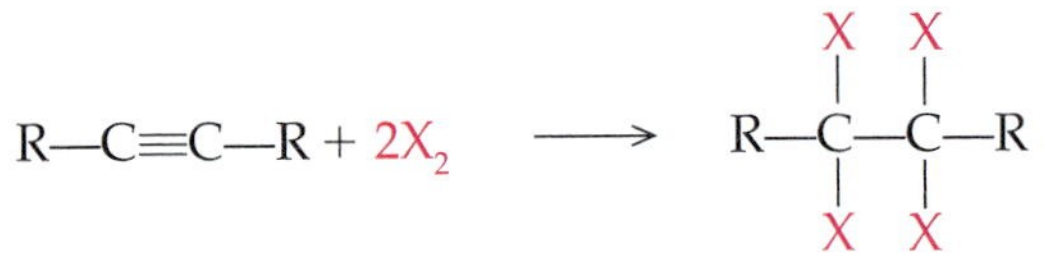

$$R{-}C{\equiv}C{-}R + 2X_2 \longrightarrow \begin{matrix} & X & X & \\ & | & | & \\ R{-}C{-}&C{-}R \\ & | & | & \\ & X & X & \end{matrix}$$

Alkyne	Halogen (2 mol)	Tetrahaloalkane

QUESTION 11.15

Write balanced equations for the complete chlorination of each of the following alkynes:

  a. $H_3CC{\equiv}CCH_3$
  b. $H_3CC{\equiv}CCH_2CH_3$

QUESTION 11.16

Using the I.U.P.A.C. Nomenclature System, name each of the products and reactants in the reactions described in Question 11.15.

Below we see an equation representing the bromination of 1-pentene. Notice that the solution of reactants is red because of the presence of bromine. However, the product is colorless (Figure 11.3).

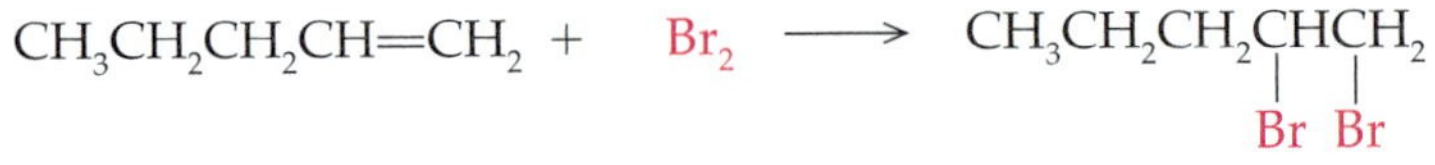

$$CH_3CH_2CH_2CH{=}CH_2 + Br_2 \longrightarrow CH_3CH_2CH_2\underset{\underset{Br}{|}}{C}H\underset{\underset{Br}{|}}{C}H_2$$

1-Pentene (colorless)	Bromine (red)	1,2-Dibromopentane (colorless)

This bromination reaction can be used to show the presence of double or triple bonds in an organic compound. The reaction mixture is red because of the presence of dissolved bromine. If the red color is lost, the bromine has been consumed. Thus bromination has occurred, and the compound must have had a carbon–carbon double or triple bond. The greater the amount of bromine that must be added to the reaction, the more unsaturated the compound. For instance, a diene or an alkyne would consume twice as much bromine as an alkene with a single double bond.

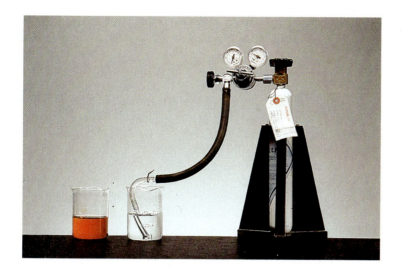

Figure 11.3

Bromination of an alkene. The solution on the left is red because of the presence of bromine and absence of an alkene or alkyne. The solution on the right is colorless because the bromine has completely reacted with the unsaturated hydrocarbon.

Hydration: Addition of H_2O

A water molecule can be added to an alkene. This reaction, termed **hydration,** requires a trace of strong acid (H^+) as a catalyst. The product is an alcohol, as shown in the following reaction:

$$
\underset{\text{Alkene}}{\overset{\displaystyle R \diagdown \;\; \diagup R}{\underset{\displaystyle R \diagup \;\; \diagdown R}{\overset{\displaystyle C}{\underset{\displaystyle C}{\|}}}}} \;+\; \underset{\text{Water}}{\overset{\displaystyle H}{\underset{\displaystyle OH}{|}}} \;\xrightarrow{\;H^+\;}\; \underset{\text{Alcohol}}{\overset{\displaystyle R}{\underset{\displaystyle R}{\overset{\displaystyle R-C-H}{\underset{\displaystyle R-C-OH}{|}}}}}
$$

The following equation shows the hydration of ethene to produce ethanol.

$$
\underset{\text{Ethene}}{\overset{\displaystyle H \diagdown \;\; \diagup H}{\underset{\displaystyle H \diagup \;\; \diagdown H}{\overset{C}{\underset{C}{\|}}}}} \;+\; \underset{\text{Water}}{\overset{\color{red}H}{\underset{\color{red}OH}{|}}} \;\xrightarrow{\;H^+\;}\; \underset{\substack{\text{Ethanol}\\(\text{ethyl alcohol})}}{\overset{\displaystyle H}{\underset{\displaystyle H}{\overset{H-C-\color{red}H}{\underset{H-C-\color{red}OH}{|}}}}}
$$

With alkenes in which the groups attached to the two carbons of the double bond are different (unsymmetrical alkenes), two products are possible. For example,

$$
\underset{\substack{\text{Propene}\\(\text{propylene})}}{\overset{3 \;\; 2 \;\; 1}{H-\underset{H}{\overset{H}{C}}-\underset{H}{\overset{H}{C}}=C-H}} + H-OH \;\xrightarrow{\;H^+\;}\; \underset{\substack{\text{Major product}\\\text{2-Propanol}\\(\text{isopropyl alcohol})}}{\overset{3 \;\; 2 \;\; 1}{H-\underset{H}{\overset{H}{C}}-\underset{OH}{\overset{H}{C}}-\underset{H}{\overset{H}{C}}-H}} + \underset{\substack{\text{Minor product}\\\text{1-Propanol}\\(\text{propyl alcohol})}}{\overset{3 \;\; 2 \;\; 1}{H-\underset{H}{\overset{H}{C}}-\underset{H}{\overset{H}{C}}-\underset{OH}{\overset{H}{C}}-H}}
$$

5 **Learning Goal**

Apply Markovnikov's rule to predict the major and minor products of hydration reactions of unsymmetrical alkenes.

When hydration of an unsymmetrical alkene, such as propene, is carried out in the laboratory, one product (2-propanol) is favored over the other. The Russian chemist Vladimir Markovnikov studied many such reactions and came up with a rule that can be used to predict the major product of such a reaction. **Markovnikov's rule** tells us that the carbon of the carbon–carbon double bond that originally has more hydrogen atoms receives the hydrogen atom being added to the double bond. The remaining carbon forms a bond with the —OH. Simply stated, "the rich get richer"—the carbon with the greater number of hydrogens gets the new one as well. In the preceding example, carbon-1 has two C—H bonds originally, and carbon-2 has only one. The major product, 2-propanol, results from the new C—H bond forming on carbon-1 and the new C—OH bond on carbon-2.

Addition of water to a double bond is a reaction that we find in several biochemical pathways. For instance, the citric acid cycle is a key metabolic pathway in the complete oxidation of the sugar glucose and the release of the majority of the energy used by the body. The citric acid cycle is also the source of starting materials for the synthesis of the biological molecules needed for life. The next-to-last reaction in the citric acid cycle is the hydration of a molecule of fumarate to produce a molecule called malate.

The reactions of the citric acid cycle are described in Section 18.8.

We have seen that hydration of a double bond requires a trace of acid as a catalyst. In the cell, this reaction is catalyzed by an enzyme, or biological catalyst, called fumarase.

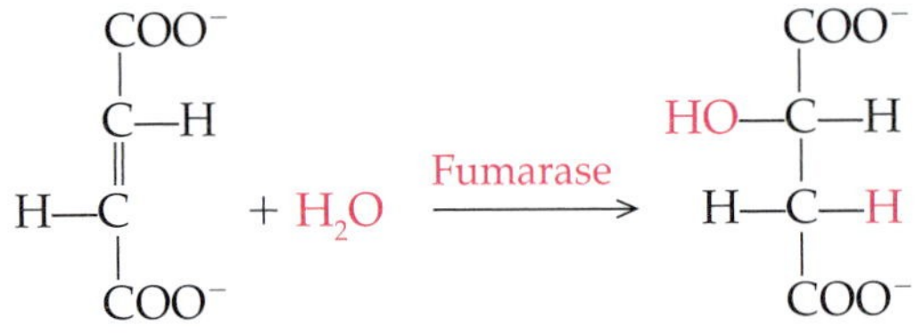

$$\text{Fumarate} + H_2O \xrightarrow{\text{Fumarase}} \text{Malate}$$

Hydration of a double bond also occurs in the β-oxidation pathway (see Section 19.2). This pathway carries out the oxidation of dietary fatty acids. Like the citric acid cycle, β-oxidation harvests the energy of the food molecules to use for body functions.

EXAMPLE 11.7 Writing Equations for the Hydration of Alkenes

Write an equation showing all the products of the hydration of 1-pentene.

5 Learning Goal
Apply Markovnikov's rule to predict the major and minor products of hydration reactions of unsymmetrical alkenes.

SOLUTION

Begin by drawing the structure of 1-pentene and water and indicating the catalyst.

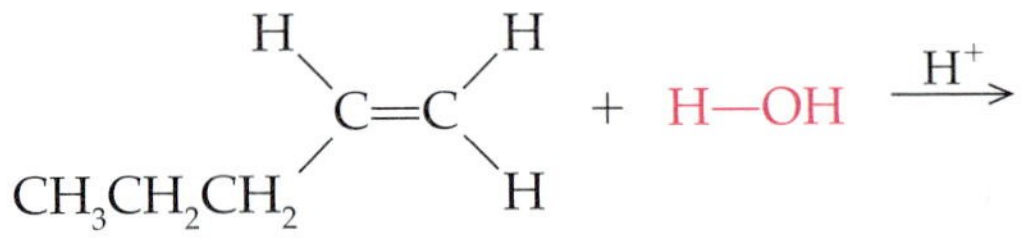

1-Pentene   Water

Markovnikov's rule tells us that the carbon atom that is already bonded to the greater number of hydrogen atoms is more likely to bond to the hydrogen atom from the water molecule. The other carbon atom is more likely to bond to the hydroxyl group. Thus, we can predict that the major product of this reaction will be 2-pentanol and that the minor product will be 1-pentanol. Now we can complete the equation by showing the products:

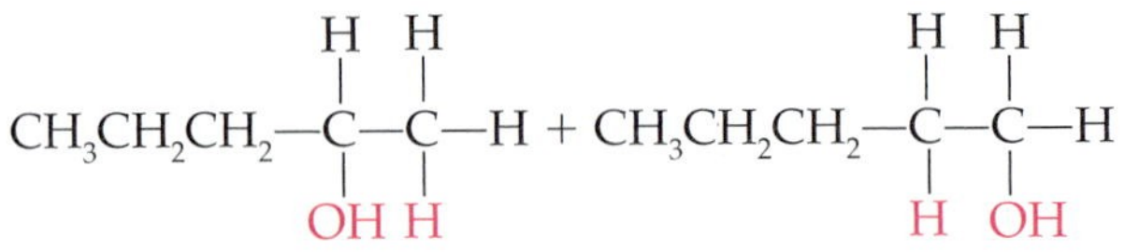

2-Pentanol (major product)   1-Pentanol (minor product)

A LIFESTYLE Connection

Life Without Polymers?

What do Nike Air-Sole shoes, Saturn automobiles, disposable diapers, tires, shampoo, and artificial joints and skin share in common? These products and a great many other items we use every day are composed of synthetic or natural polymers. The field of polymer chemistry has come a long way since the 1920s and 1930s when DuPont chemists invented nylon and Teflon.

Consider the disposable diaper. The outer, waterproof layer is composed of polyethylene. The polymerization reaction that produces polyethylene is shown in Section 11.4. The diapers have elastic to prevent leaking. The elastic is made of a natural polymer, rubber. The monomer from which natural rubber is formed is 2-methyl-1,3-butadiene. The common name of this monomer is *isoprene*. Isoprene is an important monomer in the synthesis of many natural polymers.

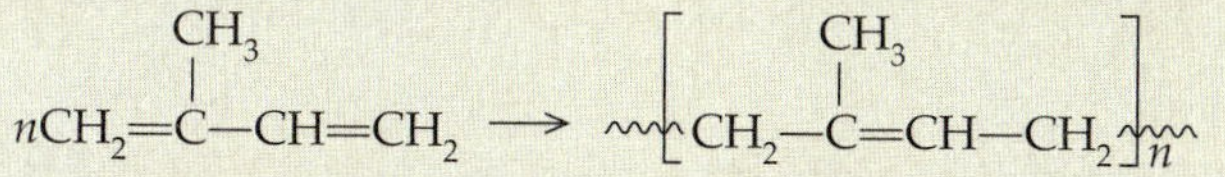

$$nCH_2{=}\overset{\overset{\displaystyle CH_3}{|}}{C}{-}CH{=}CH_2 \longrightarrow \left[CH_2{-}\overset{\overset{\displaystyle CH_3}{|}}{C}{=}CH{-}CH_2\right]_n$$

2-Methyl-1, 3-butadiene (isoprene)   Rubber polymer

The diaper is filled with a synthetic polymer called poly(acrylic acid). This polymer has the remarkable ability to absorb many times its own weight in liquid. Polymers that have this ability are called superabsorbers, but polymer chemists have no idea why they have this property! The acrylate monomer and resulting poly(acrylic acid) polymer are shown here:

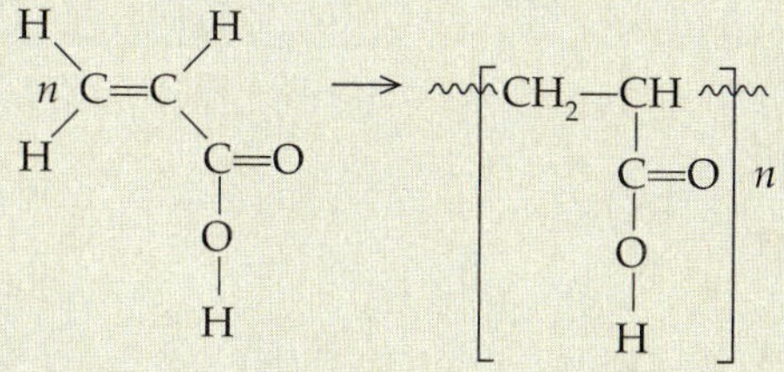

$$n\;\overset{\overset{\displaystyle H}{|}}{\underset{\underset{\displaystyle H}{|}}{C}}{=}\overset{\overset{\displaystyle H}{|}}{\underset{\underset{\displaystyle \underset{\displaystyle O{-}H}{\overset{|}{O}}}{\overset{|}{C}{=}O}}{C}} \longrightarrow \left[CH_2{-}\underset{\underset{\underset{O{-}H}{\overset{|}{O}}}{\overset{|}{C}{=}O}}{CH}\right]_n$$

Acrylate monomer   Poly(acrylic acid)

Another example of a useful polymer is Gore-Tex. This amazing polymer is made by stretching Teflon. Teflon is produced from the monomer tetrafluoroethene, as seen in the following reaction:

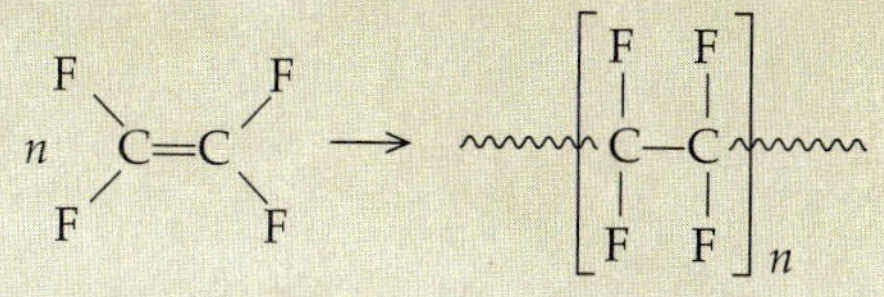

$$n\;\overset{\overset{\displaystyle F}{\diagup}}{\underset{\underset{\displaystyle F}{\diagdown}}{C}}{=}\overset{\overset{\displaystyle F}{\diagdown}}{\underset{\underset{\displaystyle F}{\diagup}}{C}} \longrightarrow \left[\overset{\overset{\displaystyle F}{|}}{\underset{\underset{\displaystyle F}{|}}{C}}{-}\overset{\overset{\displaystyle F}{|}}{\underset{\underset{\displaystyle F}{|}}{C}}\right]_n$$

Tetrafluoroethene   Teflon

Clothing made from this fabric is used to protect firefighters because of its fire resistance. Because it also insulates, Gore-Tex clothing is used by military forces and by many amateur athletes for protection during strenuous activity in the cold. In addition to its use in protective clothing, Gore-Tex has been used in millions of medical procedures for sutures, synthetic blood vessels, and tissue reconstruction.

Plastics are polymers that have changed the way we live, providing novel solutions to many problems. However, they have also created an environmental problem. Since most are not biodegradable, plastic that is discarded simply builds up in landfills. To learn more about solutions to this problem, see *An Environmental Perspective: Plastic Recycling*, online at www.mhhe.com/denniston.

FOR FURTHER UNDERSTANDING

Visit The Macrogalleria, www.pslc.ws/macrog.htm, an Internet site maintained by the Department of Polymer Science of the University of Southern Mississippi, to help you answer these questions:

Why does shrink wrap shrink?

What are optical fibers made of and how do they transmit light?

QUESTION 11.17

Write a balanced equation for the hydration of each of the following alkenes. Predict the major product of each of the reactions.

a. $CH_3CH{=}CHCH_3$

b. $CH_2{=}CHCH_2CH_2\underset{\underset{\displaystyle CH_3}{|}}{CH}CH_3$

c. $CH_3CH_2CH_2CH{=}CHCH_2CH_3$

d. $CH_3CHClCH{=}CHCHClCH_3$

QUESTION 11.18

Write a balanced equation for the hydration of each of the following alkenes. Predict the major product of each of the reactions.

a. $CH_2{=}CHCH_2CH_2CH_3$

b. $CH_3CH_2CH_2CH{=}CHCH_3$

c. $CH_3CHBrCH_2CH{=}CHCH_2Cl$

d. $CH_3CH_2CH_2CH_2CH_2CH{=}CHCH_3$

Addition Polymers of Alkenes

6 Learning Goal

Write equations representing the formation of addition polymers of alkenes.

Polymers are macromolecules composed of repeating structural units called **monomers.** A polymer may be made up of several thousand monomers. Many commercially important plastics and fibers are addition polymers made from alkenes or substituted alkenes. They are called **addition polymers** because they are made by the sequential addition of the alkene monomer. The general formula for this addition reaction follows:

$$n \quad \underset{R}{\overset{R}{\diagdown}} C = C \underset{R}{\overset{R}{\diagup}} \quad \xrightarrow[\text{Pressure}]{\substack{\text{Catalyst}\\\text{Heat}}} \quad \text{etc.} - \underset{\underset{R}{|}}{\overset{\overset{R}{|}}{C}} - \underset{\underset{R}{|}}{\overset{\overset{R}{|}}{C}} - \underset{\underset{R}{|}}{\overset{\overset{R}{|}}{C}} - \underset{\underset{R}{|}}{\overset{\overset{R}{|}}{C}} - \underset{\underset{R}{|}}{\overset{\overset{R}{|}}{C}} - \underset{\underset{R}{|}}{\overset{\overset{R}{|}}{C}} - \text{etc.}$$

Alkene monomer   Addition polymer
R = H, X, or an alkyl group

The product of the reaction is generally represented in a simplified manner:

$$\left[\underset{\underset{R}{|}}{\overset{\overset{R}{|}}{C}} - \underset{\underset{R}{|}}{\overset{\overset{R}{|}}{C}} \right]_n$$

Polyethylene is a polymer made from the monomer ethylene (ethene):

$$n CH_2 = CH_2 \longrightarrow \left[CH_2 - CH_2 \right]_n$$

Ethene   Polyethylene
(ethylene)

It is used to make bottles, injection-molded toys and housewares, and wire coverings.

Polypropylene is a plastic made from propylene (propene). It is used to make indoor-outdoor carpeting, packaging materials, toys, and housewares. When propylene polymerizes, a methyl group is located on every other carbon of the main chain:

$$n CH_2 = \underset{\underset{CH_3}{|}}{CH} \longrightarrow \left[CH_2 - \underset{\underset{CH_3}{|}}{CH} \right]_n$$

or

$$\sim CH_2 - \underset{\underset{CH_3}{|}}{CH} - CH_2 - \underset{\underset{CH_3}{|}}{CH} - CH_2 - \underset{\underset{CH_3}{|}}{CH} \sim$$

To learn more about plastics and plastic recycling, see *An Environmental Perspective: Plastic Recycling,* online at *www.mhhe.com/denniston and A Lifestyle Connection: Life Without Polymers? on the preceding page.*

Polymers made from alkenes or substituted alkenes are simply very large alkanes or substituted alkanes. Like alkanes, they are typically inert. This chemical inertness makes these polymers ideal for containers to hold juices, chemicals, and fluids used medically. They are also used to make sutures, catheters, and other indwelling devices. A variety of polymers made from substituted alkenes are listed in Table 11.1.

11.5 Aromatic Hydrocarbons

7 Learning Goal

Draw the structures and write the names of common aromatic hydrocarbons.

In the early part of the nineteenth century, chemists began to discover organic compounds with chemical properties quite distinct from alkanes, alkenes, and alkynes. They called these substances *aromatic compounds* because many of the first examples were isolated from the pleasant-smelling resins of tropical trees.

TABLE 11.1 Some Important Addition Polymers of Alkenes

Monomer Name	Formula	Polymer	Uses
Styrene	$CH_2{=}CH{-}\bigcirc$ 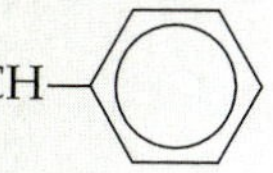	Polystyrene	Styrofoam containers
Acrylonitrile	$CH_2{=}CHCN$	Polyacrylonitrile (Orlon)	Clothing
Methyl methacrylate	$CH_2{=}C(CH_3){-}\overset{\overset{\textstyle O}{\|}}{C}OCH_3$	Polymethyl methacrylate (Plexiglas, Lucite)	Basketball backboards
Vinyl chloride	$CH_2{=}CHCl$	Polyvinyl chloride (PVC)	Plastic pipe, credit cards
Tetrafluoroethene	$CF_2{=}CF_2$	Polytetrafluoroethylene (Teflon)	Nonstick surfaces

We now define **aromatic hydrocarbons** as compounds that contain a benzene ring. The structure of the benzene ring is represented in various ways in Figure 11.4.

Structure and Properties

Benzene, discovered in 1825 by Michael Faraday, was known to have the formula C_6H_6. However, it was not until 1856 that August Kekulé proposed a model for its structure that suggested that each of the carbon atoms is bonded to a single hydrogen atom and that single and double bonds alternate around the ring. This results in two possible structures shown in Figure 11.4a.

We now know that each carbon is bonded to two others by sharing a pair of electrons. Each carbon atom also shares a pair of electrons with a hydrogen atom. The remaining six electrons are shared equally among the carbons forming a cloud of electrons above and below the ring. This structure is most accurately represented by the line formula in Figure 11.4c.

Two symbols are commonly used to represent the benzene ring. The representation in Figure 11.4b is the structure proposed by Kekulé. The structure in Figure 11.4c uses a circle to represent the electron cloud.

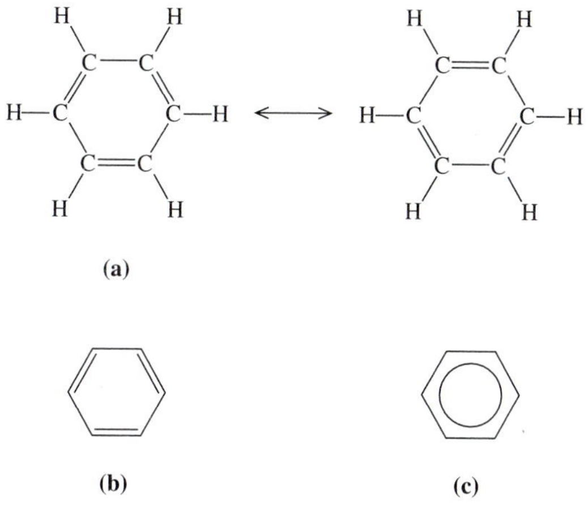

Figure 11.4

Four ways to represent the benzene molecules. The structures in (a) represent the two possible Kekulé structures. Structure (b) is a simplified diagram of one of the Kekulé structures. Structure (c) is the most commonly used representation.

Nomenclature

7 Learning Goal

Draw the structures and write the names of common aromatic hydrocarbons.

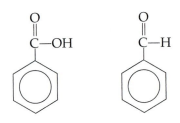

Benzoic acid

Benzaldehyde

Most simple aromatic compounds are named as derivatives of benzene. Thus, benzene is the parent compound, and the name of any atom or group bonded to benzene is used as a prefix, as in these examples:

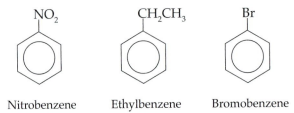

Nitrobenzene

Ethylbenzene

Bromobenzene

Other members of this family have unique names based on history rather than logic:

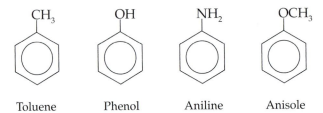

Toluene

Phenol

Aniline

Anisole

In the I.U.P.A.C Nomenclature System, when two or more groups are present on the ring, the ring carbons are numbered to give the lowest numbers to the groups.

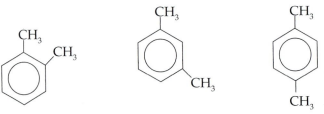

1,2-Dimethylbenzene

1,3-Dimethylbenzene

1,4-Dimethylbenzene

EXAMPLE 11.8 Naming Derivatives of Benzene

7 Learning Goal

Draw the structures and write the names of common aromatic hydrocarbons.

Name the following compounds using the I.U.P.A.C. Nomenclature System.

a.

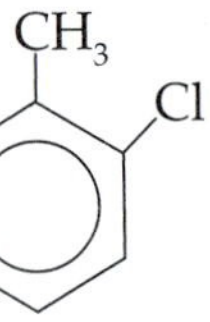

b.

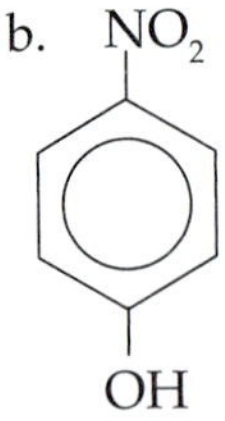

c.

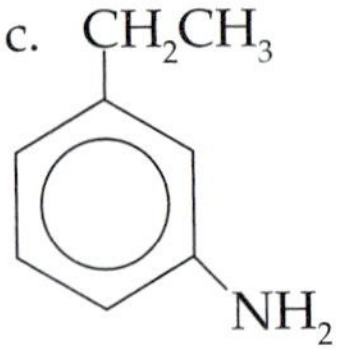

SOLUTION

Parent compound:	toluene	phenol	aniline
Substituents:	2-chloro	4-nitro	3-ethyl
Name:	2-Chlorotoluene	4-Nitrophenol	3-Ethylaniline

In I.U.P.A.C. nomenclature, the group derived by removing one hydrogen from benzene, ($—C_6H_5$), is called the **phenyl group.** An aromatic hydrocarbon with an aliphatic side chain is named as a phenyl substituted hydrocarbon. For example,

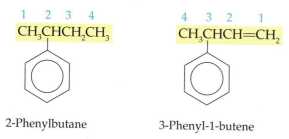

2-Phenylbutane   3-Phenyl-1-butene

QUESTION 11.19

Draw each of the following compounds:

a. 1,3,5-Trichlorobenzene
b. 2,5-Dibromophenol
c. 2-Nitroaniline

QUESTION 11.20

Draw each of the following compounds:

a. 2,3-Dichlorotoluene
b. 3-Bromoaniline
c. 1-Bromo-3-ethylbenzene

Polycyclic Aromatic Hydrocarbons

Polynuclear aromatic hydrocarbons (PAH) are composed of two or more aromatic rings joined together. Many of them have been shown to be carcinogenic, that is, they cause cancer.

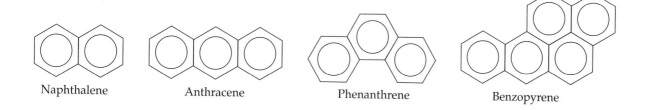

Naphthalene   Anthracene   Phenanthrene   Benzopyrene

Naphthalene has a distinctive aroma. It has been frequently used as mothballs and may cause hemolytic anemia (a condition causing breakdown of red blood cells) in humans but has not been associated with human or animal cancers. Anthracene, derived from coal tar, is the parent compound of many dyes and pigments. Phenanthrene is common in the environment, although there is no industrial use of the compound. It is a product of incomplete combustion of fossil fuels and wood. Although anthracene is a suspected carcinogen, phenanthrene has not been shown to be one. Benzopyrene is found in tobacco smoke, smokestack effluents, charcoal-grilled meat, and automobile exhaust. It is one of the most potent carcinogens known.

11.6 Heterocyclic Aromatic Compounds

Heterocyclic aromatic compounds are those having at least one atom other than carbon as part of the structure of the aromatic ring. The structures and common names of several heterocyclic aromatic compounds are shown:

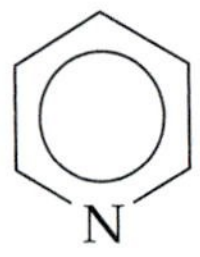

Pyridine 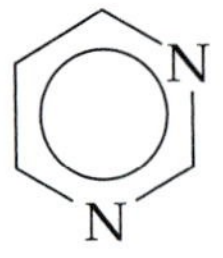   Pyrimidine 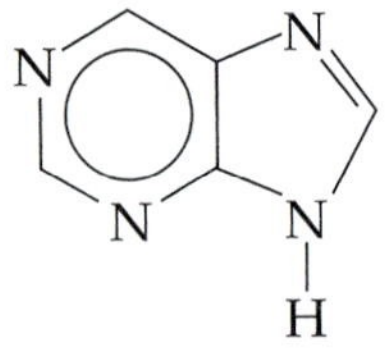   Purine

All these compounds are more similar to benzene in stability and chemical behavior than they are to alkenes. Many of these compounds are components of molecules that have significant effects on biological systems. For instance, purines and pyrimidines are components of DNA (deoxyribonucleic acid) and RNA (ribonucleic acid). DNA and RNA are the molecules responsible for storing and expressing the genetic information of an organism. The pyridine ring is found in nicotine, the addictive compound in tobacco.

See A Medical Connection: The Nicotine Patch online at www.mhhe.com/denniston.

SUMMARY OF REACTIONS

Addition Reactions of Alkenes

Hydrogenation:

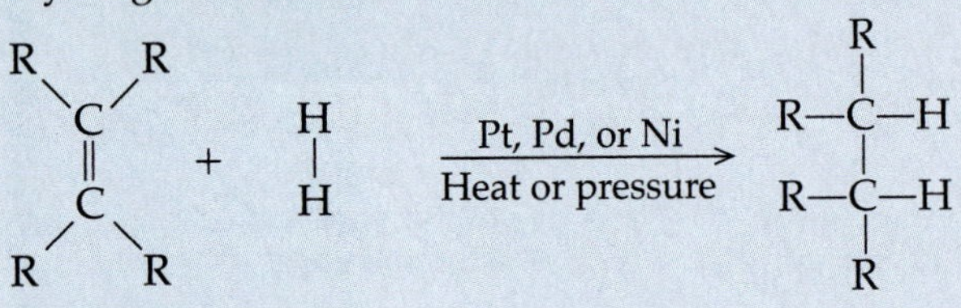

$$\text{Alkene} + \text{Hydrogen} \xrightarrow[\text{Heat or pressure}]{\text{Pt, Pd, or Ni}} \text{Alkane}$$

Hydration:

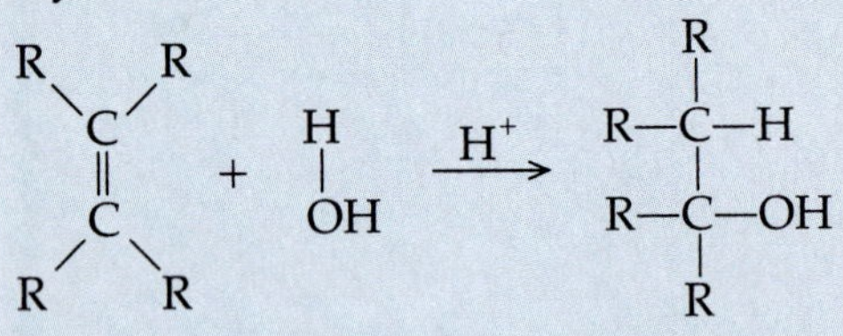

$$\text{Alkene} + \text{Water} \xrightarrow{H^+} \text{Alcohol}$$

Halogenation:

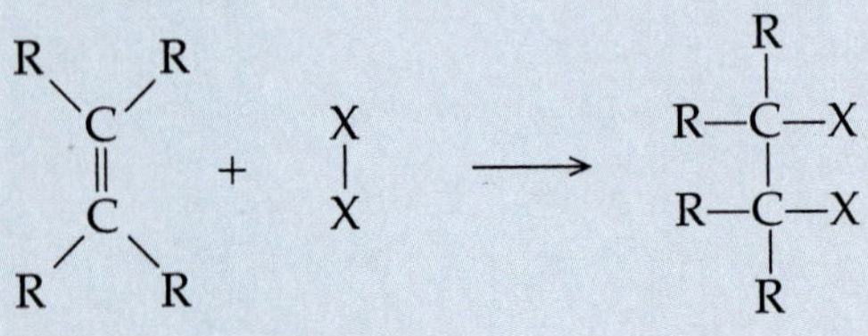

$$\text{Alkene} + \text{Halogen} \longrightarrow \text{Alkyl dihalide}$$

Addition Polymers of Alkenes

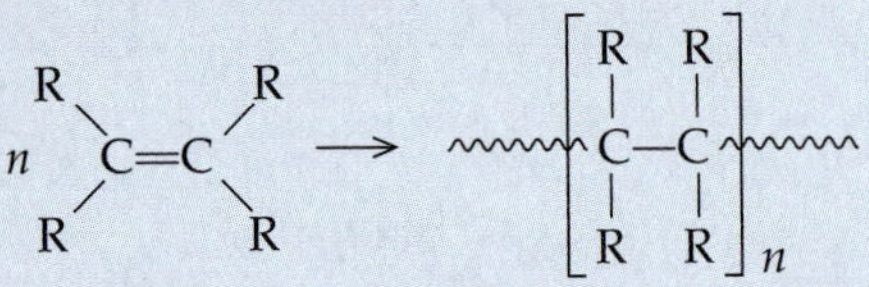

$$n \; \text{Alkene monomer} \longrightarrow \text{Addition polymer}$$

SUMMARY

11.1 Alkenes and Alkynes: Structure and Physical Properties

Alkenes and *alkynes* are *unsaturated hydrocarbons*. Alkenes are characterized by the presence of at least one carbon–carbon double bond and have the general molecular formula C_nH_{2n}. Alkynes are characterized by the presence of at least one carbon–carbon triple bond and have the general molecular formula C_nH_{2n-2}. The physical properties of alkenes and alkynes are similar to those of alkanes, but their chemical properties are quite different.

11.2 Alkenes and Alkynes: Nomenclature

Alkenes and alkynes are named by identifying the parent compound and replacing the *-ane* ending of the alkane with *-ene* (for an alkene) or *-yne* (for an alkyne). The parent chain is numbered to give the lowest number to the first of the two carbons involved in the double bond (or triple bond). Finally, all groups are named and numbered.

11.3 Geometric Isomers

The carbon–carbon double bond is rigid. This allows the formation of *geometric isomers*, or isomers that occur when two different groups are bonded to each carbon of the double bond. When similar groups are on the same side of a double bond, the prefix *cis* is used to describe the compound. When similar groups are on opposite sides of a double bond, the prefix *trans* is used.

11.4 Reactions Involving Alkenes and Alkynes

Whereas alkanes undergo *substitution reactions,* alkenes and alkynes undergo *addition reactions.* The principal addition reactions of unsaturated hydrocarbons are *halogenation, hydration,* and *hydrogenation. Polymers* can be made from alkenes or substituted alkenes.

11.5 Aromatic Hydrocarbons

Aromatic hydrocarbons contain benzene rings. Simple aromatic compounds are named as derivatives of benzene. Several members of this family have historical common names, such as aniline, phenol, and toluene. Polycyclic aromatic hydrocarbons consist of two or more aromatic molecules bonded together.

11.6 Heterocyclic Aromatic Compounds

Heterocyclic aromatic compounds are those having at least one atom other than carbon as part of the structure of the aromatic ring. They are more similar to benzene in stability and chemical behavior than they are to alkenes. Many of these compounds are components of molecules that have significant effects on biological systems, including DNA, RNA, and nicotine.

KEY TERMS

addition polymer (11.4)
addition reaction (11.4)
alkene (11.1)
alkyne (11.1)
aromatic hydrocarbon (11.5)
geometric isomers (11.3)
halogenation (11.4)
heterocyclic aromatic compound (11.6)

hydration (11.4)
hydrogenation (11.4)
Markovnikov's rule (11.4)
monomer (11.4)
phenyl group (11.5)
polymer (11.4)
unsaturated hydrocarbon (11.1)

QUESTIONS AND PROBLEMS

Alkenes and Alkynes: Structure and Physical Properties

Foundations

11.21 What is the general relationship between the carbon chain length of an alkene and its boiling point?

11.22 Describe and explain the reason for the water solubility of an alkyne.

11.23 Write the general formulas for alkanes, alkenes, and alkynes.

11.24 What are the characteristic functional groups of alkenes and alkynes?

Applications

11.25 Describe the geometry of ethene. What are the bond angles in ethene?

11.26 Describe the geometry of ethyne. What are the bond angles in ethyne?

11.27 Arrange the following groups of molecules from the highest to lowest boiling points:
    a. ethyne   propyne   2-pentyne
    b. 2-butene   3-decene   ethene

11.28 Arrange the following groups of molecules from the highest to the lowest melting points.

a.

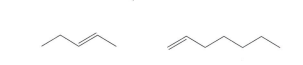

b.

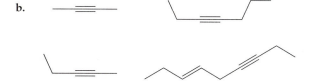

Alkenes and Alkynes: Nomenclature

Foundations

11.29 Briefly describe the rules for naming alkenes and alkynes.

11.30 What is meant by a geometric isomer?

11.31 Describe what is meant by a *cis* isomer of an alkene.

11.32 Describe what is meant by a *trans* isomer of an alkene.

Applications

11.33 Draw a condensed formula for each of the following compounds:
  a. 2-Methyl-2-hexene
  b. *trans*-3-Heptene
  c. *cis*-1-Chloro-2-pentene
  d. *cis*-2-Chloro-2-methyl-3-heptene
  e. *trans*-5-Bromo-2,6-dimethyl-3-octene

11.34 Draw a condensed formula for each of the following compounds:
  a. 2-Hexyne
  b. 4-Methyl-1-pentyne
  c. 1-Chloro-4,4,5-trimethyl-2-heptyne
  d. 2-Bromo-3-chloro-7,8-dimethyl-4-decyne

11.35 Name each of the following using the I.U.P.A.C. Nomenclature System:
  a. 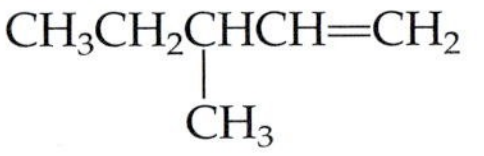 $CH_3CH_2CHCH{=}CH_2$
       |
       CH_3

  b. 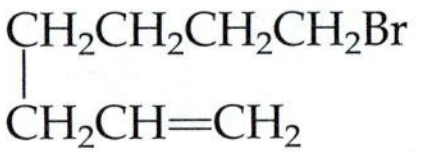 $CH_2CH_2CH_2CH_2Br$
       |
       $CH_2CH{=}CH_2$

  c. 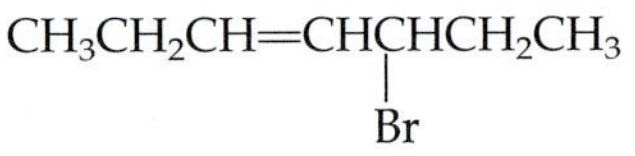 $CH_3CH_2CH{=}CHCHCH_2CH_3$
                      |
                      Br

  d.

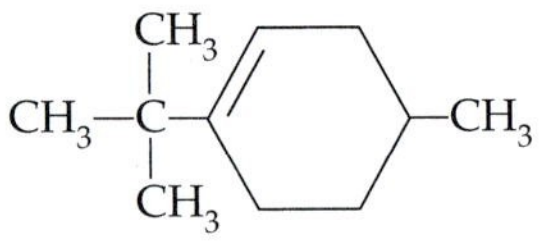

11.36 Name each of the following using the I.U.P.A.C. Nomenclature System:
  a. 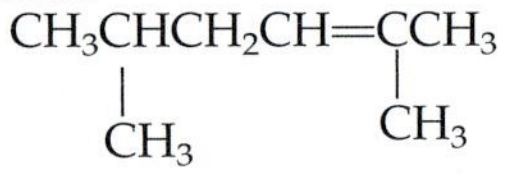 $CH_3CHCH_2CH{=}CCH_3$
       | |
       CH_3 CH_3

  b. 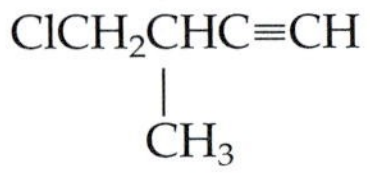 $ClCH_2CHC{\equiv}CH$
              |
              CH_3

  c. 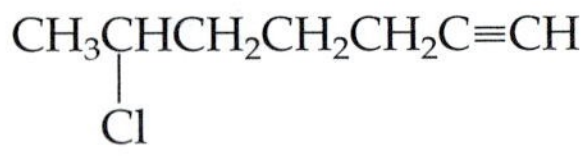 $CH_3CHCH_2CH_2CH_2C{\equiv}CH$
       |
       Cl

  d.

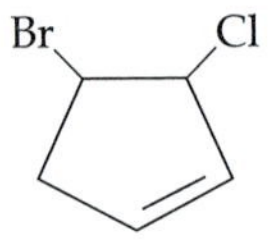

11.37 Of the following compounds, which can exist as *cis-trans* geometric isomers? Draw the two geometric isomers.
  a. 2,3-Dibromobutane
  b. 2-Heptene
  c. 2,3-Dibromo-2-butene
  d. Propene

11.38 Of the following compounds, which can exist as geometric isomers?
  a. 1-Bromo-1-chloro-2-methylpropene
  b. 1,1-Dichloroethene

  c. 1,2-Dibromoethene
  d. 3-Ethyl-2-methyl-2-hexene

11.39 Which of the following alkenes would not exhibit *cis-trans* geometric isomerism?

  a.

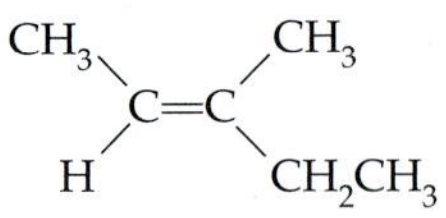

  b.

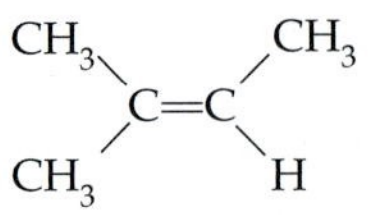

  c.

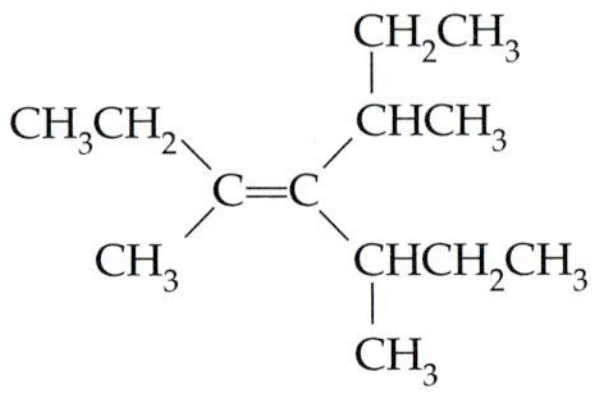

  d.

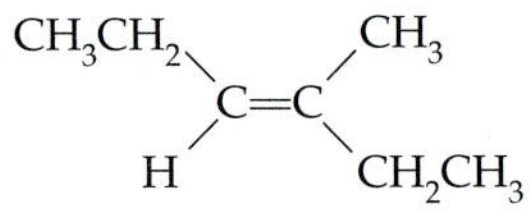

11.40 Which of the following structures have incorrect I.U.P.A.C. names? If incorrect, give the correct I.U.P.A.C. name.

  a. 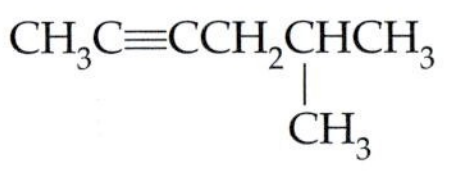 $CH_3C{\equiv}CCH_2CHCH_3$
                    |
                    CH_3

       2-Methyl-4-hexyne

  b.

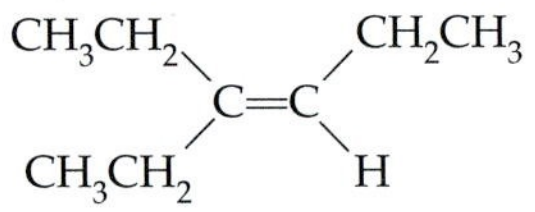

       3-Ethyl-3-hexyne

  c. 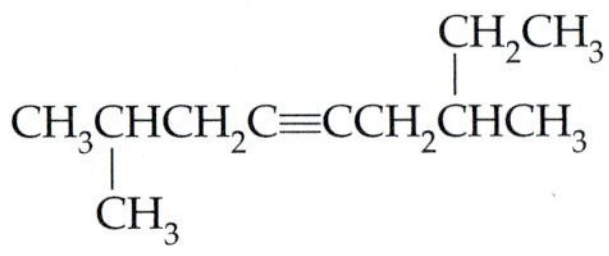 $CH_3CHCH_2C{\equiv}CCH_2CHCH_3$
       | |
       CH_3 CH_2CH_3

       2-Ethyl-7-methyl-4-octyne

  d.

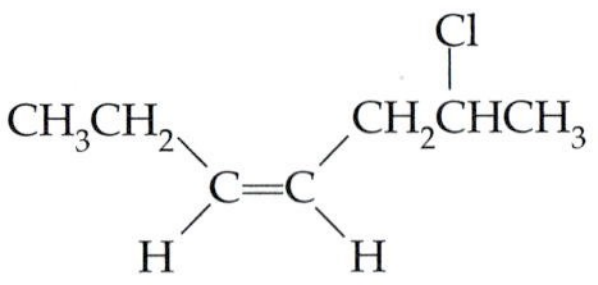

       trans-6-Chloro-3-heptene

  e.

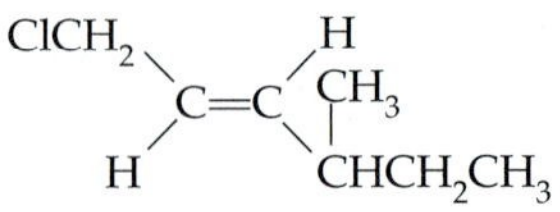

       1-Chloro-5-methyl-2-hexene

11.41 Which of the following can exist as *cis* and *trans* isomers?
  a. $H_2C{=}CH_2$
  b. $CH_3CH{=}CHCH_3$
  c. $Cl_2C{=}CBr_2$

d. ClBrC=CClBr

e. $(CH_3)_2C=C(CH_3)_2$

11.42 Draw and name all the *cis* and *trans* isomers in Question 11.41.

Reactions Involving Alkenes and Alkynes

Foundations

11.43 Write a general equation representing the hydrogenation of an alkene.

11.44 Write a general equation representing the hydrogenation of an alkyne.

11.45 Write a general equation representing the halogenation of an alkene.

11.46 Write a general equation representing the halogenation of an alkyne.

11.47 Write a general equation representing the hydration of an alkene.

11.48 What is meant by the term *addition reaction*?

Applications

11.49 How could you distinguish between a sample of cyclohexane and a sample of hexene (both C_6H_{12}) using a simple chemical test? (*Hint:* Refer to the subsection entitled "Halogenation: Addition of X_2.")

11.50 Quantitatively, 1 mol of Br_2 is consumed per mole of alkene, and 2 mol of Br_2 are consumed per mole of alkyne. How many moles of Br_2 would be consumed for 1 mol of each of the following:

a. 2-Hexyne

b. Cyclohexene

c.

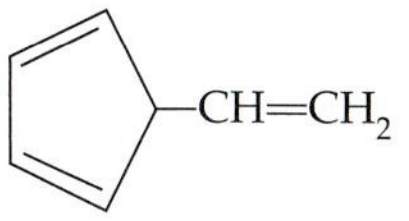

d.

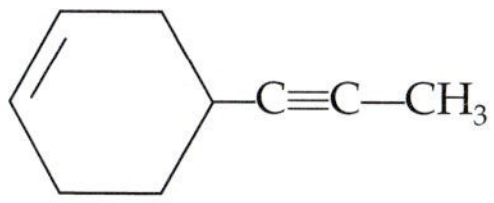

11.51 Complete each of the following reactions by supplying the missing reactant or product(s) as indicated by question marks:

a. $CH_3CH_2CH=CHCH_2CH_3 + ? \longrightarrow$
$$CH_3CH_2CH_2CH_2CH_2CH_3$$

b.

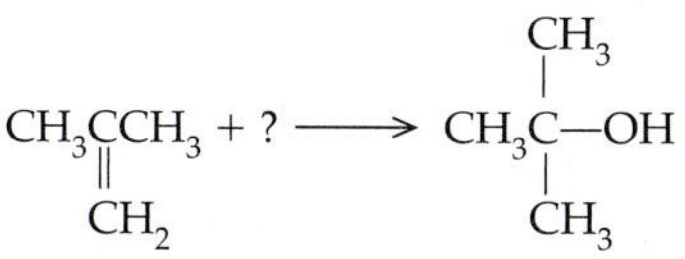

c. 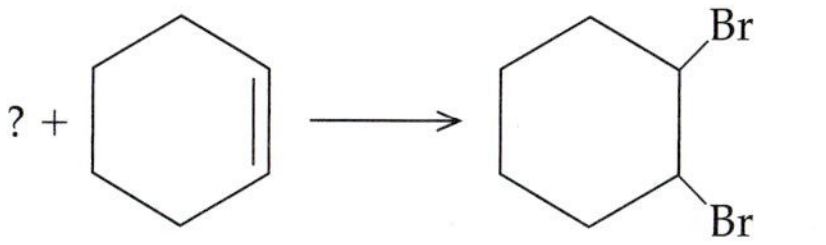 ? +

d. $2CH_3CH_2CH_2CH_2CH_2CH_3 + ?O_2 \xrightarrow{Heat} ? + ?$
(complete combustion)

e. 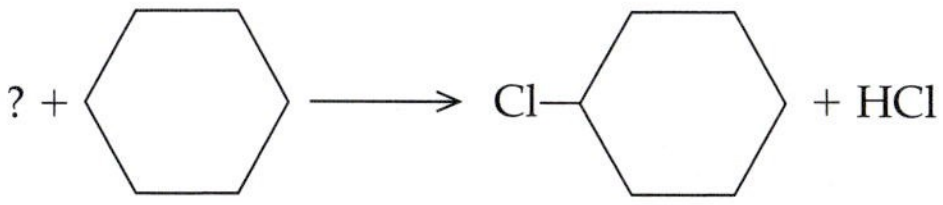 ? +

f. 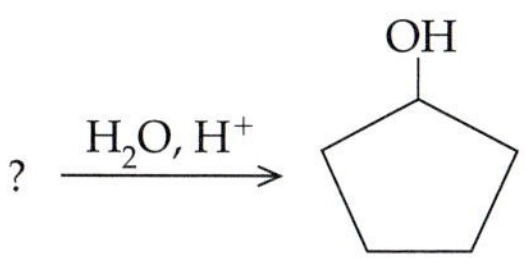 ? $\xrightarrow{H_2O, H^+}$

11.52 Draw and name the product in each of the following reactions:

a. Cyclopentene + H_2O (H^+)

b. Cyclopentene + H_2

11.53 Write a balanced equation for each of the following reactions:

a. Hydrogenation of 2-butyne

b. Halogenation of 2-pentyne

11.54 Write a balanced equation for each of the following reactions:

a. Hydration of 1-butene

b. Hydration of 2-butene

11.55 Complete each of the following by supplying the missing product indicated by the question mark:

a. 2-Butene $\xrightarrow{Br_2}$?

b. 3-Methyl-2-hexene $\xrightarrow{H_2O}$?

c.

11.56 Bromine is often used as a laboratory spot test for unsaturation in an aliphatic hydrocarbon. Bromine in CCl_4 is red. When bromine reacts with an alkene or alkyne, the alkyl halide formed is colorless; hence, a disappearance of the red color is a positive test for unsaturation. A student tested the contents of two vials, A and B, both containing compounds with a molecular formula, C_6H_{12}. Vial A decolorized bromine, but vial B did not. How may the results for vial B be explained? What class of compound would account for this?

11.57 What is meant by the term *polymer*?

11.58 What is meant by the term *monomer*?

11.59 Write an equation representing the synthesis of Teflon from tetrafluoroethene. (*Hint:* Refer to Table 11.1.)

11.60 Write an equation representing the synthesis of polystyrene. (*Hint:* Refer to Table 11.1.)

11.61 Provide the I.U.P.A.C. name for each of the following molecules. Write a balanced equation for the hydration of each.

a. $CH_3CH=CHCH_2CH_3$

b. 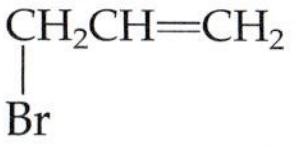 $CH_2CH=CH_2$ with Br substituent

c.

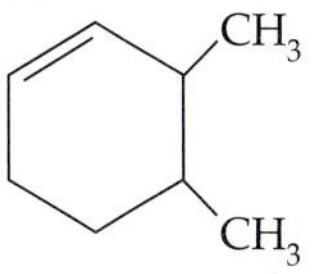

11.62 Provide the I.U.P.A.C. name for each of the following molecules. Write a balanced equation for the hydrogenation of each.

a.

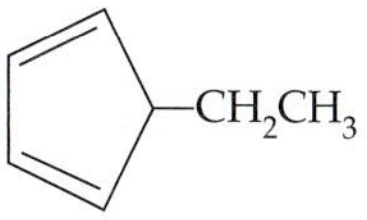

b. $CH_3CH=CHCH_2CH=CHCH_2CH=CHCH_3$

c. 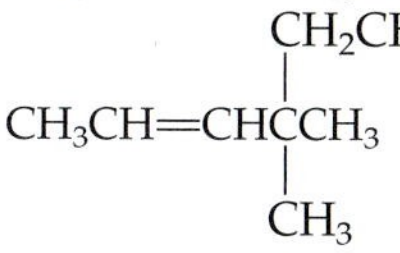 $CH_3CH=CHCCH_3$ with CH_2CH_3 and CH_3 substituents

11.63 Write an equation for the addition reaction that produced each of the following molecules:

a. 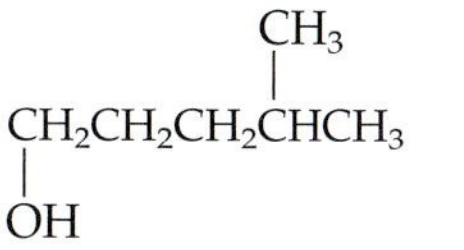 $CH_2CH_2CH_2CHCH_3$ with CH_3 and OH substituents

b. 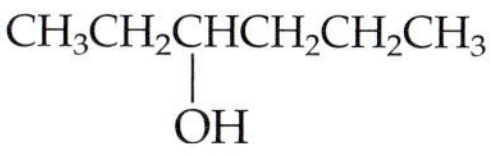 $CH_3CH_2CHCH_2CH_2CH_3$ with OH substituent

c.

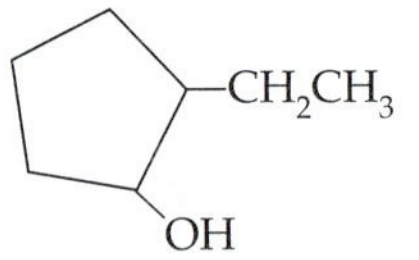

11.64 Write an equation for the addition reaction that produced each of the following molecules:

a. CH₃CH₂CHCHCH₃
with OH and CH₂CH₃ substituents

$$CH_3CH_2\underset{\underset{\displaystyle CH_2CH_3}{|}}{\overset{\overset{\displaystyle OH}{|}}{C}}HCHCH_3$$

b.

$$CH_3\underset{\underset{\displaystyle OH}{|}}{C}HCH_2CH_3$$

c.

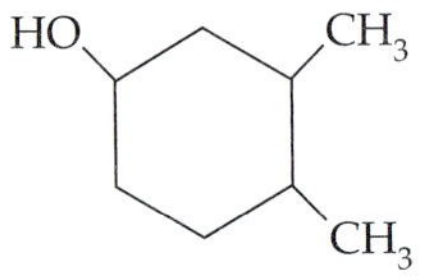

11.65 Draw the structure of each of the following compounds and write a balanced equation for the complete hydrogenation of each:
   a. 1,4-Hexadiene
   b. 2,4,6-Octatriene
   c. 1,3-Cyclohexadiene
   d. 1,3,5-Cyclooctatriene

11.66 Draw the structure of each of the following compounds and write a balanced equation for the bromination of each:
   a. 3-Methyl-1,4-hexadiene
   b. 4-Bromo-1,3-pentadiene
   c. 3-Chloro-2,4-hexadiene
   d. 3-Bromo-1,3-Cyclohexadiene

Aromatic Hydrocarbons

Foundations

11.67 Where did the term *aromatic hydrocarbon* first originate?

11.68 What chemical characteristic of the aromatic hydrocarbons is most distinctive?

Applications

11.69 Draw the structure for each of the following compounds:
   a. 2,4-Dibromotoluene
   b. 1,2,4-Triethylbenzene
   c. Isopropylbenzene
   d. 2-Bromo-5-chlorotoluene

11.70 Name each of the following compounds, using the I.U.P.A.C. system.
   a.

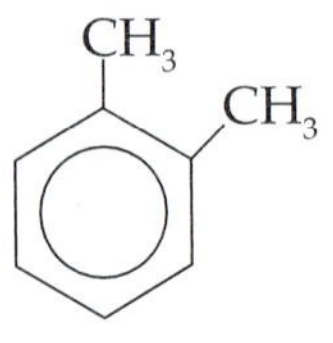

   b.

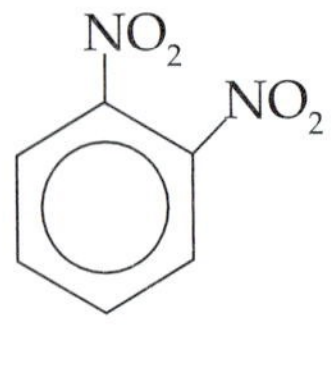

c.

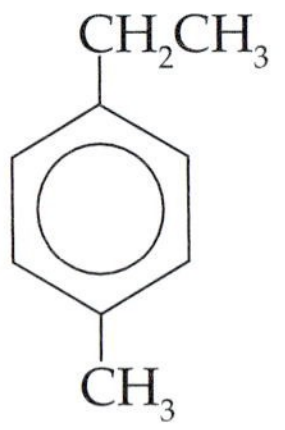

d.

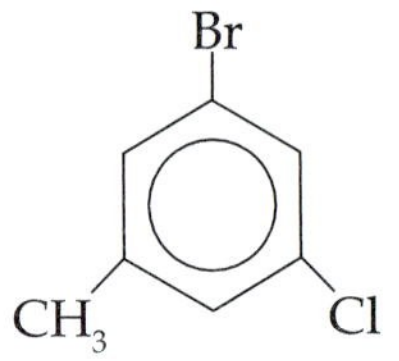

e.

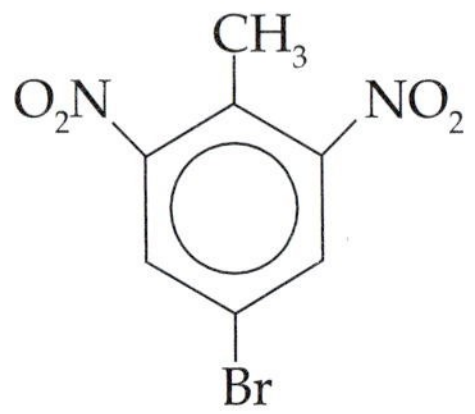

11.71 Draw each of the following compounds, using condensed formulas:
   a. 1,2-Dimethylbenzene
   b. Propylbenzene
   c. 1,3,5-Trinitrobenzene
   d. 3-Chlorotoluene

11.72 Draw each of the following compounds, using condensed formulas:
   a. 1-Bromo-3-chlorobenzene
   b. Isopropylbenzene
   c. 1,2,3-Trimethylbenzene
   d. 1-Bromo-2,3-dichlorobenzene

Heterocyclic Aromatic Compounds

11.73 Draw the general structure of a pyrimidine.
11.74 Which biological molecules contain pyrimidine rings?
11.75 Draw the general structure of a purine.
11.76 Which biological molecules contain purine rings?

FOR FURTHER UNDERSTANDING

1. There is a plastic polymer called polyvinylidene difluoride (PVDF) that can be used to sense a baby's breath and thus be used to prevent sudden infant death syndrome (SIDS). The secret is that this polymer can be specially processed so that it becomes piezoelectric (produces an electrical current when it is physically deformed) and pyroelectric (develops an electrical potential when its temperature changes). When a PVDF film is placed beside a sleeping baby, it will set off an alarm if the baby stops breathing. The structure of this polymer is shown here:

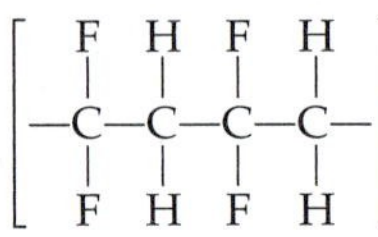

   Go to the library and investigate some of the other amazing uses of PVDF. Draw the structure of the alkene from which this compound is produced.

2. Isoprene is the repeating unit of the natural polymer, rubber. It is also the starting material for the synthesis of cholesterol and several lipid-soluble vitamins, including vitamin A and vitamin K. The structure of isoprene is shown below.

$$CH_2{=}\overset{\overset{\displaystyle CH_3}{|}}{C}{-}CH{=}CH_2$$

What is the I.U.P.A.C. name for isoprene?

3. When polyacrylonitrile is burned, toxic gases are released. In airplane fires, more passengers die from inhalation of toxic fumes than from burns. Refer to Table 11.1 for the structure of acrylonitrile. What toxic gas would you predict would be the product of the combustion of these polymers?

4. If a molecule of polystyrene consists of 25,000 monomers, what is the molar mass of the molecule?

5. A factory produces one million tons of polypropylene. How many moles of propene would be required to produce this amount? What is the volume of this amount of propene at 25°C and 1 atm?

12

LEARNING GOALS

1 Draw the structures and be able to rank selected alcohols by relative water solubility, boiling points, or melting points.

2 Write the common and I.U.P.A.C. names for common alcohols.

3 Discuss the biological, medical, or environmental significance of several alcohols.

4 Classify alcohols as primary, secondary, or tertiary.

5 Write equations representing the preparation of alcohols by the hydration of an alkene.

6 Write equations showing the dehydration of an alcohol.

7 Write equations representing the oxidation of alcohols.

8 Discuss the use of phenols as germicides.

9 Write names and draw structures for common ethers and discuss their use in medicine.

10 Write equations representing formation of an ether by a dehydration reaction between two alcohol molecules.

11 Write names and draw structures for simple thiols and discuss their biological significance.

12 Draw the structures and discuss the physical properties of aldehydes and ketones.

13 Write the common and I.U.P.A.C. names for common aldehydes and ketones.

14 Discuss the biological, medical, or environmental significance of several aldehydes and ketones.

15 Write equations representing the oxidation and reduction of aldehydes and ketones.

Oxygen- and Sulfur-Containing Organic Compounds

Research tells us that even as babies, we prefer sweet tastes over all others, and the sugar molecules in our diet that satisfy this sweet tooth all contain two of the functional groups that we will study in this chapter: the hydroxyl group and the carbonyl group. Glucose (blood sugar) and fructose (fruit sugar), sweet individually, can be chemically combined to produce sucrose, or table sugar. The United States Department of Agriculture reports that the average American consumes about 152 pounds of sucrose each year! It is little wonder that physicians, nutritionists, and dentists are concerned about the obesity and tooth decay caused by so much sugar in our diets.

As a result of these concerns, the food chemistry industry has invested billions of dollars in the synthesis of non-nutritive sugar substitutes, such as aspartame (Equal) and Splenda. You can also find a variety of candies, soft drinks, and gums that are labeled "sugar-free." A quick check of the nutritional label reveals that, although they are sugar-free, they are not calorie-free! These sweets contain sugar alcohols, such as sorbitol or mannitol, instead of sucrose. Sorbitol and mannitol look very much like glucose and fructose, except that they are missing the carbonyl group.

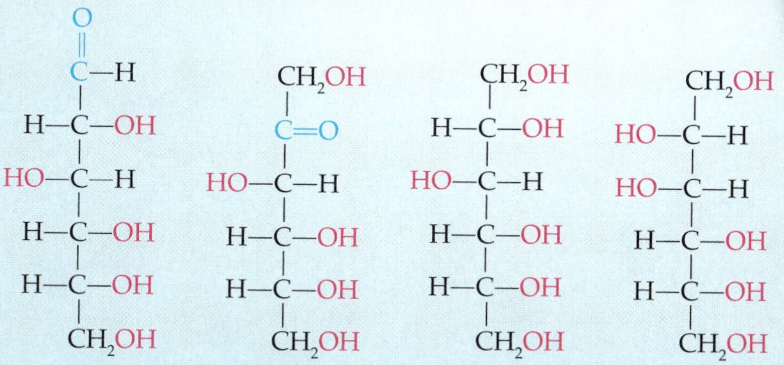

The sugar that sweetens the candies shown above is sucrose, composed of D-fructose and D-glucose bonded together. A key step in preparing such delightful confections is dissolving the sugar in water. Examine the structures of D-fructose and D-glucose at left and the structure of sucrose in Figure 14.11 and explain why these sugars are easily dissolved in water.

D-Glucose	D-Fructose	D-Sorbitol	D-Mannitol

Compared to sucrose, they range in sweetness from about half to nearly the same. They also have fewer calories than table sugar (about one-third to one-half the calories). Sugar alcohols actually absorb heat from the surroundings when they dissolve. As a result, they

—*Continued next page*

cause a cooling sensation in the mouth. You may have noticed this when eating certain breath freshening mints and gums.

In biological systems, the hydroxyl and carbonyl groups are important to the structures of sugars, fats, and proteins. They allow these biological molecules to undergo a variety of reactions such as oxidation, reduction, hydration, and dehydration that are essential for life.

The thiol group is found in the structure of some amino acids and is essential for keeping proteins in the proper three-dimensional shape required for their biological function.

Figure 12.1

Ball-and-stick model of the simple alcohol, ethanol.

 Learning Goal

Draw the structures and be able to rank selected alcohols by relative water solubility, boiling points, or melting points.

Electronegativity is discussed in Section 3.1. Hydrogen bonding is described in detail in Section 6.2.

12.1 Alcohols

Structure and Physical Properties

An **alcohol** is an organic compound that contains a **hydroxyl group** (—OH) attached to an alkyl group (Figure 12.1).

The hydroxyl groups of alcohols are very polar because the oxygen and hydrogen atoms have significantly different electronegativities. Because the two atoms involved in this polar bond are oxygen and hydrogen, hydrogen bonds can form between alcohol molecules (Figure 12.2). As a result of this hydrogen bonding, alcohols boil at much higher temperatures than hydrocarbons of similar molecular weight.

Alcohols of one to four carbon atoms are very soluble in water, and those with five or six carbons are moderately soluble in water. This is due to the ability of the alcohol to form hydrogen bonds with water molecules (see Figure 12.2b). As the nonpolar, or hydrophobic, portion of an alcohol (the carbon chain) becomes larger relative to the polar, hydrophilic, region (the hydroxyl group), the water solubility of an alcohol decreases. As a result, alcohols of seven carbon atoms or more are nearly insoluble in water.

An increase in the number of hydroxyl groups along a carbon chain will increase the influence of the polar hydroxyl group. It follows, then, that diols and triols are more water soluble than alcohols with only a single hydroxyl group.

Figure 12.2

(a) Hydrogen bonding between alcohol molecules. (b) Hydrogen bonding between alcohol molecules and water molecules.

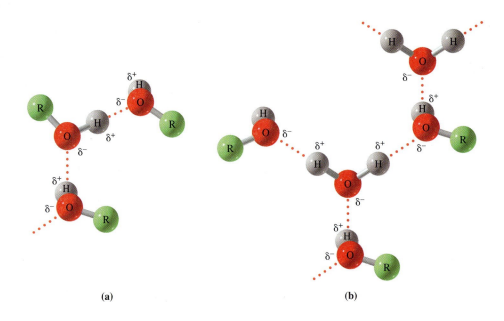

(a) (b)

The presence of polar hydroxyl groups in large biological molecules—for instance, proteins and nucleic acids—allows intramolecular hydrogen bonding that keeps these molecules in the shapes needed for biological function.

The term hydrophobic, which literally means "water fearing," is used to describe a molecule or a region of a molecule that is nonpolar and, thus, more soluble in nonpolar solvents than in water. Similarly, the term hydrophilic, meaning water loving, is used to describe a polar molecule or region of a molecule that is more soluble in the polar solvent, water, than in a nonpolar solvent.

Nomenclature

I.U.P.A.C. Names

In the I.U.P.A.C. Nomenclature System, alcohols are named according to the following steps:

- Determine the name of the *parent compound*, the longest continuous carbon chain containing the —OH group.
- Replace the *-e* ending of the alkane chain with the *-ol* ending of the alcohol. Following this pattern, an alkane becomes an alkanol. For instance, etha*ne* becomes ethan*ol*, and propa*ne* becomes propan*ol*.
- Number the parent chain to give the carbon bearing the hydroxyl group the lowest possible number.
- Name and number all substituents, and add them as prefixes to the "alkanol" name.
- Alcohols containing two hydroxyl groups are named *-diols*. Those bearing three hydroxyl groups are called *-triols*. A number giving the position of each of the hydroxyl groups is needed in these cases.

2 Learning Goal
Write the common and I.U.P.A.C. names for common alcohols.

The way to determine the parent compound was described in Section 10.2.

EXAMPLE 12.1 Using I.U.P.A.C. Nomenclature to Name an Alcohol

Name the following alcohol using I.U.P.A.C. nomenclature:

SOLUTION

Parent compound: heptane (becomes heptanol)

Position of —OH: carbon-2 (*NOT* carbon-6)

Substituents: 6-methyl

Name: 6-Methyl-2-heptanol

2 Learning Goal
Write the common and I.U.P.A.C. names for common alcohols.

Common Names

The common names for alcohols are derived from the alkyl group corresponding to the parent compound. The name of the alkyl group is followed by the word *alcohol*. For some alcohols, such as ethylene glycol and glycerol, historical names are used. The following examples provide the I.U.P.A.C. and common names of several alcohols:

2 Learning Goal
Write the common and I.U.P.A.C. names for common alcohols.

See Section 10.2 for the names of the common alkyl groups.

2-Propanol
(isopropyl
alcohol)

1,2-Ethanediol
(ethylene glycol)

Ethanol
(ethyl alcohol)
(grain alcohol)

QUESTION 12.1

Use the I.U.P.A.C. Nomenclature System to name each of the following compounds.

a. $CH_3CHCH_2CH_2CH_2OH$
   |
   CH_3

b. $CH_3CHCH_2CHCH_3$
   | |
   OH CH_2CH_3

c. $CH_2—CH—CH_2$
   | | |
   OH OH OH
(Common name: Glycerol)

d. $CH_3CH_2CH—CHCH_2CH_2OH$
       | |
       Cl CH_3

QUESTION 12.2

Give the common name and the I.U.P.A.C. name for each of the following compounds.

a. $CH_3CH_2CH_2CH_2CH_2CH_2CH_2OH$

b. CH_3CHCH_3
   |
   OH

c. CH_3
      |
   CH_3CHCH_2OH

Medically Important Alcohols

3 Learning Goal

Discuss the biological, medical, or environmental significance of several alcohols.

Figure 12.3

Champagne, a sparkling wine, results when fermentation is carried out in a sealed bottle. Under these conditions, the CO_2 produced during fermentation is trapped in the wine.

Methanol

Methanol (methyl alcohol), CH_3OH, is a colorless and odorless liquid that is used as a solvent and as the starting material for the synthesis of methanal (formaldehyde). Methanol is often called *wood alcohol* because it can be made by heating wood in the absence of air. In fact, ancient Egyptians produced methanol by this process and, mixed with other substances, used it for embalming. It was not until 1661 that Robert Boyle first isolated pure methanol, which he called *spirit of box*, because he purified it by distillation from boxwood. Methanol is toxic and can cause blindness and perhaps death if ingested. Methanol may also be used as fuel, especially for "formula" racing cars.

Ethanol

Ethanol (ethyl alcohol), CH_3CH_2OH, is a colorless and odorless liquid and is the alcohol in alcoholic beverages. It is also widely used as a solvent and as a raw material for the preparation of other organic chemicals.

The ethanol used in alcoholic beverages comes from the **fermentation** of carbohydrates (sugars and starches). The beverage produced depends on the starting material and the fermentation process: scotch (grain), bourbon (corn), burgundy wine (grapes and grape skins), and chablis wine (grapes without red skins) (Figure 12.3). The following equation summarizes the fermentation process:

$$C_6H_{12}O_6 \xrightarrow[\text{enzyme action}]{\text{Several steps involving}} 2CH_3CH_2OH + 2CO_2$$

Sugar   Ethanol
(glucose)   (ethyl alcohol)

The alcoholic beverages listed have quite different alcohol concentrations. Wines are generally 12–13% alcohol because the yeasts that produce the ethanol are killed by ethanol concentrations of 12–13%. To produce bourbon or scotch with an alcohol concentration of 40–45% ethanol (80 or 90 proof), the original fermentation products must be distilled.

2-Propanol

2-Propanol (isopropyl alcohol) was commonly called *rubbing alcohol* because patients with high fevers were often given alcohol baths to reduce body temperature. Rapid evaporation of the alcohol results in skin cooling. This practice is no longer commonly followed.

It is also used as a disinfectant (Figure 12.4), an astringent (skin-drying agent), an industrial solvent, and a raw material in the synthesis of organic chemicals. It is colorless, has a very slight odor, and is toxic when ingested.

1,2-Ethanediol

1,2-Ethanediol (ethylene glycol) is used as automobile antifreeze. When added to water in the radiator, the ethylene glycol solute lowers the freezing point and raises the boiling point of the water. Ethylene glycol has a sweet taste but is extremely poisonous. For this reason, color additives are used in antifreeze to ensure that it is properly identified.

1,2,3-Propanetriol

1,2,3-Propanetriol (glycerol) is a viscous, sweet-tasting, nontoxic liquid. It is very soluble in water and is used in cosmetics, pharmaceuticals, and lubricants. Glycerol is obtained as a by-product of the hydrolysis of fats.

Classification of Alcohols

Alcohols are classified as **primary (1°), secondary (2°), or tertiary (3°),** depending on the number of alkyl groups attached to the **carbinol carbon,** the carbon bearing the hydroxyl (—OH) group. If no alkyl groups are attached, the alcohol is methyl alcohol; if there is a single alkyl group, the alcohol is a primary alcohol; an alcohol with two alkyl groups bonded to the carbon bearing the hydroxyl group is a secondary alcohol, and if three alkyl groups are attached, the alcohol is a tertiary alcohol.

Methyl alcohol   1° Alcohol   2° Alcohol   3° Alcohol

Distillation is the separation of compounds in a mixture based on differences in boiling points.

 To learn about the effects of alcohol consumption during pregnancy, see A Medical Connection: Fetal Alcohol Syndrome online at www.mhhe.com/denniston.

$$CH_3CHCH_3$$
$$|$$
$$OH$$

2-Propanol
(isopropyl alcohol)

$$CH_2—CH_2$$
$$||$$
$$OHOH$$

1,2-Ethanediol
(ethylene glycol)

$$CH_2—CH—CH_2$$
$$|||$$
$$OHOHOH$$

1,2,3-Propanetriol
(glycerol)

4   **Learning Goal**
Classify alcohols as primary, secondary, or tertiary.

Figure 12.4

Isopropyl alcohol, or rubbing alcohol, is used as a disinfectant before and after an injection or blood test.

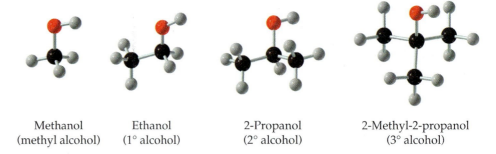

Methanol	Ethanol	2-Propanol	2-Methyl-2-propanol
(methyl alcohol)	(1° alcohol)	(2° alcohol)	(3° alcohol)

QUESTION 12.3

Classify each of the following alcohols as 1°, 2°, or 3°.

a. $CH_3CH_2CH_2CH_2OH$

b. $CH_3CH_2CHCH_2CH_3$
$\qquad\qquad\quad |$
$\qquad\qquad\ \ OH$

c.

    CH_3

    OH

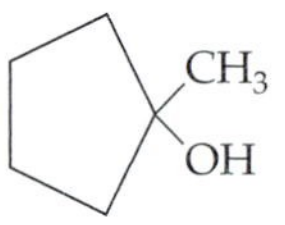

QUESTION 12.4

Classify each of the following alcohols as 1°, 2°, or 3°.

a. $CH_3CH_2CHCH_3$
$\qquad\qquad\quad |$
$\qquad\qquad\ \ OH$

b. $\qquad\qquad\quad CH_3$
$\qquad\qquad\qquad\ \ |$
$\ \ CH_3CH_2CH_2CCH_3$
$\qquad\qquad\qquad\ \ |$
$\qquad\qquad\qquad\ OH$

c. CH_3CH_2OH

d. $CH_2{-}CHOH$
$\quad\ |\qquad\ |$
$\ \ CH_2{-}CH_2$

Reactions Involving Alcohols

Preparation of Alcohols

5 Learning Goal

Write equations representing the preparation of alcohols by the hydration of an alkene.

Hydration of alkenes is described in Section 11.4.

The most important reactions of alkenes are *addition reactions*. Addition of a water molecule to the carbon–carbon double bond of an alkene produces an alcohol. This reaction, called **hydration**, requires a trace of acid (H^+) as a catalyst, as shown in the following equation:

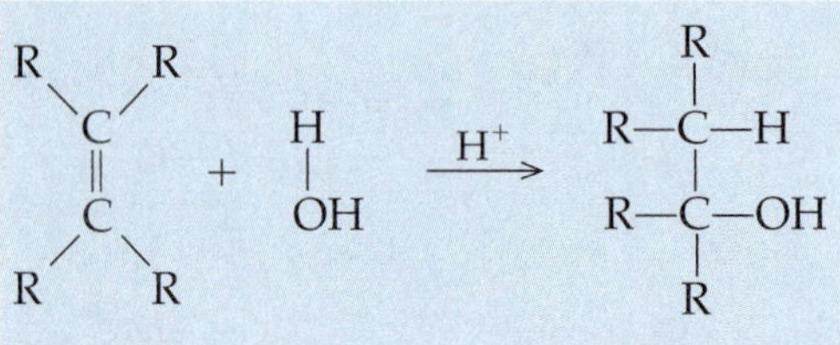

$$
\underset{\text{Alkene}}{\underset{R}{\overset{R}{}}\begin{array}{c} R\diagdown\quad\diagup R \\ C \\ \| \\ C \\ R\diagup\quad\diagdown R \end{array}} + \underset{\text{Water}}{\begin{array}{c} H \\ | \\ OH \end{array}} \xrightarrow{H^+} \underset{\text{Alcohol}}{\begin{array}{c} R \\ | \\ R{-}C{-}H \\ | \\ R{-}C{-}OH \\ | \\ R \end{array}}
$$

CHEMISTRY at the Crime Scene | Fingerprinting

In 1880, Henry Faulds published a paper in the prestigious British journal *Nature*. The paper, entitled "On the Skin-Furrows of the Hand," began with Faulds' observation of the fingerprint of a prehistoric pottery maker in the clay of a pot. This initial observation led him to begin a study of fingerprints in humans and monkeys. In his conclusion, he suggested that "bloody finger-marks" might lead to "scientific identification of criminals." He noted that the Chinese had been using fingerprints to identify criminals since early times and that the Egyptians made their criminals seal confessions with a thumbprint.

Criminal investigators worldwide were quick to realize the value of this unique signature in identifying criminals. Soon, a variety of techniques were available for observing fingerprints on solid surfaces, such as glass or metal. But how can a criminalist view latent prints on porous materials, such as a piece of paper? The solution to this problem came in 1954 when it was reported that the reagent ninhydrin would react with amino acids secreted from sweat glands.

Ninhydrin was discovered in 1910 by English chemist, Siegfried Ruhemann. He observed that a reaction between ninhydrin and amino acids produces a deep purple substance called Ruhemann's purple, or RP, in honor of his discovery.

However, it was not until 1954 that two Swedish scientists, Oden and von Hoffsten, recognized that this reaction could be used to detect the small amounts of amino acids deposited on porous materials.

When ninhydrin is applied to paper, the hydroxyl groups of the ninhydrin react with the amino acids in the perspiration deposited on the paper when it is handled by a person. The result is a bright purple image of the fingerprints.

Because fingerprints are unique to each individual, they are an invaluable line of evidence in the criminalist's toolkit. The ability of ninhydrin to reveal invisible prints on porous surfaces, along with its affordability and ease of use, makes it one of the most powerful reagents in crime scene investigation.

FOR FURTHER UNDERSTANDING

What other functional groups can you identify in the ninhydrin molecule? Aside from RP and CO_2 gas, what other molecule studied in this chapter is produced in the reaction between ninhydrin and an amino acid?

What types of crimes might be exposed using ninhydrin?

$$H_2N-CH-COOH + 2 \quad \text{(Ninhydrin)} \longrightarrow \text{(Ruhemann's purple)} + CO_2\uparrow + R-\overset{O}{\underset{}{C}}-H$$

Amino acid   Ninhydrin   Ruhemann's purple

QUESTION 12.5

Write a balanced equation showing the hydration of each of the following alkenes. Remember that Markovnikov's rule must be applied in the case of unsymmetrical alkenes.

a. $CH_3CH{=}CH_2$

b. $CH_2{=}CH_2$

c. $CH_3CH_2CH{=}CHCH_2CH_3$

QUESTION 12.6

Using the I.U.P.A.C. Nomenclature System, name the reactants and products for each of the reactions in Question 12.5.

QUESTION 12.7

Classify each of the alcohols produced in the reactions of Question 12.5 as primary (1°), secondary (2°), or tertiary (3°).

QUESTION 12.8

Which of the alkenes in Question 12.5 can be found as both *cis* and *trans* isomers?

Dehydration of Alcohols

Alcohols undergo **dehydration** (lose water) when heated with concentrated sulfuric acid (H_2SO_4) or phosphoric acid (H_3PO_4). We have just seen that alkenes can be hydrated to give alcohols. Dehydration is simply the reverse process: the conversion of an alcohol back to an alkene. This is seen in the following general equation and the examples that follow:

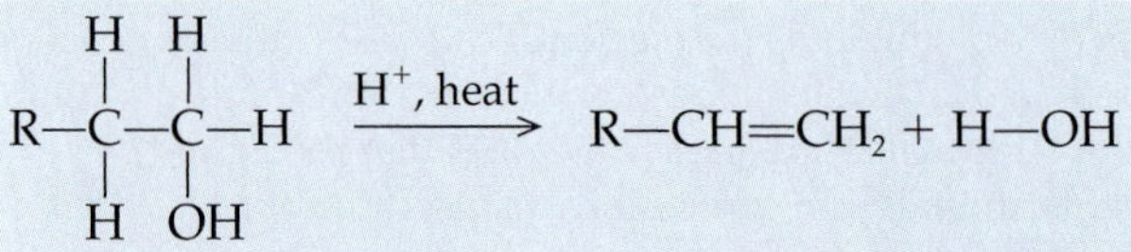

$$\underset{\text{Alcohol}}{R-\overset{\overset{\displaystyle H}{|}}{\underset{\underset{\displaystyle H}{|}}{C}}-\overset{\overset{\displaystyle H}{|}}{\underset{\underset{\displaystyle OH}{|}}{C}}-H} \xrightarrow{H^+,\ \text{heat}} \underset{\text{Alkene}}{R-CH=CH_2} + \underset{\text{Water}}{H-OH}$$

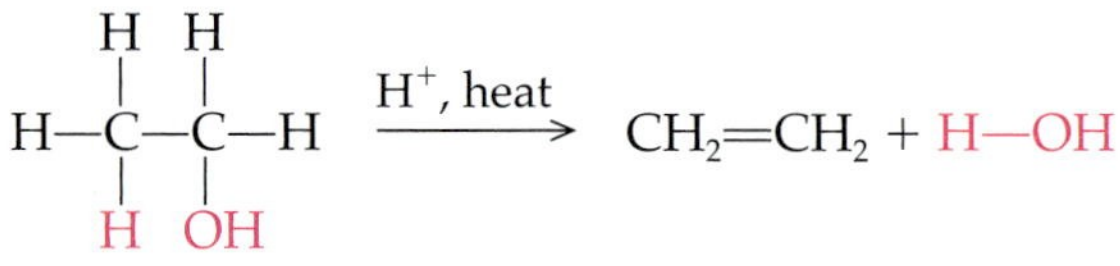

$$\underset{\substack{\text{Ethanol}\\ \text{(ethyl alcohol)}}}{H-\overset{\overset{\displaystyle H}{|}}{\underset{\underset{\displaystyle H}{|}}{C}}-\overset{\overset{\displaystyle H}{|}}{\underset{\underset{\displaystyle OH}{|}}{C}}-H} \xrightarrow{H^+,\ \text{heat}} \underset{\substack{\text{Ethene}\\ \text{(ethylene)}}}{CH_2=CH_2} + H-OH$$

In some cases, dehydration of alcohols produces a mixture of products, as seen in the following example:

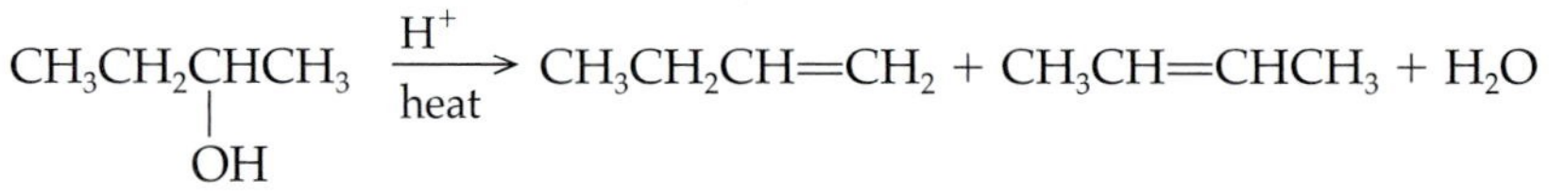

$$\underset{\text{2-Butanol}}{\underset{\underset{\displaystyle OH}{|}}{CH_3CH_2CHCH_3}} \xrightarrow[\text{heat}]{H^+} \underset{\substack{\text{1-Butene}\\ \text{(minor product)}}}{CH_3CH_2CH=CH_2} + \underset{\substack{\text{2-Butene}\\ \text{(major product)}}}{CH_3CH=CHCH_3} + H_2O$$

Notice in the equation shown above, the major product is the more highly substituted alkene. In 1875, the Russian chemist Alexander Zaitsev developed a rule to describe such reactions. **Zaitsev's rule** states that in a dehydration reaction, the alkene with the greatest number of alkyl groups on the double bonded carbons (the more highly substituted alkene) is the major product of the reaction.

EXAMPLE 12.2 Predicting the Products of Alcohol Dehydration

Predict the products of the dehydration of 3-pentanol.

SOLUTION

Because 3-pentanol is a symmetrical alcohol, the only product of dehydration is 2-pentene:

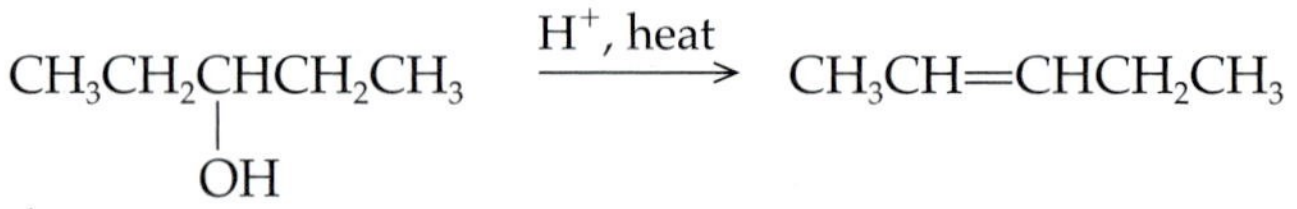

$$\underset{\text{3-Pentanol}}{\underset{\underset{\displaystyle OH}{|}}{CH_3CH_2CHCH_2CH_3}} \xrightarrow{H^+,\ \text{heat}} \underset{\text{2-Pentene}}{CH_3CH=CHCH_2CH_3}$$

QUESTION 12.9

Predict the products obtained by reacting each of the following alkenes with water and a trace of acid:

a. Ethene
b. Propene
c. 2-Butene

QUESTION 12.10

Name the products of the reactions in Question 12.9 using the I.U.P.A.C. Nomenclature System, and classify each of the product alcohols as primary (1°), secondary (2°), or tertiary (3°).

QUESTION 12.11

Predict the products obtained by reacting each of the following alkenes with water and a trace of acid:

a. 1-Butene
b. 2-Methylpropene

QUESTION 12.12

Name the products of the reactions in Question 12.11 using the I.U.P.A.C. Nomenclature System, and classify each of the product alcohols as primary (1°), secondary (2°), or tertiary (3°).

7 Learning Goal
Write equations representing the oxidation of alcohols.

Oxidation Reactions

Alcohols may be oxidized to aldehydes, ketones, and carboxylic acids. An **oxidation** reaction involves a gain of oxygen or the loss of hydrogen. A **reduction** reaction involves the loss of oxygen or gain of hydrogen. If two hydrogens are gained or lost for every oxygen gained or lost, the reaction is neither an oxidation nor a reduction. The most commonly used oxidizing agents are solutions of basic potassium permanganate ($KMnO_4/OH^-$) and chromic acid (H_2CrO_4). The symbol [O] over the reaction arrow is used throughout this book to designate any general oxidizing agent, as in the following reactions:

Oxidation of methanol produces the aldehyde, methanal:

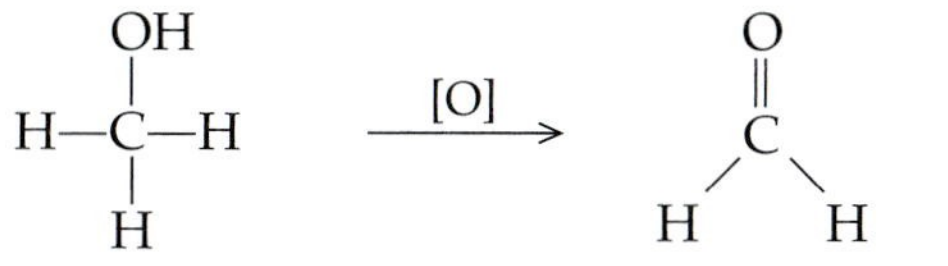

Methanol
(methyl alcohol)
An alcohol

Methanal
(formaldehyde)
An aldehyde

Note that the symbol [O] is used throughout this book to designate any oxidizing agent.

Oxidation of a primary alcohol produces an aldehyde:

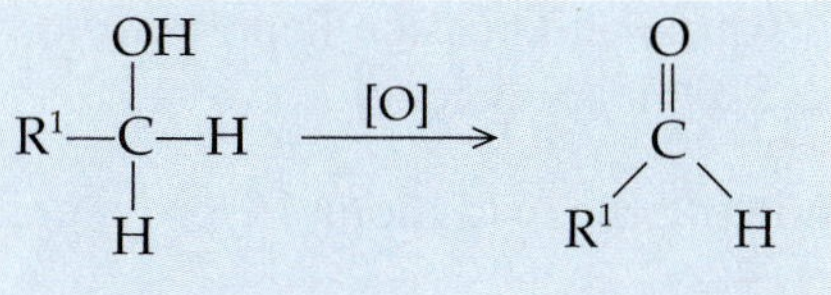

1° Alcohol

An aldehyde

As we will see in Section 12.5, aldehydes can undergo further oxidation to produce carboxylic acids.

7 Learning Goal

Write equations representing the oxidation of alcohols.

EXAMPLE 12.3 Writing an Equation Representing the Oxidation of a Primary Alcohol

Write an equation showing the oxidation of 2,2-dimethylpropanol to produce 2,2-dimethylpropanal.

SOLUTION

Begin by writing the structure of the reactant, 2,2-dimethylpropanol, and indicate the need for an oxidizing agent by placing the designation [O] over the reaction arrow. Then show the oxidation of the hydroxyl group to the aldehyde carbonyl group.

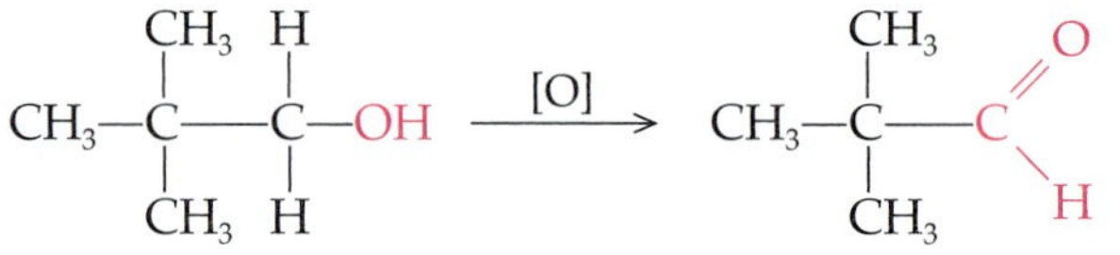

2,2-Dimethylpropanol   2,2-Dimethylpropanal

QUESTION 12.13

Write an equation showing the oxidation of the following primary alcohols:

a. 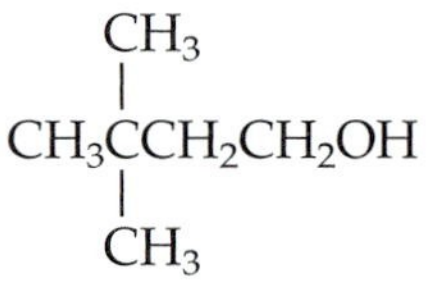
$$CH_3CCH_2CH_2OH$$
with CH_3 above and CH_3 below the second carbon

b. CH_3CH_2OH

QUESTION 12.14

Name each of the reactant alcohols and product aldehydes in Question 12.13 using the I.U.P.A.C. Nomenclature System. *Hint:* Refer to the example above, as well as to Section 12.5 to name the aldehyde products.

Oxidation of a secondary alcohol produces a ketone:

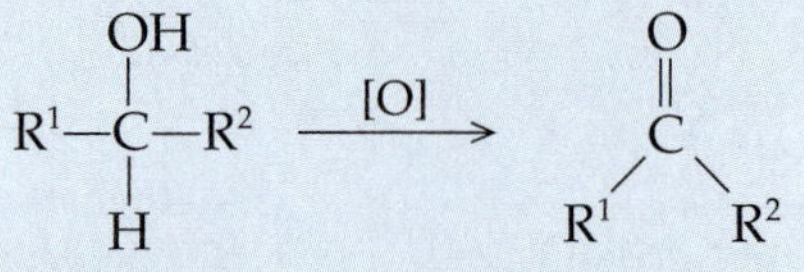

2° Alcohol   A ketone

EXAMPLE 12.4 Writing an Equation Representing the Oxidation of a Secondary Alcohol

7 Learning Goal

Write equations representing the oxidation of alcohols.

Write an equation showing the oxidation of 2-propanol to produce propanone.

Continued—

SOLUTION

Begin by writing the structure of the reactant, 2-propanol, and indicate the need for an oxidizing agent by placing the designation [O] over the reaction arrow. Then show the oxidation of the hydroxyl group to the ketone carbonyl group.

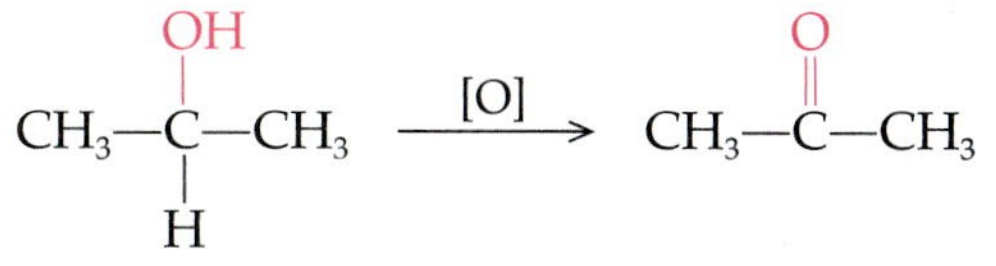

$$CH_3-\overset{\overset{\displaystyle OH}{|}}{\underset{\underset{\displaystyle H}{|}}{C}}-CH_3 \xrightarrow{[O]} CH_3-\overset{\overset{\displaystyle O}{\|}}{C}-CH_3$$

2-Propanol   Propanone

QUESTION 12.15

Write an equation showing the oxidation of the following secondary alcohols:

a. 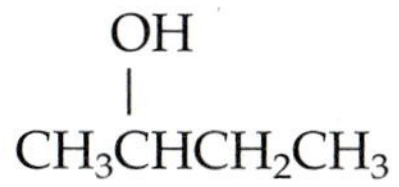 $\overset{\overset{\displaystyle OH}{|}}{CH_3CHCH_2CH_3}$

b. 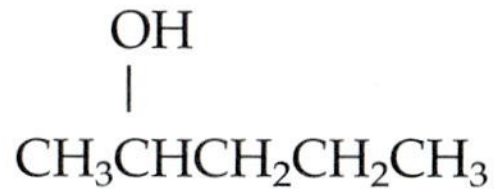 $\overset{\overset{\displaystyle OH}{|}}{CH_3CHCH_2CH_2CH_3}$

QUESTION 12.16

Name each of the reactant alcohols and product ketones in Question 12.15 using the I.U.P.A.C. Nomenclature System. *Hint:* Refer to the example above, as well as to Section 12.5, to name the ketone products.

Tertiary alcohols cannot be oxidized:

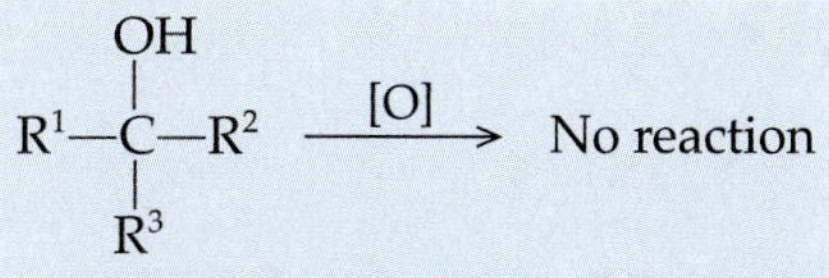

$$R^1-\overset{\overset{\displaystyle OH}{|}}{\underset{\underset{\displaystyle R^3}{|}}{C}}-R^2 \xrightarrow{[O]} \text{No reaction}$$

3° Alcohol

For the oxidation reaction to occur, the carbon bearing the hydroxyl group must contain at least one C—H bond. Because tertiary alcohols contain three C—C bonds to the carbinol carbon, they cannot undergo oxidation.

When ethanol is metabolized in the liver, it is oxidized to ethanal (acetaldehyde). If too much ethanol is present in the body, an overabundance of ethanal is formed, which causes many of the adverse effects of the "morning-after hangover." Continued oxidation of ethanal produces ethanoic acid (acetic acid), which is used as an energy source by the cell and eventually oxidized to CO_2 and H_2O. These reactions, summarized as follows, are catalyzed by liver enzymes.

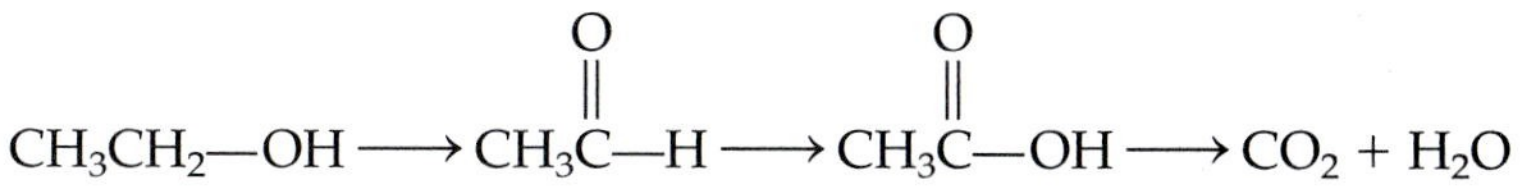

$$CH_3CH_2-OH \longrightarrow CH_3\overset{\overset{\displaystyle O}{\|}}{C}-H \longrightarrow CH_3\overset{\overset{\displaystyle O}{\|}}{C}-OH \longrightarrow CO_2 + H_2O$$

Ethanol   Ethanal   Ethanoic acid
(ethyl alcohol)   (acetaldehyde)   (acetic acid)

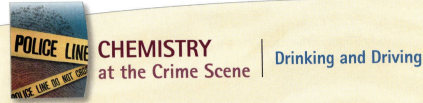

CHEMISTRY at the Crime Scene | Drinking and Driving

Nearly 41,000 people are killed each year in alcohol-related automobile accidents that result from the effects of ethanol on behavior, reflexes, and coordination. Blood alcohol levels of 0.05–0.15% seriously inhibit coordination. Blood levels in excess of 0.08% are considered evidence of intoxication in most states. Blood alcohol levels in the range of 0.30–0.50% produce unconsciousness and the risk of death.

One tool in the prevention of alcohol-related traffic deaths and, perhaps, in gathering evidence at the scene of an accident,

is the breathalyzer test. The suspect is required to exhale into a solution that will react with the unmetabolized alcohol in the breath. The partial pressure of the alcohol in the exhaled air has been demonstrated to be proportional to the blood alcohol level. The solution is an acidic solution of dichromate ion, which is yellow-orange. The alcohol reduces the chromium in the dichromate ion from +6 to +3, the Cr^{3+} ion, which is green. The intensity of the green color is measured, and it is proportional to the amount of ethanol that was oxidized. The reaction is

$$16H^+ + 2Cr_2O_7^{2-} + 3CH_3CH_2OH \longrightarrow$$

Yellow-orange

$$3CH_3COOH + 4Cr^{3+} + 11H_2O$$

Green

Of course, those with a positive breathalyzer test are given a more accurate blood test to ultimately establish their guilt or innocence. But the breathalyzer technology continues to provide a field-level screen for those endangering the lives of others by drinking and driving.

Driving under the influence of alcohol impairs coordination and reflexes.

FOR FURTHER UNDERSTANDING

Explain why the intensity of the green color in the reaction solution is proportional to the level of alcohol in the breath.

Construct a graph that represents this relationship.

---

8 Learning Goal

Discuss the use of phenols as germicides.

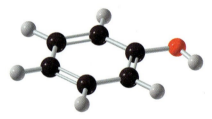

Figure 12.5

Ball-and-stick model of phenol. Keep in mind that this model provides accurate geometry, but does not show the cloud of shared electrons above and below the benzene ring. Review Section 11.5 for a more accurate description of the benzene ring.

12.2 Phenols

Phenols are compounds in which the hydroxyl group is attached to a benzene ring (Figure 12.5). Like alcohols, they are polar compounds because of the polar hydroxyl group. Thus, the simpler phenols are somewhat soluble in water. They are found in flavorings and fragrances (mint and savory) and are used as preservatives (butylated hydroxytoluene, BHT). Examples include

Thymol (mint)   Carvacrol (savory)   Butylated hydroxytoluene, BHT (food preservative)

Phenols are also widely used in health care as germicides. Carbolic acid, a dilute solution of phenol, was used as an antiseptic and disinfectant by Joseph Lister in his early work to decrease postsurgical infections. He used carbolic acid to bathe surgical wounds and to "sterilize" his instruments. Other derivatives of phenol that are used as antiseptics and disinfectants include hexachlorophene, hexylresorcinol, and 2-phenylphenol. The structures of these compounds are shown below:

Phenol
(carbolic acid;
phenol dissolved
in water;
antiseptic)

Hexachlorophene
(antiseptic)

Hexylresorcinol
(antiseptic)

2-Phenylphenol
(antiseptic)

12.3 Ethers

Ethers have the general formula R—O—R, and thus they are structurally related to alcohols (R—O—H) and water (H—O—H). As with alcohols and water, the lone pairs of electrons on the oxygen atom cause ethers to have an angular structure. The C—O bonds of ethers are polar, so ether molecules are polar (Figure 12.6). However, ethers do not form hydrogen bonds to one another because there is no —OH group. Therefore, they have much lower boiling points than alcohols of similar molar mass but higher boiling points than alkanes of similar molar mass. Compare the following examples:

9 **Learning Goal**
Write names and draw structures for common ethers and discuss their use in medicine.

$CH_3CH_2CH_2CH_3$ CH_3—O—CH_2CH_3 $CH_3CH_2CH_2OH$

Butane
(butane)
M.W. = 58
b.p. = −0.5°C

Methoxyethane
(ethyl methyl ether)
M.W. = 60
b.p. = 7.9°C

1-Propanol
(propyl alcohol)
M.W. = 60
b.p. = 97.2°C

In the I.U.P.A.C. system of naming ethers, the —OR substituent is named as an alkoxy group. This is analogous to the name *hydroxy* for the —OH group. Thus, CH_3—O— is methoxy, CH_3CH_2—O— is ethoxy, and so on.

An alkoxy group is an alkyl group bonded to an oxygen atom (—OR).

EXAMPLE 12.5   Using I.U.P.A.C. Nomenclature to Name an Ether

Name the following ether using I.U.P.A.C. nomenclature:

$$O—CH_3$$
$$|$$
$$CH_3CH_2CHCH_2CH_2CH_2CH_2CH_2CH_3$$
$$1 \quad 2 \quad 3 \quad 4 \quad 5 \quad 6 \quad 7 \quad 8 \quad 9$$

SOLUTION

Parent compound: nonane

Position of alkoxy group: carbon-3 (*NOT* carbon-7)

Substituents: 3-methoxy

Name: 3-Methoxynonane

In the common system of nomenclature, ethers are named by placing the names of the two alkyl groups attached to the ether oxygen as prefixes in front of

Figure 12.6

Ball-and-stick model of the ether, methoxymethane (dimethyl ether).

the word *ether.* The names of the two groups can be placed either alphabetically or by size (smaller to larger), as seen in the following examples:

$$CH_3—O—CH_3 \qquad CH_3—O—CH_2CH_3 \qquad CH_3CH_2—O—CH(CH_3)_2$$

Dimethyl ether
or
methyl ether

Ethyl methyl ether
or
methyl ethyl ether

Ethyl isopropyl ether

9 Learning Goal
Write names and draw structures for common ethers and discuss their use in medicine.

EXAMPLE 12.6 Naming Ethers Using the Common Nomenclature System

Write the common name for each of the following ethers:

$$CH_3CH_2—O—CH_2CH_3 \qquad CH_3—O—CH_2CH_2CH_3$$

SOLUTION

Alkyl Groups:	two ethyl groups	methyl and propyl
Name:	Diethyl ether	Methyl propyl ether

Notice that there is only one correct name for methyl propyl ether because the methyl group is smaller than the propyl group and it would be first in an alphabetical listing also.

QUESTION 12.17

Write the common and I.U.P.A.C. names for each of the following ethers:

a. $CH_3CH_2—O—CH_2CH_2CH_3$
b. $CH_3—O—CH_2CH_2CH_3$

QUESTION 12.18

Write the common and I.U.P.A.C. names for each of the following ethers:

a. $CH_3CH_2—O—CH_2CH_2CH_2CH_2CH_3$
b. $CH_3CH_2CH_2CH_2—O—CH_2CH_2CH_3$

Chemically, ethers are moderately inert. They do not react with reducing agents or bases under normal conditions. However, they are extremely volatile and highly flammable (easily oxidized in air) and hence must always be treated with great care.

Ethers may be prepared by a dehydration reaction (removal of water) between two alcohol molecules, as shown in the following general reaction. The reaction requires heat and acid.

10 Learning Goal
Write equations representing formation of an ether by a dehydration reaction between two alcohol molecules.

$$R^1—OH + R^2—OH \xrightarrow[\text{heat}]{H^+} R^1—O—R^2 + H_2O$$

Alcohol   Alcohol   Ether   Water

EXAMPLE 12.7 Writing an Equation Representing the Synthesis of an Ether via a Dehydration Reaction

Write an equation showing the synthesis of dimethyl ether.

— Continued—

SOLUTION

Write the structure of the product. Note that the alkyl substituents of this ether are two methyl groups. Thus, the alcohol that must undergo dehydration to produce dimethyl ether is methanol.

$$CH_3OH + CH_3OH \xrightarrow{H^+} CH_3{-}O{-}CH_3 + H_2O$$

Methanol   Methanol   Dimethyl ether   Water

QUESTION 12.19

Write an equation showing the dehydration reaction that would produce diethyl ether. Provide structures and names for all reactants and products.

QUESTION 12.20

Write an equation showing the dehydration reaction between two molecules of 2-propanol. Provide structures and names for all reactants and products.

Diethyl ether was the first general anesthetic used. The dentist, Dr. William Morton, is credited with its introduction in the 1800s. Diethyl ether functions as an anesthetic by interacting with the central nervous system. It appears that diethyl ether (and many other general anesthetics) functions by accumulating in the lipid material of the nerve cells, thereby interfering with nerve impulse transmission. This results in analgesia, a lessened perception of pain.

Halogenated ethers are also routinely used as general anesthetics (Figure 12.7). They are less flammable than diethyl ether and are therefore safer to store and work with. *Penthrane* and *Enthrane* (trade names) are two of the more commonly used members of this family:

$$
\begin{array}{cc}
\overset{\displaystyle F}{\underset{\displaystyle F}{CH_3{-}O{-}C}}{-}\overset{\displaystyle Cl}{\underset{\displaystyle Cl}{C}}{-}H &
\overset{\displaystyle F}{\underset{\displaystyle F}{H{-}C}}{-}O{-}\overset{\displaystyle F}{\underset{\displaystyle F}{C}}{-}\overset{\displaystyle F}{\underset{\displaystyle Cl}{C}}{-}H
\end{array}
$$

Penthrane   Enthrane

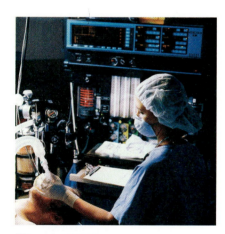

Figure 12.7

An anesthesiologist administers Penthrane to a surgical patient.

12.4 Thiols

Compounds that contain the sulfhydryl group (—SH) are called **thiols.** They are similar to alcohols in structure, but the sulfur atom replaces the oxygen atom.

Thiols and many other sulfur compounds have nauseating aromas. They are found in substances as different as the defensive spray of the North American striped skunk, onions, and garlic. The structures of the two most common compounds in the defense spray of the striped skunk, *trans*-2-butene-1-thiol and 3-methyl-1-butanethiol, are seen in Figure 12.8. These structures are contrasted with the structures of the two molecules that make up the far more pleasant

11 Learning Goal

Write names and draw structures for simple thiols and discuss their biological significance.

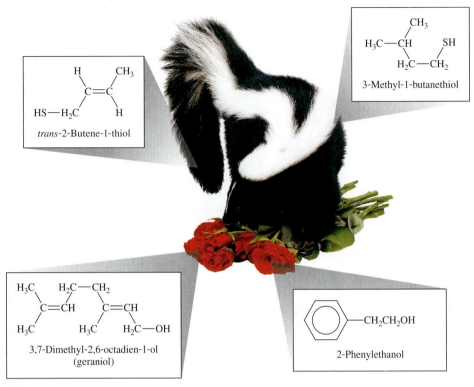

Figure 12.8

This skunk on a bed of roses is surrounded by scent molecules. The two most common compounds in the defense spray of the striped skunk are the thiols, *trans*-2-butene-1-thiol and 3-methyl-1-butanethiol. Two alcohols, 2-phenylethanol and geraniol, are the major components of the scent of roses.

scent of roses: geraniol, an unsaturated alcohol, and 2-phenylethanol, an aromatic alcohol.

The I.U.P.A.C. rules for naming thiols are similar to those for naming alcohols, except that the full name of the alkane is retained. The suffix *-thiol* follows the name of the parent compound.

11 Learning Goal

Write names and draw structures for simple thiols and discuss their biological significance.

EXAMPLE 12.8 Naming Thiols Using the I.U.P.A.C. Nomenclature System

Write the I.U.P.A.C. name for the thiols CH_3CH_2SH and $HSCH_2CH_2SH$.

SOLUTION

Retain the full name of the parent compound and add the suffix *-thiol*.

$$CH_3CH_2—SH \qquad HS—CH_2CH_2—SH$$

Parent compound:	ethane	ethane
Position of —SH:	carbon-1 (*must be!*)	carbon-1 and carbon-2
Name:	Ethanethiol	1,2-Ethanedithiol

Amino acids are the subunits from which proteins are made. A protein is a long polymer, or chain, of many amino acids bonded to one another.

The amino acid cysteine is a thiol that has an important role in the structure and shape of many proteins. Two cysteine molecules can undergo oxidation to form cystine. The new bond is called a **disulfide** (—S—S—) bond.

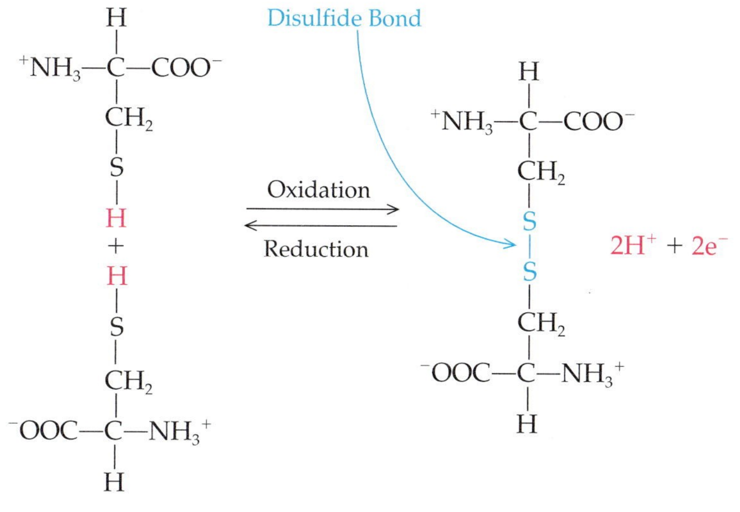

2-Cysteine   Cystine

If the two cysteines are in different protein chains, the disulfide bond between them forms a bridge joining them together. If the two cysteines are in the same protein chain, a loop is formed. An example of the importance of disulfide bonds is seen in the production and structure of the protein hormone insulin, which controls blood sugar levels in the body (Figure 12.9).

Coenzyme A is a thiol that serves as a "carrier" of acetyl groups ($CH_3CO—$) in biochemical reactions. It plays a central role in metabolism by shuttling acetyl groups from one reaction to another. When the two-carbon acetate group is attached to coenzyme A, the product is acetyl coenzyme A (acetyl CoA). The bond between coenzyme A and the acetyl group is a high-energy *thioester bond*. Formation of the high-energy thioester bond "energizes" the acetyl group so that it can participate in other biochemical reactions.

The reactions involving coenzyme A are discussed in detail in Chapters 18 and 19.

A high-energy bond is one that releases a great deal of energy when it is broken.

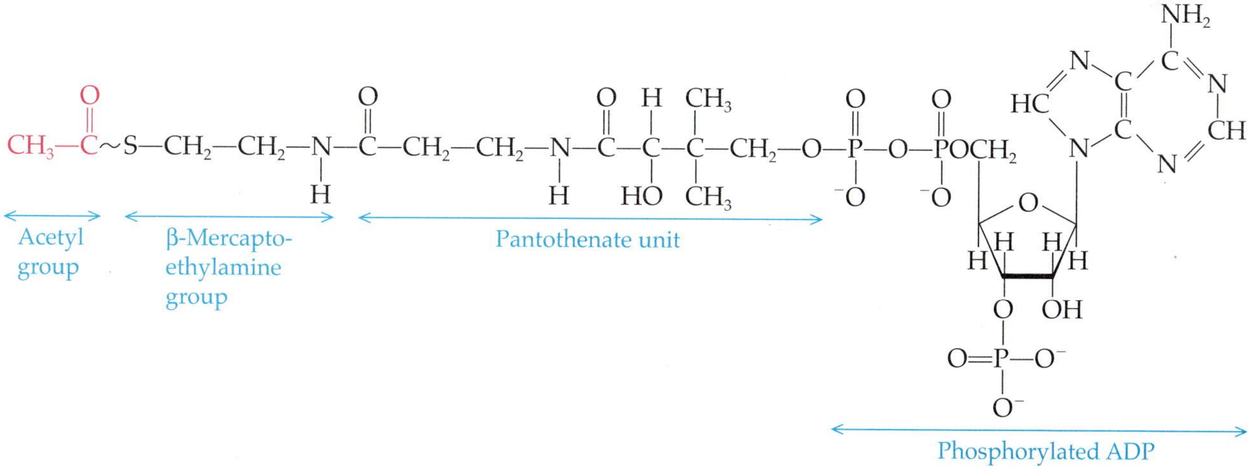

Acetyl coenzyme A
(acetyl CoA)

Acetyl CoA is made and used in the energy-producing reactions that provide most of the energy for life processes. It is also required for the biosynthesis of many biological molecules.

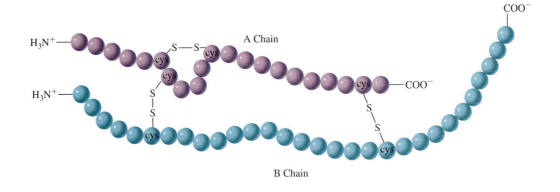

Insulin

Figure 12.9

Disulfide bonds hold the A and B chains together and form a hairpin loop in the A chain.

12.5 Aldehydes and Ketones

Aldehydes and ketones are carbonyl-containing compounds distinguished by the location of the **carbonyl group** ($>C=O$) within the carbon chain. In aldehydes, the carbonyl group is always located at the end of the carbon chain (carbon-1). In ketones, the carbonyl group is located within the carbon chain of the molecule. Thus, in ketones, the carbonyl carbon is attached to two other carbon atoms. However, in aldehydes, the carbonyl carbon is attached to at least one hydrogen atom; the second atom attached to the carbonyl carbon of an aldehyde may be another hydrogen or a carbon atom (Figure 12.10).

Structure and Physical Properties

Aldehydes and **ketones** are polar compounds because of the polar carbonyl group. Because of the dipole–dipole attractions between molecules, they boil at higher temperatures than hydrocarbons or ethers that have the same number of carbon atoms or are of equivalent molecular weight. Because they cannot form intermolecular hydrogen bonds, their boiling points are lower than those of alcohols of comparable molecular weight.

Carbonyl group

12 Learning Goal

Draw the structures and discuss the physical properties of aldehydes and ketones.

Dipole–dipole attraction

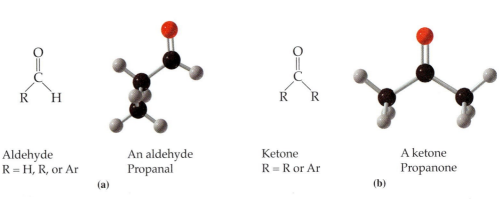

Aldehyde
R = H, R, or Ar

An aldehyde
Propanal

(a)

Ketone
R = R or Ar

A ketone
Propanone

(b)

Figure 12.10

The structures of aldehydes and ketones. (a) The general structure of an aldehyde and a ball-and-stick model of the aldehyde propanal. (b) The general structure of a ketone and a ball-and-stick model of the ketone propanone.

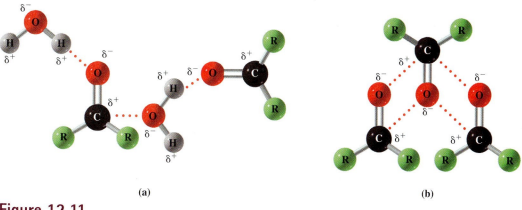

(a) **(b)**

Figure 12.11

(a) Hydrogen bonding between the carbonyl group of an aldehyde or ketone and water. (b) Polar interactions between carbonyl groups of aldehydes or ketones.

Aldehydes and ketones can form intermolecular hydrogen bonds with water (Figure 12.11). As a result, the smaller members of the two families (five or fewer carbon atoms) are reasonably soluble in water. However, as the carbon chain length increases, the compounds become less polar and more hydrocarbonlike. These larger compounds are soluble in nonpolar organic solvents.

QUESTION 12.21

Which member in each of the following pairs will be more water soluble?

a. $CH_3(CH_2)_2CH_3$ or $CH_3\overset{\overset{\displaystyle O}{\|}}{C}CH_3$

b. $CH_3\overset{\underset{\displaystyle O}{\|}}{C}CH_2CH_2CH_3$ or $CH_3\overset{\underset{\displaystyle OH}{|}}{CH}CH_2CH_2CH_3$

QUESTION 12.22

Which member in each of the following pairs would have a higher boiling point?

a. $CH_3CH_2\overset{\underset{\displaystyle O}{\|}}{C}OH$ or $CH_3CH_2\overset{\underset{\displaystyle O}{\|}}{C}H$

b. $CH_3\overset{\underset{\displaystyle O}{\|}}{C}OH$ or $CH_3\overset{\overset{\displaystyle O}{\|}}{C}CH_3$

Nomenclature

Naming Aldehydes

In the I.U.P.A.C. system, aldehydes are named according to the following set of rules:

- Determine the parent compound, that is, the longest continuous carbon chain containing the carbonyl group.
- Replace the final -*e* of the parent alkane with -*al*.

13 **Learning Goal**
Write the common and I.U.P.A.C. names for common aldehydes and ketones.

- Number the chain beginning with the carbonyl carbon (or aldehyde group) as carbon-1.
- Number and name all substituents as usual. No number is used for the position of the carbonyl group because it is always at the end of the parent chain. Therefore, it must be carbon-1.

Several examples are provided here with common names given in parentheses:

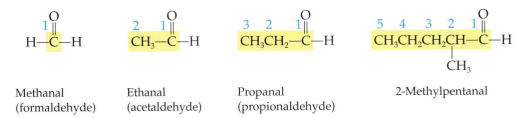

Methanal
(formaldehyde)

Ethanal
(acetaldehyde)

Propanal
(propionaldehyde)

2-Methylpentanal

EXAMPLE 12.9 **Using the I.U.P.A.C. Nomenclature System to Name Aldehydes**

13 Learning Goal

Write the common and I.U.P.A.C. names for common aldehydes and ketones.

Name the aldehydes represented by the following condensed formulas:

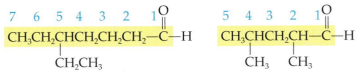

SOLUTION

Parent compound:	heptane (becomes heptanal)	pentane (becomes pentanal)
Position of carbonyl group:	carbon-1 (*must be!*)	carbon-1 (*must be!*)
Substituents:	5-ethyl	2,4-dimethyl
Name:	5-Ethylheptanal	2,4-Dimethylpentanal

Notice that the position of the carbonyl group is not indicated by a number. By definition, the carbonyl group is located at the end of the carbon chain of an aldehyde. The carbonyl carbon is defined as carbon-1; thus, it is not necessary to include the position of the carbonyl group in the name of the compound.

---

13 Learning Goal

Write the common and I.U.P.A.C. names for common aldehydes and ketones.

Carboxylic acid nomenclature is described in Section 13.1.

The common names of the aldehydes are derived from the same Latin roots as the corresponding carboxylic acids. The common names of the first five aldehydes are presented in Table 12.1.

In the common system of nomenclature, substituted aldehydes are named as derivatives of the straight-chain parent compound (see Table 12.1). Greek letters are used to indicate the positions of the substituents. The carbon atom bonded to the carbonyl group is the α-carbon, the next is the β-carbon, and so on.

$$
\begin{array}{ccccc}
\delta & \gamma & \beta & \alpha & O \\
| & | & | & | & || \\
-C-&C-&C-&C-&C-H \\
| & | & | & | &
\end{array}
$$

TABLE 12.1 I.U.P.A.C. and Common Names and Formulas for Several Aldehydes

I.U.P.A.C. Name	Common Name	Formula
Methanal	Formaldehyde	$\overset{\displaystyle O}{\underset{\displaystyle \parallel}{}}$ H—C—H
Ethanal	Acetaldehyde	$CH_3\overset{O}{\overset{\parallel}{C}}$—H
Propanal	Propionaldehyde	$CH_3CH_2\overset{O}{\overset{\parallel}{C}}$—H
Butanal	Butyraldehyde	$CH_3CH_2CH_2\overset{O}{\overset{\parallel}{C}}$—H
Pentanal	Valeraldehyde	$CH_3CH_2CH_2CH_2\overset{O}{\overset{\parallel}{C}}$—H

Consider the following examples:

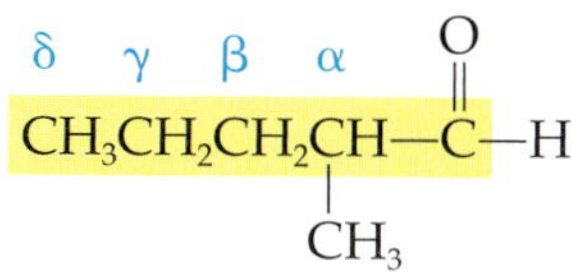

δ   γ   β   α   O

CH₃CH₂CH₂CH—C—H

             |

             CH₃

2-Methylpentanal
(α-methylvaleraldehyde)

δ   γ   β   α   O

CH₃CH₂CHCH₂—C—H

         |

         CH₃

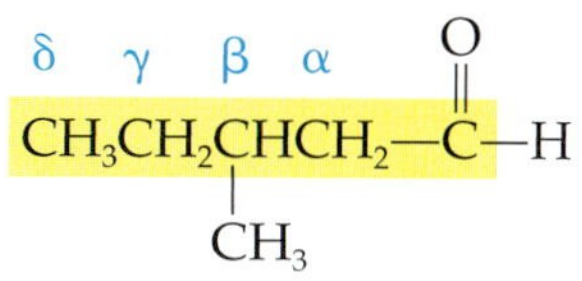

3-Methylpentanal
(β-methylvaleraldehyde)

EXAMPLE 12.10 Using the Common Nomenclature System to Name Aldehydes

Name the aldehydes represented by the following condensed formulas:

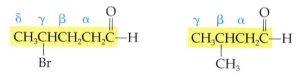

13 Learning Goal

Write the common and I.U.P.A.C. names for common aldehydes and ketones.

SOLUTION

Parent compound:	pentane (becomes valeraldehyde)	butane (becomes butyraldehyde)
Position of carbonyl group:	carbon-1 (*must be!*)	carbon-1 (*must be!*)
Substituents:	γ-bromo	β-methyl
Name:	γ-Bromovaleraldehyde	β-Methylbutyraldehyde

In the common system of nomenclature for aldehydes, the carbon atom bonded to the carbonyl group is called the α-carbon, the next is the β-carbon, etc. Also remember that by definition, the carbonyl group is located at the beginning of the carbon chain of an aldehyde. Thus, it is not necessary to include the position of the carbonyl group in the name of the compound.

QUESTION 12.23

Use the I.U.P.A.C. and common nomenclature systems to name each of the following compounds.

a. 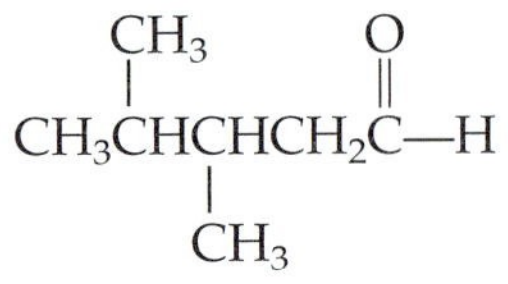 $CH_3CHCHCH_2\overset{\overset{\displaystyle O}{\|}}{C}{-}H$
   with CH_3 on the second carbon and CH_3 below the third carbon

b. 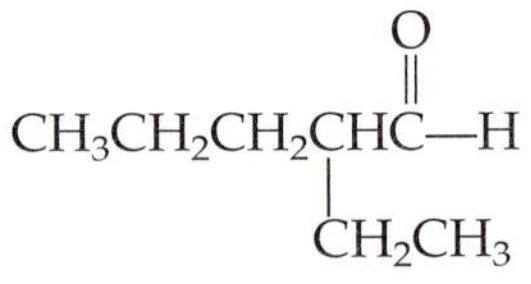 $CH_3CH_2CH_2\overset{\overset{\displaystyle O}{\|}}{C}H{-}H$ with CH_2CH_3 below

QUESTION 12.24

Write the condensed structural formula for each of the following compounds.

a. 4-Fluorohexanal
b. α,β-Dimethylbutyraldehyde

Naming Ketones

13 Learning Goal
Write the common and I.U.P.A.C. names for common aldehydes and ketones.

The rules for naming ketones in the I.U.P.A.C. Nomenclature System are directly analogous to those for naming aldehydes. In ketones, however, the *-e* ending of the parent alkane is replaced with the *-one* suffix of the ketone family, and the location of the carbonyl carbon is indicated by a number. The longest carbon chain is numbered to give the carbonyl carbon the lowest possible number. For example,

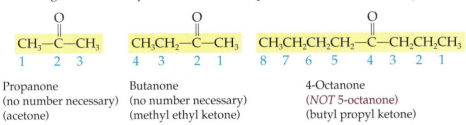

Propanone
(no number necessary)
(acetone)

Butanone
(no number necessary)
(methyl ethyl ketone)

4-Octanone
(*NOT* 5-octanone)
(butyl propyl ketone)

13 Learning Goal
Write the common and I.U.P.A.C. names for common aldehydes and ketones.

EXAMPLE 12.11 | **Using the I.U.P.A.C Nomenclature System to Name Ketones**

Name the ketones represented by the following condensed formulas.

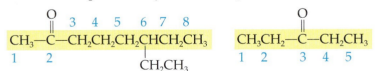

SOLUTION

Parent compound:	octane (becomes octanone)	pentane (becomes pentanone)
Position of carbonyl group:	carbon-2 (*NOT* carbon-7)	carbon-3
Substituents:	6-ethyl	none
Name:	6-Ethyl-2-octanone	3-Pentanone

The common names of ketones are derived by naming the alkyl groups that are bonded to the carbonyl carbon. These are used as prefixes followed by the word *ketone*. The alkyl groups may be arranged alphabetically or by size (smaller to larger).

EXAMPLE 12.12 **Using the Common Nomenclature System to Name Ketones**

13 **Learning Goal**

Write the common and I.U.P.A.C. names for common aldehydes and ketones.

Name the ketones represented by the following condensed formulas:

$$CH_3CH_2CH_2-\overset{\overset{\displaystyle O}{\|}}{C}-CH_3 \qquad CH_3CH_2-\overset{\overset{\displaystyle O}{\|}}{C}-CH_2CH_2CH_2CH_2CH_3$$

SOLUTION

Identify the alkyl groups that are bonded to the carbonyl carbon.

Alkyl groups:	propyl and methyl	ethyl and pentyl
Name:	Methyl propyl ketone	Ethyl pentyl ketone

QUESTION 12.25

Use the I.U.P.A.C. Nomenclature System to name each of the following compounds.

a. $CH_3\overset{\overset{\displaystyle }{}}{\underset{\underset{\displaystyle I}{|}}{CH}}\overset{\overset{\displaystyle O}{\|}}{C}CH_3$

b. $CH_3\overset{}{\underset{\underset{\displaystyle CH_2CH_2CH_2CH_3}{|}}{CH}}CH_2\overset{\overset{\displaystyle O}{\|}}{C}CH_3$

QUESTION 12.26

Use the I.U.P.A.C. Nomenclature System to name each of the following compounds.

a. $CH_3\overset{}{\underset{\underset{\displaystyle CH_3}{|}}{CH}}\overset{\overset{\displaystyle O}{\|}}{C}CH_3$

b. $CH_3\overset{}{\underset{\underset{\displaystyle CH_3}{|}}{CH}}\overset{\overset{\displaystyle O}{\|}}{C}CH_2CH_3$

Important Aldehydes and Ketones

Methanal (formaldehyde) is a gas (b.p. −21°C). It is available commercially as an aqueous solution called *formalin*. Formalin has been used as a preservative for tissues and as an embalming fluid. See A Medical Connection: Formaldehyde and Methanol Poisoning for more information on methanal.

14 **Learning Goal**

Discuss the biological, medical, or environmental significance of several aldehydes and ketones.

Ethanal (acetaldehyde) is produced from ethanol in the liver. Ethanol is oxidized in this reaction, which is catalyzed by the liver enzyme alcohol dehydrogenase. The ethanal that is produced in this reaction is responsible for the symptoms of a hangover.

Propanone (acetone), the simplest ketone, is an important and versatile solvent for organic compounds. It has the ability to dissolve organic compounds and is also miscible with water. As a result, it has a number of industrial applications and is used as a solvent in adhesives, paints, cleaning solvents, nail polish, and nail polish remover. Propanone is flammable and should therefore be treated with appropriate care. *Butanone*, a four-carbon ketone, is also an important industrial solvent.

Many aldehydes and ketones are produced industrially as food and fragrance chemicals, medicinals, and agricultural chemicals. They are particularly important to the food industry, where they are used as artificial and/or natural additives to food. Vanillin, a principal component of natural vanilla, is shown in Figure 12.12. Artificial vanilla flavoring is a dilute solution of synthetic vanillin

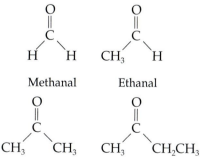

Methanal   Ethanal

Propanone   Butanone

Figure 12.12

Important aldehydes and ketones.

A MEDICAL Connection | Formaldehyde and Methanol Poisoning

Most aldehydes have irritating, unpleasant odors, and formaldehyde is no exception. Formaldehyde is also an extremely toxic substance. As an aqueous solution, called formalin, it has been used to preserve biological tissues and for embalming. It has also been used to disinfect environmental surfaces, body fluids, and feces. Under no circumstances is it used as an antiseptic on human tissue because of its toxic fumes and the skin irritations that it causes.

Formaldehyde is used in the production of some killed-virus vaccines. When a potentially deadly virus, such as polio virus, is treated with heat and formaldehyde, the genetic information (RNA) is damaged beyond repair. The proteins of the virus also react with formaldehyde. However, the shape of the proteins, which is critical for a protective immune response against the virus, is not changed. Thus, when a child is injected with the Salk killed-virus polio vaccine, the immune system recognizes viral proteins and produces antibodies that protect against polio virus infection.

Formaldehyde can also be produced in the body! As we saw in this chapter, methanol can be oxidized to produce formaldehyde. In the body, the liver enzyme alcohol dehydrogenase, whose function it is to detoxify alcohols by oxidizing them, catalyzes the conversion of methanol to formaldehyde (methanal). The formaldehyde then reacts with cellular macromolecules, including proteins, causing severe damage (remember, it is used as an embalming agent!). As a result, methanol poisoning can cause blindness, respiratory failure, convulsions, and death.

Clever physicians have devised a treatment for methanol poisoning that is effective if administered soon enough after ingestion. Since the same enzyme that oxidizes methanol to formaldehyde (methanal) also oxidizes ethanol to acetaldehyde (ethanal), doctors reasoned that administering an intravenous solution of ethanol to the patient could protect against the methanol poisoning. If the ethanol concentration in the body is higher than the methanol concentration, most of the alcohol dehydrogenase enzymes of the liver will be carrying out the oxidation of the ethanol. This is called *competitive inhibition* because the methanol and ethanol molecules are competing for binding to the enzymes. The molecule that is in the higher concentration will more frequently bind to the enzyme and undergo reaction. In this case, the result is that the alcohol dehydrogenase enzymes are kept busy oxidizing ethanol and producing the less-toxic (not nontoxic) product, acetaldehyde. This gives the body time to excrete the methanol before it is oxidized to the potentially deadly formaldehyde.

Patient receiving an intravenous solution of ethanol to treat methanol poisoning.

FOR FURTHER UNDERSTANDING

Acetaldehyde is described as less toxic than formaldehyde. Do some background research on the effects of these two aldehydes on biological systems.

You have studied enzymes in previous biology courses. Using what you learned in those classes with information from Section 16.10 in this text, put together an explanation of the way in which competitive inhibition works. Can you think of other types of poisoning for which competitive inhibition might be used to develop an effective treatment?

dissolved in ethanol. Figure 12.12 also shows other examples of important aldehydes and ketones.

QUESTION 12.27

Draw the structure of the aldehyde synthesized from ethanol in the liver.

QUESTION 12.28

Draw the structure of a ketone that is an important, versatile solvent for organic compounds.

 The complex aldehyde, retinal, produced from vitamin A, is essential for vision. To discover more about the process of vision and the role of retinal, *see A Medical Connection: The Chemistry of Vision* online at www.mhhe.com/denniston.

The ketone dihydroxyacetone is the active ingredient in many self-tanning lotions. To learn more about ketones and the chemistry of self-tanning lotions, see *A Lifestyle Connection: That Golden Tan Without the Fear of Skin Cancer* online at www.mhhe.com/denniston.

7 Learning Goal

Write equations representing the oxidation of alcohols.

Reactions Involving Aldehydes and Ketones

Preparation of Aldehydes and Ketones

Aldehydes and ketones are prepared primarily by the oxidation of the corresponding alcohol. As we saw in Section 12.1, the oxidation of methyl alcohol gives methanal (formaldehyde). The oxidation of a primary alcohol produces an aldehyde, and the oxidation of a secondary alcohol yields a ketone. Tertiary alcohols do not undergo oxidation under the conditions normally used.

QUESTION 12.29

Write an equation showing the oxidation of 1-propanol.

QUESTION 12.30

Write an equation showing the oxidation of 2-butanol.

15 Learning Goal

Write equations representing the oxidation and reduction of aldehydes and ketones.

Oxidation Reactions

Aldehydes are easily oxidized further to carboxylic acids, whereas ketones do not generally undergo further oxidation. Aldehydes are so easily oxidized that it is often very difficult to prepare them because they continue to react to give the carboxylic acid rather than the desired aldehyde. Aldehydes are susceptible to air oxidation, even at room temperature, and cannot be stored for long periods. The following example shows a general equation for the oxidation of an aldehyde to a carboxylic acid:

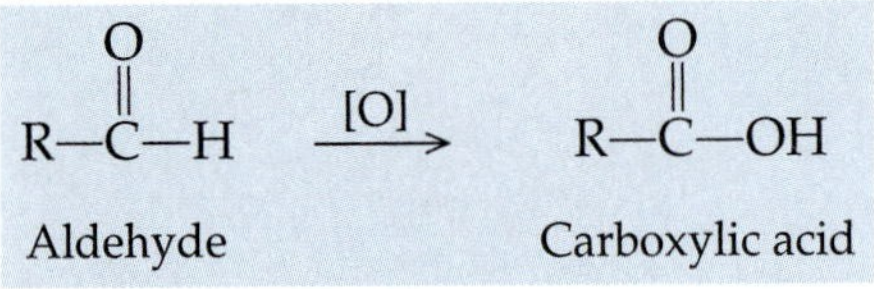

$$\underset{\text{Aldehyde}}{\text{R}-\overset{\displaystyle \overset{O}{\|}}{\text{C}}-\text{H}} \xrightarrow{[O]} \underset{\text{Carboxylic acid}}{\text{R}-\overset{\displaystyle \overset{O}{\|}}{\text{C}}-\text{OH}}$$

Many oxidizing agents can be used. Both basic potassium permanganate and chromic acid are good oxidizing agents, as the following specific example shows:

In basic solution, the product is the carboxylic acid anion:

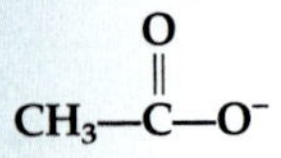

$$\text{CH}_3-\overset{\displaystyle \overset{O}{\|}}{\text{C}}-\text{O}^-$$

The rules for naming carboxylic acid anions are described in Section 13.1.

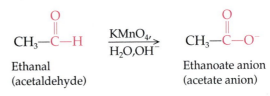

Ethanal
(acetaldehyde)

Ethanoate anion
(acetate anion)

Aldehydes and ketones can be distinguished on the basis of differences in their reactivity. The most common laboratory test for aldehydes is the **Tollens' test.** When exposed to the Tollens' reagent, a basic solution of $Ag(NH_3)_2^+$, an aldehyde undergoes oxidation. The silver ion (Ag^+) is reduced to silver metal (Ag^0) as the aldehyde is oxidized to a carboxylic acid anion.

$$R-\overset{\overset{O}{\|}}{C}-H + Ag(NH_3)_2^+ \longrightarrow R-\overset{\overset{O}{\|}}{C}-O^- + Ag^0$$

Aldehyde   Silver ammonia complex— Tollens' reagent   Carboxylate anion   Silver metal mirror

Silver metal precipitates from solution and coats the flask, producing a smooth silver mirror, as seen in Figure 12.13. The test is therefore often called the Tollens' silver mirror test. The commercial manufacture of silver mirrors uses a similar process. Ketones cannot be oxidized to carboxylic acids and do not react with the Tollens' reagent.

EXAMPLE 12.13 **Writing Equations for the Reaction of an Aldehyde and of a Ketone with Tollens' Reagent**

Write equations for the reaction of propanal and 2-pentanone with Tollens' reagent.

SOLUTION

$$CH_3CH_2\overset{\overset{O}{\|}}{C}-H + Ag(NH_3)_2^+ \longrightarrow CH_3CH_2\overset{\overset{O}{\|}}{C}-O^- + Ag^0$$

Propanal   Propanoate anion

$$CH_3CH_2CH_2\overset{\overset{O}{\|}}{C}CH_3 + Ag(NH_3)_2^+ \longrightarrow \text{No reaction}$$

2-Pentanone

QUESTION 12.31

Write an equation for the reaction of ethanal with Tollens' reagent.

QUESTION 12.32

Write an equation for the reaction of propanone with Tollens' reagent.

Another test that is used to distinguish between aldehydes and ketones is Benedict's test. Here, a buffered aqueous solution of copper(II) hydroxide and sodium citrate reacts to oxidize aldehydes but does not generally react with ketones. Cu^{2+} is reduced to Cu^+ in the process. Cu^{2+} is soluble and gives a blue solution, whereas Cu^+ precipitates as the red solid copper(I) oxide, Cu_2O.

All simple sugars (monosaccharides) are either aldehydes or ketones. Glucose is an aldehyde sugar that is commonly called *blood sugar* because it is the sugar transported in the blood and used for energy by many cells. In uncontrolled diabetes, glucose may be found in the urine. One early method used to determine the

Silver ions are very mild oxidizing agents. They will oxidize aldehydes but not alcohols.

Animation
The Silver Mirror

15 Learning Goal
Write equations representing the oxidation and reduction of aldehydes and ketones.

Cu(II) is an even milder oxidizing agent than silver ion.

Figure 12.13

The silver precipitate produced by the Tollens' reaction is deposited on glass. The progress of the reaction is visualized in panels (a) through (d). Silver mirrors are made in a similar process.

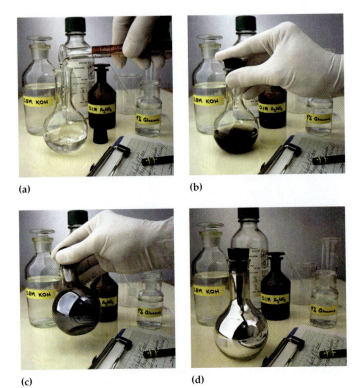

(a)   (b)

(c)   (d)

amount of glucose in the urine was to observe the color change of the Benedict's test. The amount of precipitate formed is directly proportional to the amount of glucose in the urine (Figure 12.14). The reaction of glucose with the Benedict's reagent is represented in the following equation:

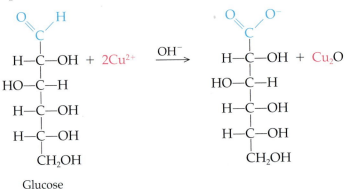

Glucose

Figure 12.14

The amount of precipitate formed and thus the color change observed in the Benedict's test are directly proportional to the amount of reducing sugar in the sample.

15 Learning Goal

Write equations representing the oxidation and reduction of aldehydes and ketones.

Hydrogenation was first discussed in Section 11.4 for the hydrogenation of alkenes.

We should also note that when the carbonyl group of a ketone is bonded to a —CH_2OH group, the molecule will give a positive Benedict's test. This occurs because such ketones are converted to aldehydes under basic conditions. In Chapter 14, we will see that this applies to the ketone sugars, as well. They are converted to aldehyde sugars and react with Benedict's reagent.

Reduction Reactions

Aldehydes and ketones are both readily reduced to the corresponding alcohol by a variety of reducing agents. Throughout the text, the symbol [H] over the reaction arrow represents a reducing agent.

The classical method of aldehyde or ketone reduction is **hydrogenation.** The carbonyl compound is reacted with hydrogen gas and a catalyst (nickel, platinum, or palladium metal) in a pressurized reaction vessel. Heating may also be necessary. The carbon–oxygen double bond (the carbonyl group) is reduced to a carbon–oxygen

A MEDICAL Connection | Alcohol Abuse and Antabuse

According to a recent study carried out by the Centers for Disease Control and Prevention,[1] 75,000 Americans die each year as a result of alcohol abuse. Of these, 34,833 people died of cirrhosis of the liver, cancer, or other drinking-related diseases. The remaining 40,933 died in alcohol-related automobile accidents. Of those who died, 72% were men and 6% were under the age of 21. In fact, a separate study has estimated that 1400 college-age students die each year of alcohol-related causes.

These numbers are striking. Alcohol abuse is now the third leading cause of preventable death in the United States, outranked only by tobacco use and poor diet and exercise habits. As the study concluded, "These results emphasize the importance of adopting effective strategies to reduce excessive drinking, including increasing alcohol excise taxes and screening for alcohol misuse in clinical settings."

Tetraethylthiuram disulfide
(disulfiram)
Antabuse

One approach to treatment of alcohol abuse, the drug tetraethylthiuram disulfide or disulfiram, has been used since 1951. The activity of this drug generally known by the trade name Antabuse was discovered accidentally by a group of Danish researchers who were testing it for antiparasitic properties. They made the observation that those who had taken disulfiram became violently ill after consuming any alcoholic beverage. Further research revealed that this compound inhibits one of the liver enzymes in the pathway for the oxidation of alcohols.

We have seen that ethanol is oxidized to ethanal (acetaldehyde) in the liver. This reaction is catalyzed by the enzyme alcohol dehydrogenase. Acetaldehyde, which is more toxic than ethanol, is responsible for many of the symptoms of a hangover. The enzyme acetaldehyde dehydrogenase oxidizes acetaldehyde into ethanoic acid (acetic acid), which then is used in biochemical pathways that harvest energy for cellular work or that synthesize fats.

Antabuse inhibits acetaldehyde dehydrogenase. This inhibition occurs within one to two hours of taking the drug and continues up to fourteen days. When a person who has taken Antabuse drinks an alcoholic beverage, the level of acetaldehyde quickly reaches levels that are five to ten times higher than would normally occur after a drink. Within just a few minutes, the symptoms of a severe hangover are experienced and may continue for several hours.

The drug Antabuse may be useful in treating alcohol abuse.

Experts in drug and alcohol abuse have learned that drugs such as Antabuse are generally not effective on their own. However, when used in combination with support groups and/or psychotherapy to solve underlying behavioral or psychological problems, Antabuse is an effective deterrent to alcohol abuse.

1. Alcohol-Attributable Deaths and Years of Potential Life Lost—United States, 2001, Morbidity and Mortality Weekly Report, 53 (37): 866–870, September 24, 2004, also available at http://www.cdc.gov/mmwr/preview/mmwrhtml/mm5337a2.htm.

FOR FURTHER UNDERSTANDING

Antabuse alone is not a cure for alcoholism. Consider some of the reasons why this is so.

Write equations showing the oxidation of ethanal to ethanoic acid as a pathway with the product of the first reaction serving as the reactant for the second. Explain the physiological effects of Antabuse in terms of these chemical reactions.

single bond. This is similar to the reduction of an alkene to an alkane (the reduction of a carbon–carbon double bond to a carbon–carbon single bond). The addition of hydrogen to a carbon–oxygen double bond is shown in the following general equation:

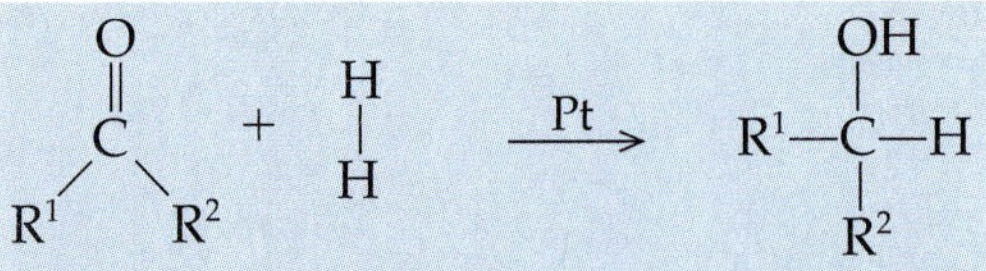

$$
\underset{\substack{\text{Aldehyde} \\ \text{or ketone}}}{\overset{\overset{\displaystyle O}{\|}}{\underset{R^1}{C}\underset{R^2}{}}} + \underset{\text{Hydrogen}}{\overset{\displaystyle H}{\underset{\displaystyle H}{|}}} \xrightarrow{\text{Pt}} \underset{\text{Alcohol}}{\overset{\displaystyle OH}{R^1-\underset{\underset{R^2}{|}}{C}-H}}
$$

The hydrogenation (reduction) of a ketone produces a secondary alcohol, as seen in the following equation showing the reduction of the ketone, 3-octanone:

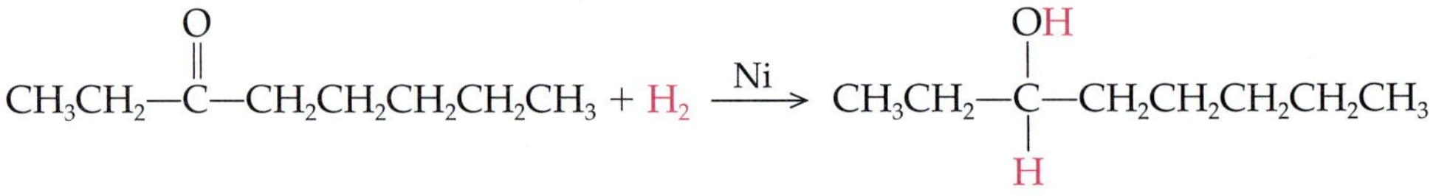

$$
\underset{\substack{\text{3-Octanone} \\ \text{(A ketone)}}}{CH_3CH_2-\overset{\overset{\displaystyle O}{\|}}{C}-CH_2CH_2CH_2CH_2CH_3} + H_2 \xrightarrow{\text{Ni}} \underset{\substack{\text{3-Octanol} \\ \text{(A secondary alcohol)}}}{CH_3CH_2-\overset{\overset{\displaystyle OH}{|}}{\underset{\underset{\displaystyle H}{|}}{C}}-CH_2CH_2CH_2CH_2CH_3}
$$

EXAMPLE 12.14   **Writing an Equation Representing the Hydrogenation of a Ketone**

15 **Learning Goal**
Write equations representing the oxidation and reduction of aldehydes and ketones.

Write an equation showing the hydrogenation of 3-pentanone.

SOLUTION

The product of the reduction of a ketone is a secondary alcohol, in this case, 3-pentanol.

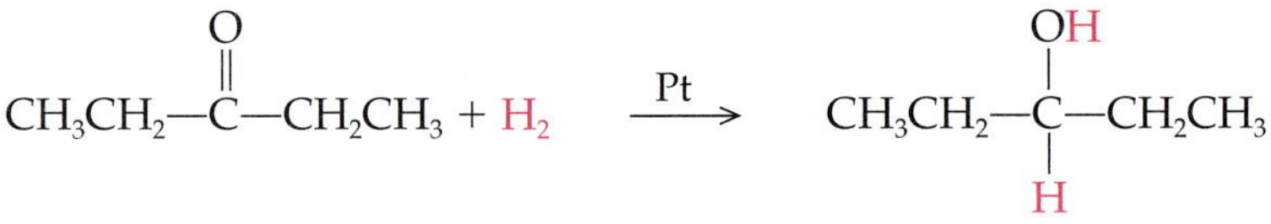

$$
\underset{\text{3-Pentanone}}{CH_3CH_2-\overset{\overset{\displaystyle O}{\|}}{C}-CH_2CH_3} + H_2 \xrightarrow{\text{Pt}} \underset{\text{3-Pentanol}}{CH_3CH_2-\overset{\overset{\displaystyle OH}{|}}{\underset{\underset{\displaystyle H}{|}}{C}}-CH_2CH_3}
$$

The hydrogenation of an aldehyde results in the production of a primary alcohol, as seen in the following equation showing the reduction of the aldehyde, butanal:

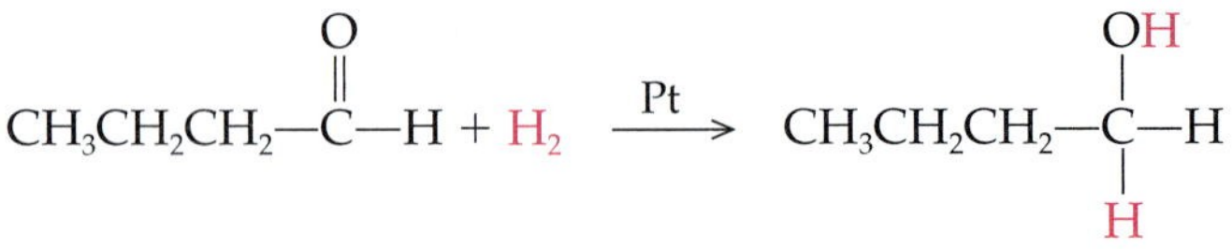

$$
\underset{\substack{\text{Butanal} \\ \text{(An aldehyde)}}}{CH_3CH_2CH_2-\overset{\overset{\displaystyle O}{\|}}{C}-H} + H_2 \xrightarrow{\text{Pt}} \underset{\substack{\text{1-Butanol} \\ \text{(A primary alcohol)}}}{CH_3CH_2CH_2-\overset{\overset{\displaystyle OH}{|}}{\underset{\underset{\displaystyle H}{|}}{C}}-H}
$$

EXAMPLE 12.15   **Writing an Equation Representing the Hydrogenation of an Aldehyde**

Write an equation showing the hydrogenation of 3-methylbutanal.

Continued—

SOLUTION

Recall that the reduction of an aldehyde results in the production of a primary alcohol, in this case, 3-methyl-1-butanol.

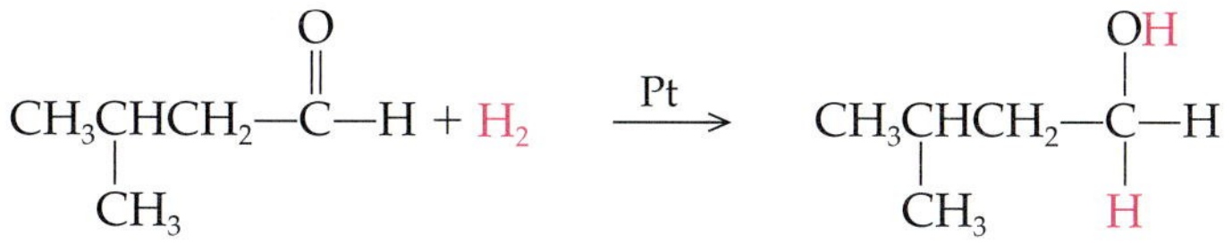

$$\underset{\underset{CH_3}{|}}{CH_3CHCH_2}-\overset{\overset{O}{\|}}{C}-H + H_2 \xrightarrow{Pt} \underset{\underset{CH_3}{|}}{CH_3CHCH_2}-\overset{\overset{OH}{|}}{\underset{\underset{H}{|}}{C}}-H$$

3-Methylbutanal   3-Methyl-1-butanol

QUESTION 12.33

Label each of the following as an oxidation or a reduction reaction.

a. Ethanal to ethanol
b. Benzoic acid to benzaldehyde
c. Cyclohexanone to cyclohexanol
d. 2-Propanol to propanone
e. 2,3-Butanedione (found in butter) to 2,3-butanediol

QUESTION 12.34

Write an equation for each of the reactions in Question 12.33.

A biological example of the reduction of a ketone occurs in the body, particularly during strenuous exercise when the lungs and circulatory system may not be able to provide enough oxygen to the muscles. Under these circumstances, lactate fermentation begins. In this reaction, the enzyme *lactate dehydrogenase* reduces pyruvate, the product of glycolysis, a pathway for the breakdown of glucose, into lactate. The source of hydrogen ions for this reaction is nicotinamide adenine dinucleotide (NADH), which is oxidized in the course of the reaction.

The role of the lactate fermentation in exercise is discussed in greater detail in Section 18.4.

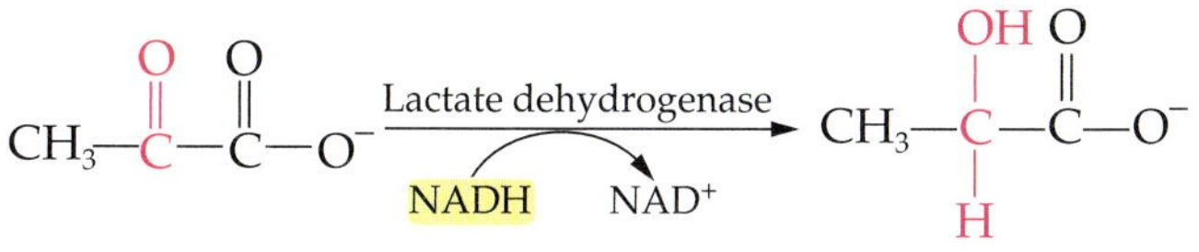

$$CH_3-\overset{\overset{O}{\|}}{C}-\overset{\overset{O}{\|}}{C}-O^- \xrightarrow[\underset{NADH \quad NAD^+}{}]{Lactate\ dehydrogenase} CH_3-\overset{\overset{OH}{|}}{\underset{\underset{H}{|}}{C}}-\overset{\overset{O}{\|}}{C}-O^-$$

Pyruvate   Lactate

SUMMARY OF REACTIONS

Preparation of Alcohols

Hydration of alkenes:

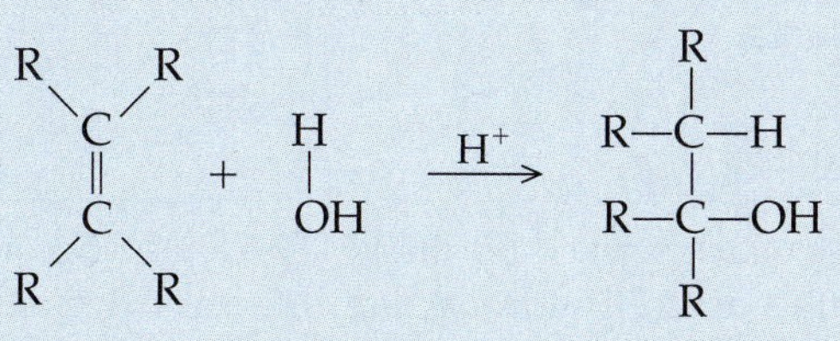

$$\underset{\underset{R}{|}}{\overset{\overset{R}{|}}{C}}=\underset{\underset{R}{|}}{\overset{\overset{R}{|}}{C}} + \underset{OH}{\overset{H}{|}} \xrightarrow{H^+} \begin{array}{c} R-\overset{|}{C}-H \\ R-\overset{|}{\underset{\underset{R}{|}}{C}}-OH \end{array}$$

Alkene   Water   Alcohol

Dehydration of Alcohols

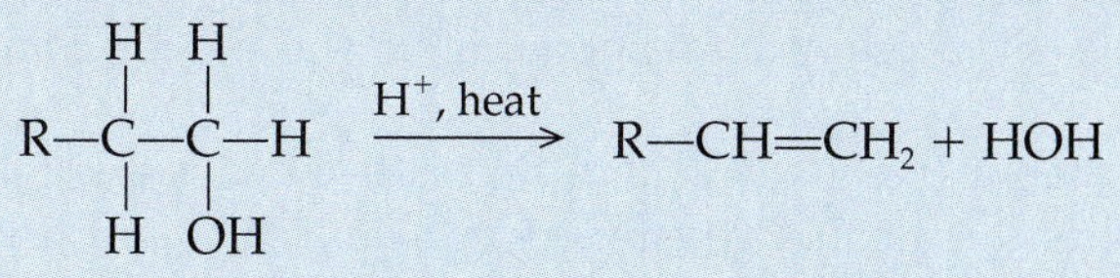

$$\underset{\underset{H}{|}\ \underset{OH}{|}}{\overset{\overset{H}{|}\ \overset{H}{|}}{R-C-C-H}} \xrightarrow{H^+,\ heat} R-CH=CH_2 + HOH$$

Alcohol   Alkene   Water

—Continued

Continued—

Oxidation Reactions

Oxidation of a primary alcohol:

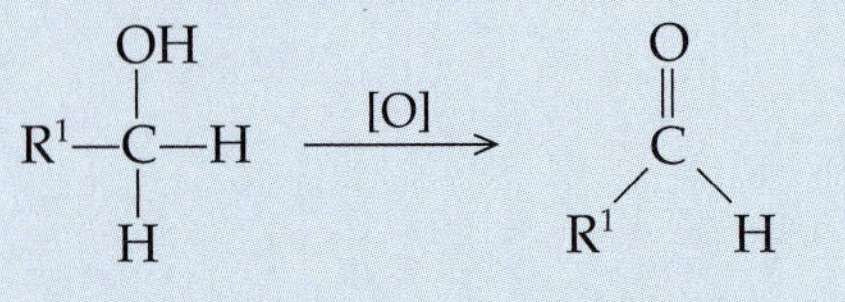

1° Alcohol   An aldehyde

Oxidation of a secondary alcohol:

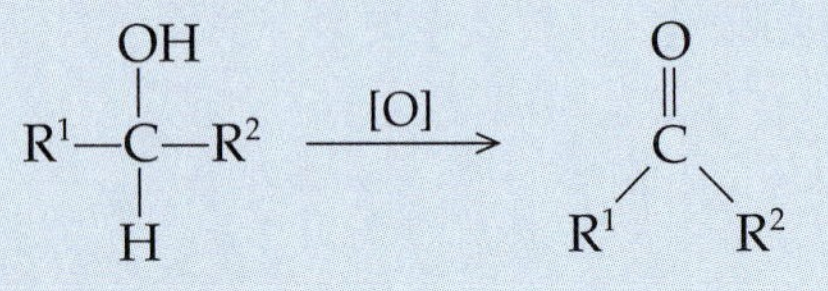

2° Alcohol   A ketone

Oxidation of a tertiary alcohol:

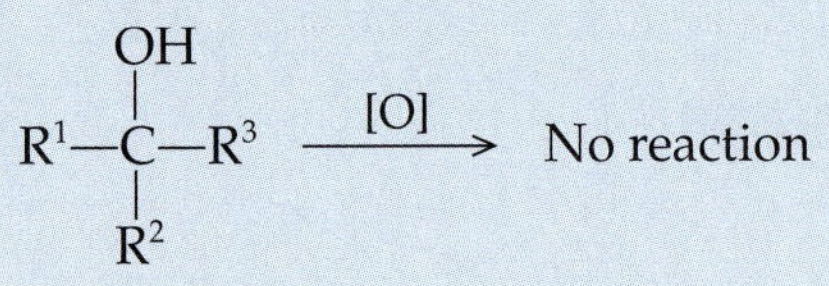

3° Alcohol

Dehydration Synthesis of an Ether

$$R^1\text{—OH} + R^2\text{—OH} \xrightarrow[\text{heat}]{H^+} R^1\text{—O—}R^2 + H_2O$$

Alcohol   Alcohol   Ether   Water

Oxidation of an Aldehyde

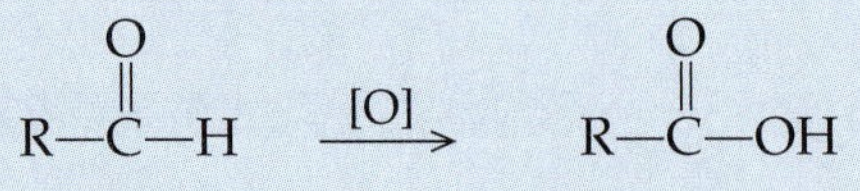

Aldehyde   Carboxylic acid

Reduction of Aldehydes and Ketones

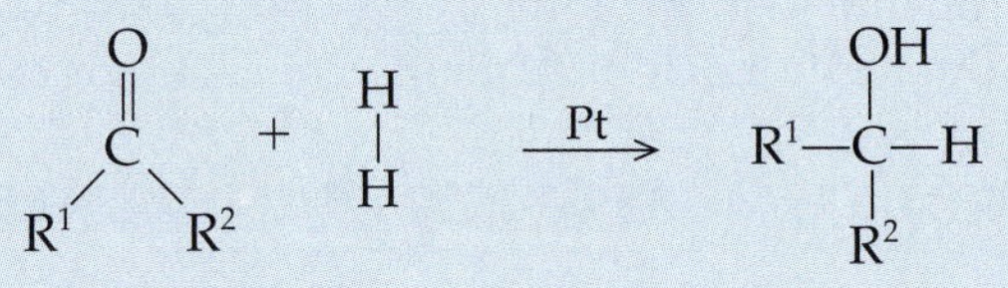

Aldehyde   Hydrogen   Alcohol
or Ketone

SUMMARY

12.1 Alcohols

Alcohols are characterized by the *hydroxyl group (—OH)* and have the general formula R—OH. They are very polar, owing to the polar hydroxyl group, and are able to form intermolecular hydrogen bonds. Because of hydrogen bonding between alcohol molecules, they have higher boiling points than hydrocarbons of comparable molecular weight. Smaller alcohols are very water soluble.

In the I.U.P.A.C. system, alcohols are named by determining the parent compound and replacing the *-e* ending with *-ol*. The chain is numbered to give the hydroxyl group the lowest possible number. Common names are derived from the alkyl group corresponding to the parent compound.

Methanol is a toxic alcohol that is used as a solvent. Ethanol is the alcohol consumed in beer, wine, and distilled liquors. Isopropanol is used as a disinfectant. Ethylene glycol (1,2-ethanediol) is used as antifreeze, and glycerol (1,2,3-propanetriol) is used in cosmetics and pharmaceuticals.

Alcohols may be classified as *primary, secondary,* or *tertiary,* depending on the number of alkyl groups attached to the *carbinol carbon*, the carbon bearing the hydroxyl group. A primary alcohol has a single alkyl group bonded to the *carbinol carbon*. Secondary and tertiary alcohols have two and three alkyl groups, respectively.

Alcohols can be prepared by *hydration* of alkenes or reduction of aldehydes and ketones. Alcohols can undergo *dehydration* to yield alkenes. Primary and secondary alcohols undergo oxidation reactions to yield aldehydes and ketones, respectively. Tertiary alcohols do not undergo oxidation.

12.2 Phenols

Phenols are compounds in which the hydroxyl group is attached to a benzene ring; they have the general formula Ar—OH. Many phenols are important as antiseptics and disinfectants.

12.3 Ethers

Ethers are characterized by the R—O—R functional group. Ethers are generally nonreactive but are extremely flammable. Diethyl ether was the first general anesthetic used in medical practice. It has since been replaced by Penthrane and Enthrane, which are less flammable.

12.4 Thiols

Thiols are characterized by the sulfhydryl group (—SH). The amino acid, cysteine, is a thiol that is extremely important for maintaining the correct shapes of proteins. Coenzyme A is a thiol that serves as a "carrier" of acetyl groups in biochemical reactions.

12.5 Aldehydes and Ketones

The *carbonyl group* ($\rangle$C$=$O) is characteristic of *aldehydes* and *ketones.* The carbonyl group and the two groups attached to it are coplanar. In ketones, the carbonyl carbon is attached to two carbon-containing groups, whereas in aldehydes, the carbonyl carbon is attached to at least one hydrogen; the second group attached to the carbonyl carbon in aldehydes may be another hydrogen or a carbon atom. Owing to the polar carbonyl group, aldehydes and ketones are polar compounds. Their boiling points are higher than those of comparable hydrocarbons but lower than those of comparable alcohols. Small aldehydes and ketones are reasonably soluble in water because of the hydrogen bonding between the carbonyl group and water molecules. Larger carbonyl-containing compounds are less polar and thus are more soluble in nonpolar organic solvents.

In the I.U.P.A.C. Nomenclature System, aldehydes are named by determining the parent compound and replacing the final *-e* of the parent alkane with *-al.* The chain is numbered beginning with the carbonyl carbon as carbon-1. Ketones are named by determining the parent compound and replacing the *-e* ending of the parent alkane with the *-one* suffix of the ketone family. The longest carbon chain is numbered to give the carbonyl carbon the lowest possible number. In the common system of nomenclature, substituted aldehydes are named as derivatives of the parent compound. Greek letters indicate the positions of substituents. Common names of ketones are derived by naming the alkyl groups bonded to the carbonyl carbon. These names are followed by the word *ketone.*

Many members of the aldehyde and ketone families are important as food and fragrance chemicals, medicinals, and agricultural chemicals. Methanal (formaldehyde) is used to preserve tissue. Ethanal causes the symptoms of a hangover and is oxidized to produce acetic acid commercially. Propanone is a useful and versatile solvent for organic compounds.

In the laboratory, aldehydes and ketones are prepared by the oxidation of alcohols. Oxidation of a primary alcohol produces an aldehyde; oxidation of a secondary alcohol yields a ketone. Tertiary alcohols do not react under these conditions. Aldehydes and ketones can be distinguished from one another on the basis of their ability to undergo oxidation reactions. The *Tollens' test* and *Benedict's test* are the most common such tests. Aldehydes are easily oxidized to carboxylic acids. Ketones do not undergo further oxidation reactions. Aldehydes and ketones are readily reduced to alcohols by *hydrogenation.*

KEY TERMS

alcohol (12.1)	ketone (12.5)
aldehyde (12.5)	oxidation (12.1)
Benedict's test (12.5)	phenol (12.2)
carbinol carbon (12.1)	primary (1°) alcohol (12.1)
carbonyl group (12.5)	reduction (12.1)
dehydration (12.1)	secondary (2°) alcohol
disulfide (12.4)	(12.1)
ether (12.3)	tertiary (3°) alcohol (12.1)
fermentation (12.1)	thiol (12.4)
hydration (12.1)	Tollens' test (12.5)
hydrogenation (12.5)	Zaitsev's rule (12.1)
hydroxyl group (12.1)	

QUESTIONS AND PROBLEMS

Alcohols: Structure and Physical Properties

Foundations

12.35 Describe the relationship between the water solubility of alcohols and their hydrocarbon chain length.

12.36 Explain the relationship between the water solubility of alcohols and the number of hydroxyl groups in the molecule.

Applications

12.37 Arrange the following compounds in order of increasing boiling point, beginning with the lowest:
  a. $CH_3CH_2CH_2CH_2CH_3$
  b. $CH_3CHCH_2CHCH_3$
          $\quad\quad\;\;|\quad\quad\;\;|$
          $\quad\quad\;$OH$\quad\;$OH
  c. $CH_3CHCH_2CH_2CH_3$
          $\quad\quad\;\;|$
          $\quad\quad\;$OH
  d. $CH_3CH_2CH_2$—O—CH_2CH_3

12.38 Why do alcohols have higher boiling points than alkanes?

Alcohols: Nomenclature

Foundations

12.39 Summarize the I.U.P.A.C. rules for naming alcohols.

12.40 Summarize the rules for determining the common names for alcohols.

Applications

12.41 Give the I.U.P.A.C. name for each of the following compounds:
  a. $CH_3CH_2CH_2CH_2CH_2CH_2CH_2OH$

  b. CH_3CHCH_3
          $\quad\quad\;|$
          $\quad\quad$OH

  c. CH_3—$\underset{\underset{CH_2OH}{|}}{\overset{\overset{CH_3}{|}}{C}}$—$CH_3$

12.42 Give the I.U.P.A.C. name for each of the following compounds:

a.  $CH_3CH_2CHCH_2CH_2CH_2OH$
    with Br on the third carbon

b. 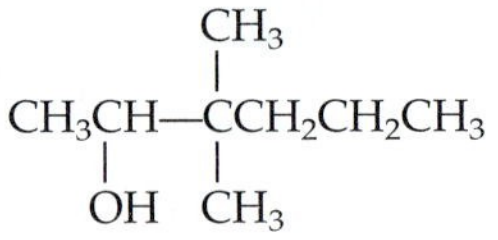 $CH_3CH—CCH_2CH_2CH_3$
    with CH_3 above, OH and CH_3 below

c. 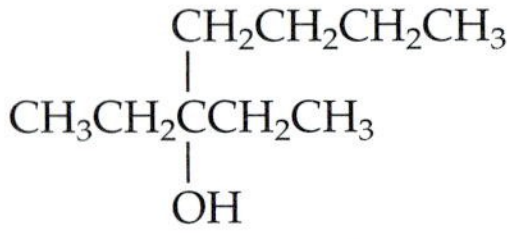 $CH_3CH_2CCH_2CH_3$
    with $CH_2CH_2CH_2CH_3$ above and OH below

12.43 Draw each of the following, using complete structural formulas:
a. 3-Hexanol
b. 1,2,3-Pentanetriol
c. 2-Methyl-2-pentanol
d. Cyclohexanol
e. 3,4-Dimethyl-3-heptanol

12.44 Draw the structure of each of the following compounds:
a. Pentyl alcohol
b. Isopropyl alcohol
c. Octyl alcohol
d. Propyl alcohol
e. 4-Methyl-2-hexanol
f. Isobutyl alcohol
g. 1,5-Pentanediol
h. 2-Nonanol

Medically Important Alcohols

12.45 What are the principal uses of methanol, ethanol, and isopropyl alcohol?

12.46 What is fermentation?

12.47 Why must fermentation products be distilled to produce liquors such as scotch?

12.48 If a bottle of distilled alcoholic spirits—for example, scotch whiskey—is labeled as 80 proof, what is the percentage of alcohol in the scotch?

Classification of Alcohols

Foundations

12.49 Define the term *carbinol carbon.*

12.50 Define the terms *primary, secondary,* and *tertiary alcohol,* and draw a general structure for each.

Applications

12.51 Classify each of the following as a 1°, 2°, or 3° alcohol:
a. 3-Methyl-1-butanol
b. 2-Methylcyclopentanol
c. *t*-Butyl alcohol
d. 1-Methylcyclopentanol
e. 2-Methyl-2-pentanol

12.52 Classify each of the following as a 1°, 2°, or 3° alcohol:
a. $CH_3CH_2CH_2CH_2CH_2CH_2CH_2OH$

b. 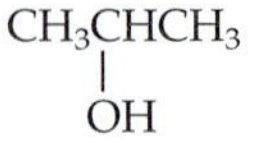 CH_3CHCH_3
    with OH below

c. 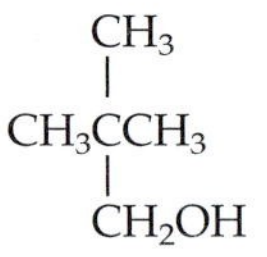 CH_3CCH_3
    with CH_3 above and CH_2OH below

d.  $CH_3CH_2CHCH_2CH_2CH_2OH$
    with Br above

e. 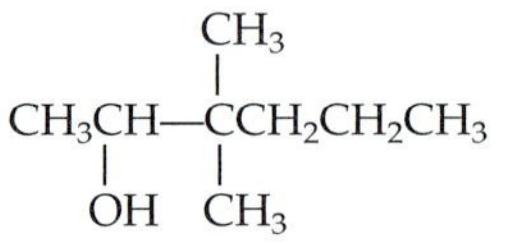 $CH_3CH—CCH_2CH_2CH_3$
    with CH_3 above, OH and CH_3 below

Reactions Involving Alcohols

Foundations

12.53 Write a general equation representing each of the following reactions:
a. The preparation of an alcohol by hydration of an alkene.
b. The dehydration of an alcohol.

12.54 Write general equations representing the oxidation of a 1° alcohol and a 2° alcohol.

Applications

12.55 Predict the products formed by the hydration of the following alkenes:
a. 1-Pentene
b. 2-Pentene
c. 3-Methyl-1-butene
d. 3,3-Dimethyl-1-butene

12.56 Draw the alkene products of the dehydration of the following alcohols:
a. 2-Pentanol
b. 3-Methyl-1-pentanol
c. 2-Butanol
d. 4-Chloro-2-pentanol
e. 1-Propanol

12.57 Which product(s) would result from the oxidation of each of the following alcohols with, for example, potassium permanganate? If no reaction occurs, write N.R.
a. 2-Butanol
b. 2-Methyl-2-hexanol
c. Cyclohexanol
d. 1-Methyl-1-cyclopentanol

12.58 Give the oxidation products of the following alcohols. If no reaction occurs, write N.R.

a. 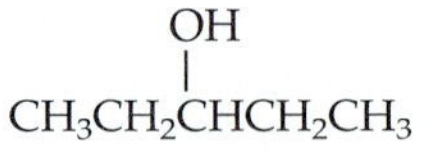 $CH_3CH_2CHCH_2CH_3$
    with OH above

b. $CH_3CH_2CH_2OH$

c. 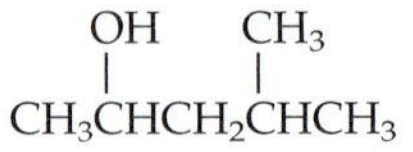 $CH_3CHCH_2CHCH_3$
    with OH and CH_3 above

d. 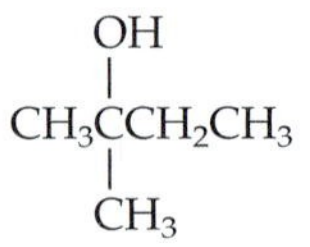 $CH_3CCH_2CH_3$
    with OH above and CH_3 below

e. 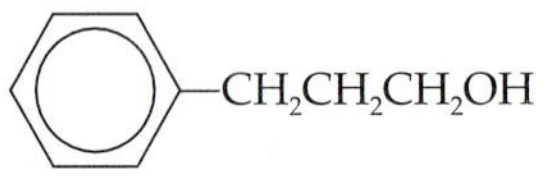 (benzene ring)—$CH_2CH_2CH_2OH$

Phenols

12.59 What are phenols? Describe the water solubility of phenols.

12.60 List some phenol compounds that are commonly used as antiseptics or disinfectants.

Ethers

12.61 Draw all of the alcohols and ethers of molecular formula $C_4H_{10}O$.

12.62 Name each of the isomers drawn for Problem 12.61.

12.63 Ethers may be prepared by the removal of water (dehydration) between two alcohols, as shown. Give the structure(s) of the ethers formed by the reaction of the following alcohol(s) under acidic conditions with heat.

    a. $2CH_3CH_2OH \longrightarrow$?

    b. $CH_3OH + CH_3CH_2OH \longrightarrow$?

12.64 Draw the structural formula for each of the following ethers:

    a. Methyl propyl ether

    b. 2-Methoxyoctane

    c. Diisopropyl ether

    d. 3-Ethoxypentane

Thiols

12.65 Cystine is an amino acid formed from the oxidation of two cysteine molecules to form a disulfide bond. The molecular formula of cystine is $C_6H_{12}O_4N_2S_2$. Draw the structural formula of cystine. (*Hint:* For the structure of cysteine, see Figure 16.3.)

12.66 Give the I.U.P.A.C. name for each of the following thiols.

    a. $CH_3CH_2CH_2$—SH

    b. $CH_3\underset{\underset{SH}{|}}{C}HCH_2CH_3$

    c. $\underset{\underset{SH}{|}}{C}H_2\underset{\underset{SH}{|}}{C}HCH_3$

Aldehydes and Ketones: Structure and Physical Properties

Foundations

12.67 Explain the relationship between carbon chain length and water solubility of aldehydes or ketones.

12.68 Explain the dipole–dipole interactions that occur between molecules containing carbonyl groups.

Applications

12.69 Draw intermolecular hydrogen bonding between ethanal and water.

12.70 Explain briefly why simple (containing fewer than five carbon atoms) aldehydes and ketones exhibit appreciable solubility in water.

Aldehydes and Ketones: Nomenclature

Foundations

12.71 Summarize the rules of the I.U.P.A.C. Nomenclature System for naming aldehydes.

12.72 Summarize the rules of the I.U.P.A.C. Nomenclature System for naming ketones.

Applications

12.73 Use the I.U.P.A.C. Nomenclature System to name each of the following compounds:

    a. $CH_3\overset{\overset{O}{||}}{C}CH_2CH_3$

    b. $H\overset{\overset{O}{||}}{C}\underset{\underset{CH_2CH_2CH_3}{|}}{C}HCH_2CH_3$

    c. $CH_3CH_2CH_2\overset{\overset{O}{||}}{C}H$

    d. $CH_3\underset{\underset{CH_3}{|}}{\overset{\overset{Br}{|}}{C}}CH_2CH_2\overset{\overset{O}{||}}{C}H$

12.74 Give the I.U.P.A.C. name for each of the following compounds:

    a. $CH_3\underset{\underset{Br}{|}}{C}HCH_2\overset{\overset{O}{||}}{C}H$

    b. $CH_3\underset{\underset{Cl}{|}}{\overset{\overset{CH_3}{|}}{C}}CH_2\overset{\overset{O}{||}}{C}CH_2CH_2CH_3$

    c. $CH_3\overset{\overset{O}{||}}{C}CH_2\underset{\underset{CH_2CH_3}{|}}{\overset{\overset{CH_2CH_3}{|}}{C}}CH_2CH_3$

    d. $CH_3\overset{\overset{O}{||}}{C}CH_2\underset{\underset{Cl}{|}}{C}HCH_2CH_3$

12.75 Draw the structure of each of the following compounds:

    a. 3-Hydroxybutanal

    b. 2-Methylpentanal

    c. 4-Bromohexanal

    d. 3-Iodopentanal

    e. 2-Hydroxy-3-methylheptanal

12.76 Draw the structure of each of the following compounds:

    a. Dimethyl ketone

    b. Methyl propyl ketone

    c. Ethyl butyl ketone

    d. Diisopropyl ketone

Important Aldehydes and Ketones

12.77 Why is acetone a good solvent for many organic compounds?

12.78 List several aldehydes and ketones that are used as food or fragrance chemicals.

Reactions Involving Aldehydes and Ketones

Foundations

12.79 Indicate whether each of the following statements is true or false.

    a. Aldehydes and ketones can be oxidized to produce carboxylic acids.

   b. Oxidation of a primary alcohol produces an aldehyde.
   c. Oxidation of a tertiary alcohol produces a ketone.
   d. Alcohols can be produced by the oxidation of an aldehyde or ketone.

12.80 Indicate whether each of the following statements is true or false.
   a. Ketones, but not aldehydes, react in the Tollens' silver mirror test.
   b. Oxidation of 1-butanol produces a mixture of butanal and butanone.
   c. Tertiary alcohols are oxidized to become carboxylic acids.
   d. Most aldehydes are readily oxidized to become carboxylic acids.

Applications

12.81 Draw the structures of each of the following compounds. Then draw and name the product that you would expect to produce by oxidizing each of these alcohols:
   a. 2-Butanol
   b. 2-Methyl-1-propanol
   c. Cyclopentanol
   d. 2-Methyl-2-propanol
   e. 2-Nonanol
   f. 1-Decanol

12.82 An unknown has been determined to be one of the following three compounds:

$$CH_3CH_2CCH_2CH_3$$

3-Pentanone

$$CH_3CH_2CH_2CH_2CH$$

Pentanal

$$CH_3CH_2CH_2CH_2CH_3$$

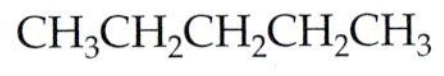

Pentane

The unknown is fairly soluble in water and produces a silver mirror when treated with the silver ammonia complex. A red precipitate appears when it is treated with the Benedict's reagent. Which of the compounds is the correct structure for the unknown? Explain your reasoning.

12.83 An aldehyde can be oxidized to produce a carboxylic acid. Draw the carboxylic acid that would be produced by the oxidation of each of the following aldehydes:
   a. Methanal
   b. Ethanal
   c. Propanal
   d. Butanal

12.84 An alcohol can be oxidized to produce an aldehyde or a ketone. What aldehyde or ketone is produced by the oxidation of each of the following alcohols?
   a. Methanol
   b. 1-Propanol
   c. 3-Pentanol
   d. 2-Methyl-2-butanol

FOR FURTHER UNDERSTANDING

1. You are provided with two solvents: water (H_2O) and hexane ($CH_3CH_2CH_2CH_2CH_2CH_3$). You are also provided with two biological molecules whose structures are shown here:

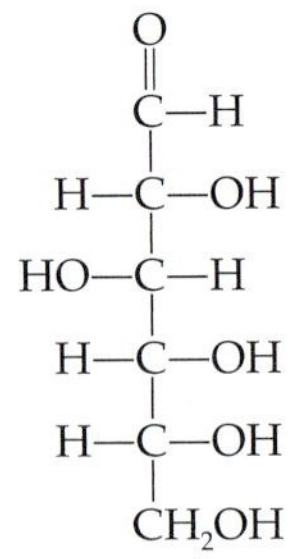

Predict which biological molecule would be more soluble in water and which would be more soluble in hexane. Defend your prediction. Design a careful experiment to test your hypothesis.
   Consider the digestion of dietary molecules in the digestive tract. Which of the two biological molecules shown in this problem would be more easily digested under the conditions present in the digestive tract?

2. Cholesterol is an alcohol and a steroid (Chapter 15). Diets that contain large amounts of cholesterol have been linked to heart disease and atherosclerosis, hardening of the arteries. The narrowing of the artery, caused by plaque buildup, is very apparent. Cholesterol is directly involved in this buildup. Describe the various functional groups and principal structural features of the cholesterol molecule. Would you use a polar or nonpolar solvent to dissolve cholesterol? Explain your reasoning.

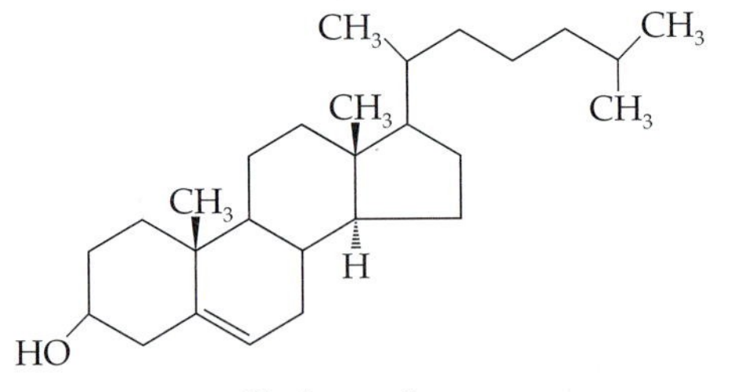

Cholesterol

3. An unknown compound A is known to be an alcohol with the molecular formula $C_4H_{10}O$. When dehydrated, compound A gave only one alkene product, C_4H_8, compound B. Compound A could not be oxidized. What are the identities of compound A and compound B?

4. Sulfides are the sulfur analogs of ethers, that is, ethers in which oxygen has been substituted by a sulfur atom. They are named in an analogous manner to the ethers with the term *sulfide* replacing *ether*. For example, CH_3—S—CH_3 is dimethyl sulfide. Draw the sulfides that correspond to the following ethers and name them:
   a. Diethyl ether
   b. Methyl propyl ether
   c. Dibutyl ether
   d. Ethyl phenyl ether

5. Dimethyl sulfoxide (DMSO) has been used by many sports enthusiasts as a linament for sore joints; it acts as an

anti-inflammatory agent and a mild analgesic (pain killer). However, it is no longer recommended for this purpose because it carries toxic impurities into the blood. DMSO is a sulfoxide—it contains the $S{=}O$ functional group. DMSO is prepared from dimethyl sulfide by mild oxidation, and it has the molecular formula C_2H_6SO. Draw the structure of DMSO.

6. Design a synthesis for each of the following compounds, using any inorganic reagent of your choice and any hydrocarbon or alkyl halide of your choice:
   a. Octanal
   b. Cyclohexanone
   c. 2-Phenylethanoic acid

7. When alkenes react with ozone, O_3, the double bond is cleaved, and an aldehyde and/or a ketone is produced. The reaction, called *ozonolysis*, is shown in general as:

$$\begin{array}{c}\diagup\\C{=}C\\\diagup\end{array} + O_3 \longrightarrow \begin{array}{c}\\C{=}O + O{=}C\\\end{array}$$

Predict the ozonolysis products that are formed when each of the following alkenes is reacted with ozone:
   a. 1-Butene
   b. 2-Hexene
   c. *cis*-3,6-Dimethyl-3-heptene

13

OUTLINE

LEARNING GOALS

1. Write structures and describe the physical properties of carboxylic acids.

2. Determine the common and I.U.P.A.C. names of carboxylic acids.

3. Describe the biological, medical, or environmental significance of several carboxylic acids.

4. Write equations representing acid-base reactions of carboxylic acids.

5. Write structures and describe the physical properties of esters.

6. Determine the common and I.U.P.A.C. names of esters.

7. Write equations representing the synthesis and hydrolysis of an ester.

8. Define the term saponification and describe how soap works in the emulsification of grease and oil.

9. Classify amines as primary, secondary, or tertiary.

10. Describe the physical properties of amines.

11. Draw and name simple amines using common and systematic nomenclature.

12. Write equations representing the basicity and neutralization of amines.

13. Describe the structure of quaternary ammonium salts and discuss their use as antiseptics and disinfectants.

14. Describe the physical properties of amides.

15. Draw the structure and write the common and I.U.P.A.C. names of amides.

16. Write equations representing the hydrolysis of amides.

Carboxylic Acids, Esters, Amines, and Amides

Carboxylic acids and their derivatives the esters are a common part of our daily lives. You may encounter many of them in the Chef's Salad you have for lunch. The two-carbon carboxylic acid, acetic acid in aqueous solution, or vinegar, adds that tartness to the Italian dressing on your Chef's Salad. Propionic acid, with three carbons, gives that tangy flavor to the Swiss cheese on your salad.

Long-chain carboxylic acids are called fatty acids and they form esters when they react with an alcohol, such as glycerol. The olive oil in your salad dressing and the solid fat in the meat and cheese on your salad are examples of such triglycerides.

Perhaps you had a fruit salad for breakfast. If so, you may have enjoyed a number of sweet, fruity-tasting esters, such as 3-methylbutyl ethanoate (banana) or methyl thiobutanoate (strawberries).

Some carboxylic acids are key elements in exercise physiology and others are used in the treatment of diseases. For instance, lactic acid is a product of our metabolism that builds up in muscles and blood when we are exercising so strenuously that we cannot provide sufficient oxygen to working muscle. We will learn more about the lactate fermentation and its role in exercising muscle in Chapter 18. A four-carbon carboxylic acid, butanoic acid (butyric acid), is being used to treat sickle cell anemia. Although we don't yet understand how, this compound "turns on" the gene for fetal hemoglobin, protecting some patients from the harmful effects of this genetic disorder. To learn more about this therapy, see A Medical Connection: Wake Up Sleeping Gene online at www.mhhe.com/denniston.

In this chapter, we will also add a new element to the structure of organic molecules. That element is nitrogen, the fourth most common atom in living systems. It is an important component of the structure of the nucleic acids, DNA and RNA, which are the molecules that carry the genetic information for living systems. It is also essential to the structure and function of proteins and is found in many plant products, including nicotine and opiate drugs.

The salad in this figure is part of a healthy diet. Using the Internet, make a list of the organic molecules found in these foods and their role in diet and health.

13.1 Carboxylic Acids

Carboxylic acids (Figure 13.1a) have the following general structure:

$$\underset{\text{Aromatic carboxylic acid}}{Ar-\overset{\displaystyle O}{\overset{\|}{C}}-OH} \qquad\qquad \underset{\text{Aliphatic carboxylic acid}}{R-\overset{\displaystyle O}{\overset{\|}{C}}-OH}$$

They are characterized by the carboxyl group, shown in red, which may also be written in condensed form as —COOH. The name carboxylic acid describes this family of compounds quite well. It is taken from the terms carbonyl and hydroxyl, the two structural units that make up the carboxyl group. The word acid in the name tells us one of the more important properties of these molecules: they dissociate in water to release protons. Thus, they are acids.

Structure and Physical Properties

The **carboxyl group** consists of two very polar functional groups, the carbonyl group and the hydroxyl group. Thus, **carboxylic acids** are very polar compounds. In addition, they can hydrogen bond to one another and to molecules of a polar solvent such as water. As a result of intermolecular hydrogen bonding, they boil at higher temperatures than aldehydes, ketones, or alcohols of comparable molecular weight.

As with alcohols, smaller carboxylic acids are soluble in water (Figure 13.2). However, solubility falls off dramatically as the carbon content of the carboxylic acid increases because the molecules become more hydrocarbonlike and less polar. For example, acetic acid (the two-carbon carboxylic acid found in vinegar) is completely soluble in water, but hexadecanoic acid (a sixteen-carbon carboxylic acid found in palm oil) is insoluble in water.

The lower molecular weight carboxylic acids have sharp, sour tastes and unpleasant aromas. Formic acid, HCOOH, is used as a chemical defense by ants and causes the burning sensation of the ant bite. Acetic acid, CH_3COOH, is found in vinegar; propionic acid, CH_3CH_2COOH, is responsible for the tangy flavor of Swiss cheese; and butyric acid, $CH_3CH_2CH_2COOH$, causes the stench associated with rancid butter and gas gangrene.

The longer chain carboxylic acids are generally called **fatty acids** and are important components of biological membranes and triglycerides, the major lipid storage form in the body.

QUESTION 13.1

Assuming that each of the following pairs of molecules has the same carbon chain length, which member of each of the following pairs has the lower boiling point?

a. a carboxylic acid or a ketone
b. a ketone or an alcohol
c. an alcohol or an alkane

Learning Goal

1 Learning Goal

Write structures and describe the physical properties of carboxylic acids.

Figure 13.1

Ball-and-stick models of (a) a carboxylic acid, propanoic acid, and (b) an ester, methyl ethanoate.

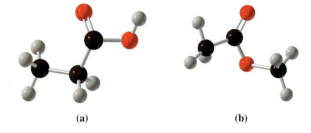

(a)   (b)

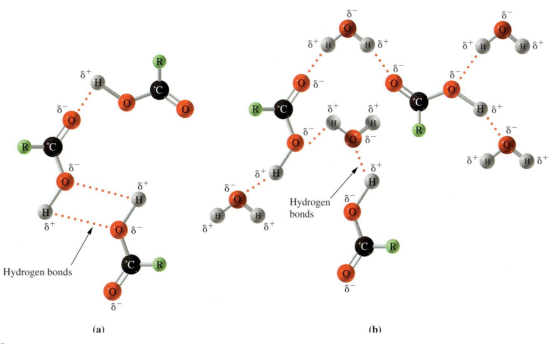

(a)   (b)

Figure 13.2

Hydrogen bonding (a) between carboxylic acid molecules and (b) between carboxylic acid molecules and water molecules.

QUESTION 13.2

Assuming that each of the following pairs of molecules has the same carbon chain length, which member of each of the following pairs has the lower boiling point?

a. an ether or an aldehyde
b. an aldehyde or a carboxylic acid
c. an ether or an alcohol

QUESTION 13.3

Why would you predict that a carboxylic acid would be more polar and have a higher boiling point than an aldehyde of comparable molecular weight?

QUESTION 13.4

Why would you predict that a carboxylic acid would be more polar and have a higher boiling point than an alcohol of comparable molecular weight?

Nomenclature

In the I.U.P.A.C. Nomenclature System, carboxylic acids are named according to the following set of rules:

- Determine the parent compound, the longest continuous carbon chain bearing the carboxyl group.
- Number the chain so that the carboxyl carbon is carbon-1.

2 **Learning Goal**

Determine the common and I.U.P.A.C. names of carboxylic acids.

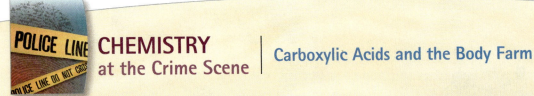

CHEMISTRY at the Crime Scene | Carboxylic Acids and the Body Farm

Dr. Arpad Vass of Oak Ridge National Laboratory explains his unusual vocation in the following way, "Each year there are about 90,000 homicides in this country and for every one, you need to know when the person died. Our research can help answer the where and when, and that should help law enforcement officials solve crimes." Time and location of death are key pieces of information in a criminal investigation. Whether a suspect is convicted or goes free may rest on an accurate estimate of the time since death (TSD). Dr. Vass and graduate student Jennifer Love study decomposing bodies on a research plot in Tennessee. Although its official name is University of Tennessee Anthropological Research Facility, it has been called the "Body Farm" since Patricia Cornwall published a novel of that name in 1994.

Vass and Love are trying to develop an instrument that can sample the air around a corpse to determine the concentrations of certain carboxylic acids, also referred to as volatile fatty acids. What they envision is a "tricorder" like device that will sample the air for valeric acid, propionic acid, and the straight and branched types of butyric acid. Because the ratio of these carboxylic acids changes in a predictable way following death, measurement of their relative concentrations, from air or from the soil under the corpse, could provide an accurate TSD.

The research is done by taking daily measurements of the air around decomposing bodies at the "Body Farm." At death, the proteins and lipids of the body begin to break down. This decomposition produces, among other substances, the carboxylic acids being studied. The bodies may be placed in various environments so that the researchers can study the effects of different temperature and moisture levels on the process of decomposition. The more that is learned about the amounts and types of organic compounds produced during decomposition under different conditions, the more precise the TSD determinations will be. In fact, Vass considers that these processes are a chemical "clock" and is working to refine the accuracy of that clock so that, eventually, TSD may be measured accurately in hours, rather than days.

FOR FURTHER UNDERSTANDING

Draw the structures of the carboxylic acids described in this article and write their I.U.P.A.C. names.

The bones of a murdered 9-year-old boy were found lying on his father's property. Immediately the father was suspected of having killed the child. A sample of the soil under the boy's bones was sent to Dr. Vass for analysis. No volatile fatty acids were found in the soil sample. What can you conclude from these data?

- Replace the -e ending of the parent alkane with the suffix -oic acid. If there are two carboxyl groups, the suffix -dioic acid is used.
- Name and number substituents in the usual way.

The following examples illustrate the naming of carboxylic acids with one carboxyl group:

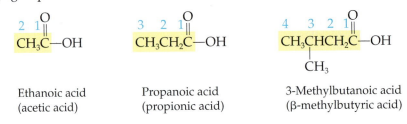

Ethanoic acid
(acetic acid)

Propanoic acid
(propionic acid)

3-Methylbutanoic acid
(β-methylbutyric acid)

The following examples illustrate naming carboxylic acids with two carboxyl groups:

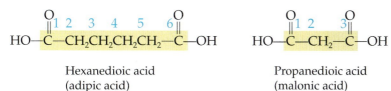

Hexanedioic acid
(adipic acid)

Propanedioic acid
(malonic acid)

EXAMPLE 13.1 Use the I.U.P.A.C. Nomenclature System to Name a Carboxylic Acid

Name the following carboxylic acid using the I.U.P.A.C. Nomenclature System.

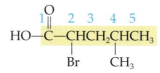

2 Learning Goal
Determine the common and I.U.P.A.C. names of carboxylic acids.

SOLUTION

Parent compound: pentane (becomes pentanoic acid)
Position of —COOH: carbon-1 (*must be!*)
Substituents: 2-bromo and 4-methyl
Name: 2-Bromo-4-methylpentanoic acid

QUESTION 13.5

Determine the I.U.P.A.C. name for each of the following structures. Remember that —COOH is an alternative way to represent the carboxyl group.

a. 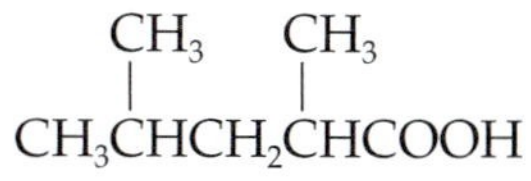 CH₃CHCH₂CHCOOH (with CH₃ and CH₃ substituents)

b. 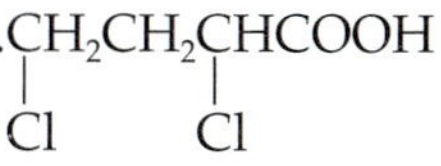 CH₂CH₂CHCOOH (with Cl and Cl substituents)

QUESTION 13.6

Write the structure for each of the following carboxylic acids.

a. 2,3-Dihydroxybutanoic acid   b. 2-Bromo-3-chloro-4-methylhexanoic acid

---

As we have seen so often, the use of common names, rather than systematic names, still persists. Often these names have evolved from the source of a given compound. This is certainly true of the carboxylic acids. Table 13.1 shows the I.U.P.A.C. and common names of several carboxylic acids, as well as their sources and the Latin or Greek words that gave rise to the common names. Not only are the prefixes different from those used in the I.U.P.A.C. system, the suffix is different as well. Common names end in *-ic acid* rather than *-oic acid*.

In the common system of nomenclature, substituted carboxylic acids are named as derivatives of the parent compound (see Table 13.1). Greek letters are used to indicate the position of the substituent. The carbon atom bonded to the carboxyl group is the α-carbon, the next is the β-carbon, and so on.

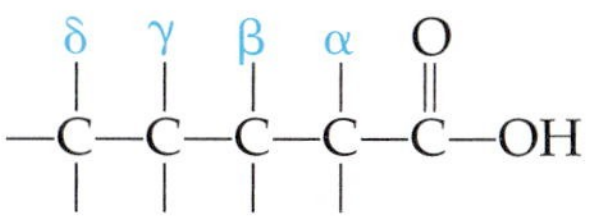

2 Learning Goal
Determine the common and I.U.P.A.C. names of carboxylic acids.

$$\overset{\delta}{-C}-\overset{\gamma}{C}-\overset{\beta}{C}-\overset{\alpha}{C}-\overset{O}{C}-OH$$

TABLE 13.1 Names and Sources of Some Common Carboxylic Acids

Name	Structure	Source	Root
Formic acid (methanoic acid)	HCOOH	Ants	L: *formica*, ant
Acetic acid (ethanoic acid)	CH$_3$COOH	Vinegar	L: *acetum*, vinegar
Propionic acid (propanoic acid)	CH$_3$CH$_2$COOH	Swiss cheese	Gk: *protos*, first; *pion*, fat
Butyric acid (butanoic acid)	CH$_3$(CH$_2$)$_2$COOH	Rancid butter	L: *butyrum*, butter
Valeric acid (pentanoic acid)	CH$_3$(CH$_2$)$_3$COOH	Valerian root	
Caproic acid (hexanoic acid)	CH$_3$(CH$_2$)$_4$COOH	Goat fat	L: *caper*, goat
Caprylic acid (octanoic acid)	CH$_3$(CH$_2$)$_6$COOH	Goat fat	L: *caper*, goat
Capric acid (decanoic acid)	CH$_3$(CH$_2$)$_8$COOH	Goat fat	L: *caper*, goat
Palmitic acid (hexadecanoic acid)	CH$_3$(CH$_2$)$_{14}$COOH	Palm oil	
Stearic acid (octadecanoic acid)	CH$_3$(CH$_2$)$_{16}$COOH	Tallow (beef fat)	Gk: *stear*, tallow

Note: I.U.P.A.C. names are shown in parentheses.

Some examples of common names are

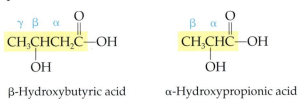

β-Hydroxybutyric acid   α-Hydroxypropionic acid

EXAMPLE 13.2 **Naming Carboxylic Acids Using the Common System of Nomenclature**

Write the common name for each of the following carboxylic acids.

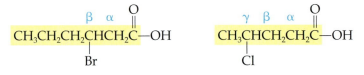

SOLUTION

Parent compound:	caproic acid	valeric acid
Substituents:	β-bromo	γ-chloro
Name:	β-Bromocaproic acid	γ-Chlorovaleric acid

QUESTION 13.7

Provide the common name for each of the following molecules. Keep in mind that the carboxyl group can be represented as —COOH.

a. 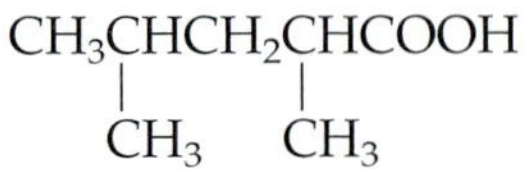 CH$_3$CHCH$_2$CHCOOH
         | |
       CH$_3$ CH$_3$

b. 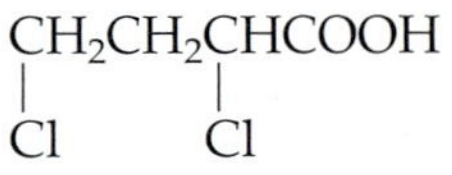 CH$_2$CH$_2$CHCOOH
     | |
    Cl   Cl

QUESTION 13.8

Provide the common name for each of the following molecules.

a. 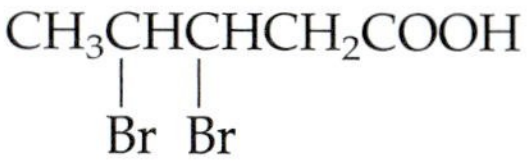 CH₃CHCHCH₂COOH
     | |
    Br Br

a. $CH_3CHCHCH_2COOH$ with Br, Br substituents

b. 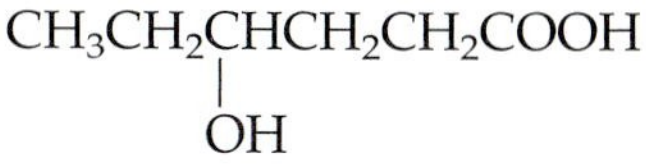 $CH_3CH_2CHCH_2CH_2COOH$ with OH substituent

---

Benzoic acid is the simplest aromatic carboxylic acid.

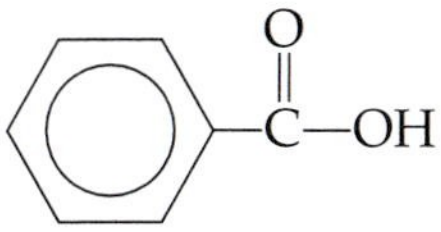

Benzoic acid

When naming aromatic carboxylic acids, assume that the carboxyl group is bonded to carbon-1. Then number the ring to give the lowest numbers to the remaining groups.

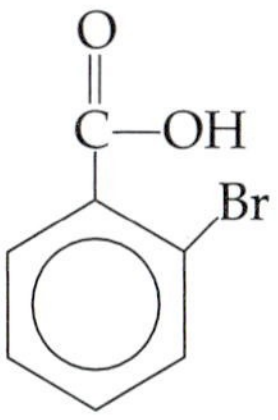

2-Bromobenzoic acid 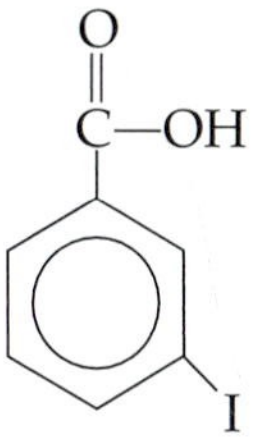   3-Iodobenzoic acid

QUESTION 13.9

Draw structures for each of the following compounds.

a. 2,4,6-Tribromobenzoic acid
b. 2,3-Dibromobenzoic acid

QUESTION 13.10

Draw structures for each of the following compounds.

a. 3,4-Dichlorobenzoic acid
b. 2-Bromo-3-chlorobenzoic acid

Some Important Carboxylic Acids

As Table 13.1 shows, many carboxylic acids occur in nature. Fatty acids can be isolated from a variety of sources including palm oil, coconut oil, butter, milk, lard, and tallow (beef fat). More complex carboxylic acids are also found in a variety of foodstuffs. For example, citric acid is found in citrus fruits and is often used to give the sharp taste to sour candies. It is also added to foods as a preservative and antioxidant. Adipic acid (hexanedioic acid) gives tartness to soft drinks and helps to retard spoilage.

3   **Learning Goal**

Describe the biological, medical, or environmental significance of several carboxylic acids.

AN ENVIRONMENTAL Perspective | Garbage Bags from Potato Peels?

One problem facing society is its enormous accumulation of trash. To try to control the mountains of garbage that we produce, many institutions and towns practice recycling of aluminum, paper, and plastics. One problem that remains, however, is the plastic trash bag. We stuff the trash bag full of biodegradable garbage and bury it in a landfill, but soil bacteria can't break down the plastic to get to the biodegradable materials inside. Imagine a twenty-fourth century archeologist excavating one of these monuments to our society!

The good news is that laboratory research and products of bacterial metabolism are providing new materials that have the properties of plastics, but are readily biodegradable. For instance, sheets of plastic can be made by making polymers of lactic acid, which is a natural carboxylic acid produced by fermentation of sugars, particularly in milk and working muscle. Because many common soil bacteria can break down polylactic acid (PLA), trash bags made from this polymer would be quickly broken down in landfill soil.

Making plastic from lactic acid requires a huge supply of this carboxylic acid. As it turns out, we can produce an enormous quantity of lactic acid from garbage. When French fries are produced, nearly half of the potato is wasted. That amounts to about ten billion pounds of potato waste each year. When cheese is made, the curds are separated from the whey, and several billion gallons of whey are poured down the drain each year. Potato waste and whey can easily be broken down to produce glucose, which, in turn, can be converted into lactic acid used to make biodegradable plastics. PLA plastics have been available since the early 1990s and have been used successfully for sutures, medical implants, and drug delivery systems.

Biodegradable plastic can be made from garbage, such as potato peels left over from making French fries.

$$n \; H_3C-\overset{\overset{\displaystyle H}{|}}{\underset{\underset{\displaystyle OH}{|}}{C}}-\overset{\overset{\displaystyle O}{\|}}{C}-OH \longrightarrow \left[-O-\overset{\overset{\displaystyle CH_3}{|}}{\underset{\underset{\displaystyle H}{|}}{C}}-\overset{\overset{\displaystyle O}{\|}}{C}-O-\overset{\overset{\displaystyle CH_3}{|}}{\underset{\underset{\displaystyle H}{|}}{C}}-\overset{\overset{\displaystyle O}{\|}}{C}-O- \right]_n$$

Lactic acid   Polylactic acid (PLA)

Nature has provided other biodegradable plastics that are produced by bacteria. The most common of these in nature is polyhydroxybutyrate (PHB) made by the bacterium *Alcaligenes eutrophus*. PHB is a homopolymer (made up of a

Bacteria in milk produce lactic acid as a product of fermentation of sugars. Lactic acid contributes a tangy flavor to yogurt and buttermilk. It is also used as a food preservative to lower the pH to a level that retards the microbial growth that causes food spoilage. Lactic acid is produced in muscle cells when an individual is exercising strenuously. If the level of lactic acid in the muscle and bloodstream becomes high enough, the muscle can't continue to work.

Remember, the carboxyl group can be represented as —COOH, CO₂H, or

$$-\overset{\overset{\displaystyle O}{\|}}{C}-OH$$

COOH
|
H—C—H
|
HO—C—COOH
|
H—C—H
|
COOH

Citric acid
(citrus fruit)

COOH
|
H—C—OH
|
CH₃

Lactic acid
(yogurt)

COOH
|
H—C—H
|
H—C—H
|
H—C—H
|
H—C—H
|
COOH

Adipic acid
(beet juice)

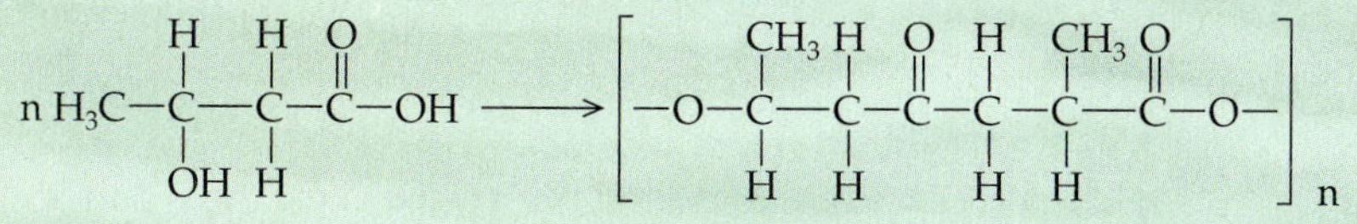

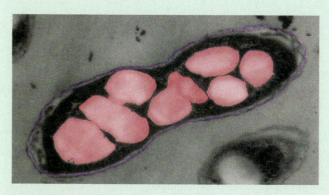

β-Hydroxybutyric acid → Polyhydroxybutyric acid (PHB)

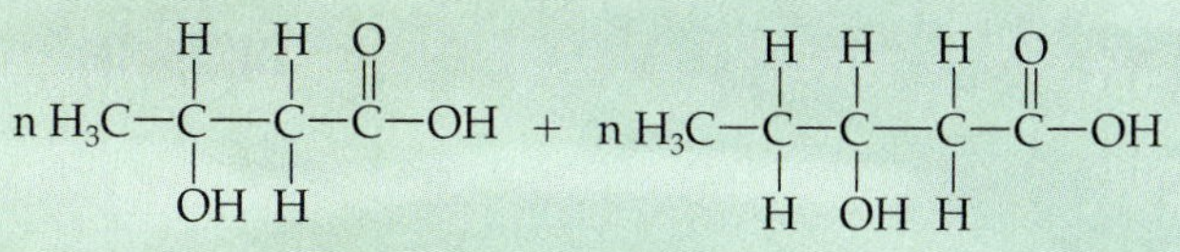

β-Hydroxybutyric acid + β-Hydroxyvaleric acid

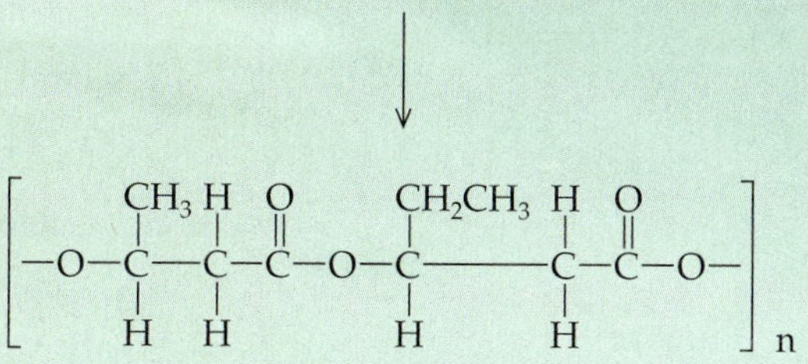

Biopol—a heteropolymer

single monomer) of 3-hydroxybutanoic acid (β-hydroxybu-tyric acid).

Unfortunately, this natural polymer does not have physical properties that would allow it to be a commercially successful plastic. This led to interest in making heteropolymers, polymers composed of two or more monomers. One such polymer with useful properties is a heteropolymer of β-hydroxybutyric acid and β-hydroxyvaleric acid, which has been given the name Biopol.

Biopol has properties that make it commercially useful, and it is completely broken down into carbon dioxide and water by microorganisms in the soil. Thus, it is completely biodegradable. Because the bacteria produced a low yield of Biopol, scientists produced transgenic plants in hopes that they would produce high yields of the polymer. This approach also encountered problems.

For the time being, biodegradable plastics cannot outcompete their nonbiodegradable counterparts. Future research and development will be required to reduce the cost of commercial production and fulfill the promise of an "environmentally friendly" garbage bag.

Computer colorized inclusions of polyhydroxybutyrate in *Alcaligenes eutrophus.*

FOR FURTHER UNDERSTANDING

Why are biodegradable plastics useful as sutures?

Two hydrogen atoms are lost in each reaction that adds a carboxylic acid to the polymers described here. What type of chemical reaction is this?

Reactions Involving Carboxylic Acids

Preparation of Carboxylic Acids

Many small carboxylic acids are prepared on a commercial scale by oxidation of the corresponding alcohol or aldehyde. The general reaction is

These reactions were discussed in Sections 12.1 and 12.5.

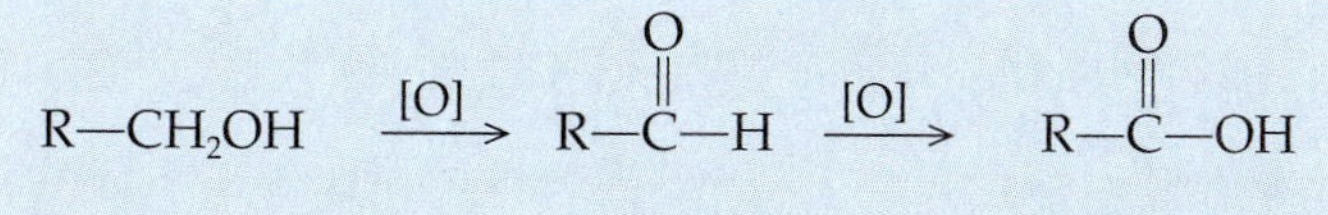

$$R-CH_2OH \xrightarrow{[O]} R-\overset{O}{\overset{\parallel}{C}}-H \xrightarrow{[O]} R-\overset{O}{\overset{\parallel}{C}}-OH$$

Primary alcohol   Aldehyde   Carboxylic acid

EXAMPLE 13.3 Writing Equations for the Oxidation of a Primary Alcohol to a Carboxylic Acid

Write an equation showing the oxidation of 1-propanol to propanoic acid.

— *Continued*—

EXAMPLE 13.3 —*Continued*

SOLUTION

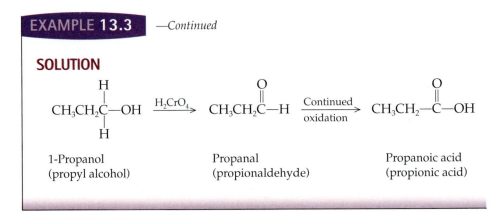

1-Propanol
(propyl alcohol)

Propanal
(propionaldehyde)

Propanoic acid
(propionic acid)

Acid–Base Reactions

4 Learning Goal
Write equations representing acid-base reactions of carboxylic acids.

Carboxylic acids behave as acids because they are proton donors. They are weak acids that dissociate to form carboxylate ions and hydrogen ions, as shown in the following equation:

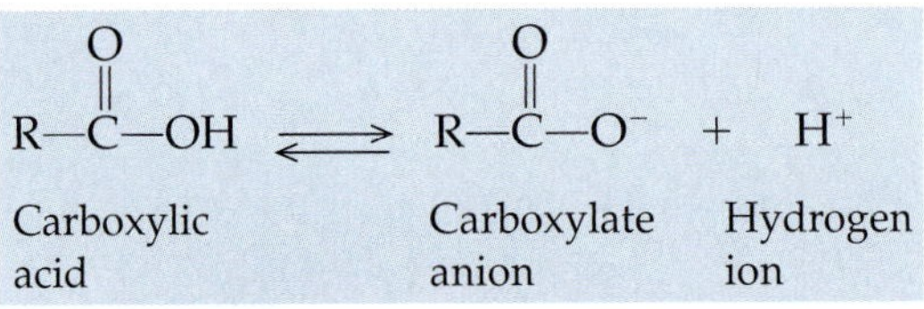

$$R-\overset{\overset{\displaystyle O}{\|}}{C}-OH \rightleftharpoons R-\overset{\overset{\displaystyle O}{\|}}{C}-O^- + H^+$$

Carboxylic acid   Carboxylate anion   Hydrogen ion

The properties of weak acids are described in Section 8.1.

Carboxylic acids are weak acids because they dissociate only slightly in solution. The majority of the acid remains in solution in the undissociated form. Typically, less than 5% of the acid is ionized (approximately five carboxylate ions to every ninety-five carboxylic acid molecules).

When a strong base is added to a carboxylic acid, neutralization occurs. The acid protons are removed by the OH^- to form water and the carboxylate ion. The equilibrium shown in the reaction above is shifted to the right, owing to removal of H^+. This is an illustration of LeChatelier's principle.

LeChatelier's principle is described in Section 5.4.

$$R-\overset{\overset{\displaystyle O}{\|}}{C}-OH + NaOH \longrightarrow R-\overset{\overset{\displaystyle O}{\|}}{C}-O^-Na^+ + H_2O$$

Carboxylic acid   Strong base   Carboxylic acid salt   Water

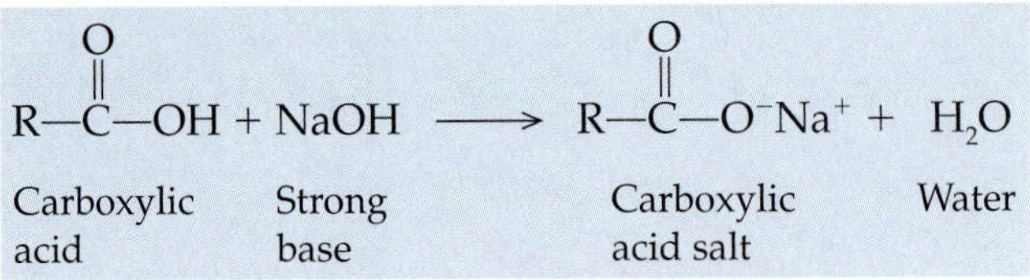

The carboxylate anion and the cation of the base form the carboxylic acid salt.

The following examples show the neutralization of acetic acid and benzoic acid in solutions of the strong base NaOH.

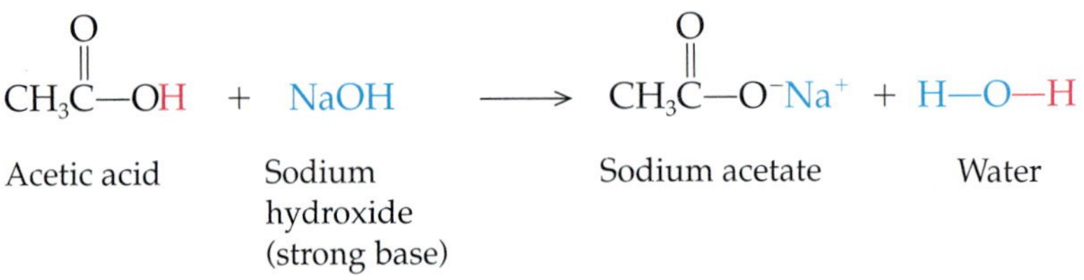

$$CH_3\overset{\overset{\displaystyle O}{\|}}{C}-OH + NaOH \longrightarrow CH_3\overset{\overset{\displaystyle O}{\|}}{C}-O^-Na^+ + H-O-H$$

Acetic acid   Sodium hydroxide (strong base)   Sodium acetate   Water

Sodium benzoate is commonly used as a food preservative.

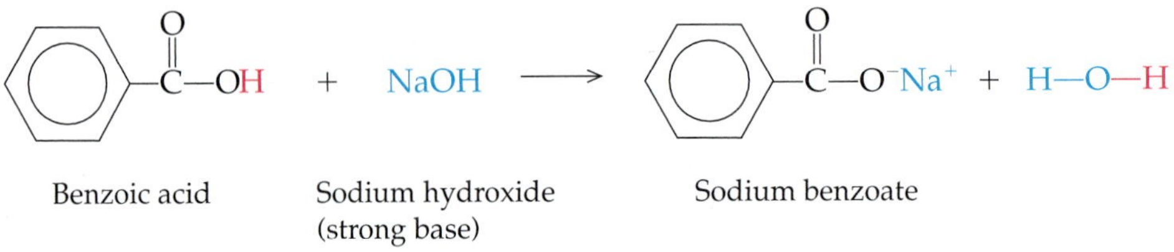

Benzoic acid   Sodium hydroxide (strong base)   Sodium benzoate

EXAMPLE 13.4   Writing an Equation to Show the Neutralization of
a Carboxylic Acid by a Strong Base

Write an equation showing the neutralization of propanoic acid by potassium
hydroxide.

4   **Learning Goal**

Write equations representing acid-base
reactions of carboxylic acids.

SOLUTION

The protons of the acid are removed by the OH^- of the base. This produces
water. The cation of the base, in this case potassium ion, forms the salt of the
carboxylic acid.

$$CH_3CH_2\!-\!\overset{\displaystyle O}{\overset{\|}{C}}\!-\!OH \;+\; KOH \;\longrightarrow\; CH_3CH_2\!-\!\overset{\displaystyle O}{\overset{\|}{C}}\!-\!O^-K^+ \;+\; H_2O$$

| Propanoic acid | Potassium hydroxide | Potassium salt of propanoic acid | Water |

QUESTION 13.11

Write the formula of the organic product obtained through each of the following
reactions.

a. $CH_3CH_2CH_2OH \xrightarrow{\;H_2CrO_4\;}$?

b. $H\overset{\displaystyle O}{\overset{\|}{C}}CH_2CH_2CH_2CH_3 \xrightarrow{\;H_2CrO_4\;}$?

QUESTION 13.12

Write the formula of the organic product obtained through each of the following
reactions.

a. $CH_3CH_2COOH + KOH \longrightarrow$?
b. $CH_3CH_2CH_2COOH + Ba(OH)_2 \longrightarrow$?

The salt of a carboxylic acid is named by replacing the *-ic acid* suffix with *-ate*. Thus
acetic acid becomes acetate, and benzoic acid becomes benzoate. This name is preceded
by the name of the appropriate cation, sodium in the examples of the neutralization of
acetic acid and benzoic acid on the previous page.

EXAMPLE 13.5   Naming the Salt of a Carboxylic Acid

Write the common and I.U.P.A.C. names of the salt produced in the reaction
shown in Example 13.4.

SOLUTION

$$CH_3CH_2\!-\!\overset{\displaystyle O}{\overset{\|}{C}}\!-\!O^-K^+$$

I.U.P.A.C. name of the parent carboxylic acid: propanoic acid
Replace the *-ic acid* ending with *-ate:*   propanoate
Name of the cation of the base: potassium

Continued—

EXAMPLE 13.5 —*Continued*

Name of the carboxylic acid salt: potassium propanoate

Common name of the parent carboxylic acid: propionic acid
Replace the *-ic acid* ending with *-ate:* propionate
Name of the cation of the base: potassium
Name of the carboxylic acid salt: potassium propionate

Carboxylic acid salts are ionic substances. As a result, they are very soluble in water. Long-chain carboxylic acid salts (fatty acid salts) are called *soaps*.

13.2 Esters

In this chapter, we will also study esters (Figure 13.1b), which have the following general structures:

$$R-\overset{\overset{\displaystyle O}{\|}}{C}-O-R \qquad Ar-\overset{\overset{\displaystyle O}{\|}}{C}-O-Ar \qquad Ar-\overset{\overset{\displaystyle O}{\|}}{C}-O-R$$

Examples of aliphatic and aromatic esters

The group shown in red is called the **acyl group.** The acyl group is part of the functional group of carboxylic acid derivatives, including esters and amides.

Structure and Physical Properties

> **5 Learning Goal**
> Write structures and describe the physical properties of esters.

Esters are mildly polar and have pleasant aromas. Many esters are found in natural foodstuffs; banana oil (3-methylbutyl ethanoate; common name, isoamyl acetate), pineapples (ethyl butanoate; common name, ethyl butyrate), and raspberries (isobutyl methanoate; common name, isobutyl formate) are but a few examples.

See An Environmental Perspective: The Chemistry of Flavor and Fragrance later in this chapter.

Esters boil at approximately the same temperature as aldehydes or ketones of comparable molecular weight. The simpler ones are somewhat soluble in water.

Nomenclature

> **6 Learning Goal**
> Determine the common and I.U.P.A.C. names of esters.

Esters are **carboxylic acid derivatives,** organic compounds derived from carboxylic acids. They are formed from the reaction of a carboxylic acid with an alcohol, and both of these reactants are reflected in the naming of the ester. They are named according to the following set of rules:

- Use the *alkyl* or *aryl* portion of the alcohol name as the first name.
- The *-ic acid* ending of the name of the carboxylic acid is replaced with *-ate* and follows the first name.

For example, in the following reaction, ethanoic acid reacts with methanol to produce methyl ethanoate:

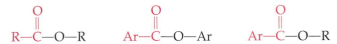

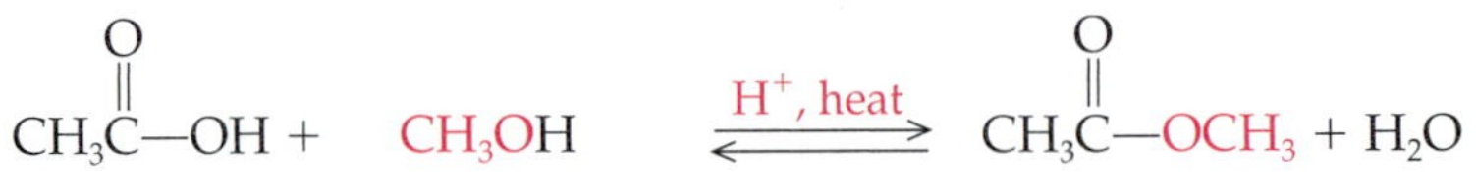

$$CH_3\overset{\overset{\displaystyle O}{\|}}{C}-OH + CH_3OH \xrightleftharpoons{H^+,\ heat} CH_3\overset{\overset{\displaystyle O}{\|}}{C}-OCH_3 + H_2O$$

Ethano*ic acid*   *Methanol*   *Methyl* ethano*ate*
(acetic acid)   (methyl alcohol)   (methyl acetate)

Similarly, acetic *acid* and *ethanol* react to produce *ethyl* acet*ate,* and the product of the reaction between benz*oic acid* and *isopropyl* alcohol is *isopropyl* benz*oate.*

Naming esters is analogous to naming the salts of carboxylic acids. Consider the following comparison:

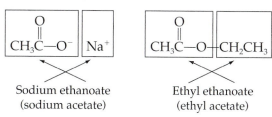

| Sodium ethanoate | Ethyl ethanoate |
| (sodium acetate) | (ethyl acetate) |

As shown in this example, the alkyl group of the alcohol, rather than Na^+, has displaced the acidic hydrogen of the carboxylic acid.

EXAMPLE 13.6 **Naming Esters Using the I.U.P.A.C. and Common Nomenclature Systems**

Write the I.U.P.A.C. and common names for the following ester:

$$CH_3CH_2CH_2C\!-\!OCH_2CH_3$$
(with O double-bonded to C)

6 **Learning Goal**
Determine the common and I.U.P.A.C. names of esters.

SOLUTION

I.U.P.A.C. and common names of parent carboxylic acid:	butanoic acid	butyric acid
Replace the *ic acid* ending of the carboxylic acid with *-ate:*	butanoate	butyrate
Name of the alkyl portion of the alcohol:	ethyl	ethyl
I.U.P.A.C. and common names of the ester:	Ethyl butanoate	Ethyl butyrate

---

QUESTION 13.13

Name each of the following esters using both the I.U.P.A.C. and common nomenclature systems.

a.
$$CH_3CH_2CH_2C\!-\!OCH_2CH_2CH_3$$
(with O double-bonded to C)

b.
$$CH_3CH_2CH_2C\!-\!OCH_2CH_3$$
(with O double-bonded to C)

QUESTION 13.14

Name each of the following esters using both the I.U.P.A.C. and common nomenclature systems.

a.
$$CH_3C\!-\!OCH_2CH_2CH_3$$
(with O double-bonded to C)

b.
$$CH_3CH_2C\!-\!OCH_2CH_2CH_2CH_3$$
(with O double-bonded to C)

7 Learning Goal

Write equations representing the synthesis and hydrolysis of an ester.

Esterification is reversible. The direction of the reaction is determined by the conditions chosen. Excess alcohol favors ester formation. The carboxylic acid is favored when excess water is present.

Reactions Involving Esters

Preparation of Esters

The conversion of a carboxylic acid to an ester requires heat and is catalyzed by a trace of acid (H^+). When esters are prepared directly from a carboxylic acid and an alcohol, a water molecule is lost, as in the reaction:

$$R^1{-}\overset{\displaystyle O}{\overset{\|}{C}}{-}OH + R^2OH \underset{\longleftarrow}{\overset{H^+,\ heat}{\longrightarrow}} R^1{-}\overset{\displaystyle O}{\overset{\|}{C}}{-}OR^2 + H_2O$$

Carboxylic acid   Alcohol   Ester   Water

$$CH_3CH_2\overset{\displaystyle O}{\overset{\|}{C}}{-}OH + CH_3OH \underset{\longleftarrow}{\overset{H^+,\ heat}{\longrightarrow}} CH_3CH_2\overset{\displaystyle O}{\overset{\|}{C}}{-}OCH_3 + H{-}O{-}H$$

Propanoic acid   Methanol   Methyl propanoate
(propionic acid)   (methyl alcohol)   (methyl propionate)

Esterification is a *dehydration* reaction, so called because a water molecule is eliminated during the reaction.

EXAMPLE 13.7   **Writing Equations Representing Esterification Reactions**

Write an equation showing the esterification reaction that would produce ethyl butanoate.

SOLUTION

The name, ethyl butanoate, tells us that the alcohol used in the reaction is ethanol and the carboxylic acid is butanoic acid. We must remember that a trace of acid and heat are required for the reaction and that the reaction is reversible. With this information, we can write the following equation representing the reaction:

$$CH_3CH_2CH_2\overset{\displaystyle O}{\overset{\|}{C}}{-}OH + CH_3CH_2OH \underset{\longleftarrow}{\overset{H^+,\ heat}{\longrightarrow}} CH_3CH_2CH_2\overset{\displaystyle O}{\overset{\|}{C}}{-}OCH_2CH_3 + H_2O$$

Butanoic acid   Ethanol   Ethyl butanoate
(butyric acid)   (ethyl butyrate)

---

7 Learning Goal

Write equations representing the synthesis and hydrolysis of an ester.

QUESTION 13.15

Write equations showing the esterification reactions that would produce butyl ethanoate and ethyl propanoate.

QUESTION 13.16

Write equations showing the esterification reactions that would produce methyl butanoate and propyl methanoate.

Hydrolysis of Esters

Acid Hydrolysis. **Hydrolysis,** sometimes also referred to as *hydration,* refers to the cleavage of any bond by the addition of a water molecule. Esters undergo hydrolysis reactions in water, as shown in the general reaction:

7 **Learning Goal**
Write equations representing the synthesis and hydrolysis of an ester.

$$R^1-\overset{\overset{\displaystyle O}{\|}}{C}-OR^2 + H_2O \underset{}{\overset{H^+,\, heat}{\rightleftharpoons}} R^1-\overset{\overset{\displaystyle O}{\|}}{C}-OH + R^2OH$$

| Ester | Water | Carboxylic acid | Alcohol |

This reaction requires heat. A small amount of acid (H^+) may be added to catalyze the reaction, as in the following example:

$$CH_3CH_2\overset{\overset{\displaystyle O}{\|}}{C}-OCH_2CH_2CH_3 + H_2O \underset{}{\overset{H^+,\, heat}{\rightleftharpoons}} CH_3CH_2\overset{\overset{\displaystyle O}{\|}}{C}-OH + CH_3CH_2CH_2OH$$

Propyl propanoate   Propanoic acid   1-Propanol
(propyl propionate)   (propionic acid)   (propanol)

Base Hydrolysis. The base-catalyzed hydrolysis of an ester is called **saponification.**

$$R^1-\overset{\overset{\displaystyle O}{\|}}{C}-OR^2 + H_2O \overset{NaOH,\, heat}{\longrightarrow} R^1-\overset{\overset{\displaystyle O}{\|}}{C}-O^-Na^+ + R^2OH$$

| Ester | Water | Carboxylic acid salt | Alcohol |

Under basic conditions, the acid cannot exist. Thus, the reaction yields the salt of the carboxylic acid having the cation of the basic catalyst.

$$CH_3\overset{\overset{\displaystyle O}{\|}}{C}-OCH_2CH_2CH_2CH_3 \overset{NaOH,\, heat}{\longrightarrow} CH_3\overset{\overset{\displaystyle O}{\|}}{C}-O^-Na^+ + CH_3CH_2CH_2CH_2OH$$

Butyl ethanoate   Sodium ethanoate   1-Butanol
(butyl acetate)   (sodium acetate)   (butyl alcohol)

The carboxylic acid is formed when the reaction mixture is neutralized with an acid such as HCl.

$$CH_3\overset{\overset{\displaystyle O}{\|}}{C}-O^-Na^+ + HCl \longrightarrow CH_3\overset{\overset{\displaystyle O}{\|}}{C}-OH + NaCl$$

Sodium ethanoate   Ethanoic acid
(sodium acetate)   (acetic acid)

QUESTION 13.17

Complete each of the following reactions by drawing the structures of the missing products.

a. $CH_3\overset{\overset{\displaystyle O}{\|}}{C}-OCH_2CH_2CH_3 + H_2O \underset{}{\overset{H^+,\, heat}{\rightleftharpoons}} ?$

b. $CH_3CH_2CH_2CH_2CH_2\overset{\overset{\displaystyle O}{\|}}{C}-OCH_2CH_2CH_3 + H_2O \overset{KOH,\, heat}{\longrightarrow} ?$

AN ENVIRONMENTAL Perspective | The Chemistry of Flavor and Fragrance

Carboxylic acids are often foul smelling. For instance, butyric acid is one of the worst smelling compounds imaginable—the smell of rancid butter.

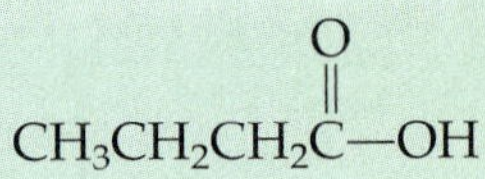

$$CH_3CH_2CH_2\overset{\overset{\displaystyle O}{\|}}{C}-OH$$

Butanoic acid
(butyric acid)

Butyric acid is also a product of fermentation reactions carried out by *Clostridium perfringens*. This organism is the most common cause of gas gangrene. Butyric acid contributes to the notable foul smell accompanying this infection.

By forming esters of butyric acid, one can generate compounds with pleasant smells. Ethyl butyrate is the essence of pineapple oil.

Pineapple

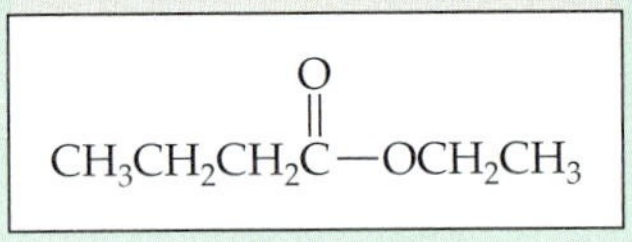

$$CH_3CH_2CH_2\overset{\overset{\displaystyle O}{\|}}{C}-OCH_2CH_3$$

Ethyl butanoate
(ethyl butyrate)

Volatile esters are often pleasant in both aroma and flavor. Natural fruit flavors are complex mixtures of many esters and other organic compounds. Chemists can isolate these mixtures and identify the chemical components. With this information, they are able to synthesize artificial fruit flavors, using just a few of the esters found in the natural fruit. As a result, the artificial flavors rarely have the full-bodied flavor of nature's original blend.

Raspberries

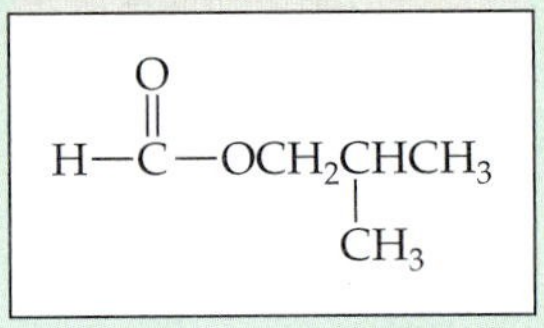

$$H-\overset{\overset{\displaystyle O}{\|}}{C}-OCH_2\underset{\underset{\displaystyle CH_3}{|}}{C}HCH_3$$

Isobutyl methanoate
(isobutyl formate)

Bananas

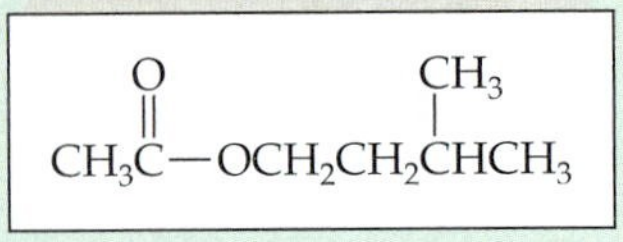

$$CH_3\overset{\overset{\displaystyle O}{\|}}{C}-OCH_2CH_2\underset{\underset{\displaystyle CH_3}{|}}{C}HCH_3$$

3-Methylbutyl ethanoate
(isoamyl acetate)

Oranges

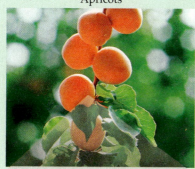

$$CH_3C\overset{O}{\overset{\|}{-}}OCH_2CH_2CH_2CH_2CH_2CH_2CH_2CH_3$$

Octyl ethanoate
(octyl acetate)

Apples

$$CH_3CH_2CH_2C\overset{O}{\overset{\|}{-}}OCH_3$$

Methyl butanoate
(methyl butyrate)

Apricots

$$CH_3CH_2CH_2C\overset{O}{\overset{\|}{-}}OCH_2CH_2CH_2CH_2CH_3$$

Pentyl butanoate
(pentyl butyrate)

Strawberries

$$CH_3CH_2CH_2C\overset{O}{\overset{\|}{-}}SCH_3$$

Methyl thiobutanoate
(methyl thiobutyrate)
(a thioester in which sulfur replaces oxygen)

FOR FURTHER UNDERSTANDING

Draw the structure of methyl salicylate, which is found in oil of wintergreen.

Write an equation for the synthesis of each of the esters shown in this connection.

QUESTION 13.18

Use the I.U.P.A.C. Nomenclature System to name each of the products in Question 13.17.

Triesters of glycerol are more commonly referred to as triglycerides. We know them as solid fats, generally from animal sources, and liquid oils, typically from plants. We will study triglycerides in detail in Section 15.3.

8 Learning Goal

Define the term saponification and describe how soap works in the emulsification of grease and oil.

A more detailed diagram of a micelle is shown in Figure 19.1.

An emulsion is a suspension of very fine droplets of one liquid in another. In this case, it is oil in water.

Fats and oils are triesters of the alcohol glycerol. When they are hydrolyzed by saponification, the products are **soaps,** which are the salts of long-chain carboxylic acids (fatty acid salts). According to Roman legend, soap was discovered by washerwomen following a heavy rain on Mons Sapo ("Mount Soap"). An important sacrificial altar was located on the mountain. The rain mixed with the remains of previous animal sacrifices—wood ash and animal fat—at the base of the altar. Thus, the three substances required to make soap accidentally came together—water, fat, and alkali (potassium carbonate and potassium hydroxide, called *potash,* leached from the wood ash). The soap mixture flowed down the mountain and into the Tiber River, where the washerwomen quickly realized its value.

We still use the old Roman recipe to make soap from water, a strong base, and natural fats and oils obtained from animals or plants. The synthesis of a soap is shown in Figure 13.3.

When soap is dissolved in water, the carboxylate end actually dissolves. The hydrocarbon part is repelled by the water molecules so that a thin film of soap is formed on the surface of the water with the hydrocarbon chains protruding outward. When soap solution comes in contact with oil or grease, the hydrocarbon part dissolves in the oil or grease, but the polar carboxylate group remains dissolved in water. When particles of oil or grease are surrounded by soap molecules, the resulting "units" formed are called *micelles.* A simplified view of this phenomenon is shown in Figure 13.4.

Micelles repel one another because they are surrounded on the surface by negatively charged carboxylate ions. Mechanical action (for example, scrubbing or tumbling in a washing machine) causes oil or grease to be surrounded by soap molecules and broken into small droplets so that relatively small micelles are formed. These small micelles are then washed away. Careful examination of this solution shows that it is an *emulsion* containing suspended micelles.

Condensation Polymers

As we saw in Chapter 11, *polymers* are macromolecules, very large molecules. They result from the combination of many smaller molecules, usually in a repeating pattern, to give molecules whose formula weight may be 10,000 g/mol or greater. The small molecules that make up the polymer are called *monomers.*

A polymer may be made from a single type of monomer (A). Such a polymer, called a homopolymer, would have the following general structure:

chain continues~A-A-A-A-A-A-A-A-A-A-A-A-A-A-A-A-A-A-A~chain continues

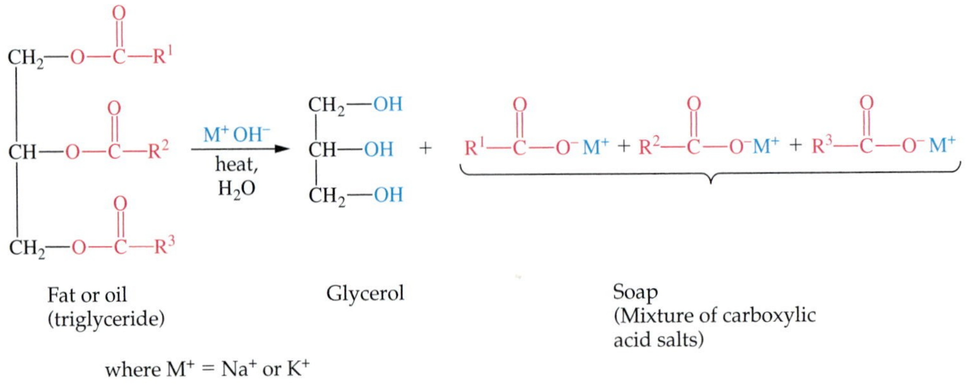

Figure 13.3

Saponification is the base-catalyzed hydrolysis of a glycerol triester.

Fat or oil
(triglyceride)

Glycerol

Soap
(Mixture of carboxylic acid salts)

where $M^+ = Na^+$ or K^+

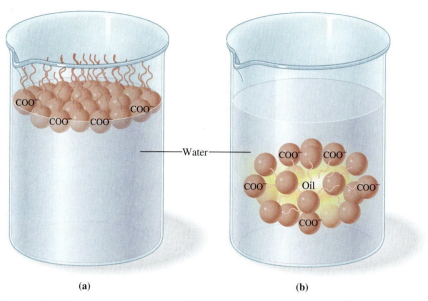

Figure 13.4

Simplified view of the action of a soap. The wiggly lines represent the long, continuous carbon chains of each soap molecule. (a) The thin film of soap molecules that forms at the water surface reduces surface tension. (b) Particles of oil and grease are surrounded by soap molecules to form a micelle.

The addition polymers of alkenes that we studied in Chapter 11 are examples of this type of polymer. Alternatively, two different monomers (A and B) may be copolymerized, producing a heteropolymer with the following structure:

chain continues~A-B-A-B-A-B-A-B-A-B-A-B-A-B-A-B~chain continues

Polyesters are heteropolymers. They are also known as condensation polymers. **Condensation polymers** are formed by the polymerization of monomers in a reaction that forms a small molecule such as water or an alcohol. Polyesters are synthesized by reacting a dicarboxylic acid and a dialcohol (diol). Notice that each of the combining molecules has two reactive functional groups, highlighted in red here:

n HOOC—⬡—COOH + n HOCH$_2$CH$_2$OH

Terephthalic acid   1,2-Ethanediol

H$^+$

O
‖
HOCH$_2$CH$_2$O—C—⬡—COOH

Another molecule of terephthalic acid can react here.

+ H$_2$O

Another molecule of 1,2-ethanediol can react here.

Reaction continues

A MEDICAL Connection | Carboxylic Acid Derivatives of Special Interest

Analgesics (pain killers) and antipyretics (fever reducers)

Aspirin (*acetylsalicylic acid*), derived from the bark of the willow, is the most widely used drug in the world. Hundreds of millions of dollars are spent annually on this compound. It is used primarily as a pain reliever (analgesic) and in the reduction of fever (antipyretic). Aspirin is among the drugs often referred to as NSAIDS, or nonsteroidal anti-inflammatory drugs. These drugs inhibit the inflammatory response by inhibiting an enzyme called cyclooxygenase, which is the first enzyme in the pathway for the synthesis of prostaglandins. Prostaglandins are responsible, in part, for pain and fever. Thus, aspirin and other NSAIDS reduce pain and fever by decreasing prostaglandin synthesis. Aspirin's side effects are a problem for some individuals. Because aspirin inhibits clotting, its use is not recommended during pregnancy, nor should it be used by individuals with ulcers. In those instances, *acetaminophen*, found in the over-the-counter pain-reliever Tylenol, is often prescribed.

The search for NSAIDS that are more effective and yet gentler on the stomach has provided two new analgesics for the over-the-counter market. These are ibuprofen (sold as Motrin, Advil, Nuprin) and naproxen (sold as Naprosyn, Naprelan, Anaprox, and Aleve).

Some common analgesics.

Many types of over-the-counter pain relievers are available.

Polyethylene terephthalate
PETE

Each time a pair of molecules reacts using one functional group from each, a new molecule is formed that still has two reactive groups. The product formed in this reaction is polyethylene terephthalate, or PETE.

When formed as fibers, polyesters are used to make fabric for clothing. These polyesters were trendy in the 1970s, during the "disco" period, but lost

Pheromones

Pheromones, chemicals secreted by animals, influence the behavior of other members of the same species. They often represent the major means of communication among simpler animals. The term *pheromone* literally means "to carry" and "to excite" (Greek, *pherein*, to carry; Greek, *horman*, to excite). They are chemicals carried or shed by one member of the species and used to alert other members of the species.

Pheromones may be involved in sexual attraction, trail marking, aggregation or recruitment, territorial marking, or signaling alarm. Others may be involved in defense or in species socialization—for example, designating various classes within the species as a whole. Among all of the pheromones, insect pheromones have been the most intensely studied. Many of the insect pheromones are carboxylic acids or acid derivatives, as seen here:

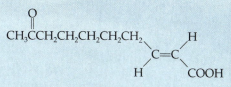

9-Keto-*trans*-2-decenoic acid
(queen bee socializing/royalty pheromone)

cis-7-Dodecenyl acetate
(cabbage looper sex pheromone)

$$CH_3CH_2CH{=}CH(CH_2)_9CH_2OCCH_3$$

Tetracecenyl acetate
(European corn borer sex pheromone)

FOR FURTHER UNDERSTANDING

A nonsteroidal anti-inflammatory drug can cause side effects if used over a long period of time. Do some research on the physiological roles of prostaglandins to develop hypotheses concerning these side effects.

How might you make use of pheromones to control pests such as the corn borer?

their popularity soon thereafter. Polyester fabrics, and a number of other synthetic polymers used in clothing, have become even more fashionable since the introduction of microfiber technology. The synthetic polymers are extruded into fibers that are only half the diameter of fine silk fibers. When these fibers are used to create fabrics, the result is a fabric that drapes freely yet retains its shape. These fabrics are generally lightweight, wrinkle resistant, and remarkably strong.

Polyester can be formed into a film called Mylar. These films, coated with aluminum foil, are used to make balloons that remain inflated for long periods. They are also used as the base for recording tapes and photographic film.

PETE can be used to make shatterproof plastic bottles, such as those used for soft drinks. However, these bottles cannot be recycled and reused directly because they cannot withstand the high temperatures required to sterilize them.

Biodegradable plastics made of both homopolymers and heteropolymers are discussed in "An Environmental Perspective: Garbage Bags from Potato Peels?" found earlier in this chapter.

PETE can't be used for any foods, such as jellies, that must be packaged at high temperatures. For these uses, a new plastic, PEN, or polyethylene naphthalate, is used.

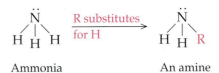

Napthalate group   Ethylene group

13.3 Amines

Structure and Physical Properties

9 Learning Goal

Classify amines as primary, secondary, or tertiary.

Amines are organic derivatives of ammonia and, like ammonia, they are basic. Amines are the most important type of organic base found in nature. We can think of them as substituted ammonia molecules in which one, two, or three of the ammonia hydrogens have been replaced by an organic group:

Ammonia   An amine

The structures drawn above and in Figure 13.5 reveal that like ammonia, amines are pyramidal. The nitrogen atom has three groups bonded to it and has a non-bonding pair of electrons.

Amines are classified according to the number of alkyl or aryl groups attached to the nitrogen. In a **primary (1°) amine,** one of the hydrogens is replaced by an organic group. In a **secondary (2°) amine,** two hydrogens are replaced. In a **tertiary (3°) amine,** three organic groups replace the hydrogens:

The geometry of ammonia is described in Section 3.5.

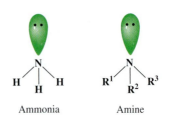

Ammonia   Amine

Figure 13.5

The pyramidal structure of amines. Note the similarities in structure between an amine and the ammonia molecule.

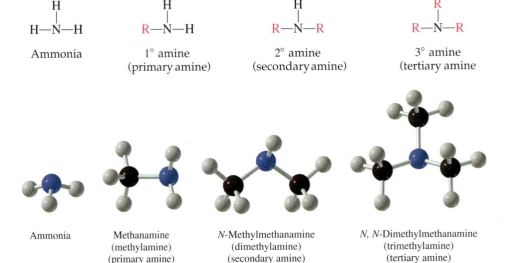

Ammonia

Methanamine
(methylamine)
(primary amine)

N-Methylmethanamine
(dimethylamine)
(secondary amine)

N, N-Dimethylmethanamine
(trimethylamine)
(tertiary amine)

EXAMPLE 13.8   **Classifying Amines as Primary, Secondary, or Tertiary**

Classify each of the following compounds as a primary, secondary, or tertiary amine.

Continued—

SOLUTION

Compare the structure of the amine with that of ammonia.

$$
\begin{array}{cc}
\overset{\displaystyle CH_3}{\underset{|}{H-N-H}} & \overset{\displaystyle H}{\underset{|}{H-N-H}}
\end{array}
\quad \text{1° amine: one hydrogen replaced}
$$

$$
\begin{array}{cc}
\overset{\displaystyle CH_3}{\underset{|}{CH_3-N-H}} & \overset{\displaystyle H}{\underset{|}{H-N-H}}
\end{array}
\quad \text{2° amine: two hydrogens replaced}
$$

$$
\begin{array}{cc}
\overset{\displaystyle CH_3}{\underset{|}{CH_3-N-CH_3}} & \overset{\displaystyle H}{\underset{|}{H-N-H}}
\end{array}
\quad \text{3° amine: three hydrogens replaced}
$$

QUESTION 13.19

Determine whether each of the following amines is primary, secondary, or tertiary.

a. $\overset{\displaystyle CH_2CH_3}{\underset{|}{CH_3CH_2NCH_3}}$

b. $CH_3CH_2CH_2NH_2$

c. $\overset{\displaystyle H}{\underset{|}{CH_3NCH_3}}$

QUESTION 13.20

Classify each of the following amines as primary, secondary, or tertiary.

a. CH_3NH_2

b. $\overset{\displaystyle CH_3}{\underset{|}{CH_3CH_2CH_2CH_2NCH_2CH_3}}$

c. $\overset{\displaystyle CH_2CH_2CH_3}{\underset{|}{HNCH_3}}$

The nitrogen atom is more electronegative than the hydrogen atoms in amines. As a result, the N—H bond is polar, and hydrogen bonding can occur between amine molecules or between amine molecules and water (Figure 13.6).

Hydrogen bonding is described in Section 6.2.

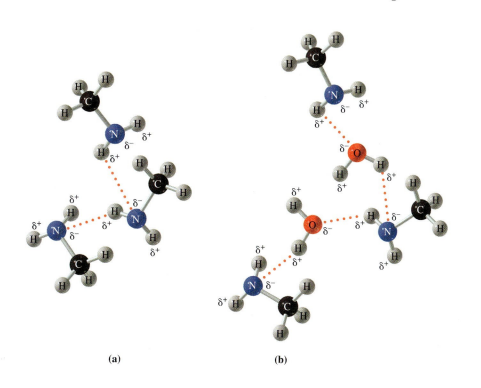

(a)   **(b)**

Figure 13.6

Hydrogen bonding (a) between methylamine molecules and (b) between methylamine molecules and water molecules. Dotted lines represent hydrogen bonds.

QUESTION 13.21

Refer to Figure 13.6 and draw a similar figure showing the hydrogen bonding that occurs between water and a 2° amine.

QUESTION 13.22

Refer to Figure 13.6 and draw hydrogen bonding between two primary amine molecules.

---

10 Learning Goal

Describe the physical properties of amines.

The ability of primary and secondary amines to form N—H⋯N hydrogen bonds is reflected in their boiling points. Primary amines have boiling points well above those of alkanes of similar molecular weight but considerably lower than those of comparable alcohols.

Tertiary amines do not have an N—H bond. As a result, they cannot form intermolecular hydrogen bonds with other tertiary amines. Consequently, their boiling points are lower than those of primary or secondary amines of comparable molecular weight as the following examples show:

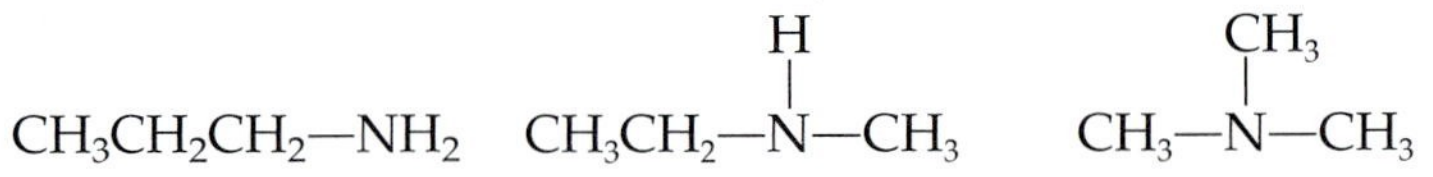

$$CH_3CH_2CH_2—NH_2 \quad CH_3CH_2—\overset{\overset{\displaystyle H}{|}}{N}—CH_3 \quad CH_3—\overset{\overset{\displaystyle CH_3}{|}}{N}—CH_3$$

Propanamine (propylamine)	N-Methylethanamine (ethylmethylamine)	N,N-Dimethylmethanamine (trimethylamine)
M.W. = 59 g/mol	M.W. = 59 g/mol	M.W. = 59 g/mol
b.p. = 48.7° C	b.p. = 36.7° C	b.p. = 2.9° C

The intermolecular hydrogen bonds formed by primary and secondary amines are not as strong as the hydrogen bonds formed by alcohols because nitrogen is not as electronegative as oxygen. For this reason, primary and secondary amines have lower boiling points than alcohols.

All amines can form intermolecular hydrogen bonds with water (O—H⋯N). As a result, small amines (six or fewer carbons) are soluble in water. As we have noted for other families of organic molecules, water solubility decreases as the length of the hydrocarbon (hydrophobic) portion of the molecule increases.

QUESTION 13.23

Which compound in each of the following pairs would you predict has a higher boiling point? Explain your reasoning.

a. Methanol or methylamine
b. Dimethylamine or water
c. Methylamine or ethylamine
d. Propylamine or butane

QUESTION 13.24

Compare the boiling points of methylamine, dimethylamine, and trimethylamine. Explain why they are different.

Nomenclature

11 Learning Goal

Draw and name simple amines using common and systematic nomenclature.

In systematic nomenclature, primary amines are named according to the following rules:

• Determine the name of the *parent compound,* the longest continuous carbon chain containing the amine group.

- Replace the *–e* ending of the alkane chain with *–amine.* Following this pattern, the alkane becomes an alkan*amine;* for instance, ethane becomes ethan*amine.*
- Number the parent chain to give the carbon bearing the amine group the lowest possible number.
- Name and number all substituents, and add them as prefixes to the "alkanamine" name.

For instance,

CH_3—NH_2 $CH_3CH_2CH_2$—NH_2 $CH_3CH_2CH_2CHCH_3$
                                                                                                      |
                                                                                                    NH_2

Methanamine   1-Propanamine   2-Pentanamine

For secondary or tertiary amines the prefix *N*-alkyl is added to the name of the parent compound. For example,

                                                                            CH_3
                                                                              |
CH_3—NH—CH_2CH_3 CH_3—N—CH_3

N-Methylethanamine   *N,N*-Dimethylmethanamine

EXAMPLE 13.9   **Writing the Systematic Name for an Amine**

Name the following amine.

$CH_3CH_2CH_2$—NH—CH_3

SOLUTION

Parent compound: propane (becomes propanamine)
Additional group on N: methyl (becomes *N*-methyl)
Name: *N*-Methylpropanamine

11 **Learning Goal**
Draw and name simple amines using common and systematic nomenclature.

Common names are often used for the simple amines. The common names of the alkyl groups bonded to the amine nitrogen are followed by the ending *-amine.* Each group is listed alphabetically in one continuous word followed by the suffix *-amine:*

                                                                            CH_3
                                                                              |
CH_3—NH_2 CH_3—NH—CH_3 CH_3—N—CH_3

Methylamine   Dimethylamine   Trimethylamine

CH_3CH_2—NH_2 CH_3CH_2—NH—CH_3

Ethylamine   Ethylmethylamine

QUESTION 13.25

Draw the complete structural formula for each of the following compounds.

a. 2-Propanamine
b. 3-Octanamine
c. *N*-Ethyl-2-heptanamine

d. 2-Methyl-2-pentanamine
e. 4-Chloro-5-iodo-1-nonanamine
f. *N,N*-Diethyl-1-pentanamine

QUESTION 13.26

Name each of the following amines using the systematic and common nomenclature systems.

a. 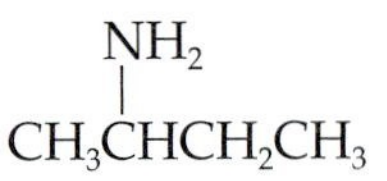
$$\begin{array}{c} NH_2 \\ | \\ CH_3CHCH_2CH_3 \end{array}$$

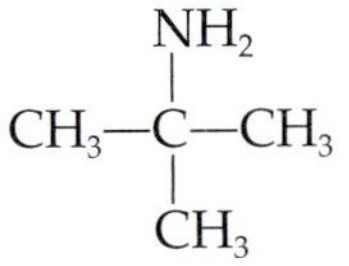

b.
$$\begin{array}{c} NH_2 \\ | \\ CH_3{-}C{-}CH_3 \\ | \\ CH_3 \end{array}$$

c.
$$\begin{array}{c} CH_3 \\ | \\ NH \\ | \\ CH_3CHCH_2CH_3 \end{array}$$

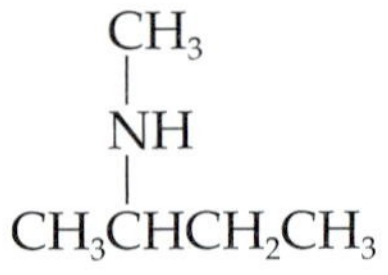

d. 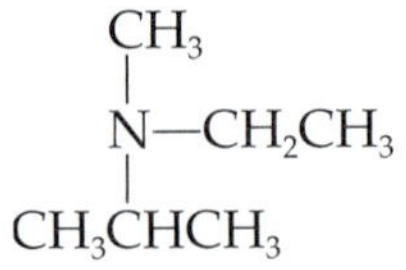
$$\begin{array}{c} CH_3 \\ | \\ N{-}CH_2CH_3 \\ | \\ CH_3CHCH_3 \end{array}$$

Medically Important Amines

Although amines play many different roles in our day-to-day lives, one important use is in medicine. A host of drugs derived from amines is responsible for improving the quality of life, whereas others, such as cocaine and heroin, are highly addictive.

Amphetamines, such as benzedrine and methedrine, stimulate the central nervous system. They elevate blood pressure and pulse rate and are often used to decrease fatigue. Medically, they have been used to treat depression and epilepsy. Amphetamines have also been prescribed as diet pills because they decrease the appetite. Their use is controlled by federal law because excess use of amphetamines can cause paranoia and mental illness.

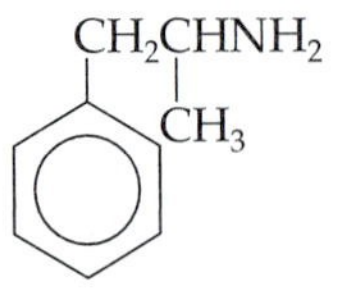

1-Phenyl-2-propanamine
(Amphetamine)

Benzedrine

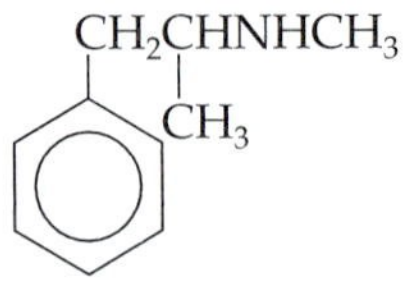

N-Methyl-1-phenyl-2-propanamine
(Methamphetamine)

Methedrine

Many of the medicinal amines are *analgesics* (pain relievers) or *anesthetics* (pain blockers). Novocaine and related compounds, for example, are used as local anesthetics. Demerol is a very strong pain reliever.

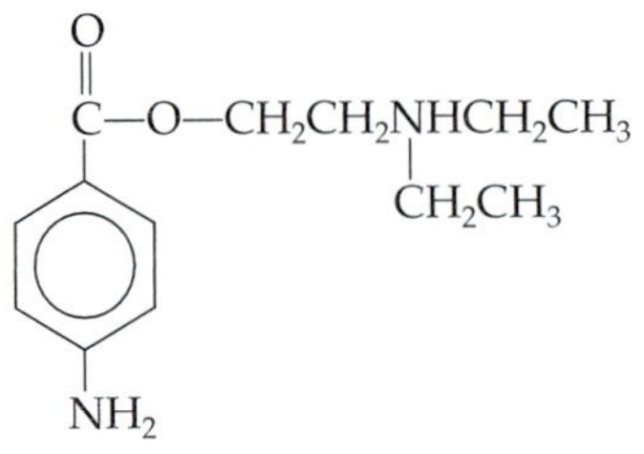

Novocaine

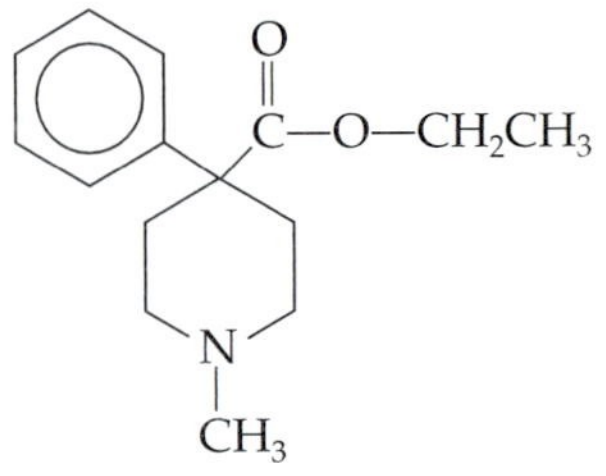

Demerol

Ephedrine, its stereoisomer pseudoephedrine, and phenylephrine (also called neosynephrine) are used as decongestants in cough syrups and nasal sprays. By shrinking the membranes that line the nasal passages, they relieve the symptoms of congestion and stuffy nose. These compounds are very closely related to L-dopa and dopamine, which are key compounds in the function of the central nervous system.

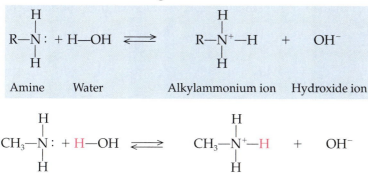

Ephedrine   Pseudoephedrine   Phenylephrine
(neosynephrine)

Recently, many states have restricted the sale of products containing ephedrine and pseudoephedrine and many drug store chains have moved these products behind the counter. The reason for these precautions is that either ephedrine or pseudoephedrine can be used as the starting material in the synthesis of methamphetamines. In response to this problem, pharmaceutical companies are replacing ephedrine and pseudoephedrine in these decongestants with phenylephrine, which cannot be used as a reactant in the synthesis of methamphetamine.

Ephedrine and pseudoephedrine are also the primary active ingredients in ephedra, a plant found in the deserts of central Asia. Ephedra is used as a stimulant in a variety of products that are sold over-the-counter as aids to boost energy, promote weight loss, and enhance athletic performance. In 2004, the Food and Drug Administration banned the use of ephedra in these over-the-counter formulations after reviewing 16,000 reports of adverse side effects including nervousness, heart irregularities, seizures, heart attacks, and 80 deaths, including that of Baltimore Orioles pitcher Steve Bechler, age 23. However, in April 2005, a Federal Judge in Texas ruled that the FDA had failed to prove that ephedra is dangerous at doses of 10 mg or lower, opening the way for the sale of ephedra-containing herbal remedies.

The *sulfa drugs*, the first chemicals used to fight bacterial infections, are synthesized from amines.

$$H_2N-\!\!\bigcirc\!\!-\overset{\overset{O}{\|}}{\underset{\underset{O}{\|}}{S}}-NH_2$$

Sulfanilamide—a sulfa drug

Methamphetamine use and abuse are discussed in "A Medical Connection: Methamphetamine" found online at www.mhhe.com/denniston.

Reactions Involving Amines

Basicity

Amines behave as weak bases, accepting H^+, when dissolved in water. The nonbonding pair (lone pair) of electrons of the nitrogen atom can be shared with a proton (H^+) from a water molecule, producing an **alkylammonium ion.** Hydroxide ions are also formed, so the resulting solution is basic.

12 **Learning Goal**
Write equations representing the basicity and neutralization of amines.

$$\underset{\substack{| \\ H}}{\overset{\substack{H \\ |}}{R-N:}} + H-OH \;\rightleftharpoons\; \underset{\substack{| \\ H}}{\overset{\substack{H \\ |}}{R-N^+-H}} + OH^-$$

Amine   Water   Alkylammonium ion   Hydroxide ion

$$\underset{\substack{| \\ H}}{\overset{\substack{H \\ |}}{CH_3-N:}} + H-OH \;\rightleftharpoons\; \underset{\substack{| \\ H}}{\overset{\substack{H \\ |}}{CH_3-N^+-H}} + OH^-$$

Methylamine   Water   Methylammonium ion   Hydroxide ion

Amine Salts

Because amines are bases, they react with acids to form alkylammonium salts.

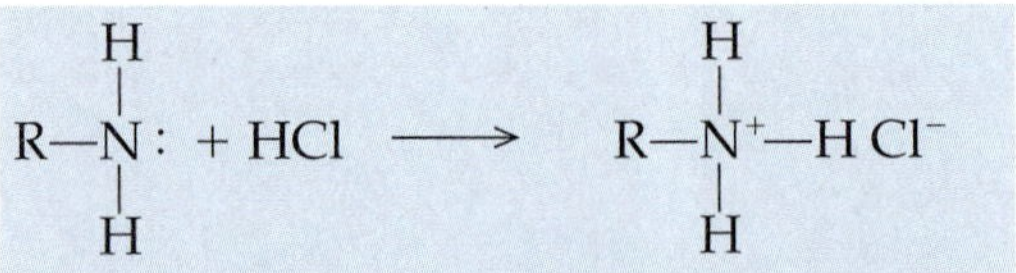

$$\underset{\text{Amine}}{\overset{\displaystyle H}{\underset{\displaystyle H}{R-N:}}} + \underset{\text{Acid}}{HCl} \longrightarrow \underset{\text{Alkylammonium salt}}{\overset{\displaystyle H}{\underset{\displaystyle H}{R-N^+-H\ Cl^-}}}$$

Recall that the reaction of an acid and a base gives a salt (Section 8.3).

The reaction of methylamine with hydrochloric acid is typical of these reactions:

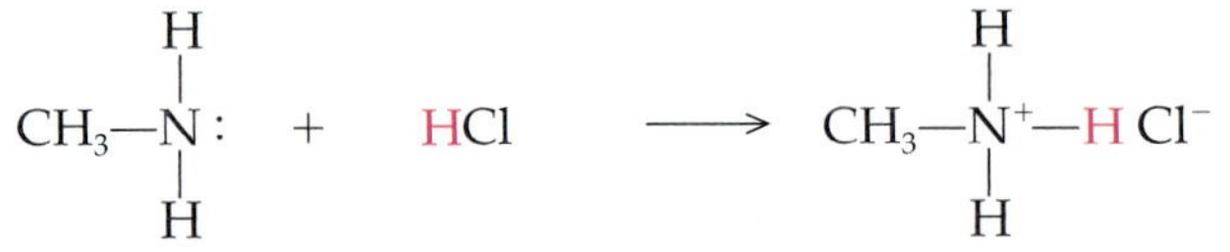

$$\underset{\text{Methylamine}}{\overset{\displaystyle H}{\underset{\displaystyle H}{CH_3-N:}}} + \underset{\substack{\text{Hydrochloric}\\\text{acid}}}{HCl} \longrightarrow \underset{\substack{\text{Methylammonium}\\\text{chloride}}}{\overset{\displaystyle H}{\underset{\displaystyle H}{CH_3-N^+-H\ Cl^-}}}$$

Alkylammonium salts are named by replacing the suffix -*amine* with *ammonium*. This is then followed by the name of the anion. The salts are ionic and hence are quite soluble in water.

A variety of important drugs are amines. They are usually administered as alkylammonium salts because the salts are much more soluble in aqueous solutions and in body fluids.

Alkylammonium salts can neutralize hydroxide ions. In this reaction, water is formed and the protonated amine cation is converted into an amine.

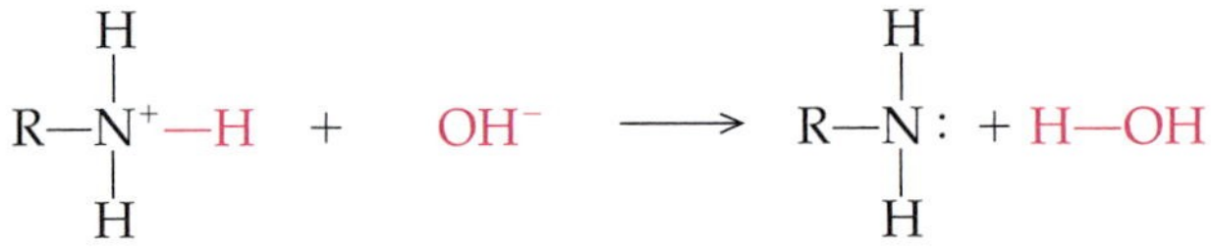

$$\underset{\substack{\text{Alkylammonium}\\\text{salt}}}{\overset{\displaystyle H}{\underset{\displaystyle H}{R-N^+-H}}} + \underset{\substack{\text{Hydroxide}\\\text{ion}}}{OH^-} \longrightarrow \underset{\text{Amine}}{\overset{\displaystyle H}{\underset{\displaystyle H}{R-N:}}} + \underset{\text{Water}}{H-OH}$$

The local anesthetic novocaine, which is often used in dentistry and for minor surgery, is injected as an amine salt. See Medically Important Amines earlier in this section.

Thus, by adding a strong acid to a water-insoluble amine, a water soluble alkylammonium salt can be formed. The salt can just as easily be converted back to an amine by the addition of a strong base. The ability to manipulate the solubility of physiologically active amines through interconversion of the amine and its corresponding salt is extremely important in the development, manufacture, and administration of many important drugs.

Pure cocaine is an amine and a base (structure below). This form of cocaine, referred to as "crack" or "freebase" cocaine, is generally found in the form of relatively large crystals (Figure 13.7a) that may vary in color from white to dark brown or black. As we have just seen, when an amine reacts with an acid, an alkylammonium salt is formed. When cocaine reacts with HCl, the product is cocaine hydrochloride:

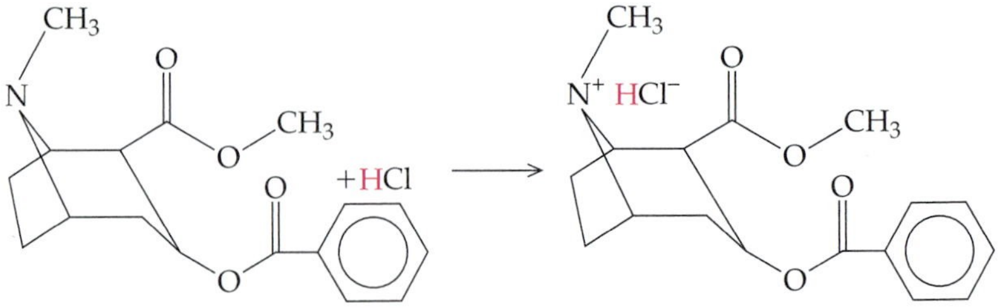

"Crack" cocaine   Cocaine hydrochloride
(a base)   (a salt)

(a)

(b)

Figure 13.7

(a) Crack cocaine is a water-insoluble base with a low melting point and a crystalline structure. (b) Powdered cocaine is a water-soluble salt of cocaine base. (c) Cocaine is extracted from the leaves of the coca plant.

(c)

The salt of cocaine is a powder (Figure 13.7b) that is soluble in water. Because it is a powder, it can be snorted, and because it is water-soluble, it dissolves in the fluids of the nasal mucous membranes and is absorbed into the bloodstream. This is a common form of cocaine because it is the direct product of the preparation from coca leaves (Figure 13.7c). A coca paste is made from the leaves and is mixed with HCl and water. After additional processing, the product is the salt of cocaine.

Cocaine hydrochloride salt can be converted into its base form by a process called "freebasing." Although the chemistry is simple, the process is dangerous because it requires highly flammable solvents. This pure cocaine is not soluble in water. It has a relatively low melting point, however, and can be smoked. Crack, so called because of the crackling noise it makes when smoked, is absorbed into the body more quickly than the snorted powder and results in a more immediate high.

Quaternary Ammonium Salts

Quaternary ammonium salts are ammonium salts that have four organic groups bonded to the nitrogen. They have the following general structure:

$$R_4N^+X^-$$
(R = any alkyl or aryl group;
X$^-$ = a halide anion, most commonly Cl$^-$)

Quaternary ammonium salts that have a very long carbon chain, sometimes called "quats," are used as disinfectants and antiseptics because they have detergent activity. Two popular quats are benzalkonium chloride (Zephiran) and cetylpyridinium chloride, found in the mouthwash Scope.

13 Learning Goal

Describe the structure of quaternary ammonium salts and discuss their use as antiseptics and disinfectants.

$$\left[\text{—CH}_2\text{—}\overset{\displaystyle CH_3}{\underset{\displaystyle CH_3}{N^+}}\text{—C}_{18}H_{37}\right]Cl^-$$

Benzalkonium
chloride

$$\left[N^+\underset{\displaystyle (CH_2)_{15}CH_3}{}\right]Cl^-$$

Cetylpyridinium
chloride

Phospholipids and biological membranes are discussed in Sections 15.3 and 15.6.

Choline is an important quaternary ammonium salt in the body. It is part of the hydrophilic "head" of the membrane phospholipid lecithin. Choline is also a precursor for the synthesis of the neurotransmitter acetylcholine.

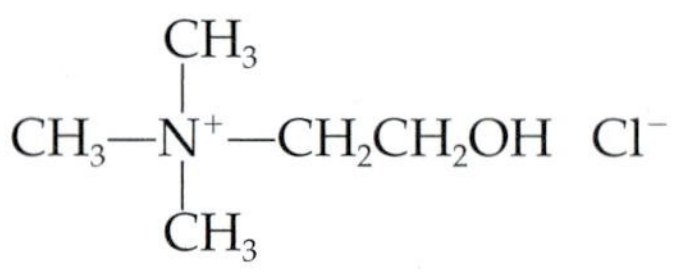

$$CH_3-N^+-CH_2CH_2OH \quad Cl^-$$

with CH_3 groups above and below the nitrogen.

Choline

13.4 Amides

Amides are the products of reactions between carboxylic acid derivatives and ammonia or amines. The general structure of an amide is shown here.

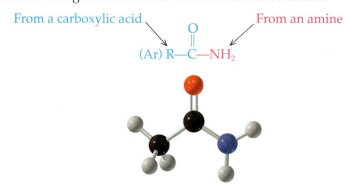

From a carboxylic acid →

$$\text{(Ar) R}-\overset{\overset{\displaystyle O}{\|}}{C}-NH_2$$

← From an amine

Ethanamide

The amide group is composed of two portions: the carbonyl group from a carboxylic acid and the amino group from ammonia or an amine. The bond between the carbonyl carbon and the nitrogen of the amine or ammonia is called the **amide bond.**

Structure and Physical Properties

14 Learning Goal

Describe the physical properties of amides.

Most amides are solids at room temperature. They have very high boiling points, and the simpler ones are quite soluble in water. Both of these properties are a result of strong intermolecular hydrogen bonding between the N—H bond of one amide and the C=O group of a second amide, as shown in Figure 13.8.

Unlike amines, amides are not bases (proton acceptors). The reason is that the highly electronegative oxygen atom of the carbonyl group causes a very strong attraction between the lone pair of nitrogen electrons and the carbonyl group. As a result, the unshared pair of electrons cannot "hold" a proton.

Nomenclature

15 Learning Goal

Draw the structure and write the common and I.U.P.A.C. names of amides.

Nomenclature of carboxylic acids is described in Section 13.1.

The common and I.U.P.A.C. names of amides are derived from the common and I.U.P.A.C. names of the carboxylic acids from which they were made. Remove the *-ic acid* ending of the common name or the *-oic acid* ending of the I.U.P.A.C. name of the carboxylic acid, and replace it with the ending *-amide.* Several examples of the common and I.U.P.A.C. nomenclature are provided in the following structures:

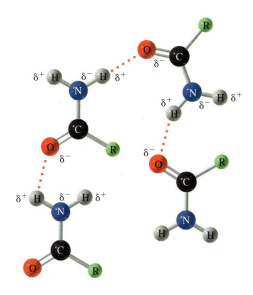

Figure 13.8

Hydrogen bonding between amide molecules.

$$CH_3\overset{\displaystyle O}{\overset{\|}{C}}-NH_2$$

Ethan*oic acid* → Ethan*amide*
or
Acet*ic acid* → Acet*amide*

$$CH_3CH_2\overset{\displaystyle O}{\overset{\|}{C}}-NH_2$$

Propan*oic acid* → Propan*amide*
or
Propion*ic acid* → Propion*amide*

Substituents on the nitrogen are placed as prefixes and are indicated by *N-* followed by the name of the substituent. There are no spaces between the prefix and the amide name. For example:

$$CH_3CH_2\overset{\displaystyle O}{\overset{\|}{C}}-NH-CH_3$$

N-Methylpropanamide

$$CH_3CH_2CH_2CH_2CH_2\overset{\displaystyle O}{\overset{\|}{C}}-NH-CH_2CH_2CH_3$$

N-Propylhexanamide

Medically Important Amides

Barbiturates, often called "downers," are derived from amides and are used as sedatives. They are also used as anticonvulsants for epileptics and for people suffering from a variety of brain disorders that manifest themselves in neurosis, anxiety, and tension.

Barbital—a barbiturate

Phenacetin and acetaminophen are also amides. Acetaminophen is an aromatic amide that is commonly used in place of aspirin, particularly by people who are allergic to aspirin or who suffer stomach bleeding from the use of aspirin. It was first synthesized in 1893 and is the active ingredient in Tylenol and Datril. Like aspirin, acetaminophen relieves pain and reduces fever. However, unlike aspirin, it is not an anti-inflammatory drug.

A MEDICAL Connection | Semisynthetic Penicillins

The antibacterial properties of penicillin were discovered by Alexander Fleming in 1929. These natural penicillins produced by several species of the mold *Penicillium*, had a number of drawbacks. They were effective only against a type of bacteria referred to as gram-positive because of a staining reaction based on their cell wall structure. They were also very susceptible to destruction by bacterial enzymes called β-lactamases, and some were destroyed by stomach acid and had to be administered by injection.

To overcome these problems, chemists have produced semisynthetic penicillins by modifying the core structure. The core of penicillins is 6-aminopenicillanic acid, which consists of a thiazolidine ring fused to a β-lactam ring. In addition, there is an R group bonded via an amide bond to the core structure.

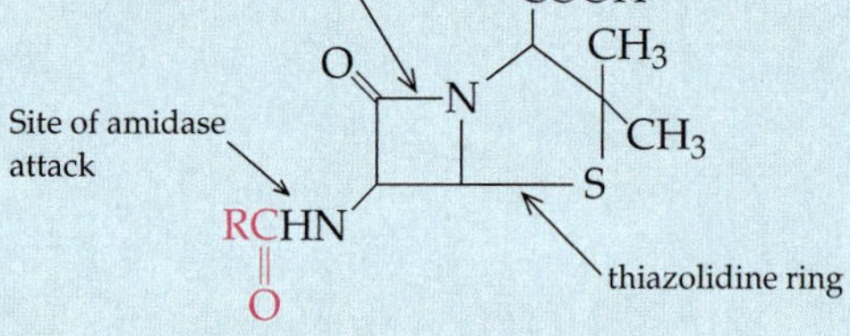

6-Aminopenicillanic acid

The β-lactam ring confers antimicrobial properties. However, the R group determines the degree of antibacterial activity, the pharmacological properties, including the types of bacteria against which it is active and the degree of resistance to the β-lactamases exhibited by any particular penicillin antibiotic. These are the properties that must be modified to produce penicillins that are acid resistant, effective with a broad spectrum of bacteria, and β-lactamase resistant.

Chemists simply remove the natural R group by cleaving the amide bond with an enzyme called an amidase. They then replace the R group and test the properties of the "new" antibiotic. Among the resulting semisynthetic penicillins are ampicillin, methicillin, and oxacillin.

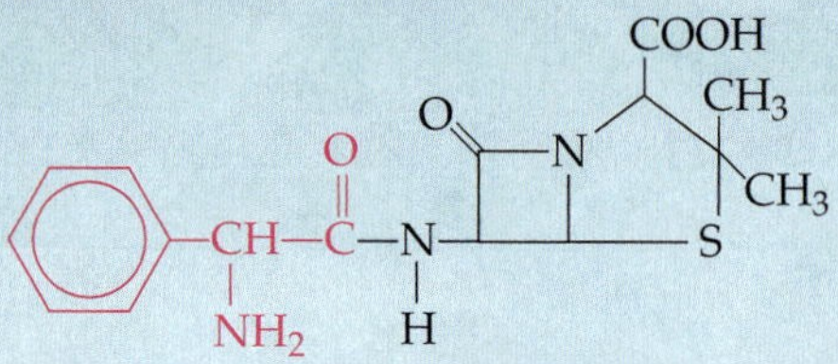

Ampicillin

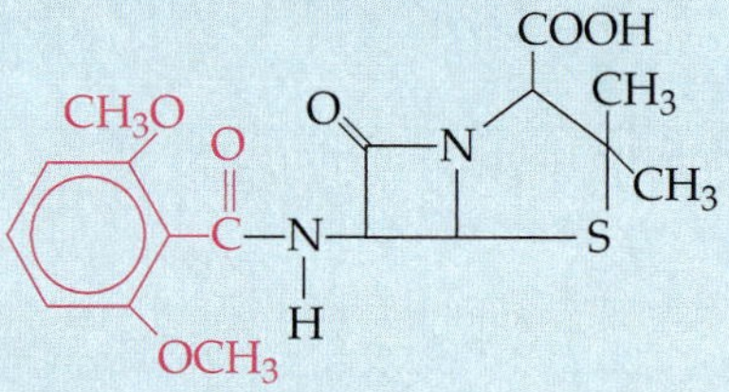

Methicillin

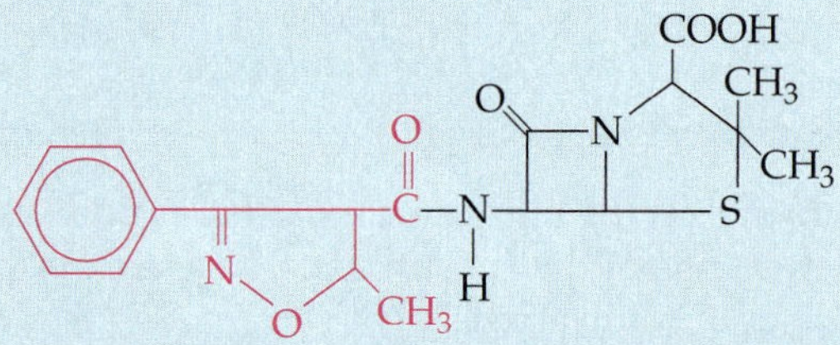

Oxacillin

FOR FURTHER UNDERSTANDING

Using the Internet and other resources, investigate and describe the properties of some new penicillins and the bacteria against which they are effective.

Why does changing the R group of a penicillin result in altered chemical and physiological properties?

Phenacetin was synthesized in 1887 and used as an analgesic for almost a century. Its structure and properties are similar to those of acetaminophen. However, it was banned by the U.S. Food and Drug Administration in 1983 because of the kidney damage and blood disorders that it causes.

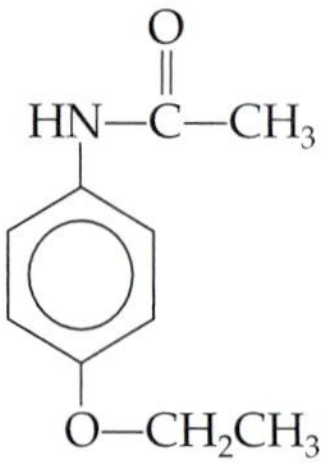

Phenacetin

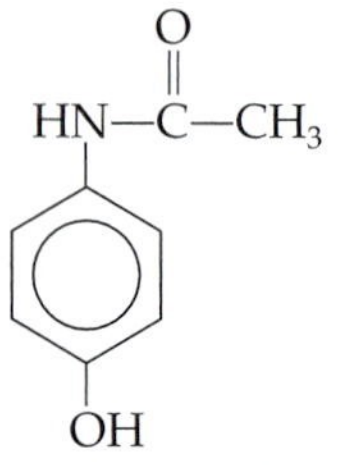

Acetaminophen

Reactions Involving Amides

Hydrolysis of Amides

Hydrolysis of an amide results in breaking the amide bond to produce a carboxylic acid and ammonia or an amine. It is very difficult to hydrolyze the amide bond. The reaction requires heating the amide in the presence of a strong acid or base.

16 Learning Goal

Write equations representing the hydrolysis of amides.

$$R-\overset{\overset{\displaystyle O}{\|}}{C}-NH-R^1 + H_3O^+ \longrightarrow R-\overset{\overset{\displaystyle O}{\|}}{C}-OH + R^1-N^+H_3$$

Amide   Strong acid   Carboxylic acid   Alkylammonium ion or ammonium ion

Ammoniumion is formed if $R^1 = H$

$$CH_3CH_2CH_2\overset{\overset{\displaystyle O}{\|}}{C}-NH_2 + H_3O^+ \longrightarrow CH_3CH_2CH_2\overset{\overset{\displaystyle O}{\|}}{C}-OH + N^+H_4$$

Butanamide (butyramide)   Butanoic acid (butyric acid)

If a strong base is used, the products are the amine and the salt of the carboxylic acid:

$$R-\overset{\overset{\displaystyle O}{\|}}{C}-NH-R^1 + NaOH \longrightarrow R-\overset{\overset{\displaystyle O}{\|}}{C}-O^-Na^+ + R^1-NH_2$$

Amide   Strong base   Carboxylic acid salt   Amine or ammonia

$$CH_3CH_2\overset{\overset{\displaystyle O}{\|}}{C}-NHCH_3 + NaOH \longrightarrow CH_3CH_2\overset{\overset{\displaystyle O}{\|}}{C}-O^-Na^+ + CH_3NH_2$$

N-Methylpropanamide (*N*-methylpropionamide)   Sodium propanoate (sodium propionate)   Methanamine (methylamine)

SUMMARY OF REACTIONS

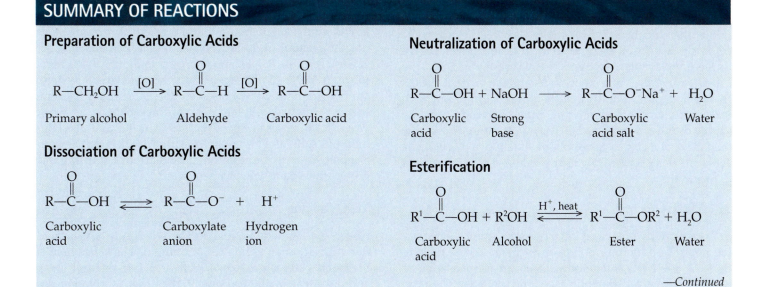

Preparation of Carboxylic Acids

$$R-CH_2OH \xrightarrow{[O]} R-\overset{\overset{\displaystyle O}{\|}}{C}-H \xrightarrow{[O]} R-\overset{\overset{\displaystyle O}{\|}}{C}-OH$$

Primary alcohol   Aldehyde   Carboxylic acid

Dissociation of Carboxylic Acids

$$R-\overset{\overset{\displaystyle O}{\|}}{C}-OH \rightleftharpoons R-\overset{\overset{\displaystyle O}{\|}}{C}-O^- + H^+$$

Carboxylic acid   Carboxylate anion   Hydrogen ion

Neutralization of Carboxylic Acids

$$R-\overset{\overset{\displaystyle O}{\|}}{C}-OH + NaOH \longrightarrow R-\overset{\overset{\displaystyle O}{\|}}{C}-O^-Na^+ + H_2O$$

Carboxylic acid   Strong base   Carboxylic acid salt   Water

Esterification

$$R^1-\overset{\overset{\displaystyle O}{\|}}{C}-OH + R^2OH \underset{}{\overset{H^+, heat}{\rightleftharpoons}} R^1-\overset{\overset{\displaystyle O}{\|}}{C}-OR^2 + H_2O$$

Carboxylic acid   Alcohol   Ester   Water

—Continued

Continued—

Acid Hydrolysis of Esters

$$R^1-\overset{\overset{\displaystyle O}{\|}}{C}-OR^2 + H_2O \underset{\longleftarrow}{\overset{H^+,\ heat}{\rightleftharpoons}} R^1-\overset{\overset{\displaystyle O}{\|}}{C}-OH + R^2OH$$

Ester　　　Water　　　　　Carboxylic　　Alcohol
　　　　　　　　　　　　　　acid

Saponification

$$R^1-\overset{\overset{\displaystyle O}{\|}}{C}-OR^2 + H_2O \overset{NaOH,\ heat}{\longrightarrow} R^1-\overset{\overset{\displaystyle O}{\|}}{C}-O^-Na^+ + R^2OH$$

Ester　　　Water　　　　　Carboxylic　　Alcohol
　　　　　　　　　　　　　　acid salt

Basicity of Amines

$$R-NH_2 + H-OH \rightleftharpoons R-\overset{\overset{\displaystyle H}{|}}{\underset{\underset{\displaystyle H}{|}}{N^+}}-H\ + OH^-$$

Amine　　Water　　　Alkylammonium　Hydroxide
　　　　　　　　　　ion　　　　　　ion

Neutralization of Amines

$$R-NH_2 + HCl \longrightarrow R-\overset{\overset{\displaystyle H}{|}}{\underset{\underset{\displaystyle H}{|}}{N^+}}-H\ Cl^-$$

Amine　　Acid　　Alkylammonium salt

Hydrolysis of Amides

$$R-\overset{\overset{\displaystyle O}{\|}}{C}-NH-R^1 + H_3O^+ \longrightarrow R-\overset{\overset{\displaystyle O}{\|}}{C}-OH + R^1-N^+H_3$$

Amide　　　　　Strong　　　Carboxylic　Alkyl-
　　　　　　　acid　　　　acid　　　ammonium
　　　　　　　　　　　　　　　　　ion

$$R-\overset{\overset{\displaystyle O}{\|}}{C}-NH-R^1 + NaOH \longrightarrow R-\overset{\overset{\displaystyle O}{\|}}{C}-O^-Na^+ + R^1-NH_2$$

Amide　　　　　Strong　　　Carboxylic　Amine
　　　　　　　base　　　　acid salt　or
　　　　　　　　　　　　　　　　　ammonia

SUMMARY

13.1 Carboxylic Acids

The functional group of the *carboxylic acids* is the *carboxyl group* (—COOH). Because the carboxyl group is extremely polar and carboxylic acids can form intermolecular hydrogen bonds, they have higher boiling points and melting points than alcohols. Lower molecular weight carboxylic acids are water-soluble and tend to taste sour and have unpleasant aromas. Longer chain carboxylic acids are called *fatty acids*. Carboxylic acids are named (I.U.P.A.C.) by replacing the *-e* ending of the parent compound with *-oic acid*. Common names are often derived from the source of the carboxylic acid. They are synthesized by the oxidation of primary alcohols or aldehydes. Carboxylic acids are weak acids. They are neutralized by strong bases to form salts.

13.2 Esters

Esters are mildly polar and have pleasant aromas. The boiling points and melting points of esters are comparable to those of aldehydes and ketones. Esters are formed from the reaction between carboxylic acids and alcohols. They can undergo hydrolysis back to the parent carboxylic acid and alcohol. The base-catalyzed hydrolysis of an ester is called *saponification*. *Soaps* are salts of long-chain carboxylic acids (fatty acids).

13.3 Amines

Amines are a family of organic compounds that contain an amino group or substituted amino group. A *primary amine* has the general formula RNH_2, a *secondary amine* has the general formula R_2NH, and a *tertiary amine* has the general formula R_3N. In the systematic nomenclature system, amines are named alkanamines. In the common system, they are named alkylamines. Amines behave as weak bases, forming *alkylammonium ions* in water and alkylammonium salts when they react with acids. *Quaternary ammonium salts* are ammonium salts that have four organic groups bonded to the nitrogen atom.

13.4 Amides

Amides are formed in a reaction between a carboxylic acid derivative and an amine (or ammonia). The *amide bond* is the bond between the carbonyl carbon of the *acyl group* and the nitrogen of the amine. In the I.U.P.A.C. Nomenclature System, they are named by replacing the *-oic acid* ending of the carboxylic acid with the *-amide* ending. In the common system of nomenclature the *-ic acid* ending of the carboxylic acid is replaced by the *-amide* ending. Hydrolysis of an amide produces a carboxylic acid and an amine (or ammonia).

KEY TERMS

acyl group (13.2)
alkylammonium ion (13.3)
amide (13.4)
amide bond (13.4)
amine (13.3)
carboxyl group (13.1)
carboxylic acid (13.1)
carboxylic acid
  derivative (13.2)
condensation polymer (13.2)

ester (13.2)
fatty acid (13.1)
hydrolysis (13.2)
primary (1°) amine (13.3)
quaternary ammonium
  salt (13.3)
saponification (13.2)
secondary (2°) amine (13.3)
soap (13.2)
tertiary (3°) amine (13.3)

QUESTIONS AND PROBLEMS

Carboxylic Acids: Structure and Properties

Foundations

13.27 The functional group is largely responsible for the physical and chemical properties of various chemical families. Explain why a carboxylic acid is more polar and has a higher boiling point than an alcohol or an aldehyde of comparable molecular weight.

13.28 Explain why carboxylic acids are weak acids.

Applications

13.29 Which member of each of the following pairs has the lower boiling point?
  a. Hexanoic acid or 3-hexanone
  b. 3-Hexanone or 3-hexanol
  c. 3-Hexanol or hexane
  d. Dipropyl ether or hexanal
  e. Hexanal or hexanoic acid
  f. Ethanol or ethanoic acid

13.30 Arrange the following from highest to lowest melting points:

Draw the condensed structural formula for each of these line drawings and provide the I.U.P.A.C. name for each.

Carboxylic Acids: Structure and Nomenclature

Foundations

13.31 Summarize the I.U.P.A.C. nomenclature rules for naming carboxylic acids.

13.32 Describe the rules for determining the common names of carboxylic acids.

Applications

13.33 Write the complete structural formulas for each of the following carboxylic acids:
  a. 2-Bromopentanoic acid
  b. 2-Bromo-3-methylbutanoic acid
  c. 2,4,6-Trimethylstearic acid
  d. Propenoic acid

13.34 Name each of the following carboxylic acids, using both the common and the I.U.P.A.C. Nomenclature Systems:

a. $H-\overset{\overset{\displaystyle O}{\|}}{C}-OH$

b. $CH_3\overset{\overset{\displaystyle CH_3}{|}}{C}HCH_2\overset{\overset{\displaystyle O}{\|}}{C}-OH$

c. $CH_3CH_2\overset{\overset{\displaystyle Br}{|}}{C}H\overset{}{C}H\underset{\underset{\displaystyle CH_3}{|}}{C}H_2\overset{\overset{\displaystyle O}{\|}}{C}-OH$

d. $CH_3CH_2\overset{\overset{\displaystyle CH_2CH_3}{|}}{C}HCH_2CH_2\overset{\overset{\displaystyle O}{\|}}{C}-OH$

13.35 Provide the common and I.U.P.A.C. names for each of the following compounds:

a. $CH_3\overset{\overset{\displaystyle OH}{|}}{C}H\overset{\overset{\displaystyle O}{\|}}{C}-OH$

b. $CH_3\overset{\overset{\displaystyle OH}{|}}{C}HCH_2\overset{\overset{\displaystyle O}{\|}}{C}-OH$

c. $CH_3\overset{\overset{\displaystyle CH_3}{|}}{\underset{\underset{\displaystyle CH_3}{|}}{C}}CH_2CH_2\overset{\overset{\displaystyle O}{\|}}{C}-OH$

d. $CH_3CH_2\overset{\overset{\displaystyle Cl}{|}}{\underset{\underset{\displaystyle Cl}{|}}{C}}CH_2\overset{\overset{\displaystyle O}{\|}}{C}-OH$

13.36 Draw the structure of each of the following carboxylic acids:
  a. β-Chlorobutyric acid
  b. α, β-Dibromovaleric acid
  c. β, γ-Dihydroxybutyric acid
  d. δ-Bromo-γ-chloro-β-methylcaproic acid

Carboxylic Acids: Reactions

Foundations

13.37 How is a soap prepared?

13.38 How do soaps assist in the removal of oil and grease from clothing?

Applications

13.39 Complete each of the following reactions by supplying the missing part(s) indicated by the question mark(s):

a. $CH_3CH_2CH_2OH \xrightarrow{?(1)} CH_3CH_2\overset{O}{\underset{\|}{C}}-OH \underset{?(3)}{\overset{NaOH}{\rightleftarrows}} ?(4)$

$\downarrow ?(2)$

$CH_3CH_2\overset{O}{\underset{\|}{C}}-\underset{\underset{CH_3}{|}}{O}CHCH_3$

b. $CH_3COOH + NaOH \longrightarrow ?$

c. $CH_3CH_2CH_2CH_2CH_2COOH + NaOH \longrightarrow ?$

d. $? + CH_3CH_2\overset{\overset{CH_3}{|}}{C}HOH \xrightarrow{H^+} CH_3\overset{O}{\underset{\|}{C}}-O\underset{\underset{CH_3}{|}}{C}HCH_2CH_3$

13.40 Complete each of the following reactions by supplying the missing products:
   a. $CH_3CH_2CH_2CH_2COOH + KOH$
   b. Benzoic acid + sodium hydroxide

Esters: Structure, Physical Properties, and Nomenclature

Foundations

13.41 Explain why esters are described as mildly polar.
13.42 Compare the boiling points of esters to those of aldehydes or ketones of similar molecular weight.

Applications

13.43 Write each of the following, using condensed formulas:
   a. Methyl benzoate
   b. Butyl decanoate
   c. Methyl propionate
   d. Ethyl propionate
13.44 Use the I.U.P.A.C. Nomenclature System to name each of the following esters:

a. $CH_3\overset{O}{\underset{\|}{C}}-OCH_2CH_3$

b. $CH_3CH_2\overset{O}{\underset{\|}{C}}-OCH_3$

c. $CH_3\overset{\overset{CH_3}{|}}{C}HCH_2\overset{O}{\underset{\|}{C}}-OCH_3$

d. $\underset{\underset{Br}{|}}{C}H_2\overset{}{C}H\underset{\underset{Br}{|}}{C}H_2CH_2\overset{O}{\underset{\|}{C}}-OCH_2CH_3$

Esters: Reactions

Foundations

13.45 Why is preparation of an ester referred to as a dehydration reaction?
13.46 What is meant by a hydrolysis reaction?

Applications

13.47 The structure of salicylic acid is shown. If this acid reacts with methanol, the product is an ester, methyl salicylate. Methyl salicylate is known as oil of wintergreen and is often used as a flavoring agent. Draw the structure of the product of this reaction.

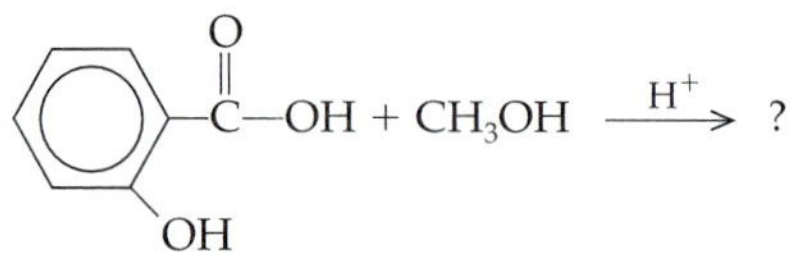

$$\text{(salicylic acid with } \overset{O}{\underset{\|}{C}}-OH \text{ and OH)} + CH_3OH \xrightarrow{H^+} ?$$

13.48 When salicylic acid reacts with acetic anhydride, one of the products is an ester, acetylsalicylic acid. Acetylsalicylic acid is the active ingredient in aspirin. Complete the equation below by drawing the structure of acetylsalicylic acid. (*Hint:* Acid anhydrides are hydrolyzed by water.)

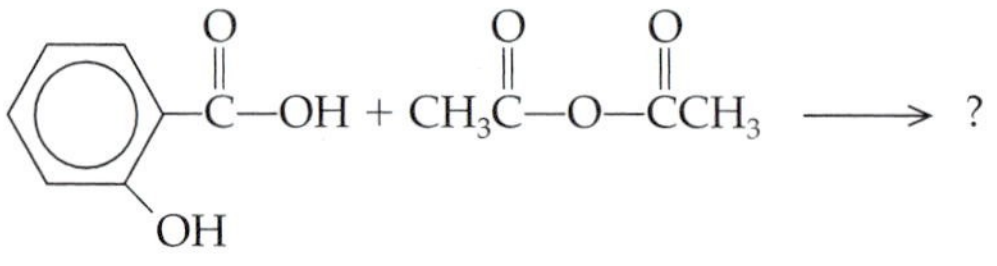

$$\text{(salicylic acid with } \overset{O}{\underset{\|}{C}}-OH \text{ and OH)} + CH_3\overset{O}{\underset{\|}{C}}-O-\overset{O}{\underset{\|}{C}}CH_3 \longrightarrow ?$$

13.49 Write an equation for the acid-catalyzed hydrolysis of each of the following esters:
   a. Propyl propanoate
   b. Butyl methanoate
   c. Ethyl methanoate
   d. Methyl pentanoate
13.50 Write an equation for the base-catalyzed hydrolysis of each of the following esters:
   a. Pentyl methanoate
   b. Hexyl propanoate
   c. Butyl hexanoate
   d. Methyl benzoate

Amines

Foundations

13.51 Describe the systematic rules for naming amines.
13.52 How are the common names of amines derived?
13.53 Describe the physiological effects of amphetamines.
13.54 Define the terms *analgesic* and *anesthetic* and list some amines that have these activities.

Applications

13.55 For each pair of compounds, predict which would have the higher boiling point. Explain your reasoning.
   a. Ethanamine or ethanol
   b. Butane or 1-propanamine
   c. Methanamine or water
   d. Ethylmethylamine or butane
13.56 Explain why a tertiary amine such as triethylamine has a significantly lower boiling point than its primary amine isomer, 1-hexanamine. Draw a diagram to illustrate your answer.
13.57 Use systematic and common nomenclature to name each of the following amines:
   a. $CH_3CH_2\overset{}{\underset{\underset{CH_3}{|}}{C}}HNH_2$

   b. $CH_3CH_2CH_2\overset{}{\underset{\underset{NH_2}{|}}{C}}HCH_2CH_3$

   c. $(CH_3)_3C-NH_2$
   d. $CH_3CH_2CH_2CH_2CH_2CH_2CH_2CH_2NH_2$
   e. $CH_3\overset{\overset{CH_3}{|}}{N}CH_2CH_3$

13.58 Draw the structure of each of the following compounds:
  a. Diethylamine
  b. Butylamine
  c. 3-Decanamine
  d. 3-Bromo-2-pentanamine

13.59 Draw the structure of each of the following compounds:
  a. Cyclohexanamine
  b. 2-Bromocyclopentanamine
  c. Tetraethylammonium iodide
  d. 3-Bromobenzenamine

13.60 Draw structural formulas for the eight isomeric amines that have the molecular formula $C_4H_{11}N$. Name each of the isomers using the systematic names and determine whether each isomer is a 1°, 2°, or 3° amine.

13.61 Classify each of the following amines as 1°, 2°, or 3°:
  a. Cyclohexanamine
  b. Dibutylamine
  c. 2-Methyl-2-heptanamine
  d. Tripentylamine

13.62 Classify each of the following amines as primary, secondary, or tertiary:
  a. Benzenamine
  b. *N*-Ethyl-2-pentanamine
  c. Ethylmethylamine
  d. Tripropylamine
  e. *m*-Chloroaniline

13.63 Complete each of the following reactions by supplying the missing reactant or product indicated by a question mark:

  a. 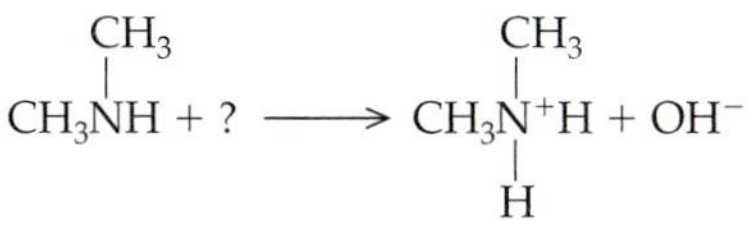 $CH_3\overset{\underset{\displaystyle |}{CH_3}}{N}H + ? \longrightarrow CH_3\overset{\underset{\displaystyle |}{CH_3}}{\underset{\underset{\displaystyle |}{H}}{N}}{}^+H + OH^-$

  b. 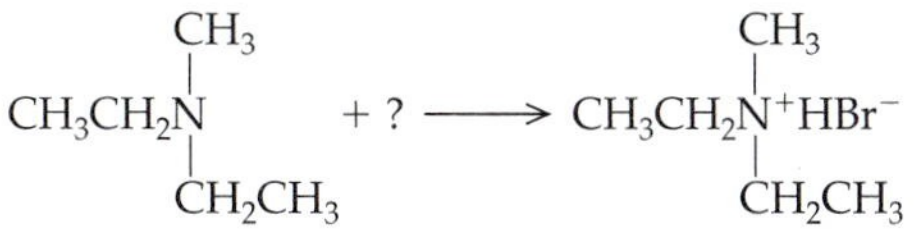 $CH_3CH_2\overset{\underset{\displaystyle |}{CH_3}}{\underset{\underset{\displaystyle |}{CH_2CH_3}}{N}} + ? \longrightarrow CH_3CH_2\overset{\underset{\displaystyle |}{CH_3}}{\underset{\underset{\displaystyle |}{CH_2CH_3}}{N}}{}^+HBr^-$

  c. $CH_3CH_2CH_2NH_2 + H_2O \longrightarrow ? + OH^-$

  d. 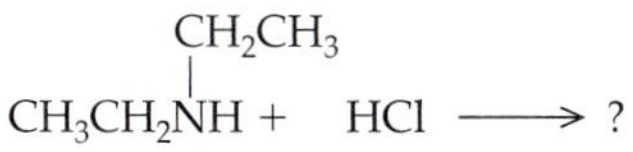 $CH_3CH_2\overset{\underset{\displaystyle |}{CH_2CH_3}}{N}H + HCl \longrightarrow ?$

13.64 Complete each of the following reactions by supplying the missing reactant or product indicated by a question mark:
  a. $CH_3CH_2NH_2 + H_2O \longrightarrow ? + OH^-$

  b. 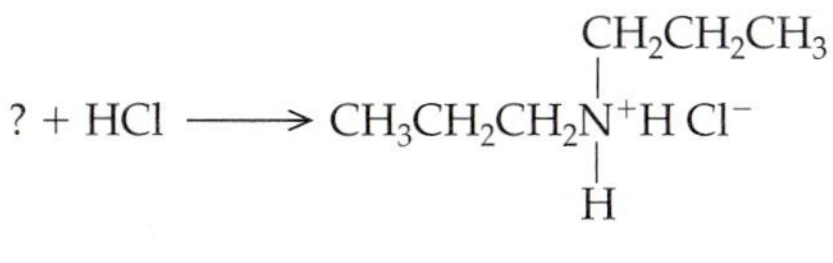 $? + HCl \longrightarrow CH_3CH_2CH_2\overset{\underset{\displaystyle |}{CH_2CH_2CH_3}}{\underset{\underset{\displaystyle |}{H}}{N}}{}^+H\,Cl^-$

  c. 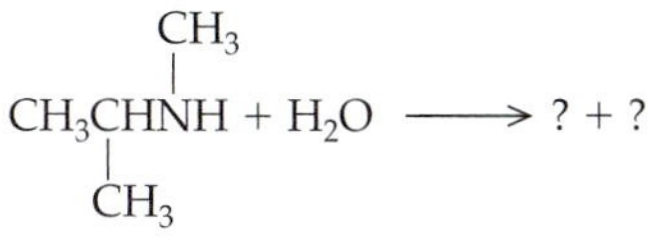 $CH_3\overset{\underset{\displaystyle |}{CH_3}}{\underset{\underset{\displaystyle |}{CH_3}}{C}}HNH + H_2O \longrightarrow ? + ?$

  d. $NH_3 + HBr \longrightarrow ?$

13.65 Most drugs containing amine groups are not administered as the amine but rather as the ammonium salt. Can you suggest a reason why?

13.66 How would you quickly convert an alkylammonium salt into a water-insoluble amine? Explain the rationale for your answer.

Amides

Foundations

13.67 Why do amides have very high boiling points?

13.68 Describe the water solubility of amides in relation to their carbon chain length.

Applications

13.69 Use the I.U.P.A.C. and common systems of nomenclature to name the following amides:

  a. $CH_3CH_2\overset{\overset{\displaystyle O}{\|}}{C}NH_2$

  b. $CH_3CH_2CH_2CH_2\overset{\overset{\displaystyle O}{\|}}{C}NH_2$

  c. $CH_3\overset{\overset{\displaystyle O}{\|}}{C}N(CH_3)_2$

13.70 Use the I.U.P.A.C. Nomenclature System to name each of the following amides:

  a. 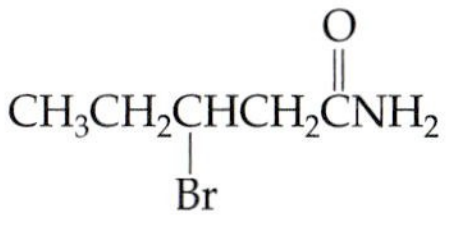 $CH_3CH_2\overset{\underset{\displaystyle |}{\underset{\displaystyle Br}{}}}{C}HCH_2\overset{\overset{\displaystyle O}{\|}}{C}NH_2$

  b. (structure of bromobenzene ring with $-\overset{\overset{\displaystyle O}{\|}}{C}NH_2$ group; Br substituent)

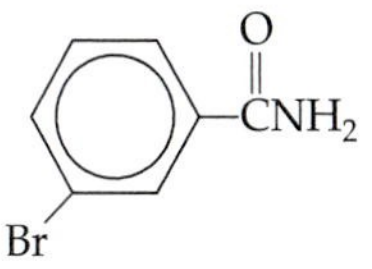

  c. $CH_3\overset{\underset{\displaystyle |}{CH_3}}{C}H\overset{\overset{\displaystyle O}{\|}}{C}NH_2$

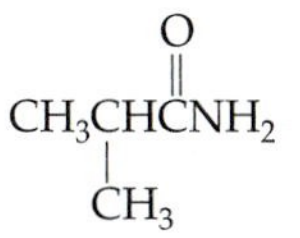

13.71 Draw the condensed structural formula of each of the following amides:
  a. Ethanamide
  b. *N*-Methylpropanamide
  c. *N,N*-Diethylbenzamide
  d. 3-Bromo-4-methylhexanamide
  e. *N,N*-Dimethylacetamide

13.72 Draw the condensed formula of each of the following amides:
  a. Acetamide
  b. 4-Methylpentanamide
  c. *N,N*-Dimethylpropanamide
  d. Formamide
  e. *N*-Ethylpropionamide

13.73 Explain why amides are neutral in the acid-base sense.

13.74 Lidocaine is often used as a local anesthetic. For medicinal purposes, it is often used in the form of its hydrochloride salt because the salt is water soluble. In the structure of lidocaine hydrochloride shown, locate the amide functional group.

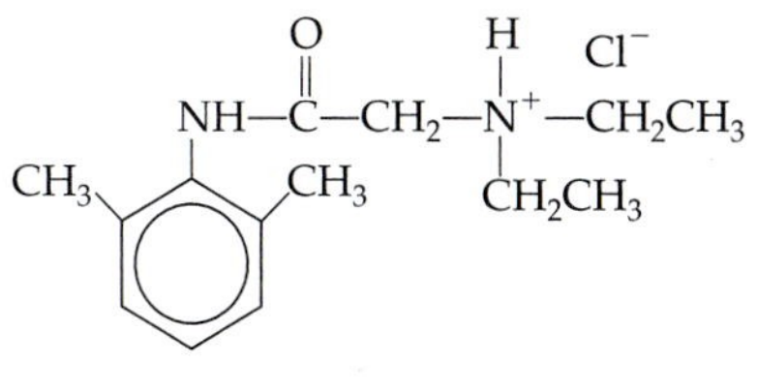

Lidocaine hydrochloride

13.75 The antibiotic penicillin BT contains functional groups discussed in this chapter. In the structure of penicillin BT shown, locate and name as many functional groups as you can.

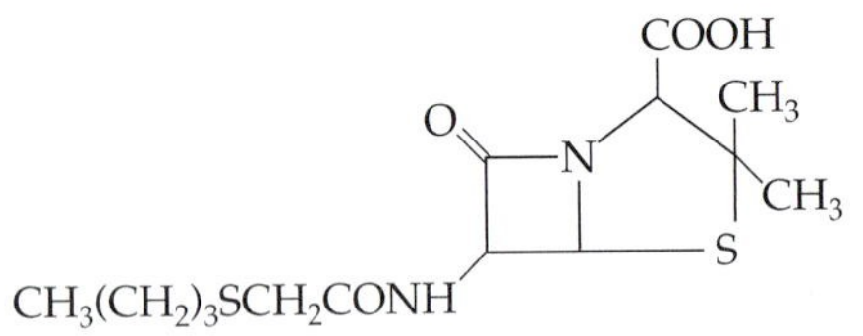

$$\text{CH}_3(\text{CH}_2)_3\text{SCH}_2\text{CONH}$$ ... Penicillin BT structure with COOH, CH₃, CH₃, O, N, S

Penicillin BT

13.76 The structure of saccharin, an artificial sweetener, is shown. Circle the amide group.

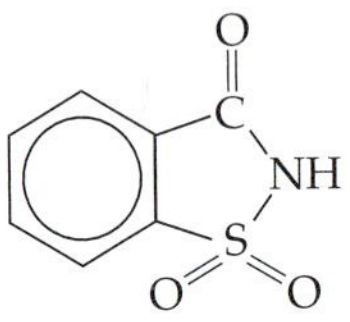

Saccharin

13.77 Complete each of the following reactions by supplying the missing reactant(s) or product(s) indicated by a question mark. Provide the systematic name for all the reactants and products.

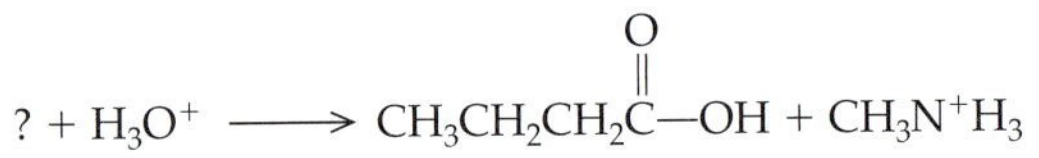

a. 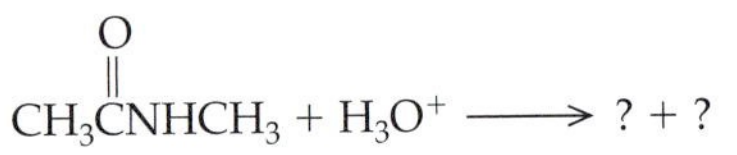 $\text{CH}_3\overset{\text{O}}{\underset{\|}{\text{C}}}\text{NHCH}_3 + \text{H}_3\text{O}^+ \longrightarrow ? + ?$

b. $? + \text{H}_3\text{O}^+ \longrightarrow \text{CH}_3\text{CH}_2\text{CH}_2\overset{\text{O}}{\underset{\|}{\text{C}}}-\text{OH} + \text{CH}_3\text{N}^+\text{H}_3$

c. $\text{CH}_3\underset{\underset{\text{CH}_3}{|}}{\text{CH}}\text{CH}_2\overset{\text{O}}{\underset{\|}{\text{C}}}\text{NHCH}_2\text{CH}_3 + ? \longrightarrow$

$\text{CH}_3\underset{\underset{\text{CH}_3}{|}}{\text{CH}}\text{CH}_2\overset{\text{O}}{\underset{\|}{\text{C}}}-\text{OH} + ?$

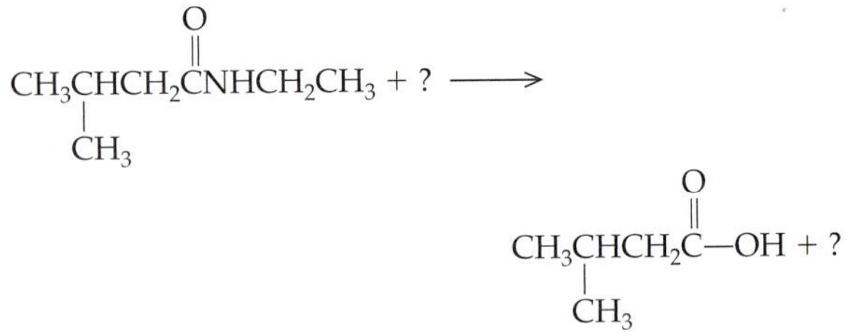

13.78 Complete each of the following by supplying the missing reagents. Draw the structures of each of the reactants and products.
a. *N*-Methylpropanamide + ? $\longrightarrow$ propanoic acid + ?
b. *N,N*-Dimethylacetamide + strong acid $\longrightarrow$? + ?
c. Formamide + strong acid $\longrightarrow$? + ?

FOR FURTHER UNDERSTANDING

1. Radioactive isotopes of an element behave chemically in exactly the same manner as the nonradioactive isotopes. As a result, they can be used as tracers to investigate the details of chemical reactions. A scientist is curious about the origin of the bridging oxygen atom in an ester molecule. She has chosen to use the radioactive isotope oxygen-18 to study the following reaction:

$$\text{CH}_3\text{CH}_2\text{OH} + \text{CH}_3\overset{\text{O}}{\underset{\|}{\text{C}}}-\text{OH} \xrightarrow{\text{H}^+,\text{ heat}} \text{CH}_3\overset{\text{O}}{\underset{\|}{\text{C}}}-\text{O}-\text{CH}_2\text{CH}_3 + \text{H}_2\text{O}$$

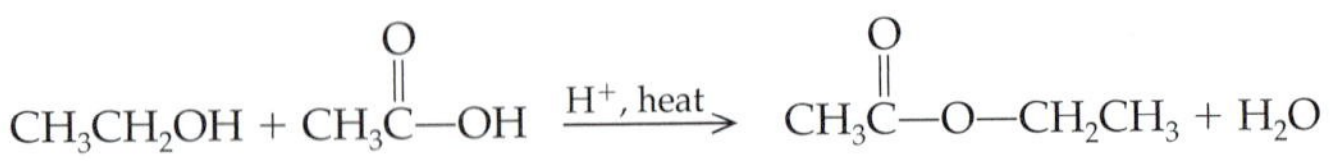

Design experiments using oxygen-18 that will demonstrate whether the oxygen in the water molecule came from the —OH of the alcohol or the —OH of the carboxylic acid.

2. Triglycerides are the major lipid storage form in the human body. They are formed in an esterification reaction between glycerol (1,2,3-propanetriol) and three fatty acids (long-chain carboxylic acids). Write a balanced equation for the formation of a triglyceride formed in a reaction between glycerol and three molecules of decanoic acid.

3. Chloramphenicol is a very potent, broad-spectrum antibiotic. It is reserved for life-threatening bacterial infections because it is quite toxic. It is also a very bitter tasting chemical. As a result, children had great difficulty taking the antibiotic. A clever chemist found that the taste could be improved considerably by producing the palmitate ester. Intestinal enzymes hydrolyze the ester, producing chloramphenicol, which can then be absorbed. The following structure is the palmitate ester of chloramphenicol. Draw the structure of chloramphenicol.

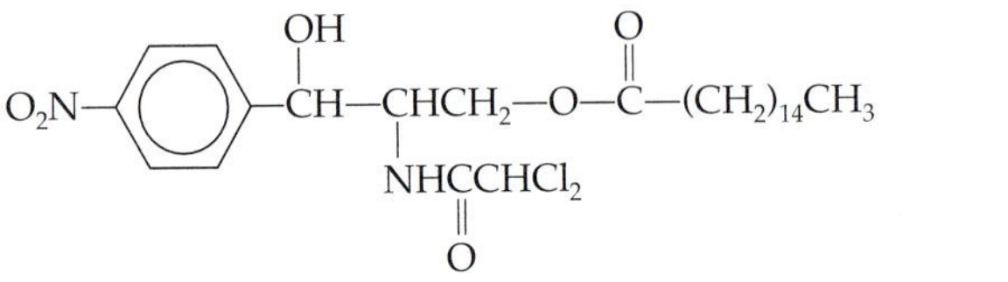

$$\text{O}_2\text{N}-\overset{\overset{\text{OH}}{|}}{\bigcirc}-\text{CH}-\underset{\underset{\underset{\underset{\|}{\text{O}}}{\text{NHCCHCl}_2}}{|}}{\text{CH}}\text{CH}_2-\text{O}-\overset{\text{O}}{\underset{\|}{\text{C}}}-(\text{CH}_2)_{14}\text{CH}_3$$

Chloramphenicol palmitate

4. Acetyl coenzyme A (acetyl CoA) can serve as a donor of acetate groups in biochemical reactions. One such reaction is the formation of acetylcholine, an important neurotransmitter involved in nerve signal transmission at neuromuscular junctions. The structure of choline is shown below. Draw the structure of acetylcholine.

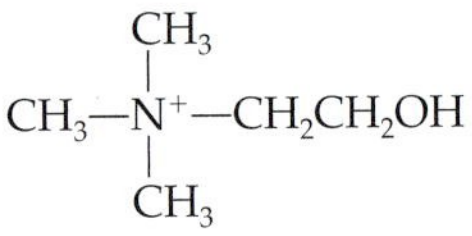

$$\text{CH}_3-\overset{\overset{\text{CH}_3}{|}}{\underset{\underset{\text{CH}_3}{|}}{\text{N}^+}}-\text{CH}_2\text{CH}_2\text{OH}$$

Choline

5. Histamine is made and stored in blood cells called *mast cells*. Mast cells are involved in the allergic response. Release of histamine in response to an allergen causes dilation of capillaries. This, in turn, allows fluid to leak out of the capillary resulting in local swelling. It also causes an increase in the volume of the vascular system. If this increase is great enough, a severe drop in blood pressure may cause shock. Histamine is produced by decarboxylation (removal of the carboxylate group as CO_2) of the amino acid histidine shown below. Draw the structure of histamine.

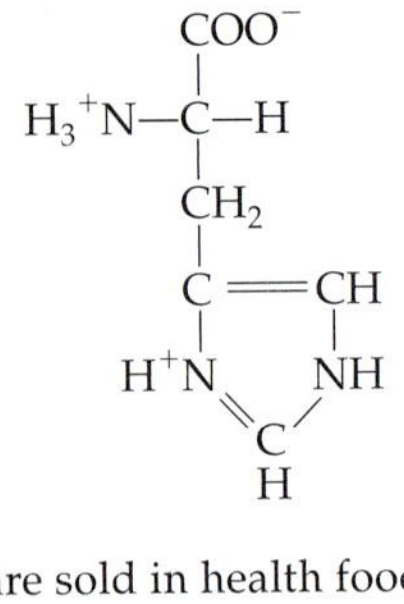

Histidine structure with COO⁻, H₃⁺N—C—H, CH₂, C=CH, H⁺N, NH, C, H

6. Carnitine tablets are sold in health food stores. It is claimed that carnitine will enhance the breakdown of body fat. Carnitine is a tertiary amine found in mitochondria, cell organelles in which food molecules are completely oxidized and ATP is produced. Carnitine is involved in transporting the acyl groups of fatty acids from the cytoplasm into the mitochondria. The fatty acyl group is transferred from a fatty

acyl CoA molecule and esterified to carnitine. Inside the mitochondria, the reaction is reversed and the fatty acid is completely oxidized. The structure of carnitine is shown here:

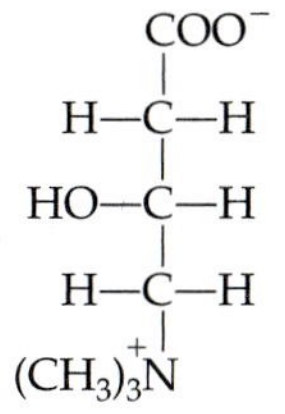

$$
\begin{array}{c}
COO^- \\
| \\
H-C-H \\
| \\
HO-C-H \\
| \\
H-C-H \\
| \\
(CH_3)_3\overset{+}{N}
\end{array}
$$

Draw the acyl carnitine molecule that is formed by esterification of palmitic acid with carnitine.

7. Bulletproof vests are made of the polymer called Kevlar. It is produced by the copolymerization of the following two monomers:

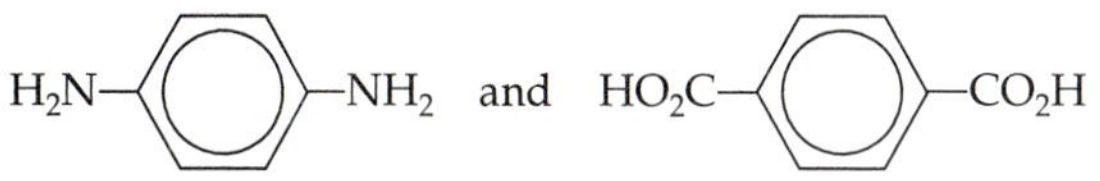

$H_2N-\bigcirc-NH_2$ and $HO_2C-\bigcirc-CO_2H$

Draw the structure of a portion of Kevlar polymer.

14

LEARNING GOALS

1 Explain the difference between complex and simple carbohydrates and know the amounts of each recommended in the daily diet.

2 Apply the systems of classifying and naming monosaccharides according to the functional group and number of carbons in the chain.

3 Explain stereoisomerism.

4 Determine whether a molecule has a chiral center.

5 Identify monosaccharides as either D- or L-.

6 Draw and name the common monosaccharides using structural formulas.

7 Given the linear structure of a monosaccharide, draw the Haworth projection of its α- and β-cyclic forms and vice versa.

8 By inspection of the structure, predict whether a sugar is reducing or nonreducing.

9 Discuss the use of Benedict's reagent to measure the level of glucose in urine.

10 Draw and name the common disaccharides and discuss their significance in biological systems.

11 Describe the difference between galactosemia and lactose intolerance.

12 Discuss the structural, chemical, and biochemical properties of starch, glycogen, and cellulose.

Carbohydrates

E mil Fischer's father, a wealthy businessman, once said that Emil was too stupid to be a businessman and had better be a student. Lucky for the field of biochemistry, Emil did just that. Originally he wanted to study physics, but his cousin Otto Fischer convinced him to study chemistry. Beginning as an organic chemist, Fischer launched a career that eventually led to groundbreaking research in biochemistry. By following his career, we get a glimpse of the entire field.

Early on, Fischer discovered the active ingredients in coffee and tea, caffeine and theobromine. Eventually he discovered their structures and synthesized them in the laboratory. In work that he carried out between 1882 and 1906, Fischer demonstrated that adenine and guanine belong to a family of compounds he called purines. All purines have the same core structure and differ from one another by the functional groups attached to the ring.

We will study the purines in Chapter 17 where we will learn that they and another type of nitrogenous base, the pyrimidines, are essential components of the genetic molecules DNA and RNA. By reading the order of these nitrogenous bases, we can decipher the genetic code of an organism.

In 1884, Fischer began his monumental work on sugars. In 1890, he established that sugars could exist as stereoisomers, and between 1891 and 1894, he worked out the stereochemical configuration of all known sugars found in nature. Fischer studied virtually all sugars, but one of his greatest successes was the synthesis of glucose, fructose, and mannose, three six-carbon sugars. These and other sugars are the subject of this chapter.

Between 1899 and 1908, Fischer turned his attention to proteins. He developed methods to separate and identify individual amino acids. Each of the amino acids has a common core; they differ from one another in the nature of a side chain, or R group that gives it unique properties. Fischer also worked on synthesis of proteins from amino acids. He demonstrated the nature of the peptide bond and discovered how to make small peptides in the laboratory. In Chapter 16 we will study the amino acids and the structures of some of the proteins essential for life.

Quite late in his career, Fischer even got around to studying fats, a heterogeneous group of substances characterized by their hydrophobic nature. In Chapter 15, we will study the incredibly diverse family of fats, or lipids.

In 1902, Fischer was awarded the Nobel Prize for his work on sugar and purine synthesis. But a glance at all of his accomplishments helps us to realize that this great man, who began his career as an organic chemist, established the field of biochemistry through his extraordinary studies of the molecules of life.

Carbohydrates are an essential component of a well-rounded diet. Make a list of the types of carbohydrates in your diet and the roles they play in your physiology and health.

14.1 Types of Carbohydrates

1 **Learning Goal**

Explain the difference between complex and simple carbohydrates and know the amounts of each recommended in the daily diet.

We begin our study of biochemistry with carbohydrates. Carbohydrates are produced in plants by photosynthesis (Figure 14.1).

$$6\ CO_2 + 6\ H_2O \xrightarrow{\text{Sunlight}} C_6H_{12}O_6 + 6\ O_2$$

Carbon dioxide   Water   Sugar (glucose)   Oxygen

Natural carbohydrate sources such as grains and cereals, breads, sugar cane, fruits, milk, and honey are an important source of energy for animals. **Carbohydrates** include simple sugars having the general formula $(CH_2O)_n$, as well as long polymers of these simple sugars, for instance, potato starch, and a variety of molecules of intermediate size. The simple sugar glucose, $C_6H_{12}O_6$, is the primary energy source for the brain and nervous system and can be used by many other tissues. When "burned" by cells for energy, each gram of carbohydrate releases approximately four kilocalories of energy.

A healthy diet should contain both complex carbohydrates, such as starches and cellulose, and simple sugars, such as fructose and sucrose (Figure 14.2). However, the quantity of simple sugars, especially sucrose, should be minimized because large quantities of sucrose in the diet promote obesity and tooth decay.

> A kilocalorie is the same as the calorie referred to in the "count-your-calories" books and on nutrition labels.

Complex carbohydrates are better for us than simple sugars. Starch, found in rice, potatoes, breads, and cereals, is an excellent energy source. In addition, complex carbohydrates, such as cellulose, provide us with an important supply of dietary fiber.

See A Medical Connection: Tooth Decay and Simple Sugars on page 414.

It is hard to determine exactly what percentage of the daily diet *should* consist of carbohydrates. The *actual* percentage varies widely throughout the world, from 80% in the Far East, where rice is the main component of the diet, to 40–50% in the United States. Currently, it is recommended that 45–65% of the calories in the diet should come from carbohydrates and that no more than 10% of the daily caloric intake should be sucrose.

QUESTION 14.1

What is the current recommendation for the amount of carbohydrates that should be included in the diet? Of the daily intake of carbohydrates, what percentage should be simple sugar?

QUESTION 14.2

Distinguish between simple and complex sugars. What are some sources of complex carbohydrates?

Figure 14.1

Carbohydrates are produced by plants such as this potato in the process of photosynthesis, which uses the energy of sunlight to produce hexoses from CO_2 and H_2O.

Monosaccharides such as glucose and fructose are the simplest carbohydrates because they contain a single (*mono-*) sugar (*saccharide*) unit. **Disaccharides,** including sucrose and lactose, consist of two monosaccharide units joined through bridging oxygen atoms. Such a bond is called a **glycosidic bond. Oligosaccharides** consist of three to ten monosaccharide units joined by glycosidic bonds. The largest and most complex carbohydrates are **polysaccharides,** which are long, often highly branched, chains of monosaccharides. Starch, glycogen, and cellulose are examples of polysaccharides.

Figure 14.2

Carbohydrates from a variety of foods are an essential component of the diet.

14.2 Monosaccharides

Monosaccharides are composed of carbon, hydrogen, and oxygen, and most are characterized by the general formula $(CH_2O)_n$, in which n is any whole number from 3 to 7. They can be named on the basis of the functional groups they contain. A monosaccharide with a ketone (carbonyl) group is a **ketose.** If an aldehyde (carbonyl) group is present, it is called an **aldose.** Because monosaccharides also contain many hydroxyl groups, they are sometimes called *polyhydroxyaldehydes* or *polyhydroxyketones.*

This general formula is an oversimplification because several biologically important monosaccharides, such as the blood group antigens, are chemically modified; and deoxyribose, the monosaccharide found in DNA, has one fewer oxygen atom than the general formula above would predict.

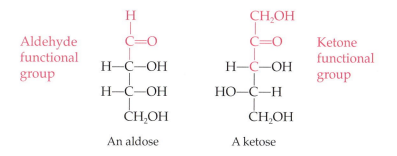

An aldose   A ketose

Another system of nomenclature tells us the number of carbon atoms in the main skeleton. A three-carbon monosaccharide is a *triose,* a four-carbon sugar is a *tetrose,* a five-carbon sugar is a *pentose,* a six-carbon sugar is a *hexose,* and so on. Combining the two naming systems gives even more information about the sugar. For example, an aldotetrose is a four-carbon sugar that is also an aldehyde.

2 Learning Goal

Apply the systems of classifying and naming monosaccharides according to the functional group and number of carbons in the chain.

A MEDICAL Connection | Tooth Decay and Simple Sugars

How many times have you heard the lecture from parents or your dentist about brushing your teeth after a sugary snack? Annoying as this lecture might be, it is based on sound scientific data that demonstrate that the cause of tooth decay is plaque and acid formed by the bacterium *Streptococcus mutans* using sucrose as its substrate.

Saliva is teeming with bacteria in concentrations up to one hundred million (10^8) per milliliter of saliva! Within minutes after you brush your teeth, sticky glycoproteins in the saliva adhere to tooth surfaces. Then millions of oral bacteria immediately bind to this surface.

Although many oral bacteria stick to the tooth surface, as the accompanying diagram shows, only *S. mutans* causes cavities. The reason for this is that this organism alone can make the enzyme *glucosyl transferase*. This enzyme acts only on the disaccharide sucrose, breaking it down into glucose and fructose. The glucose is immediately added to a growing polysaccharide called *dextran*, the glue that allows the bacteria to adhere to the tooth surface, contributing to the formation of plaque.

Now the bacteria embedded in the dextran take in the fructose and use it in the lactic acid fermentation. The lactic acid that is produced lowers the pH on the tooth surface and begins to dissolve calcium from the tooth enamel. Even though we produce about one liter of saliva each day, the acid cannot be washed away from the tooth surface because the dextran plaque is not permeable to saliva.

So what can we do to prevent tooth decay? Of course, brushing after each meal and flossing regularly reduce plaque buildup. Eating a diet rich in calcium also helps build strong tooth enamel. Foods rich in complex carbohydrates, such as fruits and vegetables, help prevent cavities in two ways. Glucosyl transferase can't use complex carbohydrates in its cavity-causing chemistry, and eating fruits and vegetables helps to remove plaque mechanically.

Perhaps the most effective way to prevent tooth decay is to avoid sucrose-containing snacks between meals. Studies have shown that eating sucrose-rich foods doesn't cause much tooth decay if followed immediately by brushing. However, even small amounts of sugar eaten between meals actively promote cavity formation.

FOR FURTHER UNDERSTANDING

It has been suggested that tooth decay could be prevented by a vaccine that would rid the mouth of *Streptococcus mutans*. Explain this from the point of view of the chemical reactions that are described above.

What steps could you take following a sugary snack to help prevent tooth decay, even when it is not possible to brush your teeth?

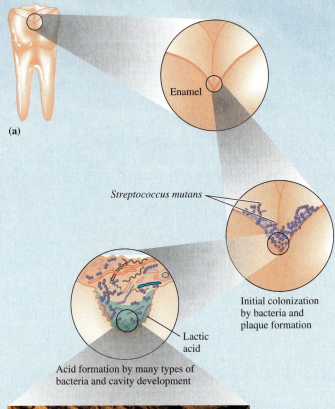

(a)

Enamel

Streptococcus mutans

Initial colonization by bacteria and plaque formation

Lactic acid

Acid formation by many types of bacteria and cavity development

(b)

(a) The complex process of tooth decay. (b) Electron micrograph of dental plaque.

A MEDICAL Connection

Chemistry Through the Looking Glass

In his children's story *Through the Looking Glass*, Lewis Carroll's heroine Alice wonders whether "looking-glass milk" would be good to drink. Many biological molecules, such as the sugars, exist as two stereoisomers, *enantiomers*, that are mirror images of one another. Because two mirror-image forms occur, it is rather remarkable that in our bodies, and in most of the biological world, only one of the two is found. For instance, the common sugars are members of the D-family, whereas all the common amino acids that make up our proteins are members of the L-family. It is not too surprising, then, that the enzymes in our bodies that break down the sugars and proteins we eat are *stereospecific*, that is, they recognize only one mirror-image isomer. Knowing this, we can make an educated guess that "looking-glass milk" could not be digested by our enzymes and therefore would not be a good source of food for us. It is even possible that it might be toxic to us!

Pharmaceutical chemists are becoming more and more concerned with the stereochemical purity of the drugs that we take. Consider a few examples. In 1960, the drug thalidomide was commonly prescribed in Europe as a sedative. However, during that year, hundreds of women who took thalidomide during pregnancy gave birth to babies with severe birth defects. Thalidomide, it turned out, was a mixture of two enantiomers. One is a sedative; the other is a teratogen, a chemical that causes birth defects.

One of the common side effects of taking antihistamines for colds or allergies is drowsiness. Again, this is the result of the fact that antihistamines are mixtures of enantiomers. One causes drowsiness; the other is a good decongestant.

One enantiomer of the compound carvone is associated with the smell of spearmint; the other produces the aroma of caraway seeds or dill. One mirror-image form of limonene smells like lemons; the other has the aroma of oranges.

The pain reliever ibuprofen is currently sold as a mixture of enantiomers, but one is a much more effective analgesic than the other.

Taste, smell, and the biological effects of drugs in the body all depend on the stereochemical form of compounds and their interactions with cellular enzymes or receptors. As a result, chemists are actively working to devise methods of separating the isomers in pure form. Alternatively, methods of conducting stereospecific syntheses that produce only one stereoisomer are being sought. By preparing pure stereoisomers, the biological activity of a compound can be much more carefully controlled. This will lead to safer medications.

FOR FURTHER UNDERSTANDING

The substrate for an enzyme must fit precisely into the active site of an enzyme. That means that the shape of the active site must be complementary to the shape of the substrate. It also means that the charge distribution must be complementary. For instance, if there is a partial positive charge on the substrate, there must be a complementary partial negative charge on the surface of the enzyme. Draw the structures of D- and L-glyceraldehyde and show why an enzyme that binds to D-glyceraldehyde will not be able to bind tightly to L-glyceraldehyde.

Pharmaceutical companies are investing large sums of money in the development of drugs that are stereochemically pure. Explain the importance of this line of research.

Each monosaccharide also has a unique name. These names are shown in blue for the following structures. Because monosaccharides can exist in several different isomeric forms, it is important to provide the complete name. Thus the complete names of the following structures are D-glyceraldehyde, D-glucose, and D-fructose. These names tell us that the structure represents one particular sugar and also identifies the sugar as one of two possible isomeric forms (D- or L-).

$$
\begin{array}{ccc}
H & H & CH_2OH \\
| & | & | \\
C=O & C=O & C=O \\
| & | & | \\
H-C-OH & H-C-OH & HO-C-H \\
| & | & | \\
CH_2OH & HO-C-H & H-C-OH \\
 & | & | \\
 & H-C-OH & H-C-OH \\
 & | & | \\
 & H-C-OH & CH_2OH \\
 & | & \\
 & CH_2OH &
\end{array}
$$

Aldose	Aldose	Ketose
Triose	Hexose	Hexose
Aldotriose	Aldohexose	Ketohexose
D-Glyceraldehyde	D-Glucose	D-Fructose

QUESTION 14.3

What is the structural difference between an aldose and a ketose?

QUESTION 14.4

Explain the difference between:

a. A ketohexose and an aldohexose
b. A triose and a pentose

14.3 Stereoisomers and Stereochemistry

Stereoisomers

The prefixes D- and L- found in the complete name of a monosaccharide are used to identify one of two possible isomeric forms called **stereoisomers.** By definition, each member of a pair of stereoisomers must have the same molecular formula and the same bonding. How then do isomers of the D-family differ from those of the L-family? D- and L-isomers differ in the spatial arrangements of atoms in the molecule.

Stereochemistry is the study of the different spatial arrangements of atoms. A general example of a pair of stereoisomers is shown in Figure 14.3. In this example, the general molecule C-abcd is formed from the bonding of a central carbon to four different groups: a, b, c, and d. This results in two molecules rather than one. Each isomer is bonded together through the exact *same* bonding pattern, yet they are *not* identical. If they were identical, they would be superimposable one upon the other; *they are not.* They are therefore stereoisomers. These two stereoisomers have a mirror-image relationship that is analogous to the mirror-image relationship of the left and right hands (see Figure 14.3b).

Two stereoisomers that are nonsuperimposable mirror images of one another are called a pair of **enantiomers.** Molecules that can exist in enantiomeric forms are called **chiral molecules.** The term simply means that as a result of different three-dimensional arrangements of atoms, the molecule can exist in two mirror-image forms. For any pair of nonsuperimposable mirror-image forms (enantiomers), one is always designated D- and the other L-.

3 Learning Goal

Explain stereoisomerism.

Build models of these compounds using toothpicks and gumdrops of five different colors to prove this to yourself.

Animation

Chiral Molecules
Chiral Molecules (B)

You can find further information online at www.mhhe.com/denniston in Stereochemistry and Stereoisomers Revisited.

Mirror

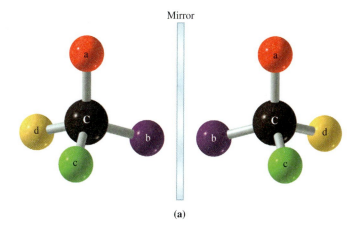

(a)

Nonsuperimposable mirror images:
enantiomers

Mirror

(b)

Figure 14.3

(a) A pair of enantiomers for the general molecule C–abcd. (b) Mirror-image right and left hands.

A carbon atom that has four different groups bonded to it is called a **chiral carbon** atom. Any molecule containing a chiral carbon can exist as a pair of enantiomers. Consider the simplest carbohydrate, **glyceraldehyde,** which is shown in Figure 14.4. Note that the second carbon is bonded to four different groups. It is therefore a chiral carbon. As a result, we can draw two enantiomers of glyceraldehyde that are nonsuperimposable mirror images of one another. Larger biological molecules typically have more than one chiral carbon.

4   **Learning Goal**
Determine whether a molecule has a chiral center.

Rotation of Plane–Polarized Light

Stereoisomers can be distinguished from one another by their different optical properties. Each member of a pair of stereoisomers will rotate plane-polarized light in a different direction.

White light is a form of electromagnetic radiation that consists of many different wavelengths (colors) vibrating in *planes* that are all perpendicular to the direction of the light beam (Figure 14.5). To measure optical properties of enantiomers, scientists use special light sources to produce *monochromatic light,* that is, light of a single wavelength. The monochromatic light is passed through a polarizing material, like a Polaroid lens, so that only waves in one plane can pass through. The light that emerges from the lens is *plane-polarized light* (Figure 14.5).

Applying these principles, scientists have developed the *polarimeter* to measure the ability of a compound to change the angle of the plane of plane-polarized light (Figure 14.5). The polarimeter allows the determination of the specific rotation of a compound, that is, the measure of its ability to rotate plane-polarized light.

 The polarimeter, measurement of the rotation of plane-polarized light, and the calculation of specific rotation are discussed in detail online at www.mhhe.com/denniston in "Stereochemistry and Stereoisomers Revisited."

Figure 14.4

(a) Structural formulas of D- and L-glyceraldehyde. The end of the molecule with the carbonyl group is the most oxidized end. The D- or L-configuration of a monosaccharide is determined by the orientation of the functional groups attached to the chiral carbon farthest from the oxidized end. In the D-enantiomer, the —OH is to the right. In the L-enantiomer, the —OH is to the left. (b) A three-dimensional representation of D- and L-glyceraldehyde.

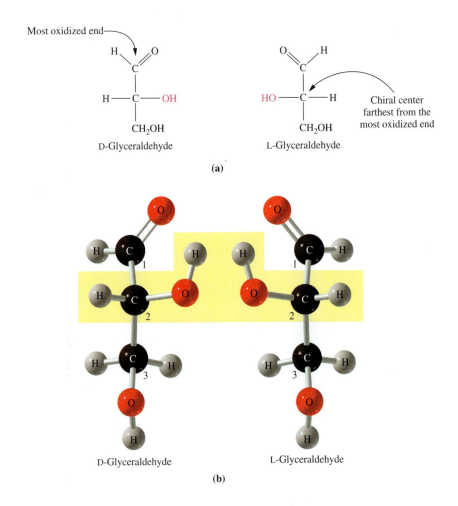

(a)

(b)

Figure 14.5

Schematic drawing of a polarimeter.

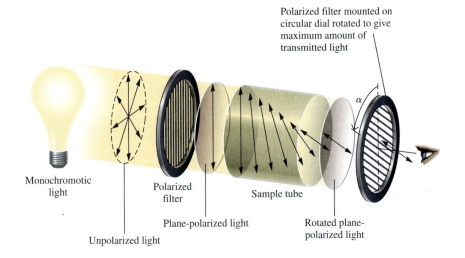

Some compounds rotate light in a clockwise direction. These are said to be *dextrorotatory* and are designated by a plus sign (+) before the specific rotation value. Other substances rotate light in a counterclockwise direction. These are called *levorotatory* and are indicated by a minus sign (−) before the specific rotation value.

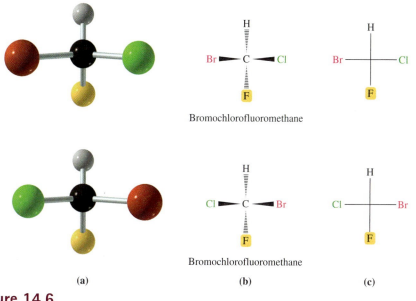

Figure 14.6

Drawing a Fischer projection. (a) The ball-and-stick models for the stereoisomers of bromochlorofluoromethane. (b) The wedge-and-dash and (c) Fischer projections of these molecules.

The Relationship Between Molecular Structure and Optical Activity

In 1848, Louis Pasteur was the first to see a relationship between the structure of a compound and the effect of that compound on plane-polarized light. In his studies of winemaking, Pasteur noticed that salts of tartaric acid formed in the wine. He also noticed that two types of crystals, mirror images of one another, were produced. When he dissolved each type of crystal in water, he found that they rotated plane-polarized light in opposite directions. From this and the work of others, scientists realized that two enantiomers, which are identical to one another in all other chemical and physical properties, will rotate plane-polarized light to the same degree, but in opposite directions.

Fischer Projection Formulas

Emil Fischer devised a simple way to represent the structure of stereoisomers. The **Fischer projection** is a two-dimensional drawing of a molecule that shows a chiral carbon at the intersection of two lines. The horizontal lines represent bonds projecting out of the page, and the vertical lines represent bonds that project into the page. Figure 14.6 shows how to draw Fischer projections for the stereoisomers of bromochlorofluoromethane.

EXAMPLE 14.1 Drawing Fischer Projections for a Sugar

Draw the Fischer projections for the stereoisomers of glyceraldehyde.

SOLUTION

Review the structures of the two stereoisomers of glyceraldehydes (Figure 14.4b). The ball-and-stick models can be represented using three-dimensional wedge

Continued—

5 Learning Goal

Identify monosaccharides as either D- or L-.

EXAMPLE 14.1 —*Continued*

drawings. Remember that the most oxidized carbon for sugars (the aldehyde or ketone group) is always drawn at the top of the structure.

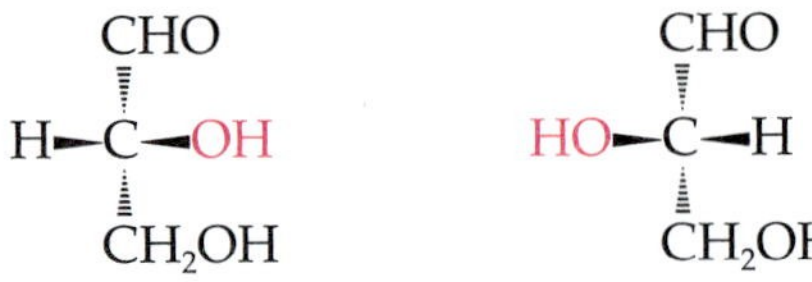

```
        CHO                    CHO
         |                      |
    H — C — OH          HO — C — H
         |                      |
      CH₂OH                  CH₂OH
   D-Glyceraldehyde      L-Glyceraldehyde
```

Remember that in the wedge diagram, the solid wedges represent bonds directed toward the reader. The dashed wedges represent bonds directed away from the reader and into the page. In these molecules, the center carbon is the only chiral carbon in the structure. To convert these wedge representations to a Fischer projection, simply use a horizontal line in place of each solid wedge and use a vertical line to represent each dashed wedge. The chiral carbon is represented by the point at which the vertical and horizontal lines cross, as shown below.

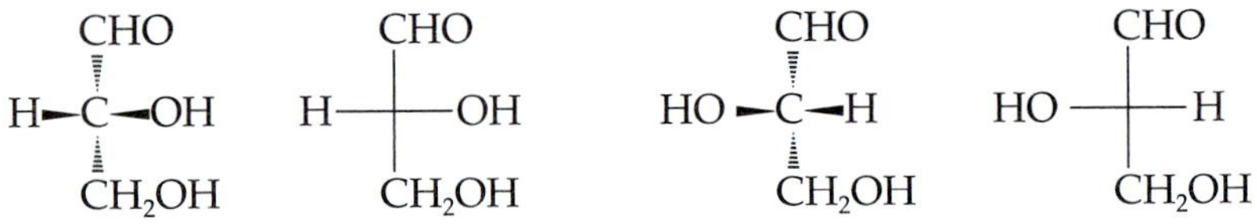

```
   CHO         CHO              CHO          CHO
    |           |                |            |
H—C—OH      H——OH         HO—C—H      HO——H
    |           |                |            |
  CH₂OH       CH₂OH            CH₂OH        CH₂OH

   D-Glyceraldehyde              L-Glyceraldehyde
```

QUESTION 14.5

Indicate whether each of the following molecules is an aldose or a ketose.

a.
```
   CH₃
    |
    C=O
    |
H—C—OH
    |
  CH₂OH
```

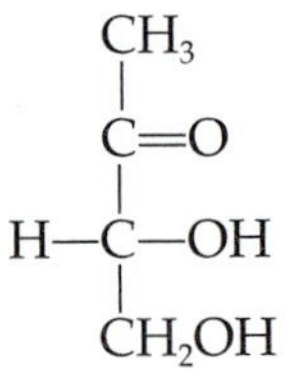

b.
```
    H
    |
    C=O
    |
H—C—OH
    |
H—C—OH
    |
HO—C—H
    |
  CH₂OH
```

c.
```
  CH₂OH
    |
    C=O
    |
HO—C—H
    |
H—C—OH
    |
H—C—OH
    |
  CH₂OH
```

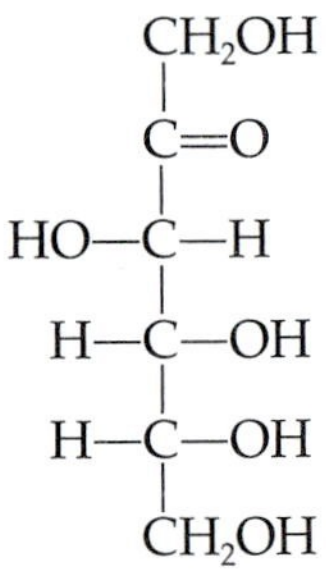

d.
```
    H
    |
    C=O
    |
H—C—OH
    |
  CH₂OH
```

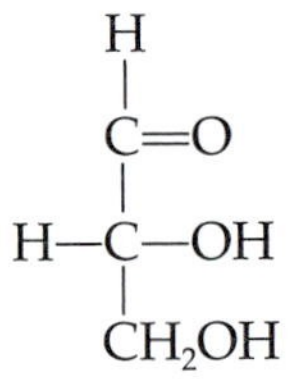

e.
```
   CH₃
    |
    C=O
    |
H—C—OH
    |
H—C—OH
    |
H—C—OH
    |
  CH₂OH
```

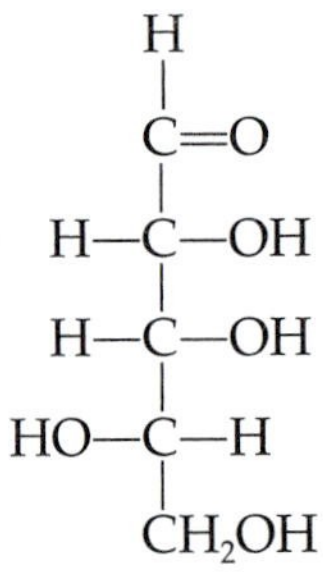

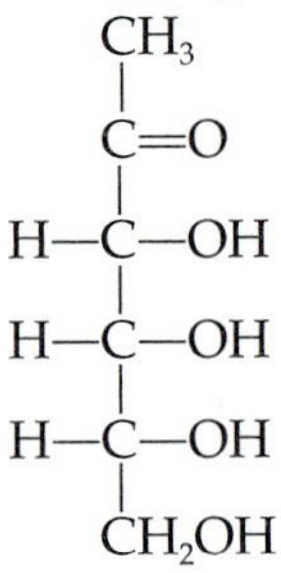

f.
```
    H
    |
    C=O
    |
HO—C—H
    |
H—C—OH
    |
HO—C—H
    |
  CH₂OH
```

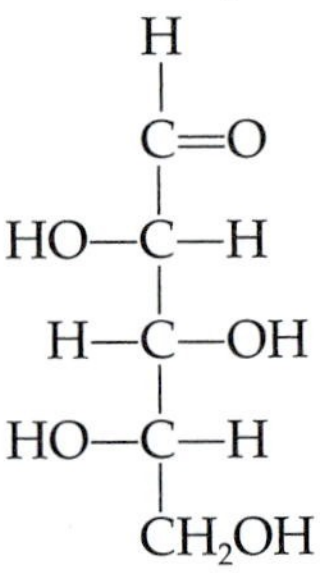

QUESTION 14.6

Draw the Fischer projection for each molecule in Question 14.5.

QUESTION 14.7

Place an asterisk beside each chiral carbon in the Fischer projections you drew for Question 14.6.

QUESTION 14.8

In the Fischer projections you drew for Question 14.6, indicate which bonds project toward you and which project into the page.

The D- and L-System of Nomenclature

In 1891 Emil Fischer came up with a nomenclature system that would allow scientists to distinguish between enantiomers.

Although specific rotation is an experimental value that must be measured, the D- and L-designations of all other monosaccharides are determined by comparison of their structures with D- and L-glyceraldehyde. Sugars with more than three carbons will have more than one chiral carbon. *By convention*, the position of the hydroxyl group on the chiral carbon farthest from the carbonyl group (the most oxidized end of the molecule) determines whether a monosaccharide is in the D- or L-configuration. If the —OH group is on the right, the molecule is in the D-configuration. If the —OH group is on the left, the molecule is in the L-configuration. Almost all carbohydrates in living systems are members of the D-family.

> Fischer named the molecule with the —OH group on the right D-glyceraldehyde and the molecule with the —OH on the left L-glyceraldehyde.

> The structures and designations of D- and L-glyceraldehyde are defined by convention. The D- and L-terminology is generally applied only to carbohydrates and amino acids. For organic molecules, the D- and L-convention has been replaced by a new system that provides the absolute configuration of a chiral carbon. This system, called the (R) and (S) system, is described in Stereochemistry and Stereoisomers Revisited, online at www.mhhe.com/denniston.

```
        H                    H                  CH₂OH
        |                    |                    |
        C=O                  C=O                  C=O
        |                    |                    |
   H—C—OH             H—C—OH            HO—C—H
        |                    |                    |
      CH₂OH           HO—C—H             H—C—OH
                             |                    |
                        H—C—OH            H—C—OH
                             |                    |
                        H—C—OH              CH₂OH
                             |
                          CH₂OH

  D-Glyceraldehyde      D-Glucose           D-Fructose
```

QUESTION 14.9

Determine the configuration (D- or L-) for each of the molecules in Question 14.5.

QUESTION 14.10

Explain the difference between the D- and L-designation and the (+) and (−) designation.

14.4 Biologically Important Monosaccharides

Monosaccharides, the simplest carbohydrates, have backbones of from three to seven carbons. There are many monosaccharides, but we will focus on those that are most common in biological systems. These include the five- and six-carbon sugars: glucose, fructose, galactose, ribose, and deoxyribose.

> **6** **Learning Goal**
> Draw and name the common monosaccharides using structural formulas.

Glucose

Glucose is the most important sugar in the human body. It is found in numerous foods and has several common names, including dextrose, grape sugar, and blood sugar. Glucose is broken down in glycolysis and other pathways to release energy for body functions.

Figure 14.7

Cyclization of glucose to give α- and β-D-glucose. Note that the carbonyl carbon (C-1) becomes chiral in this process, yielding the α- and β-forms of glucose.

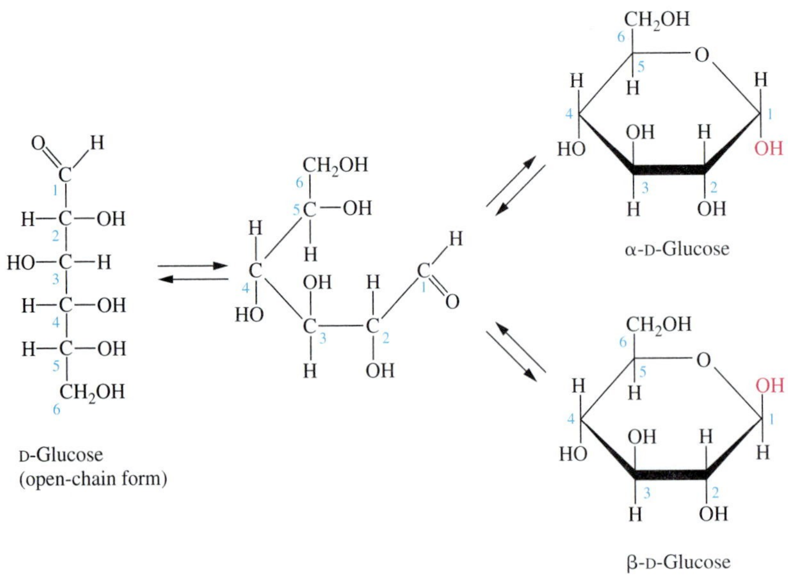

D-Glucose
(open-chain form)

α-D-Glucose

β-D-Glucose

The concentration of glucose in the blood is critical to normal body function. As a result, it is carefully controlled by the hormones insulin and glucagon. Normal blood glucose levels are 100–120 mg/100 mL, with the highest concentrations appearing after a meal. Insulin stimulates the uptake of the excess glucose by most cells of the body, and after one to two hours, levels return to normal. If blood glucose concentrations drop too low, the individual feels lightheaded and shaky. When this happens, glucagon stimulates the liver to release glucose into the blood, reestablishing normal levels. We will take a closer look at this delicate balancing act in Sections 18.11 and 19.7.

The molecular formula of glucose, an aldohexose, is $C_6H_{12}O_6$. The structure of glucose is shown in Figure 14.7, and the method used to draw this structure is described in Example 14.2.

6 Learning Goal

Draw and name the common monosaccharides using structural formulas.

EXAMPLE 14.2 | Drawing the Structure of a Monosaccharide

Draw the structure for D-glucose.

SOLUTION

Glucose is an aldohexose.

Step 1. Draw six carbons in a straight vertical line; each carbon is separated from the ones above and below it by a bond:

1 C
|
2 C
|
3 C
|
4 C
|
5 C
|
6 C

Continued—

Step 2. The most highly oxidized carbon is, by convention, drawn as the uppermost carbon (carbon-1). In this case, carbon-1 is an aldehyde carbon:

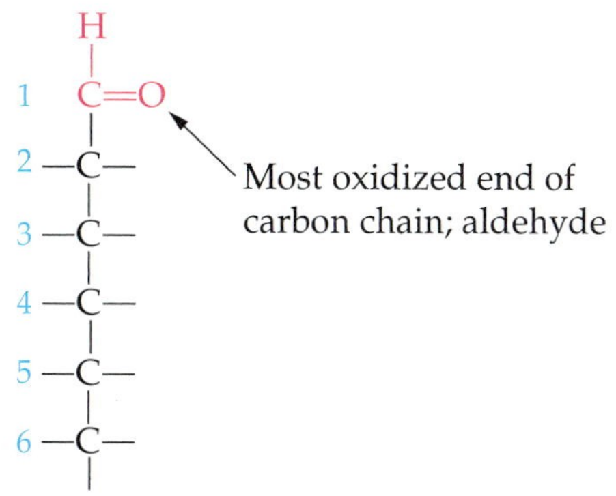

$$
\begin{array}{rl}
 & \text{H} \\
 & | \\
1 & \text{C}{=}\text{O} \\
 & | \\
2 & \text{—C—} \\
 & | \\
3 & \text{—C—} \\
 & | \\
4 & \text{—C—} \\
 & | \\
5 & \text{—C—} \\
 & | \\
6 & \text{—C—} \\
 & |
\end{array}
$$

Most oxidized end of carbon chain; aldehyde

Step 3. The atoms are added to the next to the last carbon atom, at the bottom of the chain, to give either the D- or L-configuration as desired. Remember, when the —OH group is to the right, you have D-glucose. When in doubt, compare your structure to D-glyceraldehyde!

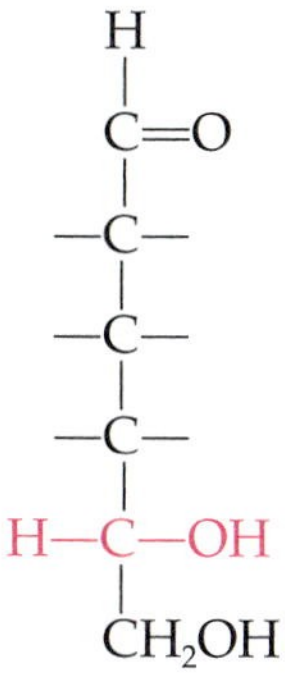

$$
\begin{array}{c}
\text{H} \\
| \\
\text{C}{=}\text{O} \\
| \\
\text{—C—} \\
| \\
\text{—C—} \\
| \\
\text{—C—} \\
| \\
\text{H—C—OH} \\
| \\
\text{CH}_2\text{OH}
\end{array}
\qquad
\begin{array}{c}
\text{H} \\
| \\
\text{C}{=}\text{O} \\
| \\
\text{H—C—OH} \\
| \\
\text{CH}_2\text{OH}
\end{array}
$$

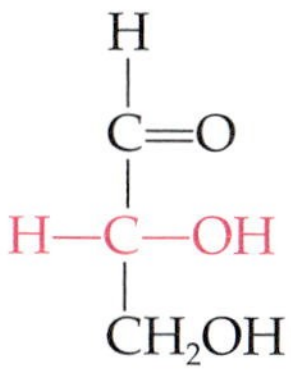

Compare chiral carbons farthest from the carbonyl group

D-Isomer   D-Glyceraldehyde

Step 4. All remaining atoms are then added to give the desired carbohydrate. For example, one would draw the following structure for D-glucose.

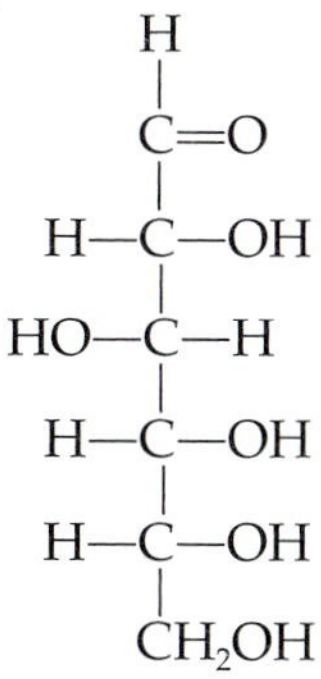

$$
\begin{array}{c}
\text{H} \\
| \\
\text{C}{=}\text{O} \\
| \\
\text{H—C—OH} \\
| \\
\text{HO—C—H} \\
| \\
\text{H—C—OH} \\
| \\
\text{H—C—OH} \\
| \\
\text{CH}_2\text{OH}
\end{array}
$$

D-Glucose

The positions for the hydrogen atoms and the hydroxyl groups on the remaining carbons must be learned for each sugar.

QUESTION 14.11

Draw the structure of D-ribose. (Information on the structure of D-ribose is found later in this chapter.)

QUESTION 14.12

Draw the structure of D-galactose. (Information on the structure of D-galactose is found later in this chapter.)

QUESTION 14.13

Draw the structure of L-ribose.

QUESTION 14.14

Draw the structure of L-galactose.

In actuality, the open-chain form of glucose is present in very small concentrations in cells. It exists in cyclic form under physiological conditions because the carbonyl group at C-1 of glucose reacts with the hydroxyl group at C-5 to give a six-membered ring. For D-glucose, two isomers can be formed in this reaction (see Figure 14.7). These isomers are called α- and β-D-glucose. In the α-isomer, the C-1 hydroxyl group is below the ring, and in the β-isomer, the C-1 hydroxyl group is above the ring. Like the stereoisomers discussed previously, the α and β forms can be distinguished from one another because they rotate plane-polarized light differently.

In Figure 14.7 a new type of structural formula, called a **Haworth projection,** is presented. Although on first inspection it appears complicated, it is quite simple to derive a Haworth projection from a structural formula, as Example 14.3 shows.

EXAMPLE 14.3 | Drawing the Haworth Projection of a Monosaccharide from the Structural Formula

7 Learning Goal

Given the linear structure of a monosaccharide, draw the Haworth projection of its α- and β-cyclic forms and vice versa.

Draw the Haworth projections of α- and β-D-glucose.

SOLUTION

1. Before attempting to draw a Haworth projection, look at the first steps of ring formation shown here:

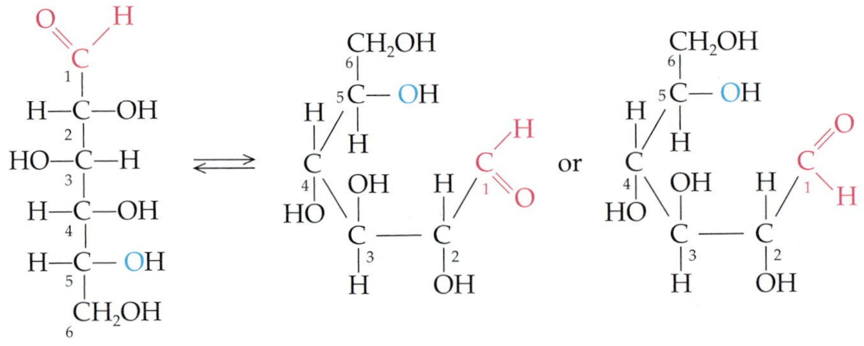

Glucose
(open chain)

Glucose
(intermediates in ring formation)

Try to imagine that you are seeing the molecules shown above in three dimensions. Some of the substituent groups on the molecule will be above the ring, and some will be beneath it. The question then becomes: How do you determine which groups to place above the ring and which to place beneath the ring?

Continued—

2. Look at the two-dimensional structural formula. Note the groups (drawn in blue) to the left of the carbon chain. These are placed above the ring in the Haworth projection.

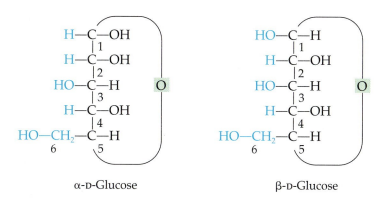

α-D-Glucose   β-D-Glucose

3. Now note the groups (drawn in red) to the right of the carbon chain. These will be located beneath the carbon ring in the Haworth projection.

α-D-Glucose   β-D-Glucose

4. Thus in the Haworth projection of the cyclic form of any D-sugar, the —CH₂OH group is always "up." When the —OH group at C-1 is also "up," *cis* to the —CH₂OH group, the sugar is β-D-glucose. When the —OH group at C-1 is "down," *trans* to the —CH₂OH group, the sugar is α-D-glucose.

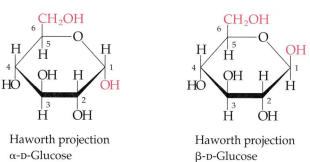

Haworth projection   Haworth projection
α-D-Glucose   β-D-Glucose

QUESTION 14.15

Refer to the linear structure of D-galactose. Draw the Haworth projections of α- and β-D-galactose.

QUESTION 14.16

Refer to the linear structure of D-ribose. Draw the Haworth projections of α- and β-D-ribose. Note that D-ribose is a pentose.

Fructose

Fructose, also called levulose and fruit sugar, is the sweetest of all sugars. It is found in large amounts in honey, corn syrup, and sweet fruits. The structure of fructose is similar to that of glucose. When there is a —CH_2OH group instead of a —CHO group at carbon-1 and a —C=O group instead of CHOH at carbon-2, the sugar is a ketose. In this case, it is D-fructose.

Cyclization of fructose produces α- and β-D-fructose:

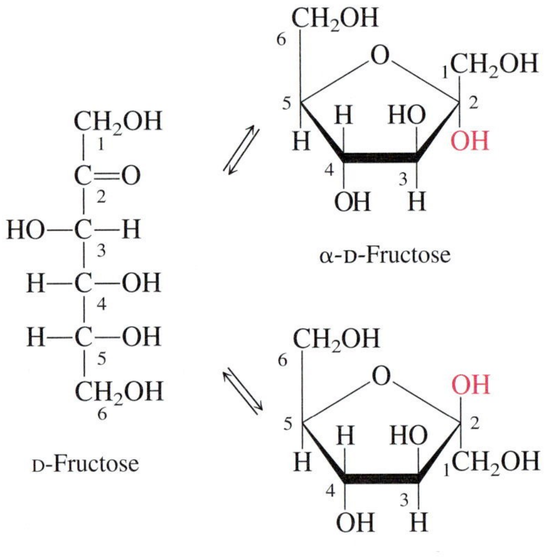

α-D-Fructose

D-Fructose

β-D-Fructose

Galactose

Galactose is another important hexose. The linear structure of D-galactose and the Haworth projections of α-D-galactose and β-D-galactose are shown here:

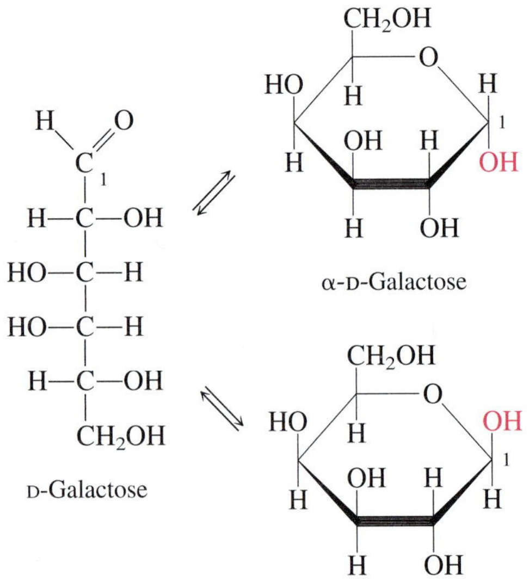

D-Galactose

α-D-Galactose

β-D-Galactose

Galactose is found in biological systems as a component of the disaccharide lactose, or milk sugar. This is the principal sugar found in the milk of most mammals. β-D-Galactose and a modified form, β-D-*N*-acetylgalactosamine, are also components of blood group antigens.

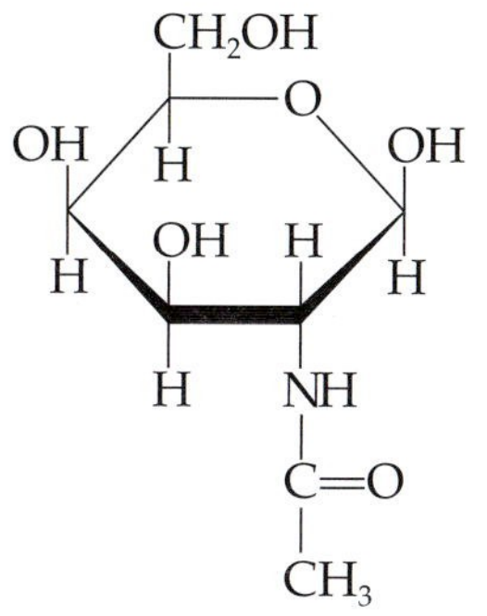

β-D-*N*-Acetylgalactosamine

Ribose and Deoxyribose, Five-Carbon Sugars

Ribose is a component of many biologically important molecules, including RNA and various coenzymes that are required by many of the enzymes that carry out biochemical reactions in the body. The structure of the five-carbon sugar D-ribose is shown in its open-chain form and in the α- and β-cyclic forms.

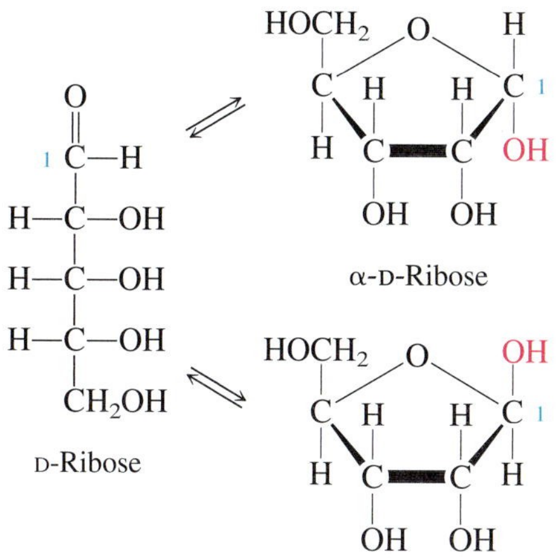

D-Ribose

α-D-Ribose

β-D-Ribose

DNA, the molecule, that carries the genetic information of the cell, contains 2-deoxyribose. In this molecule, the —OH group at C-2 has been replaced by a hydrogen, hence the designation "2-deoxy," indicating the absence of an oxygen.

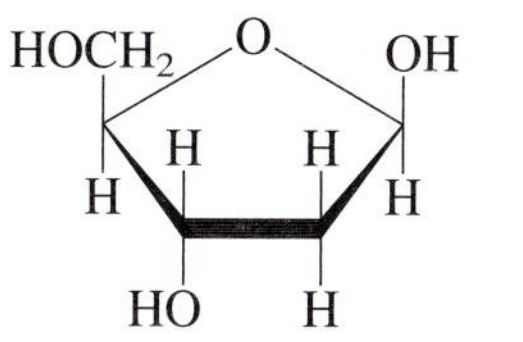

β-D-2-Deoxyribose

Reducing Sugars

The aldehyde group of aldoses is readily oxidized by Benedict's reagent. Recall that **Benedict's reagent** is a basic buffer solution that contains Cu^{2+} ions. The Cu^{2+}

8 Learning Goal

By inspection of the structure, predict whether a sugar is reducing or nonreducing.

ions are reduced to Cu^+ ions, which, in basic solution, precipitate as brick-red Cu_2O. The aldehyde group of the aldose is oxidized to a carboxylic acid, which undergoes an acid-base reaction to produce a carboxylate anion.

$$\underset{\substack{|\\CH_2OH}}{\overset{\substack{O\quad H\\\diagdown\!\!\diagup\\C\\|}}{H-C-OH}} + 2Cu^{2+} \text{ (buffer)} + 5OH^- \longrightarrow \underset{\substack{|\\CH_2OH}}{\overset{\substack{O\quad O^-\\\diagdown\!\!\diagup\\C\\|}}{H-C-OH}} + Cu_2O(s) + 3H_2O$$

Although ketones generally are not easily oxidized, ketoses, such as fructose, can be because they are readily converted to aldoses in Benedict's reagent. Because the metal ions in the solution are reduced, the sugars are reducing agents and are called **reducing sugars.** All monosaccharides and all common disaccharides, except sucrose, are reducing sugars.

For many years, Benedict's reagent was used to test for *glucosuria,* the presence of excess glucose in urine. Individuals suffering from *Type I insulin-dependent diabetes mellitus* do not produce the hormone insulin, which controls the uptake of glucose from the blood. When the blood glucose level rises above 160 mg/100 mL, the kidney is unable to reabsorb the excess, and glucose is found in urine. Although the level of blood glucose could be controlled by the injection of insulin, urine glucose levels were monitored to ensure that the amount of insulin injected was correct. Benedict's reagent was a useful tool because the amount of Cu_2O formed, and hence the degree of color change in the reaction, is directly proportional to the amount of reducing sugar in urine. A brick-red color indicates a very high concentration of glucose in urine. Yellow, green, and blue-green solutions indicate decreasing amounts of glucose in urine, and a blue solution indicates an insignificant concentration.

Use of Benedict's reagent to test urine glucose levels has largely been replaced by chemical tests that provide more accurate results. The most common technology is based on a test strip that is impregnated with the enzyme glucose oxidase and other agents that will cause a measurable color change. In one such kit, the compounds that result in color development include the enzyme peroxidase, a compound called orthotolidine, and a yellow dye. When a drop of urine is placed on the strip, the glucose oxidase catalyzes the conversion of glucose into gluconic acid and hydrogen peroxide.

See A Medical Connection: Diabetes Mellitus and Ketone Bodies in Chapter 19.

9 **Learning Goal**

Discuss the use of Benedict's reagent to measure the level of glucose in urine.

$$\underset{\text{D-Glucose}}{\overset{\substack{O\\||\\C-H\\|\\H-C-OH\\|\\HO-C-H\\|\\H-C-OH\\|\\H-C-OH\\|\\CH_2OH}}{}} + O_2 \quad\xrightarrow{\substack{\text{Glucose}\\\text{oxidase}}}\quad \underset{\text{D-Gluconic acid}}{\overset{\substack{O\\||\\C-OH\\|\\H-C-OH\\|\\HO-C-H\\|\\H-C-OH\\|\\H-C-OH\\|\\CH_2OH}}{}} + H_2O_2$$

The enzyme peroxidase catalyzes a reaction between the hydrogen peroxide and orthotolidine. This produces a blue product. The yellow dye on the test strip simply serves to "dilute" the blue end product, thereby allowing greater accuracy of the test over a wider range of glucose concentrations. The test strip remains yellow if there is no glucose in the sample. It will vary from a pale green to a dark blue, depending on the concentration of glucose in the urine sample.

Frequently, doctors recommend that diabetics monitor their *blood* glucose levels multiple times each day because this provides a more accurate indication of how well the diabetic is controlling his or her diet. Many small, inexpensive glucose meters are available that couple the oxidation of glucose by glucose oxidase with an appropriate color change system. As with the urine test, the intensity of the color change is proportional to the amount of glucose in the blood. A photometer within the device reads the color change and displays the glucose concentration. An even newer technology uses a device that detects the electrical charge generated by the oxidation of glucose. In this case, the amount of electrical charge is proportional to the glucose concentration.

14.5 Biologically Important Disaccharides

Recall that disaccharides consist of two monosaccharides joined through an "oxygen bridge." These carbon–oxygen bonds are called *glycosidic bonds.*

10 Learning Goal
Draw and name the common disaccharides and discuss their significance in biological systems.

Maltose

If an α-D-glucose and a second glucose are linked, as shown in Figure 14.8, the disaccharide is **maltose,** or malt sugar. This is one of the intermediates in the hydrolysis of starch. Because the C-1 hydroxyl group of α-D-glucose is attached to C-4 of another glucose molecule, the disaccharide is linked by an α(1 → 4) glycosidic bond.

Maltose is a reducing sugar. Any disaccharide that has a free —OH group at C-1 is a reducing sugar because the cyclic structure can open at this position to form a free aldehyde. Disaccharides that do not contain a free —OH group on C-1 do not react with Benedict's reagent and are called **nonreducing sugars.**

Lactose

Milk sugar, or **lactose,** is a disaccharide made up of one molecule of β-D-galactose and one of either α- or β-D-glucose. Galactose differs from glucose only in the configuration of the hydroxyl group at C-4 (Figure 14.9). In the cyclic form of glucose, the C-4 hydroxyl group is "down," and in galactose it is "up." In lactose, the C-1 hydroxyl group of β-D-galactose is bonded to the C-4 hydroxyl group of either an α- or β-D-glucose. The bond between the two monosaccharides is therefore a β(1 → 4) glycosidic bond (Figure 14.10).

11 Learning Goal
Describe the difference between galactosemia and lactose intolerance.

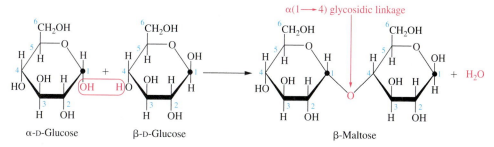

Figure 14.8

Glycosidic bond formed between the C-1 hydroxyl group of α-D-glucose and the C-4 hydroxyl group of β-D-glucose. The disaccharide is called β-maltose because the hydroxyl group at the reducing end of the disaccharide has the β-configuration.

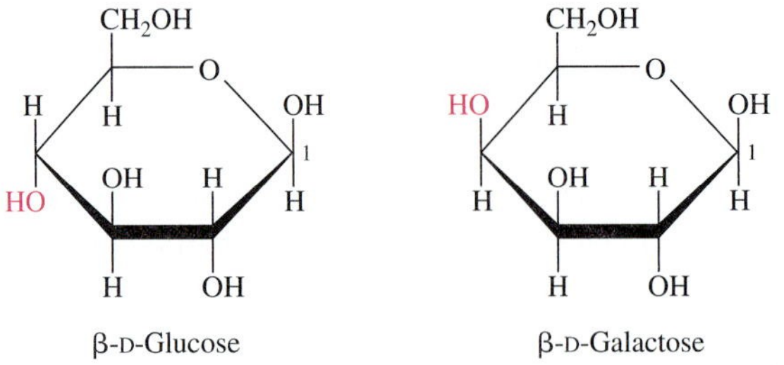

Figure 14.9

Comparison of the cyclic forms of glucose and galactose. Note that galactose is identical to glucose except in the position of the C-4 hydroxyl group.

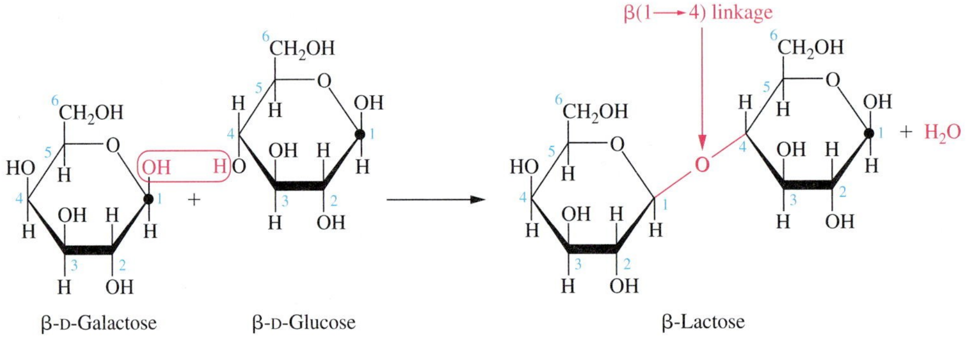

Figure 14.10

Glycosidic bond formed between the C-1 hydroxyl group of β-D-galactose and the C-4 hydroxyl group of β-D-glucose. The disaccharide is called β-lactose because the hydroxyl group at the reducing end of the disaccharide has the β-configuration.

Glycolysis is discussed in Chapter 18.

Lactose is the principal sugar in the milk of most mammals. To be used by the body as an energy source, lactose must be hydrolyzed to produce glucose and galactose. Note that this is simply the reverse of the reaction shown in Figure 14.10. Glucose liberated by the hydrolysis of lactose is used directly in the energy-harvesting reactions of glycolysis. However, a series of reactions is necessary to convert galactose into a phosphorylated form of glucose that can be used in cellular metabolic reactions. In humans, the genetic disease **galactosemia** is caused by the absence of one or more of the enzymes needed for this conversion. A toxic compound formed from galactose accumulates in people who suffer from galactosemia. If the condition is not treated, galactosemia leads to severe mental retardation, cataracts, and early death. However, the effects of this disease can be avoided entirely by providing galactosemic infants with a diet that does not contain galactose. Such a diet, of course, cannot contain lactose and therefore must contain no milk or milk products.

Many adults, and some children, are unable to hydrolyze lactose because they do not make the enzyme *lactase*. This condition, which affects 20% of the population of the United States, is known as **lactose intolerance.** Undigested lactose remains in the intestinal tract and causes cramping and diarrhea that can eventually lead to dehydration. Some of the lactose is metabolized by intestinal bacteria that release organic acids and CO_2 gas into the intestines, causing further discomfort. Lactose intolerance is unpleasant, but its effects can be avoided by a diet that excludes milk and milk products. Alternatively, the enzyme that hydrolyzes lactose is available in tablet form. When ingested with dairy products, it breaks down the lactose, preventing symptoms.

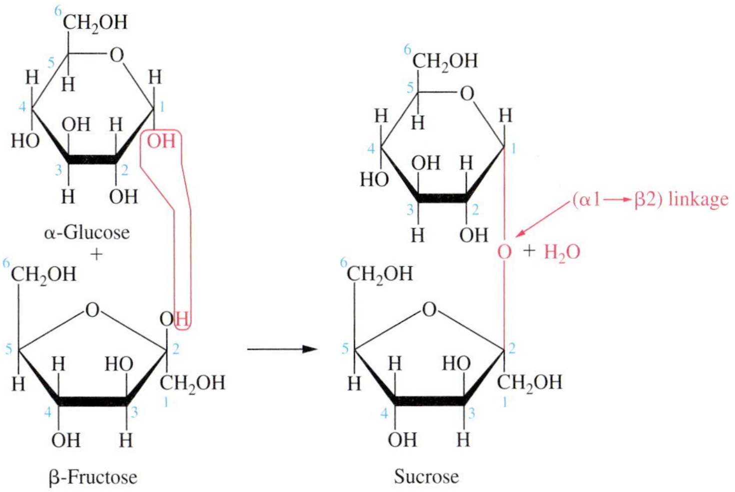

Figure 14.11

Glycosidic bond formed between the C-1 hydroxyl of α-D-glucose and the C-2 hydroxyl of β-D-fructose. This bond is called an (α1 → β2) glycosidic linkage. The disaccharide formed in this reaction is sucrose.

Sucrose

Sucrose is also called table sugar, cane sugar, or beet sugar. Sucrose is an important carbohydrate in plants. It is water-soluble and can easily be transported through the circulatory system of the plant. It cannot be synthesized by animals. High concentrations of sucrose produce high osmotic pressure, which inhibits the growth of microorganisms, so it is used as a preservative. Of course, it is also widely used as a sweetener. It is estimated that the average American consumes 100–125 pounds of sucrose each year. It has been suggested that sucrose in the diet is undesirable because it represents a source of empty calories; it contains no vitamins or minerals. However, the only negative association that has been scientifically verified is the link between sucrose in the diet and dental caries, or cavities (see A Medical Connection: Tooth Decay and Simple Sugars earlier in this chapter.)

Sucrose is a disaccharide of α-D-glucose joined to β-D-fructose (Figure 14.11). The glycosidic linkage between α-D-glucose and β-D-fructose is quite different from those that we have examined for lactose and maltose. This bond is called an (α1 → β2) glycosidic linkage, because it involves the C-1 carbon of glucose and the C-2 carbon of fructose (noted in red in Figure 14.11). Sucrose will not react with Benedict's reagent and is not a reducing sugar.

14.6 Polysaccharides

Starch

Most carbohydrates that are found in nature are large polymers of glucose. Thus a polysaccharide is a large polymer composed of many monosaccharide units (monomers) joined in one or more chains.

Plants have the ability to use the energy of sunlight to produce monosaccharides, principally glucose, from CO_2 and H_2O. Although sucrose is the major transport form of sugar in the plant, starch (a polysaccharide) is the principal storage form in most plants. These plants store glucose in starch granules. Nearly all plant cells contain some starch granules, but in some seeds, such as corn, as much as 80% of the cell's dry weight is starch.

 Learning Goal

Discuss the structural, chemical, and biochemical properties of starch, glycogen, and cellulose.

A polymer (Section 11.4) is a large molecule made up of many small units, monomers, held together by chemical bonds.

CHEMISTRY at the Crime Scene | Blood Group Antigens

Before DNA fingerprinting was available, blood typing was often used to analyze blood from a crime scene. Such evidence could exonerate a suspect or provide further evidence of guilt. But with only four human blood types, you can imagine how inexact conclusions could be drawn from this evidence!

In 1904, Dr. Karl Landsteiner performed a series of experiments on the blood of workers in his laboratory. He separated the blood cells from the serum, the liquid component of the blood, and mixed these samples in test tubes. When he mixed serum from one individual with blood cells of another, Landsteiner observed that, in some instances, the serum samples caused clumping, or *agglutination*, of red blood cells (RBC). (See figure below.) As a result of many such experiments, Landsteiner showed that there are four human blood groups, designated A, B, AB, and O.

We now know that differences among blood groups reflect differences among oligosaccharides attached to the proteins and lipids of the RBC membranes. The oligosaccharides on the RBC surface have a common core, as shown in the accompanying figure, consisting of β-D-N-acetylgalactosamine, galactose, N-acetylneuraminic acid (sialic acid), and L-fucose. It is the terminal monosaccharide of this oligosaccharide that distinguishes the cells and governs the compatibility of the blood types.

The A blood group antigen has β-D-N-acetylgalactosamine at its end, whereas the B blood group antigen has α-D-galactose. In type O blood, neither of these sugars is found on the cell surface; only the core oligosaccharide is present. Some oligosaccharides on type AB blood cells have a terminal β-D-N-acetylgalactosamine, whereas others have a terminal α-D-galactose.

Why does agglutination occur? The clumping reaction that occurs when incompatible bloods are mixed is an antigen–antibody reaction. Antigens are large molecules, often portions of bacteria or viruses, that stimulate the immune defenses of the body to produce protective antibodies. Antibodies bind to the foreign antigens and help to destroy them.

People with type A blood also have antibodies against type B blood (anti-B antibodies) in the blood serum. If the person with type A blood receives a transfusion of type B blood, the anti-B antibodies bind to the type B blood cells, causing clumping and destruction of those cells that can result in death. Individuals with type B blood also produce anti-A antibodies and therefore cannot receive a transfusion from a type A individual. Those with type AB blood have neither anti-A nor anti-B antibodies in their blood. (If they did, they would destroy their own red blood cells!) Type O blood has no A or B antigens on the RBC but has both anti-A and anti-B antibodies. Because of the presence of both types of antibodies, type O individuals can receive transfusions only from a person who is also type O.

In the forensic laboratory, this same agglutination reaction can be used to determine the blood type of a sample from a crime scene for comparison with the blood of the victim and that of a suspect. The results may clear the suspect or further incriminate him or her. Because there are only four human blood types, significant additional evidence is needed to prove the guilt of a suspect. In fact, DNA fingerprinting is currently being used to reevaluate guilty verdicts obtained with this less reliable evidence.

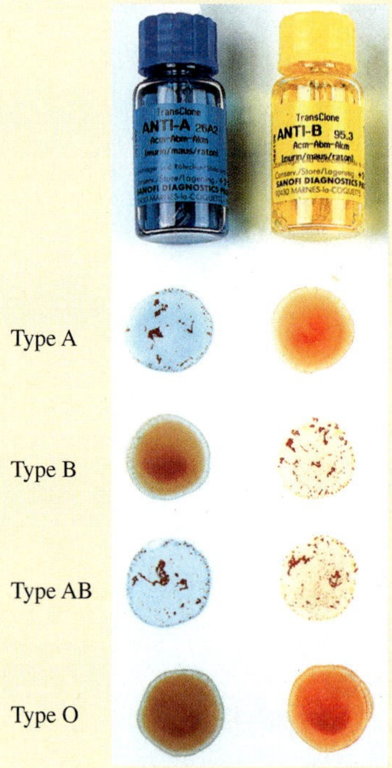

Type A

Type B

Type AB

Type O

ABO blood typing kit. When antibodies bind to antigens on the cell surface, clumping occurs.

FOR FURTHER UNDERSTANDING

People with type AB blood are referred to as universal recipients. Explain why these people can receive blood of any of the four ABO types.

Why is blood typing insufficient to prove the guilt of a suspect?

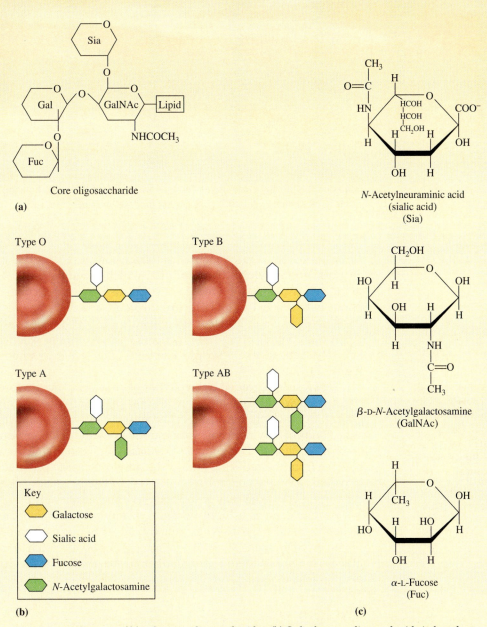

(a) Schematic diagram of blood group oligosaccharides. (b) Only the core oligosaccharide is found on the surface of type O red blood cells. On type A red blood cells, β-D-N-acetylgalactosamine is linked to the galactose (Gal) of the core oligosaccharide. On type B red blood cells, a galactose molecule is found attached to the galactose of the core oligosaccharide. (c) The structures of some of the unusual monosaccharides found in blood group oligosaccharides.

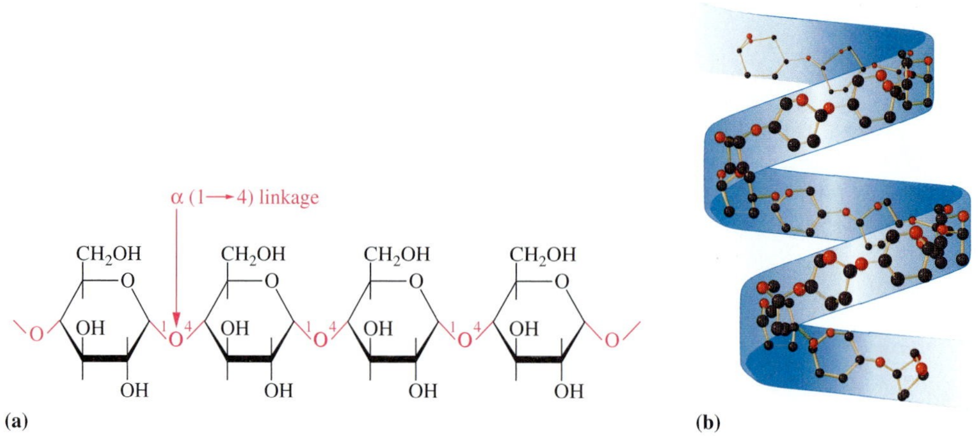

(a)   (b)

Figure 14.12

Structure of amylose. (a) A linear chain of α-D-glucose joined in α(1 → 4) glycosidic linkage makes up the primary structure of amylose. (b) Owing to hydrogen bonding, the amylose chain forms a left-handed helix that contains six glucose units per turn.

Starch is a heterogeneous material composed of the glucose polymers, **amylose** and **amylopectin.** Amylose, which accounts for 20% of the starch of a plant cell, is a linear polymer of α-D-glucose molecules connected by glycosidic bonds between C-1 of one glucose molecule and C-4 of a second glucose. Thus, the glucose units in amylose are joined by α(1 → 4) glycosidic bonds. A single chain can contain up to four thousand glucose units. Amylose coils up into a helix that repeats every six glucose units. The structure of amylose is shown in Figure 14.12.

Amylose is degraded by two types of enzymes. They are produced in the pancreas, from which they are secreted into the small intestine, and in the salivary glands, from which they are secreted into the saliva. α-*Amylase* cleaves the glycosidic bonds of amylose chains at random along the chain, producing shorter polysaccharide chains. The enzyme β-*amylase* sequentially cleaves the disaccharide maltose from the reducing end of the amylose chain. The maltose is hydrolyzed into glucose by the enzyme *maltase.* The glucose is quickly absorbed by intestinal cells and used by the cells of the body as a source of energy.

Amylopectin is a highly branched amylose in which the branches are attached to the C-6 hydroxyl groups by α(1 → 6) glycosidic bonds (Figure 14.13). The main chains consist of α(1 → 4) glycosidic bonds. Each branch contains 20–25 glucose units, and there are so many branches that the main chain can scarcely be distinguished.

> Enzymes are proteins that serve as biological catalysts. They speed up biochemical reactions so that life processes can function. These enzymes are called α(1 → 4) glycosidases because they cleave α(1 → 4) glycosidic bonds.

Glycogen

Glycogen is the major glucose storage molecule in animals. The structure of glycogen is similar to that of amylopectin. The "main chain" is linked by α(1 → 4) glycosidic bonds, and it has numerous α(1 → 6) glycosidic bonds, which provide many branch points along the chain. Glycogen differs from amylopectin only by having more and shorter branches. Otherwise, the two molecules are virtually identical. The structure of glycogen is shown in Figure 14.13.

Glycogen is stored in the liver and skeletal muscle. Glycogen synthesis and degradation in the liver are carefully regulated. As we will see in Sections 18.11 and 19.7. these two processes are intimately involved in keeping blood glucose levels constant.

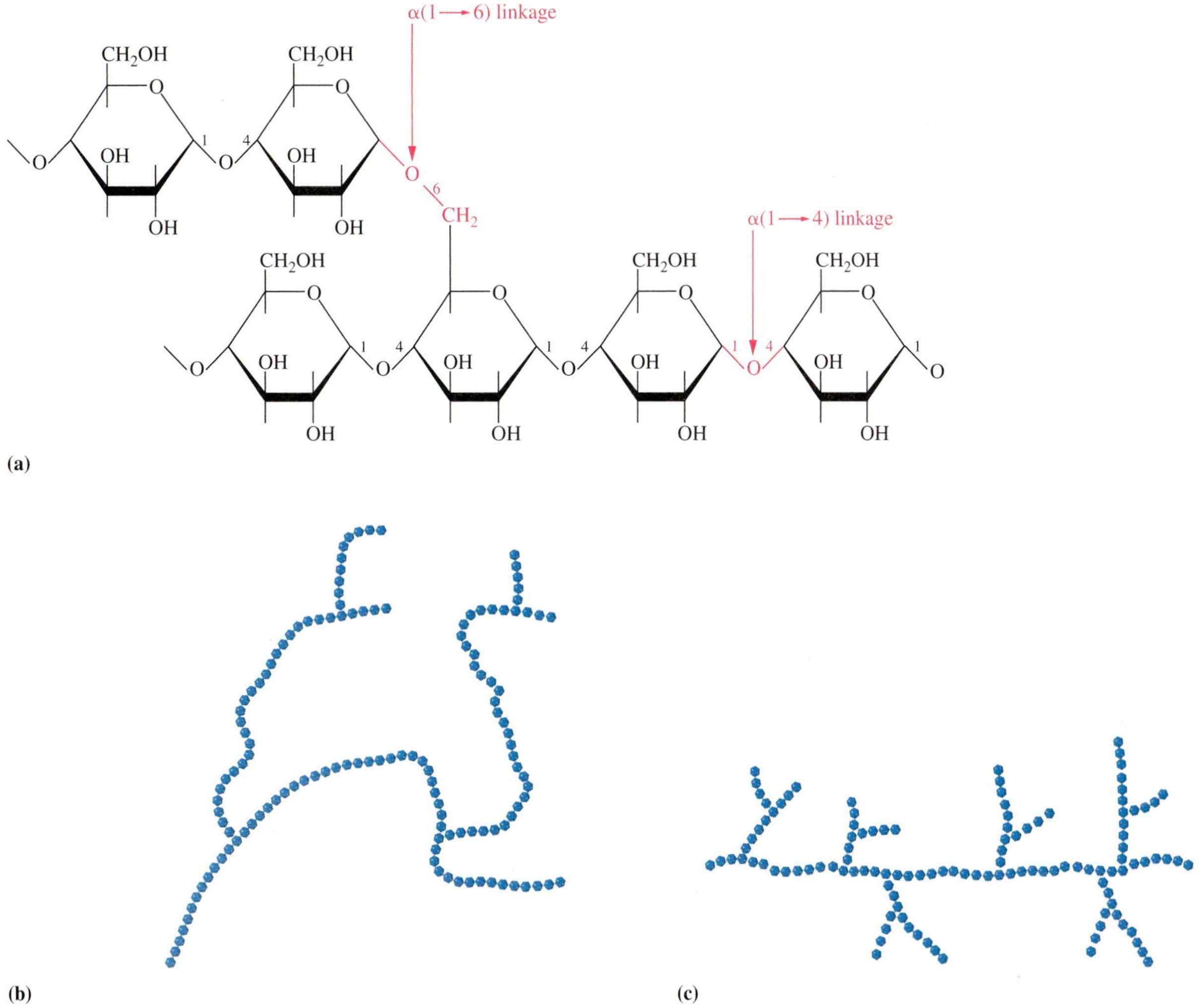

$\alpha(1 \longrightarrow 6)$ linkage

$\alpha(1 \longrightarrow 4)$ linkage

(a)

(b)

(c)

Figure 14.13

Structure of amylopectin and glycogen. (a) Both amylopectin and glycogen consist of chains of α-D-glucose molecules joined in α(1 → 4) glycosidic linkages. Branching from these chains are other chains of the same structure. Branching occurs by formation of α(1 → 6) glycosidic bonds between glucose units. (b) A representation of the branched-chain structure of amylopectin. (c) A representation of the branched-chain structure of glycogen. Glycogen differs from amylopectin only in that the branches are shorter and there are more of them.

Cellulose

The most abundant polysaccharide, in fact the most abundant organic molecule in the world, is **cellulose,** a polymer of β-D-glucose units linked by β(1 → 4) glycosidic bonds (Figure 14.14). A molecule of cellulose typically contains about 3000 glucose units, but the largest known cellulose, produced by the alga *Valonia,* contains 26,000 glucose molecules.

Cellulose is a structural component of the plant cell wall. The unbranched structure of the cellulose polymer and the β(1 → 4) glycosidic linkages allow cellulose molecules to form long, straight chains of parallel cellulose molecules called *fibrils.* These fibrils are quite rigid and are held together tightly by hydrogen bonds; thus, it is not surprising that cellulose is a cell wall structural element.

In contrast to glycogen, amylose, and amylopectin, cellulose *cannot* be digested by humans. The reason is that we cannot synthesize the enzyme *cellulase,*

To learn about some unusual monosaccharides and polysaccharides important to human health, see A Medical Connection: Monosaccharide Derivatives and Heteropolysaccharides of Medical Interest online at www.mhhe.com/denniston

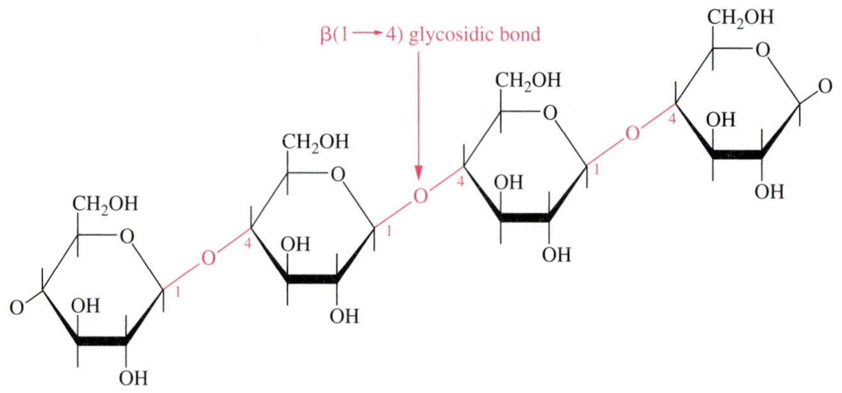

Figure 14.14

The structure of cellulose.

which can hydrolyze the β(1 → 4) glycosidic linkages of the cellulose polymer; only a few animals, such as termites, cows, and goats, are able to digest cellulose. These animals have, within their digestive tracts, microorganisms that produce the enzyme cellulase. The sugars released by this microbial digestion can then be absorbed and used by these animals. In humans, cellulose from fruits and vegetables serves as fiber in the diet.

QUESTION 14.17

Which chemical reactions are catalyzed by α-amylase and β-amylase?

QUESTION 14.18

What is the function of cellulose in the human diet? How does this relate to the structure of cellulose?

SUMMARY

14.1 Types of Carbohydrates

Carbohydrates are found in a wide variety of naturally occurring substances and serve as principal energy sources for the body. Dietary carbohydrates include complex carbohydrates, such as starch in potatoes, and simple carbohydrates, such as sucrose.

Carbohydrates are classified as *monosaccharides* (one sugar unit), *disaccharides* (two sugar units), *oligosaccharides* (three to ten sugar units), or *polysaccharides* (many sugar units).

14.2 Monosaccharides

Monosaccharides that have an aldehyde as their most oxidized functional group are *aldoses*, and those having a ke-

tone group as their most oxidized functional group are *ketoses*. They may be classified as *trioses*, *tetroses*, *pentoses*, and so forth, depending on the number of carbon atoms in the carbohydrate.

14.3 Stereoisomers and Stereochemistry

Stereoisomers of monosaccharides exist because of the presence of *chiral carbon* atoms. They are classified as D- or L- depending on the arrangement of the atoms on the chiral carbon farthest from the aldehyde or ketone group. If the —OH on this carbon is to the right, the stereoisomer is of the D-family. If the —OH group is to the left, the stereoisomer is of the L-family.

Each member of a pair of stereoisomers will rotate plane-polarized light in a different direction. A polarimeter is used to measure the direction of rotation of plane-polarized light. Compounds that rotate light in a clockwise direction are termed dextrorotatory and are designated by a plus sign (+). Compounds that rotate light in a counterclockwise direction are called levorotatory and are indicated by a minus sign (−).

The *Fischer projection* is a two-dimensional drawing of a *chiral molecule* that shows a chiral carbon at the intersection of two lines. Horizontal lines represent bonds projecting out of the page and vertical lines represent bonds that project into the page. The most oxidized carbon is always represented at the "top" of the structure.

14.4 Biologically Important Monosaccharides

Important monosaccharides include *glyceraldehyde, glucose, fructose, galactose,* and *ribose.* Monosaccharides containing five or six carbon atoms can exist as five-membered or six-membered rings. Formation of a ring produces a new chiral carbon at the original carbonyl carbon, which is designated either α or β depending on the orientation of the groups. A *Haworth projection* is a means of representing the orientation of groups around a cyclic sugar molecule.

Reducing sugars are oxidized by *Benedict's reagent.* All monosaccharides and all common disaccharides, except sucrose, are reducing sugars. At one time, Benedict's reagent was used to determine the concentration of glucose in urine.

14.5 Biologically Important Disaccharides

Important disaccharides include *lactose* and *sucrose.* Lactose is a disaccharide of β-D-galactose bonded β(1 → 4) with D-glucose. In *galactosemia,* defective metabolism of galactose leads to accumulation of a toxic by-product. The ill effects of galactosemia are avoided by exclusion of milk and milk products from the diet of affected infants. Sucrose is a dimer composed of α-D-glucose bonded (α1 → β2) with β-D-fructose.

14.6 Polysaccharides

Starch, the storage polysaccharide of plant cells, is composed of approximately 80% amylose and 20% amylopectin. *Amylose* is a polymer of α-D-glucose units bonded α(1 → 4). Amylose forms a helix. *Amylopectin* has many branches. Its main chain consists of α-D-glucose units bonded α(1 → 4). The branches are connected by α(1 → 6) glycosidic bonds.

Glycogen, the major storage polysaccharide of animal cells, resembles amylopectin, but it has more, shorter branches. The liver reserve of glycogen is used to regulate blood glucose levels.

Cellulose is a major structural molecule of plants. It is a β(1 → 4) polymer of D-glucose that can contain thousands of glucose monomers. Cellulose cannot be digested by animals because they do not produce an enzyme capable of cleaving the β(1 → 4) glycosidic linkage.

KEY TERMS

aldose (14.2)	hexose (14.2)
amylopectin (14.6)	ketose (14.2)
amylose (14.6)	lactose (14.5)
Benedict's reagent (14.4)	lactose intolerance
carbohydrate (14.1)	(14.5)
cellulose (14.6)	maltose (14.5)
chiral carbon (14.3)	monosaccharide (14.1)
chiral molecule (14.3)	nonreducing sugar (14.5)
disaccharide (14.1)	oligosaccharide (14.1)
enantiomers (14.3)	pentose (14.2)
Fischer projection (14.3)	polysaccharide (14.1)
fructose (14.4)	reducing sugar (14.4)
galactose (14.4)	ribose (14.4)
galactosemia (14.5)	saccharide (14.1)
glucose (14.4)	stereochemistry (14.3)
glyceraldehyde (14.3)	stereoisomers (14.3)
glycogen (14.6)	sucrose (14.5)
glycosidic bond (14.1)	tetrose (14.2)
Haworth projection (14.4)	triose (14.2)

QUESTIONS AND PROBLEMS

Types of Carbohydrates

Foundations

14.19 What is the difference between a monosaccharide and a disaccharide?

14.20 What is a polysaccharide?

14.21 What is the general molecular formula for a simple sugar?

14.22 What biochemical process is ultimately responsible for the synthesis of sugars?

Applications

14.23 Read the labels on some of the foods in your kitchen, and see how many products you can find that list one or more carbohydrates among the ingredients in the package. Make a list of these compounds, and attempt to classify them as monosaccharides, disaccharides, or polysaccharides.

14.24 Some disaccharides are often referred to by their common names. What are the chemical names of (a) milk sugar, (b) beet sugar, and (c) cane sugar?

14.25 How many kilocalories of energy are released when 1 g of carbohydrate is "burned" or oxidized?

14.26 List some natural sources of carbohydrates.

14.27 Draw and provide the names of an aldohexose and a ketohexose.

14.28 Draw and provide the name of an aldotriose.

Monosaccharides

Foundations

14.29 Define the term *aldose.*

14.30 Define the term *ketose.*

14.31 What is a ketopentose?

14.32 What is an aldotriose?

Applications

14.33 Identify each of the following sugars.

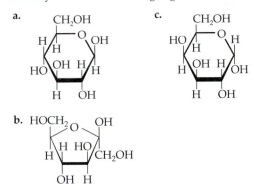

14.34 Draw the open-chain form of the sugars in Problem 14.33.
14.35 Draw all of the different possible aldotrioses of molecular formula $C_3H_6O_3$.
14.36 Draw all of the different possible aldotetroses of molecular formula $C_4H_8O_4$.

Stereoisomers and Stereochemistry

Foundations

14.37 Define the term *stereoisomer*.
14.38 Define the term *enantiomer*.
14.39 Define the term *chiral carbon*.
14.40 Draw an aldotetrose. Note each chiral carbon with an asterisk (*).
14.41 Explain how a polarimeter works.
14.42 What is plane-polarized light?
14.43 What is a Fischer projection?
14.44 How would you produce a Fischer projection beginning with a three-dimensional model of a sugar?

Applications

14.45 Is there any difference between dextrose and D-glucose?
14.46 The linear structure of D-glucose is shown in Figure 14.7. Draw its mirror image.
14.47 How are D- and L-glyceraldehyde related?
14.48 Determine whether each of the following is a D- or L-sugar:

a.
```
      O
      ||
      CH
      |
 H—C—OH
      |
 H—C—OH
      |
   CH₂OH
```

b.
```
      O
      ||
      CH
      |
 H—C—OH
      |
 H—C—OH
      |
 HO—C—H
      |
   CH₂OH
```

c.
```
      O
      ||
      CH
      |
 H—C—OH
      |
 H—C—OH
      |
 HO—C—H
      |
   CH₂OH
```

14.49 Draw a Fischer projection formula for each of the following compounds. Indicate each of the chiral carbons with asterisks (*).

a.
```
      O
      ||
      C—H
      |
 HO—C—H
      |
  H—C—OH
      |
 HO—C—H
      |
 HO—C—H
      |
   CH₂OH
```

b.
```
      O
      ||
      C—H
      |
  H—C—OH
      |
  H—C—OH
      |
   CH₂OH
```

c.
```
      O
      ||
      C—H
      |
 HO—C—H
      |
  H—C—OH
      |
 HO—C—H
      |
  H—C—OH
      |
 HO—C—H
      |
   CH₂OH
```

14.50 Draw a Fischer projection formula for each of the following compounds. Indicate each of the chiral carbons with asterisks (*).

a.
```
      O
      ||
      C—H
      |
  H—C—H
      |
 HO—C—H
      |
 HO—C—H
      |
 HO—C—H
      |
   CH₂OH
```

b.
```
      O
      ||
      C—H
      |
  H—C—H
      |
  H—C—OH
      |
   CH₂OH
```

c.
```
      O
      ||
      C—H
      |
 HO—C—H
      |
 HO—C—H
      |
 HO—C—H
      |
  H—C—OH
      |
 HO—C—H
      |
   CH₂OH
```

Biologically Important Monosaccharides

Foundations

14.51 What is a Haworth projection?
14.52 Why does cyclization of D-glucose give two isomers, α- and β-D-glucose?

Applications

14.53 Draw the structure of the open-chain form of D-fructose, and show how it cyclizes to form α- and β-D-fructose.
14.54 Draw the structure of the open chain form of D-ribose, and show how it cyclizes to form α- and β-D-ribose.
14.55 Which of the following would give a positive Benedict's test?
  a. Sucrose
  b. Glycogen
  c. β-Maltose
  d. α-Lactose
14.56 Why was Benedict's reagent useful for determining the amount of glucose in urine?
14.57 Describe what is meant by a pair of enantiomers. Draw an example of a pair of enantiomers.
14.58 What is a chiral carbon atom?

Biologically Important Disaccharides

Foundations

14.59 What is a glycosidic bond?
14.60 Define the term *disaccharide*.

Applications

14.61 Maltose is a disaccharide isolated from amylose that consists of two glucose units linked α(1 → 4). Draw the structure of this molecule.
14.62 Sucrose is a disaccharide formed by linking α-D-glucose and β-D-fructose by an (α1 → β2) bond. Draw the structure of this disaccharide.
14.63 What is the major biological source of lactose?
14.64 What metabolic defect causes galactosemia? What simple treatment prevents most of the ill effects of galactosemia?
14.65 What are the major physiological effects of galactosemia?
14.66 What is lactose intolerance? What is the difference between lactose intolerance and galactosemia?

Polysaccharides

Foundations

14.67 What is a polymer?
14.68 What form of sugar is used as the major transport sugar in a plant?
14.69 What is the major storage form of sugar in a plant?
14.70 What is the major structural form of sugar in a plant?

Applications

14.71 What is the difference between the structure of cellulose and the structure of amylose?

14.72 How does the structure of amylose differ from that of amylopectin and glycogen?

14.73 What is the major physiological purpose of glycogen? Where in the body do you find glycogen stored?

14.74 Where are α-amylase and β-amylase produced? Where do α-amylase and β-amylase carry out their enzymatic functions?

FOR FURTHER UNDERSTANDING

1. The following is the structure of salicin, a bitter-tasting compound found in the bark of willow trees:

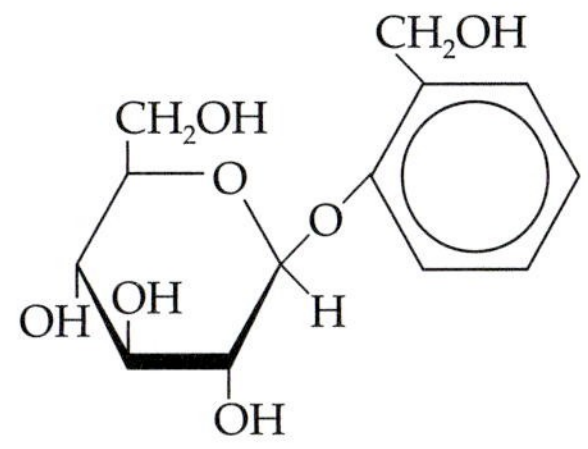

Salicin

The aromatic ring portion of this structure is quite insoluble in water. How would forming a glycosidic bond between the aromatic ring and β-D-glucose alter the solubility? Explain your answer.

2. Ancient peoples used salicin to reduce fevers. Write an equation for the acid-catalyzed hydrolysis of the glycosidic bond of salicin. Compare the aromatic product with the structure of acetylsalicylic acid (aspirin). Use this information to develop a hypothesis explaining why ancient peoples used salicin to reduce fevers.

3. Chitin is a modified cellulose in which the C-2 hydroxyl group of each glucose is replaced by

This nitrogen-containing polysaccharide makes up the shells of lobsters, crabs, and the exoskeletons of insects. Draw a portion of a chitin polymer consisting of four monomers.

4. Pectins are polysaccharides obtained from fruits and berries and used to thicken jellies and jams. Pectins are α(1 → 4) linked D-galacturonic acid. D-Galacturonic acid is D-galactose in which the C-6 hydroxyl group has been oxidized to a carboxyl group. Draw a portion of a pectin polymer consisting of four monomers.

5. Peonin is a red pigment found in the petals of peony flowers. Consider the structure of peonin:

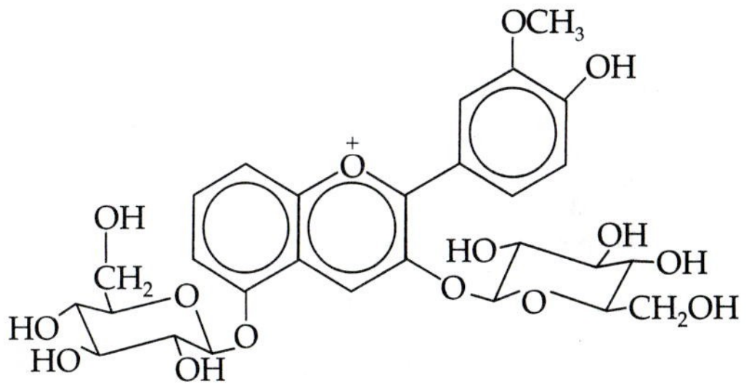

Why do you think peonin is bonded to two hexoses? Which monosaccharide(s) would be produced by acid-catalyzed hydrolysis of peonin?

15

LEARNING GOALS

1. Discuss the physical and chemical properties and biological function of each of the families of lipids.

2. Write the structures of saturated and unsaturated fatty acids.

3. Compare and contrast the structure and properties of saturated and unsaturated fatty acids.

4. Write equations representing the reactions that fatty acids undergo.

5. Describe the functions of prostaglandins.

6. Discuss the mechanism by which aspirin reduces pain.

7. Draw the structure of a phospholipid and discuss its amphipathic nature.

8. Discuss the general classes of sphingolipids and their functions.

9. Draw the structure of the steroid nucleus and discuss the functions of steroid hormones.

10. Describe the function of lipoproteins in triglyceride and cholesterol transport in the body.

11. Describe the structure of the cell membrane and discuss its functions.

12. Discuss passive and facilitated diffusion of materials through a cell membrane.

13. Describe the mechanism of action of the Na^+-K^+ ATPase.

Lipids and Their Functions in Biochemical Systems

Lipids seem to be the most controversial group of biological molecules, particularly in the fields of medicine and nutrition. We are concerned about the use of anabolic steroids by athletes. Although these hormones increase muscle mass and enhance performance, we are just beginning to understand the damage they cause to the body.

We worry about what types of dietary fat we should consume. We hear frequently about the amounts of saturated fats and cholesterol in our diets because a strong correlation has been found between these lipids and heart disease. Large quantities of dietary saturated fats may also predispose an individual to colon, esophageal, stomach, and breast cancers. As a result, we are advised to reduce our intake of cholesterol and saturated fats.

Nonetheless, lipids serve a wide variety of functions essential to living systems and are required in our diet. In this chapter we will study the diverse collection of molecules referred to as lipids. We will see that triglycerides are both a dietary source of energy and a, sometimes unwanted, storage form of energy. Other lipids serve as structural components of the cell; for instance, phospholipids and cholesterol are components of the membranes around each of our cells. Some of the chemical messengers of our bodies are lipids. These include the steroid hormones and the hormonelike prostaglandins. Even some of the vitamins that are required in our diet are lipids and any diet that is completely fat-free will result in deficiencies of these vitamins.

This quick tour through a few of the potential hazards of lipids and their essential roles in our bodies makes it easy to understand why lipids are so controversial and why so much literature has been published about the lipids in our diets. But, in fact, standards of fat intake have not been experimentally determined. The most recent U.S. Dietary Guidelines recommend that our daily fat intake should be between 20 and 35% of our daily caloric intake. Fewer than 10% of the calories we consume should be saturated fats and cholesterol intake should be less than 300 mg/day. Overall, the intake of fats and oils that are high in saturated fats or in trans-fatty acids should be limited.

The foxglove plant is the source of digitoxin and other cardiotonic steroids. Read A Medical Connection: Steroids and the Treatment of Heart Disease in this chapter and explain why the ingestion of foxglove can have deadly consequences.

15.1 Biological Functions of Lipids

The term **lipids** refers to a collection of organic molecules of varying chemical composition. They are grouped together on the basis of their solubility in nonpolar solvents. Lipids are commonly subdivided into four main groups:

1. *Fatty acids* (saturated and unsaturated)
2. *Glycerides* (glycerol-containing lipids)
3. *Nonglyceride lipids* (sphingolipids, steroids, waxes)
4. *Complex lipids* (lipoproteins)

In this chapter, we examine the structure, properties, chemical reactions, and biological functions of each of the lipid groups shown in Figure 15.1.

As a result of differences in their structures, lipids serve many different functions in the human body. The following brief list will give you an idea of the importance of lipids in biological processes:

- *Energy source.* When oxidized, each gram of fat releases 9 kcal of energy, or more than twice the energy released by oxidation of a gram of carbohydrate.
- *Energy storage.* Most of the energy stored in the body is in the form of lipids (triglycerides) in fat cells called *adipocytes.*
- *Cell membrane structural components.* Phosphoglycerides, sphingolipids, and steroids make up the basic structure of all cell membranes which control the flow of molecules into and out of cells.
- *Hormones.* The steroid hormones are critical chemical messengers that allow tissues of the body to communicate with one another. The hormonelike prostaglandins also exert strong biological effects.
- *Vitamins.* The lipid-soluble vitamins, A, D, E, and K, play a major role in critical biological processes, including blood clotting and vision.
- *Vitamin absorption.* Dietary fat "carries" lipid-soluble vitamins. All are transported into cells of the small intestine. Therefore, a diet that is too low in fat (less than 20% of calories) can result in a deficiency of these four vitamins.
- *Protection.* Fats serve as a shock absorber, or protective layer, for the vital organs. About 4% of the total body fat is reserved for this critical function.
- *Insulation.* Fat stored beneath the skin (subcutaneous fat) insulates the body from extremes of cold temperatures.

Lipid-soluble vitamins are discussed in detail online at www.mhhe.com/denniston in Lipid-Soluble Vitamins.

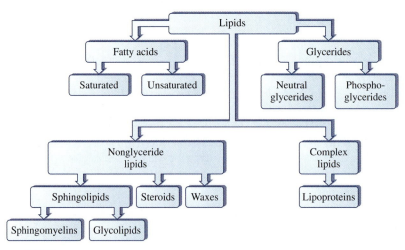

Figure 15.1

The classification of lipids.

A MEDICAL Connection | Lifesaving Lipids

In the intensive-care nursery, the premature infant struggles for life. Born three and a half months early, the baby weighs only 1.6 pounds, and the lungs labor to provide enough oxygen to keep the tiny body alive. Premature infants often have respiratory difficulties because they have not yet begun to produce *pulmonary surfactant*.

Pulmonary surfactant is a combination of phospholipids and proteins that reduces surface tension (Section 6.2) in the alveoli of the lungs. (Alveoli are the small, thin-walled air sacs in the lungs.) This allows efficient gas exchange across the membranes of the alveolar cells; oxygen can more easily diffuse from the air into the tissues and carbon dioxide can easily diffuse from the tissues into the air. Without pulmonary surfactant, gas exchange in the lungs is very poor.

Pulmonary surfactant is not produced until early in the sixth month of pregnancy. Premature babies born before they have begun secretion of natural surfactant suffer from *respiratory distress syndrome (RDS)*, which is caused by the severe difficulty they have obtaining enough oxygen from the air that they breathe.

Until recently, RDS was a major cause of death among premature infants, but now a lifesaving treatment is available. A fine aerosol of an artificial surfactant is administered directly into the trachea. The Glaxo-Wellcome Company product EXO-SURF Neonatal contains the phospholipid lecithin to reduce surface tension; 1-hexadecanol, which spreads the lecithin; and a polymer called tyloxapol, which disperses both the lecithin and the 1-hexadecanol.

Artificial pulmonary surfactant therapy has dramatically reduced premature infant death caused by RDS and appears to have reduced overall mortality for all babies born weighing less than 700 g (about 1.5 pounds). Advances such as this have come about as a result of research on the makeup of body tissues and

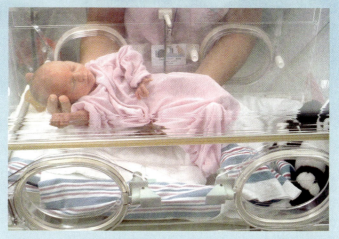

A premature baby in an incubator.

secretions in both healthy and diseased individuals. Often, such basic research provides the information needed to develop effective therapies.

FOR FURTHER UNDERSTANDING

Intramuscular doses of dexamethasone given to the mother before the birth of the child have been used to induce fetal surfactant production and lung maturation. This treatment has been shown to reduce RDS and mortality by 40%. Use the Internet to investigate the use of dexamethasone in treatment of infant respiratory distress syndrome. To what class of lipids does dexamethasone belong?

Draw the structures of 1-hexadecanol and lecithin. Use the structures to explain how they interact on the surface of the alveoli.

15.2 Fatty Acids

Structure and Properties

Fatty acids are long-chain monocarboxylic acids. As a consequence of their biosynthesis, fatty acids generally contain an *even number* of carbon atoms. The general formula for a **saturated fatty acid** is $CH_3(CH_2)_nCOOH$, in which n in biological systems is an even integer. If $n = 16$, the result is an 18-carbon saturated fatty acid, stearic acid, having the following structural formula:

2 **Learning Goal**
Write the structures of saturated and unsaturated fatty acids.

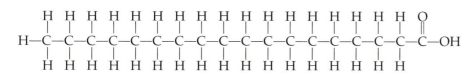

TABLE 15.1 Common Saturated and Unsaturated Fatty Acids

Common Saturated Fatty Acids

Common Name	I.U.P.A.C. Name	Melting Point (°C)	RCOOH Formula	Condensed Formula
Capric	Decanoic	32	$C_9H_{19}COOH$	$CH_3(CH_2)_8COOH$
Lauric	Dodecanoic	44	$C_{11}H_{23}COOH$	$CH_3(CH_2)_{10}COOH$
Myristic	Tetradecanoic	54	$C_{13}H_{27}COOH$	$CH_3(CH_2)_{12}COOH$
Palmitic	Hexadecanoic	63	$C_{15}H_{31}COOH$	$CH_3(CH_2)_{14}COOH$
Stearic	Octadecanoic	70	$C_{17}H_{35}COOH$	$CH_3(CH_2)_{16}COOH$
Arachidic	Eicosanoic	77	$C_{19}H_{39}COOH$	$CH_3(CH_2)_{18}COOH$

Common Unsaturated Fatty Acids

Common Name	I.U.P.A.C. Name	Melting Point (°C)	RCOOH Formula	Number of Double Bonds	Position of Double Bonds
Palmitoleic	cis-9-Hexadecenoic	0	$C_{15}H_{29}COOH$	1	9
Oleic	cis-9-Octadecenoic	16	$C_{17}H_{33}COOH$	1	9
Linoleic	cis,cis-9,12-Octadecadienoic	5	$C_{17}H_{31}COOH$	2	9, 12
Linolenic	All cis-9,12,15-Octadecatrienoic	−11	$C_{17}H_{29}COOH$	3	9, 12, 15
Arachidonic	All cis-5,8,11,14-Eicosatetraenoic	−50	$C_{19}H_{31}COOH$	4	5, 8, 11, 14

Condensed Formula

Palmitoleic	$CH_3(CH_2)_5CH{=}CH(CH_2)_7COOH$
Oleic	$CH_3(CH_2)_7CH{=}CH(CH_2)_7COOH$
Linoleic	$CH_3(CH_2)_4CH{=}CH-CH_2-CH{=}CH(CH_2)_7COOH$
Linolenic	$CH_3CH_2CH{=}CH-CH_2-CH{=}CH-CH_2-CH{=}CH(CH_2)_7COOH$
Arachidonic	$CH_3(CH_2)_4CH{=}CH-CH_2-CH{=}CH-CH_2-CH{=}CH-CH_2-CH{=}CH-(CH_2)_3COOH$

The saturated fatty acids may be thought of as derivatives of alkanes, the saturated hydrocarbons described in Chapter 10.

Note that each of the carbons in the chain is bonded to the maximum number of hydrogen atoms. To help remember the structure of a saturated fatty acid, you might think of each carbon in the chain being "saturated" with hydrogen atoms. Examples of common saturated fatty acids are given in Table 15.1. An example of an **unsaturated fatty acid** is the 18-carbon unsaturated fatty acid oleic acid, which has the following structural formula:

Unsaturated fatty acids may be thought of as derivatives of alkenes, unsaturated hydrocarbons discussed in Chapter 11.

A discussion of trans-fatty acids is found in Section 11.3.

In the case of unsaturated fatty acids, there is at least one carbon-to-carbon double bond. The double bonds found in almost all naturally occurring unsaturated fatty acids are in the *cis* configuration. In addition, the double bonds are not randomly located in the hydrocarbon chain. Both the placement and the geometric configuration of the double bonds are dictated by the enzymes that catalyze the biosynthesis of unsaturated fatty acids. Examples of common unsaturated fatty acids are also given in Table 15.1.

EXAMPLE 15.1 **Writing the Structural Formula of an Unsaturated Fatty Acid**

Draw the structural formula for palmitoleic acid.

2 **Learning Goal**

Write the structures of saturated and unsaturated fatty acids.

SOLUTION

Referring to Table 15.1, we see that the I.U.P.A.C. name of palmitoleic acid is *cis*-9-hexadecenoic acid. The name tells us that this is a 16-carbon fatty acid having a carbon-to-carbon double bond between carbons 9 and 10. The name also reveals that this is the *cis* isomer.

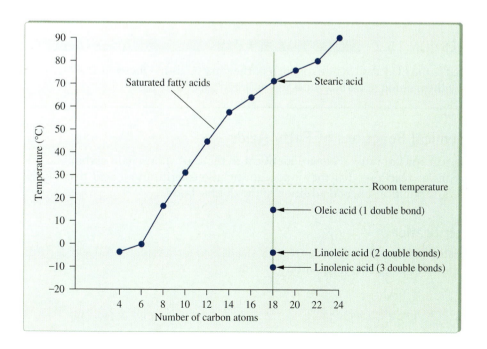

Examination of Table 15.1 and Figure 15.2 reveals several interesting and important points about the physical properties of fatty acids.

3 **Learning Goal**

Compare and contrast the structure and properties of saturated and unsaturated fatty acids.

- The melting points of saturated fatty acids increase with increasing carbon number. Those with ten or more carbons are solids at room temperature.
- The melting points of unsaturated fatty acids increase with increasing hydrocarbon chain length.
- The melting point of a saturated fatty acid is greater than that of an unsaturated fatty acid of the same chain length. The reason is that the *cis* double bond produces "kinked" molecules that cannot stack in an organized arrangement and thus have lower intermolecular attractions and lower melting points.

Figure 15.2

The melting points of fatty acids. Melting points of both saturated and unsaturated fatty acids increase as the number of carbon atoms in the chain increases. The melting points of unsaturated fatty acids are lower than those of the corresponding saturated fatty acid with the same number of carbon atoms. Also, as the number of double bonds in the chain increases, the melting points decrease.

3 Learning Goal

Compare and contrast the structure and properties of saturated and unsaturated fatty acids.

EXAMPLE 15.2 **Examining the Similarities and Differences Between Saturated and Unsaturated Fatty Acids**

Construct a table comparing the structure and properties of saturated and unsaturated fatty acids.

SOLUTION

Property	Saturated Fatty Acid	Unsaturated Fatty Acid
Chemical composition	Carbon, hydrogen, oxygen	Carbon, hydrogen, oxygen
Chemical structure	Hydrocarbon chain with a terminal carboxyl group	Hydrocarbon chain with a terminal carboxyl group
Carbon-carbon bonds within the hydrocarbon chain	Only C—C single bonds	At least one C—C double bond
"Shape" of hydrocarbon chain	Linear, fully extended	Bend in carbon chain at site of C—C double bond
Physical state at room temperature	Solid	Liquid
Melting point for two fatty acids of the same hydrocarbon chain length	Higher	Lower
Relationship between melting point and chain length	Longer chain length, higher melting point	Longer chain length, higher melting point

---

QUESTION 15.1

Draw formulas for each of the following fatty acids:

a. Oleic acid
b. Lauric acid
c. Linoleic acid
d. Stearic acid

QUESTION 15.2

What is the I.U.P.A.C. name for each of the fatty acids in Question 15.1? (*Hint:* Review the naming of carboxylic acids in Section 13.1 and Table 15.1.)

Chemical Reactions of Fatty Acids

4 Learning Goal

Write equations representing the reactions that fatty acids undergo.

The reactions of fatty acids are identical to those of short-chain carboxylic acids. The major reactions that they undergo include esterification, acid hydrolysis of esters, saponification, and addition at the double bond.

Esterification

Esterification is described in Section 13.2.

In **esterification,** fatty acids react with alcohols to form esters and water according to the following general equation:

$$\underset{\text{Fatty acid}}{R^1 - \overset{\overset{\text{O}}{\|}}{C} - OH} + \underset{\text{Alcohol}}{HOR^2} \xrightarrow{H^+, \text{heat}} \underset{\text{Ester}}{R^1 - \overset{\overset{\text{O}}{\|}}{C} - OR^2} + \underset{\text{Water}}{H - OH}$$

Writing Equations Representing the Esterification of Fatty Acids

Write an equation representing the esterification of capric acid with propyl alcohol and write the I.U.P.A.C. names of each of the organic reactants and products.

4 **Learning Goal**
Write equations representing the reactions that fatty acids undergo.

SOLUTION

From Table 15.1 we see that capric acid is also known as decanoic acid. Knowing that decanoic acid is a 10-carbon fatty acid and propyl alcohol (propanol) is a 3-carbon alcohol, we can predict that the product of this reaction will be propyl decanoate. Now we can write an equation for this reaction, remembering to note the need for an acid catalyst and heat:

$$
\underset{\text{Decanoic acid}}{CH_3(CH_2)_8-\overset{\overset{\displaystyle O}{\|}}{C}-OH} + \underset{\text{Propanol}}{CH_3CH_2CH_2OH} \xrightarrow{H^+,\ heat}
$$

$$
\underset{\text{Propyl decanoate}}{CH_3(CH_2)_8-\overset{\overset{\displaystyle O}{\|}}{C}-O-CH_2CH_2CH_3} + H_2O
$$

QUESTION 15.3

Write the complete equation for the esterification of lauric acid and ethyl alcohol. Write the I.U.P.A.C. name for each of the organic reactants and products.

QUESTION 15.4

Write the complete equation for the esterification of capric acid and 1-pentanol. Write the I.U.P.A.C. name for each of the organic reactants and products.

Acid Hydrolysis

Recall that hydrolysis is the reverse of esterification, producing fatty acids from esters:

Acid hydrolysis is discussed in Section 13.2.

$$
\underset{\text{Ester}}{R^1-\overset{\overset{\displaystyle O}{\|}}{C}-OR^2} + \underset{\text{Water}}{HO-H} \xrightarrow{H^+,\ heat} \underset{\text{Fatty acid}}{R^1-\overset{\overset{\displaystyle O}{\|}}{C}-OH} + \underset{\text{Alcohol}}{R^2OH}
$$

Writing Equations Representing the Acid Hydrolysis of a Fatty Acid Ester

Write an equation representing the acid hydrolysis of methyl decanoate and write the I.U.P.A.C. names of each of the organic reactants and products.

4 **Learning Goal**
Write equations representing the reactions that fatty acids undergo.

SOLUTION

From the name of the ester, methyl decanoate, we know that it was prepared from methanol and decanoic acid. Now we can write an equation representing

Continued—

EXAMPLE 15.4 —*Continued*

the hydrolysis of methyl decanoate into methanol and decanoic acid (capric acid). Remember to note the need for an acid catalyst and heat over the reaction arrow:

$$CH_3(CH_2)_8-\overset{\overset{\displaystyle O}{\|}}{C}-O-CH_3 + H_2O \xrightarrow{H^+, \text{ heat}} CH_3(CH_2)_8-\overset{\overset{\displaystyle O}{\|}}{C}-OH + CH_3OH$$

Methyl decanoate   Decanoic acid   Methanol

QUESTION 15.5

Write a complete equation for the acid hydrolysis of butyl propionate. Write the I.U.P.A.C. name for each of the organic reactants and products.

QUESTION 15.6

Write a complete equation for the acid hydrolysis of ethyl butyrate. Write the I.U.P.A.C. name for each of the organic reactants and products.

Saponification

Saponification is described in Section 13.2.

Saponification is the base-catalyzed hydrolysis of an ester:

$$R^1-\overset{\overset{\displaystyle O}{\|}}{C}-OR^2 + NaOH \longrightarrow R^1-\overset{\overset{\displaystyle O}{\|}}{C}-O^-Na^+ + R^2OH$$

Ester   Base   Salt   Alcohol

The role of soaps in removal of dirt and grease is described in Section 13.2.

The product of this reaction, an ionized salt, is a soap. Because soaps have a long uncharged hydrocarbon tail and a negatively charged terminus (the carboxylate group), they form micelles that dissolve oil and dirt particles. Thus the dirt is emulsified and broken into small particles and can be rinsed away.

Examples of micelles are shown in Figures 13.4 and 19.1.

EXAMPLE 15.5 | **Writing Equations Representing the Base–Catalyzed Hydrolysis of a Fatty Acid Ester**

4 Learning Goal

Write equations representing the reactions that fatty acids undergo.

Write an equation representing the base-catalyzed hydrolysis of ethyl dodecanoate and write the I.U.P.A.C. names of each of the organic reactants and products.

SOLUTION

From the name of the ester, ethyl dodecanoate, we know that it was prepared from ethanol and the 12-carbon fatty acid dodecanoic acid (Refer to Table 15.1). With this information, we can write an equation representing the hydrolysis of ethyl dodecanoate into ethanol and sodium dodecanoate. Remember that in the base-catalyzed hydrolysis of an ester, the base is a reactant. Now we can write an equation representing this reaction:

$$CH_3(CH_2)_{10}-\overset{\overset{\displaystyle O}{\|}}{C}-O-CH_2CH_3 + NaOH \longrightarrow$$

Ethyl dodecanoate

$$CH_3(CH_2)_{10}-\overset{\overset{\displaystyle O}{\|}}{C}-O^-Na^+ + CH_3CH_2OH$$

Sodium dodecanoate   Ethanol

CHEMISTRY at the Crime Scene | Adipocere and Mummies of Soap

One November evening in 1911, widower Patrick Higgins stepped into his local pub in Abercorn, Scotland. To the surprise of his drinking companions, he did not have his two young boys with him. Neighbors knew that the boys had been a great burden on Patrick since the death of his wife; so they believed Patrick's story that he left William, age 6, and John, age 4, with two women in Edinburgh who had offered to adopt the boys.

More than eighteen months had passed when an object was seen floating in the Hopetoun Quarry, an unused, flooded quarry near town. When the object was fished out, it was obvious that it was the body of a young boy; the rescuers were stunned to find another small body tied to the first by a rope. How were these bodies preserved after such a long time and why did they float? The answer is that their bodies had almost completely turned into adipocere, or more simply, soap.

Forensic scientists are trying to understand the nature of the reaction that creates *adipocere*, the technical term for the yellowish-white, greasy, wax-like substance, which results from the saponification of fatty tissue. Some researchers hope that this information may allow determination of the postmortem interval (length of time since death). Others simply value the process because it helps preserve the body so well that even after long periods of time, it can be easily recognized and any wounds or injuries can be observed.

It is known that adipocere is produced when body fat is hydrolyzed (water is needed) to release fatty acids. Because the fatty acids lower the pH in the tissues, they inhibit many of the bacteria that would begin the process of decay. Certain other bacteria, particularly *Clostridium welchii*, an organism that cannot grow in the presence of oxygen, is known to speed up the formation of adipocere in moist, warm, anaerobic (oxygenless) environments. Adipocere forms first in subcutaneous tissues, including the cheeks, breasts, and buttocks. Given appropriate warmth and damp conditions, it may be seen as early as three to four weeks after death; but more commonly it is not observed until five to six months after death.

Adipocere formation in John and William Higgins was so extensive that their former neighbors had no trouble recognizing them. At the post mortem, another advantage of adipocere formation became obvious—it had preserved the stomach contents of the boys! From this, the coroner learned that the boys had eaten Scotch broth about an hour before they died. Investi-

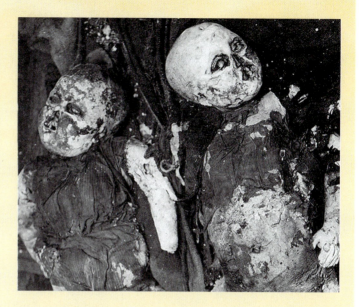

gators were able to find the woman who had given the broth to the boys and, from her testimony, learned that she had fed them on the last day they were seen in the village. Clearly their father had lied about the adoption by two Edinburgh women! In under one and one-half hours, a jury convicted the father of murdering his sons and he was hanged in October 1913.

FOR FURTHER UNDERSTANDING

Adipocere is the technical term for "soap mummification." It comes from the Latin words *adipis* or fat, as in adipose tissue, and *cera*, which means wax. Draw a triglyceride composed of the fatty acids myristic acid, stearic acid, and oleic acid. Write a balanced equation showing a possible reaction that would lead to the formation of adipocere.

Forensic scientists are studying adipocere formation as a possible source of information to determine postmortem interval (length of time since death) of bodies of murder or accident victims. Among the factors being studied are the type of soil, including pH, moisture, temperature, and presence or absence of lime. How might each of these factors influence the rate of adipocere formation and hence the determination of the postmortem interval?

QUESTION 15.7

Write a complete equation for the reaction of butyl propionate with KOH. Write the I.U.P.A.C. name for each of the organic reactants and products.

QUESTION 15.8

Write a complete equation for the reaction of ethyl butyrate with NaOH. Write the I.U.P.A.C. name for each of the organic reactants and products.

Problems can arise when "hard" water is used for cleaning because the high concentrations of Ca^{2+} and Mg^{2+} in such water cause fatty acid salts to precipitate. Not only does this interfere with the emulsifying action of the soap, it also leaves a hard scum on the surface of sinks and tubs.

$$2R-\overset{\overset{\displaystyle O}{\|}}{C}-O^- + Ca^{2+} \longrightarrow (R-\overset{\overset{\displaystyle O}{\|}}{C}-O^-)_2Ca^{2+}(s)$$

Reaction at the Double Bond (Unsaturated Fatty Acids)

Hydrogenation is described in Section 11.4.

Hydrogenation is an example of an addition reaction. The following is a typical example of the addition of hydrogen to the double bonds of a fatty acid:

$$CH_3(CH_2)_4CH=CHCH_2CH=CH(CH_2)_7COOH \xrightarrow{2H_2,\ Ni} CH_3(CH_2)_{16}COOH$$

Linoleic acid   Stearic acid

4   **Learning Goal**
Write equations representing the reactions that fatty acids undergo.

EXAMPLE 15.6   **Writing Equations for the Hydrogenation of a Fatty Acid**

Write an equation representing the hydrogenation of oleic acid and write the I.U.P.A.C. names of each of the organic reactants and products.

SOLUTION

From Table 15.1 we see that the I.U.P.A.C. name for oleic acid is *cis*-9-octadecenoic acid. The name tells us that this is an 18-carbon unsaturated fatty acid with a double bond in the *cis*-configuration between carbons 9 and 10. The hydrogenation reaction, which requires a catalyst such as Ni, will produce the saturated fatty acid octadecanoic acid (stearic acid), as represented in the following equation:

$$CH_3(CH_2)_7CH=CH(CH_2)_7-\overset{\overset{\displaystyle O}{\|}}{C}-OH \xrightarrow{H_2,\ Ni} CH_3(CH_2)_{16}-\overset{\overset{\displaystyle O}{\|}}{C}-OH$$

cis-9-Octadecenoic acid   Octadecanoic acid

QUESTION 15.9

Write a balanced equation showing the hydrogenation of *cis*-9-hexadecenoic acid.

QUESTION 15.10

Write a balanced equation showing the hydrogenation of arachidonic acid.

Hydrogenation of vegetable oils produces a mixture of *cis* and *trans* unsaturated fatty acids. The *trans* unsaturated fatty acids are thought to contribute to atherosclerosis (hardening of the arteries).

Hydrogenation is used in the food industry to convert polyunsaturated vegetable oils into saturated solid fats. *Partial hydrogenation* is carried out to add hydrogen to some, but not all, double bonds in polyunsaturated oils. In this way, liquid vegetable oils are converted into solid form. Crisco is one example of a hydrogenated vegetable oil.

Margarine is also produced by partial hydrogenation of vegetable oils, such as corn oil or soybean oil. The extent of hydrogenation is carefully controlled so that the solid fat will be spreadable and have the consistency of butter when eaten. If too many double bonds were hydrogenated, the resulting product would have the undesirable consistency of animal fat. Artificial color is added to the product, and it may be mixed with milk to produce a butterlike appearance and flavor.

Eicosanoids: Prostaglandins, Leukotrienes, and Thromboxanes

Some of the unsaturated fatty acids containing more than one double bond cannot be synthesized by the body. For many years, it has been known that linolenic acid and linoleic acid, called **essential fatty acids,** are necessary for specific biochemical functions and must be supplied in the diet (see Table 15.1). The function of linoleic acid became clear in the 1960s when it was discovered that linoleic acid is required for the biosynthesis of **arachidonic acid,** the precursor of a class of hormonelike molecules known as **eicosanoids.** The name is derived from the Greek word *eikos,* meaning "twenty," because they are all derivatives of twenty-carbon fatty acids. The eicosanoids include three groups of structurally related compounds: prostaglandins, leukotrienes, and thromboxanes.

A hormone is a chemical signal that is produced by a specialized tissue and is carried by the bloodstream to target tissues. Eicosanoids are referred to as hormonelike because they affect the cells that produce them, as well as other target tissues.

5 Learning Goal

Describe the functions of prostaglandins.

Prostaglandins are extremely potent biological molecules with hormonelike activity. They got their name because they were originally isolated from seminal fluid produced in the prostate gland. More recently they also have been isolated from most animal tissues. Prostaglandins are unsaturated carboxylic acids consisting of a twenty-carbon skeleton that contains a five-carbon ring. Several general classes of prostaglandins are grouped under the designations A, B, E, and F, among others. The examples in Figure 15.3 illustrate the general structure of prostaglandins and the current nomenclature system.

Prostaglandins are made in most tissues and exert their biological effects on the cells that produce them and on other cells in the immediate vicinity. Because prostaglandins and closely related leukotrienes and thromboxanes affect so many body processes and because they often cause opposing effects in different tissues, it can be difficult to keep track of their many regulatory functions. The following is a brief summary of some of the biological processes that are thought to be regulated by prostaglandins, leukotrienes, and thromboxanes.

1. **Blood clotting.** Blood clots form when a blood vessel is damaged, yet such clotting along the walls of undamaged vessels could result in heart attack or stroke. *Thromboxane A_2* (Figure 15.4) stimulates constriction of blood vessels and aggregation of platelets. PGI_2 (prostacyclin) inhibits platelet aggregation and causes dilation of blood vessels and thus prevents the untimely production of blood clots.
2. **The inflammatory response.** When tissue is damaged by mechanical injury, burns, or invasion by microorganisms, a variety of white blood cells descend on the damaged site. The result is swelling, redness, fever, and pain. Prostaglandins are thought to promote aspects of the inflammatory response, especially pain and fever. Drugs such as aspirin block prostaglandin synthesis and help to relieve the symptoms.

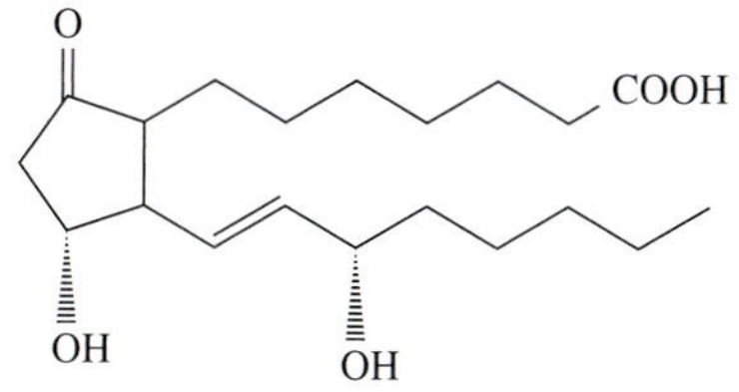

Prostaglandin E_1

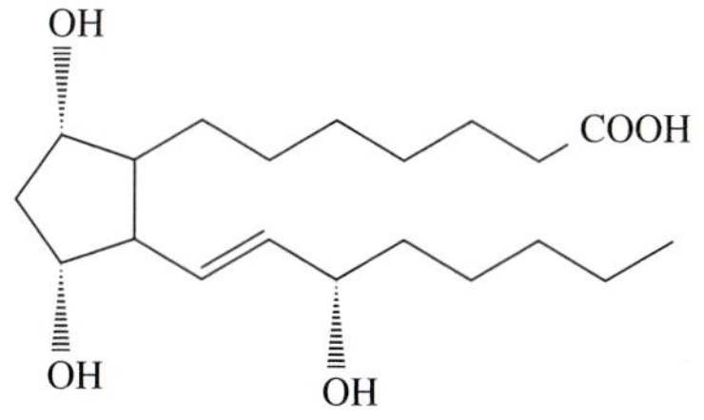

Prostaglandin F_1

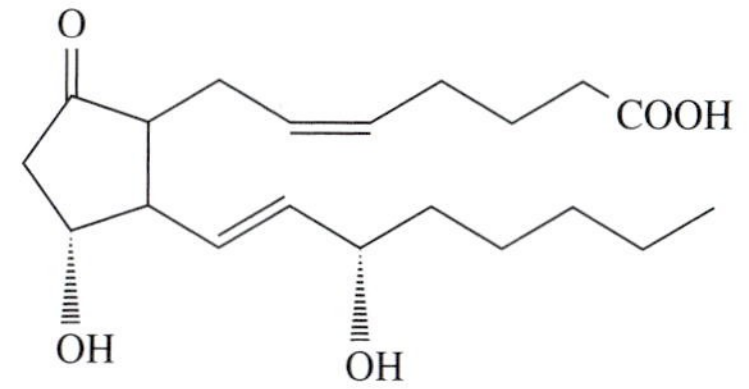

Prostaglandin E_2

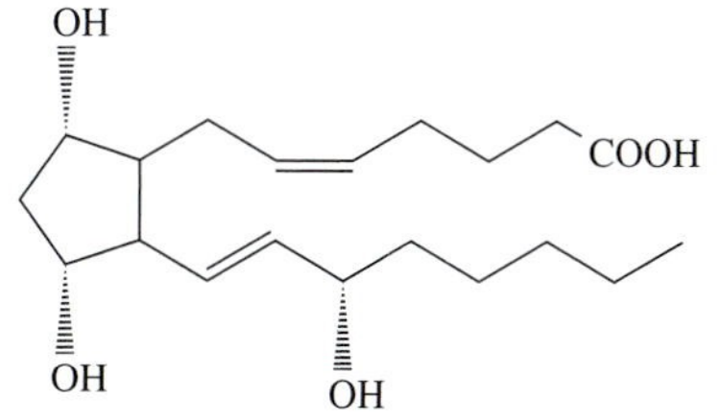

Prostaglandin F_2

Figure 15.3

The structures of four prostaglandins. The nomenclature of prostaglandins is based on the arrangement of the carbon skeleton and the number and orientation of double bonds, hydroxyl groups, and ketone groups. For example, in the name PGF_2, PG stands for prostaglandin, F indicates a particular group of prostaglandins with a hydroxyl group bonded to carbon-9, and 2 indicates that there are two carbon–carbon double bonds in the compound.

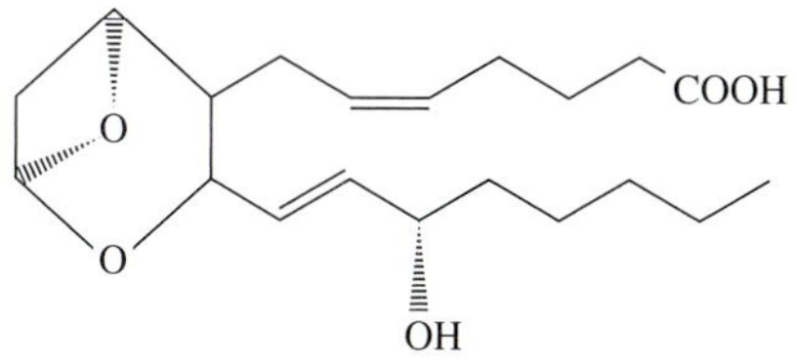

Thromboxane A_2

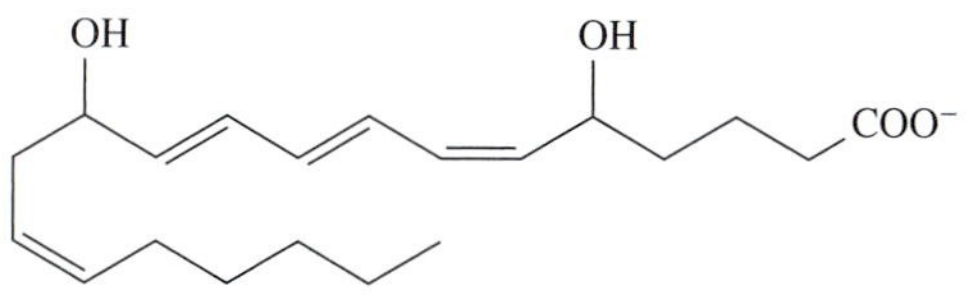

Leukotriene B_4

Figure 15.4

The structures of thromboxane A_2 and leukotriene B_4.

3. **Reproductive system.** PGE$_2$ stimulates smooth muscle contraction, particularly uterine contractions. There is strong evidence that dysmenorrhea (painful menstruation) may be the result of an excess of two prostaglandins. Drugs, such as ibuprofen, that inhibit prostaglandin synthesis are found to provide relief from these symptoms.

4. **Gastrointestinal tract.** Prostaglandins have been shown both to inhibit the secretion of acid and increase the secretion of a protective mucus layer into the stomach. Because aspirin inhibits prostaglandin synthesis, it may encourage stomach ulcers by inhibiting the formation of the normal protective mucus layer and allowing increased secretion of stomach acid.

5. **Kidneys.** Prostaglandins produced in the kidneys cause the renal blood vessels to dilate increasing blood flow and, thus, increasing water and electrolyte excretion.

6. **Respiratory tract.** Eicosanoids produced by certain white blood cells, *leukotrienes* (see Figure 15.4), promote the constriction of the bronchi associated with asthma. Other prostaglandins promote bronchodilation.

As mentioned, prostaglandins stimulate the inflammatory response and, as a result, are partially responsible for the cascade of events that cause pain. Aspirin has long been known to alleviate such pain, and we now know that it does so by inhibiting the synthesis of prostaglandins (Figure 15.5).

The first two steps of prostaglandin synthesis (Figure 15.6), the release of arachidonic acid from the membrane and its conversion to PGH$_2$ by the enzyme,

Figure 15.5

Aspirin inhibits the synthesis of prostaglandins by acetylating the enzyme, cyclooxygenase. The acetylated enzyme is no longer functional.

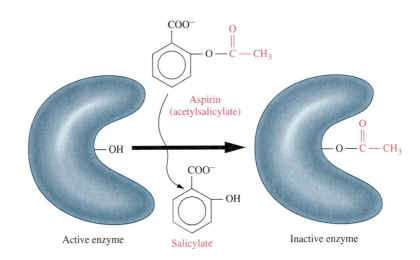

Figure 15.6

A summary of the synthesis of several prostaglandins from arachidonic acid.

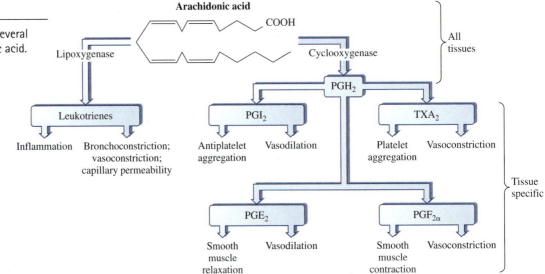

cyclooxygenase, occur in all tissues that produce prostaglandins. The conversion of PGH_2 into the other biologically active forms is tissue specific and requires enzymes found only in certain tissues.

Aspirin works by inhibiting cyclooxygenase, which catalyzes the first step in the pathway leading from arachidonic acid to PGH_2. The acetyl group of aspirin becomes covalently bound to the enzyme, thereby inactivating it (Figure 15.5). Because the reaction catalyzed by cyclooxygenase occurs in all cells, aspirin effectively inhibits synthesis of all prostaglandins.

6 **Learning Goal**
Discuss the mechanism by which aspirin reduces pain.

15.3 Glycerides

Neutral Glycerides

Glycerides are lipid esters that contain the glycerol (1,2,3-propanetriol) molecule and fatty acids. They may be subdivided into two classes: neutral glycerides and phosphoglycerides. Neutral glycerides are nonionic and nonpolar. Phosphoglyceride molecules have a polar region, the phosphoryl group, in addition to nonpolar fatty acid tails. The structures of each of these types of glycerides are critical to their function.

The esterification of glycerol with a fatty acid produces a **neutral glyceride.** Esterification may occur at one, two, or all three positions, producing **mono-glycerides, diglycerides,** or **triglycerides.** You will also see these referred to as *mono-, di-,* or *triacylglycerols.*

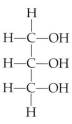

Glycerol
(1,2,3-propanetriol)

> **EXAMPLE 15.7**   **Writing an Equation for the Synthesis of a Monoglyceride**
>
> Write a general equation for the esterification of glycerol and one fatty acid.
>
> **SOLUTION**
> The fatty acid $R-\overset{O}{\underset{}{C}}-OH$ will react with one of the —OH groups of glycerol, forming the ester and releasing a water molecule.
>
> $$\text{Glycerol} \qquad \text{Fatty acid} \qquad \text{Monoglyceride} \qquad \text{Water}$$

4 **Learning Goal**
Write equations representing the reactions that fatty acids undergo.

QUESTION 15.11

Write an equation for the esterification of glycerol with two molecules of stearic acid.

QUESTION 15.12

Write a balanced equation for the esterification of glycerol with one molecule of myristic acid.

The most important neutral glycerides are triglycerides, the major component of fat cells. A triglyceride consists of a glycerol backbone (shown in black) joined to three fatty acid units through ester bonds (shown in red). These long molecules readily stack with one another and constitute the majority of the lipids stored in the body's fat cells.

The principal function of triglycerides in biochemical systems is the storage of energy. If more energy-rich nutrients are consumed than are required for

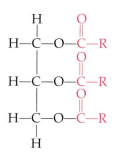

Triglyceride

Lipid metabolism is discussed in Chapter 19.

See A Lifestyle Connection: Losing Those Unwanted Pounds of Adipose Tissue in Chapter 19.

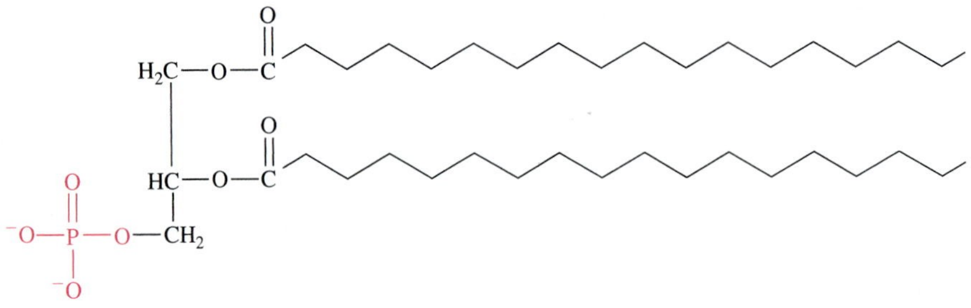

Phosphatidate

(a)

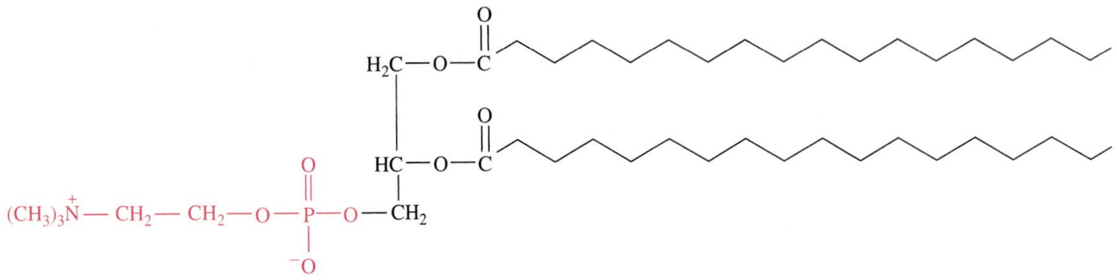

Phosphatidylcholine (lecithin)

(b)

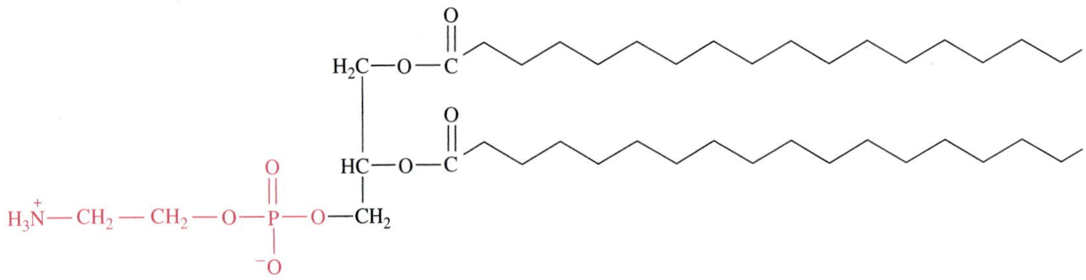

Phosphatidylethanolamine (cephalin)

(c)

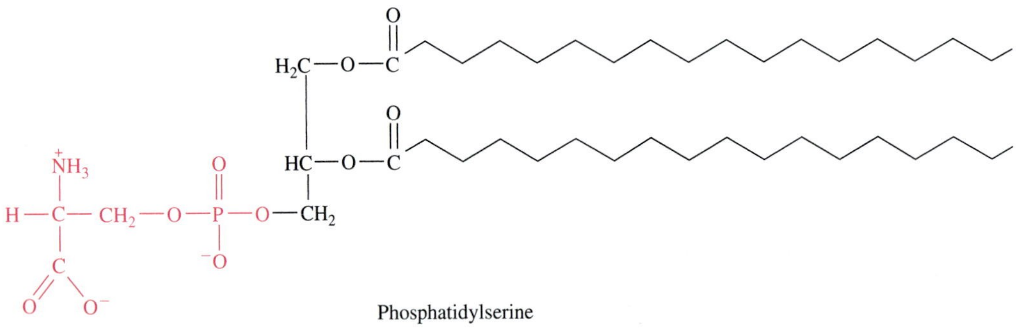

Phosphatidylserine

(d)

Figure 15.7

The structures of (a) phosphatidate and the common membrane phospholipids, (b) phosphatidylcholine (lecithin), (c) phosphatidylethanolamine (cephalin), and (d) phosphatidylserine.

metabolic processes, much of the excess is converted to neutral glycerides and stored as triglycerides in fat cells of *adipose tissue*. When energy is needed, triglycerides are metabolized by the body, and energy is released. For this reason, exercise, along with moderate reduction in caloric intake, is recommended for overweight individuals.

Phosphoglycerides

Phospholipids are a group of lipids that are phosphate esters. The presence of the phosphoryl group results in a molecule with a polar head (the phosphoryl group) and a nonpolar tail (the alkyl chain of the fatty acid). Because the phosphoryl group ionizes in solution, a charged lipid results.

The most abundant membrane lipids are derived from glycerol-3-phosphate and are known as **phosphoglycerides.** Phosphoglycerides contain acyl groups derived from long-chain fatty acids at C-1 and C-2 of glycerol-3-phosphate. At C-3, the phosphoryl group is joined to glycerol by a phosphoester bond. The simplest phosphoglyceride contains a free phosphoryl group and is known as a **phosphatidate** (Figure 15.7). When the phosphoryl group is attached to another hydrophilic molecule, a more complex phosphoglyceride is formed. For example, *phosphatidylcholine (lecithin)* and *phosphatidylethanolamine (cephalin)* are found in the membranes of most cells (Figure 15.7).

Lecithin possesses a polar "head" and a nonpolar "tail." Thus, it is an *amphipathic* molecule. The ionic "head" is hydrophilic and interacts with water molecules, whereas the nonpolar "tail" is hydrophobic and interacts with nonpolar molecules. This amphipathic nature is central to the structure and function of cell membranes.

Lecithin is also the major phospholipid in pulmonary surfactant. It is also found in egg yolks and soybeans and is used as an emulsifying agent in ice cream. An **emulsifying agent** aids in the suspension of triglycerides in water. Emulsification occurs because the hydrophilic head of lecithin dissolves in water and its hydrophobic tail dissolves in triglycerides.

Cephalin is similar in general structure to lecithin; the amine group bonded to the phosphoryl group is the only difference.

 Learning Goal
Draw the structure of a phospholipid and discuss its amphipathic nature.

A phosphoester is the product of the reaction between phosphoric acid and an alcohol.

See A Medical Connection: Lifesaving Lipids, earlier in this chapter.

QUESTION 15.13

Using condensed formulas, draw the mono-, di-, and triglycerides that would result from the esterification of glycerol with each of the following fatty acids.

  a. Oleic acid   b. Capric acid

QUESTION 15.14

Using condensed formulas, draw the mono-, di-, and triglycerides that would result from the esterification of glycerol with each of the following fatty acids.

  a. Palmitic acid   b. Lauric acid

15.4 Nonglyceride Lipids

Sphingolipids

Sphingolipids are lipids that are not derived from glycerol. Like phospholipids, sphingolipids are amphipathic, having a polar head group and two nonpolar fatty acid tails, and are structural components of cellular membranes. They are derived from sphingosine, a long-chain, nitrogen-containing (amino) alcohol:

Learning Goal
Discuss the general classes of sphingolipids and their functions.

$$CH_3(CH_2)_{12}CH=CH-\overset{\overset{\displaystyle OH}{|}}{\underset{\underset{\displaystyle CH_2OH}{|}}{\overset{|}{\underset{|}{C}}-H}}$$
$$H_2N-C-H$$

Sphingosine

Sphingolipids include sphingomyelins and glycosphingolipids. **Sphingomyelins** are the only class of sphingolipids that are also phospholipids:

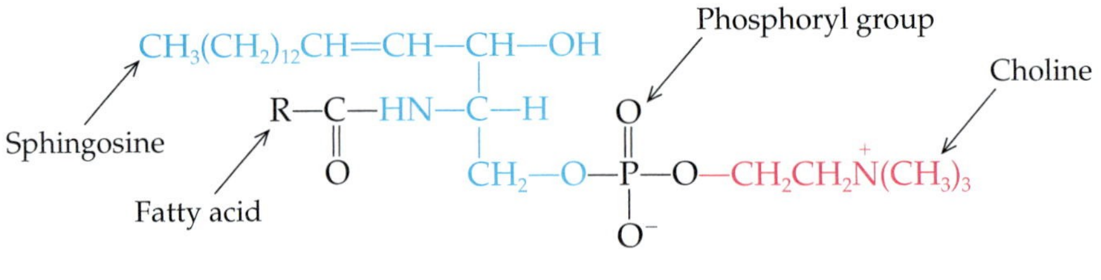

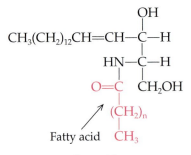

Sphingomyelin

Sphingomyelins are located throughout the body, but are particularly important structural lipid components of nerve cell membranes. They are found in abundance in the myelin sheath that surrounds and insulates cells of the central nervous system. In humans, about 25% of the lipids of the myelin sheath are sphingomyelins. Their role is essential to proper cerebral function and nerve transmission.

Glycosphingolipids, or *glycolipids,* include cerebrosides, sulfatides, and gangliosides and are built on a ceramide backbone structure, which is a fatty acid amide derivative of sphingosine:

Ceramide

To learn about several genetic defects in sphingolipid breakdown, see A Medical Connection: Disorders of Sphingolipid Metabolism online at www.mhhe.com/denniston

Cerebrosides are characterized by the presence of a single monosaccharide head group. Two common cerebrosides are glucocerebroside, found in the membranes of macrophages (cells that protect the body by ingesting and destroying foreign microorganisms) and galactocerebroside, found almost exclusively in the membranes of brain cells. Glucocerebroside consists of ceramide bonded to the hexose glucose; galactocerebroside consists of ceramide joined to the monosaccharide, galactose.

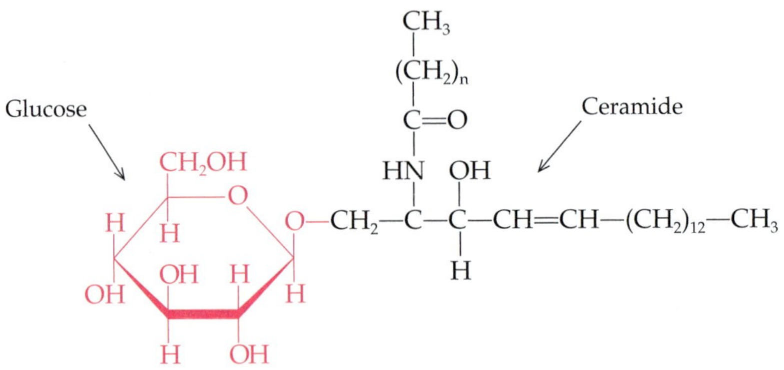

Glucocerebroside

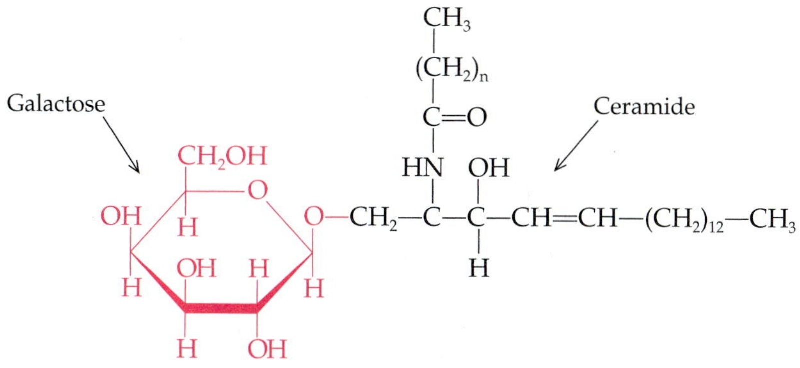

Galactocerebroside

Sulfatides are derivatives of galactocerebroside that contain a sulfate group. Notice that they carry a negative charge at physiological pH.

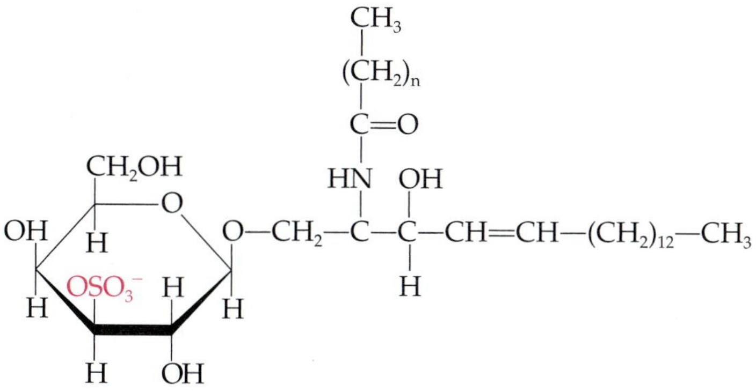

A sulfatide of galactocerebroside

Gangliosides are glycolipids that possess oligosaccharide groups, including one or more molecules of *N*-acetylneuraminic acid (sialic acid). First isolated from membranes of nerve tissue, gangliosides are found in most tissues of the body.

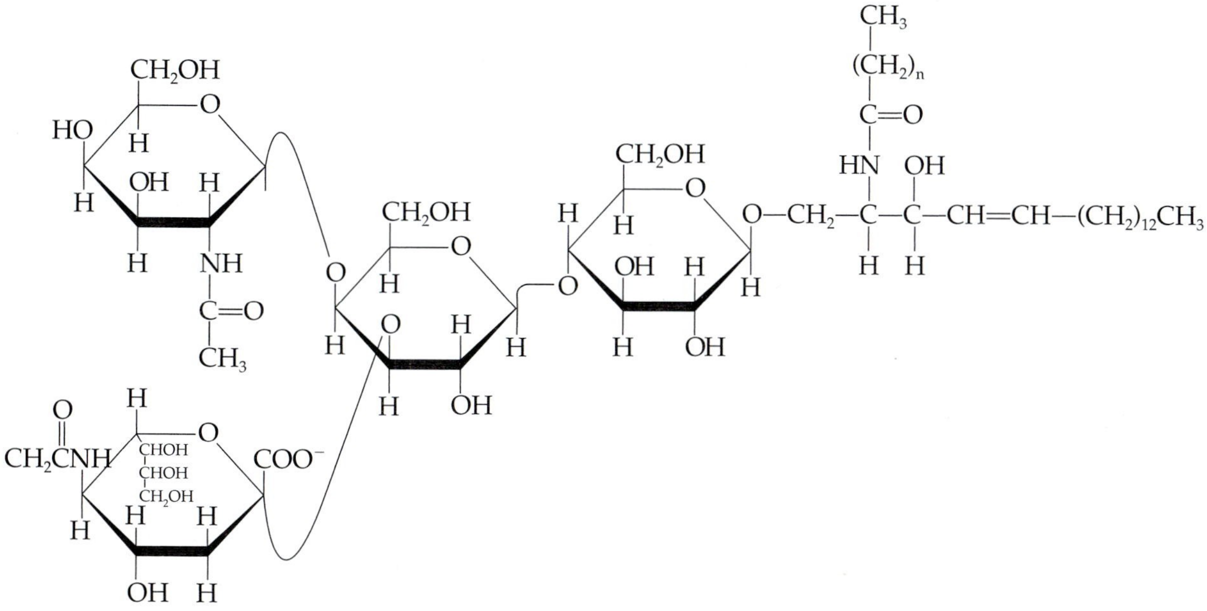

A ganglioside associated with Tay-Sachs disease, a fatal genetic disorder in which harmful levels of this ganglioside build up in nerve cells of the brain.

A MEDICAL Connection | Steroids and the Treatment of Heart Disease

The foxglove plant *(Digitalis purpurea)* is an herb that produces one of the most powerful known stimulants of heart muscle. The active ingredients of the foxglove plant (digitalis) are so-called cardiac glycosides, or *cardiotonic steroids*, which include digitoxin, digosin, and gitalin.

Digitoxin

The structure of digitoxin, one of the cardiotonic steroids produced by the foxglove plant.

Digitalis purpurea, the foxglove plant.

These drugs are used clinically in the treatment of congestive heart failure, which results when the heart is not beating with strong, efficient strokes. When the blood is not propelled through the cardiovascular system efficiently, fluid builds up in the lungs and lower extremities (edema). The major symptoms of congestive heart failure are an enlarged heart, weakness, edema, shortness of breath, and fluid accumulation in the lungs.

This condition was originally described in 1785 by a physician, William Withering, who found a peasant woman whose folk medicine was famous as a treatment for chronic heart problems. Her potion contained a mixture of more than twenty herbs, but Dr. Withering, a botanist as well as physician, quickly discovered that foxglove was the active ingredient in the mixture. Withering used *Digitalis purpurea* successfully to treat congestive heart failure and even described some cautions in its use.

Cardiotonic steroids are extremely strong heart stimulants. A dose as low as 1 mg increases the stroke volume of the heart (volume of blood per contraction), increases the strength of the contraction, and reduces the heart rate. When the heart is pumping more efficiently because of stimulation by digitalis, edema disappears.

Digitalis can be used to control congestive heart failure, but the dose must be carefully determined and monitored because the therapeutic dose is close to the dose that causes toxicity. The symptoms that result from high body levels of cardiotonic steroids include vomiting, blurred vision and lightheadedness, increased water loss, convulsions, and death. Only a physician can determine the initial dose and maintenance schedule for an individual to control congestive heart failure and yet avoid the toxic side effects.

FOR FURTHER UNDERSTANDING

Foxglove is a perennial plant, that is, it is a plant that will grow back each year for at least three years. Occasionally, foxglove first-year growth has been mistaken for comfrey, another plant with medical applications. Greeks and Romans used comfrey to treat wounds and to stop heavy bleeding, as well as for bronchial problems. Explain why the use of foxglove in place of comfrey might have fatal consequences.

Drugs such as digitalis are referred to as cardiac glycosides and as cardiotonic steroids. Explain why both names are valid.

Steroids

9 Learning Goal

Draw the structure of the steroid nucleus and discuss the functions of steroid hormones.

Steroids are a naturally occurring family of organic molecules of biochemical and medical interest. A great deal of controversy has surrounded various steroids. We worry about the amount of cholesterol in the diet and the possible health effects. We are concerned about the use of anabolic steroids by athletes wishing to build muscle mass and improve their performance. However, members of this family of molecules derived from cholesterol have many important functions in the body.

The bile salts that aid in the emulsification and digestion of lipids are steroid molecules, as are the sex hormones, testosterone and estrone.

All steroids contain the steroid nucleus (steroid carbon skeleton) as shown here:

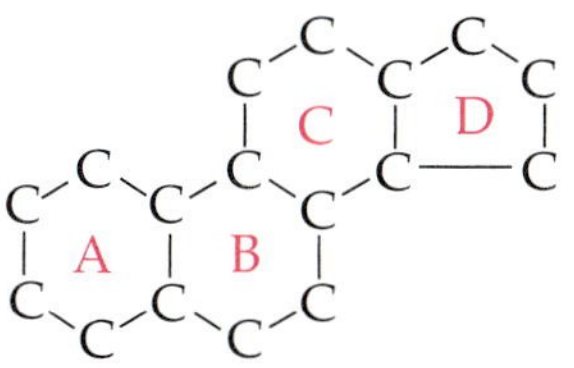

Lipid digestion is described in Section 19.1.

Carbon skeleton of
the steroid nucleus

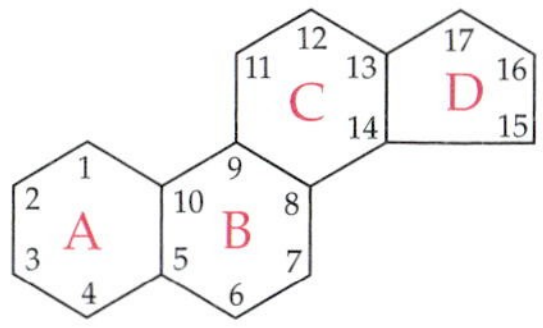

Steroid nucleus

The steroid carbon skeleton consists of four fused rings. Each ring pair has two carbons in common. Thus two fused rings share one or more common bonds as part of their ring backbones. For example, rings A and B, B and C, and C and D are all fused in the preceding structure. Many steroids have methyl groups attached to carbons 10 and 13, as well as an alkyl, alcohol, or ketone group attached to carbon-17.

Cholesterol, a common steroid, is found in the membranes of most animal cells. It is an amphipathic molecule and is readily soluble in the hydrophobic region of membranes. It is involved in regulation of the fluidity of the membrane as a result of the nonpolar fused ring. However, the hydroxyl group is polar and functions like the polar heads of sphingolipids and phospholipids.

There is a strong correlation between the concentration of cholesterol found in the blood plasma and heart disease, particularly atherosclerosis (hardening of the arteries). Cholesterol, in combination with other substances, contributes to a narrowing of the artery passageway. As narrowing increases, more pressure is necessary to ensure adequate blood flow, and high blood pressure (hypertension) develops. Hypertension is also linked to heart disease.

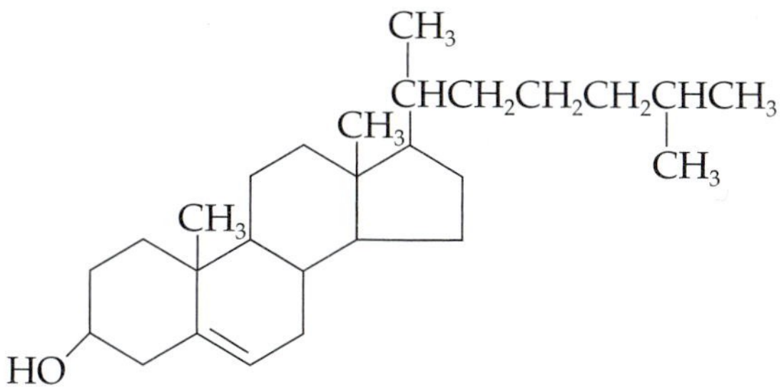

Cholesterol

Egg yolks contain a high concentration of cholesterol, as do many dairy products and animal fats. As a result, it has been recommended that the amounts of these products in the diet be regulated to moderate the dietary intake of cholesterol.

Bile salts are amphipathic derivatives of cholesterol that are synthesized in the liver and stored in the gallbladder. The principal bile salts in humans are cholate and chenodeoxycholate. Bile salts are emulsifying agents whose polar hydroxyl and carboxylate groups interact with water and whose hydrophobic regions bind to lipids.

Following a meal, bile flows from the gallbladder to the duodenum (the uppermost region of the small intestine). Here the bile salts emulsify dietary fats into small droplets that can be more readily digested by lipases (lipid digesting enzymes) also found in the small intestine. Bile salts are described in greater detail in Section 19.1.

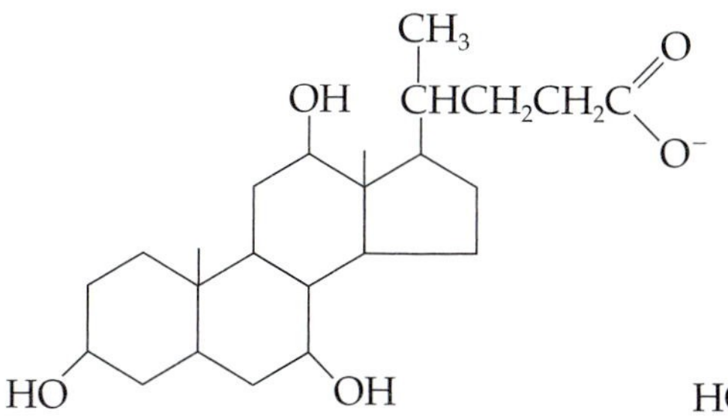

Cholate

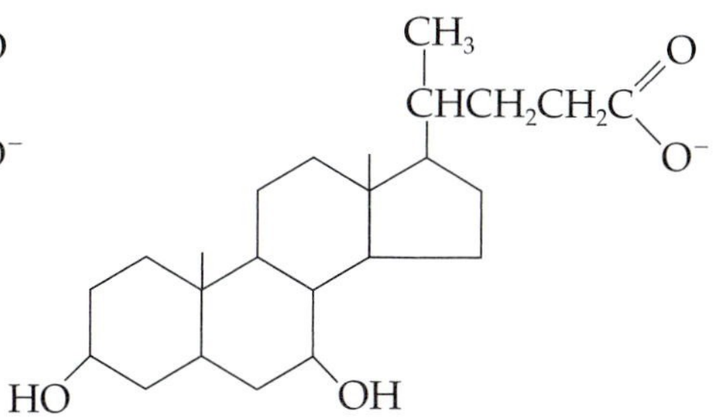

Chenodeoxycholate

Steroids play a role in the reproductive cycle. In a series of chemical reactions, cholesterol is converted to the steroid *progesterone,* the most important hormone associated with pregnancy. Produced in the ovaries and in the placenta, progesterone is responsible for both the successful initiation and the successful completion of

Animation

Mechanism of Steroid Hormone Action

pregnancy. It prepares the lining of the uterus to accept the fertilized egg. Once the egg is attached, progesterone is involved in the development of the fetus and plays a role in the suppression of further ovulation during pregnancy.

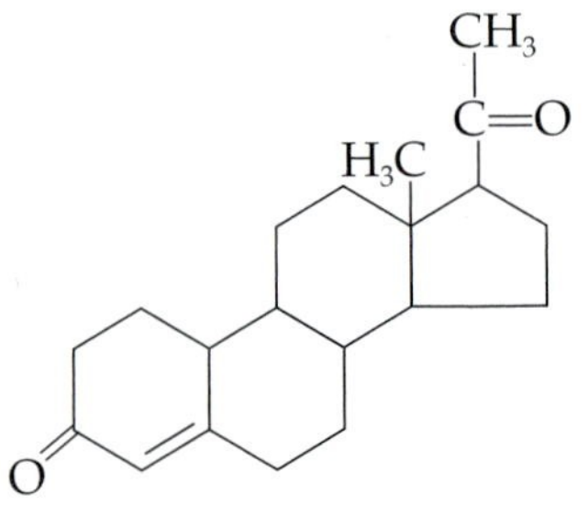

19-Norprogesterone

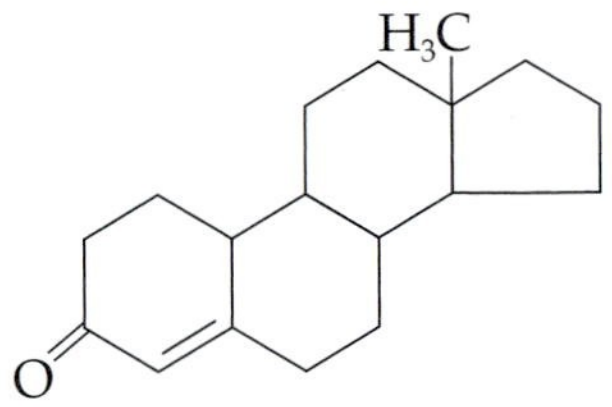

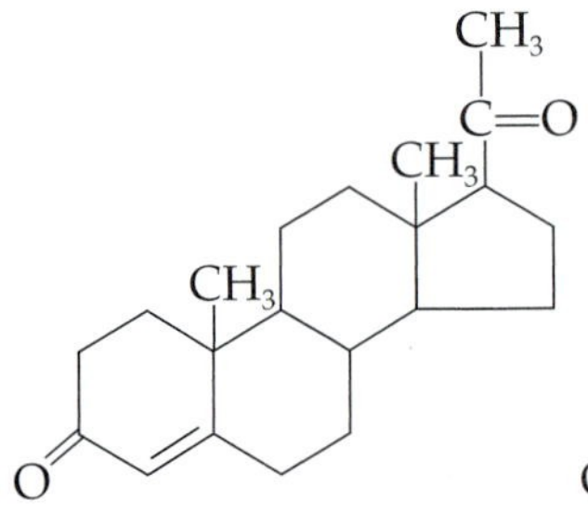

Progesterone

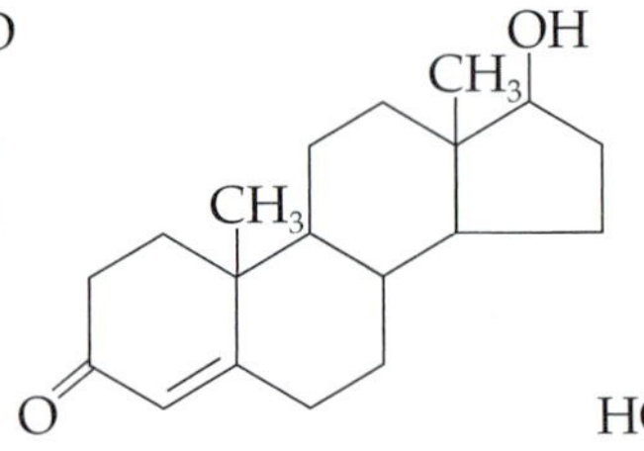

Testosterone

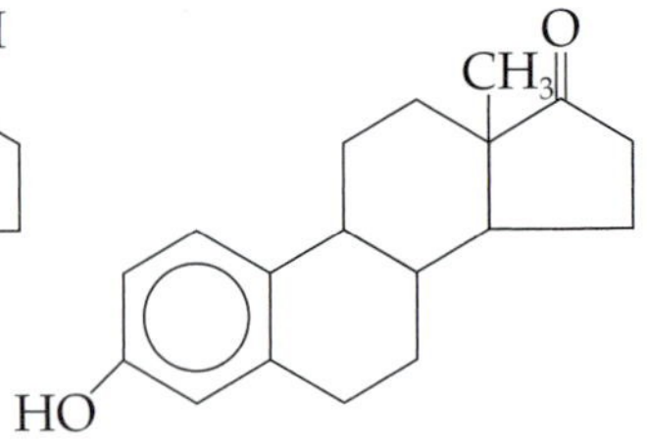

Estrone

Both *testosterone*, a male sex hormone found in the testes, and *estrone*, a female sex hormone, are produced by the chemical modification of progesterone. These hormones are involved in the development of male and female sex characteristics.

Many steroids have played important roles in the development of birth control agents. 19-Norprogesterone was one of the first synthetic birth control agents. However, its use was severely limited because it had to be taken by injection. A related compound, norlutin (chemical name: 17-α-ethynyl-19-nortestosterone), was found to provide both the strength and the effectiveness of 19-norprogesterone and could be taken orally.

Norlutin

Currently, "combination" oral contraceptives are prescribed most frequently. These include a progesterone and an estrogen. They are also used to regulate menstruation in patients with heavy menstrual bleeding. There are at least thirty combination pills currently available. All of these compounds act by inducing a false pregnancy, which prevents ovulation. When oral contraception is discontinued, ovulation usually returns within three menstrual cycles.

Cortisone is important to the proper regulation of a number of biochemical processes. Cortisone is also used in the treatment of rheumatoid arthritis, asthma, gastrointestinal disorders, and many skin conditions. However, treatment with cortisone is not without risk. Possible side effects of cortisone therapy include fluid retention, sodium retention, and potassium loss that can lead to congestive heart failure. Other side effects include muscle weakness, osteoporosis, gastrointestinal upsets including peptic ulcers, and neurological symptoms, including vertigo, headaches, and convulsions.

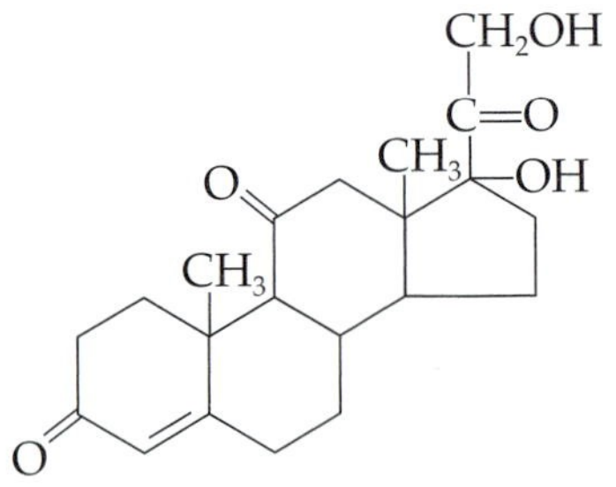

Cortisone

QUESTION 15.15

Draw structure of the steroid nucleus. Note the locations of the A, B, C, and D steroid rings.

QUESTION 15.16

What is meant by the term *fused ring*?

Waxes

Waxes are derived from many different sources and have a variety of chemical compositions, depending on the source. Paraffin wax, for example, is composed of a mixture of solid hydrocarbons (usually straight-chain compounds). Natural waxes generally are composed of a long-chain fatty acid esterified to a long-chain alcohol. Because the long hydrocarbon tails are extremely hydrophobic, waxes are completely insoluble in water. Waxes are also solid at room temperature, owing to their high molecular weights. Two examples of waxes are myricyl palmitate, a major component of beeswax, and whale oil (spermaceti wax), from the head of the sperm whale, which is composed of cetyl palmitate.

Naturally occurring waxes have a variety of uses. Lanolin, which serves as a protective coating for hair and skin, is used in skin creams and ointments. Carnauba wax is used in automobile polish. Whale oil was once used as a fuel, in ointments, and in candles. However, synthetic waxes have replaced whale oil to a large extent, because of efforts to ban the hunting of whales.

$$CH_3(CH_2)_{14}-\overset{\overset{\displaystyle O}{\|}}{C}-O-(CH_2)_{29}CH_3$$

Myricyl palmitate
(beeswax)

$$CH_3(CH_2)_{14}-\overset{\overset{\displaystyle O}{\|}}{C}-O-(CH_2)_{15}CH_3$$

Cetyl palmitate
(whale oil)

15.5 Complex Lipids

Complex lipids are lipids that are bonded to other types of molecules. The most common and important complex lipids are **plasma lipoproteins,** which are responsible for the transport of other lipids in the body. Lipoprotein particles consist of a core of hydrophobic lipids surrounded by amphipathic proteins, phospholipids, and cholesterol (Figure 15.8).

There are four major classes of human plasma lipoproteins:

- **Chylomicrons,** which have a density of less than 0.95 g/mL, carry dietary triglycerides from the intestine to other tissues. The remaining lipoproteins are classified by their densities.
- **Very low density lipoproteins (VLDL)** have a density of 0.95–1.019 g/mL. They bind triglycerides synthesized in the liver and carry them to adipose and other tissues for storage or use as an energy source.
- **Low-density lipoproteins (LDL)** are characterized by a density of 1.019–1.063 g/mL. They carry cholesterol to peripheral tissues and help regulate cholesterol levels in those tissues. These are richest in cholesterol, frequently carrying 40% of the plasma cholesterol.
- **High-density lipoproteins (HDL)** have a density of 1.063–1.210 g/mL. They are bound to plasma cholesterol; however, they transport cholesterol from peripheral tissues to the liver.

A summary of the composition of each of the plasma lipoproteins is presented in Figure 15.9.

There is evidence that high levels of HDL in the blood help reduce the incidence of atherosclerosis, perhaps because HDL carries cholesterol from the peripheral tissues back to the liver. In the liver, some of the cholesterol is used for bile synthesis and secreted into the intestines, from which it is excreted.

A final correlation has been made between diet and atherosclerosis. People whose diet is high in saturated fats tend to have high levels of cholesterol in the blood. Although the relationship between saturated fatty acids and cholesterol metabolism is unclear, it is known that a diet rich in unsaturated fats results in decreased cholesterol levels. In fact, the use of unsaturated fat in the diet results in a decrease in the level of LDL and an increase in the level of HDL. With the positive correlation between heart disease and high cholesterol levels, the current dietary recommendations include a diet that is low in fat and the substitution of unsaturated fats (vegetable oils) for saturated fats (animal fats).

10 **Learning Goal**
Describe the function of lipoproteins in triglyceride and cholesterol transport in the body.

Recently an inflammatory protein, the C-reactive protein (CRP), has been implicated in atherosclerosis. A test for the level of this protein in the blood is being suggested as a way to predict the risk of heart attack. A high sensitivity CRP test is now widely available.

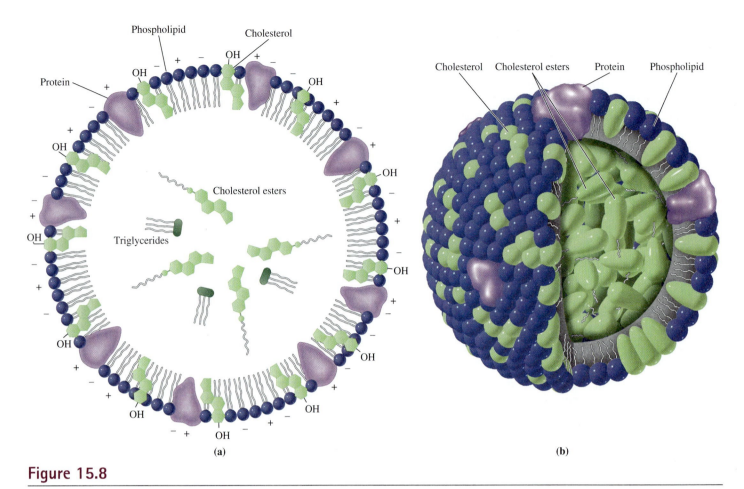

Figure 15.8

A model for the structure of a plasma lipoprotein. The various lipoproteins are composed of a shell of protein, cholesterol, and phospholipids surrounding more hydrophobic molecules such as triglycerides or cholesterol esters (cholesterol esterified to fatty acids). (a) Cross section, (b) three-dimensional view.

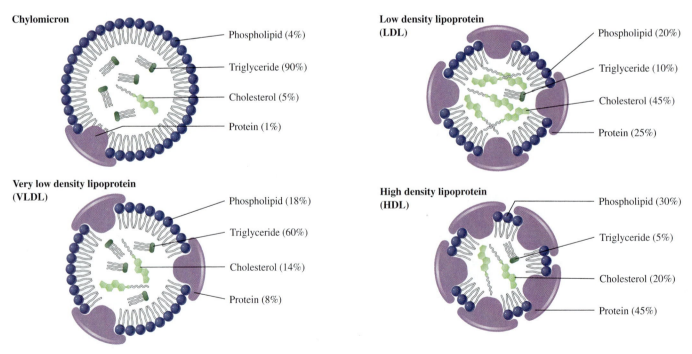

Figure 15.9

A summary of the relative amounts of cholesterol, phospholipid, protein, and triglycerides in the four classes of lipoproteins.

15.6 The Structure of Biological Membranes

Biological membranes are *lipid bilayers* in which hydrophobic hydrocarbon tails are packed in the center of the bilayer and ionic head groups are exposed on the surface to interact with water (Figure 15.10). The hydrocarbon tails of membrane phospholipids provide a thin shell of nonpolar material that prevents mixing of molecules on either side. The nonpolar tails of membrane phospholipids thus provide a barrier between the interior of the cell and its surroundings. The polar heads of lipids are exposed to water, and they are highly solvated.

11 Learning Goal

Draw the structure of the cell membrane and discuss its functions.

Fluid Mosaic Structure of Biological Membranes

Membranes are not static; they are composed of molecules in motion. The fluidity of biological membranes is determined by the proportions of saturated and unsaturated fatty acid groups in the membrane phospholipids. About half of the fatty acids that are isolated from membrane lipids from all sources are unsaturated.

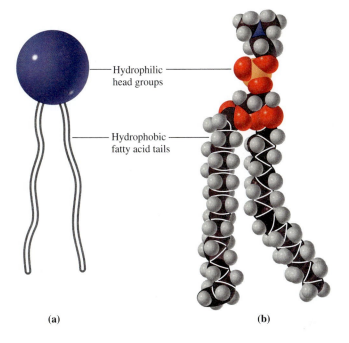

(a)

(b)

Hydrophilic head groups

Hydrophobic fatty acid tails

Hydrophilic surface

Bilayer

Hydrophobic interior

(c)

Figure 15.10

(a) Representation of a phospholipid.
(b) Space-filling model of a phospholipid.
(c) Representation of a phospholipid bilayer membrane. (d) Line formula structure of a bilayer membrane composed of phospholipids, cholesterol, and sphingolipids.

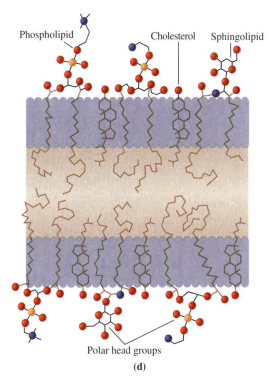

Phospholipid   Cholesterol   Sphingolipid

Polar head groups

(d)

To learn about antibiotics that destroy membrane structure, see A Medical Connection: Antibiotics That Destroy Membrane Integrity online at www.mhhe.com/denniston.

The unsaturated fatty acid tails of phospholipids contribute to membrane fluidity because of the bends introduced into the hydrocarbon chain by the double bonds. Because of these "kinks," the fatty acid tails do not pack together tightly.

We also find that the percentage of unsaturated fatty acid groups in membrane lipids is inversely proportional to the temperature of the environment. Generally, the body temperatures of mammals are quite constant, and the fatty acid composition of their membrane lipids is therefore usually very uniform. One interesting exception is the reindeer. Much of the year the reindeer must travel through ice and snow. Thus the hooves and lower legs must function at much colder temperatures than the rest of the body. Because of this, the percentage of unsaturation in the membranes varies along the length of the reindeer leg. We find that the proportion of unsaturated fatty acids increases closer to the hoof, permitting the membranes to function in the low temperatures of ice and snow to which the lower leg is exposed.

Thus, membranes are fluid, regardless of the environmental temperature conditions. In fact, it has been estimated that membranes have the consistency of olive oil.

Although the hydrophobic barrier created by the fluid lipid bilayer is an important feature of membranes, the proteins embedded within the lipid bilayer are equally important and are responsible for critical cellular functions. The presence of these membrane proteins was revealed by an electron microscopic technique called *freeze-fracture*. Cells are frozen to very cold temperatures and then fractured with a very fine diamond knife. Some of the cells are fractured between the two layers of the lipid bilayer. When viewed with the electron microscope, the membrane appeared to be a mosaic, studded with proteins. Because of the fluidity of membranes and the appearance of the proteins seen by electron microscopy, our concept of membrane structure is called the **fluid mosaic model** (Figure 15.11).

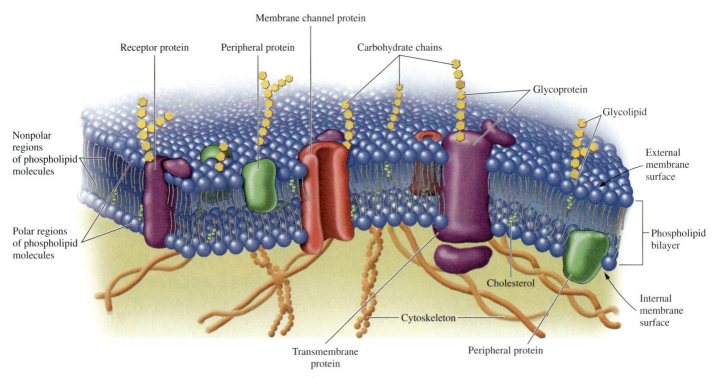

Figure 15.11

The fluid mosaic model of membrane structure.

Some of the observed proteins, called **peripheral membrane proteins,** are bound only to one of the surfaces of the membrane by interactions between ionic head groups of the membrane lipids and ionic amino acids on the surface of the peripheral protein. Other membrane proteins, called **transmembrane proteins,** are embedded within the membrane and extend completely through it, being exposed both inside and outside the cell. The region exposed to the outside of the cell typically has oligosaccharides covalently attached. Hence these proteins are *glycoproteins.* Typically the part of the transmembrane protein that extends into the cell is attached to filaments of the cytoplasmic skeleton.

Membrane Transport

The cell membrane allows the cell to interact with its environment and control passage of material into and out of the cell. The external cell membrane controls the entrance of fuel and the exit of waste products. Most of these transport processes are controlled by transmembrane transport proteins. These transport proteins are the cellular gatekeepers, whose function in membrane transport is analogous to the function of enzymes in carrying out cellular chemical reactions. However, some molecules pass through the membranes unassisted, by the **passive transport** processes of diffusion and osmosis. These are referred to as passive processes because they do not require any energy expenditure by the cell.

12 Learning Goal

Discuss passive and facilitated diffusion of materials through a cell membrane.

Passive Diffusion: The Simplest Form of Membrane Transport

Diffusion (Section 7.4) is the *net* movement of a solute with the gradient (from an area of high concentration to an area of low concentration). Because of the lipid bilayer structure of the membrane, only a few molecules are able to diffuse freely across a membrane. These include small molecules such as O_2 and CO_2. Any large or highly charged molecules or ions are not able to pass through the lipid bilayer directly (Figure 15.12).

Animation
Diffusion Through Cell Membranes

Animation
How Diffusion Works

Facilitated Diffusion: Specificity of Molecular Transport

Most molecules are transported across biological membranes by specific protein carriers known as *permeases.* When a solute diffuses through a membrane from an area of high concentration to an area of low concentration by passing through a channel within a permease, the process is known as **facilitated diffusion.** No energy is consumed by facilitated diffusion; thus it is another means of passive transport, and the direction of transport depends upon the concentrations of metabolite on each side of the membrane.

Animation
How Facilitated Diffusion Works

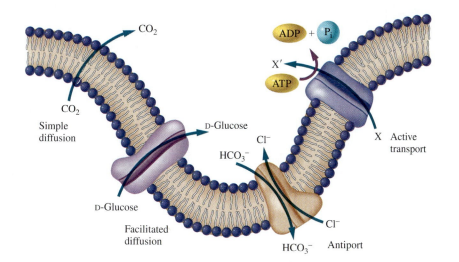

Figure 15.12

Transport across a cell membrane may occur by simple diffusion or facilitated diffusion. Transport of Cl^- and HCO_3^- occurs in opposite directions across cell membranes, a process called antiport. Energy for active transport mechanisms is required for many important substances.

A MEDICAL Connection | Liposome Delivery Systems

Liposomes were discovered by Dr. Alec Bangham in 1961. During his studies on phospholipids and blood clotting, he found that if he mixed phospholipids and water, tiny phospholipid bilayer sacs, called liposomes, would form spontaneously. Since that first observation, liposomes have been developed as efficient delivery systems for everything from antitumor and antiviral drugs, to the hair-loss therapy minoxidil!

If a drug is included in the solution during formation of liposomes, the phospholipids will form a sac around the solution. In this way, the drug becomes encapsulated within the phospholipid sphere. These liposomes can be injected intravenously or applied to body surfaces. Sometimes scientists include hydrophilic molecules in the surface of the liposome. This increases the length of time that they will remain in circulation in the bloodstream. These so-called stealth liposomes are being used to carry anticancer drugs, such as doxorubicin and mitoxantrone. Liposomes are also being used as carriers for the antiviral drugs, such as AZT and ddC, that are used to treat human immunodeficiency virus infection.

A clever trick to help target the drug-carrying liposome is to include an antibody on the surface of the liposome. These antibodies are proteins designed to bind specifically to the surface of a tumor cell. Upon attaching to the surface of the tumor cell, the liposome "membrane" fuses with the cell membrane. In this way, the deadly chemicals are delivered only to those cells targeted for destruction. This helps to avoid many of the unpleasant side effects of chemotherapy that occur when normal healthy cells are killed by the drug.

Another application of liposomes is in the cosmetics industry. Liposomes can be formed that encapsulate a vitamin, herbal agent, or other nutritional element. When applied to the skin, the liposomes pass easily through the outer layer of dead skin, delivering their contents to the living skin cells beneath. As with the pharmaceutical liposomes, these liposomes, sometimes called cosmeceuticals, fuse with skin cells. Thus, they directly deliver the beneficial cosmetic agent directly to the cells that can benefit the most.

Since their accidental discovery forty years ago, much has been learned about the formation of liposomes and ways to engineer them for more efficient delivery of their contents. This is another example of the marriage of serendipity, an accidental discovery, with scientific research and technological application. As the development of new types of liposomes continues, we can expect that even more ways will be found to improve the human condition.

FOR FURTHER UNDERSTANDING

From what you know of the structure of phospholipids, explain the molecular interactions that cause liposomes to form.

Could you use liposome technology to deliver a hydrophobic drug? Explain your answer.

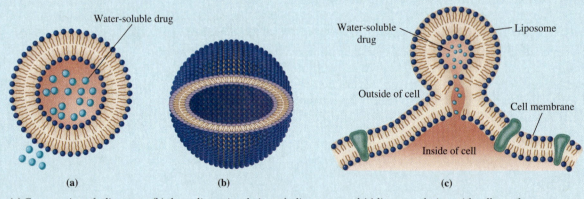

(a) Cross section of a liposome, (b) three-dimensional view of a liposome, and (c) liposome fusing with cell membrane.

Red blood cells use a single permease to transport Cl^- into the cell in exchange for HCO_3^- ions (Figure 15.12). This two-way transport, known as *antiport*, occurs by facilitated diffusion so that Cl^- flows from a high exterior concentration to a low interior one, and bicarbonate flows from a high interior concentration to a low exterior one.

Energy Requirements for Transport

Simple diffusion and facilitated diffusion involve the spontaneous flow of materials from a region of higher concentration to an area of lower concentration

13 Learning Goal
Describe the mechanism of action of the Na^+-K^+ ATPase.

Animation
Primary Active Transport

(a concentration gradient). To survive, cells often must move substances "uphill," against a concentration gradient. This phenomenon, called **active transport,** requires energy (Figure 15.13). Many ions and food molecules are imported through the cell membrane by active transport. The energy used for this process may consume more than half of the total energy harvested by cellular metabolism.

A good example of active transport is the Na^+-K^+ ATPase, which moves these ions into and out of the cell against their gradients (Figure 15.13). Cells must maintain a high concentration of Na^+ outside the cell and a high concentration of K^+ inside the cell. This requires a continuous supply of cellular energy in the form of adenosine triphosphate (ATP). Over one-third of the total ATP produced by the cell is used to maintain these Na^+ and K^+ concentration gradients across the cell membrane. Thus, the name *Na^+-K^+ ATPase* refers to the enzymatic activity that hydrolyzes ATP. The hydrolysis of ATP releases the energy needed to move Na^+ and K^+ ions across the cell membrane. For each ATP molecule hydrolyzed, three Na^+ are moved out of the cell and two K^+ are transported into the cell.

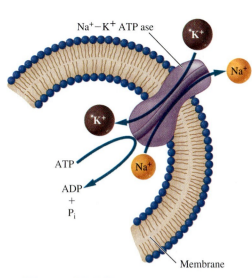

Figure 15.13

Schematic diagram of the active transport mechanism of the Na^+-K^+ ATPase.

Animation

Sodium-Potassium Exchange Pump 2
How the Sodium Potassium Pump Works
Sodium Potassium Exchange Pump

QUESTION 15.17

Define the term *diffusion*.

QUESTION 15.18

What is the difference between active and passive transport?

SUMMARY

15.1 Biological Functions of Lipids

Lipids are organic molecules characterized by their solubility in nonpolar solvents. Lipids are subdivided into classes based on structural characteristics: *fatty acids, glycerides,* nonglycerides, and *complex lipids*. Lipids serve many functions in the body, including energy storage, protection of organs, insulation, and absorption of vitamins. Other lipids are energy sources, hormones, or vitamins. Cells store chemical energy in the form of lipids, and the cell membrane is a lipid bilayer.

15.2 Fatty Acids

Fatty acids are *saturated* and *unsaturated* carboxylic acids containing between twelve and twenty-four carbon atoms. Fatty acids with even numbers of carbon atoms occur most frequently in nature. The reactions of fatty acids are identical to those of carboxylic acids. They include *esterification,* production by acid hydrolysis of esters, *saponification,* and addition at the double bond. *Prostaglandins,* thromboxanes, and leukotrienes are derivatives of twenty-carbon fatty acids that have a variety of physiological effects.

15.3 Glycerides

Glycerides are the most abundant lipids. The triesters of glycerol (*triglycerides*) are of greatest importance. *Neutral*

glycerides are important because of their ability to store energy. Ionic *phospholipids* are important components of all biological membranes.

15.4 Nonglyceride Lipids

Nonglyceride lipids consist of *sphingolipids, steroids,* and *waxes*. *Sphingomyelin* is a component of the myelin sheath around cells of the central nervous system. *Steroids* are important for many biochemical functions: *cholesterol* is a membrane component; testosterone, progesterone, and estrone are sex hormones; and cortisone is an anti-inflammatory steroid that is important in the regulation of many biochemical pathways.

15.5 Complex Lipids

Plasma lipoproteins are complex lipids that transport other lipids through the bloodstream. *Chylomicrons* carry dietary triglycerides from the intestine to other tissues. *Very low density lipoproteins* carry triglycerides synthesized in the liver to other tissues for storage. *Low-density lipoproteins* carry cholesterol to peripheral tissues and help regulate blood cholesterol levels. *High-density lipoproteins* transport cholesterol from peripheral tissues to the liver.

15.6 The Structure of Biological Membranes

The *fluid mosaic model* of membrane structure pictures biological membranes that are composed of lipid bilayers in which proteins are embedded. Membrane lipids contain polar head groups and nonpolar hydrocarbon tails. The hydrocarbon tails of phospholipids are derived from saturated and unsaturated long-chain fatty acids containing an even number of carbon atoms.

The simplest type of membrane transport is passive diffusion of a substance across the lipid bilayer from the region of higher concentration to that of lower concentration. Many metabolites are transported across biological membranes by permeases that form pores through the membrane. The conformation of the pore is complementary to that of the substrate to be transported. Cells use energy to transport molecules across the plasma membrane against their concentration gradients, a process known as *active transport.*

Na^+-K^+ ATPase hydrolyzes one molecule of ATP to provide the driving force for pumping three Na^+ out of the cell in exchange for two K^+.

KEY TERMS

active transport (15.6)
arachidonic acid (15.2)
atherosclerosis (15.4)
cholesterol (15.4)
chylomicron (15.5)
complex lipid (15.5)
diglyceride (15.3)
eicosanoid (15.2)
emulsifying agent (15.3)
essential fatty acid (15.2)
esterification (15.2)
facilitated diffusion (15.6)
fatty acid (15.2)
fluid mosaic model (15.6)
glyceride (15.3)
high-density lipoprotein (HDL) (15.5)
hydrogenation (15.2)
lipid (15.1)
low-density lipoprotein (LDL) (15.5)

monoglyceride (15.3)
neutral glyceride (15.3)
passive transport (15.6)
peripheral membrane protein (15.6)
phosphatidate (15.3)
phosphoglyceride (15.3)
phospholipid (15.3)
plasma lipoprotein (15.5)
prostaglandin (15.2)
saponification (15.2)
saturated fatty acid (15.2)
sphingolipid (15.4)
sphingomyelin (15.4)
steroid (15.4)
transmembrane protein (15.6)
triglyceride (15.3)
unsaturated fatty acid (15.2)
very low density lipoprotein (VLDL) (15.5)
wax (15.4)

QUESTIONS AND PROBLEMS

Biological Functions of Lipids

Foundations

15.19 List the four main groups of lipids.
15.20 List the biological functions of lipids.

Applications

15.21 In terms of solubility, explain why a diet that contains no lipids can lead to a deficiency in lipid-soluble vitamins.

15.22 Why are lipids (triglycerides) such an efficient molecule for the storage of energy in the body?

Fatty Acids

Foundations

15.23 What is the difference between a saturated and an unsaturated fatty acid? Write the structures for a saturated and an unsaturated fatty acid.
15.24 As the length of the hydrocarbon chain of saturated fatty acids increases, what is the effect on melting points? As the number of carbon-carbon double bonds in fatty acids increases, what is the effect on melting points?

Applications

15.25 Explain the relationship between fatty-acid-chain length and melting points that you described in answer to Question 15.24.
15.26 Explain the relationship that you described in answer to Question 15.24 for the effect of the number of carbon–carbon double bonds in fatty acids on their melting points.
15.27 Draw the structures of each of the following fatty acids:
   a. Decanoic acid
   b. Stearic acid
   c. *trans*-5-Decenoic acid
   d. *cis*-5-Decenoic acid
15.28 What are the common and I.U.P.A.C. names of each of the following fatty acids?
   a. $C_{15}H_{31}COOH$
   b. $C_{11}H_{23}COOH$
   c. $CH_3(CH_2)_5CH{=}CH(CH_2)_7COOH$
   d. $CH_3(CH_2)_7CH{=}CH(CH_2)_7COOH$
15.29 Write an equation for each of the following reactions:
   a. Esterification of glycerol with three molecules of myristic acid
   b. Acid hydrolysis of a triglyceride containing three stearic acid molecules
   c. Reaction of decanoic acid with KOH
   d. Hydrogenation of linoleic acid
15.30 Write an equation for each of the following reactions:
   a. Esterification of glycerol with three molecules of palmitic acid
   b. Acid hydrolysis of a triglyceride containing three oleic acid molecules
   c. Reaction of stearic acid with KOH
   d. Hydrogenation of oleic acid
15.31 What is the function of essential fatty acids? What molecules are formed from arachidonic acid?
15.32 What is the biochemical basis for the effectiveness of aspirin in decreasing the inflammatory response? What is the role of prostaglandins in the inflammatory response?
15.33 List four effects of prostaglandins.
15.34 What are the functions of thromboxane A_2 and leukotrienes?

Glycerides

Foundations

15.35 What are emulsifying agents and what are their practical uses?
15.36 Why are triglycerides also referred to as triacylglycerols?

Applications

15.37 What do you predict would be the physical state of a triglyceride with three saturated fatty acid tails? Explain your reasoning.
15.38 What do you predict would be the physical state of a triglyceride with three unsaturated fatty acid tails? Explain your reasoning.

15.39 Draw the structure of the triglyceride molecule formed by esterification at C-1, C-2, and C-3 with hexadecanoic acid, *trans*-9-hexadecenoic acid, and *cis*-9-hexadecenoic acid, respectively.

15.40 Draw the structure of the phosphatidate formed between glycerol-3-phosphate that is esterified at C-1 and C-2 with capric and lauric acids, respectively.

Nonglyceride Lipids

Foundations

15.41 Define the term *sphingolipid*. What are the two major types of sphingolipids?

15.42 Define the term *glycosphingolipid*. Distinguish among the three types of glycosphingolipids, cerebrosides, sulfatides, and gangliosides.

Applications

15.43 What is the biological function of sphingomyelin?

15.44 Why are sphingomyelins amphipathic?

15.45 What is the role of cholesterol in biological membranes?

15.46 How does cholesterol contribute to atherosclerosis?

15.47 What are the biological functions of progesterone, testosterone, and estrone?

15.48 How has our understanding of the steroid sex hormones contributed to the development of oral contraceptives?

15.49 What is the medical application of cortisone?

15.50 What are the possible side effects of cortisone treatment?

15.51 A wax found in beeswax is myricyl palmitate. Which fatty acid and which alcohol are used to form this compound?

15.52 A wax found in the head of sperm whales is cetyl palmitate. Which fatty acid and which alcohol are used to form this compound?

Complex Lipids

Foundations

15.53 What are the four major types of plasma lipoproteins?

15.54 What is the function of each of the four types of plasma lipoproteins?

Applications

15.55 Distinguish among the four plasma lipoproteins in terms of their composition and their function.

15.56 What is the correlation between saturated fats in the diet and atherosclerosis?

The Structure of Biological Membranes

Foundations

15.57 What is the basic structure of a biological membrane?

15.58 Describe the fluid mosaic model of membrane structure.

15.59 Describe peripheral membrane proteins.

15.60 Describe transmembrane proteins and list some of their functions.

Applications

15.61 What is the major effect of cholesterol on the properties of biological membranes?

15.62 Why do the hydrocarbon tails of membrane phospholipids provide a barrier between the inside and outside of the cell?

15.63 How will the properties of a biological membrane change if the fatty acid tails of the phospholipids are converted from saturated to unsaturated chains?

15.64 What is the function of unsaturation in the hydrocarbon tails of membrane lipids?

Biological Membranes: Transport

Foundations

15.65 Explain the difference between simple diffusion across a membrane and facilitated diffusion.

15.66 Define the term *active transport*.

Applications

15.67 How does active transport differ from facilitated diffusion?

15.68 By what mechanism are Cl^- and HCO_3^- ions transported across the red blood cell membrane?

15.69 What is the meaning of the term *antiport*?

15.70 Why is the function of the Na^+-K^+ ATPase an example of active transport?

15.71 What is the stoichiometry of the Na^+-K^+ ATPase?

15.72 How will the Na^+ and K^+ concentrations of a cell change if Na^+-K^+ ATPase is inhibited?

FOR FURTHER UNDERSTANDING

1. Olestra is a fat substitute that provides no calories, yet has all the properties of a naturally occurring fat. It has a creamy, tongue-pleasing consistency. Unlike other fat substitutes, olestra can withstand heating. Thus, it can be used to prepare foods such as potato chips and crackers. Olestra is a sucrose polyester and is produced by esterification of six, seven, or eight fatty acids to molecules of sucrose. Draw the structure of one such molecule having eight stearic acid acyl groups attached.

2. Liposomes can be made by vigorously mixing phospholipids (like phosphatidylcholine) in water. When the mixture is allowed to settle, spherical vesicles form that are surrounded by a phospholipid bilayer "membrane." Pharmaceutical chemists are trying to develop liposomes as a targeted drug delivery system. By adding the drug of choice to the mixture described above, liposomes form around the solution of drug. Specific proteins can be incorporated into the mixture that will end up within the phospholipid bilayers of the liposomes. These proteins are able to bind to targets on the surface of particular kinds of cells in the body. Explain why injection of liposome encapsulated pharmaceuticals might be a good drug delivery system.

3. "Cholesterol is bad and should be eliminated from the diet." Do you agree or disagree? Defend your answer.

4. Why would a phospholipid such as lecithin be a good emulsifying agent for ice cream?

5. When a plant becomes cold-adapted, the composition of the membranes changes. What changes would you predict in fatty acid and cholesterol composition? Explain your reasoning.

16

LEARNING GOALS

1   List the functions of proteins.

2   Draw the general structure of an amino acid and classify amino acids based on their R groups.

3   Describe the primary structure of proteins and draw the structure of the peptide bond.

4   Describe the types of secondary structure of a protein and discuss the forces that maintain secondary structure.

5   Describe the structure and functions of fibrous proteins.

6   Describe the tertiary and quaternary structure of a protein and list the R group interactions that maintain protein shape.

7   Describe the roles of hemoglobin and myoglobin.

8   Describe how extremes of pH and temperature cause denaturation of proteins.

9   Classify enzymes according to the type of reaction catalyzed.

10   Describe the effect that enzymes have on the activation energy of a reaction.

11   Discuss the role of the active site and the importance of enzyme specificity and describe the difference between the lock-and-key model and the induced fit model of enzyme-substrate complex formation.

12   Discuss the roles of cofactors and coenzymes in enzyme activity.

13   Explain how pH and temperature affect the rate of an enzyme-catalyzed reaction.

14   Discuss the mechanisms by which certain chemicals inhibit enzyme activity.

15   Provide examples of medical uses of enzymes.

Protein Structure and Enzymes

I magine the earth about four billion years ago: it was young then, not even a billion years old. Beginning as a red-hot molten sphere, the earth's surface had cooled slowly and become solid rock. But the interior, still extremely hot, erupted through the crust spewing hot gases and lava. Eventually these eruptions produced craggy land masses and an atmosphere composed of gases such as hydrogen, carbon dioxide, ammonia, and water vapor. As the water vapor cooled, it condensed into liquid water, forming ponds and shallow seas.

At the dawn of biological life, the surface of the earth was still very hot and covered with rocky peaks and hot shallow oceans. The atmosphere was not very inviting either—filled with noxious gases and containing no molecular oxygen. Yet this is the environment where life on our planet began.

Some scientists think that they have found bacteria—living fossils—that may be very closely related to the first inhabitants of earth. These bacteria thrive at temperatures higher than the boiling point of water. Some need only H_2, CO_2, and H_2O for their metabolic processes and they quickly die in the presence of molecular oxygen.

But this lifestyle raises some uncomfortable questions. For instance, how do these bacteria survive at these extreme temperatures that would cook the life-forms with which we are more familiar? Researcher Mike Adams of the University of Georgia has found some of the answers. Adams and his students have studied the structure of an enzyme, a protein that acts as a biological catalyst, from one of these extraordinary bacteria. He has found that the three-dimensional structure of the super-hot enzymes is held together by many more attractive forces than the structure of the low-temperature version of the same enzyme. Thus these proteins are stable and functional even at temperatures above the boiling point of water!

In this chapter, we will study the structure and function of proteins, some of which serve as enzymes, speeding up critical chemical reactions in the body. But, as we will see, proteins have many other roles in the structure and metabolism of all living organisms.

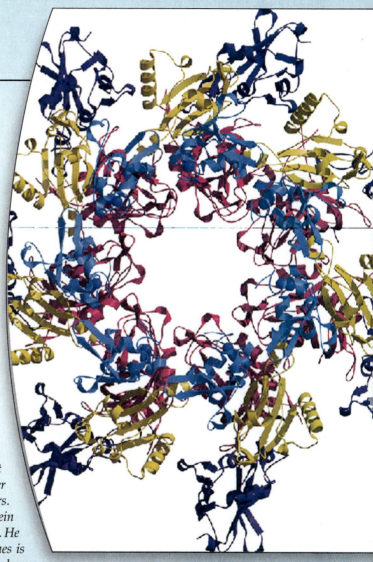

Anthrax reached national attention in the 2001 anthrax attack that killed five people. Scientists have determined the three-dimensional structure of an anthrax toxin called protective antigen. In its active form, the protein has a quaternary structure consisting of seven subunits that form a channel in cell membranes. Describe the various weak interactions that maintain quaternary structure and that allow a protein to become embedded within a membrane.

471

16.1 Cellular Functions of Proteins

In the 1800s, Johannes Mulder came up with the name **protein,** a term derived from a Greek word that means "of first importance." Proteins are a very important class of food molecules because they provide an organism with carbon and hydrogen and also with nitrogen and sulfur. These latter two elements are unavailable from fats and carbohydrates, the other major classes of food molecules.

Proteins have many biological functions, as the following short list suggests.

Learning Goal

List the functions of proteins.

Animation

The Immune Response

Antibodies are discussed in greater detail in A Medical Connection: Immunoglobulins: Proteins That Defend the Body online at www.mhhe.com/denniston.

To learn about the use of protein hormones in the battle against cancer, see A Chemistry Connection: Angiogenesis Inhibitors: Proteins That Inhibit Tumor Growth online at www.mhhe.com/denniston.

- Enzymes are biological catalysts. The majority of the enzymes that have been studied are proteins. Reactions that would take days or weeks or require extremely high temperatures without enzymes are completed in an instant. For example, the digestive enzymes *pepsin, trypsin,* and *chymotrypsin* break down proteins in our diet so that subunits can be absorbed for use by our cells.
- **Defense proteins** include **antibodies** (also called *immunoglobulins*) which are specific protein molecules produced by specialized cells of the immune system in response to foreign **antigens.** These foreign invaders include bacteria and viruses that infect the body. Each antibody has regions that precisely fit and bind to a single antigen. It helps to end the infection by binding to the antigen and helping to destroy it or remove it from the body.
- **Transport proteins** carry materials from one place to another in the body. The protein *transferrin* transports iron from the liver to the bone marrow, where it is used to synthesize the heme group for hemoglobin. The proteins *hemoglobin* and *myoglobin* are responsible for transport and storage of oxygen in higher organisms, respectively.
- **Regulatory proteins** control many aspects of cell function, including metabolism and reproduction. We can function only within a limited set of conditions. For life to exist, body temperature, the pH of the blood, and blood glucose levels must be carefully regulated. Many of the hormones that regulate body function, such as *insulin* and *glucagon,* are proteins.
- **Structural proteins** provide mechanical support to large animals and provide them with their outer coverings. Our hair and fingernails are largely composed of the protein, *keratin.* Other proteins provide mechanical strength for our bones, tendons, and skin. Without such support, large, multicellular organisms like ourselves could not exist.
- **Movement proteins** are necessary for all forms of movement. Our muscles, including that most important muscle, the heart, contract and expand through the interaction of *actin* and *myosin* proteins. Sperm can swim because they have long flagella made up of proteins.
- **Storage proteins** serve as sources of amino acids for embryos and infants. Egg *albumin* and *casein* in milk are examples of nutrient storage proteins.

16.2 The α-Amino Acids

Structure of Amino Acids

Learning Goal

Draw the general structure of an amino acid and classify amino acids based on their R groups.

Conjugate acids and bases are described in detail in Chapter 8.

The proteins of the body are made up of some combination of twenty different subunits called **α-amino acids.** The general structure of an α-amino acid is shown in Figure 16.1. We find that nineteen of the twenty amino acids that are commonly isolated from proteins have this same general structure; they are primary amines on the α-carbon. The remaining amino acid, proline, is a secondary amine.

Notice that the α-carbon in the general structure is attached to a carboxylate group (a carboxyl group that has lost a proton, —COO$^-$) and a protonated amino group (an amino group that has gained a proton, —NH$_3^+$). At pH 7, conditions required for life functions, you will not find amino acids in which the carboxylate group is protonated (—COOH) and the amino group is unprotonated (—NH$_2$).

Under these conditions, the carboxyl group is in the conjugate base form (—COO⁻), and the amino group is in its conjugate acid form (—NH₃⁺). Any neutral molecule with equal numbers of positive and negative charges is called a *zwitterion*. Thus, amino acids in water exist as dipolar ions called zwitterions.

The α-carbon of each amino acid is also bonded to a hydrogen atom and a side chain, or R group. In a protein, the R groups interact with one another through a variety of weak attractive forces. These interactions participate in folding the protein chain into a precise three-dimensional shape that determines its ultimate function. They also maintain that three-dimensional conformation.

Stereoisomers of Amino Acids

The α-carbon is attached to four different groups in all amino acids except glycine. The α-carbon of most α-amino acids is therefore chiral, allowing mirror-image forms, enantiomers, to exist. Glycine has two hydrogen atoms attached to the α-carbon and is the only amino acid commonly found in proteins that is not chiral.

The configuration of α-amino acids isolated from proteins is L-. This is based on comparison of amino acids with D-glyceraldehyde (Figure 16.2). In Figure 16.2a, we see a comparison of D- and L-glyceraldehyde with D- and L-alanine. Notice that the most oxidized end of the molecule, in each case the carbonyl group, is drawn at the top of the molecule. In the D-isomer of glyceraldehyde, the —OH group is on the right. Similarly, in the D-isomer of alanine, the —N⁺H₃ is on the right. In the L-isomers of the two compounds, the —OH and —N⁺H₃ groups are on the left. By this comparison with the enantiomers of glyceraldehyde, we can define the D- and L-enantiomers of the amino acids. Figure 16.2b shows the ball-and-stick models of the D- and L-isomers of alanine.

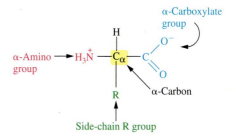

Figure 16.1

General structure of an α-amino acid. All amino acids isolated from proteins, with the exception of proline, have this general structure.

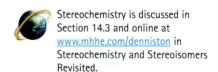

Stereochemistry is discussed in Section 14.3 and online at www.mhhe.com/denniston in Stereochemistry and Stereoisomers Revisited.

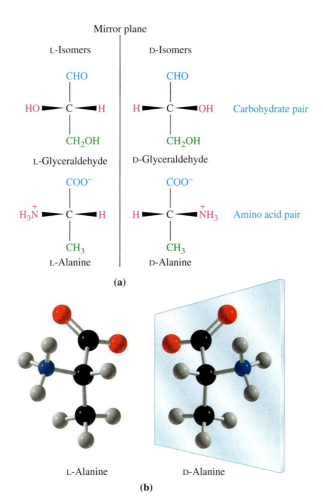

Figure 16.2

(a) Structure of D- and L-glyceraldehyde and their relationship to D- and L-alanine. (b) Ball-and-stick models of D- and L-alanine.

A MEDICAL Connection | Proteins in the Blood

The blood plasma of a healthy individual typically contains 60–80 g/L of protein. This protein can be separated into five classes designated α through γ. The separation is based on the overall surface charge on each of the types of protein.

The most abundant protein in the blood is albumin, making up about 55% of the blood protein. Albumin contributes to the osmotic pressure of the blood simply because it is a dissolved molecule. It also serves as a nonspecific transport molecule for important metabolites that are otherwise poorly soluble in water. Among the molecules transported through the blood by albumin are bilirubin (a waste product of the breakdown of hemoglobin), Ca^{2+}, and fatty acids (organic anions).

α-Globulins (α₁ and α₂) make up 13% of the plasma proteins. They include glycoproteins (proteins with sugar groups attached), high-density lipoproteins, haptoglobin (a transport protein for free hemoglobin), ceruloplasmin (a copper transport protein), prothrombin (a protein involved in blood clotting), and very low density lipoproteins. The most abundant is

α₁-globulin α₁-antitrypsin. Although the name leads us to believe that this protein inhibits a digestive enzyme, trypsin, the primary function of α₁-antitrypsin is inactivation of an enzyme that causes damage in the lungs (see also, A Medical Connection: α₁-Antitrypsin and Familial Emphysema in this chapter). α₁-Antichymotrypsin is another inhibitor found in the bloodstream. This protein, along with amyloid proteins, is found in the amyloid plaques characteristic of Alzheimer's disease (AD). As a result, it has been suggested that an overproduction of this protein may contribute to AD. In the blood, α₁-antichymotrypsin is also found complexed to prostate specific antigen (PSA), the protein antigen that is measured as an indicator of prostate cancer. Elevated PSA levels are observed in those with the disease. It is interesting to note that PSA is a chymotrypsin-like proteolytic enzyme.

β-Globulins represent 13% of the blood plasma proteins and include transferrin (an iron transport protein) and fibrinogen, a protein involved in coagulation of blood that comprises 7% of plasma protein. Finally, γ-globulins, IgG, IgM, IgA, IgD, and IgE, make up the remaining 11% of plasma proteins. γ-Globulins are synthesized by B lymphocytes, but most of the remaining plasma proteins are synthesized in the liver. A frequent hallmark of liver disease is reduced amounts of one or more plasma proteins.

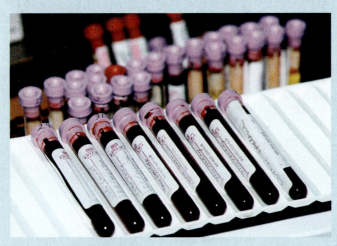

Blood samples drawn from patients.

FOR FURTHER UNDERSTANDING

Develop a hypothesis to explain why albumin in the blood can serve as a nonspecific carrier for such diverse substances as bilirubin, Ca^{2+}, and fatty acids. (Hint: Consider what you know about the structures of amino acid R groups.)

Both fibrinogen and prothrombin are involved in formation of blood clots when they are converted into proteolytic enzymes. However, they are normally found in the blood in an inactive form. Develop an explanation for this observation.

In Chapter 14, we learned that almost all monosaccharides found in nature are in the D-family. Just the opposite is true of α-amino acids. Almost all α-amino acids isolated from proteins in nature are members of the L-family. In other words, the orientation of the four groups around the chiral carbon of these α-amino acids resembles the orientation of the four groups around the chiral carbon of L-glyceraldehyde.

Classes of Amino Acids

The hydrophobic interaction between nonpolar R groups is one of the forces that helps maintain the proper three-dimensional shape of a protein.

Because all amino acids have a carboxyl group and an amino group, all differences among amino acids depend upon their side-chain R groups. Amino acids are grouped in Figure 16.3 according to the polarity of their side chains.

The side chains of some amino acids are nonpolar. They prefer contact with one another to contact with water and are said to be **hydrophobic** ("water-fearing")

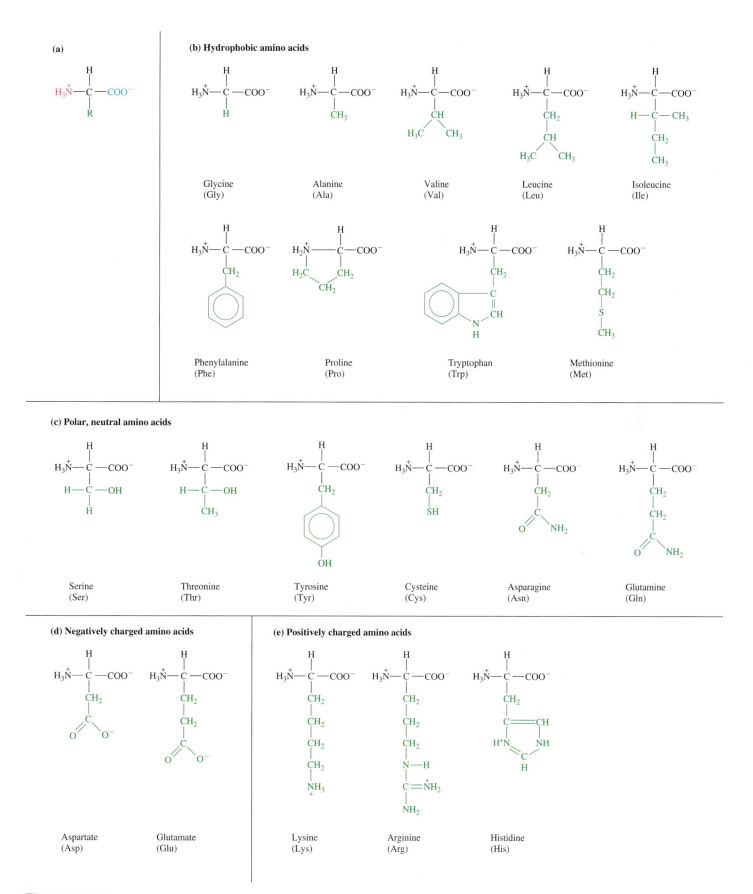

Figure 16.3

Structures of the amino acids at pH 7.0. (a) The general structure of an amino acid. Structures of (b) hydrophobic; (c) polar, neutral; (d) negatively charged; and (e) positively charged amino acids.

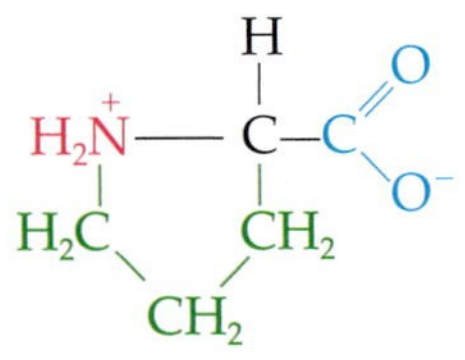

Proline (Pro)

Hydrogen bonding (Section 6.2) is another weak interaction that helps maintain the proper three-dimensional structure of a protein. The positively and negatively charged amino acids within a protein interact with one another to form ionic bridges that also help to keep the protein chain folded in a precise way.

amino acids. They are generally found buried in the interior of proteins, where they can associate with one another and remain isolated from water. The R group of proline is unique; it is actually bonded to the α-amino group, forming a secondary amine.

The side chains of the remaining amino acids are polar. Because they are attracted to polar water molecules, they are said to be **hydrophilic** ("water-loving") **amino acids.** Hydrophilic side chains are often found on the surfaces of proteins. Polar amino acids can be subdivided into three classes.

- *Polar, neutral amino acids* have R groups that have high affinity for water but are not ionic at pH 7. Serine, threonine, tyrosine, cysteine, asparagine, and glutamine fall into this category. Most of these amino acids associate with one another by hydrogen bonding, but cysteine molecules form disulfide bonds with one another.
- *Negatively charged amino acids* have ionized carboxyl groups in their side chains. At pH 7, these amino acids have a net charge of -1. Aspartate and glutamate are the two amino acids in this category. They are acidic amino acids because ionization of the carboxylic acid releases a proton.
- *Positively charged amino acids.* At pH 7, lysine, arginine, and histidine have a net positive charge because their side chains contain positive groups. These amino groups are basic because the side chain reacts with water, picking up a proton and releasing a hydroxide anion.

The names of the amino acids can be abbreviated by a three-letter code. These abbreviations are shown in Figure 16.3.

QUESTION 16.1

Write the three-letter abbreviation and draw the structure of each of the following amino acids.

   a. Glycine   b. Proline   c. Threonine   d. Aspartate   e. Lysine

QUESTION 16.2

Indicate whether each of the amino acids listed in Question 16.1 is polar, nonpolar, basic, or acidic.

16.3 The Peptide Bond

3 Learning Goal

Describe the primary structure of proteins and draw the structure of the peptide bond.

Proteins are linear polymers of L-α-amino acids in which the carboxyl group of one amino acid is linked to the amino group of another amino acid. The **peptide bond** is an *amide bond* formed between the $-COO^-$ group of one amino acid and the $\alpha\text{-}N^+H_3$ group of another amino acid. The reaction, shown below for the amino acids glycine and alanine, is a dehydration reaction because a water molecule is lost as the amide bond is formed.

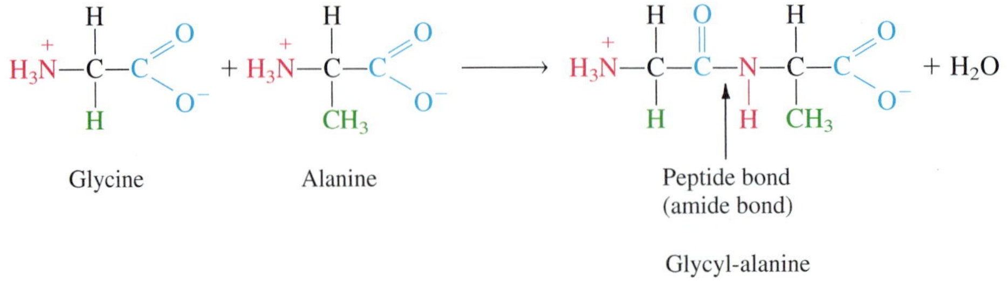

      Glycine   Alanine   Peptide bond
                                                        (amide bond)

                                                              Glycyl-alanine

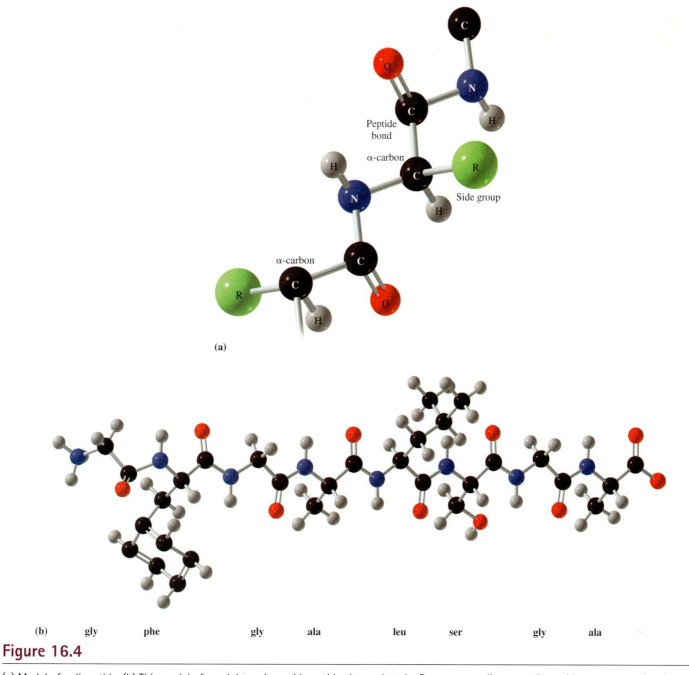

Peptide bond

α-carbon

R

Side group

H

H

N

α-carbon

R

H

O

(a)

(b) gly phe gly ala leu ser gly ala

Figure 16.4

(a) Model of a dipeptide. (b) This model of an eight amino acid peptide shows that the R groups on adjacent amino acids are on opposite sides of the chain because of the rigid peptide bond.

The molecule formed by condensing two amino acids is called a *dipeptide* (Figure 16.4). The amino acid with a free α-N^+H_3 group is known as the amino terminal, or simply the **N-terminal amino acid,** and the amino acid with a free —COO^- group is known as the carboxyl, or **C-terminal amino acid.** Structures of proteins are conventionally written with their N-terminal amino acid on the left.

The number of amino acids in small peptides is indicated by the prefixes *di-* (two units), *tri-* (three units), *tetra-* (four units), and so forth. Peptides are named as derivatives of the C-terminal amino acid, which receives its entire name. For all other amino acids, the ending *-ine* is changed to *-yl.* Thus, the dipeptide

To understand why the N-terminal amino acid is placed first and the C-terminal amino acid is placed last, we need to look at the process of protein synthesis. As we will see in Chapter 17, the N-terminal amino acid is the first amino acid of the protein. It forms a peptide bond involving its carboxyl group and the amino group of the second amino acid in the protein. Thus, a free amino group literally projects from the "left" end of the protein. Similarly, the C-terminal amino acid is the last amino acid added to the protein during protein synthesis. Because the peptide bond is formed between the amino group of this amino acid and the carboxyl group of the previous amino acid, a free carboxyl group projects from the "right" end of the protein chain.

alanyl-glycine has glycine as its C-terminal amino acid, as indicated by its full name, *glycine*:

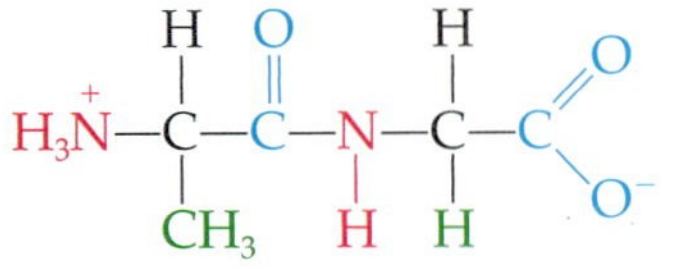

Alanyl-glycine
(ala-gly)

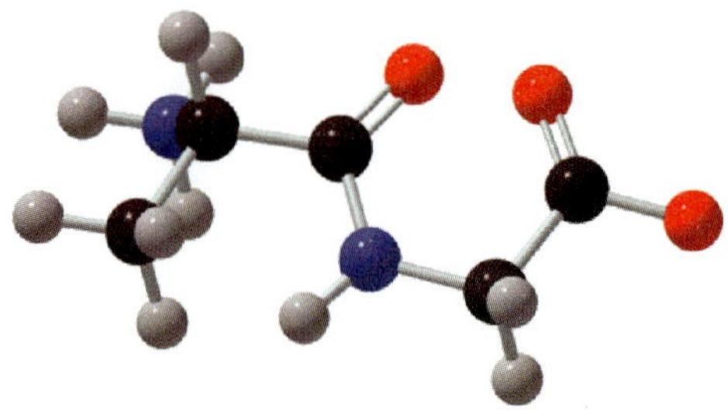

Alanyl-glycine

The dipeptide formed from alanine and glycine that has alanine as its C-terminal amino acid, glycyl-alanine, is the product of the reaction shown above. These two dipeptides have the same amino acid composition, but different amino acid sequences.

The structures of small peptides can easily be drawn with practice if certain rules are followed. First note that the backbone of the peptide contains the repeating sequence

$$N-C-C-N-C-C-N-C-C$$
$$121212$$

in which N is the α-amino group, carbon-1 is the α-carbon, and carbon-2 is the carboxyl group. Carbon-1 is always bonded to a hydrogen atom and to the R group side chain that is unique to each amino acid. Continue drawing as outlined in Example 16.1.

3 **Learning Goal**
Describe the primary structure of proteins and draw the structure of the peptide bond.

EXAMPLE 16.1 **Writing the Structure of a Tripeptide**

Draw the structure of the tripeptide alanyl-glycyl-valine.

SOLUTION

Step 1. Write the backbone for a tripeptide. It will contain three sets of three atoms, or nine atoms in all. Remember that the N-terminal amino acid is written to the left.

$$N-C-C \qquad N-C-C \qquad N-C-C$$
$$\text{Set 1} \qquad\quad \text{Set 2} \qquad\quad \text{Set 3}$$

Step 2. Add oxygens to the carboxyl carbons and hydrogens to the amino nitrogens:

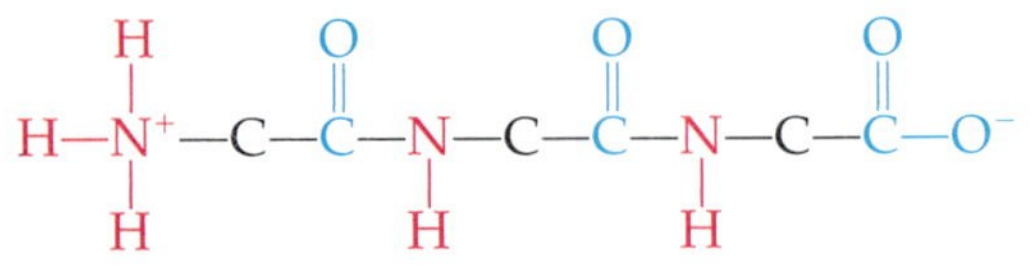

Step 3. Add hydrogens to the α-carbons:

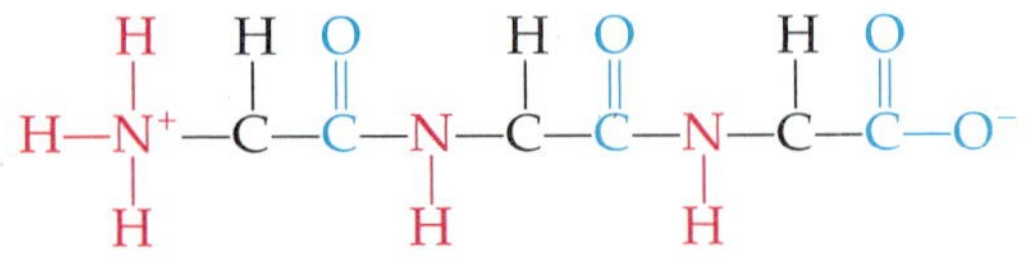

Continued—

Step 4. Add the side chains. In this example (ala-gly-val) they are, from left to right, —CH₃, H, and —CH(CH₃)₂:

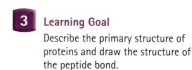

QUESTION 16.3

Write the structure of each of the following peptides at pH 7:

a. Alanyl-phenylalanine
b. Lysyl-alanine
c. Phenylalanyl-tyrosyl-leucine

QUESTION 16.4

Write the structure of each of the following peptides at pH 7:

a. Glycyl-valyl-serine
b. Threonyl-cysteine
c. Isoleucyl-methionyl-aspartate

16.4 The Primary Structure of Proteins

The **primary structure** of a protein is the amino acid sequence of the protein chain (Figure 16.4b). It results from covalent bonding between the amino acids in the chain (peptide bonds). The primary structures of proteins are translations of information contained in genes. Each protein has a different primary structure with different amino acids in different places along the chain.

Ultimately, the primary structure of a protein will determine its biologically active form. The interactions among the R groups of the amino acids in the protein chain depend on the location of those R groups along the chain. These interactions will govern how the protein chain folds, which, in turn, dictates its final three-dimensional structure and its biological function.

3 **Learning Goal**
Describe the primary structure of proteins and draw the structure of the peptide bond.

The genetic code and the process of protein synthesis are described in Sections 17.5 and 17.6.

16.5 The Secondary Structure of Proteins

The primary sequence of a protein, the chain of covalently linked amino acids, folds into regularly repeating structures that resemble designs in a tapestry. These repeating structures define the **secondary structure** of the protein. The secondary structure is the result of hydrogen bonding between the amide hydrogens and carbonyl oxygens of the peptide bonds. Many hydrogen bonds are needed to maintain secondary structure. The two most common types of secondary structure are the α-helix and the β-pleated sheet.

4 **Learning Goal**
Describe the types of secondary structure of a protein and discuss the forces that maintain secondary structure.

 See also A Chemistry Connection: Collagen: A Protein That Holds Us Together online at www.mhhe.com/denniston.

A LIFESTYLE Connection | The Opium Poppy and Peptides in the Brain

The seeds of the oriental poppy contain morphine. *Morphine* is a narcotic that has a variety of effects on the body and the brain, including drowsiness, euphoria, mental confusion, and chronic constipation. Although morphine was first isolated in 1805, not until the 1850s and the advent of the hypodermic was it effectively used as a painkiller. During the American Civil War, morphine was used extensively to relieve the pain of wounds and amputations. It was at this time that the addictive properties were noticed. By the end of the Civil War, over 100,000 soldiers were addicted to morphine.

As a result of the Harrison Act (1914), morphine came under government control and was made available only by prescription. Although morphine is addictive, *heroin*, a derivative of morphine, is much more addictive and induces a greater sense of euphoria that lasts for a longer time.

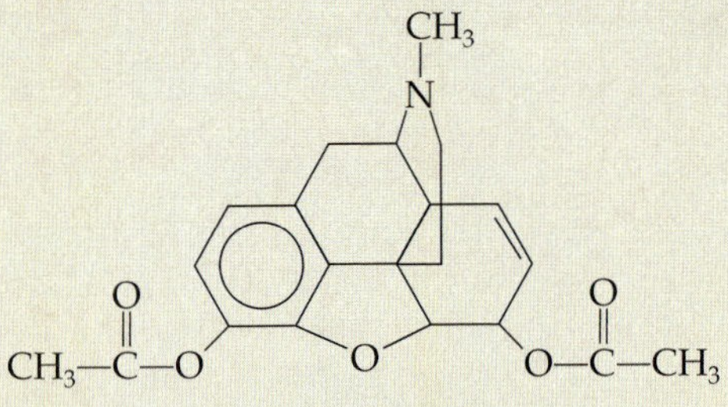

Heroin

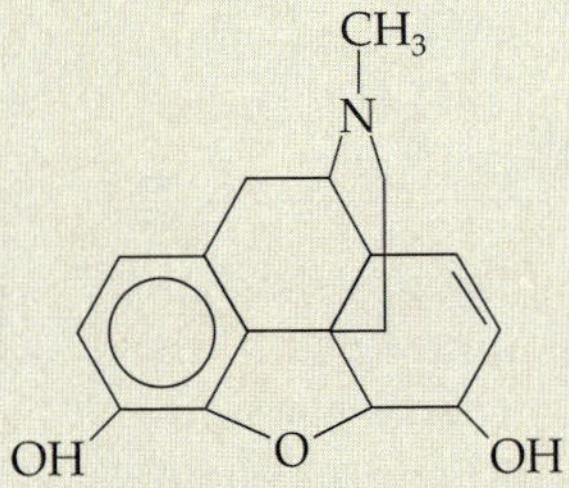

Morphine

The structures of heroin and morphine.

Why do heroin and morphine have such powerful effects on the brain? Both drugs have been found to bind to *receptors* on the surface of the cells of the brain. The function of these receptors is to bind specific chemical signals and to direct the brain cells to respond. Yet it seemed odd that the cells of our brain should have receptors for a plant chemical. This mystery was solved in 1975, when John Hughes discovered that the brain itself synthesizes small peptide hormones with a morphinelike structure. Two of these opiate peptides are called *methionine enkephalin*, or met-enkephalin, and *leucine enkephalin*, or leu-enkephalin.

These neuropeptide hormones have a variety of effects. They inhibit intestinal motility and blood flow to the gastrointestinal tract. This explains the chronic constipation of morphine users. In addition, it is thought that these *enkephalins* play a role in pain perception, perhaps serving as a pain blockade. This is supported by the observation that they are found in higher concentrations in the bloodstream following painful stimulation. It is further suspected that they may play a role in mood and mental health. The so-called runner's high is thought to be a euphoria brought about by an excessively long or strenuous run!

Unlike morphine, the action of enkephalins is short-lived. They bind to the cellular receptor and thereby induce the cells to respond. Then they are quickly destroyed by enzymes in the brain that hydrolyze the peptide bonds of the enkephalin. Once destroyed, they are no longer able to elicit a cellular response. Morphine and heroin bind to these same receptors and induce the cells to respond. However, these drugs are not destroyed and therefore persist in the brain for long periods at concentrations high enough to continue to cause biological effects.

Many researchers are working to understand why drugs like morphine and heroin are addictive. Studies with cells in culture have suggested one mechanism for morphine tolerance and addiction. Normally, when the cell receptors bind to enkephalins, this signals the cell to decrease the production of a chemical

Poppies seen here growing in the wild are the source of the natural opiate drugs morphine and codeine.

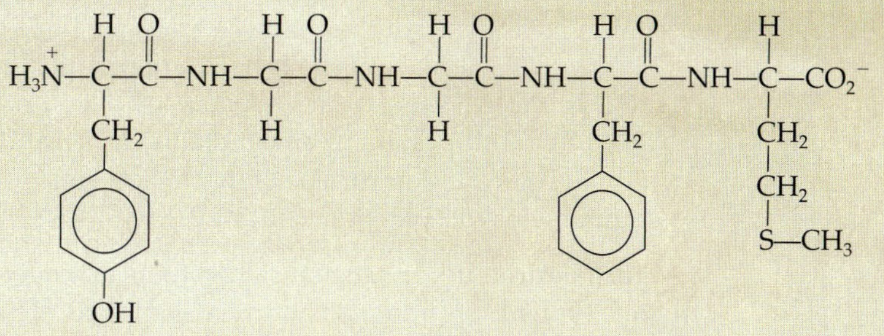

Tyr-Gly-Gly-Phe-Met
Methionine enkephalin

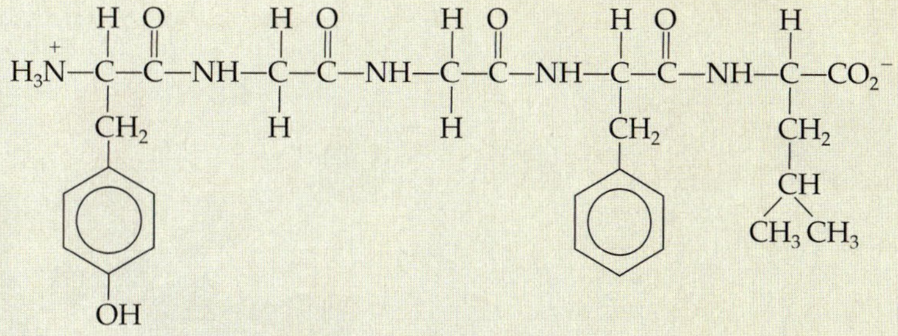

Tyr-Gly-Gly-Phe-Leu
Leucine enkephalin

Structures of the peptide opiates leucine enkephalin and methionine enkephalin. These are the body's own opiates.

messenger called *cyclic AMP*, or simply cAMP. (This compound is very closely related to the nucleotide adenosine-5'-monophosphate.) The decrease in cAMP level helps to block pain and elevate one's mood. When morphine is applied to these cells, they initially respond by decreasing cAMP levels. However, with chronic use of morphine, the cells become desensitized; that is, they do not decrease cAMP production and thus behave as though no morphine were present. However, a greater amount of morphine will once again cause the decrease in cAMP levels. Thus addiction and the progressive need for more of the drug seem to result from biochemical reactions in cells.

This logic can be extended to understand withdrawal symptoms. When an addict stops using the drug, he or she exhibits withdrawal symptoms that include excessive sweating, anxiety, and tremors. The cause may be that the high levels of morphine were keeping the cAMP levels low, thus reducing pain and causing euphoria. When morphine is removed completely, the cells overreact and produce huge quantities of cAMP. The result is all of the unpleasant symptoms known collectively as the *withdrawal syndrome*.

Clearly, morphine and heroin have demonstrated the potential for misuse and are a problem for society in several respects. Nonetheless, morphine remains one of the most effective painkillers known. Certainly, for people suffering from cancer, painful burns, or serious injuries, the risk of addiction is far outweighed by the benefits of relief from excruciating pain.

FOR FURTHER UNDERSTANDING

Compare the structures of leucine and methionine enkephalin with those of heroin and morphine. What similarities do you see that might cause them to bind to the same receptors on the surfaces of nerve cells?

What characteristics might you look for in a nonaddictive drug that could be used both to combat heroin addiction and to treat the symptoms of withdrawal?

α–Helix

The most common type of secondary structure is a coiled, helical conformation known as the **α-helix** (Figure 16.5). The α-helix has several important features.

- Every amide hydrogen and carbonyl oxygen is involved in a hydrogen bond. These hydrogen bonds lock the α-helix into place.
- Every carbonyl oxygen is hydrogen-bonded to an amide hydrogen four amino acids away in the chain.
- The hydrogen bonds of the α-helix are parallel to the long axis of the helix (see Figure 16.5).
- The polypeptide chain in an α-helix is right-handed. It is oriented like a normal screw. If you turn a screw clockwise it goes into the wall; turned counterclockwise, it comes out of the wall.

5 Learning Goal

Describe the structure and functions of fibrous proteins.

Fibrous proteins are structural proteins arranged in fibers or sheets that have only one type of secondary structure. **α-Keratins** are fibrous proteins that form the covering (hair, wool, nails, hooves, and fur) of most land animals. Human hair is a typical example of the structure of α-keratins. The proteins of hair consist almost

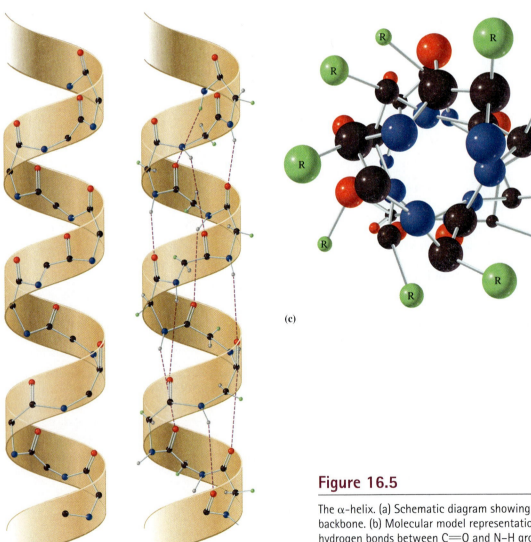

(a) (b) (c)

Figure 16.5

The α-helix. (a) Schematic diagram showing only the helical backbone. (b) Molecular model representation. Note that all of the hydrogen bonds between C=O and N–H groups are parallel to the long axis of the helix. (c) Top view of an α-helix. The side chains of the helix point away from the long axis of the helix. The view is into the barrel of the helix.

exclusively of polypeptide chains coiled into α-helices. A single α-helix is coiled in a bundle with two other helices to give a three-stranded superstructure called a *protofibril* that is part of an array known as a *microfibril* (Figure 16.6). These structures, which resemble "molecular pigtails," possess great mechanical strength, and they are virtually insoluble in water.

The major structural property of a coiled coil superstructure of α-helices is its great mechanical strength. This property is applied very efficiently in both the fibrous proteins of skin and those of muscle. As you can imagine, these proteins must be very strong to carry out their functions of mechanical support and muscle contraction.

β-Pleated Sheet

The second common secondary structure in proteins resembles the pleated folds of drapery and is known as **β-pleated sheet** (Figure 16.7a). All of the carbonyl oxygens and amide hydrogens in a β-pleated sheet are involved in hydrogen bonds, and the polypeptide chain is nearly completely extended.

Some fibrous proteins are composed of β-pleated sheets. For example, the silkworm produces *silk fibroin,* a protein whose structure is entirely β-pleated sheet (Figure 16.7). The polypeptide chains of a β-pleated sheet are almost completely extended, and silk does not stretch easily. Glycine accounts for nearly half of the amino acids of silk fibroin. Alanine and serine account for most of the others. The methyl groups of alanines and the hydroxymethyl groups of serines lie on opposite sides of the sheet. Thus the stacked sheets nestle comfortably, like sheets of corrugated cardboard, because the R groups are small enough to allow the stacked-sheet superstructure.

16.6 The Tertiary Structure of Proteins

Most fibrous proteins, such as silk, collagen, and α-keratins, are almost completely insoluble in water. (Our skin would do us very little good if it dissolved in the rain.) The majority of cellular proteins, however, are soluble in the cell cytoplasm. Soluble proteins are usually **globular proteins.** Globular proteins have three-dimensional structures called the **tertiary structure** of the protein, which are distinct from their secondary structure. The polypeptide chain with its regions of secondary structure, α-helix and β-pleated sheet, further folds on itself to achieve the tertiary structure.

The tertiary structure is maintained by interactions among the side chains, the R groups, of the amino acids. The structure is maintained by the following molecular interactions:

- van der Waals forces between the R groups of nonpolar amino acids that are hydrophobic
- Hydrogen bonds between the polar R groups of the polar amino acids
- Ionic bonds (salt bridges) between the R groups of oppositely charged amino acids
- Covalent bonds between the thiol-containing amino acids. Two of the polar cysteines can be oxidized to a dimeric amino acid called *cystine* (Figure 16.8). The disulfide bond of cystine can be a cross-link between different proteins, or it can tie together two segments within a protein.

The bonds that maintain the tertiary structure of proteins are shown in Figure 16.9. Globular proteins are extremely compact. "Hinge" regions of random coil connect regions of α-helix and β-pleated sheet. These roughly spherical proteins serve as enzymes, carrying out metabolic reactions. Others are transport or storage proteins, such as hemoglobin or albumin.

α-Helix

Protofibril

Microfibril

Microfibril

Macrofibril

Cell

One hair

Figure 16.6

Structure of α-keratins. These proteins are assemblies of triple-helical protofibrils that are assembled in an array known as a *microfibril.* These in turn are assembled into macrofibrils. Hair is a collection of macrofibrils and hair cells.

 6 **Learning Goal**

Describe the tertiary and quaternary structure of a protein and list the R group interactions that maintain protein shape.

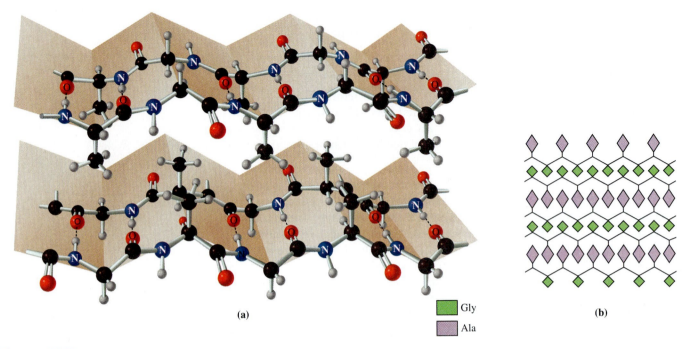

(a)

Gly
Ala

(b)

Figure 16.7

The structure of silk fibroin is almost entirely antiparallel β-pleated sheet. (a) The molecular structure of a portion of the silk fibroin protein. (b) A schematic representation of the antiparallel β-pleated sheet with nestled R groups.

Figure 16.8

Oxidation of two cysteines to give the dimer cystine. This reaction occurs in cells and is readily reversible.

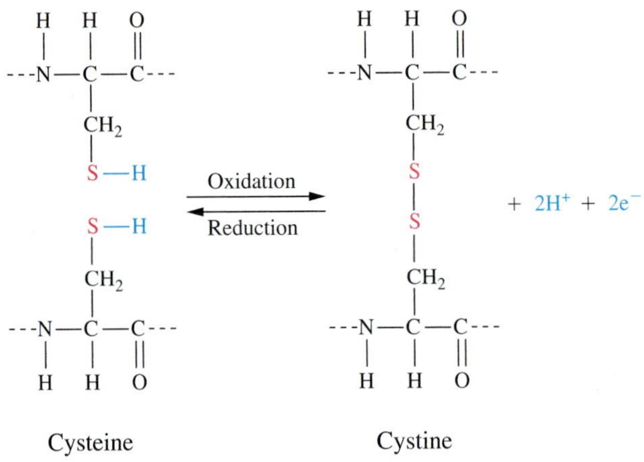

Cysteine $\xrightarrow[\text{Reduction}]{\text{Oxidation}}$ Cystine $+ 2H^+ + 2e^-$

16.7 The Quaternary Structure of Proteins

The functional form of many proteins is not a single peptide but is rather an aggregate of smaller globular peptides. For instance, the protein hemoglobin is composed of four individual globular peptide subunits: two identical α-subunits and two identical β-subunits. Only when the four peptides are bound to one another is the protein molecule functional. The association of several polypeptides to produce a functional protein defines the **quaternary structure** of a protein.

The forces that hold the quaternary structure of a protein are the same as those that hold the tertiary structure. These include hydrogen bonds between polar amino acids, ionic bridges between oppositely charged amino acids, van der Waals forces between nonpolar amino acids, and disulfide bridges.

In some cases, the quaternary structure of a functional protein involves binding to a nonprotein group. This additional group is called a **prosthetic group.** For

6 Learning Goal

Describe the tertiary and quaternary structure of a protein and list the R group interactions that maintain protein shape.

The designations α- and β- used to describe the subunits of hemoglobin do not refer to types of secondary structure.

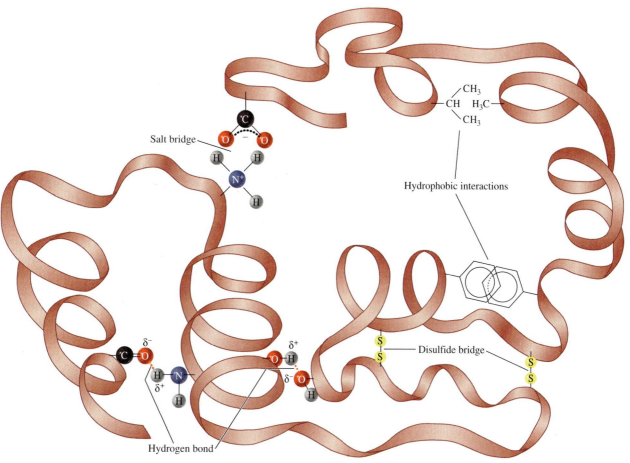

Figure 16.9

Summary of the weak interactions that help maintain the tertiary structure of a protein.

example, many of the receptor proteins on cell surfaces are **glycoproteins.** These are proteins with sugar groups covalently attached. Each of the subunits of hemoglobin is bound to an iron-containing prosthetic group called the heme group. The four levels of protein structure are depicted in Figure 16.10.

QUESTION 16.5

Describe the four levels of protein structure.

QUESTION 16.6

What are the weak interactions that maintain the tertiary structure of a protein?

16.8 Myoglobin and Hemoglobin

Myoglobin and Oxygen Storage

Most of the cells of our bodies are buried in the interior of the body and cannot directly get food molecules or eliminate waste. The circulatory system solves this problem by delivering nutrients and oxygen to body cells and carrying away wastes. Our cells require a steady supply of oxygen, but oxygen is only slightly soluble in aqueous solutions. To overcome this solubility problem, we have an

Animation
Oxygen Binding in Hemoglobin

Learning Goal
Describe the roles of hemoglobin and myoglobin.

Figure 16.10

Summary of the four levels of protein structure, using hemoglobin as an example.

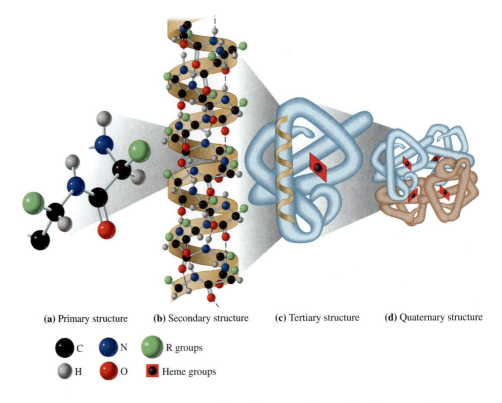

(a) Primary structure   **(b)** Secondary structure   **(c)** Tertiary structure   **(d)** Quaternary structure

C   N   R groups

H   O   Heme groups

oxygen transport protein, **hemoglobin.** Hemoglobin is found in red blood cells and is the oxygen transport protein of higher animals. **Myoglobin** is the oxygen storage protein of skeletal muscle.

Myoglobin is a globular protein shown in Figure 16.11. The Fe^{2+} ion in the heme group is the binding site for oxygen in both myoglobin and hemoglobin.

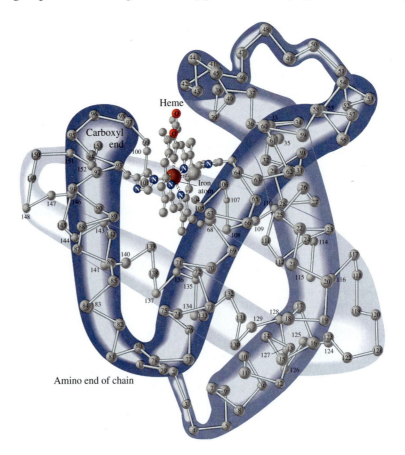

Figure 16.11

Myoglobin. The heme group has an iron atom to which oxygen binds.

Because myoglobin has a greater attraction for oxygen than does hemoglobin, oxygen from the bloodstream is efficiently transferred to the cells of the body.

Hemoglobin and Oxygen Transport

Hemoglobin (Hb) is a tetramer composed of four polypeptide subunits: two α-subunits and two β-subunits (Figure 16.12). Because each subunit of hemoglobin contains a heme group, a hemoglobin molecule can bind four molecules of oxygen:

$$Hb \; + \; 4O_2 \; \longrightarrow \; Hb(O_2)_4$$

Deoxyhemoglobin   Oxyhemoglobin

The oxygenation of hemoglobin in the lungs and the transfer of oxygen from hemoglobin to myoglobin in the tissues are very complex processes. We begin our investigation of these events with the inhalation of a breath of air.

The oxygenation of hemoglobin in the lungs is greatly favored by differences in the oxygen partial pressure (p_{O_2}) in the lungs and in the blood. The p_{O_2} in the air in the lungs is approximately 100 mm Hg; the p_{O_2} in oxygen-depleted blood is only about 40 mm Hg. Oxygen diffuses from the region of high p_{O_2} in the lungs to the region of low p_{O_2} in the blood. There it enters red blood cells and binds to the Fe^{2+} ions of the heme groups of deoxyhemoglobin, forming oxyhemoglobin. This binding helps bring more O_2 into the blood.

Sickle Cell Anemia

Sickle cell anemia is a human genetic disease that first appeared in tropical west and central Africa. It afflicts about 0.4% of African-Americans. These individuals produce a mutant hemoglobin known as sickle cell hemoglobin (Hb S). Sickle cell anemia receives its name from the sickled appearance of the red blood cells that form in this condition (Figure 16.13). The sickled cells are unable to pass through the small capillaries of the circulatory system, and circulation is hindered. This results in damage to many organs, especially bone and kidney, and can lead to death at an early age.

Sickle cell hemoglobin differs from normal hemoglobin by a single amino acid. In the β-chain of sickle cell hemoglobin, a valine (a hydrophobic amino acid) has replaced a glutamic acid (a negatively charged amino acid). This substitution provides a basis for the binding of hemoglobin S molecules to one another. When oxyhemoglobin S unloads its oxygen, individual deoxyhemoglobin S molecules bind to one another as long polymeric fibers. This occurs because the valine fits into a hydrophobic pocket on the surface of a second deoxyhemoglobin S molecule. The fibers generated in this way radically alter the shape of the red blood cell, resulting in the sickling effect.

16.9 Denaturation of Proteins

Denaturation occurs when the organized structures of a globular protein, the α-helix, the β-pleated sheet, and tertiary folds become completely disorganized. This happens because the bonds and weak attractions that maintain the secondary, tertiary, and quaternary structures of the protein are disrupted. This allows the protein to unfold, as seen in Figure 16.14. When the protein is denatured, it is no longer active.

Environmental factors that can cause denaturation include elevated temperatures and a rise or drop in pH. However organic solvents, detergents, heavy metals, and mechanical stress, such as whipping egg whites, also result in denaturation of proteins.

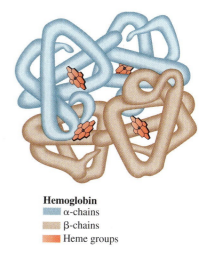

Hemoglobin
- α-chains
- β-chains
- Heme groups

Figure 16.12

Structure of hemoglobin. The protein contains four subunits, designated α and β. The α- and β-subunits face each other across a central cavity. Each subunit in the tetramer contains a heme group that binds oxygen.

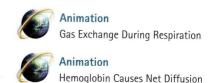

Animation
Gas Exchange During Respiration

Animation
Hemoglobin Causes Net Diffusion of Oxygen

The genetic basis of this alteration is discussed in Chapter 17.

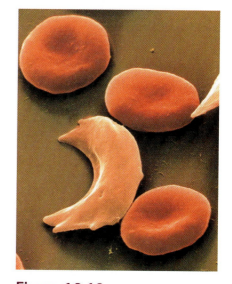

Figure 16.13

Scanning electron micrographs of normal and sickled red blood cells.

Figure 16.14

The denaturation of proteins by heat. (a) The α-helical proteins are in solution. (b) As heat is applied, the hydrogen bonds maintaining the secondary structure are disrupted, and the protein structure becomes disorganized. The protein is denatured.

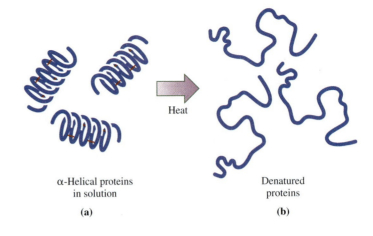

Heat

α-Helical proteins
in solution

(a)

Denatured
proteins

(b)

 8 **Learning Goal**

Describe how extremes of pH and temperature cause denaturation of proteins.

Animation
Protein Denaturation

16.10 Enzymes

The majority of the biochemical reactions required for life occur too slowly or require conditions of temperature or pH that are harmful to living systems. The solution to this problem is **enzymes,** which are biological catalysts. Without enzymes to speed up biochemical reactions, life could not exist.

The enzyme catalase provides one of the most spectacular examples of the increase in reaction rates brought about by enzymes. This enzyme is required for life in an oxygen-containing environment. In this environment, the process of aerobic (oxygen-requiring) breakdown of food molecules produces hydrogen peroxide (H_2O_2). Because H_2O_2 is toxic to the cell, it must be destroyed. One molecule of catalase converts forty million molecules of hydrogen peroxide to harmless water and oxygen every second:

$$2H_2O_2 \xrightarrow[\text{(an enzyme)}]{\text{Catalase}} 2H_2O + O_2$$

Reaction occurs forty million times every second!

This is the same reaction that you see when you pour hydrogen peroxide on a wound. The catalase released from injured cells rapidly breaks down the hydrogen peroxide. The bubbles that you see are oxygen gas released as a product of the reaction.

Classification of Enzymes

9 **Learning Goal**

Classify enzymes according to the type of reaction catalyzed.

Enzymes may be classified according to the type of reaction that they catalyze. The six classes are as follows.

Oxidoreductases

Oxidoreductases are enzymes that catalyze oxidation–reduction (redox) reactions. Oxidases catalyze oxidation reactions. Reductases catalyze reduction reactions, and dehydrogenases remove hydrogen atoms from a reactant. *Lactate dehydrogenase* is an oxidoreductase that removes hydrogen from a molecule of lactate.

Recall that redox reactions involve electron transfer from one substance to another (Section 4.3).

Transferases

The significance of phosphate group transfers in energy metabolism is discussed in Section 18.3.

Transferases are enzymes that catalyze the transfer of functional groups from one molecule to another. For example, a *transaminase* catalyzes the transfer of an amino functional group, and a *kinase* catalyzes the transfer of a phosphate group. Kinases play a major role in energy-harvesting processes involving ATP.

Hydrolases

Hydrolases catalyze hydrolysis reactions, that is, the addition of a water molecule to a bond resulting in bond breakage. These reactions are important in the digestive process. For example, *lipases* catalyze the hydrolysis of the ester bonds in triglycerides, proteases catalyze the hydrolysis of peptide bonds, and carbohydrases catalyze the hydrolysis of glycosidic bonds.

Lyases

Lyases catalyze the addition of a group to a double bond or the removal of a group to form a double bond. Some examples include decarboxylases, which remove carboxyl groups, and carboxylases, which add a carboxyl group.

Isomerases

Isomerases rearrange the functional groups within a molecule and catalyze the conversion of one isomer into another.

Ligases

Ligases are enzymes that catalyze a reaction in which a C—C, C—S, C—O, or C—N bond is made or broken. This is accompanied by an ATP-ADP interconversion.

EXAMPLE 16.2 **Classifying Enzymes According to the Type of Reaction That They Catalyze**

9 **Learning Goal**
Classify enzymes according to the type of reaction catalyzed.

Classify the enzyme that catalyzes each of the following reactions, and explain your reasoning.

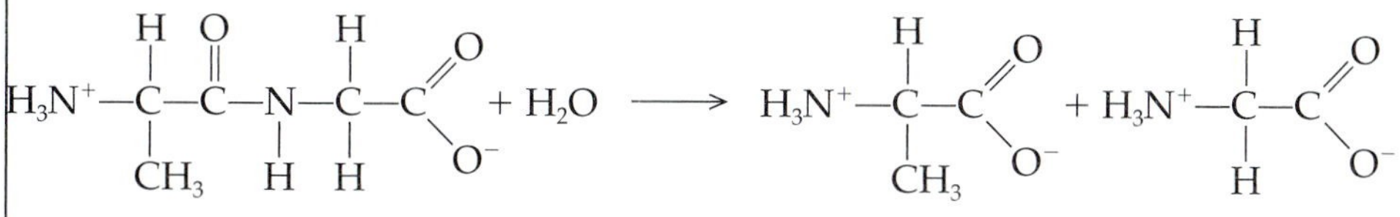

Alanyl-glycine → Alanine + Glycine

SOLUTION

The reaction occurring here involves breaking a bond, in this case a peptide bond, by adding a water molecule. The enzyme is classified as a *hydrolase*, specifically a *peptidase*.

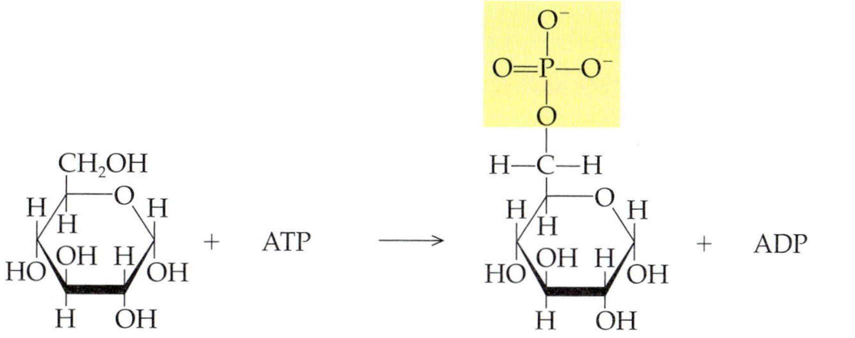

Glucose + Adenosine triphosphate → Glucose-6-phosphate + Adenosine diphosphate

Continued—

EXAMPLE 16.2 —Continued

SOLUTION

This is the first reaction in the biochemical pathway called *glycolysis*. A phosphoryl group is transferred from a donor molecule, adenosine triphosphate, to the recipient molecule, glucose. The products are glucose-6-phosphate and adenosine diphosphate. This enzyme, called *hexokinase*, is an example of a *transferase*.

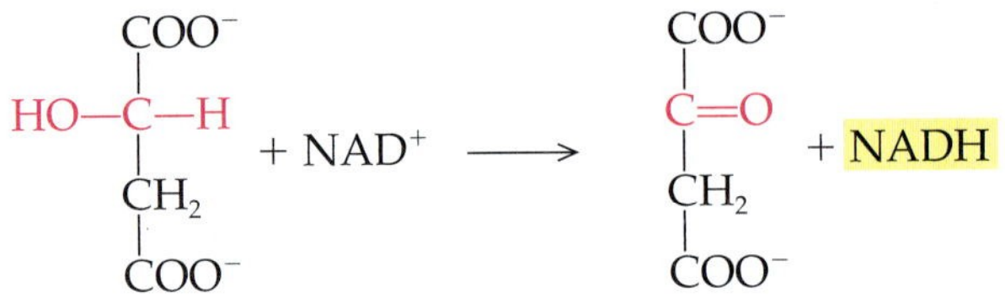

Malate   Oxaloacetate

SOLUTION

In this reaction, the reactant malate is oxidized and the coenzyme NAD^+ is reduced. The enzyme that catalyzes this reaction, *malate dehydrogenase*, is an *oxidoreductase*.

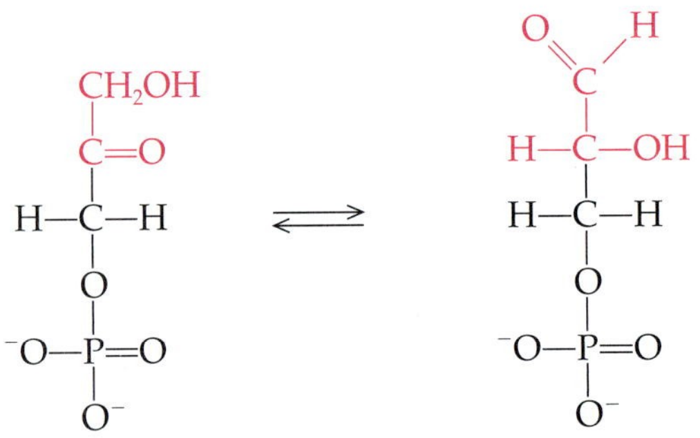

Dihydroxyacetone   Glyceraldehyde-3-phosphate
phosphate

SOLUTION

Careful inspection of the structure of the reactant and the product reveals that they each have the same number of carbon, hydrogen, oxygen, and phosphorus atoms; thus, they must be structural isomers. The enzyme must be an *isomerase*. Its name is *triose phosphate isomerase*.

QUESTION 16.7

To which class of enzymes does each of the following belong?

a. Pyruvate kinase
b. Alanine transaminase
c. Triose phosphate isomerase
d. Pyruvate dehydrogenase
e. Lactase

QUESTION 16.8

To which class of enzymes does each of the following belong?

a. Phosphofructokinase
b. Lipase

c. Acetoacetate decarboxylase
d. Succinate dehydrogenase

Naming of Enzymes

The common names for some enzymes are derived from the name of the **substrate,** the reactant that binds to the enzyme and is converted into product. In many cases, the enzyme is simply named by adding the suffix -*ase* to the name of the substrate. For instance, *urease* catalyzes the hydrolysis of urea and *lactase* catalyzes the hydrolysis of the disaccharide lactose.

Other enzymes are named for the type of reaction that they catalyze. *Dehydrogenases* catalyze the removal of hydrogen atoms from a substrate, and *decarboxylases* catalyze the removal of carboxyl groups. *Hydrogenases* and *carboxylases* carry out the opposite reaction, adding hydrogen atoms or carboxyl groups to their substrates.

Some enzymes have historical names that have no relationship to either the substrates or the reactions that they catalyze. A few examples include catalase, trypsin, pepsin, and chymotrypsin.

The Effect of Enzymes on the Activation Energy of a Reaction

How does an enzyme speed up a chemical reaction? It changes the path by which the reaction occurs, providing a lower energy route for the conversion of the substrate into the **product,** the substance that results from the enzyme-catalyzed reaction. Thus enzymes speed up reactions by lowering the activation energy of the reaction (Figure 16.15).

10 Learning Goal
Describe the effect that enzymes have on the activation energy of a reaction.

The Enzyme–Substrate Complex

The following series of reversible reactions represents the steps in an enzyme-catalyzed reaction. The first step (highlighted in blue) involves the encounter of the enzyme with its substrate and the formation of an **enzyme-substrate complex.**

11 Learning Goal
Discuss the role of the active site and the importance of enzyme specificity and describe the difference between the lock-and-key model and the induced fit model of enzyme-substrate complex formation.

$$E + S \xrightleftharpoons[\text{Step I}]{} ES \xrightleftharpoons[\text{Step II}]{} ES^* \xrightleftharpoons[\text{Step III}]{} EP \xrightleftharpoons[\text{Step IV}]{} E + P$$

| Enzyme + substrate | Enzyme– substrate complex | Transition state | Enzyme– product complex | Enzyme + product |

Animation
How Enzymes Work

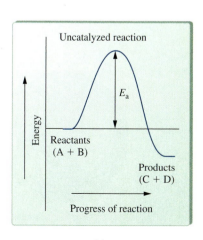

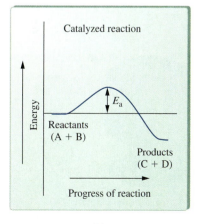

(a) (b)

Figure 16.15

Diagram of the difference in energy between the reactants (*A* and *B*) and products (*C* and *D*) for a reaction. Enzymes cannot change this energy difference but act by lowering the activation energy (E_a) for the reaction, thereby speeding up the reaction.

The part of the enzyme that binds with the substrate is called the **active site.** The characteristics of the active site that are crucial to enzyme function include the following:

- Enzyme active sites are pockets or clefts in the surface of the enzyme. The R groups in the active site that are involved in catalysis are called *catalytic groups*.
- The shape of the active site is complementary to the shape of the substrate, that is, the substrate fits neatly into the active site of the enzyme.
- An enzyme attracts and holds its substrate by weak, noncovalent interactions. The R groups involved in substrate binding, and not necessarily catalysis, make up the *binding site.*
- The conformation of the active site determines the specificity of the enzyme because only the substrate that fits into the active site will be used in a reaction.

The **lock-and-key model** of enzyme activity, shown in Figure 16.16a, was devised by Emil Fischer in 1894. At that time, it was thought that the substrate simply snapped into place like a piece of a jigsaw puzzle or a key into a lock.

Today, we know that proteins are flexible molecules. This led Daniel E. Koshland, Jr., to propose a more sophisticated model of the way enzymes and substrates interact. This model, proposed in 1958, is called the **induced fit model** (Figure 16.16b). In this model, the active site of the enzyme is not a rigid pocket into which the substrate fits precisely; rather, it is a flexible pocket that *approximates* the shape of the substrate. When the substrate enters the pocket, the active site "molds" itself around the substrate. This produces the perfect enzyme-substrate "fit."

The overall shape of a protein is maintained by many weak interactions. At any time, a few of these weak interactions may be broken by heat energy or a local change in pH. If only a few bonds are broken, they will re-form very quickly. The overall result is that there is a brief change in the shape of the enzyme. Thus, the protein or enzyme can be viewed as a flexible molecule, changing shape slightly in response to minor local changes.

QUESTION 16.9

Compare the lock-and-key and induced fit models of enzyme-substrate binding.

Figure 16.16

(a) The lock-and-key model of enzyme-substrate binding assumes that the enzyme active site has a rigid structure that is precisely complementary in shape and charge distribution to the substrate. (b) The induced fit model of enzyme-substrate binding. As the enzyme binds to the substrate, the shape of the active site conforms precisely to the shape of the substrate. The shape of the substrate may also change.

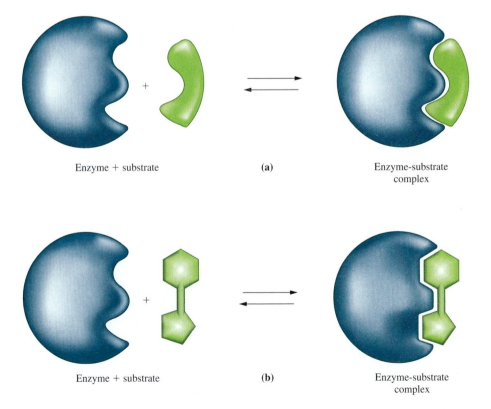

Enzyme + substrate   **(a)**   Enzyme-substrate complex

Enzyme + substrate   **(b)**   Enzyme-substrate complex

What is the relationship between an enzyme active site and its substrate?

The Transition State and Product Formation

How does enzyme-substrate binding result in a faster chemical reaction? The answer lies in step II of an enzyme-catalyzed reaction, focusing on the steps highlighted in blue:

$$E + S \underset{\text{Step I}}{\rightleftharpoons} ES \underset{\text{Step II}}{\rightleftharpoons} ES^* \underset{\text{Step III}}{\rightleftharpoons} EP \underset{\text{Step IV}}{\rightleftharpoons} E + P$$

| Enzyme + substrate | Enzyme–substrate complex | Transition state | Enzyme–product complex | Enzyme + product |

As the enzyme interacts with the substrate, it changes the shape or position of the substrate so that it is no longer energetically stable (step II). In this state, the **transition state,** the shape of the substrate is altered, because of its interaction with the enzyme, into an intermediate form having features of both the substrate and the final product. This transition state, in turn, favors the conversion of the substrate into product (step III). The product remains bound to the enzyme for a very brief time; then in step IV, the product and enzyme dissociate from one another, leaving the enzyme completely unchanged.

What kinds of transition state changes might occur in the substrate that would make a reaction proceed more rapidly?

1. The enzyme might put "stress" on a bond and thereby facilitate bond breakage. Figure 16.17 shows how this might happen when the enzyme, sucrase, catalyzes the hydrolysis of the disaccharide sucrose into the monosaccharides, glucose and fructose.
2. An enzyme may facilitate a reaction by bringing two reactants together in the proper orientation for reaction to occur. This is shown in Figure 16.18 for the condensation reaction between glucose and fructose to produce sucrose.
3. The active site of an enzyme may modify the pH of the microenvironment surrounding the substrate. To accomplish this, the enzyme may, for example,

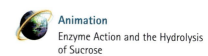

Animation
Enzyme Action and the Hydrolysis of Sucrose

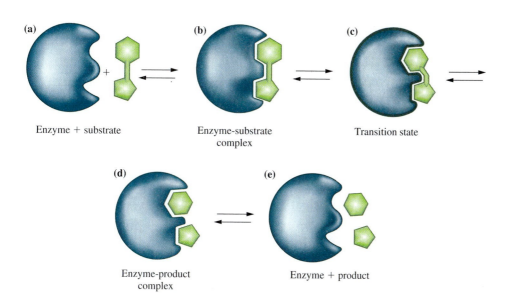

(a) Enzyme + substrate

(b) Enzyme-substrate complex

(c) Transition state

(d) Enzyme-product complex

(e) Enzyme + product

Figure 16.17

Bond breakage is facilitated by the enzyme as a result of stress on a bond. (a, b) The enzyme-substrate complex is formed. (c) In the transition state, the enzyme changes shape and thereby puts stress on the glycosidic linkage holding the two monosaccharides together. This lowers the energy of activation of this reaction. (d, e) The bond is broken, and the products are released.

Figure 16.18

An enzyme may lower the energy of activation required for a reaction by holding the substrates in close proximity and in the correct orientation. The enzyme-substrate complex forms, bringing the two monosaccharides together with the correct hydroxyl groups extended toward one another.

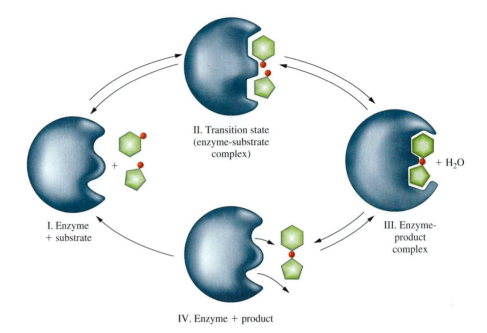

II. Transition state (enzyme-substrate complex)

I. Enzyme + substrate

+ H₂O

III. Enzyme-product complex

IV. Enzyme + product

serve as a donor or an acceptor of H^+. As a result, there would be a change in the pH in the vicinity of the substrate without disturbing the normal pH elsewhere in the cell.

QUESTION 16.11

Summarize three ways in which an enzyme might lower the energy of activation of a reaction.

QUESTION 16.12

What is the transition state in an enzyme-catalyzed reaction?

Cofactors and Coenzymes

12 Learning Goal

Discuss the roles of cofactors and coenzymes in enzyme activity.

Animation
B Vitamins

In Section 16.7, we saw that some proteins require an additional nonprotein prosthetic group to function. The same is true of some enzymes. The polypeptide portion of such an enzyme is called the **apoenzyme,** and the nonprotein prosthetic group is called the **cofactor.** Together they form the active enzyme called the **holoenzyme.** Cofactors may be metal ions, organic compounds, or organometallic compounds. They must be bound to the enzyme to maintain the correct configuration of the enzyme active site (Figure 16.19). When the cofactor is bound and the active site is in the proper conformation, the enzyme can bind the substrate and catalyze the reaction.

Other enzymes require the temporary binding of a **coenzyme.** Such binding is generally mediated by weak interactions such as hydrogen bonds. The coenzymes are organic molecules that generally serve as carriers of electrons or chemical groups. In chemical reactions, they may either donate groups to the substrate or serve as recipients of groups that are removed from the substrate. The example in Figure 16.20 shows a coenzyme accepting a functional group from one

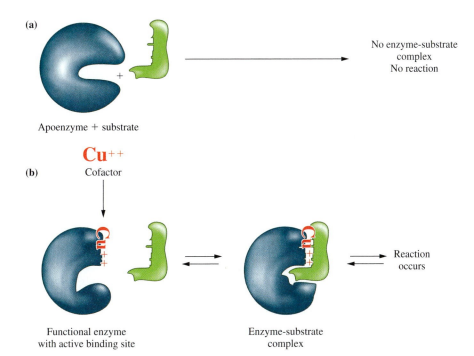

(a)

Apoenzyme + substrate

No enzyme-substrate
complex
No reaction

Cu++

(b) Cofactor

Functional enzyme
with active binding site

Enzyme-substrate
complex

Reaction
occurs

Figure 16.19

(a) The apoenzyme is unable to bind to its substrate. (b) When the required cofactor, in this case a copper ion, Cu^{2+}, is available, it binds to the apoenzyme. Now the active site takes on the correct configuration, the enzyme-substrate complex forms, and the reaction occurs.

substrate and donating it to the second substrate in a reaction catalyzed by a transferase.

Often coenzymes contain modified vitamins as part of their structure. A **vitamin** is an organic substance that is required in the diet in only small amounts. Of the water-soluble vitamins, only vitamin C has not been associated with a coenzyme. Table 16.1 is a summary of some coenzymes and the water-soluble vitamins from which they are made.

Nicotinamide adenine dinucleotide (NAD^+), shown in Figure 16.21, is an example of a coenzyme that is of critical importance in the oxidation reactions of the cellular energy-harvesting processes. NAD^+ can accept a hydride ion, a hydrogen atom with two electrons, from the substrate of these reactions. The substrate is oxidized, and the portion of NAD^+ that is derived from the vitamin, *niacin*, is reduced to produce NADH. The NADH subsequently yields the hydride ion to the first acceptor in an electron transport chain. This regenerates the NAD^+ and provides electrons for the generation of ATP, the chemical energy required by the cell. Also shown in Figure 16.21 is the hydride ion carrier $NADP^+$ and the hydrogen atom carrier FAD. Both are used in the oxidation-reduction reactions that harvest energy for the cell. Unlike NADH and $FADH_2$, NADPH serves as "reducing power" for the cell by donating hydride ions in biochemical reactions. Like NAD^+, $NADP^+$ is derived from niacin. FAD is made from the vitamin, *riboflavin*.

 Water soluble vitamins are discussed in greater detail online at www.mhhe.com/denniston in Water-Soluble Vitamins.

QUESTION 16.13

Why does the body require water-soluble vitamins?

QUESTION 16.14

What are the coenzymes formed from each of the following vitamins? What are the functions of each of these coenzymes?

a. Pantothenic acid   b. Niacin   c. Riboflavin

Figure 16.20

Some enzymes require a coenzyme to facilitate the reaction.

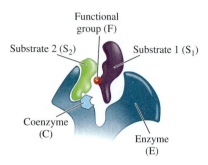

1. An enzyme with a coenzyme positioned to react with two substrates.

2. Coenzyme picks up a functional group from substrate 1.

3. Coenzyme transfers the functional group to substrate 2.

4. Products are released from enzyme.

TABLE 16.1 The Water-Soluble Vitamins and their Coenzymes

Vitamin	Coenzyme	Function
Thiamine (B_1)	Thiamine pyrophosphate	Decarboxylation reactions
Riboflavin (B_2)	Flavin mononucleotide (FMN)	Carrier of H atoms
	Flavin adenine dinucleotide (FAD)	
Niacin (B_3)	Nicotinamide adenine dinucleotide (NAD^+)	Carrier of hydride ions
	Nicotinamide adenine dinucleotide phosphate ($NADP^+$)	
Pyridoxine (B_6)	Pyridoxal phosphate	Carriers of amino and carboxyl groups
	Pyridoxamine phosphate	
Cyanocobalamin (B_{12})	Deoxyadenosyl cobalamin	Coenzyme in amino acid metabolism
Folic acid	Tetrahydrofolic acid	Coenzyme for 1-C transfer
Pantothenic acid	Coenzyme A	Acyl group carrier
Biotin	Biocytin	Coenzyme in CO_2 fixation
Ascorbic acid	Unknown	Hydroxylation of proline and lysine in collagen

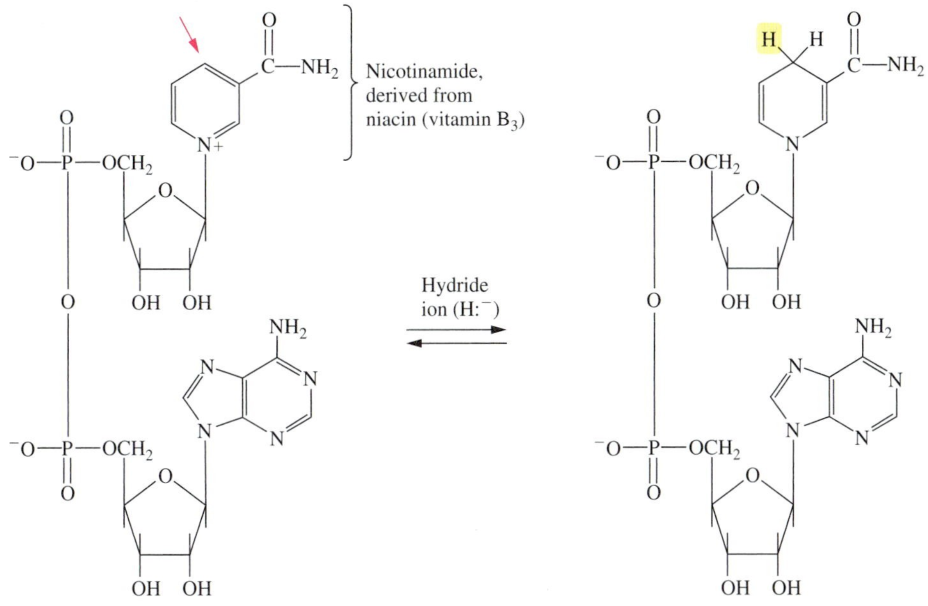

NAD$^+$
Nicotinamide adenine dinucleotide
(oxidized form)

NADH
(reduced form)

(a)

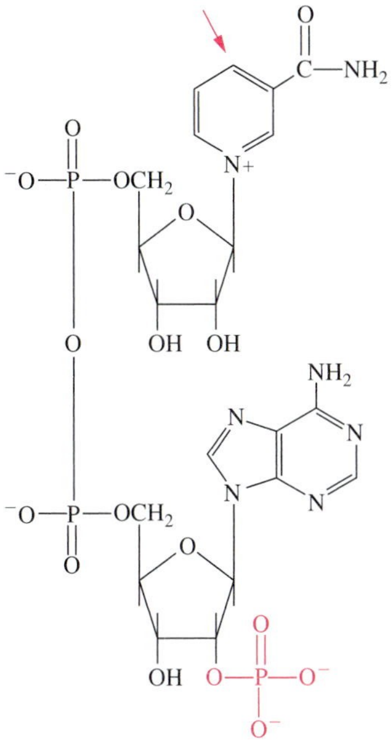

NADP$^+$
Nicotinamide adenine
dinucleotide phosphate
(oxidized form)

(b)

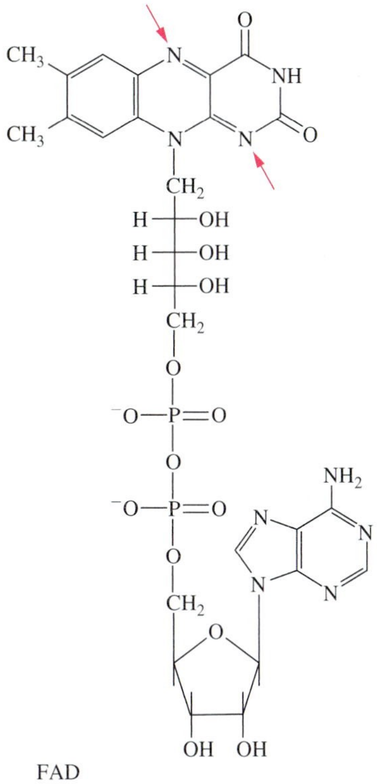

FAD
Flavin adenine dinucleotide
(oxidized form)

(c)

Figure 16.21

The structure of three coenzymes.
(a) The oxidized and reduced forms of nicotinamide adenine dinucleotide.
(b) The oxidized form of the closely related hydride ion carrier, nicotinamide adenine dinucleotide phosphate (NADP$^+$), which accepts hydride ions at the same position as NAD$^+$ (colored arrow).
(c) The oxidized form of flavin adenine dinucleotide (FAD) accepts hydrogen atoms at the positions indicated by the colored arrows.

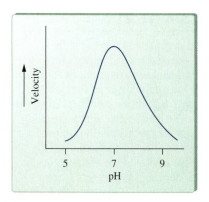

Figure 16.22

Effect of pH on the rate of an enzyme-catalyzed reaction. This enzyme functions most efficiently at pH 7. The rate of the reaction falls rapidly as the solution is made either more acidic or more basic.

13 Learning Goal

Explain how pH and temperature affect the rate of an enzyme-catalyzed reaction.

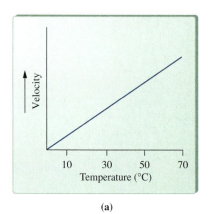

(a)

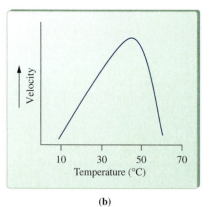

(b)

Figure 16.23

Effect of temperature on (a) uncatalyzed reactions and (b) enzyme-catalyzed reactions.

Environmental Effects

Effect of pH

Most enzymes are active only within a very narrow pH range. The cellular cytoplasm has a pH of 7, and most cytoplasmic enzymes function at a maximum efficiency at this pH. A plot of the relative rate at which a typical cytoplasmic enzyme catalyzes its specific reaction versus pH is provided in Figure 16.22.

The pH at which an enzyme functions optimally is called the **pH optimum.** Making the solution more basic or more acidic sharply decreases the rate of the reaction. At extremes of pH the enzyme loses its biologically active conformation and is *denatured.*

Although the cytoplasm of the cell and the fluids that bathe the cells have a pH that is carefully controlled so that it remains at about pH 7, there are environments within the body in which enzymes must function at a pH far from 7. Protein sequences have evolved that can maintain the proper three-dimensional structure under extreme conditions of pH. For instance, the pH of the stomach is approximately 2 as a result of the secretion of hydrochloric acid by specialized cells of the stomach lining. The proteolytic digestive enzyme, *pepsin,* must effectively degrade proteins at this extreme pH. In the case of pepsin, the enzyme has evolved so that it can maintain a stable tertiary structure at a pH of 2 and is catalytically most active in the hydrolysis of peptides that have been denatured by very low pH. Thus pepsin has a pH optimum of 2.

Effect of Temperature

Enzymes are rapidly destroyed if the temperature of the solution rises much above 37°C, but they remain stable at much lower temperatures. This is why solutions of enzymes used for clinical assays are stored in refrigerators or freezers before use. Figure 16.23 shows the effects of temperature on enzyme-catalyzed and uncatalyzed reactions. The rate of the uncatalyzed reaction steadily increases with increasing temperature because more collisions occur with sufficient energy to overcome the energy barrier for the reaction. The rate of an enzyme-catalyzed reaction also increases with modest increases in temperature because there are increasing numbers of collisions between the enzyme and the substrate. At the **temperature optimum,** the enzyme is functioning optimally and the rate of the reaction is maximal. Above the temperature optimum, increasing temperature denatures the enzyme.

Because heating enzymes and other proteins destroys their activity, a cell cannot survive very high temperatures. Thus, heat is an effective means of sterilizing medical instruments and solutions for transfusion or clinical tests. Although instruments can be sterilized by dry heat (160°C) applied for at least two hours in a dry air oven, autoclaving is a quicker, more reliable procedure. The autoclave works on the principle of the pressure cooker. The pressure causes the temperature of the steam, which would be 100°C at atmospheric pressure, to rise to 121°C. Within twenty minutes, all bacteria and viruses are killed. This is the most effective means of destroying the very heat-resistant endospores that are formed by many bacteria of clinical interest. These bacteria include the genera *Bacillus* and *Clostridium,* which are responsible for such unpleasant and deadly diseases as anthrax, gas gangrene, tetanus, and botulism food poisoning.

QUESTION 16.15

How does a decrease in pH alter the activity of an enzyme?

A MEDICAL Connection | α₁–Antitrypsin and Familial Emphysema

Nearly two million people in the United States suffer from emphysema. Emphysema is a respiratory disease caused by destruction of the alveoli, the tiny, elastic air sacs of the lung. This damage results from the irreversible destruction of a protein called elastin, which is needed for the strength and flexibility of the walls of the alveoli. When elastin is destroyed, the small air passages in the lungs, called bronchioles, become narrower or may even collapse. This severely limits the flow of air into and out of the lung, causing respiratory distress, and in extreme conditions, death.

Some people have a genetic predisposition to emphysema. This is called familial emphysema. These individuals have a genetic defect in the gene that encodes the human plasma protein α_1-antitrypsin. As the name suggests, α_1-antitrypsin is an inhibitor of the proteolytic enzyme trypsin. But, as we have seen in this chapter, trypsin is just one member of a large family of proteolytic enzymes called serine proteases. In the case of the α_1-antitrypsin activity in the lung, the inhibition of the enzyme, elastase, is the critical event.

Elastase damages or destroys elastin, which in turn promotes the development of emphysema. People with normal levels of α_1-antitrypsin are protected from familial emphysema because their α_1-antitrypsin inhibits elastase and, thus, protects the elastin. The result is healthy alveoli in the lungs. However, individuals with a genetic predisposition to emphysema have very low levels of α_1-antitrypsin. This is due to a mutation that causes a single amino acid substitution in the protein chain. Because elastase in the lungs is not effectively controlled, severe lung damage characteristic of emphysema occurs.

Emphysema is also caused by cigarette smoking. Is there a link between these two forms of emphysema? The answer is yes; research has revealed that components of cigarette smoke cause the oxidation of a methionine near the amino terminus of α_1-antitrypsin. This chemical damage destroys α_1-antitrypsin activity. There are enzymes in the lung that reduce the methionine, converting it back to its original chemical form and restoring α_1-antitrypsin activity. However, it is obvious that over a long period, smoking seriously reduces the level of α_1-antitrypsin activity. The accumulated lung damage results in emphysema in many chronic smokers.

At the current time, the standard treatment of emphysema is the use of inhaled oxygen. Studies have shown that intravenous infusion of α_1-antitrypsin isolated from human blood is both safe and effective. However, the level of α_1-antitrypsin in the blood must be maintained by repeated administration.

The α_1-antitrypsin gene has been cloned. In experiments with sheep, it was shown that the protein remains stable when administered as an aerosol. It is still functional after it has passed through the pulmonary epithelium. This research offers hope of an effective treatment for this frightful disease.

FOR FURTHER UNDERSTANDING

Draw the structure of methionine and write an equation showing the reversible oxidation of this amino acid.

Develop a hypothesis to explain why an excess of elastase causes emphysema. What is the role of elastase in this disease?

QUESTION 16.16

Heating is an effective mechanism for killing bacteria on surgical instruments. How does elevated temperature result in cellular death?

Inhibition of Enzyme Activity

Many chemicals can bind to enzymes and either eliminate or drastically reduce their catalytic ability. These chemicals, called *enzyme inhibitors,* have been used for hundreds of years. When she poisoned her victims with arsenic, Lucretia Borgia was unaware that it was binding to the thiol groups of cysteine amino acids in the proteins of her victims and thus interfering with the formation of disulfide bonds needed to stabilize the tertiary structure of enzymes. However, she was well aware of the deadly toxicity of heavy metal salts like arsenic and mercury. When you take penicillin for a bacterial infection, you are taking another enzyme inhibitor. Penicillin inhibits several enzymes that are involved in the synthesis of bacterial cell walls.

14 Learning Goal
Discuss the mechanisms by which certain chemicals inhibit enzyme activity.

Irreversible Inhibitors

Irreversible enzyme inhibitors, such as arsenic, usually bind very tightly, sometimes even covalently, to the enzyme. This generally involves binding of the inhibitor to one of the R groups of an amino acid in the active site. Inhibitor binding may block the active site binding groups so that the enzyme-substrate complex cannot form. Alternatively, an inhibitor may interfere with the catalytic groups of the active site, thereby effectively eliminating catalysis. Irreversible inhibitors, which include snake venoms and nerve gases, generally inhibit many different enzymes.

Reversible, Competitive Inhibitors

Reversible, competitive enzyme inhibitors are often referred to as **structural analogs,** that is, they are molecules that resemble the structure of the natural substrate for a particular enzyme. Because of this resemblance, the inhibitor can occupy the enzyme active site. However, no reaction can occur, and enzyme activity is inhibited (Figure 16.24). This inhibition is said to be competitive because the inhibitor and the substrate compete for binding to the enzyme active site. Thus, the degree of inhibition depends on their relative concentrations. If the inhibitor is in excess or binds more strongly to the active site, it will occupy the active site more frequently, and enzyme activity will be greatly decreased. On the other hand, if the natural substrate is present in excess, it will more frequently occupy the active site, and there will be little inhibition.

The sulfa drugs, the first antimicrobics to be discovered, are **competitive inhibitors** of a bacterial enzyme needed for the synthesis of the vitamin, folic acid. *Folic acid* is a vitamin required for the transfer of methyl groups in the biosynthesis of methionine and the nitrogenous bases required to make DNA and RNA. Humans cannot synthesize folic acid and must obtain it from the diet. Bacteria, on

See A Medical Connection: Fooling the AIDS Virus with "Look-Alike" Nucleotides in Chapter 17.

 To learn about the design of enzyme inhibitors to treat HIV infection, see also A Medical Connection: HIV Protease Inhibitors and Pharmaceutical Drug Design online at www.mhhe.com/denniston.

In addition to the folic acid supplied in the diet, we obtain folic acid from our intestinal bacteria.

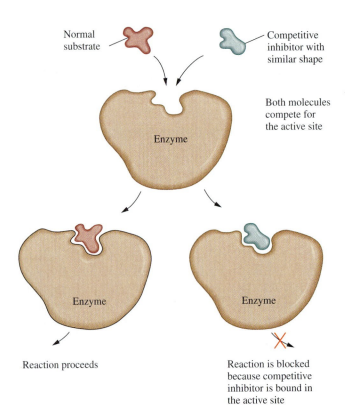

Figure 16.24

Competitive inhibition.

the other hand, must make folic acid because they cannot take it in from the environment.

para-Aminobenzoic acid (PABA) is the substrate for an early step in folic acid synthesis. The sulfa drugs, the prototype of which was discovered in the 1930s by Gerhard Domagk, are structural analogs of PABA and thus competitive inhibitors of the enzyme that uses PABA as its normal substrate.

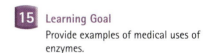

p-Aminobenzoic acid   Sulfanilamide

If the correct substrate (PABA) is bound by the enzyme, the reaction occurs, and the bacterium lives. However, if the sulfa drug is present in excess over PABA, it binds more frequently to the active site of the enzyme. No folic acid will be produced, and the bacterial cell will die.

Because we obtain our folic acid from our diets, sulfa drugs do not harm us. However, bacteria are selectively killed. Luckily, we can capitalize on this property for the treatment of bacterial infections, and as a result, sulfa drugs have saved countless lives. Although bacterial infection was the major cause of death before the discovery of sulfa drugs and other antibiotics, death caused by bacterial infection is relatively rare at present.

Uses of Enzymes in Medicine

Analysis of blood serum for levels (concentrations) of certain enzymes can provide a wealth of information about a patient's medical condition. Often, such tests are used to confirm a preliminary diagnosis based on the disease symptoms or clinical picture.

For example, when a heart attack occurs, a lack of blood supplied to the heart muscle causes some of the heart muscle cells to die. These cells release their contents, including their enzymes, into the bloodstream. Simple tests can be done to measure the amounts of certain enzymes in the blood. Such tests, called *enzyme assays*, are very precise and specific because they are based on the specificity of the enzyme-substrate complex. If you wish to test for the enzyme lactate dehydrogenase (LDH), you need only to add the appropriate substrate, in this case pyruvate and NADH. The reaction that occurs is the oxidation of NADH to NAD^+ and the reduction of pyruvate to lactate. To measure the rate of the chemical reaction, one can measure the disappearance of the substrate or the accumulation of one of the products. In the case of LDH, spectrophotometric methods (based on the light-absorbing properties of a substrate or product) are available to measure the rate of production of NAD^+. The choice of substrate determines which enzyme activity is to be measured.

Elevated blood serum concentrations of the enzymes, amylase and lipase, are indications of pancreatitis, an inflammation of the pancreas. Liver diseases such as cirrhosis and hepatitis result in elevated levels of one of the isoenzymes of lactate dehydrogenase (LDH_5), and elevated levels of alanine aminotransferase/serum glutamate–pyruvate transaminase (ALT/SGPT) and aspartate aminotransferase/serum glutamate–oxaloacetate transaminase (AST/SGOT) in blood serum. These latter two enzymes also increase in concentration following a heart attack, but the physician can differentiate between these two conditions by considering the relative increase in the two enzymes. If ALT/SGPT is elevated to a greater extent than AST/SGOT, it can be concluded that the problem is liver dysfunction.

15 **Learning Goal**
Provide examples of medical uses of enzymes.

 For more information on the use of enzymes to diagnose and treat heart attack, see A Medical Connection: Enzymes, Isoenzymes, and Myocardial Infarction online at www.mhhe.com/denniston.

 **Animation**
Transmission Across a Synapse

CHEMISTRY at the Crime Scene | Enzymes, Nerve Agents, and Poisoning

The transmission of nerve impulses at the *neuromuscular junction* involves many steps. One of them is the activity of a critical enzyme, called *acetylcholinesterase*, which catalyzes the hydrolysis of the chemical messenger, *acetylcholine*, that initiated the nerve impulse. The need for this enzyme activity becomes clear when we consider the events that begin with a message from the nerve cell and end in the appropriate response by the muscle cell. Acetylcholine is a *neurotransmitter*, that is, a chemical messenger that transmits a message from the nerve cell to the muscle cell. Acetylcholine is stored in membrane-bound bags, called *synaptic vesicles*, in the nerve cell ending.

Acetylcholinesterase comes into play in the following way. The arrival of a nerve impulse at the end plate of the nerve axon causes an influx of Ca^{2+}. This causes the acetylcholine-containing vesicles to migrate to the nerve cell membrane that is in contact with the muscle cell. This is called the *presynaptic membrane*. The vesicles fuse with the presynaptic membrane and release the neurotransmitter. The acetylcholine then diffuses across the *nerve synapse* (the space between the nerve and muscle cells) and binds to the acetylcholine receptor protein in the *postsynaptic membrane* of the muscle cell. This receptor then opens pores in the membrane through which Na^+ and K^+ ions flow into and out of the cell, respectively. This generates the nerve impulse and causes the muscle to contract. If acetylcholine remains at the neuromuscular junction, it will continue to stimulate the muscle contraction. To stop this continued stimulation, acetylcholine is hydrolyzed, and hence, destroyed by acetylcholinesterase. When this happens, nerve stimulation ceases.

$$H_3C-\overset{\overset{O}{\|}}{C}-O-CH_2CH_2-N^+-(CH_3)_3 + H_2O$$

Acetylcholine

Acetylcholinesterase

$$H_3C-\overset{\overset{O}{\diagup}}{\underset{O^-}{C}} + HO-CH_2CH_2-N^+-(CH_3)_3 + H^+$$

Acetate   Choline

Schematic diagram of the synapse at the neuromuscular junction. The nerve impulse causes acetylcholine to be released from synaptic vesicles. Acetylcholine diffuses across the synaptic cleft and binds to a specific receptor protein (R) on the postsynaptic membrane. A channel opens that allows Na^+ ions to flow into the cell and K^+ ions to flow out of the cell. This results in muscle contraction. Any acetylcholine remaining in the synaptic cleft is destroyed by acetylcholinesterase to terminate the stimulation of the muscle cell.

Inhibitors of acetylcholinesterase are used both as poisons and as drugs. Among the most important inhibitors of acetylcholinesterase are a class of compounds known as *organophos-*

Animation
Chemical Synapse

Enzymes are also used as analytical reagents in the clinical laboratory owing to their specificity. They often selectively react with one substance of interest, producing a product that is easily measured. An example of this is the clinical analysis of urea in blood. The measurement of urea levels in blood is difficult because of the complexity of blood. However, if urea is converted to ammonia using the enzyme urease, the ammonia becomes an *indicator* of urea, because it is produced from urea, and it is easily measured. This test, called the *blood urea nitrogen*

phates. One of these is the nerve agent Sarin (isopropylmethyl-fluorophosphate). Sarin forms a covalently bonded intermediate with the active site of acetylcholinesterase. Thus, it acts as an irreversible, noncompetitive inhibitor.

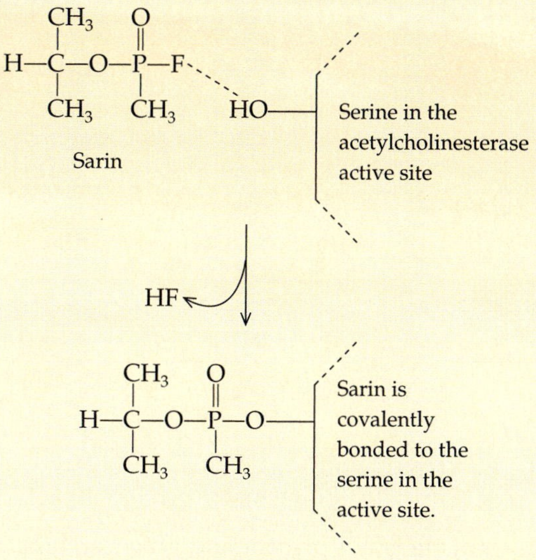

Sarin

Serine in the acetylcholinesterase active site

HF

Sarin is covalently bonded to the serine in the active site.

The covalent intermediate is stable, and acetylcholinesterase is therefore inactive, no longer able to break down acetylcholine. Nerve transmission continues, resulting in muscle spasm. Death may occur as a result of laryngeal spasm. Antidotes for poisoning by organophosphates, which include many insecticides and nerve gases, have been developed. The antidotes work by reversing the effects of the inhibitor. One of these antidotes is known as *PAM,* an acronym for *pyridine aldoxime methiodide.* This molecule displaces the organophosphate group from the active site of the enzyme, alleviating the effects of the poison.

FOR FURTHER UNDERSTANDING

Botulinum toxin inhibits release of neurotransmitters from the presynaptic membrane. What symptoms do you predict would result from this?

Why must Na^+ and K^+ enter and exit the cell through a protein channel?

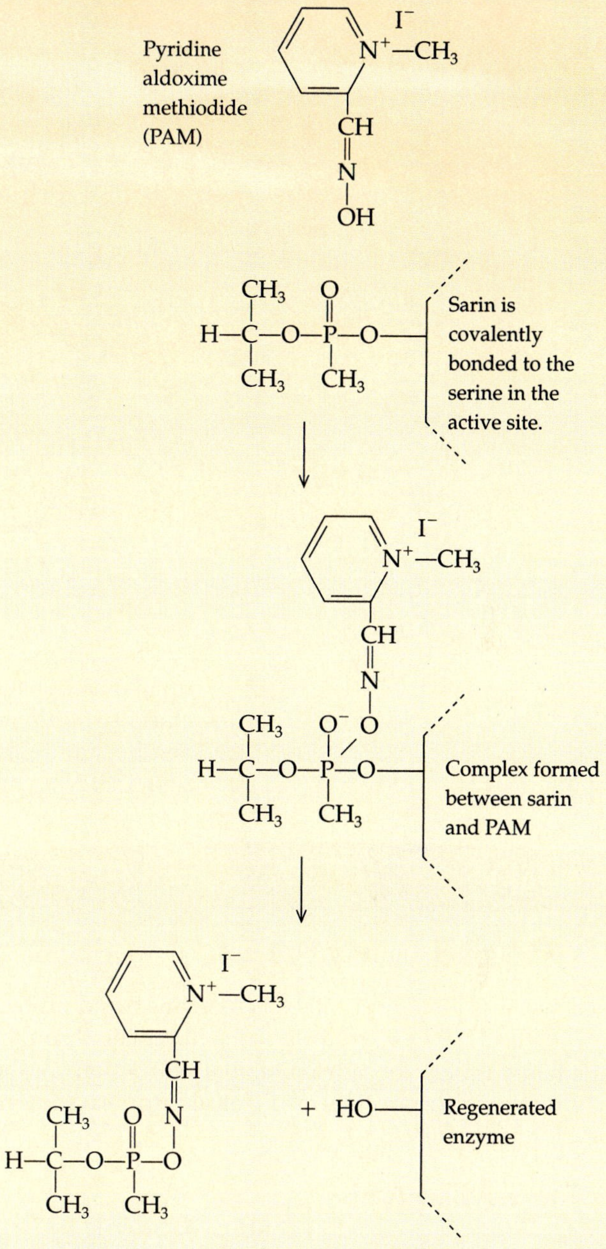

Pyridine aldoxime methiodide (PAM)

Sarin is covalently bonded to the serine in the active site.

Complex formed between sarin and PAM

Regenerated enzyme

(BUN) test, is useful in the diagnosis of kidney malfunction and serves as one example of the utility of enzymes in clinical chemistry.

Enzyme replacement therapy can also be used in the treatment of certain diseases. One such disease, Gaucher's disease, is a genetic disorder resulting in a deficiency of the enzyme *glucocerebrosidase.* In the normal situation, this enzyme breaks down a glycolipid called *glucocerebroside,* which is an important component of cell membranes.

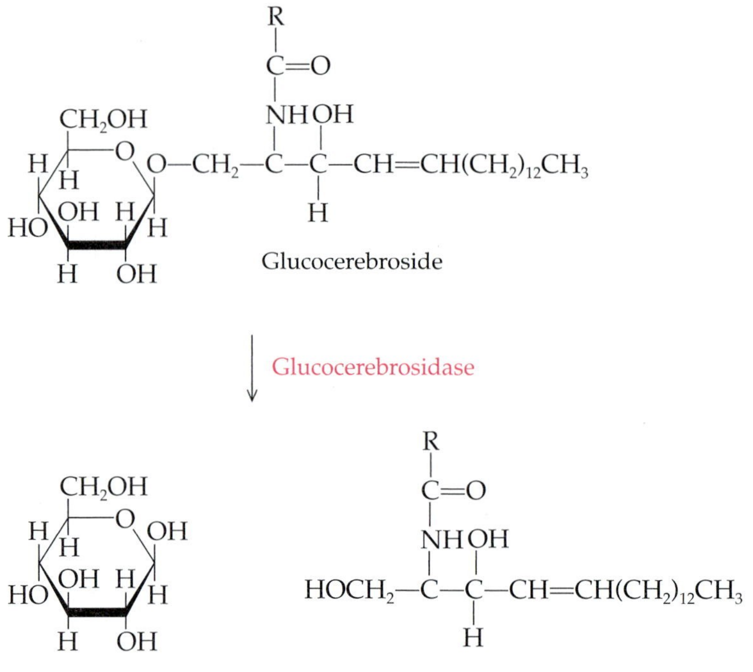

Glucocerebroside

Glucocerebrosidase

Glucose   Ceramide

In Gaucher's disease, the enzyme is not present and glucocerebroside builds up in macrophages in the liver, spleen, and bone marrow. The symptoms which include severe anemia, thrombocytopenia (reduction in the number of platelets), and hepatosplenomegaly (enlargement of the spleen and liver) can now be treated with *Cerezyme*, an enzyme produced by recombinant DNA technology.

SUMMARY

16.1 Cellular Functions of Proteins

Proteins serve as biological catalysts (enzymes) and protective *antibodies*. *Transport proteins* carry materials throughout the body. Protein hormones regulate conditions in the body. Proteins also provide mechanical support and are needed for movement.

16.2 The α–Amino Acids

Proteins are made from twenty different *amino acids*, each having an α-COO^- group and an α-N^+H_3 group. They differ only in their side-chain R groups. All α-*amino acids* are chiral except glycine. Naturally occurring amino acids have the same chirality, designated L. The amino acids are grouped according to the polarity of their R groups.

16.3 The Peptide Bond

Amino acids are joined by *peptide bonds* to produce peptides and proteins. The peptide bond is an amide bond formed in the reaction between the carboxyl group of one amino acid and the amino group of another. The peptide bond is planar and relatively rigid.

16.4 The Primary Structure of Proteins

Proteins are linear polymers of amino acids. The linear sequence of amino acids defines the *primary structure* of the protein. Evolutionary relationships between species of organisms can be deduced by comparing the primary structures of their proteins.

16.5 The Secondary Structure of Proteins

The *secondary structure* of a protein is the folding of the primary sequence into an α-*helix* or β-*pleated sheet*. These structures are maintained by hydrogen bonds between the amide hydrogen and the carbonyl oxygen of the peptide bond. Usually, *structural proteins*, such as α-*keratins* and silk fibroin, are composed entirely of α-helix or β-pleated sheet.

16.6 The Tertiary Structure of Proteins

Globular proteins contain varying amounts of α-helix and β-pleated sheet folded into higher levels of structure called the *tertiary structure*. The tertiary structure of a protein is maintained by attractive forces between the R groups of amino acids. These forces include hydrophobic interactions, hydrogen bonds, ionic bridges, and disulfide bonds.

16.7 The Quaternary Structure of Proteins

Some proteins are composed of more than one peptide. They are said to have a *quaternary structure.* Weak attractions between amino acid R groups hold the peptide subunits of the protein together. Some proteins require an attached, nonprotein *prosthetic group.*

16.8 Myoglobin and Hemoglobin

Myoglobin, the oxygen storage protein of skeletal muscle, has a prosthetic group called the *heme group.* The heme group is the site of oxygen binding. *Hemoglobin* consists of four peptides. It transports oxygen from the lungs to the tissues. Myoglobin has greater affinity for oxygen than does hemoglobin, and so oxygen is efficiently transferred from hemoglobin in the blood to myoglobin in tissues. A mutant hemoglobin is responsible for the genetic disease *sickle cell anemia.*

16.9 Denaturation of Proteins

Heat disrupts the hydrogen bonds and hydrophobic interactions that maintain protein structure. As a result, the protein unfolds and the organized structure is lost. The protein is said to be *denatured.*

16.10 Enzymes

Enzymes are most frequently named by using the common system of nomenclature. The names are useful because they are often derived from the name of the substrate and/or the reaction of the substrate that is catalyzed by the enzyme. Enzymes are classified according to function. The six general classes are *oxidoreductases, transferases, hydrolases, lyases, isomerases,* and *ligases.*

Enzymes are the biological catalysts of cells. They lower the activation energies of the reactions they catalyze.

Formation of an *enzyme-substrate complex* is the first step in an enzyme-catalyzed reaction. This involves binding the substrate to the active site of the enzyme. The lock-and-key model of substrate binding describes the enzyme as a rigid structure into which the substrate fits precisely. The newer induced fit model describes the enzyme as a flexible molecule. The shape of the active site approximates the shape of the substrate and then "molds" itself around the substrate.

An enzyme-catalyzed reaction is mediated through an unstable *transition state.* This may involve the enzyme putting "stress" on a bond, bringing reactants into close proximity and in the correct orientation, or altering the local pH.

Cofactors are metal ions, organic compounds, or organometallic compounds that bind to an enzyme and help maintain the correct configuration of the active site. The term *coenzyme* refers specifically to an organic group that binds transiently to the enzyme during the reaction. It accepts or donates chemical groups.

Enzymes are sensitive to pH and temperature. High temperatures or extremes of pH rapidly inactivate most enzymes by denaturing them.

Enzyme activity can be destroyed by a variety of inhibitors. *Irreversible inhibitors,* or poisons, bind tightly to enzymes and destroy their activity permanently. *Competitive inhibitors* are generally *structural analogs* of the natural substrate for the enzyme. They compete with the normal substrate for binding to the active site. When the competitive inhibitor is bound by the active site, the reaction cannot occur, and no product is produced.

Analysis of blood serum for unusually high levels of certain enzymes provides valuable information on a patient's condition. Such analysis is used to diagnose heart attack, liver disease, and pancreatitis. Enzymes are also used as analytical reagents, as in the blood urea nitrogen (BUN) test, and in the treatment of disease.

KEY TERMS

α-amino acid (16.2)
active site (16.10)
antibody (16.1)
antigen (16.1)
apoenzyme (16.10)
coenzyme (16.10)
cofactor (16.10)
competitive inhibitor (16.10)
C-terminal amino acid (16.3)
defense proteins (16.1)
denaturation (16.9)
enzyme (16.10)
enzyme-substrate complex (16.10)
fibrous protein (16.5)
globular protein (16.6)
glycoprotein (16.7)
α-helix (16.5)
hemoglobin (16.8)
holoenzyme (16.10)
hydrolase (16.10)
hydrophilic amino acid (16.2)
hydrophobic amino acid (16.2)
induced fit model (16.10)
irreversible enzyme inhibitor (16.10)
isomerase (16.10)
α-keratin (16.5)
ligase (16.10)
lock-and-key model (16.10)
lyase (16.10)

movement protein (16.1)
myoglobin (16.8)
N-terminal amino acid (16.3)
oxidoreductase (16.10)
peptide bond (16.3)
pH optimum (16.10)
β-pleated sheet (16.5)
primary structure (of a protein) (16.4)
product (16.10)
prosthetic group (16.7)
protein (16.1)
quaternary structure (of a protein) (16.7)
regulatory protein (16.1)
reversible, competitive enzyme inhibitor (16.10)
secondary structure (of a protein) (16.5)
sickle cell anemia (16.8)
storage protein (16.1)
structural analog (16.10)
structural protein (16.1)
substrate (16.10)
temperature optimum (16.10)
tertiary structure (of a protein) (16.6)
transferase (16.10)
transition state (16.10)
transport protein (16.1)
vitamin (16.10)

QUESTIONS AND PROBLEMS

Cellular Functions of Proteins

Foundations

16.17 Define the term *enzyme*.
16.18 Define the term *antibody*.

Applications

16.19 Of what significance are enzymes in the cell?
16.20 How do antibodies protect us against infection?

The α–Amino Acids

Foundations

16.21 Write the basic general structure of an L-α-amino acid.
16.22 Draw the D- and L-isomers of serine. Which would you expect to find in nature?

Applications

16.23 What is a chiral carbon?
16.24 Why are all α-amino acids chiral except glycine?
16.25 What is the importance of the R groups of amino acids?
16.26 Describe the classification of the R groups of amino acids, and provide an example of each class.

The Peptide Bond

Foundations

16.27 Define the term *peptide bond*.
16.28 What type of bond is the peptide bond?

Applications

16.29 Write the structure of each of the following peptides:
  a. His-trp-cys
  b. Gly-leu-ser
  c. Arg-ile-val
16.30 Write the structure of each of the following peptides:
  a. Ile-leu-phe
  b. His-arg-lys
  c. Asp-glu-ser

The Primary Structure of Proteins

Foundations

16.31 Define the *primary structure* of a protein.
16.32 What type of bond joins amino acids to one another in the primary structure of a protein?

Applications

16.33 How does the primary structure of a protein determine its three-dimensional shape?
16.34 How does the primary structure of a protein ultimately determine its biological function?

The Secondary Structure of Proteins

Foundations

16.35 Define the secondary structure of a protein.
16.36 What are the two most common types of secondary structure?

Applications

16.37 What type of secondary structure is characteristic of
  a. α-keratins?
  b. silk fibroin?
16.38 Describe the forces that maintain the two types of secondary structure: α-helix and β-pleated sheet.

The Tertiary Structure of Proteins

Foundations

16.39 Define the tertiary structure of a protein.
16.40 Use examples of specific amino acids to show the variety of weak interactions that maintain tertiary protein structure.

Applications

16.41 What is the role of cystine in maintaining protein structure?
16.42 Explain the relationship between the secondary and tertiary protein structures.

The Quaternary Structure of Proteins

Foundations

16.43 Describe the quaternary structure of proteins.
16.44 What weak interactions are responsible for maintaining quaternary protein structure?

Applications

16.45 What is a glycoprotein?
16.46 What is a prosthetic group?

Myoglobin and Hemoglobin

Foundations

16.47 What is the structure and function of hemoglobin?
16.48 What is the structure and function of myoglobin?

Applications

16.49 Carbon monoxide binds tightly to the heme groups of hemoglobin and myoglobin. How does this affinity reflect the toxicity of carbon monoxide?
16.50 The blood of the horseshoe crab is blue because of the presence of a protein called hemocyanin. What is the function of hemocyanin?
16.51 Why does replacement of glutamic acid with valine alter hemoglobin and ultimately result in sickle cell anemia?
16.52 How do sickled red blood cells hinder circulation?

Denaturation of Proteins

Foundations

16.53 Define the term *denaturation*.
16.54 What environmental factors denature proteins?

Applications

16.55 Why is heat an effective means of sterilization?
16.56 Why is it important that blood have several buffering mechanisms to avoid radical pH changes?

Enzymes

Foundations

16.57 How are the common names of enzymes often derived?
16.58 What is the most common characteristic used to classify enzymes?
16.59 Define the term *substrate*.
16.60 Define the term *product*.
16.61 Define the term *enzyme-substrate complex*.
16.62 Define the term *active site*.
16.63 What is the role of a cofactor in enzyme activity?
16.64 How does a coenzyme function in an enzyme-substrate complex.

Applications

16.65 Describe the function implied by the name of each of the following enzymes:
  a. Citrate decarboxylase
  b. Adenosine diphosphate phosphorylase

c. Oxalate reductase

d. Nitrite oxidase

e. *cis-trans* Isomerase

16.66 List the six classes of enzymes based on the type of reaction catalyzed. Briefly describe the function of each class, and provide an example of each.

16.67 What is the activation energy of a reaction?

16.68 What is the effect of an enzyme on the activation energy of a reaction?

16.69 What is the lock-and-key model of enzyme-substrate binding?

16.70 Why is the induced fit model of enzyme-substrate binding a much more accurate model than the lock-and-key model?

16.71 Outline the four general stages in an enzyme-catalyzed reaction.

16.72 What types of transition states might be envisioned that would decrease the energy of activation of an enzyme?

16.73 What is the function of NAD^+? What class of enzymes would require a coenzyme of this sort?

16.74 What is the function of FAD? What class of enzymes would require this coenzyme?

16.75 How will each of the following changes in conditions alter the rate of an enzyme-catalyzed reaction?

a. Decreasing the temperature from 37°C to 10°C

b. Increasing the pH of the solution from 7 to 11

c. Heating the enzyme from 37°C to 100°C

16.76 Why are enzymes that are used for clinical assays in hospitals stored in refrigerators?

16.77 How do the sulfa drugs selectively kill bacteria while causing no harm to humans?

16.78 Why are irreversible enzyme inhibitors often called *poisons?*

16.79 Suppose that a certain drug company manufactured a compound that had nearly the same structure as a substrate for a certain enzyme but that could not be acted upon chemically by the enzyme. What type of interaction would the compound have with the enzyme?

16.80 The addition of phenylthiourea to a preparation of the enzyme, polyphenoloxidase, completely inhibits the activity of the enzyme.

a. Knowing that phenylthiourea binds all copper ions, what conclusion can you draw about whether polyphenoloxidase requires a cofactor?

b. What kind of inhibitor is phenylthiourea?

16.81 List the enzymes whose levels are elevated in blood serum following a myocardial infarction.

16.82 List the enzymes whose levels are elevated as a result of hepatitis or cirrhosis of the liver.

FOR FURTHER UNDERSTANDING

1. Proteins involved in transport of molecules or ions into or out of cells are found in the membranes of all cells. They are classified as transmembrane proteins because some regions are embedded within the lipid bilayer, whereas other regions protrude into the cytoplasm or outside the cell. Review the classification of amino acids based on the properties of their R groups. What type of amino acids would you expect to find in the regions of the proteins embedded within the membrane? What type of amino acids would you expect to find on the surface of the regions in the cytoplasm or that protrude outside the cell?

2. A biochemist is trying to purify the enzyme hexokinase from a bacterium that normally grows in the Arctic Ocean at 5°C. In the next lab, a graduate student is trying to purify the same protein from a bacterium that grows in the vent of a volcano at 98°C. To maintain the structure of the protein from the Arctic bacterium, the first biochemist must carry out all her purification procedures at refrigerator temperatures. The second biochemist must perform all his experiments in a warm room incubator. In molecular terms, explain why the same kind of enzyme from organisms with different optimal temperatures for growth can have such different thermal properties.

3. The α-keratin of hair is rich in the amino acid cysteine. The location of these cysteines in the protein chain is genetically determined; as a result of the location of the cysteines in the protein, a person may have curly, wavy, or straight hair. How can the location of cysteines in α-keratin result in these different styles of hair? Propose a hypothesis to explain how a "perm" causes straight hair to become curly.

4. Calculate the number of different pentapeptides you can make in which the amino acids phenylalanine, glycine, serine, leucine, and histidine are each found. Imagine how many proteins could be made from the twenty amino acids commonly found in proteins.

5. Ethylene glycol is a poison that causes about fifty deaths a year in the United States. Treating people who have drunk ethylene glycol with massive doses of ethanol can save their lives. Suggest a reason for the effect of ethanol.

6. L-1-(*p*-Toluenesulfonyl)-amido-2-phenylethylchloromethyl ketone (TPCK, shown below) inhibits chymotrypsin, but not trypsin. Propose a hypothesis to explain this observation.

7. A graduate student is trying to make a "map" of a short peptide so that she can eventually determine the amino acid sequence. She digested the peptide with several proteases and determined the sizes of the resultant digestion products.

Enzyme	M.W. of Digestion Products
Trypsin	2000, 3000
Chymotrypsin	500, 1000, 3500
Elastase	500, 1000, 1500, 2000

Suggest experiments that would allow the student to map the order of the enzyme digestion sites along the peptide.

17

LEARNING GOALS

1 Draw the general structure of DNA and RNA nucleotides.

2 Describe the structure of DNA and compare it with RNA.

3 Explain DNA replication.

4 List three classes of RNA molecules and describe their functions.

5 Explain the process of transcription.

6 List and explain the three types of post-transcriptional modifications of eukaryotic mRNA.

7 Describe the essential elements of the genetic code and develop a "feel" for its elegance.

8 Describe the process of translation.

9 Define mutation and understand how mutations can cause cancer and cell death.

Introduction to Molecular Genetics

These twin girls are identical. Explain why this is so.

Look around at the students in your chemistry class. They all share many traits: upright stance, a head with two eyes, a nose, and a mouth facing forward, one ear on each side of the head, and so on. You would have no difficulty listing the similarities that define you and your classmates as Homo sapiens.

As you look more closely at the individuals, you begin to notice many differences. Eye color, hair color, skin color, the shape of the nose, height, body build: all these traits, and many more, show amazing variety from one person to the next. Even within one family, in which the similarities may be more pronounced, each individual has a unique appearance. Only identical twins look exactly alike—well, most of the time.

The molecule responsible for all these similarities and differences is deoxyribonucleic acid (DNA). Tightly wound up in structures called chromosomes in the nucleus of the cell, DNA carries the genetic code to produce the thousands of different proteins that make us who we are. These proteins include enzymes that are responsible for production of the pigment melanin. The more melanin we are genetically programmed to make, the darker our hair, eyes, and skin will be. Others are structural proteins. The gene for α-keratin that makes up hair determines whether our hair will be wavy, straight, or curly. Thousands of genes carry the genetic information for thousands of proteins that dictate our form and, some believe, our behavior.

Genetic traits are passed from one generation to the next. When a sperm fertilizes an ovum, a zygote is created from a single set of maternal chromosomes and a single set of paternal chromosomes. As this fertilized egg divides, each daughter cell will receive one copy of each of these chromosomes. The genes on these chromosomes will direct fetal development from that fertilized cell to a newborn with all the characteristics we recognize as human.

In this chapter, we will explore the structure of DNA and the molecular events that translate the genetic information of a gene into the structure of a protein.

17.1 The Structure of the Nucleotide

Learning Goal

Draw the general structure of DNA and RNA nucleotides.

In 1950, the genetic information was demonstrated to be deoxyribonucleic acid, or DNA. In 1953, just over fifty years ago, James Watson and Francis Crick published a paper describing the structure of DNA molecules.

Nucleotide Structure

From the work of Watson and Crick and many others, we now know that there are two types of nucleic acids essential for life. These are **deoxyribonucleic acid (DNA)** and **ribonucleic acid (RNA)** which are long polymers of **nucleotides.** Every nucleotide is composed of a nitrogenous base, a five-carbon sugar, and at least one phosphoryl group.

Nitrogenous bases are cyclic compounds with at least one nitrogen atom in the ring structure. There are two types of nitrogenous bases: purines and pyrimidines. **Purines,** which include adenine and guanine, consist of a six-member ring fused to a five-member ring. **Pyrimidines,** which include thymine, cytosine, and uracil, consist of a single six-member ring. Structures of these molecules are shown in Figure 17.1.

The five-carbon sugar in RNA is ribose, and the sugar in DNA is 2′-deoxyribose. The only difference between these two sugars is found at the 2′-carbon. Ribose has a hydroxyl group attached to this carbon, and deoxyribose has a hydrogen atom.

Each nucleotide consists of either ribose or deoxyribose, one of the five nitrogenous bases, and one or more phosphoryl groups (Figure 17.2). A nucleotide with the sugar ribose is a **ribonucleotide,** and one having the sugar 2′-deoxyribose is a **deoxyribonucleotide.** Because there are two cyclic molecules in a nucleotide, the sugar, and the base, we need an easy way to describe the ring atoms of each. For this reason the ring atoms of the sugar are designated with a prime to distinguish them from the atoms of the base (Figures 17.1 and 17.2).

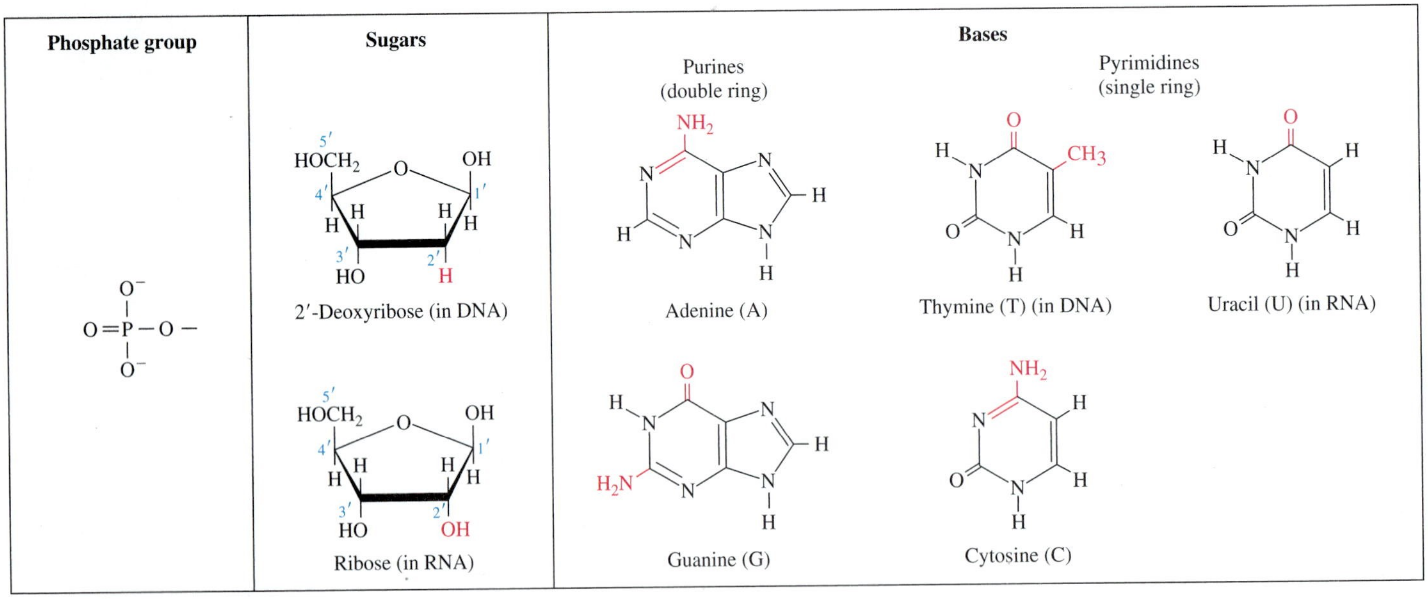

Figure 17.1

The components of nucleic acids include phosphate groups, the five-carbon sugars ribose and deoxyribose, and purine and pyrimidine nitrogenous bases. The ring positions of the sugars are designated with primes (′) to distinguish them from the ring positions of the bases.

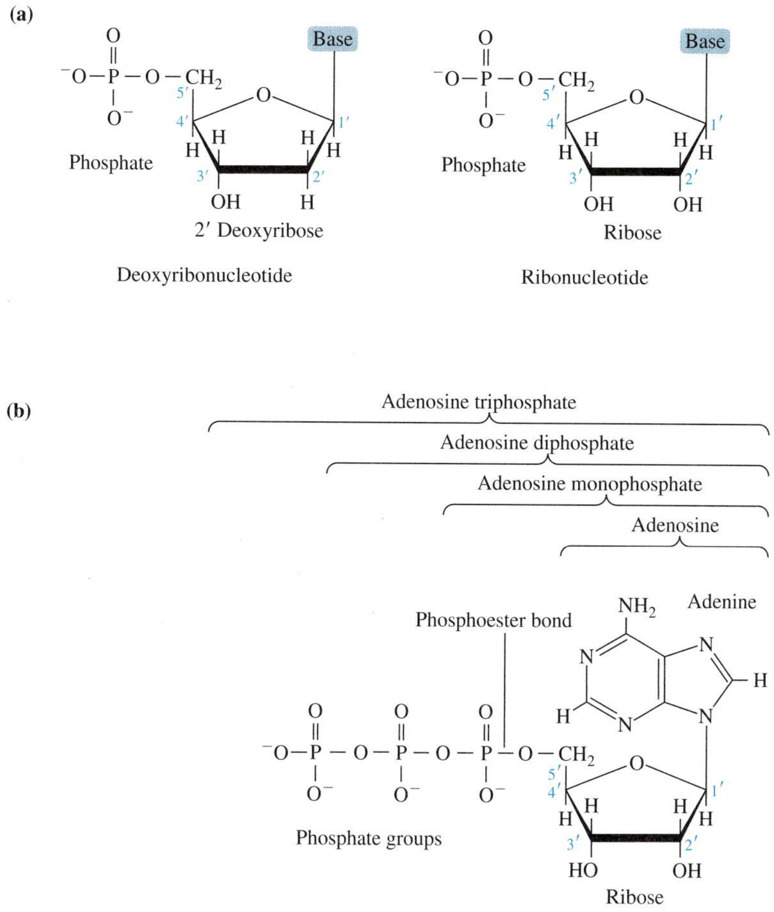

(a)

Deoxyribonucleotide   Ribonucleotide

(b)

Figure 17.2

(a) The general structures of a deoxyribonucleotide and a ribonucleotide. (b) A specific example of a ribonucleotide, adenosine triphosphate.

QUESTION 17.1

Referring to the structures in Figures 17.1 and 17.2, draw the structures for nucleotides consisting of the following units.

a. Ribose, adenine, two phosphoryl groups
b. 2′-Deoxyribose, guanine, three phosphoryl groups

QUESTION 17.2

Referring to the structures in Figures 17.1 and 17.2, draw the structures for nucleotides consisting of the following units.

a. 2′-Deoxyribose, thymine, one phosphoryl group
b. Ribose, cytosine, three phosphoryl groups
c. Ribose, uracil, one phosphoryl group

17.2 The Structure of DNA and RNA

A single strand of DNA is a polymer of nucleotides bonded to one another by 3′–5′ **phosphodiester bonds.** The backbone of the polymer is called the *sugar-phosphate backbone* because it is composed of alternating units of the five-carbon

2 Learning Goal

Describe the structure of DNA and compare it with RNA.

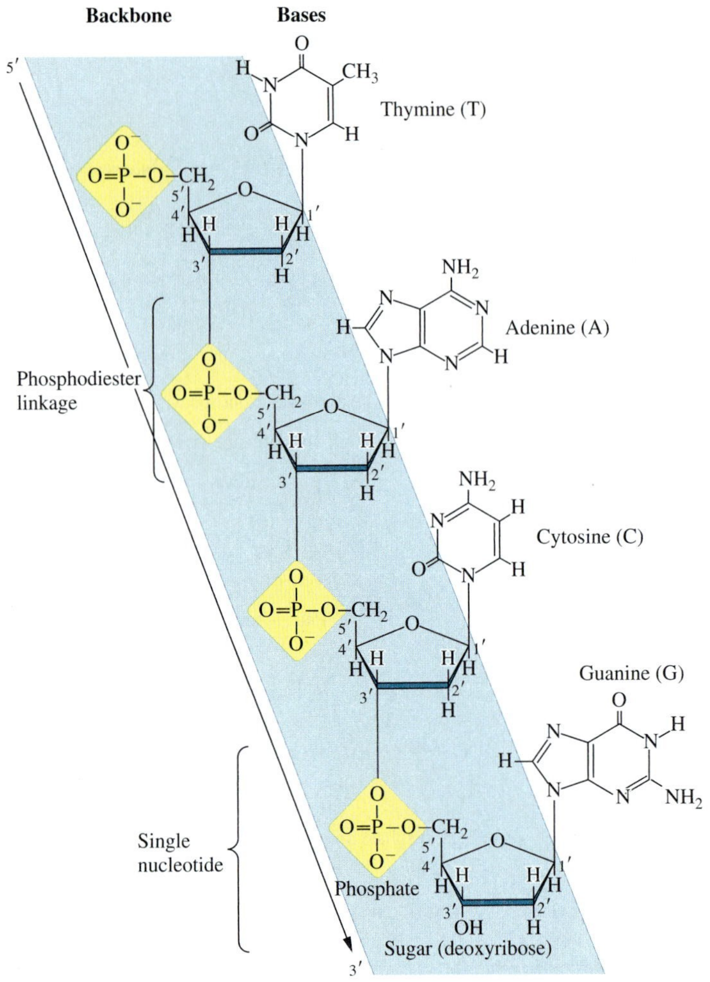

Figure 17.3

The covalent, primary structure of DNA.

sugar 2'-deoxyribose and phosphoryl groups in a phosphodiester linkage. A nitrogenous base is bonded to each sugar (Figure 17.3).

DNA Structure: The Double Helix

Animation
Molecular Structure of DNA

James Watson and Francis Crick were the first to describe the three-dimensional structure of DNA in 1953. By building models based on the experimental results of others, Watson and Crick concluded that DNA is a **double helix** of two strands of DNA wound around one another. The structure of the double helix is often compared to a spiral staircase. The sugar-phosphate backbones of the two strands of DNA spiral around the outside of the helix like the handrails on a spiral staircase. The nitrogenous bases extend into the center at right angles to the axis of the helix. You can imagine the nitrogenous bases forming the steps of the staircase. The structure of this elegant molecule is shown in Figure 17.4.

Hydrogen bonding between the nitrogenous bases in the center of the helix helps hold the double helix together. Adenine forms two hydrogen bonds with thymine, and cytosine forms three hydrogen bonds with guanine (Figure 17.4). These are called **base pairs.** The two strands of DNA are **complementary strands** because the sequence of bases on one automatically determines the sequence of bases on the other. When there is an adenine on one strand, there will always be a thymine in the same location on the opposite strand.

Key Features

- Two strands of DNA form a right-handed double helix.

- The bases in opposite strands hydrogen bond according to the AT/GC rule.

- The two strands are antiparallel with regard to their 5′ to 3′ directionality.

- There are ~10.0 nucleotides in each strand per complete 360° turn of the helix.

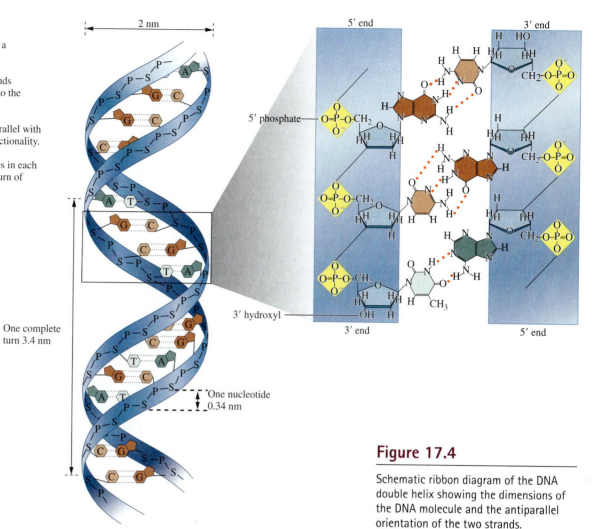

Figure 17.4

Schematic ribbon diagram of the DNA double helix showing the dimensions of the DNA molecule and the antiparallel orientation of the two strands.

One last important feature of the DNA double helix is that the two strands are **antiparallel strands,** as this example shows:

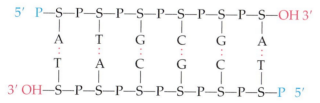

In other words, the two strands of the helix run in opposite directions (see Figure 17.4). Only when the two strands are antiparallel can the base pairs form the hydrogen bonds that hold the two strands together.

In simple organisms, such as bacteria, the DNA is a circular molecule that is supercoiled very tightly into a structure called a nucleoid. In the cells of more complex organisms, such as humans, the DNA is wrapped around small disks of histone proteins. At this level, the DNA looks like beads on a string. This string of beads then coils into the larger structures we call chromosomes.

RNA Structure

The sugar-phosphate backbone of RNA consists of ribonucleotides, also linked by 3′–5′ phosphodiester bonds. These phosphodiester bonds are identical to those

A MEDICAL Connection | Fooling the AIDS Virus with "Look–Alike" Nucleotides

The virus that is responsible for the acquired immune deficiency syndrome (AIDS) is called the *human immunodeficiency virus*, or *HIV*. HIV is a member of a family of viruses called *retroviruses*, all of which have single-stranded RNA as their genetic material. The RNA is copied into a double-stranded DNA molecule by a viral enzyme called *reverse transcriptase*. This process is the opposite of the central dogma, which states that the flow of genetic information is from DNA to RNA. But these viruses reverse that flow, RNA to DNA. For this reason, these viruses are called retroviruses, which literally means "backward viruses." The process of producing a DNA copy of the RNA is called *reverse transcription*.

Because our genetic information is DNA and it is expressed by the classical DNA → RNA → protein pathway, our cells have no need for a reverse transcriptase enzyme. Thus, the HIV reverse transcriptase is a good target for antiviral chemotherapy because inhibition of reverse transcription should kill the virus but have no effect on the human host. Many drugs have been tested for the ability to selectively inhibit HIV reverse transcription. Among these is the DNA chain terminator 3'-azido-2', 3'-dideoxythymidine, commonly called *AZT* or *zidovudine*.

How does AZT work? It is one of many drugs that looks like one of the normal nucleosides. These are called *nucleoside analogs*. A nucleoside is just a nucleotide without any phosphate groups attached. The analog is phosphorylated by the cell and then tricks a polymerase, in this case viral reverse transcriptase, into incorporating it into the growing DNA chain in place of the normal phosphorylated nucleoside. AZT is a nucleoside analog that looks like the nucleoside thymidine except that in the 3' position of the deoxyribose sugar there is an azido group ($-N_3$) rather

than the 3'-OH group. Compare the structures of thymidine and AZT shown in the accompanying figure. The 3'-OH group is necessary for further DNA polymerization because there the phosphoester linkage must be made between the growing DNA strand and the next nucleotide. If an azido group or some other group is present at the 3' position, the nucleotide analog can be incorporated into the growing DNA strand, but further chain elongation is blocked, as shown in the following figure. If the viral RNA cannot be reverse transcribed into the DNA form, the virus will not be able to replicate and can be considered dead.

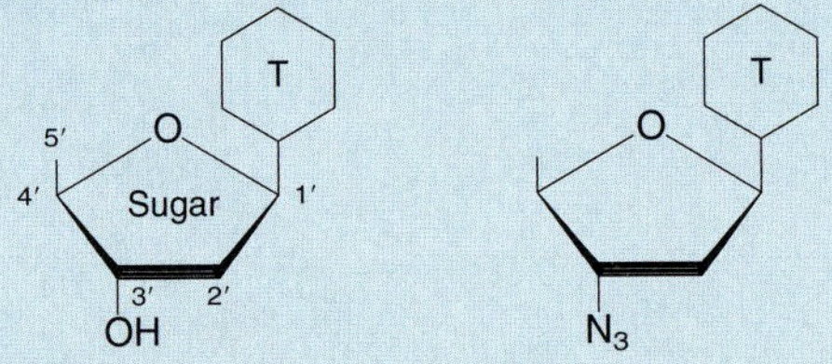

Comparison of the structures of the normal nucleoside, 2'-deoxythymidine, and the nucleoside analog, 3'-azido-2', 3'-dideoxythymidine.

AZT is particularly effective because the HIV reverse transcriptase actually prefers it to the normal nucleotide, thymidine. Nonetheless, AZT is not a cure. At best, it prolongs the life of a person with AIDS for a year or two. Eventually, however, AZT has a negative effect on the body. The cells of our bone

---

found in DNA. However, RNA molecules differ from DNA molecules in three basic properties.

- RNA molecules are usually *single-stranded*.
- The sugar-phosphate backbone of RNA consists of *ribonucleotides* linked by 3'–5' phosphodiester bonds. Thus, the sugar *ribose* is found in place of 2'-deoxyribose.
- The nitrogenous base *uracil* (U) replaces thymine (T).

Although RNA molecules are single-stranded, base pairing can still occur between uracil and adenine and between guanine and cytosine. We will show the importance of this property as we examine the way in which RNA molecules are involved in the expression of the genetic information in DNA.

17.3 DNA Replication

3 Learning Goal
Explain DNA replication.

DNA must be replicated before a cell divides so that each daughter cell inherits a copy of each gene. A cell that is missing a critical gene will die, just as an individual with a genetic disease, a defect in an important gene, may die early in life.

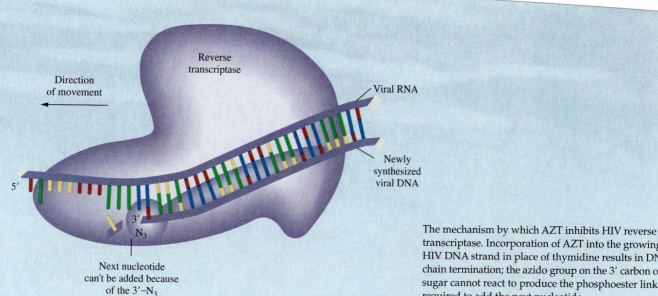

Direction of movement

Reverse transcriptase

Viral RNA

Newly synthesized viral DNA

5'

3'
N_3

Next nucleotide can't be added because of the 3'–N_3

The mechanism by which AZT inhibits HIV reverse transcriptase. Incorporation of AZT into the growing HIV DNA strand in place of thymidine results in DNA chain termination; the azido group on the 3' carbon of the sugar cannot react to produce the phosphoester linkage required to add the next nucleotide.

marrow are constantly dividing to produce new blood cells: red blood cells to carry oxygen to the tissues, white blood cells of the immune system, and platelets for blood clotting. For cells to divide, they must replicate their DNA. The DNA polymerases of these dividing cells also accidentally incorporate AZT into the growing DNA chains with the result that cells of the bone marrow begin to die. This can result in anemia and even further depression of the immune response.

Another problem that has arisen with prolonged use of AZT is that AZT-resistant mutants of the virus appear. It is well known that HIV is a virus that mutates rapidly. Some of these mutant forms of the virus have an altered reverse transcriptase that will no longer use AZT. When these mutants appear, AZT is no longer useful in treating the infection.

Fortunately, research with other nucleoside analogs, alternative types of antiviral treatments, and combinations of drugs has provided a more effective means of treating HIV infection.

FOR FURTHER UNDERSTANDING

AZT is a very effective competitive inhibitor of reverse transcriptase. Review what you have learned about competitive inhibition and structural analogs in Chapter 16 and develop a hypothesis to explain why AZT is so effective.

As noted in this connection, resistance to AZT is a common problem. This has led to the simultaneous use of two or more drugs in the treatment of HIV AIDS, for instance, AZT and a protease inhibitor. Explain why multiple drug therapy reduces the problem of viral drug resistance.

Thus, it is essential that the process of DNA replication produces an absolutely accurate copy of the original genetic information. If mistakes are made in critical genes, the result may be lethal mutations.

The structure of the DNA molecule suggested the mechanism for its accurate replication. Because adenine can base pair only with thymine and cytosine with guanine, Watson and Crick first suggested that an enzyme could "read" the nitrogenous bases on one strand of a DNA molecule and add complementary bases to a strand of DNA being synthesized. The product of this mechanism would be a new DNA molecule in which one strand is the original, or parent, strand and the second strand is a newly synthesized, or daughter, strand. This mode of DNA replication is called **semiconservative replication** (Figure 17.5).

Experimental evidence for this mechanism of DNA replication was provided by an experiment designed by Matthew Meselson and Franklin Stahl in 1958. *Escherichia coli* cells were grown in a medium in which $^{15}NH_4^+$ was the sole nitrogen source. ^{15}N is a nonradioactive, heavy isotope of nitrogen. Thus, growing the cells in this medium resulted in all of the cellular DNA containing this heavy isotope.

The cells containing only $^{15}NH_4^+$ were then added to a medium containing only the abundant isotope of nitrogen, $^{14}NH_4^+$, and were allowed to grow for one cycle of cell division. When the daughter DNA molecules were isolated and

Animation
DNA Replication
Bidirectional DNA Replication

Isotopes are atoms of the same element having the same number of protons but different numbers of neutrons and, therefore, different mass numbers.

Animation
Meselson and Stahl Experiment

Figure 17.5

In semiconservative DNA replication, each parent strand serves as a template for the synthesis of a new daughter strand.

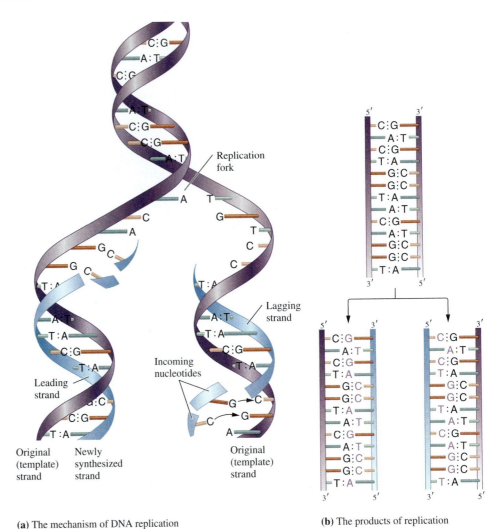

(a) The mechanism of DNA replication

(b) The products of replication

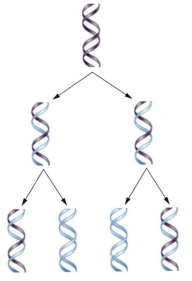

Figure 17.6

Representation of the Meselson and Stahl experiment. The DNA from cells grown in medium containing $^{15}NH_4^+$ is shown in purple. After a single generation in a medium containing $^{14}NH_4^+$, the daughter DNA molecules have one ^{15}N-labeled parent strand and one ^{14}N-labeled daughter strand (blue). After a second generation in a $^{14}NH_4^+$ containing medium, there are equal numbers of $^{14}N/^{15}N$ DNA molecules and $^{14}N/^{14}N$ DNA molecules.

analyzed, it was found that each was made up of one strand of "heavy" DNA, the parental strand, and one strand of "light" DNA, the new daughter strand. After a second round of cell division, half of the isolated DNA contained no ^{15}N and half contained a 50/50 mixture of ^{14}N and ^{15}N-labeled DNA (Figure 17.6). This demonstrated conclusively that each parental strand of the DNA molecule serves as the template for the synthesis of a daughter strand and that each newly synthesized DNA molecule is composed of one parental strand and one newly synthesized daughter strand.

17.4 Information Flow in Biological Systems

The **central dogma** of molecular biology states that in cells the flow of genetic information contained in DNA is a one-way street that leads from DNA to RNA to protein. The process by which a single strand of DNA serves as a template for the synthesis of an RNA molecule is called **transcription**. The word *transcription* is derived from the Latin word *transcribere* and simply means "to make a copy." Thus, in this process, part of the information in the DNA is copied into a strand of RNA. The process by which the message is converted into protein is called **translation**. Unlike transcription, the process of translation involves converting the information from one language to another. In this case, the genetic information in the linear sequence of nucleotides is being translated into a protein, a linear sequence of amino acids. The expression of the information contained in DNA is fundamental to the growth, development, and maintenance of all organisms.

Classes of RNA Molecules

Three classes of RNA molecules are produced by transcription: messenger RNA, transfer RNA, and ribosomal RNA.

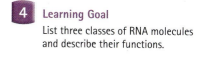

Learning Goal

List three classes of RNA molecules and describe their functions.

1. **Messenger RNA (mRNA)** carries the genetic information for a protein from DNA to the ribosomes. It is a complementary RNA copy of a gene on the DNA.
2. **Ribosomal RNA (rRNA)** is a structural and functional component of the ribosomes, which are "platforms" on which protein synthesis occurs. There are three types of rRNA molecules in bacterial ribosomes and four in the ribosomes of eukaryotes.
3. **Transfer RNA (tRNA)** translates the genetic code of the mRNA into the primary sequence of amino acids in the protein. In addition to the primary structure, tRNA molecules have a cloverleaf-shaped secondary structure resulting from base pair hydrogen bonding (A—U and G—C) and a roughly L-shaped tertiary structure (Figure 17.7). The sequence CCA is found at the 3' end of the tRNA. The 3'—OH group of the terminal nucleotide, adenosine, can

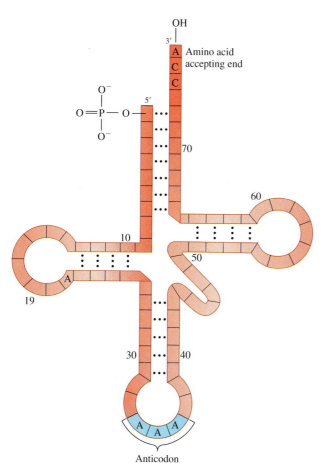

Figure 17.7

Structure of tRNA. The primary structure of a tRNA is the linear sequence of ribonucleotides. Here we see the hydrogen-bonded secondary structure of a tRNA showing the three loops and the amino acid accepting end.

be covalently attached to an amino acid. Three nucleotides at the base of the cloverleaf structure form the **anticodon.** This triplet of bases forms hydrogen bonds to a **codon** (complementary sequence of bases) on a messenger RNA (mRNA) molecule on the surface of a ribosome during protein synthesis. This hydrogen bonding of codon and anticodon brings the correct amino acid to the site of protein synthesis at the appropriate location in the growing peptide chain.

Transcription

5 Learning Goal

Explain the process of transcription.

Animation
mRNA Synthesis (Transcription)

Transcription, shown in Figure 17.8, is catalyzed by the enzyme, **RNA polymerase.** The process occurs in three stages. The first, called *initiation*, involves binding RNA polymerase to a specific nucleotide sequence, the **promoter,** at the beginning of a gene. This interaction of RNA polymerase with specific promoter DNA sequences allows RNA polymerase to recognize the starting point for transcription. It also determines which DNA strand will be transcribed. Unlike DNA replication, transcription produces a complementary copy of only one of the two strands of DNA. As it binds to the DNA, RNA polymerase separates the two strands of DNA so that it can "read" the base sequence of the DNA.

The second stage, chain elongation, begins as the RNA polymerase "reads" the DNA template strand and catalyzes the polymerization of a complementary RNA copy. With each catalytic step, RNA polymerase transfers a complementary ribonucleotide to the end of the growing RNA chain and catalyzes the formation of

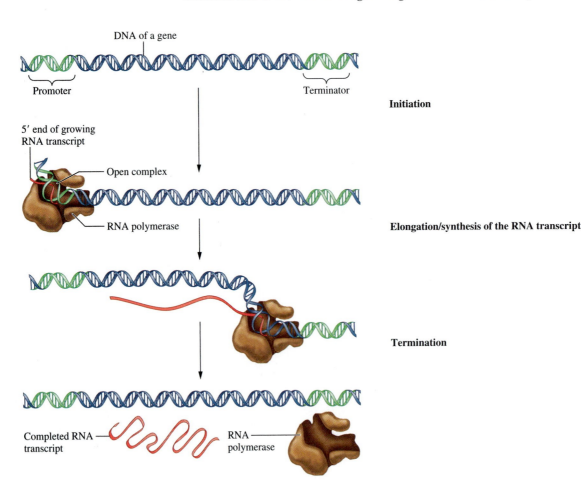

Figure 17.8

The stages of transcription.

a 3′–5′ phosphodiester bond between the 5′ phosphoryl group of the incoming ribonucleotide and the 3′ hydroxyl group of the last ribonucleotide of the growing RNA chain.

The final stage of transcription is termination. The RNA polymerase finds a termination sequence at the end of the gene and releases the newly formed RNA molecule.

QUESTION 17.3

What is the function of RNA polymerase in the process of transcription?

QUESTION 17.4

What is the function of the promoter sequence in the process of transcription?

Post-Transcriptional Processing of RNA

In bacteria, termination releases a mature mRNA for translation. But in more complex organisms, such as humans, transcription produces a **primary transcript** that must undergo extensive **post-transcriptional modification** before it is exported from the nucleus for translation in the cytoplasm.

Primary transcripts undergo three post-transcriptional modifications. These are the addition of a 5′ cap structure and a 3′ poly(A) tail, and RNA splicing.

In the first modification, a **cap structure** is enzymatically added to the 5′ end of the primary transcript. The cap structure (Figure 17.9) consists of 7-methylguanosine

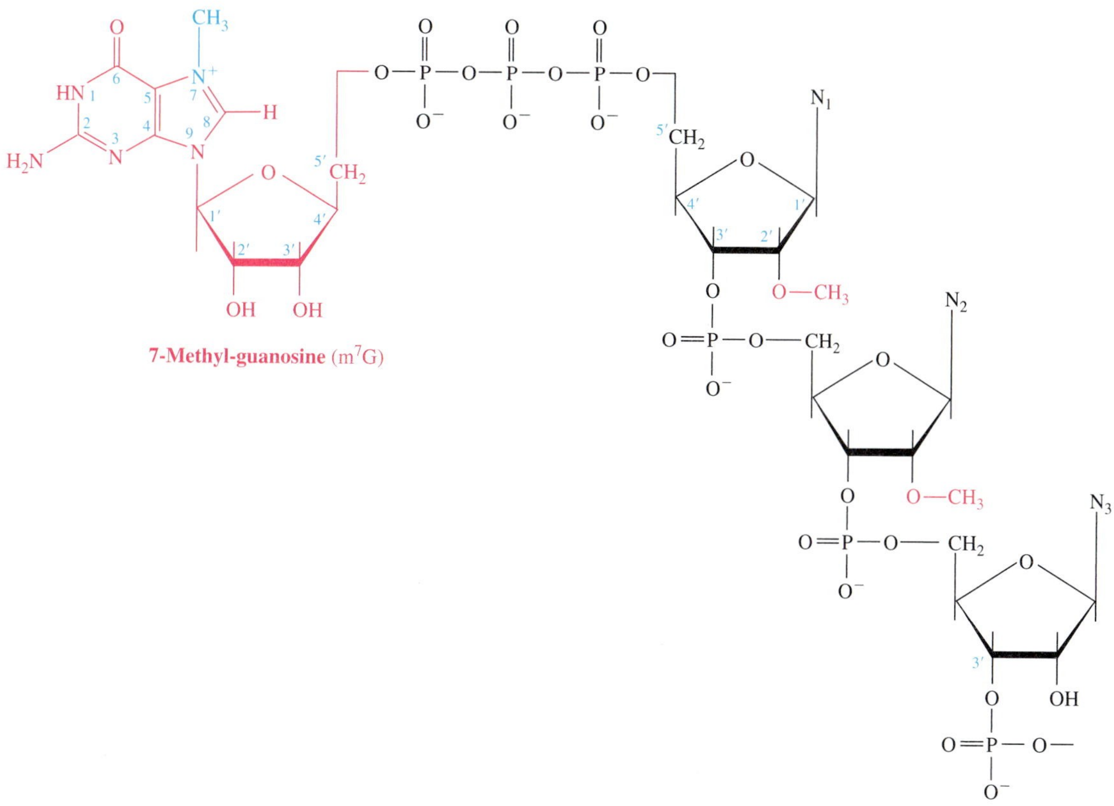

6 **Learning Goal**

List and explain the three types of post-transcriptional modifications of eukaryotic mRNA.

7-Methyl-guanosine (m^7G)

Figure 17.9

The 5′-methylated cap structure of eukaryotic mRNA.

attached to the 5′ end of the RNA by a 5′–5′ triphosphate bridge. The first two nucleotides of the mRNA are also methylated. The cap structure is required for efficient translation of the final mature mRNA.

The second modification is the enzymatic addition of a **poly(A)** tail to the 3′ end of the transcript. *Poly(A) polymerase* uses ATP and catalyzes the stepwise polymerization of one hundred to two hundred adenosine nucleotides on the 3′ end of the RNA. The poly(A) tail protects the 3′ end of the mRNA from enzymatic degradation and thus prolongs the lifetime of the mRNA.

The third modification, **RNA splicing,** involves the removal of portions of the primary transcript that are not protein coding. Bacterial genes are continuous; all nucleotide sequences of the gene are found in the mRNA. However, study of the gene structure of eukaryotes revealed a fascinating difference. Eukaryotic genes are discontinuous; there are *extra* DNA sequences within these genes that do not encode any amino acid sequences for the protein. These sequences are called *intervening sequences* or **introns.** The primary transcript contains both the introns and the protein coding sequences, called **exons.** The presence of introns in the mRNA would make it impossible for the process of translation to synthesize the correct protein. Therefore they must be removed, which is done by RNA splicing.

One of the first genes shown to contain introns was the gene for the β subunit of adult hemoglobin (Figure 17.10). On the DNA, the gene for β-hemoglobin is 1200 nucleotides long, but only 438 nucleotides carry the genetic information for protein. The remaining sequences are found in two introns of 116 and 646 nucleotides that are removed by splicing before translation. It is interesting that the larger intron is longer than the final β-hemoglobin mRNA! In the genes that have been studied, introns have been found that range in size from 50 to 20,000 nucleotides in length, and there may be many throughout a gene. Thus, a typical human gene might be ten to thirty times longer than the final mRNA.

Animation
How Spliceosomes Process RNA

Eukaryotes are organisms having cells with a true nucleus surrounded by a nuclear membrane.

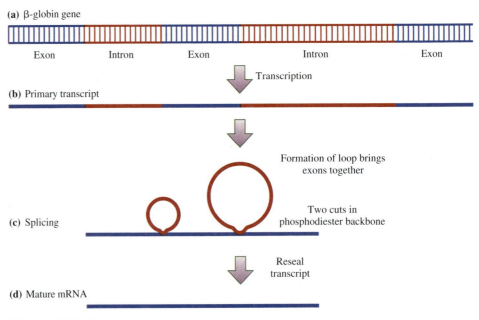

(a) β-globin gene

Exon   Intron   Exon   Intron   Exon

Transcription

(b) Primary transcript

(c) Splicing

Formation of loop brings exons together

Two cuts in phosphodiester backbone

Reseal transcript

(d) Mature mRNA

Figure 17.10

Schematic diagram of mRNA splicing. (a) The β-globin gene contains protein coding exons, as well as noncoding sequences called introns. (b) The primary transcript of the DNA carries both the introns and the exons. (c) The introns are looped out, the phosphodiester backbone of the mRNA is cut twice, and the pieces are tied together. (d) The final mature mRNA now carries only the coding sequences (exons) of the gene.

17.5 The Genetic Code

The mRNA carries the genetic code for a protein. But what is the nature of this code? In 1954, George Gamow proposed that because there are only four "letters" in the DNA alphabet (A, T, G, and C) and because there are twenty amino acids, the genetic code must contain words made of at least three letters taken from the four letters in the DNA alphabet. How did he come to this conclusion? He reasoned that a code of two-letter words constructed from any combination of the four letters has a "vocabulary" of only sixteen words (4^2). In other words, there are only sixteen different ways to put A, T, C, and G together two bases at a time (AA, AT, AC, AG, TT, TA, etc.). That is not enough to encode all twenty amino acids. A code of four-letter words gives 256 words (4^4), far more than are needed. A code of three-letter words, however, has a possible vocabulary of sixty-four words (4^3), sufficient to encode the twenty amino acids but not too excessive.

A series of elegant experiments proved that Gamow was correct by demonstrating that the genetic code is a triplet code. Mutations were introduced into the DNA of a bacterial virus. These mutations inserted (or deleted) one, two, or three nucleotides into a gene. The researchers then looked for the protein encoded by that gene. When one or two nucleotides were inserted, no protein was produced. However, when a third base was inserted, the sense of the mRNA was restored, and the protein was made. You can imagine this experiment by using a sentence composed of only three-letter words. For instance,

THE CAT RAN OUT

What happens to the "sense" of the sentence if we insert one letter?

THE FCA TRA NOU T

The reading frame of the sentence has been altered, and the sentence is now nonsense. Can we now restore the sense of the sentence by inserting a second letter?

THE FAC ATR ANO UT

No, we have not restored the sense of the sentence. Once again, we have altered the reading frame, but because our code has only three-letter words, the sentence is still nonsense. If we now insert a third letter, it should restore the correct reading frame:

THE FAT CAT RAN OUT

By inserting three new letters, we have restored the sense of the message by restoring the reading frame. This is exactly the way in which the message of the mRNA is interpreted. Each group of three nucleotides in the sequence of the mRNA is called a *codon,* and each codes for a single amino acid. If the sequence is interrupted or changed, it can change the amino acid composition of the protein that is produced or even result in the production of no protein at all.

As we noted, a three-letter genetic code contains sixty-four words, called *codons,* but there are only twenty amino acids. Thus, there are forty-four more codons than are required to specify all of the amino acids found in proteins. Three of the codons—UAA, UAG, and UGA—specify termination signals for the process of translation. But this still leaves us with forty-one additional codons. What is the function of the "extra" code words? Francis Crick (recall Watson and Crick and the double helix) proposed that the genetic code is a **degenerate code.** The term *degenerate* is used to indicate that different triplet codons may serve as code words for the same amino acid.

The complete genetic code is shown in Figure 17.11. We can make several observations about the genetic code. First, methionine and tryptophan are the only amino acids that have a single codon. All others have at least two codons, and serine

7 **Learning Goal**
Describe the essential elements of the genetic code and develop a "feel" for its elegance.

FIRST BASE	SECOND BASE				THIRD BASE
	U	C	A	G	
U	UUU Phenylalanine	UCU Serine	UAU Tyrosine	UGU Cysteine	U
U	UUC Phenylalanine	UCC Serine	UAC Tyrosine	UGC Cysteine	C
U	UUA Leucine	UCA Serine	UAA STOP	UGA STOP	A
U	UUG Leucine	UCG Serine	UAG STOP	UGG Tryptophan	G
C	CUU Leucine	CCU Proline	CAU Histidine	CGU Arginine	U
C	CUC Leucine	CCC Proline	CAC Histidine	CGC Arginine	C
C	CUA Leucine	CCA Proline	CAA Glutamine	CGA Arginine	A
C	CUG Leucine	CCG Proline	CAG Glutamine	CGG Arginine	G
A	AUU Isoleucine	ACU Threonine	AAU Asparagine	AGU Serine	U
A	AUC Isoleucine	ACC Threonine	AAC Asparagine	AGC Serine	C
A	AUA Isoleucine	ACA Threonine	AAA Lysine	AGA Arginine	A
A	AUG (START) Methionine	ACG Threonine	AAG Lysine	AGG Arginine	G
G	GUU Valine	GCU Alanine	GAU Aspartic acid	GGU Glycine	U
G	GUC Valine	GCC Alanine	GAC Aspartic acid	GGC Glycine	C
G	GUA Valine	GCA Alanine	GAA Glutamic acid	GGA Glycine	A
G	GUG Valine	GCG Alanine	GAG Glutamic acid	GGG Glycine	G

Cys — ACG tRNA

Asn — UUG tRNA

Figure 17.11

The genetic code. The table shows the possible codons found in mRNA. To read the universal biological language from this chart, find the first base in the column on the left, the second base from the row across the top, and the third base from the column to the right. This will direct you to one of the sixty-four squares in the matrix. Within that square, you will find the codon and the amino acid that it specifies. In the cell, this message is decoded by tRNA molecules like those shown to the right of the table.

and leucine have six codons each. The genetic code is also somewhat mutation-resistant. For those amino acids that have multiple codons, the first two bases are often identical and thus identify the amino acid; only the third position is variable. Mutations—changes in the nucleotide sequence—in the third position therefore often have no effect on the amino acid that is incorporated into a protein.

QUESTION 17.5

Why is the genetic code said to be degenerate?

Why is the genetic code said to be mutation-resistant?

17.6 Protein Synthesis

The process of protein synthesis is called *translation*. It involves translating the genetic information from the sequence of nucleotides into the sequence of amino acids in the primary structure of a protein. Figure 17.12 shows the relationship through which the nucleotide sequence of a DNA molecule is transcribed into a complementary sequence of ribonucleotides, the mRNA molecule. Each mRNA has a short untranslated region followed by the sequences that carry the information for the order of the amino acids in the protein that will be produced in the process of translation. That genetic information is the sequence of codons along the mRNA. The decoding process is carried out by tRNA molecules.

Translation is carried out on **ribosomes,** which are complexes of ribosomal RNA (rRNA) and proteins. Each ribosome is made up of two subunits: a small and a large ribosomal subunit (Figure 17.13a). In eukaryotic cells, the small ribosomal subunit contains one rRNA molecule and thirty-three different ribosomal proteins, and the large subunit contains three rRNA molecules and about forty-nine different proteins.

Protein synthesis involves the simultaneous action of many ribosomes on a single mRNA molecule. These complexes of many ribosomes along a single mRNA are known as *polyribosomes* or **polysomes** (Figure 17.13b). Each ribosome is synthesizing one copy of the protein molecule encoded by the mRNA. Thus, many copies of a protein are produced simultaneously.

The Role of Transfer RNA

The codons of mRNA must be read if the genetic message is to be translated into protein. The molecule that decodes the information in the mRNA molecule into

8 **Learning Goal**
Describe the process of translation.

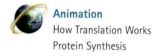

Animation
How Translation Works
Protein Synthesis

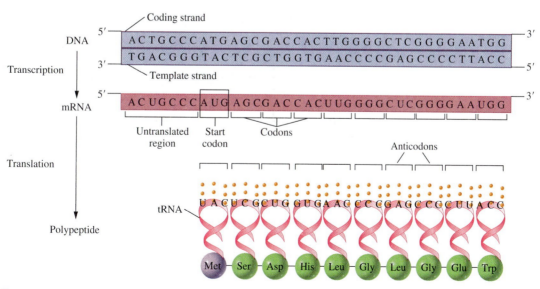

Figure 17.12

Messenger RNA (mRNA) is an RNA copy of one strand of a gene in the DNA. Each codon on the mRNA that specifies a particular amino acid is recognized by the complementary anticodon on a transfer RNA (tRNA).

Figure 17.13

Structure of the ribosome. (a) The large and small subunits form the functional complex in association with an mRNA molecule. (b) A polyribosome translating the mRNA for a β-globin chain of hemoglobin.

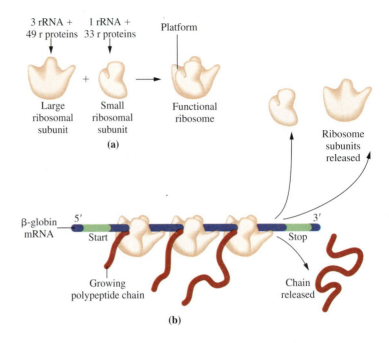

(a)

(b)

the primary structure of a protein is transfer RNA (tRNA). To decode the genetic message into the primary sequence of a protein, the tRNA must faithfully perform two functions.

First, the tRNA must covalently bind one, and only one, specific amino acid. Each tRNA is specifically recognized by the active site of an enzyme called an **aminoacyl tRNA synthetase.** This enzyme also recognizes the correct amino acid and covalently links the amino acid to the 3′ end of the tRNA molecule. The resulting structure is called an **aminoacyl tRNA.**

Second, the tRNA must be able to recognize the appropriate codon on the mRNA that calls for that amino acid. This occurs through hydrogen bonding between the anticodon of the tRNA and the codon of the mRNA.

Animation
Aminoacyl tRNA Synthetase

QUESTION 17.7

How are codons related to anticodons?

QUESTION 17.8

If the sequence of a codon on the mRNA is 5′-AUG-3′, what will be the sequence of the anticodon? Remember that hydrogen bonding rules require antiparallel strands. It is easiest to write the anticodon first 3′ → 5′ and then reverse it to the 5′ → 3′ order.

The Process of Translation

Initiation

The first stage of protein synthesis is *initiation.* Proteins called **initiation factors** assist in the formation of a translation complex composed of an mRNA molecule, the small and large ribosomal subunits, and the initiator tRNA. This initiator tRNA recognizes the codon AUG and carries the amino acid methionine.

The ribosome has two sites for binding tRNA molecules. The first site, called the **peptidyl tRNA binding site (P-site),** holds the peptidyl tRNA, the growing peptide

8 Learning Goal
Describe the process of translation.

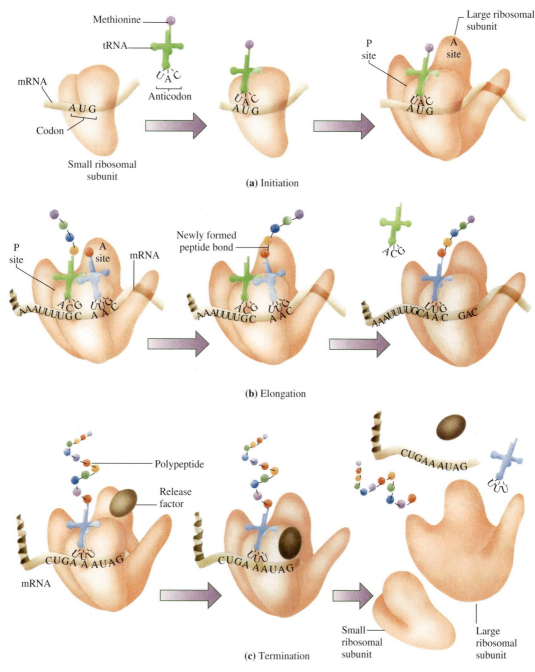

Figure 17.14

(a) Formation of an initiation complex sets protein synthesis in motion. The mRNA and proteins, called initiation factors, bind to the small ribosomal subunit. Next, a charged methionyl tRNA molecule binds, and finally, the initiation factors are released, and the large subunit binds. (b) The elongation phase of protein synthesis involves addition of new amino acids to the C-terminus of the growing peptide. An aminoacyl tRNA molecule binds at the empty A-site, and the peptide bond is formed. The uncharged tRNA molecule is released, and the peptidyl tRNA is shifted to the P-site as the ribosome moves along the mRNA. (c) Termination of protein synthesis occurs when a release factor binds the stop codon on mRNA. This leads to hydrolysis of the ester bond linking the peptide to the peptidyl tRNA molecule in the P-site. The ribosome then dissociates into its two subunits, releasing the mRNA and the newly synthesized peptide.

bound to a tRNA molecule. The second site, called the **aminoacyl tRNA binding site (A-site),** holds the aminoacyl tRNA carrying the next amino acid to be added to the peptide chain. Each of the tRNA molecules is hydrogen bonded to the mRNA molecule by codon–anticodon complementarity. The entire complex is further stabilized by the fact that the mRNA is also bound to the ribosome. Figure 17.14a shows

A MEDICAL Connection | The Ames Test for Carcinogens

Each day we come into contact with a variety of chemicals, including insecticides, food additives, hair dyes, automobile emissions, and cigarette smoke. Some of these chemicals have the potential to cause cancer. How do we determine whether these agents are harmful? More particularly, how do we determine whether they cause cancer?

If we consider the example of cigarette smoke, we see that it can be years, even centuries, before a relationship is seen between a chemical and cancer. Europeans and Americans have been smoking since Sir Walter Raleigh introduced tobacco into England in the seventeenth century. However, it was not until three centuries later that physicians and scientists demonstrated the link between smoking and lung cancer. Obviously, this epidemiological approach takes too long, and too many people die. Alternatively, we can test chemicals by treating laboratory animals, such as mice, and observing them for various kinds of cancer. However, this, too, can take years, is expensive, and requires the sacrifice of many laboratory animals. How, then, can chemicals be tested for carcinogenicity (the ability to cause cancer) quickly and inexpensively? In the 1970s, it was recognized that most carcinogens are also mutagens, that is, they cause cancer by causing mutations in the DNA, and the mutations cause the cells of the body to lose growth control. Bruce Ames, a biochemist and bacterial geneticist, developed a test using mutants of the bacterium *Salmonella typhimurium* that can demonstrate in 48–72 hours whether a chemical is a mutagen and thus a suspected carcinogen.

Ames chose several mutants of *S. typhimurium* that cannot grow unless the amino acid, histidine, is added to the growth medium. The Ames test involves subjecting these bacteria to a chemical and determining whether the chemical causes reversion of the mutation. In other words, the researcher is looking for a mutation that reverses the original mutation. When a re-version occurs, the bacteria will be able to grow in the absence of histidine.

The details of the Ames test are shown in the accompanying figure. Both an experimental and a control test are done. The control test contains no carcinogen and will show the number of spontaneous revertants that occur in the culture. If there are many colonies on the surface of the experimental plate and only a few colonies on the negative control plate, it can be concluded that the chemical tested is a mutagen. It is therefore possible that the chemical is also a carcinogen.

The Ames test has greatly accelerated our ability to test new compounds for mutagenic and possibly carcinogenic effects. However, once the Ames test identifies a mutagenic compound, testing in animals must be done to show conclusively that the compound also causes cancer.

FOR FURTHER UNDERSTANDING

A researcher carried out the Ames test in which an experimental sample was exposed to a suspected mutagen and a control sample was not. A sample from each tube was grown on a medium containing no histidine. On the experimental plate, he observed forty-three colonies and on the control plate, he observed thirty-one colonies. He concluded that the substance is a mutagen. When he reported his data and conclusion to his supervisor, she told him that his conclusions were not valid. How can the researcher modify his experimental procedure to obtain better data?

Suppose that the mutation in a strain of *S. typhimurium* produces the codon UUA instead of UUC. What is the amino acid change caused by this mutation? What base substitutions could correct the mutant codon so that once again it calls for the correct amino acid? What base substitutions would not correct the mutant codon?

---

the series of events that result in the formation of the initiation complex. The initiator methionyl tRNA occupies the P-site in this complex.

Chain Elongation

The second stage of translation is *chain elongation*. This occurs in three steps that are repeated until protein synthesis is complete. We enter the action after a tetrapeptide has already been assembled, and a peptidyl tRNA occupies the P-site (Figure 17.14b).

The first event is binding an aminoacyl-tRNA molecule to the empty A-site. Next, peptide bond formation occurs. This is catalyzed by an enzyme on the ribosome called *peptidyl transferase*. Now the peptide chain is shifted to the tRNA that occupies the A-site. Finally, the tRNA in the P-site falls away, and the ribosome changes positions so that the next codon on the mRNA occupies the A-site. This

Recent evidence indicates that the peptidyl transferase is a catalytic region of the 28S ribosomal RNA.

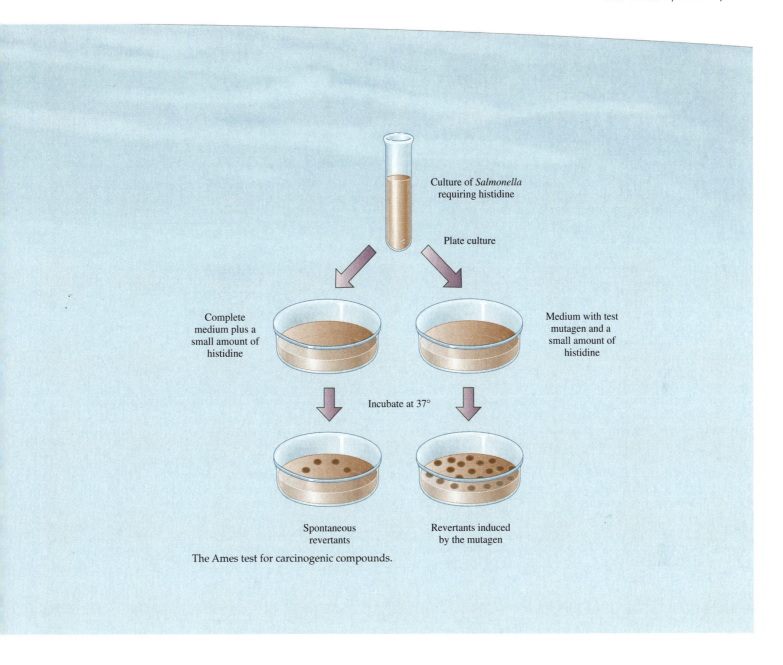

Culture of *Salmonella* requiring histidine

Plate culture

Complete medium plus a small amount of histidine

Medium with test mutagen and a small amount of histidine

Incubate at 37°

Spontaneous revertants

Revertants induced by the mutagen

The Ames test for carcinogenic compounds.

movement of the ribosome is called **translocation.** The process shifts the new peptidyl tRNA from the A-site to the P-site. The chain elongation stage of translation requires the hydrolysis of GTP to GDP and P_i. Several **elongation factors** are also involved in this process.

Termination

The last stage of translation is *termination.* There are three **termination codons**—UAA, UAG, and UGA—for which there are no corresponding tRNA molecules. When one of these "stop" codons is encountered, translation is terminated. A **release factor** binds the empty A-site. The peptidyl transferase that had previously catalyzed peptide bond formation hydrolyzes the ester bond between the peptidyl tRNA and the last amino acid of the newly synthesized protein (Figure 17.14c). At

CHEMISTRY at the Crime Scene | DNA Fingerprinting

Four U.S. Army helicopters swept over the field of illicit coca plants (*Erythroxylum* spp.) growing in a mountainous region of northern Colombia. When the soldiers were certain that the fields were unguarded, a fifth helicopter landed. From it emerged Dr. Jim Saunders, currently director of the Molecular Biology, Biochemistry, and Bioinformatics Program at Towson University, and a team of researchers from the Agricultural Research Service of the U.S. Department of Agriculture (ARS-USDA). Quickly, the scientists gathered leaves from mature plants, as well as from seedlings growing in a coca nursery, and returned to the helicopter with their valuable samples. With a final sweep over the field, the Army helicopters sprayed herbicides to kill the coca plants.

From 1997 to 2001, this scene was repeated in regions of Colombia known to have the highest coca production. The reason for these collections was to study the genetic diversity of the coca plants being grown for the illegal production of cocaine. The tool selected for this study was DNA fingerprinting.

DNA fingerprinting was developed in the 1980s by Alec Jeffries of the University of Leicester in England. The idea grew out of basic molecular genetic studies of the human genome. Scientists observed that some DNA sequences varied greatly from one person to the next. Such hypervariable regions are made up of variable numbers of repeats of short DNA sequences. They are located at many sites on different chromosomes. Each person has a different number of repeats and when his or her DNA is digested with restriction enzymes, a unique set of DNA fragments is generated. Jeffries invented DNA fingerprinting by developing a set of DNA probes that detect these

Coca nursery next to a mature field in Colombia.

variable number tandem repeats (VNTRs) when used in hybridization with Southern blots.

Although several variations of DNA fingerprinting exist, the basic technique is quite simple. DNA is digested with restriction enzymes, producing a set of DNA fragments. These are separated by electrophoresis through an agarose gel. The DNA fragments are then transferred to membrane filters and hybridized

this point, the tRNA, the newly synthesized peptide, and the two ribosomal subunits are released.

QUESTION 17.9

What is the function of the ribosomal P-site in protein synthesis?

QUESTION 17.10

What is the function of the ribosomal A-site in protein synthesis?

The quaternary structure of hemoglobin is described in Section 16.8.

The peptide that is released following translation is not necessarily in its final functional form. In some cases, the peptide is proteolytically cleaved before it becomes functional. Synthesis of digestive enzymes uses this strategy. Sometimes the protein must associate with other peptides to form a functional protein, as in the case of hemoglobin. Cellular enzymes add carbohydrate or lipid groups to some proteins, especially those that will end up on the cell surface. These final modifications

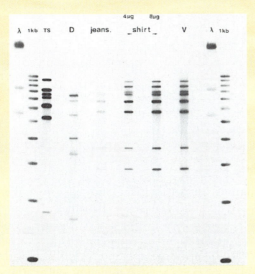

An example of a DNA fingerprint used in a criminal case. The DNA sample designated *V* is that of the victim and the sample designated *D* is that of the defendant. The samples labeled *jeans* and *shirt* were taken from the clothing of the defendant. The DNA bands from the defendant's clothing clearly match the DNA bands of the victim, providing evidence of the guilt of the defendant.

with the radioactive probe DNA. The bands that hybridize the radioactive probe are visualized by exposing the membrane to X-ray film and developing a "picture" of the gel. The result is what Jeffries calls a *DNA fingerprint,* a set of twenty-five to sixty DNA bands that are unique to an individual.

DNA fingerprinting is now routinely used for paternity testing, testing for certain genetic disorders, and identification of the dead in cases where no other identification is available. DNA fingerprints are used as evidence in criminal cases in-

volving rape and murder. In such cases, the evidence may be little more than a hair with an intact follicle on the clothing of the victim.

Less widely known is the use of DNA fingerprinting to study genetic diversity in natural populations of plants and animals. The greater the genetic diversity, the healthier the population is likely to be. Populations with low genetic diversity face a far higher probability of extinction under adverse conditions. Customs officials have used DNA fingerprinting to determine whether confiscated elephant tusks were taken illegally from an endangered population of elephants or were obtained from a legally harvested population.

In the case of the coca plants, ARS wanted to know whether the drug cartels were developing improved strains that might be hardier or more pest resistant or that have a higher concentration of cocaine. Their conclusions, which you can read in *Phytochemistry* (64: 187–197, 2003), were that the drug cartels have introduced significant genetic modification into coca plants in Colombia in the last two decades. In addition, some of these new variants, those producing the highest levels of cocaine, have been transplanted to other regions of the country. All of this indicates that the cocaine agribusiness is thriving.

FOR FURTHER UNDERSTANDING

As this sampling of applications suggests, DNA fingerprinting has become an invaluable tool in law enforcement, medicine, and basic research. What other applications of this technology can you think of?

Do some research on the development of DNA fingerprinting as a research and forensics tool. What is the probability that two individuals will have the same DNA fingerprint? How are these probabilities determined?

are specific for particular proteins and, like the sequence of the protein itself, are directed by the cellular genetic information.

17.7 Mutation, Ultraviolet Light, and DNA Repair

The Nature of Mutations

Changes can occur in the nucleotide sequence of a DNA molecule. Such a genetic change is called a **mutation.** Mutations can arise from mistakes made by DNA polymerase during DNA replication. They also result from the action of chemicals, called **mutagens,** that damage the DNA.

Mutations are classified by the kind of change that occurs in the DNA. The substitution of a single nucleotide for another is called a **point mutation:**

9 **Learning Goal**
Define mutation and understand how mutations can cause cancer and cell death.

ATGG̲ACTTC: normal DNA sequence

ATGC̲ACTTC: point mutation

Animation

Addition and Deletion Mutations
Mutation by Base Substitution

Sometimes a single nucleotide or even large sections of DNA are lost. These are called **deletion mutations:**

ATGGACTTC:	normal DNA sequence
ATGTTC:	deletion mutation

Occasionally, one or more nucleotides are added to a DNA sequence. These are called **insertion mutations:**

ATGGACTTC:	normal DNA sequence
ATGCTCGACTTC:	insertion mutation

The Results of Mutations

Some mutations are **silent mutations,** that is, they cause no change in the protein. Often, however, a mutation has a negative effect on the health of the organism. The effect of a mutation depends on how it alters the genetic code for a protein. Consider the two codons for glutamic acid: GAA and GAG. A point mutation that alters the third nucleotide of GAA to GAG will still result in the incorporation of glutamic acid at the correct position in the protein. Similarly, a GAG to GAA mutation will also be silent.

Many mutations are not silent. There are approximately four thousand human genetic disorders that result from such mutations. These occur because the mutation in the DNA changes the codon and results in incorporation of the wrong amino acid into the protein. This causes the protein to be nonfunctional or to function improperly.

Consider the human genetic disease sickle cell anemia. In the normal β-chain of hemoglobin, the sixth amino acid is glutamic acid. In the β-chain of sickle cell hemoglobin, the sixth amino acid is valine. How did this amino acid substitution arise? The answer lies in examination of the codons for glutamic acid and valine:

Glutamic acid:	GAA or GAG
Valine:	GUG, GUC, GUA, or GUU

A point mutation of A → U in the second nucleotide changes some codons for glutamic acid into codons for valine:

GAA	⟶	GUA
GAG	⟶	GUA
Glutamic acid codon		Valine codon

This mutation in a single codon leads to the change from glutamic acid to valine in the amino acid sequence at position 6 in the β-chain of human hemoglobin. The result of this seemingly minor change is sickle cell anemia in individuals who inherit two copies of the mutant gene.

QUESTION 17.11

The sequence of a gene on the mRNA is normally AUGCCCGACUUU. A point mutation in the gene results in the mRNA sequence AUGCGCGACUUU. What are the amino acid sequences of the normal and mutant proteins? Would you expect this to be a silent mutation?

QUESTION 17.12

The sequence of a gene on the mRNA is normally AUGCCCGACUUU. A point mutation in the gene results in the mRNA sequence AUGCCGGACUUU. What are the amino acid sequences of the normal and mutant proteins? Would you expect this to be a silent mutation?

A MEDICAL Connection | A Genetic Approach to Familial Emphysema

Familial emphysema is a human genetic disease resulting from the inability to produce the protein α_1-antitrypsin. See also A Medical Connection: α_1-Antitrypsin and Familial Emphysema, in Chapter 16. In individuals who have inherited one or two copies of the α_1-antitrypsin gene, this serum protein protects the lungs from the enzyme elastase. Normally, elastase fights bacteria and helps in the destruction and removal of dead lung tissue. However, the enzyme can also cause lung damage. By inhibiting elastase, α_1-antitrypsin prevents lung damage. Individuals who have inherited two defective α_1-antitrypsin genes do not produce this protein and suffer from familial, or A1AD, emphysema. In the absence of α_1-antitrypsin, the elastase and other proteases cause the severe lung damage characteristic of emphysema.

A1AD is the second most common genetic disorder in Caucasians. It is estimated that there are 100,000 sufferers in the United States and that one in five Americans carries the gene. The disorder, discovered in 1963, is often misdiagnosed as asthma or chronic obstructive pulmonary disease. In fact, it is estimated that fewer than 5% of the sufferers are diagnosed with A1AD.

The α_1-antitrypsin gene has been cloned. Early experiments with sheep showed that the protein remains stable when administered as an aerosol and remains functional after it has passed through the pulmonary epithelium. This research offers hope of an effective treatment for this disease.

The current treatment involves weekly IV injections of α_1-antitrypsin. The supply of the protein, purified from human plasma that has been demonstrated to be virus free, is rather limited. Thus, the injections are expensive. In addition, they are painful. These two factors cause some sufferers to refuse the treatment.

Recently, Dr. Terry Flotte and his colleagues at the University of Florida have taken a new approach. They have cloned the gene for α_1-antitrypsin into the DNA of an adeno-associated virus. This virus is an ideal vector for human gene replacement therapy because it replicates only in cells that are not dividing and it does not stimulate a strong immune or inflammatory response. The researchers injected the virus carrying the cloned α_1-antitrypsin gene into the muscle tissue of mice, then tested for the level of α_1-antitrypsin in the blood. The results were very promising. Effective levels of α_1-antitrypsin were produced in the muscle cells of the mice and secreted into the bloodstream. Furthermore, the level of α_1-antitrypsin remained at therapeutic levels for more than four months. The research team is planning tests with larger animals and eventually will confirm their results in human trials.

FOR FURTHER UNDERSTANDING

Of the three treatments described in this connection, which do you think has the highest probability of success in the long term? Defend your answer.

Why is it impractical to "replace" the defective gene in an adult suffering from a genetic disease such as A1AD?

Mutagens and Carcinogens

Any chemical that causes a change in the DNA sequence is called a *mutagen*. Often, mutagens are also **carcinogens,** cancer-causing chemicals. Most cancers result from mutations in a single normal cell. These mutations result in the loss of normal growth control, causing the abnormal cell to proliferate. If that growth is not controlled or destroyed, it will result in the death of the individual. We are exposed to many carcinogens during our lives. Sometimes we are exposed to a carcinogen by accident, but in some cases it is by choice. There are about three thousand chemical components in cigarette smoke, and several are potent mutagens. As a result, people who smoke have a much greater chance of lung cancer than those who don't.

Ultraviolet Light Damage and DNA Repair

Ultraviolet (UV) light is another agent that causes damage to DNA. Absorption of UV light by DNA causes adjacent pyrimidine bases to become covalently linked. The product is called a **pyrimidine dimer.** As a result of pyrimidine dimer formation, there is no hydrogen bonding between these pyrimidine molecules and the complementary bases on the other DNA strand. This stretch of DNA cannot be replicated or transcribed!

Bacteria such as *Escherichia coli* have four different mechanisms to repair ultraviolet light damage. However, even a repair process can make a mistake. Mutations occur when the UV damage repair system makes an error and causes a change in the nucleotide sequence of the DNA.

When first discovered, pyrimidine dimers were called thymine dimers.

Animation
Thymine Dimer Formation and Repair

In medicine, the pyrimidine dimerization reaction is used to advantage in hospitals where germicidal (UV) light is used to kill bacteria in the air and on environmental surfaces, such as in a vacant operating room. This cell death is caused by pyrimidine dimer formation on a massive scale. The repair systems of the bacteria are overwhelmed, and the cells die.

Of course, the same type of pyrimidine dimer formation can occur in human cells as well. Lying out in the sun all day to acquire a fashionable tan exposes the skin to large amounts of UV light. This damages the skin by formation of many pyrimidine dimers. Exposure to high levels of UV from sunlight or tanning booths has been linked to a rising incidence of skin cancer in human populations.

Consequences of Defects in DNA Repair

The human repair system for pyrimidine dimers is quite complex, requiring at least five enzymes. The first step in repair of the pyrimidine dimer is cleavage of the sugar-phosphate backbone of the DNA near the site of the damage. The enzyme that performs this is called a *repair endonuclease*. If the gene encoding this enzyme is defective, pyrimidine dimers cannot be repaired. The accumulation of mutations combined with a simultaneous decrease in the efficiency of DNA repair mechanisms leads to an increased incidence of cancer. For example, a mutation in the repair endonuclease gene, or in other genes in the repair pathway, results in the genetic skin disorder called *xeroderma pigmentosum*. People who suffer from xeroderma pigmentosum are extremely sensitive to the ultraviolet rays of sunlight and develop multiple skin cancers, usually before the age of twenty.

SUMMARY

17.1 The Structure of the Nucleotide

DNA and RNA are polymers of *nucleotides*, which are composed of a five-carbon sugar (ribose in RNA and 2′-deoxyribose in DNA), a nitrogenous base, and one, two, or three phosphoryl groups. There are two kinds of nitrogenous bases, *purines* (adenine and guanine) and *pyrimidines* (cytosine, thymine, and uracil). *Deoxyribonucleotides* are subunits of DNA. *Ribonucleotides* are subunits of RNA.

17.2 The Structure of DNA and RNA

Nucleotides are joined by 3′–5′ *phosphodiester bonds* in both DNA and RNA. DNA is a *double helix*, two strands of DNA wound around one another. The sugar-phosphate backbone is on the outside of the helix, and complementary pairs of bases extend into the center of the helix. The *base pairs* are held together by hydrogen bonds. Adenine base pairs with thymine, and cytosine base pairs with guanine. The two strands of DNA in the helix are *antiparallel strands*. RNA is single stranded.

17.3 DNA Replication

DNA replication involves synthesis of a faithful copy of the DNA molecule. It is *semiconservative;* each daughter molecule consists of one parental strand and one newly synthesized strand. DNA polymerase "reads" each parental strand and synthesizes the *complementary strand* according to the rules of base pairing.

17.4 Information Flow in Biological Systems

The *central dogma* states that the flow of biological information in cells is DNA → RNA → protein. There are three classes of RNA: *messenger RNA, transfer RNA,* and *ribosomal RNA. Transcription* is the process by which RNA molecules are synthesized. *RNA polymerase* catalyzes the synthesis of RNA. Transcription occurs in three stages: initiation, elongation, and termination. Eukaryotic genes contain *introns,* sequences that do not encode protein. These are removed from the primary transcript by *RNA splicing.* The final mRNA contains only the protein coding sequences or *exons.* This final mRNA also has an added *5′ cap structure* and a *3′ poly(A) tail.*

17.5 The Genetic Code

The genetic code is a triplet code. Each code word is called a *codon* and consists of three nucleotides. There are sixty-four codons in the genetic code. Of these, three are *termination codons* (UAA, UAG, and UGA), and the remaining sixty-one specify an amino acid. Most amino acids have several codons. As a result, the genetic code is said to be *degenerate.*

17.6 Protein Synthesis

The process of protein synthesis is called *translation.* The genetic code words on the mRNA are decoded by tRNA. Each tRNA has an *anticodon* that is complementary to a codon on the mRNA. In addition the tRNA is covalently linked to its correct amino acid. Thus, hydrogen bonding between codon and anticodon brings the correct amino acid to the site of protein synthesis. Translation also occurs in three stages called initiation, chain elongation, and termination.

17.7 Mutation, Ultraviolet Light, and DNA Repair

Any change in a DNA sequence is a *mutation*. Mutations are classified according to the type of DNA alteration, including *point mutations*, *deletion mutations*, and *insertion mutations*. Ultraviolet light (UV) causes formation of *pyrimidine dimers*. Mistakes can be made during pyrimidine dimer repair, causing UV-induced mutations. Germicidal (UV) lamps are used to kill bacteria on environmental surfaces. UV damage to skin can result in skin cancer.

KEY TERMS

aminoacyl tRNA (17.6)
aminoacyl tRNA binding site of ribosome (A-site) (17.6)
aminoacyl tRNA synthetase (17.6)
anticodon (17.4)
antiparallel strands (17.2)
base pairs (17.2)
cap structure (17.4)
carcinogen (17.7)
central dogma (17.4)
codon (17.4)
complementary strands (17.2)
degenerate code (17.5)
deletion mutation (17.7)
deoxyribonucleic acid (DNA) (17.1)
deoxyribonucleotide (17.1)
double helix (17.2)
elongation factor (17.6)
exon (17.4)
initiation factor (17.6)
insertion mutation (17.7)
intron (17.4)
messenger RNA (mRNA) (17.4)
mutagen (17.7)
mutation (17.7)
nucleotide (17.1)

peptidyl tRNA binding site of ribosome (P-site) (17.6)
phosphodiester bond (17.2)
point mutation (17.7)
poly(A) tail (17.4)
polysome (17.6)
post-transcriptional modification (17.4)
primary transcript (17.4)
promoter (17.4)
purine (17.1)
pyrimidine (17.1)
pyrimidine dimer (17.7)
release factor (17.6)
ribonucleic acid (RNA) (17.1)
ribonucleotide (17.1)
ribosomal RNA (rRNA) (17.4)
ribosome (17.6)
RNA polymerase (17.4)
RNA splicing (17.4)
semiconservative DNA replication (17.3)
silent mutation (17.7)
termination codon (17.6)
transcription (17.4)
transfer RNA (tRNA) (17.4)
translation (17.4)
translocation (17.6)

QUESTIONS AND PROBLEMS

The Structure of the Nucleotide

Foundations

17.13 Draw the structure of the purine ring, and indicate the nitrogen that is bonded to sugars in nucleotides.

17.14 a. Draw the ring structure of pyrimidines.
   b. In a nucleotide, which nitrogen atom of pyrimidine rings is bonded to the sugar?

Applications

17.15 ATP is the universal energy currency of the cell. What components make up the ATP nucleotide?

17.16 One of the energy-harvesting steps of the citric acid cycle results in the production of GTP. What is the structure of the GTP nucleotide?

The Structure of DNA and RNA

Foundations

17.17 The two strands of a DNA molecule are antiparallel. What is meant by this description?

17.18 List three differences between DNA and RNA.

Applications

17.19 How many hydrogen bonds link the adenine-thymine base pair?

17.20 How many hydrogen bonds link the guanine-cytosine base pair?

17.21 Write the structure that results when deoxycytosine-5'-monophosphate is linked to thymidine-5'-monophosphate by a $3' \rightarrow 5'$ phosphodiester bond.

17.22 Write the structure that results when adenosine-5'-monophosphate is linked to uridine-5'-monophosphate by a $3' \rightarrow 5'$ phosphodiester bond.

17.23 Describe the structure of the bacterial chromosome.

17.24 Describe the structure of the human chromosome.

DNA Replication

Foundations

17.25 What is meant by semiconservative DNA replication?

17.26 Draw a diagram illustrating semiconservative DNA replication.

Applications

17.27 If a DNA strand had the nucleotide sequence

5'-ATGCGGCTAGAATATTCCA-3'

what would be the sequence of the complementary daughter strand?

17.28 If the sequence of a double-stranded DNA is

5'-G A A T T C C T T A A G G A T C G A T C -3'
         | | | | | | | | | | | | | | | | | | |
3'-C T T A A G G A A T T C C T A G C T A G -5'

what would be the sequence of the two daughter DNA molecules after DNA replication? Indicate which strands are newly synthesized and which are parental.

Information Flow in Biological Systems

Foundations

17.29 What is the central dogma of molecular biology?

17.30 What are the roles of DNA, RNA, and protein in information flow in biological systems?

17.31 On what molecule is the anticodon found?

17.32 On what molecule is the codon found?

Applications

17.33 If a gene had the nucleotide sequence

5'-TACCTAGCTCTGGTCATTAAGGCAGTA-3'

what would be the sequence of the mRNA?

17.34 If a mRNA had the nucleotide sequence

5′-AUGCCCUUUCAUUACCCGGUA-3′

what was the sequence of the DNA strand that was transcribed?

17.35 What is meant by the term *RNA splicing?*

17.36 The following is the unspliced transcript of a eukaryotic gene:

exon 1   intron A   exon 2   intron B   exon 3   intron C   exon 4

What would the structure of the final mature mRNA look like, and which of the above sequences would be found in the mature mRNA?

17.37 List the three classes of RNA molecules.

17.38 What is the function of each of the classes of RNA molecules?

17.39 Why must introns be removed from a primary transcript?

17.40 Eukaryotic genes are described as discontinuous. What is meant by that description?

17.41 What is a poly(A) tail?

17.42 What is the purpose of the poly(A) tail on eukaryotic mRNA?

17.43 What is the cap structure?

17.44 What is the function of the cap structure on eukaryotic mRNA?

The Genetic Code

Foundations

17.45 How many codons constitute the genetic code?

17.46 What is meant by a triplet code?

17.47 What is meant by the reading frame of a gene?

17.48 What happens to the reading frame of a gene if a nucleotide is deleted?

Applications

17.49 Which two amino acids are encoded by only one codon?

17.50 Which amino acids are encoded by six codons?

17.51 An essential gene has the codon 5′-UUU-3′ in a critical position. If this codon is mutated to the sequence 5′-UUA-3′, what is the expected consequence for the cell?

17.52 An essential gene has the codon 5′-UUA-3′ in a critical position. If this codon is mutated to the sequence 5′-UUG-3′, what is the expected consequence for the cell?

Protein Synthesis

Foundations

17.53 What is the function of ribosomes?

17.54 What are the two tRNA binding sites on the ribosome?

Applications

17.55 Explain the relationship between the sequence of nucleotides of a gene in the DNA and the sequence of amino acids in the protein encoded by that gene.

17.56 Explain how a change in the sequence of nucleotides of a gene, a mutation, may alter the sequence of amino acids in the protein encoded by that gene.

17.57 Briefly describe the three stages of translation: initiation, elongation, and termination.

17.58 What peptide sequence would be formed from the mRNA

5′-AUGUGUAGUGACCAACCGAUUUCACUGUGA-3′?

The following diagram shows the reaction that produces an aminoacyl tRNA, in this case methionyl tRNA. Use the following diagram to answer Questions 17.59 and 17.60.

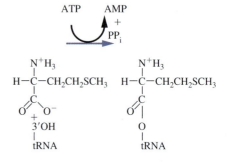

The amino acyl linkage is formed between the 3′—OH of the tRNA and the carboxylate group of the amino acid, methionine.

17.59 By what type of bond is an amino acid linked to a tRNA molecule in an aminoacyl tRNA molecule?

17.60 Draw the structure of an alanine bound to the 3′ position of adenine at the 3′ end of alanyl tRNA.

Mutation, Ultraviolet Light, and DNA Repair

Foundations

17.61 Define the term *point mutation.*

17.62 What are deletion and insertion mutations?

Applications

17.63 Why are some mutations silent?

17.64 Which is more likely to be a silent mutation, a point mutation or a deletion mutation? Explain your reasoning.

17.65 What damage does UV light cause in DNA, and how does this lead to mutations?

17.66 Explain why UV lights are effective germicides on environmental surfaces.

17.67 What is a carcinogen? Why are carcinogens also mutagens?

17.68 **a.** What causes the genetic disease xeroderma pigmentosum?
    b. Why are people who suffer from xeroderma pigmentosum prone to cancer?

FOR FURTHER UNDERSTANDING

1. It has been suggested that the triplet genetic code evolved from a two-nucleotide code. Perhaps there were fewer amino acids in the ancient proteins. Examine the genetic code in Figure 17.11. What features of the code support this hypothesis?

2. The strands of DNA can be separated by heating a DNA sample. The input heat energy breaks the hydrogen bonds between base pairs, allowing the strands to separate from one another. Suppose that you are given two DNA samples. One has a G + C content of 70% and the other has a G + C content of 45%. Which of these

samples will require a higher temperature to separate the strands? Explain your answer.

3. A mutation produces a tRNA with a new anticodon. Originally the anticodon was 5′-CCA-3′; the mutant anticodon is 5′-UCA-3′. What effect will this mutant tRNA have on cellular translation?

4. A scientist is interested in cloning the gene for blood clotting factor VIII into bacteria so that large amounts of the protein can be produced to treat hemophiliacs. Knowing that bacterial cells cannot carry out RNA splicing, she clones a complementary DNA copy of the factor VIII mRNA and introduces this into bacteria. However, there is no transcription of the cloned factor VIII gene. How could the scientist engineer the gene so that the bacterial cell RNA polymerase will transcribe it?

18

LEARNING GOALS

1 Discuss the importance of ATP in cellular energy transfer processes.

2 Describe the three stages of catabolism of dietary proteins, carbohydrates, and lipids.

3 Discuss glycolysis in terms of its two major segments.

4 Looking at an equation representing any of the chemical reactions in glycolysis, describe the kind of reaction and the significance of that reaction to the pathway.

5 Discuss the practical and metabolic roles of fermentation reactions.

6 Name the regions of the mitochondria and the function of each region.

7 Describe the reaction that results in the conversion of pyruvate to acetyl CoA, describing the location of the reaction and the components of the pyruvate dehydrogenase complex.

8 Summarize the reactions of aerobic respiration.

9 Looking at an equation representing any of the chemical reactions in the citric acid cycle, describe the kind of reaction and the significance of that reaction to the pathway.

10 Describe the process of oxidative phosphorylation.

11 Discuss the biological function of gluconeogenesis.

12 Summarize the regulation of blood glucose levels by glycogenesis and glycogenolysis.

Carbohydrate Metabolism

Just as we need energy to run, jump, and think, the cell needs a ready supply of cellular energy for the many functions that support these activities. Cells need energy for active transport, to move molecules between the environment and the cell. Energy is also needed for biosynthesis of small metabolic molecules and production of macromolecules from these intermediates. Finally, energy is required for mechanical work, including muscle contraction and motility of sperm cells. Table 18.1 lists some examples of each of these energy-requiring processes.

We need a supply of energy-rich food molecules that can be degraded, or oxidized, to provide this needed cellular energy. Our diet includes three major sources of energy: carbohydrates, fats, and proteins. Each of these types of large biological molecules must be broken down into its basic subunits—simple sugars, fatty acids and glycerol, and amino acids—before they can be taken into the cell and used to produce cellular energy. Of these classes of food molecules, carbohydrates are the most readily used. The pathway for the first stages of carbohydrate breakdown is called glycolysis. We find the same pathway in organisms as different as the simple bacterium and humans.

As we will see, anaerobic glycolysis actually releases and stores very little (2.2%) of the potential energy of glucose, but the pathway also serves as a source of biosynthetic building blocks. It also modifies the carbohydrates so that other pathways are able to release as much as 40% of the potential energy.

Aerobic catabolic pathways complete the oxidation of glucose to CO_2 and H_2O and provide most of the ATP needed by the body. This process, called aerobic respiration, produces thirty-six ATP molecules using the energy harvested from each glucose molecule that enters glycolysis. These reactions occur in metabolic pathways located in mitochondria, the cellular "power plants." Mitochondria are a type of membrane-enclosed cell organelle.

Here, in the mitochondria, the final oxidations of carbohydrates, lipids, and proteins occur. Here, also, the electrons that are harvested in these oxidation reactions are used to make ATP. In these remarkably efficient reactions, nearly 40% of the potential energy of glucose is stored as ATP.

Rock climbing requires a great deal of energy. Describe how the food molecules in an energy bar are converted into the ATP needed to sustain this high level of exercise.

Animation
Energy for Activity

TABLE 18.1 The Types of Cellular Work That Require Energy
Biosynthesis: Synthesis of Metabolic Intermediates and Macromolecules
Synthesis of glucose from CO_2 and H_2O in the process of photosynthesis in plants
Synthesis of amino acids
Synthesis of nucleotides
Synthesis of lipids
Protein synthesis from amino acids
Synthesis of nucleic acids
Synthesis of organelles and membranes
Active Transport: Movement of Ions and Molecules
Transport of H^+ to maintain constant pH
Transport of food molecules into the cell
Transport of K^+ and Na^+ into and out of nerve cells for transmission of nerve impulses
Secretion of HCl from parietal cells into the stomach
Transport of waste from the blood into the urine in the kidneys
Transport of amino acids and most hexose sugars into the blood from the intestine
Accumulation of calcium ions in the mitochondria
Motility
Contraction and flexion of muscle cells
Separation of chromosomes during cell division
Ability of sperm to swim via flagella
Movement of foreign substances out of the respiratory tract by cilia on the epithelial lining of the trachea
Translocation of eggs into the fallopian tubes by cilia in the female reproductive tract

18.1 ATP: The Cellular Energy Currency

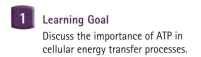

Learning Goal
Discuss the importance of ATP in cellular energy transfer processes.

Animation
A Biochemical Pathway

The degradation of fuel molecules, called **catabolism,** provides the energy for cellular energy-requiring functions, including **anabolism,** or biosynthesis. The energy of a food source can be released in one of two ways: as heat or, more important to the cell, as chemical bond energy. We can envision two alternative modes of aerobic degradation of the simple sugar glucose. Imagine that we simply set the glucose afire. This would result in its complete oxidation to CO_2 and H_2O and would release 686 kcal/mol of glucose. Yet in terms of a cell, what would be accomplished? Nothing. All of the potential energy of the bonds of glucose is lost as heat and light.

The cell uses a different strategy. With a series of enzymes, biochemical pathways in the cell carry out a step-by-step oxidation of glucose. Small amounts of energy are released at several points in the pathway and that energy is harvested and saved in the bonds of a molecule that has been called the *universal energy currency.* This molecule, shown in Figure 18.1, is **adenosine triphosphate (ATP).**

ATP serves as a "go-between" molecule that couples the *exergonic* (energy releasing) reactions of catabolism and the endergonic (energy requiring) reactions of anabolism.

The energy stored in ATP is released for cellular work when ATP is hydrolyzed to adenosine diphosphate (ADP) and an inorganic phosphate group (P_i).

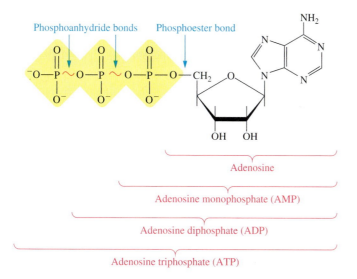

Figure 18.1

The structure of the universal energy currency, ATP. ATP is a nucleotide composed of the nitrogenous base, adenine, the five-carbon sugar, ribose, and three phosphoryl groups.

As shown in the equation below, the hydrolysis of one molecule of ATP releases 7.3 kcal/mole of energy.

$$ATP + H_2O \rightarrow ADP + P_i + 7.3 \text{ kcal/mole}$$

QUESTION 18.1

Why is ATP called the universal energy currency?

QUESTION 18.2

List five biological activities that require ATP.

18.2 Overview of Catabolic Processes

Although carbohydrates, fats, and proteins can be degraded to release energy, carbohydrates are the most readily used energy source. Any catabolic process must begin with a supply of nutrients. When we eat a meal, we are eating quantities of carbohydrates, fats, and proteins. From this point, the catabolic processes can be broken down into a series of stages. The three stages of catabolism are summarized in Figure 18.2.

Learning Goal

2 Describe the three stages of catabolism of dietary proteins, carbohydrates, and lipids.

Stage I: Hydrolysis of Dietary Macromolecules into Small Subunits

The purpose of the first stage of catabolism is to degrade large food molecules into their component subunits. These subunits—simple sugars, amino acids, fatty acids, and glycerol—are then taken into the cells of the body for use as an energy source. The process of digestion is summarized in Figure 18.3.

Polysaccharides are hydrolyzed to monosaccharides. This process begins in the mouth, where the enzyme amylase begins the hydrolysis of starch. Digestion continues in the small intestine, where pancreatic amylase further hydrolyzes the starch into maltose (a disaccharide of glucose). Maltase catalyzes the hydrolysis of maltose, producing two glucose molecules. Similarly, sucrose is hydrolyzed to glucose and fructose by the enzyme, sucrase, and lactose (milk sugar) is degraded

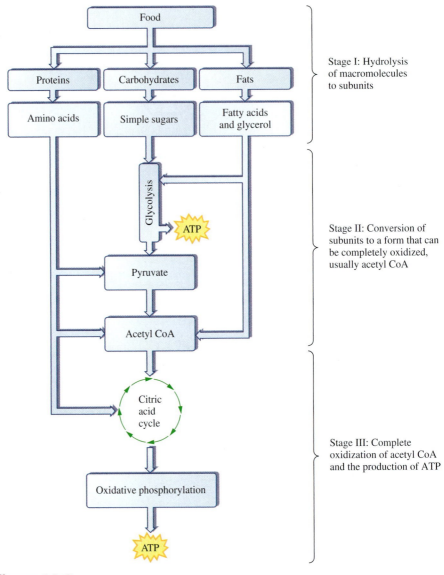

Figure 18.2

The three stages of the conversion of food into cellular energy in the form of ATP.

into the monosaccharides, glucose and galactose, by the enzyme lactase in the small intestine. The monosaccharides are taken up by the epithelial cells of the intestine in an energy-requiring process called *active transport.*

The digestion of proteins begins in the stomach, where the low pH denatures the proteins so that they are more easily hydrolyzed by the enzyme, pepsin. They are further degraded in the small intestine by trypsin, chymotrypsin, elastase, and other proteases. The products of protein digestion—amino acids and short oligopeptides—are taken up by the cells lining the intestine. This uptake also involves an active transport mechanism.

Digestion of fats does not begin until the food reaches the small intestine, even though there are lipases in both the saliva and stomach fluid. Fats arrive in the duodenum, the first portion of the small intestine, in the form of large fat globules. Bile salts produced by the liver break these up into an emulsion of tiny fat droplets. Because the small droplets have a greater surface area, the lipids are now more accessible to the action of pancreatic lipase. This enzyme hydrolyzes the fats into fatty acids and glycerol, which are taken up by intestinal cells by a transport

The digestion and transport of fats are considered in greater detail in Chapter 19.

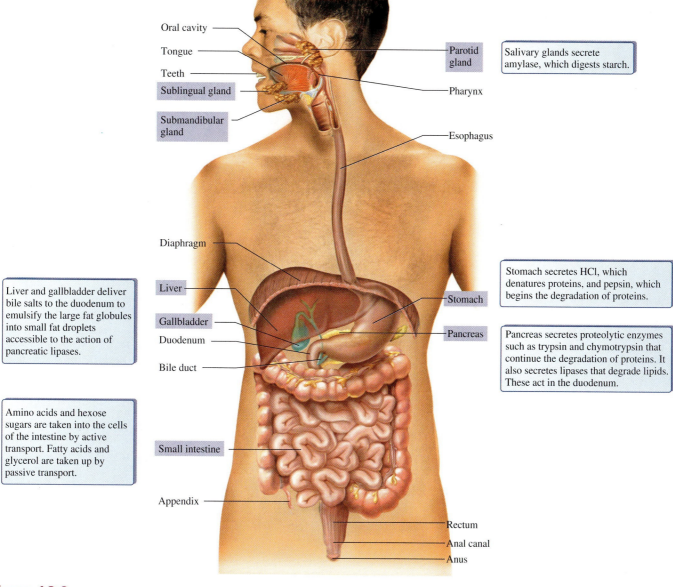

Salivary glands secrete amylase, which digests starch.

Stomach secretes HCl, which denatures proteins, and pepsin, which begins the degradation of proteins.

Pancreas secretes proteolytic enzymes such as trypsin and chymotrypsin that continue the degradation of proteins. It also secretes lipases that degrade lipids. These act in the duodenum.

Liver and gallbladder deliver bile salts to the duodenum to emulsify the large fat globules into small fat droplets accessible to the action of pancreatic lipases.

Amino acids and hexose sugars are taken into the cells of the intestine by active transport. Fatty acids and glycerol are taken up by passive transport.

Figure 18.3

An overview of the digestive processes that hydrolyze carbohydrates, proteins, and fats.

process that does not require energy. This process is called *passive transport*. A summary of these hydrolysis reactions is shown in Figure 18.4.

Active and passive transport are discussed in Section 15.6.

Stage II: Conversion of Monomers into a Form That Can Be Completely Oxidized

The monosaccharides, amino acids, fatty acids, and glycerol must now be assimilated into the pathways of energy metabolism. The two major pathways are glycolysis and the citric acid cycle (see Figure 18.2). Sugars usually enter the glycolysis pathway in the form of glucose or fructose. They are eventually converted to acetyl CoA, which is a form that can be completely oxidized in the citric acid cycle. Amino groups are removed from amino acids, and the remaining carbon skeletons enter the catabolic processes at many steps of the citric acid cycle. Fatty acids are converted to acetyl CoA and enter the citric acid cycle in that form.

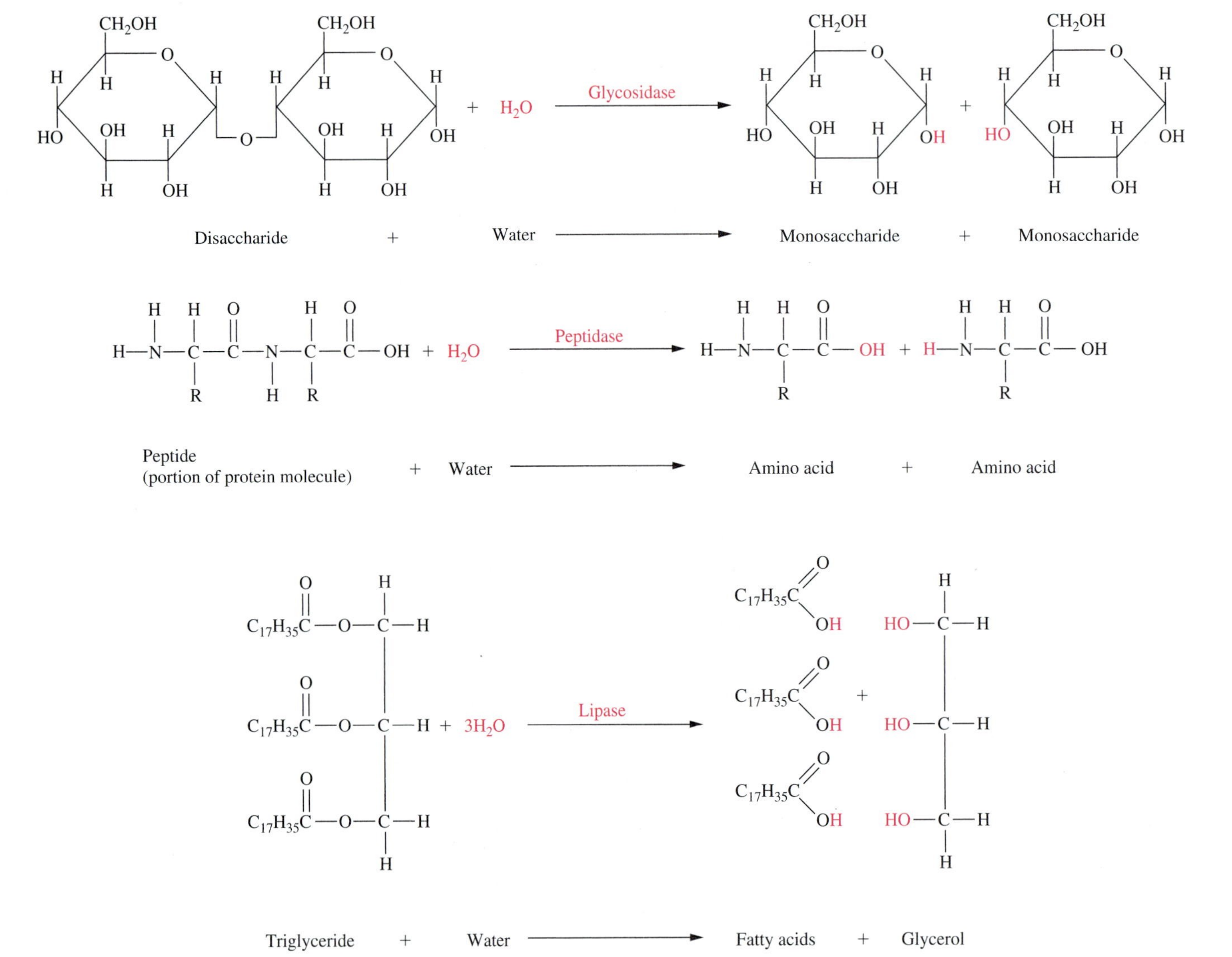

Figure 18.4

A summary of the hydrolysis reactions of carbohydrates, proteins, and fats.

Glycerol, produced by the hydrolysis of fats, is converted to glyceraldehyde-3-phosphate, one of the intermediates of glycolysis, and enters energy metabolism at that level.

Stage III: The Complete Oxidation of Nutrients and the Production of ATP

Oxidative phosphorylation is described in Section 18.9.

Acetyl CoA carries acetyl groups, two-carbon remnants of the nutrients, to the citric acid cycle. Acetyl CoA enters the cycle, and electrons and hydrogen atoms are harvested during the complete oxidation of the acetyl group to CO_2. Coenzyme A is released (recycled) to carry additional acetyl groups to the pathway. The electrons and hydrogen atoms that are harvested are used in the process of oxidative phosphorylation to produce ATP.

QUESTION 18.3

Briefly describe the three stages of catabolism.

QUESTION 18.4

Discuss the digestion of dietary carbohydrates, lipids, and proteins.

18.3 Glycolysis

An Overview

Glycolysis, also known as the Embden-Meyerhof pathway, is a pathway for carbohydrate catabolism that begins with the substrate, D-glucose. The pathway requires no oxygen; it is an anaerobic process carried out by enzymes free in the cytoplasm of the cell.

The ten steps of glycolysis, catalyzed by ten enzymes, are outlined in Figure 18.5. The first reactions of glycolysis involve an energy investment. ATP molecules are hydrolyzed, energy is released, and phosphoryl groups are added to the hexose sugars. In the remaining steps of glycolysis, energy is harvested to produce a net gain of ATP.

The three major products of glycolysis are seen in Figure 18.5. These are chemical energy in the form of ATP, chemical energy in the form of NADH, and two three-carbon pyruvate molecules. Each of these products is considered below:

Animation
How Glycolysis Works

- **Chemical energy as ATP.** Four ATP molecules are formed by the process of **substrate-level phosphorylation.** This means that a high-energy phosphoryl group from one of the substrates in glycolysis is transferred to ADP to form ATP (see Figure 18.5, steps 7 and 10). Although four ATP molecules are produced during glycolysis, the *net* gain is only two ATP molecules because two ATP molecules are used early in glycolysis (Figure 18.5, steps 1 and 3). The two ATP molecules produced represent only 2.2% of the potential energy of the glucose molecule. Thus glycolysis is not a very efficient energy-harvesting process.

- **Chemical energy in the form of reduced NAD$^+$, NADH.** Nicotinamide adenine dinucleotide (NAD$^+$) is a coenzyme derived from the vitamin, niacin. The reduced form of NAD$^+$, NADH, carries hydride anions, hydrogen atoms with two electrons (H:$^-$), removed during the oxidation of glyceraldehyde-3-phosphate (see Figure 18.5, step 6). Under aerobic conditions, the electrons and hydrogen atom are transported from the cytoplasm into the mitochondria. Here they enter an electron transport system for the generation of ATP by oxidative phosphorylation. Under anaerobic conditions, NADH is used as a source of electrons in fermentation reactions.

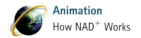

Animation
How NAD$^+$ Works

- **Two pyruvate molecules.** At the end of glycolysis, the six-carbon glucose molecule has been converted into two three-carbon pyruvate molecules. The fate of the pyruvate also depends on whether the reactions occur in the presence or absence of oxygen. Under aerobic conditions, it is used to produce acetyl CoA destined for the citric acid cycle and complete oxidation. Under anaerobic conditions, it is used as an electron acceptor in fermentation reactions.

To learn more about some rare and typically fatal genetic defects of glycolysis, see A Medical Connection: Genetic Disorders of Glycolysis online at www.mhhe.com/denniston.

In any event, these last two products must be used in some way so that glycolysis can continue to function and produce ATP. There are two reasons for this. First, if pyruvate were allowed to build up, it would cause glycolysis to stop, thereby stopping the production of ATP. Thus, pyruvate must be used in some kind of follow-up reaction, aerobic or anaerobic. Second, in step 6, glyceraldehyde-3-phosphate is oxidized and NAD$^+$ is reduced (accepts the hydride anion). The cell has only a small supply of NAD$^+$. If all the NAD$^+$ is reduced, none will be available for this reaction, and glycolysis will stop. Therefore, NADH must be reoxidized so that glycolysis can continue to produce ATP for the cell.

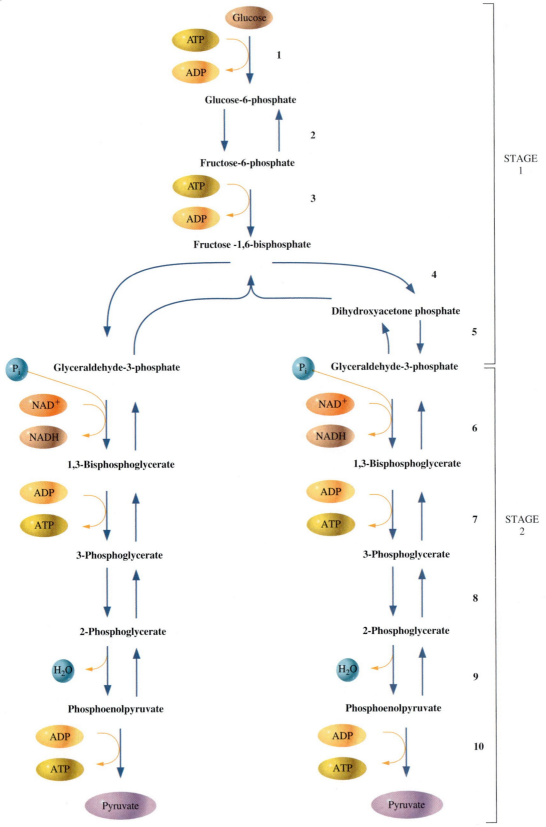

Figure 18.5

A summary of the reactions of glycolysis. These reactions occur in the cell cytoplasm.

Reactions of Glycolysis

Glycolysis can be divided into two major segments. The first is the investment of ATP energy. Without this investment, glucose would not have enough energy for glycolysis to continue, and there would be no ATP produced. This segment includes the first five reactions of the pathway. The second major segment involves the remaining reactions of the pathway (steps 6–10), those that result in a net energy yield.

Reaction 1

The substrate, glucose, is phosphorylated by the enzyme, *hexokinase,* in a coupled phosphorylation reaction. The source of the phosphoryl group is ATP. At first, this reaction seems contrary to the overall purpose of catabolism, the *production* of ATP. The expenditure of ATP in these early reactions must be thought of as an "investment." The cell actually goes into energy "debt" in these early reactions, but this is absolutely necessary to get the pathway started.

Glucose   Glucose-6-phosphate

Reaction 2

The glucose-6-phosphate formed in the first reaction is rearranged to produce the structural isomer, fructose-6-phosphate. The enzyme, *phosphoglucose isomerase,* catalyzes this isomerization. The result is that the C-1 carbon of the six-carbon sugar is exposed; it is no longer part of the ring structure. Examination of the open-chain structures reveals that this isomerization converts an aldose into a ketose.

Glucose-6-phosphate   Fructose-6-phosphate

Glucose-6-phosphate
(an aldose)   Fructose-6-phosphate
(a ketose)

---

3 Learning Goal

Discuss glycolysis in terms of its two major segments.

4 Learning Goal

Looking at an equation representing any of the chemical reactions in glycolysis, describe the kind of reaction and the significance of that reaction to the pathway.

The enzyme name can tell us a lot about the reaction (see Section 16.10). The suffix *-kinase* tells us that the enzyme is a transferase that will transfer a phosphoryl group, in this case from an ATP molecule to the substrate. The prefix *hexo-* gives us a hint that the substrate is a six-carbon sugar. Hexokinase predominantly phosphorylates the six-carbon sugar, glucose.

The enzyme name, phosphoglucose isomerase, provides clues to the reaction that is being catalyzed (Section 16.10). *Isomerase* tells us that the enzyme will catalyze the interconversion of one isomer into another. *Phosphoglucose* suggests that the substrate is a phosphorylated form of glucose.

Reaction 3

A second energy "investment" is catalyzed by the enzyme *phosphofructokinase*. The phosphoanhydride bond in ATP is hydrolyzed, and a phosphoester linkage between the phosphoryl group and the C-1 hydroxyl group of fructose-6-phosphate is formed. The product is fructose-1,6-bisphosphate.

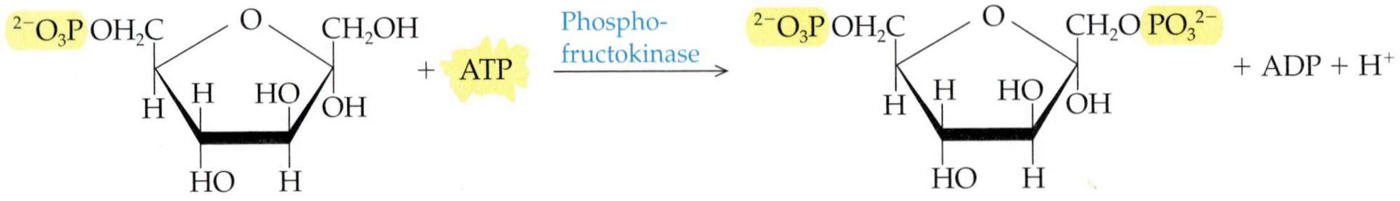

Fructose-6-phosphate   Fructose-1,6-bisphosphate

Reaction 4

Fructose-1,6-bisphosphate is split into two three-carbon intermediates in a reaction catalyzed by the enzyme *aldolase*. The products are glyceraldehyde-3-phosphate (G3P) and dihydroxyacetone phosphate.

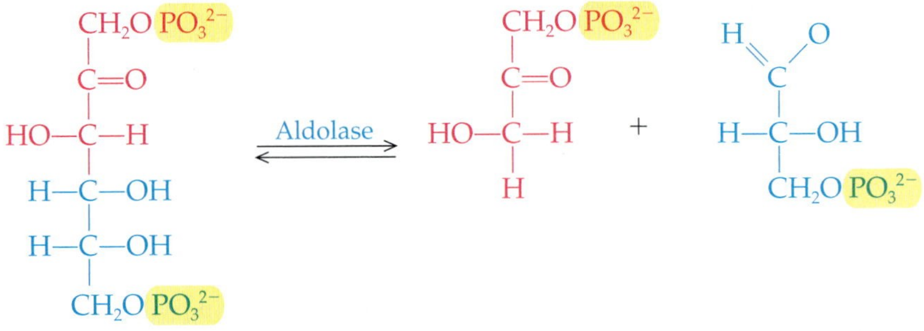

Fructose-   Dihydroxyacetone   Glyceraldehyde-
1,6-bisphosphate   phosphate   3-phosphate

Reaction 5

The enzyme name hints that two isomers of a phosphorylated three-carbon sugar are going to be interconverted (Section 16.10). The ketone dihydroxyacetone phosphate is converted into its isomeric aldehyde, glyceraldehyde-3-phosphate.

Because G3P is the only substrate that can be used by the next enzyme in the pathway, the dihydroxyacetone phosphate is rearranged to become a second molecule of G3P. The enzyme that mediates this isomerization is *triose phosphate isomerase*.

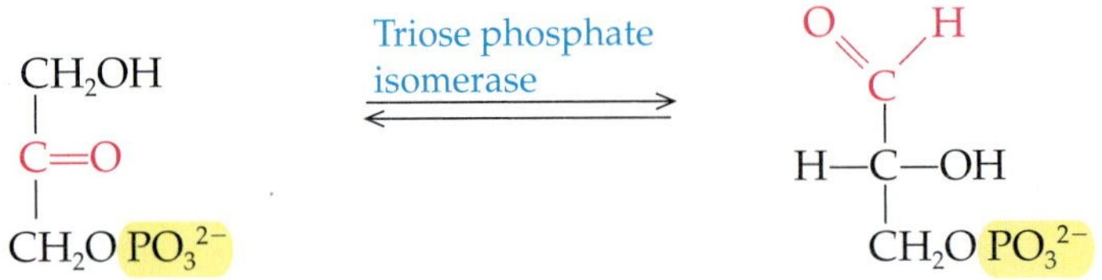

Dihydroxyacetone phosphate   Glyceraldehyde-3-phosphate

Reaction 6

The name glyceraldehyde-3-phosphate dehydrogenase tells us that the substrate, glyceraldehyde-3-phosphate, is going to be oxidized. In this reaction, we see that the aldehyde group has been oxidized to a carboxylate group (Section 12.5).

In this reaction, the aldehyde, glyceraldehyde-3-phosphate, is oxidized to a carboxylic acid in a reaction catalyzed by *glyceraldehyde-3-phosphate dehydrogenase*. This is the first step in glycolysis that harvests energy, and it involves the reduction of the coenzyme, nicotinamide adenine dinucleotide (NAD^+). This reaction occurs in two steps. First, NAD^+ is reduced to NADH as the aldehyde group of glyceraldehyde-3-phosphate is oxidized to a carboxyl group. Second, an inorganic phosphate group is transferred to the carboxyl group to give 1,3-bisphosphoglycerate. Notice that the new bond is denoted with a squiggle ($\sim$), indicating that this is a high-energy bond. This, and all remaining reactions of glycolysis, occur twice for each glucose because each glucose has been converted into two molecules of glyceraldehyde-3-phosphate.

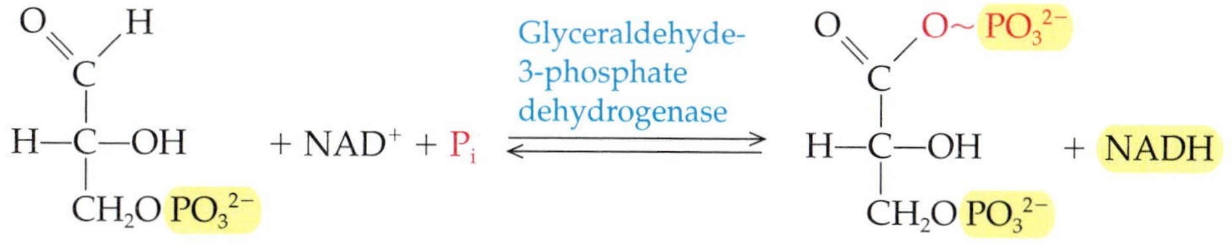

Glyceraldehyde-3-phosphate $\xrightarrow{\text{Glyceraldehyde-3-phosphate dehydrogenase}}$ 1,3-Bisphosphoglycerate

Reaction 7

In this reaction, energy is harvested in the form of *ATP*. The enzyme, *phosphoglycerate kinase*, catalyzes the transfer of the phosphoryl group of 1,3-bisphosphoglycerate to ADP. This is the first substrate-level phosphorylation of glycolysis, and it produces ATP and 3-phosphoglycerate. It is a coupled reaction in which the high-energy bond is hydrolyzed and the energy released is used to drive the synthesis of ATP.

Once again, the enzyme name reveals a great deal about the reaction. The suffix -*kinase* tells us that a phosphoryl group will be transferred. In this case, a phosphoester bond in the substrate, 1,3-bisphosphoglycerate, is hydrolyzed and ADP is phosphorylated. Note that this is a reversible reaction.

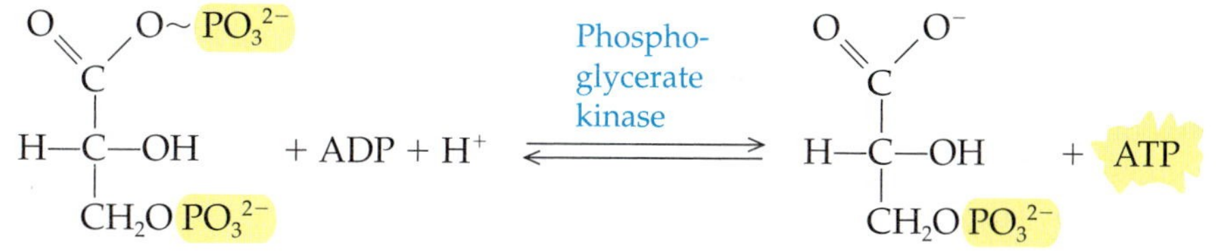

1,3-Bisphosphoglycerate $\xrightarrow{\text{Phosphoglycerate kinase}}$ 3-Phosphoglycerate

Reaction 8

3-Phosphoglycerate is isomerized to produce 2-phosphoglycerate in a reaction catalyzed by the enzyme, *phosphoglycerate mutase*. The phosphoryl group attached to the third carbon of 3-phosphoglycerate is transferred to the second carbon.

The suffix -*mutase* indicates another type of isomerase. Notice that the chemical formulas of the substrate and reactant are the same. The only difference is in the location of the phosphoryl group.

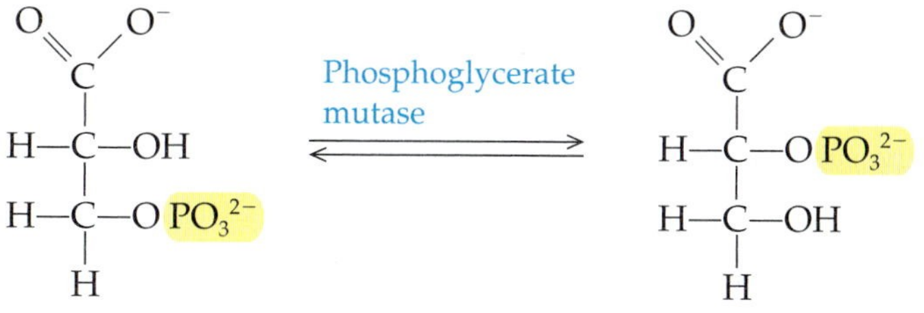

3-Phosphoglycerate $\xrightarrow{\text{Phosphoglycerate mutase}}$ 2-Phosphoglycerate

Reaction 9

In this step the enzyme, *enolase*, catalyzes the dehydration (removal of a water molecule) of 2-phosphoglycerate. The energy-rich product is phosphoenolpyruvate, the highest energy phosphorylated compound in metabolism.

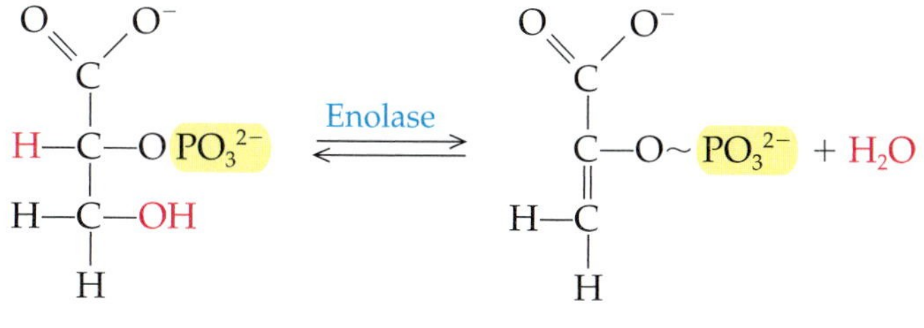

2-Phosphoglycerate $\xrightarrow{\text{Enolase}}$ Phosphoenolpyruvate

The enzyme name indicates that a phosphoryl group will be transferred (kinase) and that the product will be pyruvate.

Reaction 10

Here we see the final substrate-level phosphorylation in the pathway, which is catalyzed by *pyruvate kinase*. Phosphoenolpyruvate serves as a donor of the phosphoryl group that is transferred to ADP to produce ATP. This is another coupled reaction in which hydrolysis of the phosphoester bond in phosphoenolpyruvate provides energy for the formation of the phosphoanhydride bond of ATP. The final product of glycolysis is pyruvate.

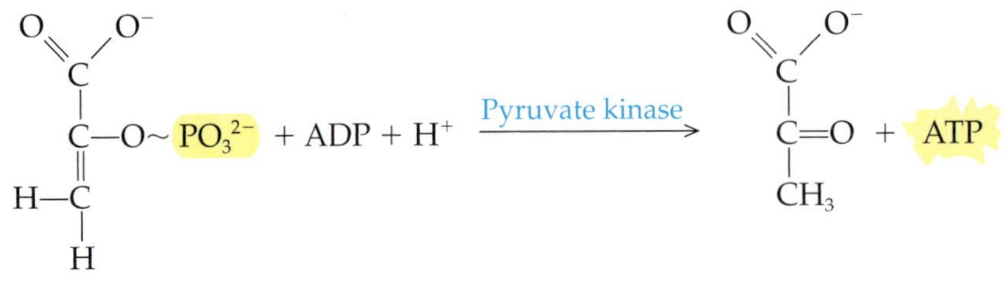

Phosphoenolpyruvate   Pyruvate

Reactions 6 through 10 occur twice per glucose molecule because the starting six-carbon sugar is split into two three-carbon molecules. Thus, in reaction 6, two NADH molecules are generated, and a total of four ATP molecules are made (steps 7 and 10). The net ATP gain from this pathway is, however, only two ATP molecules because there was an energy investment of two ATP molecules in the early steps of the pathway. This investment was paid back by the two ATP molecules produced by substrate-level phosphorylation in step 7. The actual energy yield is produced by substrate-level phosphorylation in reaction 10.

QUESTION 18.5

What is substrate-level phosphorylation?

QUESTION 18.6

What are the major products of glycolysis?

QUESTION 18.7

Describe an overview of the reactions of glycolysis.

QUESTION 18.8

How do the names of the first three enzymes of the glycolytic pathway relate to the reactions they catalyze?

18.4 Fermentations

5 Learning Goal

Discuss the practical and metabolic roles of fermentation reactions.

In the overview of glycolysis, we noted that the pyruvate produced must be used up in some way so that the pathway will continue to produce ATP. Similarly, the NADH produced by glycolysis in step 6 (see Figure 18.5) must be reoxidized at a later time, or glycolysis will grind to a halt as the available NAD^+ is used up. Under anaerobic conditions, however, different types of fermentation reactions accomplish these purposes. **Fermentations** are catabolic reactions that occur with no net oxidation. Pyruvate or an organic compound produced from pyruvate is reduced as NADH is oxidized.

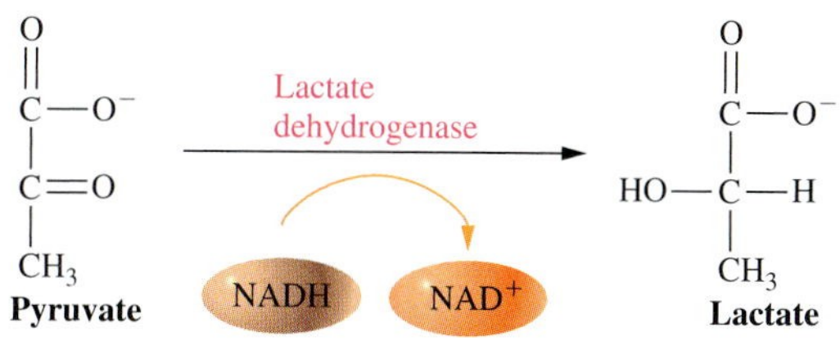

Figure 18.6

The final reaction of lactate fermentation.

Lactate Fermentation

Lactate fermentation is familiar to anyone who has performed strenuous exercise. If you exercise so hard that your lungs and circulatory system can't deliver enough oxygen to the working muscles, your aerobic (oxygen-requiring) energy-harvesting pathways are not able to supply enough ATP to your muscles. But the muscles still demand energy. Under these anaerobic conditions, lactate fermentation begins. In this reaction, the enzyme, *lactate dehydrogenase*, reduces pyruvate to lactate. NADH is the reducing agent for this process (Figure 18.6). As pyruvate is reduced, NADH is oxidized, and NAD^+ is again available, permitting glycolysis to continue.

The lactate produced in the working muscle passes into the blood. Eventually, if strenuous exercise is continued, the concentration of lactate becomes so high that this fermentation can no longer continue. Glycolysis, and thus ATP production, stops. The muscle, deprived of energy, can no longer function. This point of exhaustion is called the **anaerobic threshold.**

A variety of bacteria are able to carry out lactate fermentation under anaerobic conditions. This is of great importance in the dairy industry, because these organisms are used to produce yogurt and some cheeses. The tangy flavor of yogurt is contributed by the lactate produced by these bacteria. Unfortunately, similar organisms also cause milk to spoil.

As we saw in A Medical Connection: Tooth Decay and Simple Sugars (Chapter 14), the lactate produced by oral bacteria is responsible for the gradual removal of calcium from tooth enamel and the resulting dental cavities.

Alcohol Fermentation

Alcohol fermentation has been appreciated, if not understood, since the dawn of civilization. The fermentation process itself was discovered by Louis Pasteur during his studies of the chemistry of winemaking and "diseases of wines." Under anaerobic conditions, yeasts are able to ferment the sugars produced by fruit and grains. The sugars are broken down to pyruvate by glycolysis. This is followed by the two reactions of alcohol fermentation. First, *pyruvate decarboxylase* removes CO_2 from the pyruvate and produces acetaldehyde (Figure 18.7). Second, *alcohol dehydrogenase* catalyzes the reduction of acetaldehyde to ethanol but, more important, reoxidizes NADH in the process. The regeneration of NAD^+ allows glycolysis to continue, just as in lactate fermentation.

These applications and other fermentations are described in A Lifestyle Connection: Fermentations: The Good, the Bad, and the Ugly.

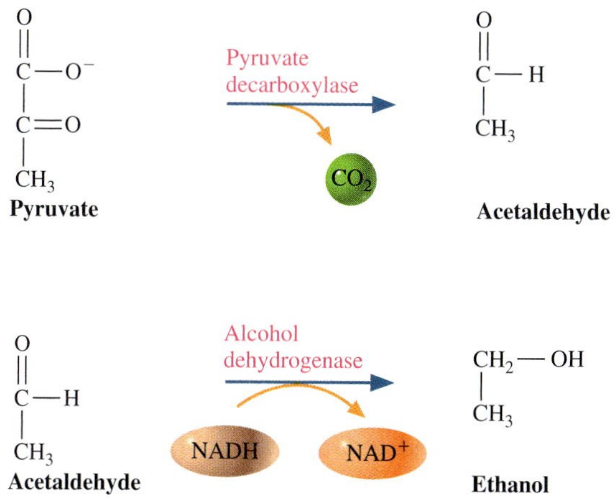

Figure 18.7

The final two reactions of alcohol fermentation.

A LIFESTYLE Connection | Fermentations: The Good, the Bad, and the Ugly

In this chapter, we have seen that fermentation is an anaerobic, cytoplasmic process that allows continued ATP generation by glycolysis. ATP production can continue because the pyruvate produced by the pathway is used in the fermentation and because NAD^+ is regenerated.

The stable end products of alcohol fermentation are CO_2 and ethanol. These have been used by humankind in a variety of ways, including the production of alcoholic beverages, bread making, and alternative fuel sources.

If alcohol fermentation is carried out by using fruit juices in a vented vat, the CO_2 will escape, and the result will be a still wine (not bubbly). But conditions must remain anaerobic; otherwise, fermentation will stop, and aerobic energy-harvesting reactions will ruin the wine. Fortunately for vintners (wine makers), when a vat is fermenting actively, enough CO_2 is produced to create a layer that keeps the oxygen-containing air away from the fermenting juice, thus maintaining an anaerobic atmosphere.

Now suppose we want to make a sparkling wine, such as champagne. To do this, we simply have to trap the CO_2 produced. In this case, the fermentation proceeds in a sealed bottle, a very strong bottle. Both the fermentation products, CO_2 and ethanol, accumulate. Under pressure within the sealed bottle the CO_2 remains in solution. When the top is "popped," the pressure is released, and the CO_2 comes out of solution in the form of bubbles.

In either case, the fermentation continues until the alcohol concentration reaches 12–13%. At that point the yeast "stews in its own juices"! That is, 12–13% ethanol kills the yeast cells that produce it. This points out a last generalization about fermentations. The stable fermentation end product, whether it is lactate or ethanol, eventually accumulates to a concentration that is toxic to the organism. Muscle fatigue is the early effect of lactate buildup in the working muscle. In the same way, continued accumulation of the fermentation product can lead to concentrations that are fatal if there is no means of getting rid of the toxic product or of getting away from it. For single-celled organisms, the result is generally death. Our bodies have evolved so that lactate buildup

The production of bread, wine, and cheese depends on fermentation processes.

The two products of alcohol fermentation, then, are ethanol and CO_2. We take advantage of this fermentation in the production of wines and other alcoholic beverages and in the process of bread making.

QUESTION 18.9

How is the alcohol fermentation in yeast similar to lactate production in skeletal muscle?

QUESTION 18.10

Why must pyruvate be used and NADH be reoxidized so that glycolysis can continue?

18.5 The Mitochondria

6 Learning Goal
Name the regions of the mitochondria and the function of each region.

Mitochondria are football-shaped organelles that are roughly the size of a bacterial cell. They are surrounded by an **outer mitochondrial membrane** and an **inner mitochondrial membrane** (Figure 18.8). The space between the two membranes is

contributes to muscle fatigue that causes the exerciser to stop the exercise. Then the lactate is removed from the blood and converted to glucose by the process of gluconeogenesis.

Another application of alcohol fermentation is the use of yeast in bread making. When we mix water, sugar, and dried yeast, the yeast cells begin to grow and carry out the process of fermentation. This mixture is then added to the flour, milk, shortening, and salt, and the dough is placed in a warm place to rise. The yeast continues to grow and ferment the sugar, producing CO_2 that causes the bread to rise. Of course, when we bake the bread, the yeast cells are killed, and the ethanol evaporates, but we are left with a light and airy loaf of bread.

Today, alcohol produced by fermentation is being considered as an alternative fuel to replace some fossil fuels. Geneticists and bioengineers are trying to develop strains of yeast that can survive higher alcohol concentrations and thus convert more of the sugar of corn and other grains into alcohol.

Bacteria perform a variety of other fermentations. The propionibacteria produce propionic acid and CO_2. The acid gives Swiss cheese its characteristic flavor, and the CO_2 gas produces the characteristic holes in the cheese. Other bacteria, the clostridia, perform a fermentation that is responsible in part for the horrible symptoms of gas gangrene. When these bacteria are inadvertently introduced into deep tissues by a puncture wound, they find a nice anaerobic environment in which to grow. In fact, these organisms are *obligate anaerobes*, that is, they are killed by even a small amount of oxygen. As they grow, they perform a fermentation called the *butyric acid, butanol, acetone fermentation*. This results in the formation of CO_2, the gas associated with gas gangrene. The CO_2 infiltrates the local tissues and helps to maintain an anaerobic environment because oxygen from the local blood supply cannot enter the area of the wound. Now able to grow well, these bacteria produce a vari-

ety of toxins and enzymes that cause extensive tissue death and necrosis. In addition, the fermentation produces acetic acid, ethanol, acetone, isopropanol, butanol, and butyric acid (which is responsible, along with the necrosis, for the characteristic foul smell of gas gangrene). Certainly, the presence of these organic chemicals in the wound enhances tissue death.

Gas gangrene is very difficult to treat. Because the bacteria establish an anaerobic region of cell death and cut off the local circulation, systemic antibiotics do not infiltrate the wound and kill the bacteria. Even our immune response is stymied. Treatment usually involves surgical removal of the necrotic tissue accompanied by antibiotic therapy. In some cases, a hyperbaric oxygen chamber is employed. The infected extremity is placed in an environment with a very high partial pressure of oxygen. The oxygen forced into the tissues is poisonous to the bacteria, and they die.

These are but a few examples of the fermentations that have an effect on humans. Regardless of the specific chemical reactions, all fermentations share the following traits:

- They use pyruvate produced in glycolysis.
- They reoxidize the NADH produced in glycolysis.
- They are self-limiting because the accumulated stable fermentation end product eventually kills the cell that produces it.

FOR FURTHER UNDERSTANDING

Write condensed structural formulas for each of the fermentation products made by clostridia in gas gangrene. Identify the functional groups and provide the I.U.P.A.C. name for each.

Explain the importance of utilizing pyruvate and reoxidizing NADH to the ability of a cell to continue producing ATP.

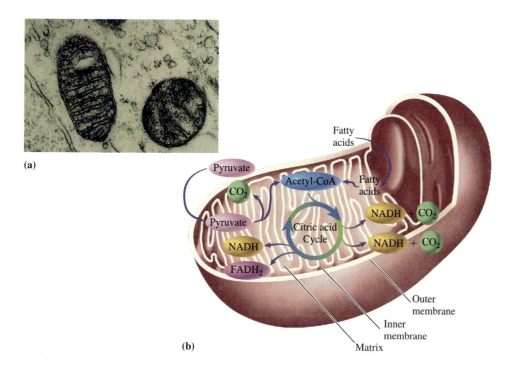

(a)

(b)

Figure 18.8

Structure of the mitochondrion.
(a) Electron micrograph of mitochondria.
(b) Schematic drawing of the mitochondrion.

A LIFESTYLE Connection | Exercise and Energy Metabolism

The Olympic sprinters get set in the blocks. The gun goes off, and roughly ten seconds later the 100-m dash is over. Elsewhere, the marathoners line up. They will run 26 miles and 385 yards in a little over two hours. Both sports involve running, but they use very different sources of energy.

Let's look at the sprinter first. The immediate source of energy for the sprinter is stored ATP. But the quantity of stored ATP is very small, only about three ounces. This allows the sprinter to run as fast as he or she can for about three seconds. Obviously, another source of stored energy must be tapped, and that energy store is *creatine phosphate:*

$$\text{O}^-\!-\!\overset{\displaystyle \text{O}}{\overset{\|}{\text{P}}}\!-\!\overset{\displaystyle \text{H}}{\overset{|}{\text{N}}}\!-\!\overset{\displaystyle \text{NH}}{\overset{\|}{\text{C}}}\!-\!\text{N}\!-\!\text{CH}_2\!-\!\text{C}\!\underset{\text{O}}{\overset{\text{O}^-}{<}}$$

The structure of creatine phosphate.

Creatine phosphate, stored in the muscle, donates its high-energy phosphate to ADP to produce new supplies of ATP.

This will keep our runner in motion for another five or six seconds before the store of creatine phosphate is also depleted. This is almost enough energy to finish the 100-m dash, but in reality, all the runners are slowing down, owing to energy depletion, and the winner is the sprinter who is slowing down the least!

Consider a longer race, the 400-m or the 800-m. These runners run at maximum capacity for much longer. When they have depleted their ATP and creatine phosphate stores, they must synthesize more ATP. Of course, the cells have been making ATP all the time, but now the demand for energy is much greater. To supply this increased demand, the anaerobic energy-generating reactions (glycolysis and lactate fermentation) and aerobic processes (citric acid cycle and oxidative phosphorylation) begin to function much more rapidly. Often, however, these athletes are running so strenuously that they cannot provide enough oxygen to the exercising muscle to allow oxidative phosphorylation to function efficiently. When this happens, the muscles must rely on glycolysis and lactate fermentation to provide *most* of the energy requirement. The chemical by-product of these anaerobic processes, lactate, builds up in the muscle and diffuses into the bloodstream. However, the concentration of lactate inevitably builds up in the working muscle and causes muscle fatigue and, eventually, muscle failure. Thus, exercise that depends primarily on anaerobic ATP production cannot continue for very long.

The marathoner presents us with a different scenario. This runner will deplete his or her stores of ATP and creatine phosphate as quickly as a short-distance runner. The anaerobic glycolytic pathway will begin to degrade glucose provided by the blood at a more rapid rate, as will the citric acid cycle and oxidative phosphorylation. The major difference in ATP production between the long-distance runner and the short- or middle-distance runner is that the muscles of the long-distance runner derive almost all the energy through aerobic pathways. These

Phosphoryl group transfer from creatine phosphate to ADP is catalyzed by the enzyme creatine kinase.

the **intermembrane space,** and the space inside the inner membrane is the **matrix space.** The enzymes of the citric acid cycle, of the β-oxidation pathway for the breakdown of fatty acids, and for the degradation of amino acids are all found in the mitochondrial matrix space.

Structure and Function

The outer mitochondrial membrane has many small pores through which small molecules (less than 10^4 g/mol) can pass. Thus, the small molecules to be oxidized for the production of ATP can easily enter the mitochondrial intermembrane space.

The inner membrane is highly folded to create a large surface area. The folded membranes are known as **cristae.** The inner mitochondrial membrane is almost completely impermeable to most substances. For this reason, it has many transport proteins to bring particular fuel molecules into the matrix space. Also embedded

Sprinters at the starting block.

needed for glycolysis and lactate fermentation. These muscle fibers fatigue rather quickly because fermentation is inefficient, quickly depleting the cell's glycogen store and causing the accumulation of lactate.

Slow twitch muscle fiber cells are about half the diameter of fast twitch muscle cells and are red. The red color is a result of the high concentrations of myoglobin in these cells. Recall that myoglobin stores oxygen for the cell (Section 16.8) and facilitates rapid diffusion of oxygen throughout the cell. In addition, slow twitch muscle fiber cells are packed with mitochondria. With this abundance of oxygen and mitochondria, these cells have the capacity for extended ATP production via aerobic pathways—ideal for endurance sports like marathon racing.

It is not surprising, then, that researchers have found that the muscles of sprinters have many more fast twitch muscle fibers and those of endurance athletes have many more slow twitch muscle fibers. One question that many researchers are trying to answer is whether the type of muscle fibers an individual has is a function of genetic makeup or training. Is a marathon runner born to be a long-distance runner, or are his or her abilities due to the type of training the runner undergoes? There is no doubt that the training regimen for an endurance runner does increase the number of slow twitch muscle fibers and that of a sprinter increases the number of fast twitch muscle fibers. But there is intriguing new evidence to suggest that the muscles of endurance athletes have a greater proportion of slow twitch muscle fibers before they ever begin training. It appears that some of us truly were born to run.

individuals continue to run long distances at a pace that allows them to supply virtually all the oxygen needed by the exercising muscle. Only aerobic pathways can provide a constant supply of ATP for exercise that goes on for hours. Theoretically, under such conditions, our runner could run indefinitely, using first his or her stored glycogen and eventually stored lipids. Of course, in reality, other factors such as dehydration and fatigue place limits on the athlete's ability to continue.

From this, we can conclude that long-distance runners must have a great capacity to produce ATP aerobically, in the mitochondria, whereas short- and middle-distance runners need a great capacity to produce energy anaerobically in the cytoplasm of the muscle cells. It is interesting to note that the muscles of these runners reflect these diverse needs.

When one examines muscle tissue that has been surgically removed, one finds two predominant types of muscle fibers. *Fast twitch muscle fibers* are large, relatively plump, pale cells. They have only a few mitochondria but contain a large reserve of glycogen and high concentrations of the enzymes that are

FOR FURTHER UNDERSTANDING

It has been said that the winner of the 100-m race is the one who is slowing down the least. Explain this observation in terms of energy-harvesting pathways.

Design an experiment to safely test whether the type of muscle fibers a runner has are the result of training or genetic makeup.

within the inner mitochondrial membrane are the protein electron carriers of the *electron transport system* and *ATP synthase*. ATP synthase is a large complex of many proteins that catalyzes the synthesis of ATP.

Origin of the Mitochondria

Mitochondria are roughly the size of bacteria; they have several other features that have led researchers to suspect that they may once have been free-living bacteria that were "captured" by eukaryotic cells. They have their own genetic information (DNA). They also make their own ribosomes that are very similar to those of bacteria. These ribosomes allow the mitochondria to synthesize some of their own proteins. Finally, mitochondria are actually self-replicating; they grow in size and divide to produce new mitochondria. All of these characteristics suggest that the mitochondria that produce the majority of the ATP for our cells evolved from bacteria "captured" perhaps as long as 1.5×10^9 years ago.

 To learn more about the genetics of mitochondria, see A Chemistry Connection: Mitochondria from Mom online at www.mhhe.com/denniston.

QUESTION 18.11

Draw a schematic diagram of a mitochondrion, and label the parts of this organelle.

QUESTION 18.12

How do mitochondria differ from the other components of eukaryotic cells?

18.6 Conversion of Pyruvate to Acetyl CoA

7 | **Learning Goal**
Describe the reaction that results in the conversion of pyruvate to acetyl CoA, describing the location of the reaction and the components of the pyruvate dehydrogenase complex.

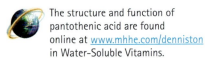

The structure and function of pantothenic acid are found online at www.mhhe.com/denniston in Water-Soluble Vitamins.

Thioester bonds are discussed in Section 12.4.

Under *aerobic* conditions, the cells can use oxygen and completely oxidize glucose to CO_2 in a metabolic pathway called the *citric acid cycle*. This pathway is often referred to as the *Krebs cycle* in honor of Sir Hans Krebs who worked out the steps of this cyclic pathway from his own experimental data and that of other researchers. It is also called the *tricarboxylic acid (TCA) cycle* because several early intermediates in the pathway have three carboxyl groups.

In the mitochondria, pyruvate is converted to a two-carbon acetyl group. This acetyl group is then "activated" to enter the reactions of the citric acid cycle. Activation occurs when the acetyl group is bonded to the thiol group of coenzyme A. **Coenzyme A** is a large thiol derived from ATP and the vitamin, pantothenic acid (Figure 18.9). It is an acceptor of acetyl groups (in red in Figure 18.9), which are bonded to it through a high-energy thioester bond. The product is acetyl coenzyme A **(acetyl CoA),** the "activated" form of the acetyl group.

Figure 18.10 shows us the reaction that converts pyruvate to acetyl CoA. First, pyruvate is decarboxylated, which means that it loses a carboxyl group that is released as CO_2. Next it is oxidized, and the hydride anion that is removed is accepted by NAD^+. Finally, the remaining acetyl group, $CH_3CO—$, is linked to coenzyme A by a thioester bond. This very complex reaction is carried out by three enzymes and five coenzymes that are organized together in a single bundle called the **pyruvate dehydrogenase complex** (see Figure 18.10). This organization allows the substrate to be passed from one enzyme to the next as each chemical reaction occurs. A schematic representation of this "disassembly line" is shown in Figure 18.10b.

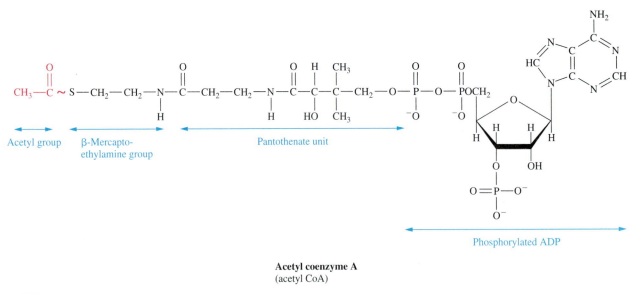

Acetyl coenzyme A
(acetyl CoA)

Figure 18.9

The structure of acetyl CoA. The bond between the acetyl group and coenzyme A is a high-energy thioester bond.

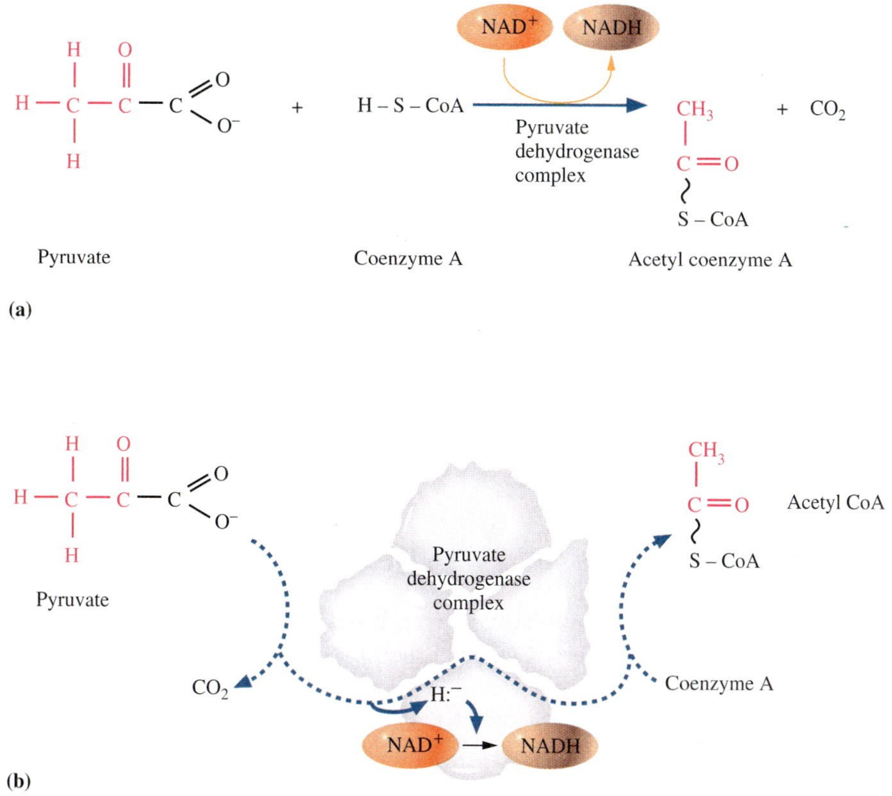

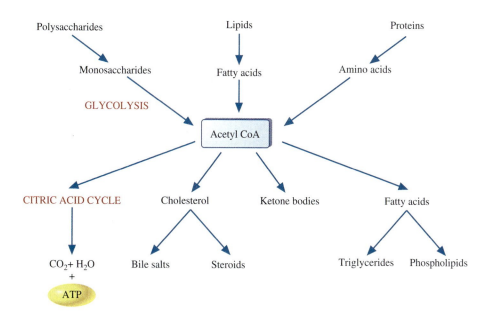

(a)

(b)

This single reaction requires four coenzymes made from four different vitamins, in addition to the coenzyme lipoamide: thiamine pyrophosphate, derived from thiamine (Vitamin B_1); FAD, derived from riboflavin (Vitamin B_2); NAD^+, derived from niacin; and coenzyme A, derived from pantothenic acid. Obviously, a deficiency in any of these vitamins would seriously reduce the amount of acetyl CoA that our cells could produce. This, in turn, would limit the amount of ATP that the body could make and would contribute to vitamin-deficiency diseases. Fortunately, a well-balanced diet provides an adequate supply of these and other vitamins.

In Figure 18.11, we see that acetyl CoA is a central character in cellular metabolism. It is produced by the degradation of glucose, fatty acids, and some amino

Figure 18.10

The decarboxylation and oxidation of pyruvate to produce acetyl CoA.
(a) The overall reaction in which CO_2 and an $H:^-$ are removed from pyruvate and the remaining acetyl group is attached to coenzyme A. This requires the concerted action of three enzymes and five coenzymes. (b) The pyruvate dehydrogenase complex that carries out this reaction is actually a cluster of enzymes and coenzymes. The substrate is passed from one enzyme to the next as the reaction occurs.

These vitamins are discussed online at www.mhhe.com/denniston in Water-Soluble Vitamins.

Figure 18.11

The central role of acetyl CoA in cellular metabolism.

acids. The major function of acetyl CoA in energy-harvesting pathways is to carry the acetyl group to the citric acid cycle, in which it will be used to produce large amounts of ATP. In addition to these catabolic duties, the acetyl group of acetyl CoA can also be used for *anabolic* or biosynthetic reactions to produce cholesterol and fatty acids. It is through this intermediate, acetyl CoA, that all energy sources (fats, proteins, and carbohydrates) are interconvertible.

QUESTION 18.13

Which vitamins are required for acetyl CoA production from pyruvate?

QUESTION 18.14

What is the major role of coenzyme A in catabolic reactions?

18.7 An Overview of Aerobic Respiration

8 Learning Goal
Summarize the reactions of aerobic respiration.

Aerobic respiration is the oxygen-requiring breakdown of food molecules and production of ATP. The different steps of aerobic respiration occur in different compartments of the mitochondria.

Under aerobic conditions cells use the citric acid cycle to completely oxidize glucose to CO_2. The enzymes for the citric acid cycle are found in the mitochondrial matrix space. The first enzyme catalyzes a reaction that joins the acetyl group of acetyl CoA (two carbons) to a four-carbon molecule (oxaloacetate) to produce citrate (six carbons). The remaining enzymes catalyze a series of rearrangements, decarboxylations (removal of CO_2), and oxidation-reduction reactions. The eventual products of this cyclic pathway are two CO_2 molecules and oxaloacetate—the molecule we began with.

Remember (Section 16.10) that it is really the hydride anion with its pair of electrons ($H:^-$) that is transferred to NAD^+ to produce NADH. Similarly, a pair of hydrogen atoms (and thus two electrons) are transferred to FAD to produce $FADH_2$.

At several steps in the citric acid cycle, a substrate is oxidized. In three of these steps, a pair of electrons is transferred from the substrate to NAD^+, producing NADH (three NADH molecules per cycle). At another step, a pair of electrons is transferred from a substrate to FAD, producing $FADH_2$ (one $FADH_2$ molecule per cycle).

The electrons are passed from NADH or $FADH_2$, through an electron transport system located in the inner mitochondrial membrane, and finally to the terminal electron acceptor, molecular oxygen (O_2). The transfer of electrons through the electron transport system causes protons (H^+) to be pumped from the mitochondrial matrix into the intermembrane compartment. The result is a high-energy H^+ reservoir.

In the final step, the energy of the H^+ reservoir is used to make ATP. This last step is carried out by the enzyme complex, ATP synthase. As protons flow back into the mitochondrial matrix through a pore in the ATP synthase complex, the enzyme catalyzes the synthesis of ATP.

This long, involved process is called *oxidative phosphorylation*, because the energy of electrons from the *oxidation* of substrates in the citric acid cycle is used to *phosphorylate* ADP and produce ATP. The details of each of these steps will be examined in upcoming sections.

Animation
How the Krebs Cycle Works

QUESTION 18.15

What is meant by the term *oxidative phosphorylation?*

QUESTION 18.16

What does the term *aerobic respiration* mean?

18.8 The Citric Acid Cycle (The Krebs Cycle)

Reactions of the Citric Acid Cycle

The **citric acid cycle,** often called the Krebs cycle in honor of the man who discovered it, is the final stage of the breakdown of carbohydrates, fats, and amino acids released from dietary proteins (Figure 18.12). To understand this important cycle, let's follow the fate of the acetyl group of an acetyl CoA as it passes through the citric acid cycle. The numbered steps listed below correspond to the steps in the citric acid cycle that are summarized in Figure 18.12.

> **9** Learning Goal
>
> Looking at an equation representing any of the chemical reactions in the citric acid cycle, describe the kind of reaction and the significance of that reaction to the pathway.

Reaction 1. This is a condensation reaction between the acetyl group of acetyl CoA and oxaloacetate. It is catalyzed by the enzyme, *citrate synthase.* The product that is formed is citrate:

Oxaloacetate   Acetyl CoA   Citrate   Coenzyme A

Reaction 2. The enzyme, *aconitase,* catalyzes the dehydration of citrate, producing *cis*-aconitate. The same enzyme, aconitase, then catalyzes addition of a water molecule to the *cis*-aconitate, converting it to isocitrate. The net effect of these two steps is the isomerization of citrate to isocitrate:

> *Notice that the conversion of citrate to cis-aconitate is a biological example of the dehydration of an alcohol to produce an alkene (Section 12.1). The conversion of cis-aconitate to isocitrate is a biochemical example of the hydration of an alkene to produce an alcohol (Sections 11.4 and 12.1).*

Citrate   *cis*-Aconitate   Isocitrate

Reaction 3. The first oxidative step of the citric acid cycle is catalyzed by *isocitrate dehydrogenase.* It is a complex reaction in which three things happen:
  a. the hydroxyl group of isocitrate is oxidized to a ketone,
  b. carbon dioxide is released, and
  c. NAD^+ is reduced to NADH.
The product of this oxidative decarboxylation reaction is α-ketoglutarate:

> *The oxidation of a secondary alcohol produces a ketone (Sections 12.1 and 12.5).*

> *The structure of NAD^+ and its reduction to NADH are shown in Figure 16.21.*

Isocitrate   α-Ketoglutarate

Figure 18.12

The reactions of the citric acid cycle.

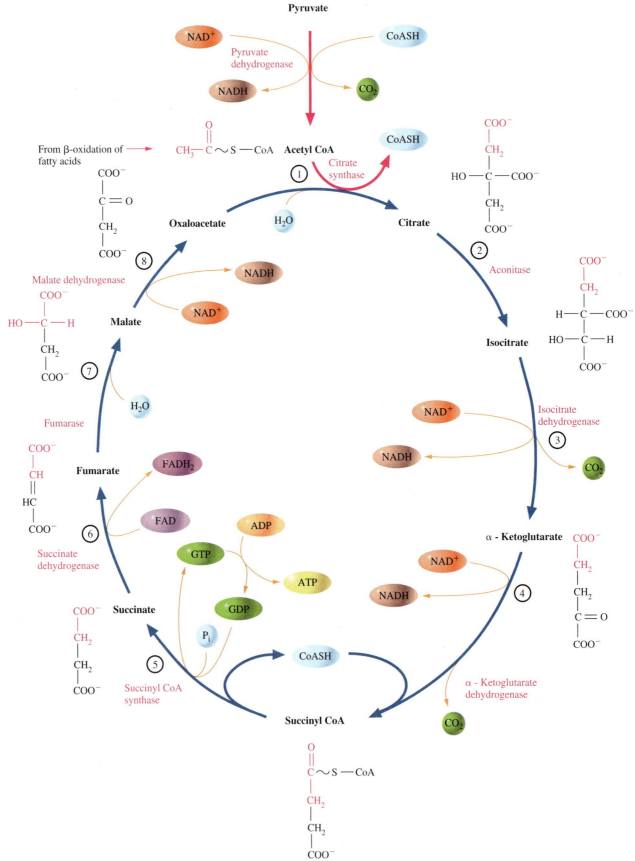

Reaction 4. Coenzyme A enters the picture again as the α-*ketoglutarate dehydrogenase* complex carries out a complex series of reactions similar to those catalyzed by the pyruvate dehydrogenase complex. The same coenzymes are required and, once again, three chemical events occur:

a. α-ketoglutarate loses a carboxylate group as CO_2,

b. it is oxidized and NAD^+ is reduced to NADH, and

c. coenzyme A combines with the product, succinate, to form succinyl CoA. The bond formed between succinate and coenzyme A is a high-energy thioester bond.

The pyruvate dehydrogenase complex was described in Section 18.6 and shown in Figure 18.10.

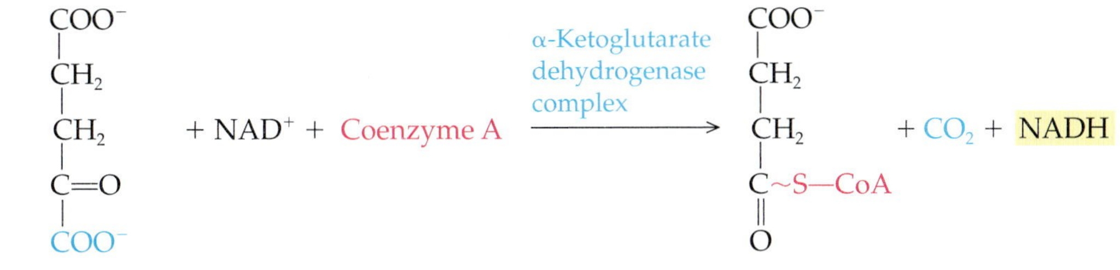

α-Ketoglutarate + NAD^+ + Coenzyme A →(α-Ketoglutarate dehydrogenase complex) Succinyl CoA + CO_2 + NADH

Reaction 5. Succinyl CoA is converted to succinate in this step, which once more is chemically very involved. The enzyme, *succinyl CoA synthase*, catalyzes a coupled reaction in which the high-energy thioester bond of succinyl CoA is hydrolyzed and an inorganic phosphate group is added to GDP to make GTP:

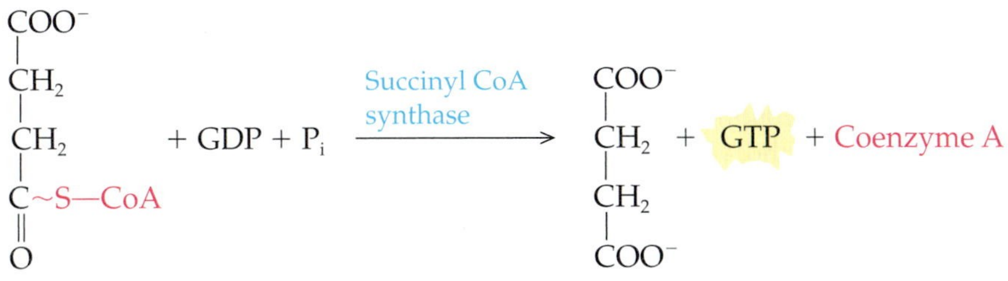

Succinyl CoA + GDP + P_i →(Succinyl CoA synthase) Succinate + GTP + Coenzyme A

Another enzyme, *dinucleotide diphosphokinase,* then catalyzes the transfer of a phosphoryl group from GTP to ADP to make ATP:

GTP + ADP →(Dinucleotide diphosphokinase) GDP + ATP

Reaction 6. Succinate dehydrogenase then catalyzes the oxidation of succinate to fumarate in the next step. The oxidizing agent, *flavin adenine dinucleotide (FAD)*, is reduced in this step:

The structure of FAD was shown in Figure 16.21.

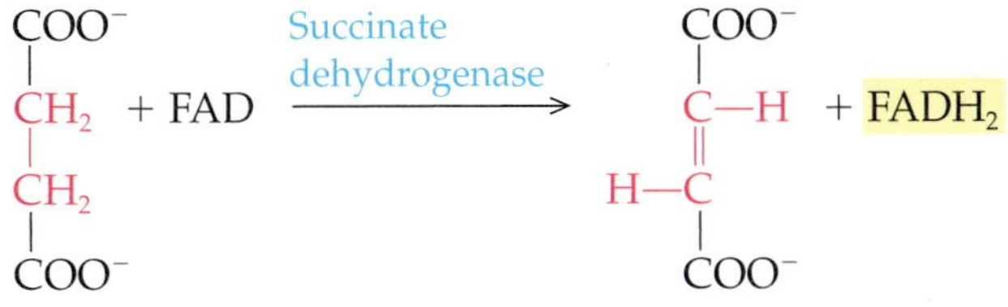

Succinate + FAD →(Succinate dehydrogenase) Fumarate + $FADH_2$

We studied hydrogenation of alkenes to produce alkanes in Section 11.4. This is simply the reverse.

This reaction is a biological example of the hydration of an alkene to produce an alcohol (Sections 11.4 and 12.1).

Reaction 7. Addition of H_2O to the double bond of fumarate gives malate. The enzyme, *fumarase,* catalyzes this reaction:

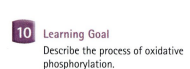

Fumarate   Malate

This reaction is a biochemical example of the oxidation of a secondary alcohol to a ketone, which we studied in Section 12.1.

Reaction 8. In the final step of the citric acid cycle, *malate dehydrogenase* catalyzes the reduction of NAD^+ to NADH and the oxidation of malate to oxaloacetate. Because the citric acid cycle "began" with the addition of an acetyl group to oxaloacetate, we have come full circle.

Malate   Oxaloacetate

18.9 Oxidative Phosphorylation

 **Learning Goal**

Describe the process of oxidative phosphorylation.

 **Animation**
Proton Pump

Animation
Electron Transport System and Formation of ATP

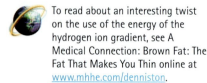 To read about an interesting twist on the use of the energy of the hydrogen ion gradient, see A Medical Connection: Brown Fat: The Fat That Makes You Thin online at www.mhhe.com/denniston.

In Section 18.7, we noted that the electrons carried by NADH can be used to produce three ATP molecules, and those carried by $FADH_2$ can be used to produce two ATP molecules. We turn now to the process by which the energy of electrons carried by these coenzymes is converted to ATP energy. It is a series of reactions called **oxidative phosphorylation,** which couples the oxidation of NADH and $FADH_2$ to the phosphorylation of ADP to generate ATP.

Electron Transport Systems and the Hydrogen Ion Gradient

Before we try to understand the mechanism of oxidative phosphorylation, let's first look at the molecules that carry out this complex process. Embedded within the mitochondrial inner membrane are **electron transport systems.** These are made up of a series of electron carriers, including coenzymes and cytochromes. All of these molecules are located within the membrane in an arrangement that allows them to pass electrons from one to the next. This array of electron carriers is called the *respiratory electron transport system* (Figure 18.13). As you would expect in such sequential oxidation-reduction reactions, the electrons lose some energy with each transfer. Some of this energy is used to make ATP.

At three sites in the electron transport system, protons (H^+) can be pumped from the mitochondrial matrix to the intermembrane space. These H^+ contribute to a high-energy H^+ reservoir. At each of the three sites, enough H^+ are pumped into the H^+ reservoir to produce one ATP molecule. The first site is NADH dehydrogenase. Because electrons from NADH enter the electron transport system by being transferred to NADH dehydrogenase, all three sites actively pump H^+, and three ATP molecules are made (see Figure 18.13). $FADH_2$ is a less "powerful" electron donor. It transfers its electrons to an electron carrier that follows NADH dehydrogenase. As a result, when $FADH_2$ is oxidized, only the second and third sites pump H^+, and only two ATP molecules are made.

The last component needed for oxidative phosphorylation is a multiprotein complex called **ATP synthase,** also called the F_0F_1 **complex** (see Figure 18.13). The

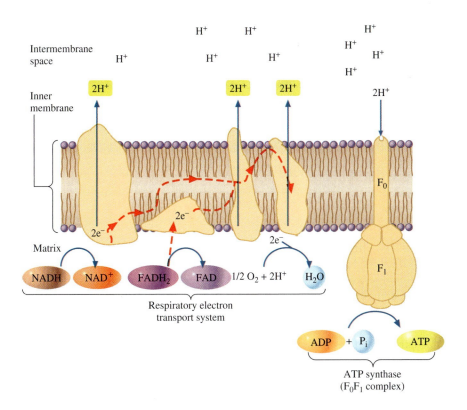

Figure 18.13

Electrons flow from NADH to molecular oxygen through a series of electron carriers embedded in the inner mitochondrial membrane. Protons are pumped from the mitochondrial matrix space into the intermembrane space. This results in a hydrogen ion reservoir in the intermembrane space. As protons pass through the channel in ATP synthase, their energy is used to phosphorylate ADP and produce ATP.

F_0 portion of the molecule is a channel through which H^+ pass. It spans the inner mitochondrial membrane, as shown in Figure 18.13. The F_1 part of the molecule is an enzyme that catalyzes the phosphorylation of ADP to produce ATP.

ATP Synthase and the Production of ATP

How does all this complicated machinery actually function? NADH carries electrons, originally from glucose, to the first carrier of the electron transport system, NADH dehydrogenase (see Figure 18.13). There, NADH is oxidized to NAD^+, which returns to the site of the citric acid cycle to be reduced again. As the dashed red line shows, the pair of electrons is passed to the next electron carrier, and H^+ are pumped to the intermembrane compartment. The electrons are passed sequentially through the electron transport system, and at two additional sites, H^+ from the matrix are pumped into the intermembrane compartment. With each transfer, the electrons lose some of their potential energy. This energy is used to transport H^+ across the inner mitochondrial membrane and into the H^+ reservoir. As mentioned above, $FADH_2$ donates its electrons to a carrier of lower energy and fewer H^+ are pumped into the reservoir.

Finally, the electrons arrive at the last carrier. They now have too little energy to accomplish any more work, but they *must* be donated to some final electron acceptor so that the electron transport system can continue to function. In aerobic organisms, the **terminal electron acceptor** is molecular oxygen, O_2, and the product is water.

As the electron transport system continues to function, a high concentration of protons builds up in the intermembrane space. This creates an H^+ gradient across the inner mitochondrial membrane. Such a gradient is an enormous energy source, like water stored behind a dam. The mitochondria use the potential energy of the gradient to synthesize ATP energy.

ATP synthase harvests the energy of this gradient by making ATP. H^+ pass through the F_0 channel back into the matrix. This causes F_1 to become an active

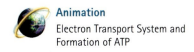

Animation

Electron Transport System and Formation of ATP

The importance of keeping the electron transport system functioning becomes obvious when we consider what occurs in cyanide poisoning. Cyanide binds to the heme group iron of cytochrome oxidase, instantly stopping electron transfers and causing death within minutes!

enzyme that catalyzes the phosphorylation of ADP to produce ATP. In this way, the energy of the H^+ reservoir is harvested to make ATP.

QUESTION 18.17

Write a balanced chemical equation for the reduction of NAD^+.

QUESTION 18.18

Write a balanced chemical equation for the reduction of FAD.

Summary of the Energy Yield

One turn of the citric acid cycle results in the production of two CO_2 molecules, three NADH molecules, one $FADH_2$ molecule, and one ATP molecule. Oxidative phosphorylation yields three ATP molecules per NADH molecule and two ATP molecules per $FADH_2$ molecule. The only exception to these energy yields is the NADH produced in the cytoplasm during glycolysis. Oxidative phosphorylation yields only two ATP molecules per cytoplasmic NADH molecule because energy must be expended to shuttle electrons from NADH in the cytoplasm to $FADH_2$ in the mitochondrion.

Knowing this information and keeping in mind that two turns of the citric acid cycle are required, we can sum up the total energy yield from the complete oxidation of one glucose molecule.

 In some tissues of the body, there is a more efficient shuttle system that results in the production of three ATP per cytoplasmic NADH. This system is described online at www.mhhe.com/denniston in Energy Yields from Aerobic Respiration: Some Alternatives.

EXAMPLE 18.1 Determining the Yield of ATP from Aerobic Respiration

Calculate the number of ATP produced by the complete oxidation of one molecule of glucose.

SOLUTION

Glycolysis:	
Substrate-level phosphorylation	2 ATP
2 NADH × 2 ATP/cytoplasmic NADH	4 ATP
Conversion of 2 pyruvate molecules to 2 acetyl CoA molecules:	
2 NADH × 3 ATP/NADH	6 ATP
Citric acid cycle (two turns):	
2 GTP × 1 ATP/GTP	2 ATP
6 NADH × 3 ATP/NADH	18 ATP
2 $FADH_2$ × 2 ATP/$FADH_2$	4 ATP
	36 ATP

This represents an energy harvest of about 40% of the potential energy of glucose.

Aerobic metabolism is very much more efficient than anaerobic metabolism. The abundant energy harvested by aerobic metabolism has had enormous consequences for the biological world. Much of the energy released by the oxidation of fuels is not lost as heat but conserved in the form of ATP. Organisms that possess abundant energy have evolved into multicellular organisms and developed specialized functions. As a consequence of their energy requirements, all multicellular organisms are aerobic.

18.10 Gluconeogenesis: The Synthesis of Glucose

Under normal conditions, we have enough glucose to satisfy our needs. However, under some conditions, the body must make glucose. This is necessary following strenuous exercise to replenish the liver and muscle supplies of glycogen. It also occurs during starvation so that the body can maintain adequate blood glucose levels to supply the brain cells and red blood cells. Under normal conditions, these two tissues use only glucose for energy.

Glucose is produced by the process of **gluconeogenesis,** the production of glucose from noncarbohydrate starting materials (Figure 18.14). Gluconeogenesis, an anabolic pathway, occurs primarily in the liver. Lactate, all amino acids except leucine and lysine, and glycerol from fats can be used to make glucose.

11 **Learning Goal**

Discuss the biological function of gluconeogenesis.

Under extreme conditions of starvation, the brain eventually switches to the use of ketone bodies. Ketone bodies are produced, under certain circumstances, from the breakdown of lipids (Section 19.3).

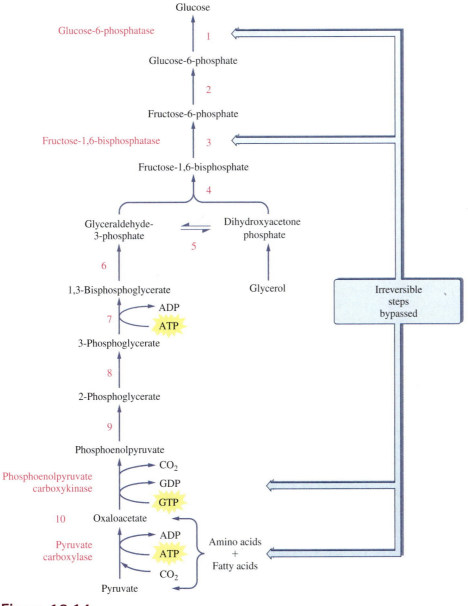

Figure 18.14

Comparison of the reactions of glycolysis and gluconeogenesis.

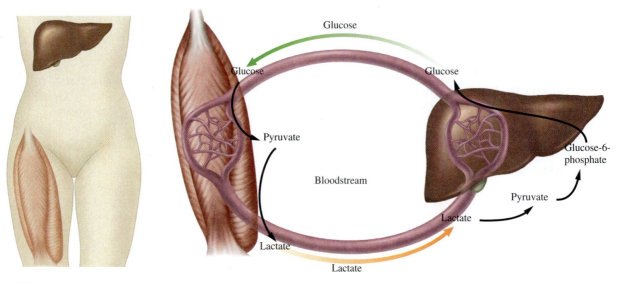

Figure 18.15

The Cori Cycle.

However, amino acids and glycerol are generally used only under starvation conditions.

At first glance, gluconeogenesis appears to be simply the reverse of glycolysis (compare Figures 18.14 and 18.5) because the intermediates of the two pathways are identical. But this is not the case because steps 1, 3, and 10 of glycolysis are irreversible, and therefore the reverse reactions must be carried out by other enzymes. These are noted in red in Figure 18.14.

As we have seen, the conversion of lactate into glucose is important in mammals. As the muscles work, they produce lactate, which is converted back to glucose in the liver. The glucose is transported into the blood and from there back to the muscle. In the muscle, it can be catabolized to produce ATP, or it can be used to replenish the muscle stores of glycogen. This cyclic process between the liver and skeletal muscles, called the **Cori Cycle,** is shown in Figure 18.15. Through this cycle, gluconeogenesis produces enough glucose to restore the depleted muscle glycogen reservoir within forty-eight hours.

18.11 Glycogen Synthesis and Degradation

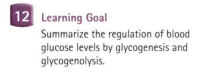

Learning Goal

Summarize the regulation of blood glucose levels by glycogenesis and glycogenolysis.

Glucose is the sole source of energy of mammalian red blood cells and the major source of energy for the brain. Neither red blood cells nor the brain can store glucose; thus a constant supply must be available as blood glucose. This is provided by dietary glucose and by the production of glucose either by gluconeogenesis or by **glycogenolysis,** the degradation of glycogen. Glycogen is a long-branched-chain polymer of glucose. Stored in the liver and skeletal muscles, it is the principal storage form of glucose.

The total amount of glucose in the blood of a 70-kg (approximately 150-lb) adult is about 20 g, but the brain alone consumes 5–6 g of glucose per hour. Breakdown of glycogen in the liver mobilizes the glucose when hormonal signals register a need for increased levels of blood glucose. Skeletal muscle also contains substantial stores of glycogen, which provide energy for rapid muscle contraction. However, this glycogen is not able to contribute to blood glucose because muscle cells do not have the enzyme, glucose-6-phosphatase. Because glucose cannot be formed from glucose-6-phosphate, it cannot be released into the bloodstream.

The Structure of Glycogen

Glycogen is a highly branched glucose polymer in which the "main chain" is linked by $\alpha\,(1 \rightarrow 4)$ glycosidic bonds. The polymer also has numerous $\alpha\,(1 \rightarrow 6)$ glycosidic bonds, which provide many branch points along the chain. This structure is shown schematically in Figure 18.16. **Glycogen granules** with a

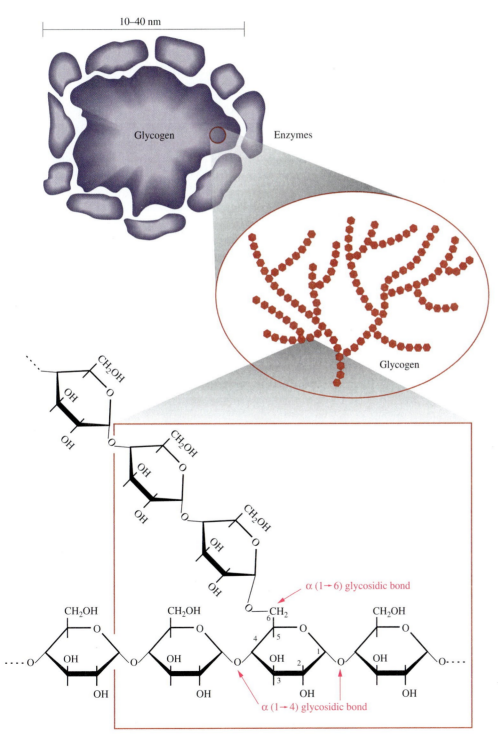

Figure 18.16

The structure of glycogen and a glycogen granule.

A MEDICAL Connection | Diagnosing Diabetes

When diagnosing diabetes, doctors consider many factors and symptoms. However, there are two primary tests to determine whether an individual is properly regulating blood glucose levels. First and foremost is the fasting blood glucose test. A person who has fasted since midnight should have a blood glucose level between 70 and 110 mg/dL in the morning. If the level is 140 mg/dL on at least two occasions, a diagnosis of diabetes is generally made.

The second commonly used test is the glucose tolerance test. For this test, the subject must fast for at least ten hours. A beginning blood sample is drawn to determine the fasting blood glucose level. This will serve as the background level for the test. The subject ingests 50–100 g of glucose (40 g/m^2 body surface), and the blood glucose level is measured at thirty minutes, and at one, two, and three hours after ingesting the glucose.

A graph is made of the blood glucose levels over time. For a person who does not have diabetes, the curve will show a peak of blood glucose at approximately one hour. There will be a reduction in the level and perhaps a slight hypoglycemia (low blood glucose level) over the next hour. Thereafter, the blood glucose level stabilizes at normal levels.

An individual is said to have impaired glucose tolerance if the blood glucose level remains between 140 and 200 mg/dL two hours after ingestion of the glucose solution. This suggests that there is a risk of the individual developing diabetes and is reason to prescribe periodic testing to allow early intervention.

If the blood glucose level remains at or above 200 mg/dL after two hours, a tentative diagnosis of diabetes is made. However, this result warrants further testing on subsequent days to rule out transient problems, such as the effect of medications on blood glucose levels.

It was recently suggested that the upper blood glucose level of 200 mg/dL should be lowered to 180 mg/dL as the standard to diagnose impaired glucose tolerance and diabetes. This would allow earlier detection and intervention. Considering the grave nature of long-term diabetic complications, it is thought to be very beneficial to begin treatment at an early stage to maintain constant blood glucose levels. For more information on diabetes, see A Medical Connection: Diabetes Mellitus and Ketone Bodies, in Chapter 19.

FOR FURTHER UNDERSTANDING

Draw a graph representing blood glucose levels for a normal glucose tolerance test.

Draw a similar graph for an individual who would be diagnosed as diabetic.

diameter of 10–40 nm are found in the cytoplasm of liver and muscle cells. These granules exist in complexes with the enzymes that are responsible for glycogen synthesis and degradation. The structure of such a granule is also shown in Figure 18.16.

Glycogenolysis: Glycogen Degradation

Two hormones control glycogenolysis, the degradation of glycogen. These are **glucagon,** a peptide hormone synthesized in the pancreas, and *epinephrine,* produced in the adrenal glands. Glucagon is released from the pancreas in response to low blood glucose, a condition called **hypoglycemia,** and epinephrine is released from the adrenal glands in response to a threat or a stress. Both situations require an increase in blood glucose, and both hormones function by altering the activity of two enzymes, glycogen phosphorylase and glycogen synthase. *Glycogen phosphorylase* is involved in glycogen degradation and is activated; *glycogen synthase* is involved in glycogen synthesis and is inactivated.

Glycogenesis: Glycogen Synthesis

The hormone **insulin,** produced by the pancreas in response to **hyperglycemia,** high blood glucose levels, stimulates the synthesis of glycogen, **glycogenesis.**

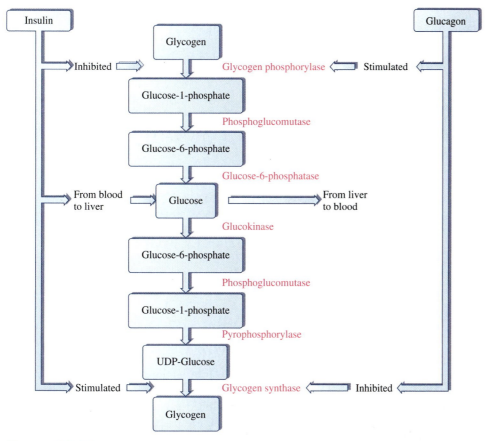

Figure 18.17

The opposing effects of the hormones insulin and glucagon on glycogen metabolism.

Insulin is perhaps one of the most influential hormones in the body because it directly alters the metabolism and uptake of glucose in all but a few cells.

When blood glucose rises, as after a meal, the beta cells of the pancreas secrete insulin. It immediately accelerates the uptake of glucose by all cells of the body except the brain and certain blood cells. In these cells, the uptake of glucose is insulin-independent. The increased uptake of glucose is especially marked in the liver, heart, skeletal muscle, and adipose tissue.

In the liver, insulin promotes glycogen synthesis and storage by inhibiting glycogen phosphorylase, thus inhibiting glycogen degradation. It also stimulates glycogen synthase and glucokinase, two enzymes that are involved in glycogen synthesis.

The opposing action of insulin and glucagon, shown in Figure 18.17, ensures that the reactions involved in glycogen degradation and synthesis do not compete with one another. In this way they provide glucose when the blood level is too low, and they cause the storage of glucose in times of excess.

Animation
Blood Sugar Regulation in Diabetics

QUESTION 18.19

Explain how glucagon affects the synthesis and degradation of glycogen.

QUESTION 18.20

How does insulin affect the storage and degradation of glycogen?

SUMMARY

18.1 ATP: The Cellular Energy Currency

Adenosine triphosphate, ATP, is a nucleotide composed of ade-nine, the sugar ribose, and a triphosphate group. The energy released by the hydrolysis of the phosphoanhydride bond between the second and third phosphoryl groups provides the energy for most cellular work.

18.2 Overview of Catabolic Processes

The body needs a supply of ATP to carry out life processes. To provide this ATP, we consume a variety of energy-rich food molecules: carbohydrates, lipids, and proteins. In the digestive tract, these large molecules are degraded into smaller molecules (monosaccharides, glycerol, fatty acids, and amino acids) that are absorbed by our cells. These molecules are further broken down to generate ATP.

18.3 Glycolysis

Glycolysis is the pathway for the *catabolism* of glucose that leads to pyruvate. It is an anaerobic process carried out by enzymes in the cytoplasm of the cell. The net harvest of ATP during glycolysis is two molecules of ATP per molecule of glucose. Two molecules of NADH are also produced. The rate of glycolysis responds to the energy demands of the cell.

18.4 Fermentations

Under anaerobic conditions, the NADH produced by glycolysis is used to reduce pyruvate to lactate in skeletal muscle (lactate *fermentation*) or to convert acetaldehyde to ethanol in yeast (alcohol fermentation).

18.5 The Mitochondria

The *mitochondria* are aerobic cell organelles that are responsible for most of the ATP production in eukaryotic cells. They are enclosed by a double membrane. The outer membrane permits low-molecular-weight molecules to pass through. The *inner mitochondrial membrane*, by contrast, is almost completely impermeable to most molecules. The inner mitochondrial membrane is the site where *oxidative phosphorylation* occurs. The enzymes of the *citric acid cycle,* of amino acid catabolism, and of fatty acid oxidation are located in the *matrix space* of the mitochondrion.

18.6 Conversion of Pyruvate to Acetyl CoA

Under aerobic conditions, pyruvate is oxidized by the *pyruvate dehydrogenase complex. Acetyl CoA,* formed in this reaction, is a central molecule in both catabolism and *anabolism.*

18.7 An Overview of Aerobic Respiration

Aerobic respiration is the oxygen-requiring degradation of food molecules and production of ATP. *Oxidative phosphorylation* is the process that uses the high-energy electrons harvested by oxidation of substrates of the citric acid cycle to produce ATP.

18.8 The Citric Acid Cycle (The Krebs Cycle)

The *citric acid cycle* is the final pathway for the degradation of carbohydrates, amino acids, and fatty acids. The citric acid cycle occurs in the matrix of the mitochondria. It is a cyclic series of biochemical reactions that accomplishes the complete oxidation of the carbon skeletons of food molecules.

18.9 Oxidative Phosphorylation

Oxidative phosphorylation is the process by which NADH and $FADH_2$ are oxidized and ATP is produced. Two molecules of ATP are produced when $FADH_2$ is oxidized, and three molecules of ATP are produced when NADH is oxidized. The complete oxidation of one glucose molecule by glycolysis, the citric acid cycle, and oxidative phosphorylation yield thirty-six molecules of ATP versus two molecules of ATP for anaerobic degradation of glucose by glycolysis and fermentation.

18.10 Gluconeogenesis: The Synthesis of Glucose

Gluconeogenesis is the pathway for glucose synthesis from noncarbohydrate starting materials. It occurs in mammalian liver. Glucose can be made from lactate, all amino acids except lysine and leucine, and glycerol. Gluconeogenesis is not simply the reversal of glycolysis. Three steps in glycolysis in which ATP is produced or consumed are bypassed by different enzymes in gluconeogenesis. All other enzymes in gluconeogenesis are shared with glycolysis.

18.11 Glycogen Synthesis and Degradation

Glycogenesis is the pathway for the synthesis of glycogen, and *glycogenolysis* is the pathway for the degradation of glycogen. The concentration of blood glucose is controlled by the liver. A high blood glucose level causes secretion of *insulin.* This hormone stimulates glycogenesis and inhibits glycogenolysis. When blood glucose levels are too low, the hormone *glucagon* stimulates gluconeogenesis and glycogen degradation in the liver.

KEY TERMS

acetyl CoA (18.6)

adenosine triphosphate
(ATP) (18.1)

aerobic respiration (18.7)

anabolism (18.1)

anaerobic threshold (18.4)

ATP synthase (18.9)

catabolism (18.1)

citric acid cycle (18.8)

coenzyme A (18.6)

Cori Cycle (18.10)

cristae (18.5)

electron transport system (18.9)

fermentation (18.4)

F_0F_1 complex (18.9)

glucagon (18.11)

gluconeogenesis (18.10)

glycogen (18.11)

glycogenesis (18.11)

glycogen granule (18.11)

glycogenolysis (18.11)

glycolysis (18.3)

hyperglycemia (18.11)

hypoglycemia (18.11)

inner mitochondrial
membrane (18.5)

insulin (18.11)

intermembrane space (18.5)

matrix space (18.5)

mitochondria (18.5)

nicotinamide adenine
dinucleotide (NAD^+) (18.3)

outer mitochondrial
membrane (18.5)

oxidative
phosphorylation (18.9)

pyruvate dehydrogenase
complex (18.6)

substrate-level
phosphorylation (18.3)

terminal electron
acceptor (18.9)

QUESTIONS AND PROBLEMS

ATP: The Cellular Energy Currency

Foundations

18.21 What molecule is primarily responsible for conserving the energy released in catabolism?

18.22 Describe the structure of ATP.

Applications

18.23 What is meant by the term *high-energy bond?*

18.24 Compare and contrast anabolism and catabolism in terms of their roles in metabolism and their relationship to ATP.

Overview of Catabolic Processes

Foundations

18.25 What is the most readily used energy source in the diet?

18.26 What is a hydrolysis reaction?

Applications

18.27 Write an equation showing the hydrolysis of maltose.

18.28 How are monosaccharides transported into a cell?

18.29 Write an equation showing the hydrolysis of a triglyceride consisting of glycerol, oleic acid, linoleic acid, and stearic acid.

18.30 How are fatty acids taken up into the cell?

Glycolysis

Foundations

18.31 What are the end products of glycolysis?

18.32 What is the net energy yield of ATP in glycolysis?

18.33 Where in the muscle cell does glycolysis occur?

18.34 Fill in the blanks:
  a. _____ molecules of ATP are produced per molecule of glucose that is converted to pyruvate.
  b. Two molecules of ATP are consumed in the conversion of _____ to fructose-1,6-bisphosphate.
  c. NAD^+ is _____ to NADH in the first energy-releasing step of glycolysis.
  d. The second substrate-level phosphorylation in glycolysis is phosphoryl group transfer from phosphoenolpyruvate to _____ .

Applications

18.35 When an enzyme has the term *kinase* in the name, what type of reaction do you expect it to catalyze?

18.36 What features do the reactions catalyzed by hexokinase and phosphofructokinase share in common?

18.37 What is the role of NAD^+ in a biochemical oxidation reaction?

18.38 Why must NADH produced in glycolysis be reoxidized to NAD^+?

Fermentations

Foundations

18.39 Write a balanced chemical equation for the conversion of acetaldehyde to ethanol.

18.40 Write a balanced chemical equation for the conversion of pyruvate to lactate.

Applications

18.41 A child was brought to the doctor's office suffering from a strange set of symptoms. When the child exercised hard, she became giddy and behaved as though drunk. What do you think is the metabolic basis of these symptoms?

18.42 A family started a batch of wine by adding yeast to grape juice and placing the mixture in a sealed bottle. Two weeks later, the bottle exploded. What metabolic reactions—and specifically, what product of those reactions—caused the bottle to explode?

The Mitochondria

Foundations

18.43 Define the term *mitochondrion.*

18.44 Define the term *cristae.*

Applications

18.45 What biochemical processes occur in the matrix space of the mitochondria?

18.46 What kinds of proteins are found in the inner mitochondrial membrane?

Conversion of Pyruvate to Acetyl CoA

Foundations

18.47 What is coenzyme A?

18.48 What is the role of coenzyme A in the reaction catalyzed by pyruvate dehydrogenase?

Applications

18.49 Under what metabolic conditions is pyruvate converted to acetyl CoA?

18.50 Write a chemical equation for the production of acetyl CoA from pyruvate. Under what conditions does this reaction occur?

18.51 How could a deficiency of riboflavin, thiamine, niacin, or pantothenic acid reduce the amount of ATP the body can produce?

18.52 In what form are the vitamins, riboflavin, thiamine, niacin, and pantothenic acid, needed by the pyruvate dehydrogenase complex?

The Citric Acid Cycle

Foundations

18.53 The pair of reactions catalyzed by aconitase results in the conversion of isocitrate to its isomer citrate. What are isomers?

18.54 The reaction catalyzed by succinate dehydrogenase is a dehydrogenation reaction. What is meant by the term *dehydrogenation reaction?*

18.55 Label each of the following statements as true or false:
   a. Both glycolysis and the citric acid cycle are aerobic processes.
   b. Both glycolysis and the citric acid cycle are anaerobic processes.
   c. Glycolysis occurs in the cytoplasm, and the citric acid cycle occurs in the mitochondria.
   d. The inner membrane of the mitochondrion is virtually impermeable to most substances.

18.56 Fill in the blanks:
   a. The proteins of the electron transport system are found in the _____, the enzymes of the citric acid cycle are found in the _____, and the hydrogen ion reservoir is found in the _____ of the mitochondria.
   b. The infoldings of the inner mitochondrial membrane are called _____.
   c. Energy released by oxidation in the citric acid cycle is conserved in the form of phosphoanhydride bonds in _____.
   d. The purpose of the citric acid cycle is the _____ of the acetyl group.

18.57 To what final products is the acetyl group of acetyl CoA converted during oxidation in the citric acid cycle?

18.58 How many molecules of ATP are produced by the complete degradation of glucose via glycolysis, the citric acid cycle, and oxidative phosphorylation?

Applications

18.59 A bacterial culture is given ^{14}C-labeled pyruvate as its sole source of carbon and energy. The following is the structure of the radiolabeled pyruvate.

$$ \overset{O}{\overset{\|}{*CH_3-C}}-\overset{O}{\overset{\|}{C}}-O^- $$

Follow the fate of the radioactive carbon through the reactions of the citric acid cycle.

18.60 A bacterial culture is given ^{14}C-labeled pyruvate as its sole source of carbon and energy. The following is the structure of the radiolabeled pyruvate.

$$ \overset{O}{\overset{\|}{CH_3-C^*}}-\overset{O}{\overset{\|}{C}}-O^- $$

Follow the fate of the radioactive carbon through the reactions of the citric acid cycle.

18.61 To what class of enzymes does dinucleotide diphosphokinase belong? Explain your answer.

18.62 To what class of enzymes does succinate dehydrogenase belong? Explain your answer.

Oxidative Phosphorylation

Foundations

18.63 Define the term *electron transport system.*
18.64 What is the terminal electron acceptor in aerobic respiration?

Applications

18.65 How many molecules of ATP are produced when one molecule of NADH is oxidized by oxidative phosphorylation?

18.66 How many molecules of ATP are produced when one molecule of $FADH_2$ is oxidized by oxidative phosphorylation?

18.67 What is the source of energy for the synthesis of ATP in mitochondria?

18.68 What is the function of the electron transport systems of mitochondria?

18.69 a. Compare the number of molecules of ATP produced by glycolysis to the number of ATP molecules produced by oxidation of glucose by aerobic respiration.
   b. Which pathway produces more ATP? Explain.

18.70 At which steps in the citric acid cycle do oxidation-reduction reactions occur?

Gluconeogenesis

18.71 Define gluconeogenesis and describe its role in metabolism.
18.72 What organ is primarily responsible for gluconeogenesis?
18.73 What is the physiological function of gluconeogenesis?
18.74 Lactate can be converted to glucose by gluconeogenesis. To what metabolic intermediate must lactate be converted so that it can be a substrate for the enzymes of gluconeogenesis?

Glycogen Synthesis and Degradation

Foundations

18.75 What organs are primarily responsible for maintaining the proper blood glucose level?
18.76 Why must the blood glucose level be carefully regulated?
18.77 What does the term *hypoglycemia* mean?
18.78 What does the term *hyperglycemia* mean?

Applications

18.79 a. What enzymes involved in glycogen metabolism are stimulated by insulin?
   b. What effect does this have on glycogen metabolism?
   c. What effect does this have on blood glucose levels?
18.80 a. What enzyme is stimulated by glucagon?
   b. What effect does this have on glycogen metabolism?
   c. What effect does this have on blood glucose levels?

FOR FURTHER UNDERSTANDING

1. An enzyme that hydrolyzes ATP (an ATPase) bound to the plasma membrane of certain tumor cells has an abnormally high activity. How will this activity affect the rate of glycolysis?

2. Explain why no net oxidation occurs during anaerobic glycolysis followed by lactate fermentation.

3. A scientist added phosphate labeled with radioactive phosphorus (^{32}P) to a bacterial culture growing anaerobically (without O_2). She then purified all the compounds produced during glycolysis. Look carefully at the steps of the pathway. Predict which of the intermediates of the pathway would be the first to contain radioactive phosphate. On which carbon of this compound would you expect to find the radioactive phosphate?

4. A one-month-old baby boy was brought to the hospital showing severely delayed development and cerebral atrophy. Blood tests showed high levels of lactate and pyruvate. By three months of age, very high levels of succinate and fumarate were found in the urine. Fumarase activity was absent in the liver and muscle tissue. The baby died at five months of age. This was the first reported case of fumarase deficiency and the

defect was recognized too late for effective therapy to be administered. What reaction is catalyzed by fumarase? How would a deficiency of this mitochondrial enzyme account for the baby's symptoms and test results?

5. A certain bacterium can grow with ethanol as its only source of energy and carbon. Propose a pathway to describe how ethanol can enter a pathway that would allow ATP production and synthesis of precursors for biosynthesis.

6. Fluoroacetate has been used as a rat poison and can be fatal when eaten by humans. Patients with fluoroacetate poisoning accumulate citrate and fluorocitrate within the cells. What enzyme is inhibited by fluoroacetate? Explain your reasoning.

7. The pyruvate dehydrogenase complex is activated by removal of a phosphoryl group from pyruvate dehydrogenase. This reaction is catalyzed by the enzyme pyruvate dehydrogenase phosphate phosphatase. A baby is born with a defect in this enzyme. What effects would this defect have on the rate of each of the following pathways: aerobic respiration, glycolysis, lactate fermentation? Explain your reasoning.

8. Pyruvate dehydrogenase phosphate phosphatase is stimulated by Ca^{2+}. In muscles, the Ca^{2+} concentration increases dramatically during muscle contraction. How would the elevated Ca^{2+} concentration affect the rate of glycolysis and the citric acid cycle?

19

LEARNING GOALS

1 Summarize the digestion and storage of lipids.

2 Describe the degradation of fatty acids by β-oxidation.

3 Explain the role of acetyl CoA in fatty acid metabolism.

4 Understand the role of ketone body production in β-oxidation.

5 Compare β-oxidation of fatty acids and fatty acid biosynthesis.

6 Describe the conversion of amino acids to molecules that can enter the citric acid cycle.

7 Explain the importance of the urea cycle and describe its essential steps.

8 Discuss the cause and effect of hyperammonemia.

9 Summarize the antagonistic effects of glucagon and insulin.

Fatty Acid and Amino Acid Metabolism

Approximately a third of all Americans are obese; that is, they are more than 20% overweight. One million are morbidly obese; they carry so much extra weight that it threatens their health. Many obese people simply eat too much and exercise too little, but others actually gain weight even though they eat fewer calories than people of normal weight. This observation led many researchers to the hypothesis that obesity in some people is a genetic disorder.

This hypothesis was supported by the 1950 discovery of an obesity mutation in mice.

In 1987, Jeffrey Friedman assembled a team of researchers to map and then clone the obesity gene that was responsible for appetite control. In 1994, after seven years of intense effort, the scientists achieved their goal.

They then injected the protein into each of ten mice that were so fat they couldn't squeeze into the feeding tunnels used for normal mice. After two weeks of treatment, each of the ten mice had lost about 30% of its weight. In addition, the mice had become more active and their metabolisms had speeded up.

Because of the dramatic results, Friedman and his colleagues called the protein leptin, from the Greek word leptos, meaning slender.

The human leptin gene also has been cloned and shown to correct genetic obesity in mice. Unfortunately, the dramatic results achieved with mice were not observed with humans. Why? It seems that nearly all of the obese volunteers already produced an abundance of leptin. In fact, fewer than ten people have been found, to date, who do not produce leptin. For these individuals, leptin injections do, indeed, reduce their appetites and lead to significant weight loss.

Clearly lipid metabolism in animals is a complex process and is not yet fully understood. The discovery of the leptin gene, and the hormone it produces, is just one part of the story. The metabolism of lipids and proteins revolves around the fate of acetyl CoA and the glycolysis and citric acid cycle pathways. In this chapter, we will see that several other metabolic pathways integrate the metabolism of fatty acids and amino acids into these other pathways and, so, complete our understanding of the breakdown and synthesis of carbohydrates, lipids, and proteins.

A healthy diet consists of carbohydrates, proteins, and lipids, all of which can be found on the picnic table in this photograph. Briefly describe the digestive processes that these foods would undergo to deliver food molecules to metabolic reactions.

19.1 Lipid Metabolism in Animals

Digestion and Absorption of Dietary Triglycerides

1 Learning Goal

Summarize the digestion and storage of lipids.

Triglycerides are highly hydrophobic ("water fearing"). Most dietary fat arrives in the duodenum, the first part of the small intestine, in the form of fat globules. These fat globules stimulate the secretion of bile from the gallbladder. **Bile** is composed of micelles of lecithin, cholesterol, protein, bile salts, inorganic ions, and bile pigments. **Micelles** (Figure 19.1) are aggregations of molecules having a polar region and a nonpolar region. The nonpolar ends of bile salts tend to bunch together when placed in water. The hydrophilic ("water loving") regions of these molecules interact with water. The major bile salts in humans are cholate and chenodeoxycholate (Figure 19.2).

After a meal is eaten, bile flows through the common bile duct into the duodenum, where bile salts emulsify the fat globules into tiny droplets. This increases the surface area of the lipid molecules, allowing them to be more easily hydrolyzed by **lipases** (Figure 19.3).

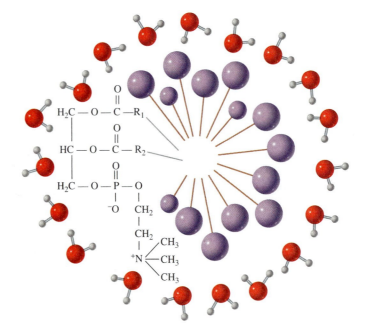

Figure 19.1

The structure of a micelle formed from the phospholipid, lecithin. The straight lines represent the long hydrophobic fatty acid tails, and the spheres represent the hydrophilic heads of the phospholipid.

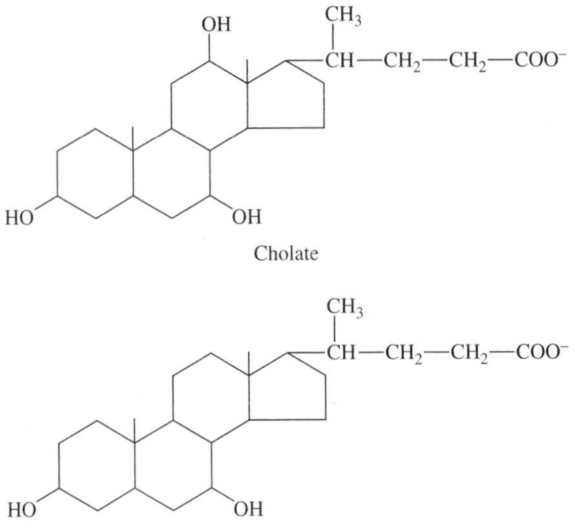

Cholate

Figure 19.2

Structures of the most common bile acids in human bile: cholate and chenodeoxycholate.

Chenodeoxycholate

Emulsification

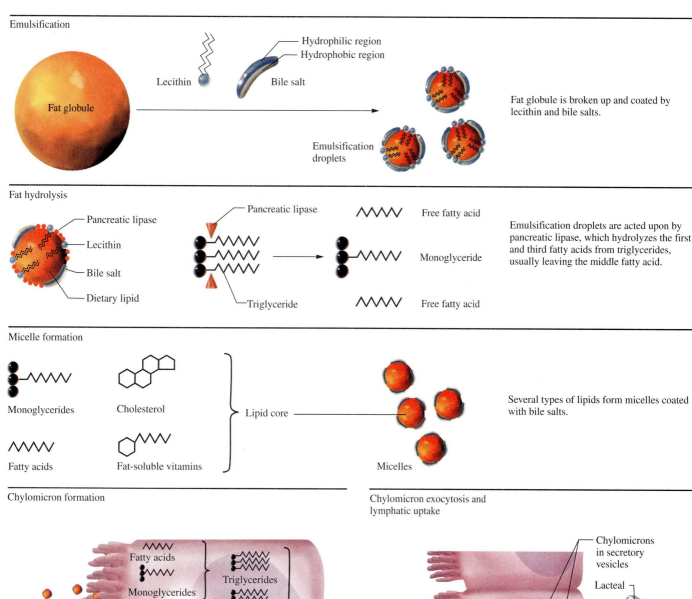

Fat globule is broken up and coated by lecithin and bile salts.

Emulsification droplets are acted upon by pancreatic lipase, which hydrolyzes the first and third fatty acids from triglycerides, usually leaving the middle fatty acid.

Several types of lipids form micelles coated with bile salts.

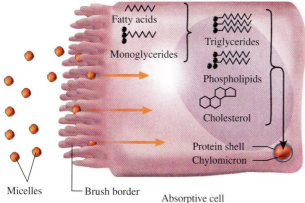

Intestinal cells absorb lipids from micelles, resynthesize triglycerides, and package triglycerides, cholesterol, and phospholipids into protein-coated chylomicrons.

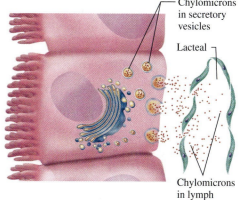

Golgi complex packages chylomicrons into secretory vesicles; chylomicrons are released from basal cell membrane by exocytosis and enter the lacteal (lymphatic capillary).

Figure 19.3

Stages of lipid digestion in the intestinal tract.

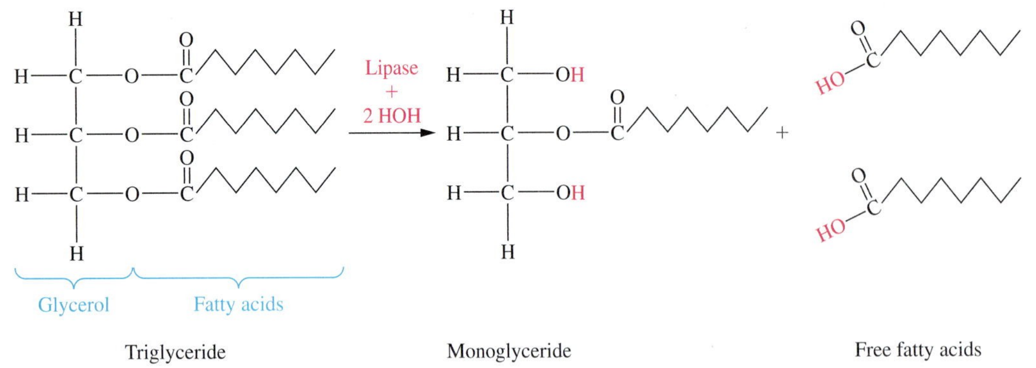

Figure 19.4

The action of pancreatic lipase in the hydrolysis of dietary lipids.

Triglycerides are described in Section 15.3.

Much of the lipid in these droplets is in the form of **triglycerides,** or triacylglycerols, which are fatty acid esters of glycerol. A protein called **colipase** binds to the surface of the lipid droplets and helps pancreatic lipases to stick to the surface and hydrolyze the ester bonds between the glycerol and fatty acids of the triglycerides (Figure 19.4). In this process, two of the three fatty acids are liberated, and the monoglycerides and free fatty acids produced mix freely with the micelles of bile. These micelles are readily absorbed through the membranes of the intestinal epithelial cells.

Plasma lipoproteins are described in Section 15.5.

Surprisingly, the monoglycerides and fatty acids are then reassembled into triglycerides that are combined with protein to produce the class of plasma lipoproteins called **chylomicrons** (Figure 19.3). These collections of lipid and protein are secreted into small lymphatic vessels and eventually arrive in the bloodstream. In the bloodstream, the triglycerides are once again hydrolyzed to produce glycerol and free fatty acids that are then absorbed by the cells. If the body needs energy, these molecules are degraded to produce ATP. If the body does not need energy, these energy-rich molecules are stored.

Lipid Storage

Fatty acids are stored in the form of triglycerides. Most of the body's triglyceride molecules are stored as fat droplets in the cytoplasm of **adipocytes** (fat cells) that make up **adipose tissue.** Each adipocyte contains a large fat droplet that accounts for nearly the entire volume of the cell. Other cells, such as those of cardiac muscle, contain a few small fat droplets. In these cells, the fat droplets are surrounded by mitochondria. When the cells need energy, triglycerides are hydrolyzed to release fatty acids that are transported into the matrix space of the mitochondria. There the fatty acids are completely oxidized, and ATP is produced.

The fatty acids provided by the hydrolysis of triglycerides are a very rich energy source for the body. The complete oxidation of fatty acids releases much more energy than the oxidation of a comparable amount of glycogen.

QUESTION 19.1

How do bile salts aid in the digestion of dietary lipids?

QUESTION 19.2

Why must dietary lipids be processed before enzymatic digestion can be effective?

19.2 Fatty Acid Degradation

An Overview of Fatty Acid Degradation

The pathway for the breakdown of fatty acids into acetyl CoA is called **β-oxidation.** The β-oxidation cycle (steps 2–5, Figure 19.5) consists of a set of four reactions whose overall form is similar to the last four reactions of the citric acid cycle. Each trip through the sequence of reactions releases acetyl CoA and returns a fatty acyl CoA molecule that has two fewer carbons. One molecule of FADH$_2$, equivalent to two ATP molecules, and one molecule of NADH, equivalent to three ATP molecules, are produced for each cycle of β-oxidation.

2 **Learning Goal**
Describe the degradation of fatty acids by β-oxidation.

3 **Learning Goal**
Explain the role of acetyl CoA in fatty acid metabolism.

This pathway is called β-oxidation because it involves the stepwise oxidation of the β-carbon of the fatty acid.

Review Section 18.9 for the ATP yields that result from oxidation of FADH$_2$ and NADH.

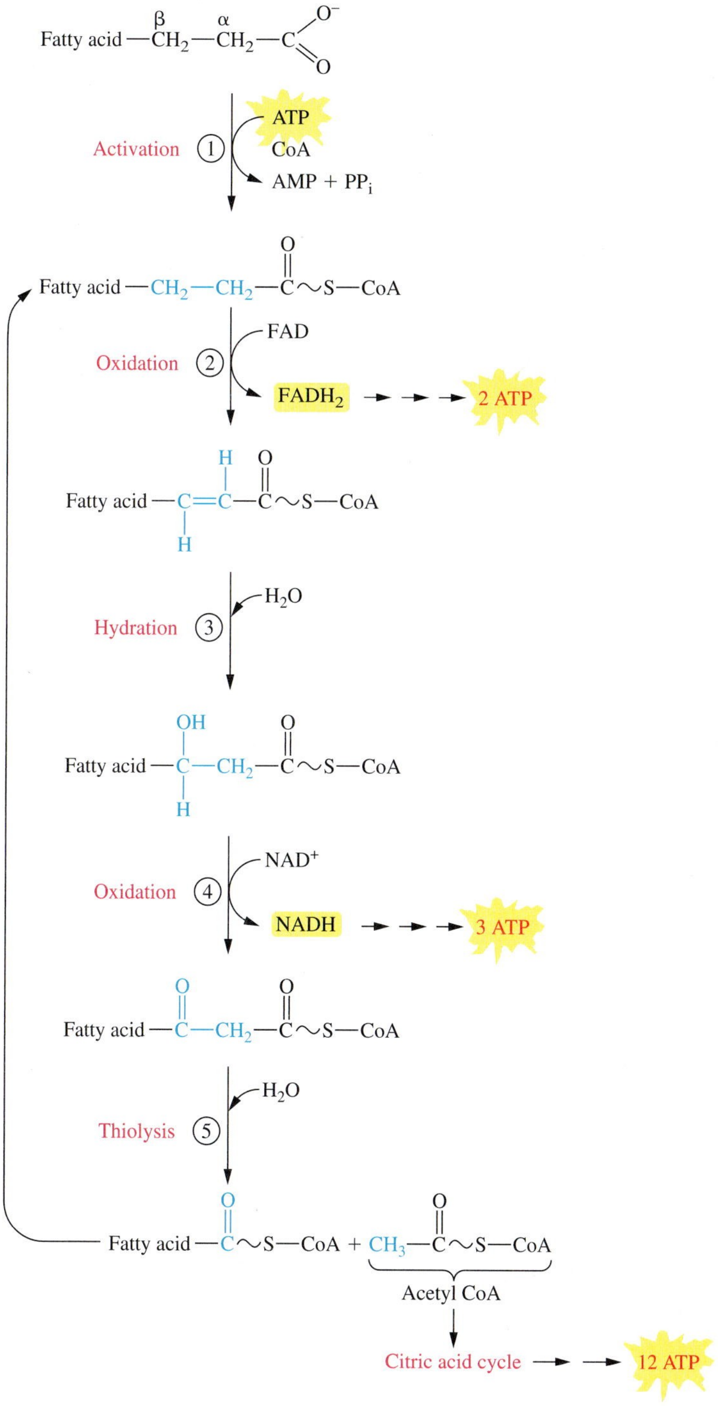

Figure 19.5

The reactions in β-oxidation of fatty acids.

A LIFESTYLE Connection

Losing Those Unwanted Pounds of Adipose Tissue

Weight, or overweight, is a topic of great concern to the American populace. A glance through almost any popular magazine quickly informs us that by today's standards, "beautiful" is synonymous with "thin." The models in all these magazines are extremely thin, and there are literally dozens of ads for weight-loss programs. Americans spend millions of dollars each year trying to attain this slim ideal of the fashion models.

Studies have revealed that this slim ideal is often below a desirable, healthy body weight. In fact, the suggested weight for a 6-foot tall male between 18 and 39 years of age is 179 pounds.

For a 5'6" female in the same age range, the desired weight is 142 pounds. For a 5'1" female, 126 pounds is recommended. Just as being too thin can cause health problems, so too can obesity.

What is obesity, and does it have disadvantages beyond aesthetics? An individual is considered to be obese if his or her body weight is more than 20% above the ideal weight for his or her height. The accompanying table lists desirable body weights, according to sex, age, height, and body frame.

Overweight carries with it a wide range of physical problems, including elevated blood cholesterol levels; high blood pressure;

Men*					Women**				
Height					**Height**				
Feet	Inches	Small Frame	Medium Frame	Large Frame	Feet	Inches	Small Frame	Medium Frame	Large Frame
5	2	128–134	131–141	138–150	4	10	102–111	109–121	118–131
5	3	130–136	133–143	140–153	4	11	103–113	111–123	120–134
5	4	132–138	135–145	142–156	5	0	104–115	113–126	122–137
5	5	134–140	137–148	144–160	5	1	106–118	115–129	125–140
5	6	136–142	139–151	146–164	5	2	108–121	118–132	128–143
5	7	138–145	142–154	149–168	5	3	111–124	121–135	131–147
5	8	140–148	145–157	152–172	5	4	114–127	124–138	134–151
5	9	142–151	148–160	155–176	5	5	117–130	127–141	137–155
5	10	144–154	151–163	158–180	5	6	120–133	130–144	140–159
5	11	146–157	154–166	161–184	5	7	123–136	133–147	143–163
6	0	149–160	157–170	164–188	5	8	126–139	136–150	146–167
6	1	152–164	160–174	168–192	5	9	129–142	139–153	149–170
6	2	155–168	164–178	172–197	5	10	132–145	142–156	152–173
6	3	158–172	167–182	176–202	5	11	135–148	145–159	155–176
6	4	162–176	171–187	181–207	6	0	138–151	148–162	158–179

*Weights at ages 25–59 based on lowest mortality. Weight in pounds according to frame (in indoor clothing weighing 5 lb, shoes with 1" heels).
**Weights at ages 25–59 based on lowest mortality. Weight in pounds according to frame (in indoor clothing weighing 3 lb, shoes with 1" heels).
Reprinted with permission of the Metropolitan Life Insurance Companies *Statistical Bulletin.*

2 Learning Goal

Describe the degradation of fatty acids by β-oxidation.

EXAMPLE 19.1 Predicting the Products of β-Oxidation of a Fatty Acid

Which products would be produced by the β-oxidation of decanoic acid?

SOLUTION

This ten-carbon fatty acid would be broken down into five acetyl CoA molecules. Because four cycles through β-oxidation are required to break down a ten-carbon fatty acid, four NADH molecules and four FADH$_2$ molecules would also be produced.

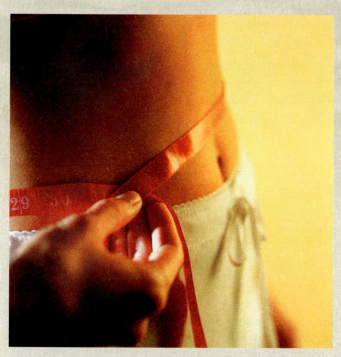

Reduced caloric intake and exercise are the keys to permanent weight loss.

increased incidence of diabetes, cancer, and heart disease; and increased probability of early death. It often causes psychological problems as well, such as guilt and low self-esteem.

Many factors may contribute to obesity. These include genetic factors, a sedentary lifestyle, and a preference for high-calorie, high-fat foods. However, the real concern is how to lose weight. How can we lose weight wisely and safely and keep the weight off for the rest of our lives? Unfortunately, the answer is *not* the answer that most people want to hear. The prevalence and financial success of the quick-weight-loss programs suggest that the majority of people want a program that is rapid and effortless. Unfortunately, most programs that promise dramatic weight reduction with little effort are usually ineffective or, worse, unsafe. The truth is that weight loss and management are best obtained by a program involving three elements.

1. *Reduced caloric intake.* A pound of body fat is equivalent to 3500 Calories (kilocalories). So if you want to lose 2 pounds each week, a reasonable goal, you must reduce your caloric intake by 1000 Calories per day. Remember that diets recommending fewer than 1200 Calories per day are difficult to maintain because they are not very satisfying and may be unsafe because they don't provide all the required vitamins and minerals. The best way to decrease Calories is to reduce fat and increase complex carbohydrates in the diet.

2. *Exercise.* Increase energy expenditures by 200–400 Calories each day. You may choose walking, running, or mowing the lawn; the type of activity doesn't matter, as long as you get moving. Exercise has additional benefits. It increases cardiovascular fitness, provides a psychological lift, and may increase the base rate at which you burn calories after exercise is finished.

3. *Behavior modification.* For some people, overweight is as much a psychological problem as it is a physical problem, and half the battle is learning to recognize the triggers that cause overeating. Several principles of behavior modification have been found very helpful.

   a. Keep a diary. Record the amount of foods eaten and the circumstances—for instance, a meal at the kitchen table or a bag of chips in the car on the way home.

   b. Identify your eating triggers. Do you eat when you feel stress, boredom, fatigue, joy?

   c. Develop a plan for avoiding or coping with your trigger situations or emotions. You might exercise when you feel that stress-at-the-end-of-the-day trigger or carry a bag of carrot sticks for the midmorning-boredom trigger.

   d. Set realistic goals, and reward yourself when you reach them. The reward should not be food related.

As you can see, there is no "quick fix" for safe, effective weight control. A commitment must be made to modify existing diet and exercise habits. Most important, those habits must be avoided forever and replaced by new, healthier behaviors and attitudes.

FOR FURTHER UNDERSTANDING

In terms of the energy-harvesting reactions we have studied in Chapters 18 and 19, explain how reduced caloric intake and increase in activity level contribute to weight loss.

If you increased your energy expenditure by 200 Calories per day and did not change your eating habits, how long would it take you to lose 10 pounds?

QUESTION 19.3

Which products would be formed by β-oxidation of each of the following fatty acids? (*Hint:* Refer to Example 19.1.)

a. Hexanoic acid   b. Octadecanoic acid

QUESTION 19.4

Which products would be formed by β-oxidation of each of the following fatty acids?

a. Octanoic acid   b. Dodecanoic acid

The Reactions of β-Oxidation

The enzymes that catalyze the β-oxidation of fatty acids are located in the matrix space of mitochondria. Special transport mechanisms are required to bring fatty acid molecules into the mitochondrial matrix. Once inside, the fatty acids are degraded by the reactions of β-oxidation. As we will see, these reactions interact with oxidative phosphorylation and the citric acid cycle to produce ATP.

Reaction 1. The first step is an *activation* reaction that results in the production of a fatty acyl CoA molecule. A thioester bond is formed between coenzyme A and the fatty acid:

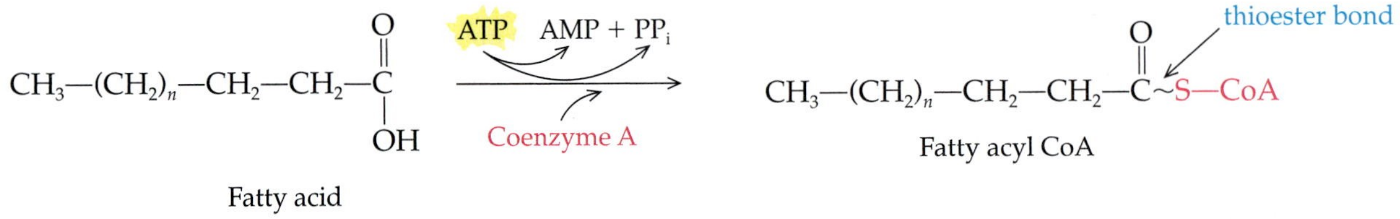

$$CH_3-(CH_2)_n-CH_2-CH_2-\underset{\underset{OH}{|}}{\overset{\overset{O}{\|}}{C}} \xrightarrow[\text{Coenzyme A}]{\text{ATP}\quad \text{AMP} + \text{PP}_i} CH_3-(CH_2)_n-CH_2-CH_2-\overset{\overset{O}{\|}}{C}{\sim}S-CoA$$

Fatty acid   thioester bond

Fatty acyl CoA

This reaction requires energy in the form of ATP, which is cleaved to AMP and pyrophosphate. This involves hydrolysis of two phosphoanhydride bonds. Here again we see the need to invest a small amount of energy so that a much greater amount of energy can be harvested later in the pathway. Coenzyme A is also required for this step. The product, a fatty acyl CoA, has a *high-energy* thioester bond between the fatty acid and coenzyme A.

Reaction 2. The next reaction is an *oxidation* reaction that removes a pair of hydrogen atoms from the fatty acid. These are used to reduce FAD to produce FADH$_2$. This *dehydrogenation* reaction is catalyzed by the enzyme *acyl-CoA dehydrogenase* and results in the formation of a carbon–carbon double bond:

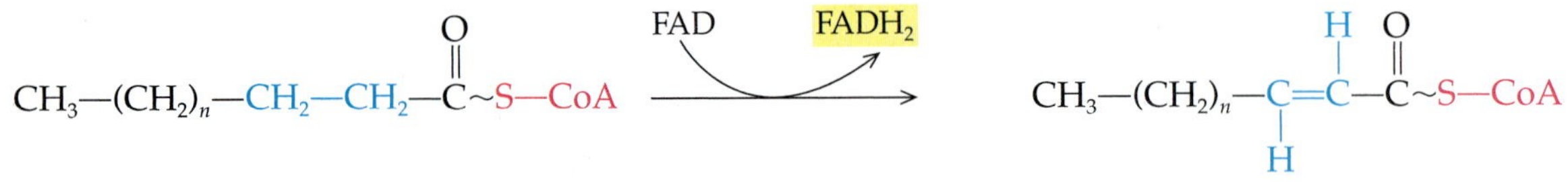

$$CH_3-(CH_2)_n-CH_2-CH_2-\overset{\overset{O}{\|}}{C}{\sim}S-CoA \xrightarrow{\text{FAD}\quad \text{FADH}_2} CH_3-(CH_2)_n-\overset{\overset{H}{|}}{C}=\underset{\underset{H}{|}}{C}-\overset{\overset{O}{\|}}{C}{\sim}S-CoA$$

Oxidative phosphorylation yields two ATP molecules for each molecule of FADH$_2$ produced by this oxidation–reduction reaction.

Reaction 3. The third reaction involves *hydration* of the double bond produced in reaction 2. As a result, the β-carbon is hydroxylated. This reaction is catalyzed by the enzyme, *enoyl-CoA hydrase.*

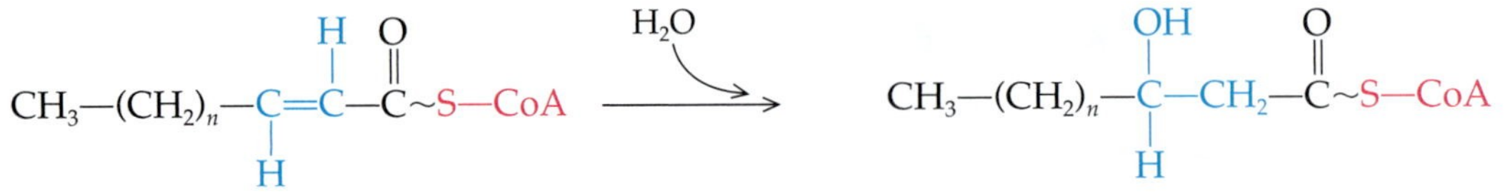

$$CH_3-(CH_2)_n-\overset{\overset{H}{|}}{C}=\underset{\underset{H}{|}}{C}-\overset{\overset{O}{\|}}{C}{\sim}S-CoA \xrightarrow{\text{H}_2\text{O}} CH_3-(CH_2)_n-\underset{\underset{H}{|}}{\overset{\overset{OH}{|}}{C}}-CH_2-\overset{\overset{O}{\|}}{C}{\sim}S-CoA$$

Reaction 4. In this *oxidation* reaction, the hydroxyl group of the β-carbon is now dehydrogenated. NAD$^+$ is reduced to form NADH that is subsequently used to produce three ATP molecules by oxidative phosphorylation. L-β-*Hydroxyacyl-CoA dehydrogenase* catalyzes this reaction.

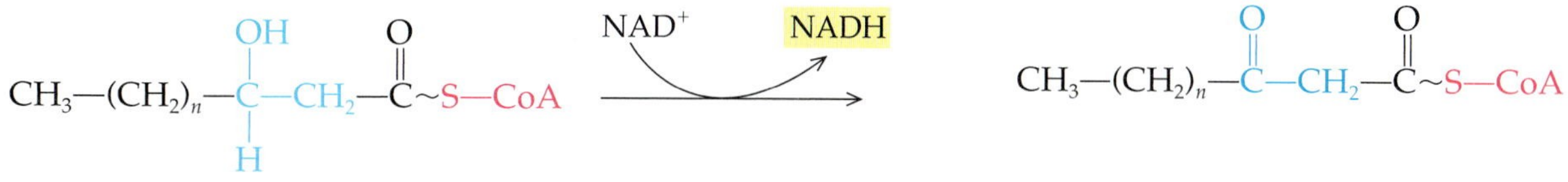

Reaction 5. The final reaction, catalyzed by the enzyme, *thiolase,* is the cleavage that releases acetyl CoA. This is accomplished by *thiolysis,* attack of a molecule of coenzyme A on the β-carbon. The result is the release of acetyl CoA and a fatty acyl CoA that is two carbons shorter than the beginning fatty acid:

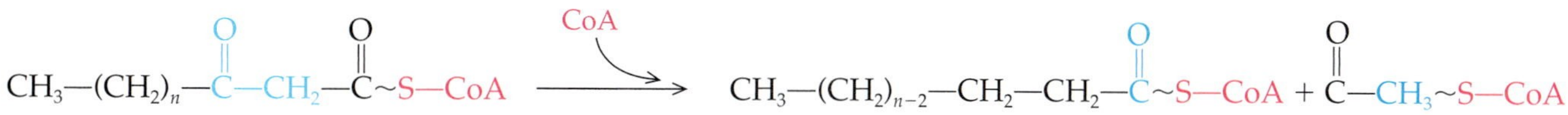

The shortened fatty acyl CoA is further oxidized by cycling through reactions 2–5 until the fatty acid carbon chain is completely degraded to acetyl CoA. The acetyl CoA produced by β-oxidation of fatty acids then enters the reactions of the citric acid cycle. Of course, this eventually results in the production of 12 ATP molecules per molecule of acetyl CoA released during β-oxidation.

As an example of the energy yield from β-oxidation, the balance sheet for ATP production when the sixteen-carbon-fatty acid palmitic acid is degraded by β-oxidation is summarized in Figure 19.6. Complete oxidation of palmitate results in production of 129 molecules of ATP, *three and one half times more energy than results from the complete oxidation of an equivalent amount of glucose.*

EXAMPLE 19.2   **Calculating the Amount of ATP Produced in the Complete Oxidation of a Fatty Acid**

How many molecules of ATP are produced in the complete oxidation of stearic acid, an eighteen-carbon saturated fatty acid?

2   Learning Goal

Describe the degradation of fatty acids by β-oxidation.

SOLUTION

Step 1 (activation)	2 ATP
Steps 2–5:	
8 FADH$_2$ × 2 ATP/FADH$_2$	16 ATP
8 NADH × 3 ATP/NADH	24 ATP
9 acetyl CoA (to citric acid cycle):	
9 × 1 GTP × 1 ATP/GTP	9 ATP
9 × 3 NADH × 3 ATP/NADH	81 ATP
9 × 1 FADH$_2$ × 2 ATP/FADH$_2$	18 ATP
	146 ATP

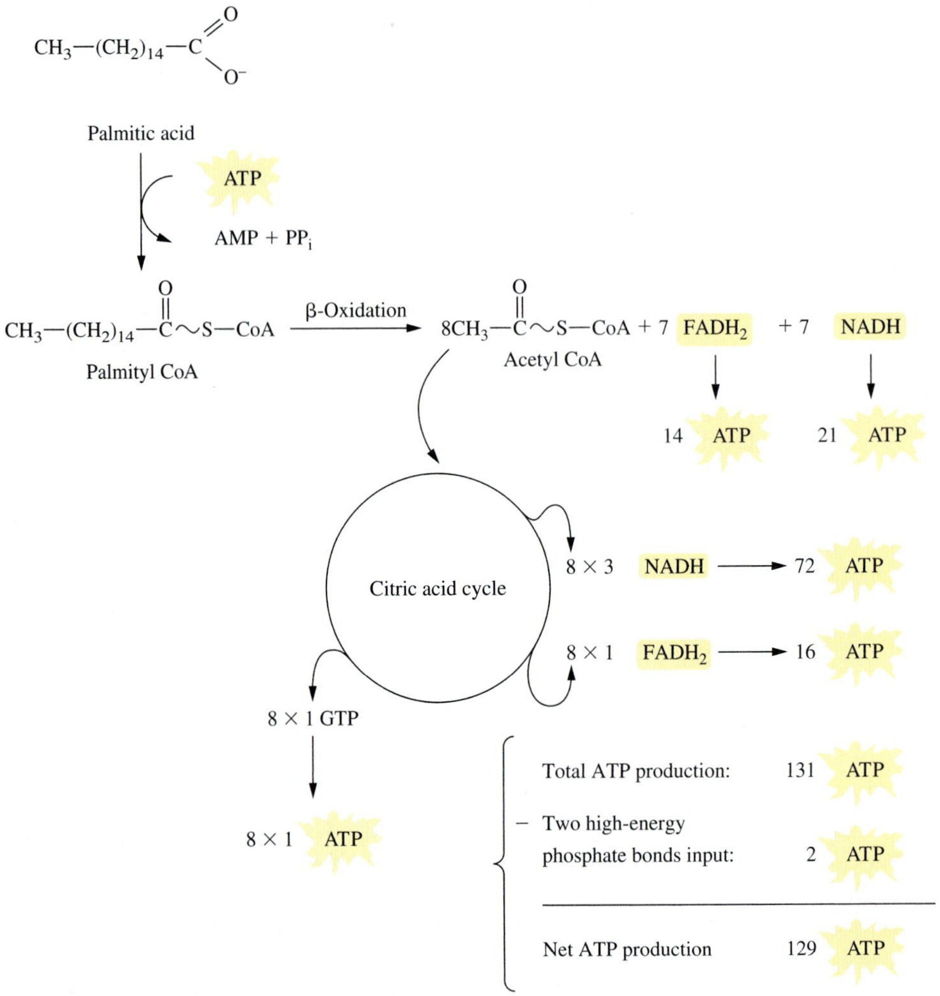

Figure 19.6

Complete oxidation of palmitic acid yields 129 molecules of ATP. Note that the activation step is considered an expenditure of two high-energy phosphoanhydride bonds because ATP is hydrolyzed to AMP + PP$_i$.

QUESTION 19.5

Write out the sequence of steps for β-oxidation of butyryl CoA.

QUESTION 19.6

What is the energy yield from the complete degradation of butyryl CoA via β-oxidation, the citric acid cycle, and oxidative phosphorylation?

19.3 Ketone Bodies

4 Learning Goal

Understand the role of ketone body production in β-oxidation.

For the acetyl CoA produced by the β-oxidation of fatty acids to efficiently enter the citric acid cycle, there must be an adequate supply of oxaloacetate. If glycolysis and β-oxidation are occurring at the same rate, there will be a steady supply of pyruvate (from glycolysis) that can be converted to oxaloacetate. But what happens if the supply of oxaloacetate is too low to allow all of the acetyl CoA to enter the citric acid cycle? Under these conditions, acetyl CoA is converted to the so-called **ketone bodies:** β-hydroxybutyrate, acetone, and acetoacetate (Figure 19.7).

Ketosis

Ketosis, abnormally high levels of blood ketone bodies, arises under some pathological conditions, such as starvation, a diet that is extremely low in carbohydrates

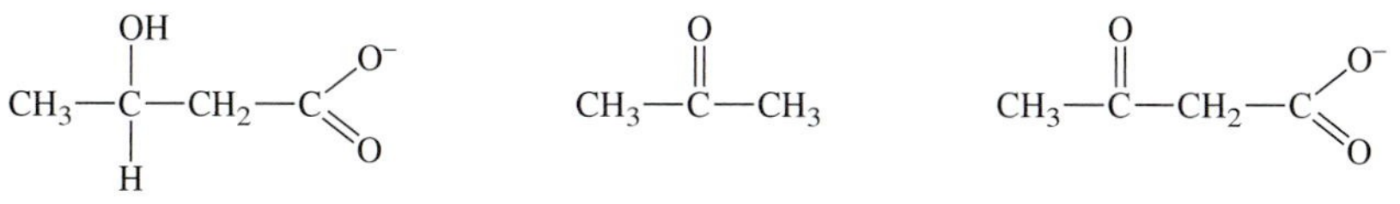

Figure 19.7

Structures of ketone bodies.

β-Hydroxybutyrate Acetone Acetoacetate

(as with the high-protein diets), or uncontrolled **diabetes mellitus.** The carbohydrate intake of a diabetic is normal, but the carbohydrates cannot get into the cell to be used as fuel. Thus diabetes amounts to starvation in the midst of plenty. In diabetes, the very high concentration of ketone acids in the blood leads to **ketoacidosis.** The ketone acids are relatively strong acids and therefore readily dissociate to release H$^+$. Under these conditions, the blood pH becomes acidic, which can lead to death.

Diabetes mellitus is a disease characterized by the appearance of glucose in the urine as a result of high blood glucose levels. The disease is usually caused by the inability to produce the hormone insulin.

Ketogenesis

The pathway for the production of ketone bodies (Figure 19.8) begins with a "reversal" of the last step of β-oxidation. When oxaloacetate levels are low, the

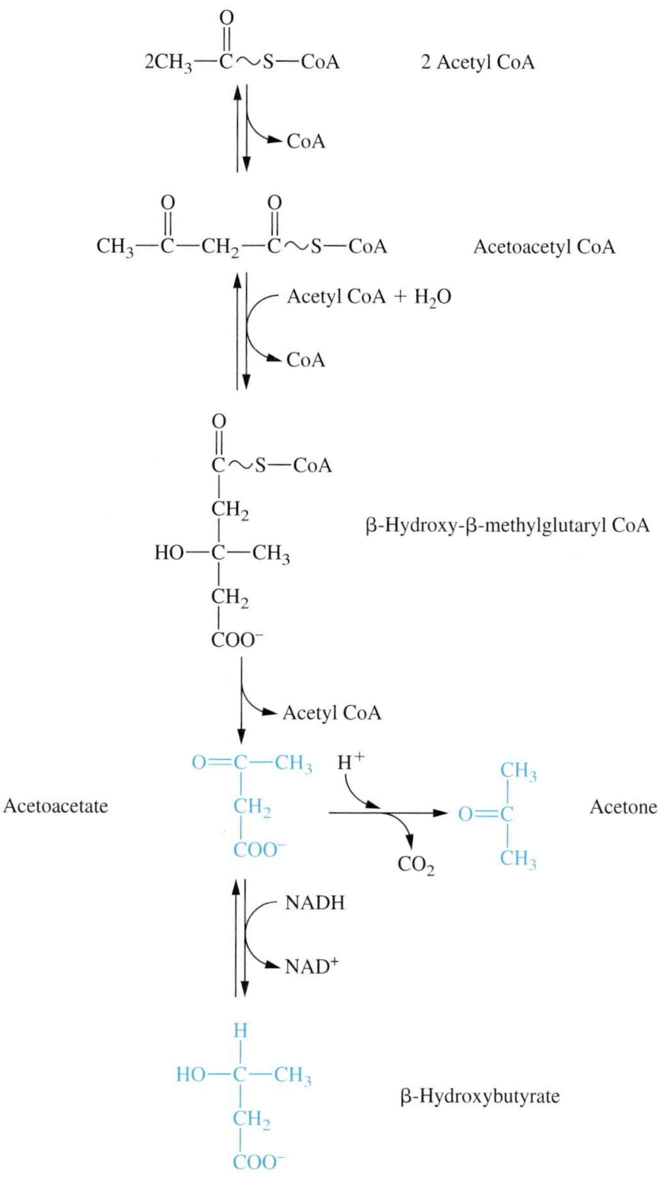

2 Acetyl CoA

Acetoacetyl CoA

β-Hydroxy-β-methylglutaryl CoA

Acetoacetate

Acetone

β-Hydroxybutyrate

Figure 19.8

Summary of the reactions involved in ketogenesis.

enzyme that normally carries out the last reaction of β-oxidation now catalyzes the fusion of two acetyl CoA molecules to produce acetoacetyl CoA.

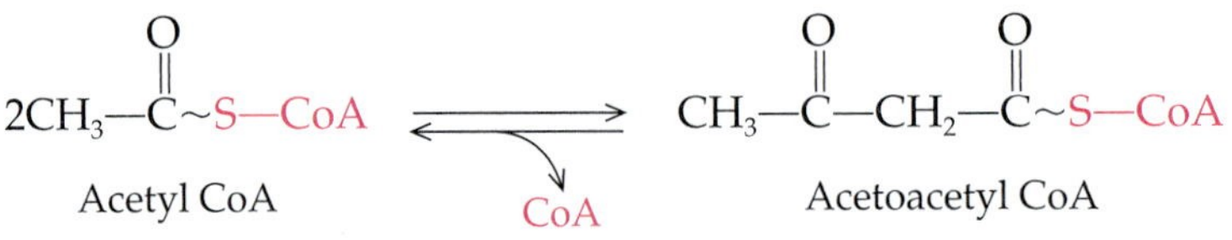

$$2CH_3-\overset{\overset{\displaystyle O}{\|}}{C}\sim S-CoA \rightleftarrows CH_3-\overset{\overset{\displaystyle O}{\|}}{C}-CH_2-\overset{\overset{\displaystyle O}{\|}}{C}\sim S-CoA$$

Acetyl CoA   CoA   Acetoacetyl CoA

Acetoacetyl CoA can react with a third acetyl CoA molecule to yield β-hydroxy-β-methylglutaryl CoA (HMG-CoA), which is then cleaved to yield acetoacetate and acetyl CoA.

In very small amounts, acetoacetate spontaneously loses carbon dioxide to give acetone. This is the reaction that causes the "acetone breath" that is often associated with uncontrolled diabetes mellitus. More frequently, it undergoes NADH-dependent reduction to produce β-hydroxybutyrate.

Acetoacetate and β-hydroxybutyrate are produced primarily in the liver. These metabolites diffuse into the blood and are circulated to other tissues, where they may be reconverted to acetyl CoA and used to produce ATP. In fact, the heart muscle derives most of its metabolic energy from the oxidation of ketone bodies, not from the oxidation of glucose. Other tissues that are best adapted to the use of glucose will increasingly rely on ketone bodies for energy when glucose becomes unavailable or limited. This is particularly true of the brain.

QUESTION 19.7

What conditions lead to excess production of ketone bodies?

QUESTION 19.8

What is the cause of the characteristic "acetone breath" that is associated with uncontrolled diabetes mellitus?

19.4 Fatty Acid Synthesis

5   **Learning Goal**

Compare β-oxidation of fatty acids and fatty acid biosynthesis.

All organisms possess the ability to synthesize fatty acids. In humans, the excess acetyl CoA produced by carbohydrate degradation is used to make fatty acids that are then stored as triglycerides.

A Comparison of Fatty Acid Synthesis and Degradation

On first examination, fatty acid synthesis appears to be simply the reverse of β-oxidation. Specifically, the fatty acid chain is constructed by the sequential addition of two-carbon acetyl groups (Figure 19.9). Although the chemistry of fatty acid synthesis and breakdown are similar, there are several major differences between β-oxidation and fatty acid biosynthesis. These are summarized as follows:

- **Intracellular location.** The enzymes responsible for fatty acid biosynthesis are located in the cytoplasm of the cell, whereas those responsible for the degradation of fatty acids are in the mitochondria.
- **Acyl group carriers.** The activated intermediates of fatty acid biosynthesis are bound to a carrier molecule called the **acyl carrier protein (ACP)** (Figure 19.10). In β-oxidation, the acyl group carrier was coenzyme A. However, there are important similarities between these two carriers. Both contain the

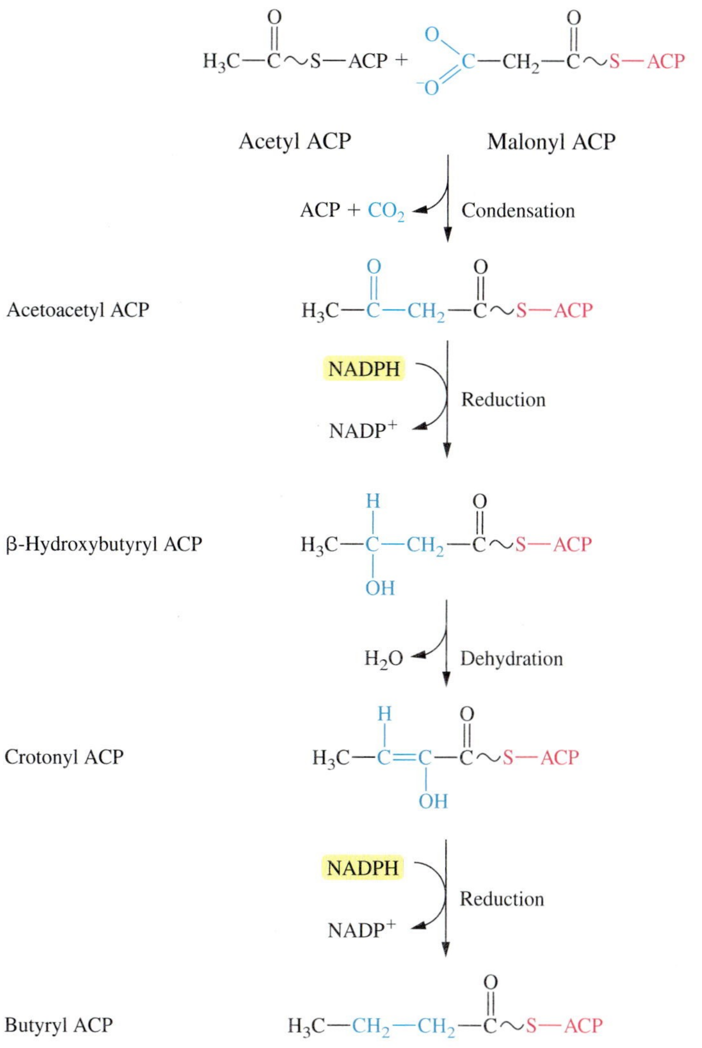

Figure 19.9

Summary of fatty acid synthesis. Malonyl ACP is produced in two reactions: carboxylation of acetyl CoA to produce malonyl CoA and transfer of the malonyl acyl group from malonyl CoA to ACP.

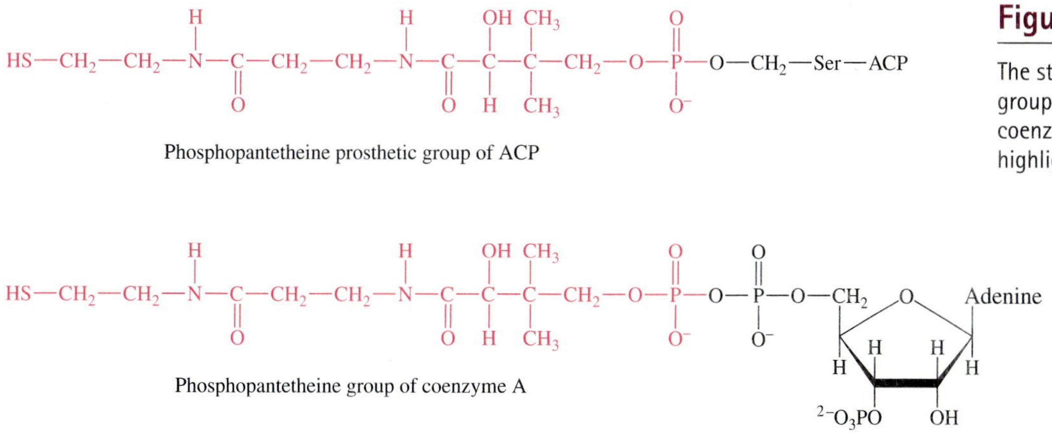

Figure 19.10

The structure of the phosphopantetheine group, the reactive group common to coenzyme A and acyl carrier protein, is highlighted in red.

phosphopantetheine group, which is made from the vitamin, pantothenic acid. In both cases, the fatty acyl group is bound by a thioester bond to the phosphopantetheine group.

- **Enzymes involved.** Fatty acid biosynthesis is carried out by a multienzyme complex known as *fatty acid synthase*. The enzymes responsible for fatty acid degradation are not physically associated in such complexes.

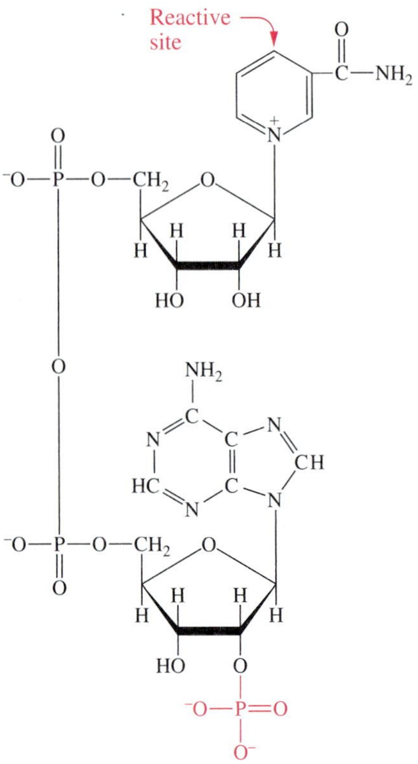

Figure 19.11

Structure of NADPH. The phosphate group shown in red is the structural feature that distinguishes NADPH from NADH.

6 **Learning Goal**

Describe the conversion of amino acids to molecules that can enter the citric acid cycle.

- **Electron carriers.** NADH and $FADH_2$ are produced by fatty acid oxidation, whereas NADPH is the reducing agent for fatty acid biosynthesis. As a general rule, *NADH is produced by catabolic reactions, and NADPH is the reducing agent of biosynthetic reactions.* These two coenzymes differ only by the presence of a phosphate group bound to the ribose ring of NADPH (Figure 19.11). The enzymes that use these coenzymes, however, are easily able to distinguish them on this basis.

QUESTION 19.9

List the four major differences between β-oxidation and fatty acid biosynthesis that reveal that the two processes are not just the reverse of one another.

QUESTION 19.10

What chemical group is part of coenzyme A and acyl carrier protein and allows both molecules to form thioester bonds to fatty acids?

19.5 Degradation of Amino Acids

Carbohydrates and lipids are not our only source of energy. As we saw in Chapter 18, dietary protein is digested to amino acids that can also be used as an energy source, although this is not their major metabolic function. Most of the amino acids used for energy come from the diet. It is only under starvation conditions, when stored glycogen has been depleted, that the body begins to burn its own protein, for instance, from muscle, as a fuel.

The degradation of amino acids occurs primarily in the liver and takes place in two stages. The first stage is the removal of the α-amino group, and the second is the degradation of the carbon skeleton. In land mammals, the amino group generally ends up in urea, which is excreted in the urine. The carbon skeletons can be converted into a variety of compounds, including citric acid cycle intermediates, pyruvate, acetyl CoA, or acetoacetyl CoA. The degradation of the carbon skeletons is summarized in Figure 19.12. Deamination reactions and the fate of the carbon skeletons of amino acids are the focus of this section.

Removal of α–Amino Groups: Transamination

The first stage of amino acid degradation, the removal of the α-amino group, is usually accomplished by a **transamination** reaction. **Transaminases** catalyze the transfer of the α-amino group from an α-amino acid to an α-keto acid:

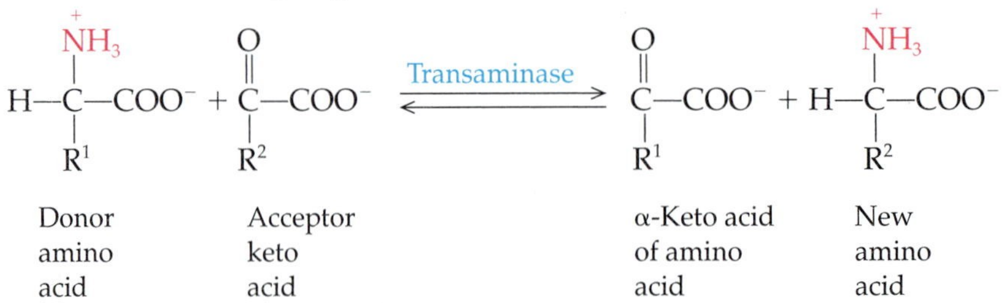

| Donor amino acid | Acceptor keto acid | | α-Keto acid of amino acid | New amino acid |

The α-amino group of a great many amino acids is transferred to α-ketoglutarate to produce the amino acid glutamate and a new keto acid. This glutamate family of transaminases is especially important because the α-keto acid corresponding to glutamate is α-ketoglutarate, a citric acid cycle intermediate. The glutamate

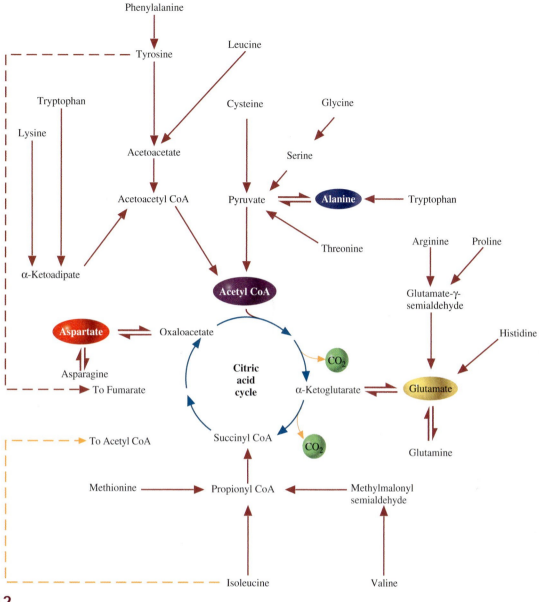

Figure 19.12

The carbon skeletons of amino acids can be converted to citric acid cycle intermediates and completely oxidized to produce ATP energy.

transaminases thus provide a direct link between amino acid degradation and the citric acid cycle.

Aspartate transaminase catalyzes the transfer of the α-amino group of aspartate to α-ketoglutarate, producing oxaloacetate and glutamate:

Pyridoxine (vitamin B$_6$)

Pyridoxal phosphate

Figure 19.13

The structure of pyridoxal phosphate, the coenzyme required for all transamination reactions, and pyridoxine, vitamin B$_6$, the vitamin from which it is derived.

For more information on these vitamins and the coenzymes that are made from them, look online at www.mhhe.com/denniston in Water-Soluble Vitamins.

Another important transaminase in mammalian tissues is *alanine transaminase,* which catalyzes the transfer of the α-amino group of alanine to α-ketoglutarate and produces pyruvate and glutamate:

Alanine   α-Ketoglutarate   Pyruvate   Glutamate

All of the more than fifty transaminases that have been discovered require the coenzyme **pyridoxal phosphate.** This coenzyme is derived from vitamin B$_6$ (pyridoxine, Figure 19.13).

Removal of α-Amino Groups: Oxidative Deamination

In the next stage of amino acid degradation, ammonium ion is liberated from the glutamate formed by the transaminase. This breakdown of glutamate, catalyzed by the enzyme, *glutamate dehydrogenase,* occurs as follows:

Glutamate   α-Ketoglutarate

This is an example of an **oxidative deamination,** an oxidation-reduction process in which NAD$^+$ is reduced to NADH and the amino acid is deaminated (the amino group is removed). A summary of the deamination reactions described is shown in Figure 19.14.

The Fate of Amino Acid Carbon Skeletons

The carbon skeletons produced by these and other deamination reactions enter glycolysis or the citric acid cycle at many steps. For instance, we have seen that transamination converts aspartate to oxaloacetate and alanine to pyruvate. The positions at which the carbon skeletons of various amino acids enter the energy-harvesting pathways are summarized in Figure 19.12.

Figure 19.14

Summary of the deamination of an α-amino acid and the fate of the ammonium ion (NH$_4^+$).

19.6 The Urea Cycle

Oxidative deamination produces large amounts of ammonium ion. Because ammonium ions are extremely toxic, they must be removed from the body, regardless of the energy expenditure required. In humans, they are detoxified in the liver by converting the ammonium ions into urea. This pathway, called the **urea cycle**, is the method by which toxic ammonium ions are kept out of the blood. The excess ammonium ions incorporated in urea are excreted in the urine (Figure 19.15).

7 **Learning Goal**
Explain the importance of the urea cycle and describe its essential steps.

Reactions of the Urea Cycle

The five reactions of the urea cycle are shown in Figure 19.15, and details of the reactions are summarized as follows:

Step 1. The first step of the cycle is a reaction in which CO_2 and NH_4^+ form carbamoyl phosphate. This reaction, which also requires ATP and H_2O, occurs in the mitochondria and is catalyzed by the enzyme *carbamoyl phosphate synthase.*

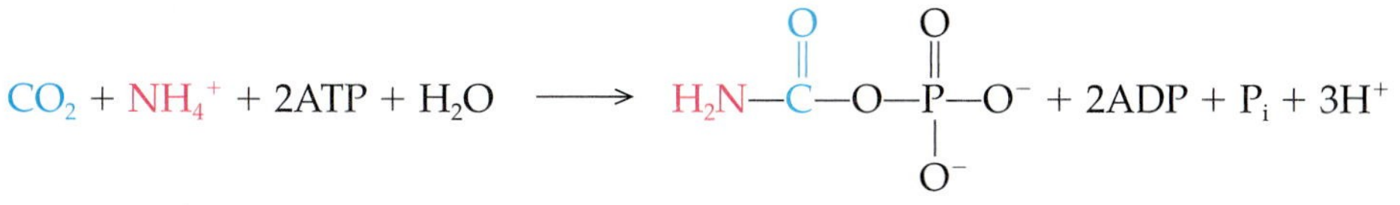

$$CO_2 + NH_4^+ + 2ATP + H_2O \longrightarrow H_2N-\overset{\overset{O}{\|}}{C}-O-\overset{\overset{O}{\|}}{\underset{\underset{O^-}{|}}{P}}-O^- + 2ADP + P_i + 3H^+$$

Carbamoyl phosphate

Step 2. The carbamoyl phosphate now condenses with the amino acid, ornithine, to produce the amino acid, citrulline. This reaction also occurs in the mitochondria and is catalyzed by the enzyme, *ornithine transcarbamoylase.*

The urea cycle involves several unusual amino acids that are not found in polypeptides.

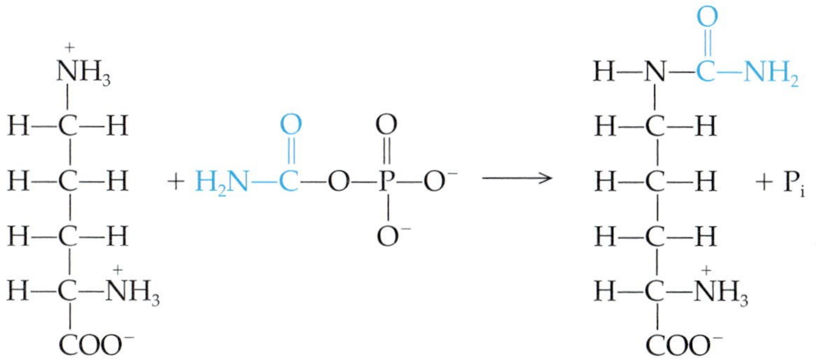

Ornithine   Carbamoyl phosphate   Citrulline

Step 3. Citrulline is transported into the cytoplasm and now condenses with aspartate to produce argininosuccinate. This reaction, which requires energy released by the hydrolysis of ATP, is catalyzed by the enzyme, *argininosuccinate synthase.*

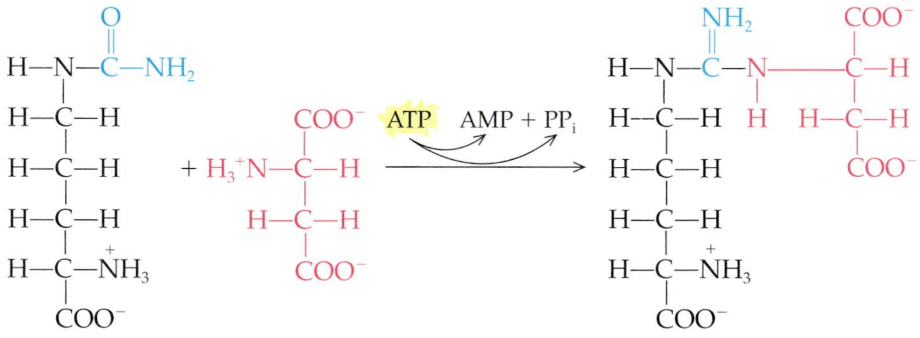

The abbreviation PP_i represents the pyrophosphate group, which consists of two phosphate groups joined by a phosphoanhydride bond:

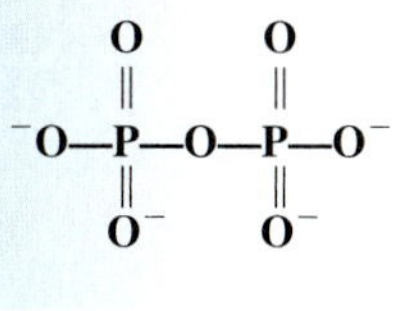

$$^-O-\overset{\overset{O}{\|}}{\underset{\underset{O^-}{\|}}{P}}-O-\overset{\overset{O}{\|}}{\underset{\underset{O^-}{\|}}{P}}-O^-$$

Citrulline   Aspartate   Argininosuccinate

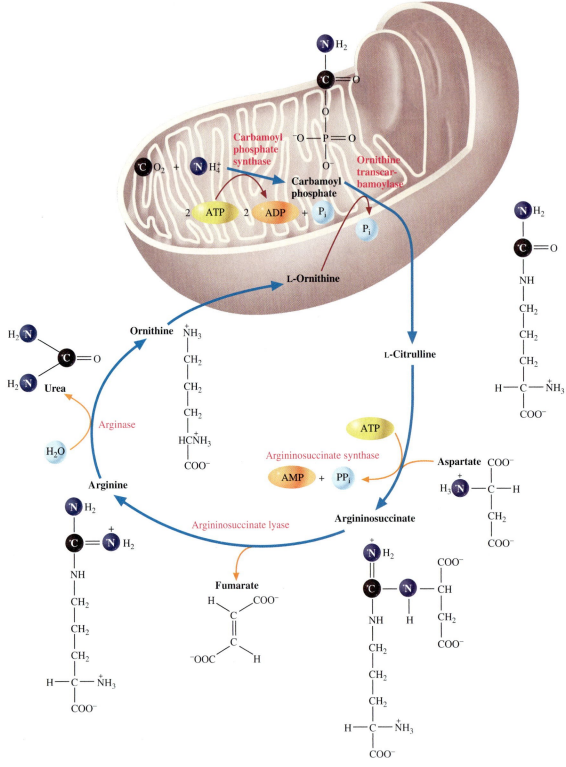

Figure 19.15

The urea cycle converts ammonium ions into urea, which is less toxic. The intracellular locations of the reactions are indicated. Citrulline, formed in the reaction between ornithine and carbamoyl phosphate, is transported out of the mitochondrion and into the cytoplasm. Ornithine, a substrate for the formation of citrulline, is transported from the cytoplasm into the mitochondrion.

Step 4. Now the argininosuccinate is cleaved to produce the amino acid, arginine, and the citric acid cycle intermediate, fumarate. This reaction is catalyzed by the enzyme, *argininosuccinate lyase.*

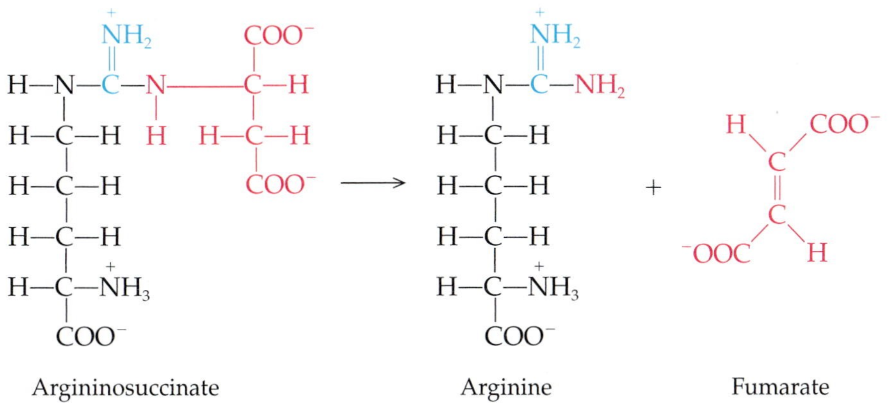

Argininosuccinate   Arginine   Fumarate

Step 5. Finally, arginine is hydrolyzed to generate urea, the product of the reaction to be excreted, and ornithine, the original reactant in the cycle. *Arginase* is the enzyme that catalyzes this reaction.

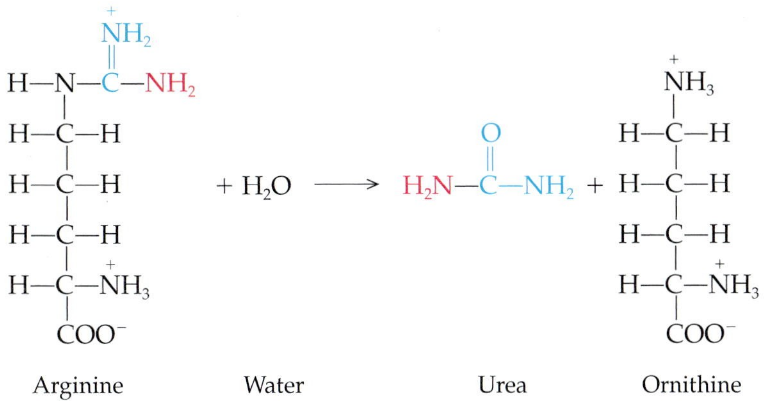

Arginine   Water   Urea   Ornithine

Note that one of the amino groups in urea is derived from the ammonium ion and the second is derived from the amino acid, aspartate.

There are genetically transmitted diseases that result from a deficiency of one of the enzymes of the urea cycle. The importance of the urea cycle is apparent when we consider the terrible symptoms suffered by afflicted individuals. A deficiency of urea cycle enzymes causes an elevation of the concentration of NH_4^+, a condition known as **hyperammonemia.** If there is a complete deficiency of one of the enzymes of the urea cycle, the result is death in early infancy. If there is a partial deficiency of one of the enzymes of the urea cycle, the result may be retardation, convulsions, and vomiting. In these milder forms of hyperammonemia, a low-protein diet leads to a lower concentration of NH_4^+ in blood and less severe clinical symptoms.

8 Learning Goal
Discuss the cause and effect of hyperammonemia.

QUESTION 19.11

What is the purpose of the urea cycle?

QUESTION 19.12

Where do the reactions of the urea cycle occur?

A MEDICAL Connection | Diabetes Mellitus and Ketone Bodies

More than one person, found unconscious on the streets of some metropolis, has been carted to jail only to die of complications arising from uncontrolled diabetes mellitus. Others are fortunate enough to arrive in hospital emergency rooms. A quick test for diabetes mellitus–induced coma is the odor of acetone on the breath of the afflicted person. Acetone is one of several metabolites produced by diabetics that are known collectively as *ketone bodies.*

The term *diabetes* was used by the ancient Greeks to designate diseases in which excess urine is produced. Two thousand years later, in the eighteenth century, the urine of certain individuals was found to contain sugar, and the name *diabetes mellitus* (Latin: *mellitus,* sweetened with honey) was given to this disease. People suffering from diabetes mellitus waste away as they excrete large amounts of sugar-containing urine.

The cause of insulin-dependent diabetes mellitus is an inadequate production of insulin by the body. Insulin is secreted in response to high blood glucose levels. It binds to the membrane receptor protein on its target cells. Binding increases the rate of transport of glucose across the membrane and stimulates glycogen synthesis, lipid biosynthesis, and protein synthesis. As a result, the blood glucose level is reduced. Clearly, the inability to produce sufficient insulin seriously impairs the body's ability to regulate metabolism.

Individuals suffering from diabetes mellitus do not produce enough insulin to properly regulate blood glucose levels. This generally results from the destruction of the β-cells of the islets of Langerhans. One theory to explain the mysterious disappearance of these cells is that a virus infection stimulates the immune system to produce antibodies that cause the destruction of the β-cells.

In the absence of insulin, the uptake of glucose into the tissues is not stimulated, and a great deal of glucose is eliminated in the urine. Without insulin, then, adipose cells are unable to take up the glucose required to synthesize triglycerides. As a result, the rate of fat hydrolysis is much greater than the rate of fat resynthesis, and large quantities of free fatty acids are liberated into the bloodstream. Because glucose is not being efficiently taken into cells, carbohydrate metabolism slows, and there is an increase in the rate of lipid catabolism. In the liver, this lipid catabolism results in the production of ketone bodies: acetone, acetoacetate, and β-hydroxybutyrate.

A similar situation can develop from improper eating, fasting, or dieting—any situation in which the body is not provided with sufficient energy in the form of carbohydrates. These ketone bodies cannot all be oxidized by the citric acid cycle, which is limited by the supply of oxaloacetate. The acetone concentration in blood rises to levels so high that acetone can be detected in the breath of untreated diabetics. The elevated concentration of ketones in the blood can overwhelm the buffering capacity of the blood, resulting in ketoacidosis. Ketones, too, will be excreted through the kidney. In fact, the presence of excess ketones in the urine can raise the osmotic concentration of the urine so that it behaves as an "osmotic diuretic," causing the excretion of enormous amounts of water. As a result, the patient may become severely dehydrated. In extreme cases, the combination of dehydration and ketoacidosis may lead to coma and death.

It has been observed that diabetics also have a higher than normal level of glucagon in the blood. As we have seen, glucagon stimulates lipid catabolism and ketogenesis. It may be that the symptoms previously described result from both the deficiency of insulin and the elevated glucagon levels. The absence of insulin may cause the elevated blood glucose and fatty acid levels, whereas glucagon, by stimulating ketogenesis, may be responsible for ketoacidosis and dehydration.

There is no cure for diabetes. However, when the problem is the result of the inability to produce active insulin, blood glucose levels can be controlled moderately well by the injection of human insulin produced from the cloned insulin gene. Unfortunately, one or even a few injections of insulin each day cannot mimic the precise control of blood glucose accomplished by the pancreas.

As a result, diabetics suffer progressive tissue degeneration that leads to early death. One primary cause of this degeneration is atherosclerosis, the deposition of plaque on the walls of blood vessels. This causes a high frequency of strokes, heart attack, and gangrene of the feet and lower extremities, often necessitating amputation. Kidney failure causes the death of about 20% of diabetics under forty years of age, and diabetic retinopathy (various kinds of damage to the retina of the eye) ranks

19.7 The Effects of Insulin and Glucagon on Cellular Metabolism

9 Learning Goal

Summarize the antagonistic effects of glucagon and insulin.

The hormone **insulin** is produced by the β-cells of the islets of Langerhans in the pancreas. It is secreted from these cells in response to an increase in the blood glucose level. Insulin lowers the concentration of blood glucose by causing a number of changes in metabolism (Table 19.1).

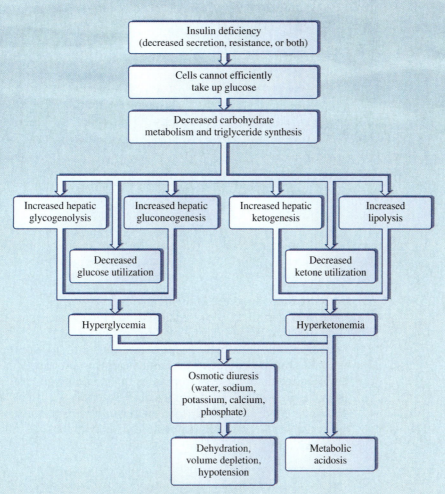

The metabolic events that occur in uncontrolled diabetes and that can lead to coma and death.

fourth among the leading causes of blindness in the United States. Nerves are also damaged, resulting in neuropathies that can cause pain or numbness, particularly of the feet.

There is no doubt that insulin injections prolong the life of diabetics, but only the presence of a fully functioning pancreas can allow a diabetic to live a life free of the complications noted here. At present, pancreas transplants do not have a good track record. Only about 50% of the transplants are functioning after one year. It is hoped that improved transplantation techniques will be developed so that diabetics can live a normal life span, free of debilitating disease.

FOR FURTHER UNDERSTANDING

The Atkins' low carbohydrate diet recommends that dieters test their urine for the presence of ketone bodies as an indicator that the diet is working. In terms of lipid and carbohydrate metabolism, explain why ketone bodies are being produced and why this is an indication that the diet is working.

An excess of ketone bodies in the blood causes ketoacidosis. Consider the chemical structure of the ketone bodies and explain why they are acids.

The simplest way to lower blood glucose levels is to stimulate storage of glucose, both as glycogen and as triglycerides. *Insulin therefore activates biosynthetic processes and inhibits catabolic processes.*

Insulin acts only on those cells, known as *target cells,* that possess a specific insulin receptor protein in their plasma membranes. The major target cells for insulin are liver, adipose, and muscle cells.

The blood glucose level is normally about 10 mM. However, a substantial meal increases the concentration of blood glucose considerably and stimulates insulin

 The effect of insulin on glycogen metabolism is described in Section 18.11.

TABLE 19.1 Comparison of the Metabolic Effects of Insulin and Glucagon

Actions	Insulin	Glucagon
Cellular glucose transport	Increased	No effect
Glycogen synthesis	Increased	Decreased
Glycogenolysis in liver	Decreased	Increased
Gluconeogenesis	Decreased	Increased
Amino acid uptake and protein synthesis	Increased	No effect
Inhibition of amino acid release and protein degradation	Decreased	No effect
Lipogenesis	Increased	No effect
Lipolysis	Decreased	Increased
Ketogenesis	Decreased	Increased

Animation

Blood Sugar Regulation in Diabetics

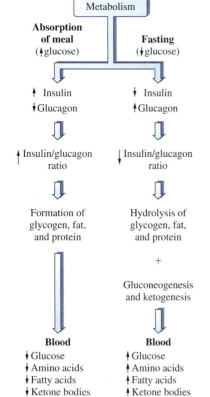

Figure 19.16

A summary of the antagonistic effects of insulin and glucagon.

secretion. Subsequent binding of insulin to the plasma membrane insulin receptor increases the rate of transport of glucose across the membrane and into cells.

Insulin exerts a variety of effects on all aspects of cellular metabolism:

- **Carbohydrate metabolism.** Insulin stimulates glycogen synthesis. At the same time, it inhibits glycogenolysis and gluconeogenesis. The overall result of these activities is the storage of excess glucose.
- **Protein metabolism.** Insulin stimulates transport and uptake of amino acids, as well as the incorporation of amino acids into proteins.
- **Lipid metabolism.** Insulin stimulates the uptake of glucose by adipose cells, as well as the synthesis and storage of triglycerides. As we have seen, storage of lipids requires a source of glucose, and insulin helps the process by increasing the available glucose. At the same time, insulin inhibits the breakdown of stored triglycerides.

As you may have already guessed, insulin is only part of the overall regulation of cellular metabolism in the body. A second hormone, **glucagon,** is secreted by the α-cells of the islets of Langerhans in response to decreased blood glucose levels. The effects of glucagon, generally the opposite of the effects of insulin, are summarized in Table 19.1. Although it has no direct effect on glucose uptake, glucagon inhibits glycogen synthesis and stimulates glycogenolysis and gluconeogenesis. It also stimulates the breakdown of fats and ketogenesis.

The antagonistic effects of these two hormones, seen in Figure 19.16, are critical for the maintenance of adequate blood glucose levels. During fasting, low blood glucose levels stimulate production of glucagon, which increases blood glucose by stimulating the breakdown of glycogen and the production of glucose by gluconeogenesis. This ensures a ready supply of glucose for the tissues, especially the brain. On the other hand, when blood glucose levels are too high, insulin is secreted. It stimulates the removal of the excess glucose by enhancing uptake and inducing pathways for storage.

QUESTION 19.13

Summarize the effects of the hormone insulin on carbohydrate, lipid, and amino acid metabolism.

QUESTION 19.14

Summarize the effects of the hormone glucagon on carbohydrate and lipid metabolism.

SUMMARY

19.1 Lipid Metabolism in Animals

Dietary lipids *(triglycerides)* are emulsified into tiny fat droplets in the intestine by the action of *bile* salts. Pancreatic *lipase* catalyzes the hydrolysis of triglycerides into monoglycerides and fatty acids. These are absorbed by intestinal epithelial cells, reassembled into triglycerides, and combined with protein to form *chylomicrons*. Chylomicrons are transported to the cells of the body through the bloodstream. Fatty acids are stored as triglycerides (triacylglycerols) in fat droplets in the cytoplasm of *adipocytes*.

19.2 Fatty Acid Degradation

Fatty acids are degraded to acetyl CoA in the mitochondria by the β-*oxidation* pathway, which involves five steps: (1) production of a fatty acyl CoA molecule, (2) oxidation of the fatty acid by an FAD-dependent dehydrogenase, (3) hydration, (4) oxidation by an NAD^+-dependent dehydrogenase, and (5) cleavage of the chain with release of acetyl CoA and a fatty acyl CoA that is two carbons shorter than the beginning fatty acid. The last four reactions are repeated until the fatty acid is completely degraded to acetyl CoA.

19.3 Ketone Bodies

Under some conditions, fatty acid degradation occurs more rapidly than glycolysis. As a result, a large amount of acetyl CoA is produced from fatty acids, but little oxaloacetate is generated from pyruvate. When oxaloacetate levels are too low, the excess acetyl CoA is converted to the *ketone bodies* acetone, acetoacetate, and β-hydroxybutyrate.

19.4 Fatty Acid Synthesis

Fatty acid biosynthesis occurs by the sequential addition of acetyl groups and, on first inspection, appears to be a simple reversal of the β-oxidation pathway. Although the biochemical reactions are similar, fatty acid synthesis differs from β-oxidation in the following ways: It occurs in the cytoplasm, utilizes *acyl carrier protein* and NADPH, and is carried out by a multienzyme complex, fatty acid synthase.

19.5 Degradation of Amino Acids

Amino acids are oxidized in the mitochondria. The first step of amino acid catabolism is deamination, the removal of the amino group. The carbon skeletons of amino acids are converted into molecules that can enter the citric acid cycle.

19.6 The Urea Cycle

In the *urea cycle*, the toxic ammonium ions released by deamination of amino acids are incorporated in urea, which is excreted in the urine.

19.7 The Effects of Insulin and Glucagon on Cellular Metabolism

Insulin stimulates biosynthetic processes and inhibits catabolism in liver, muscle, and adipose tissue. Insulin is synthesized in the β-cells of the pancreas and is secreted when the blood glucose levels become too high. The insulin receptor protein binds to the insulin. This binding mediates a variety of responses in target tissues, including the storage of glucose and lipids. *Glucagon* is secreted when blood glucose levels are too low. It has the opposite effects on metabolism, including the breakdown of lipids and glycogen.

KEY TERMS

acyl carrier protein (ACP) (19.4)
adipocyte (19.1)
adipose tissue (19.1)
bile (19.1)
chylomicron (19.1)
colipase (19.1)
diabetes mellitus (19.3)
glucagon (19.7)
hyperammonemia (19.6)
insulin (19.7)
ketoacidosis (19.3)
ketone bodies (19.3)
ketosis (19.3)
lipase (19.1)
micelle (19.1)
β-oxidation (19.2)
oxidative deamination (19.5)
phosphopantetheine (19.4)
pyridoxal phosphate (19.5)
transaminase (19.5)
transamination (19.5)
triglyceride (19.1)
urea cycle (19.6)

QUESTIONS AND PROBLEMS

Lipid Metabolism in Animals

Foundations

19.15 Describe the composition of bile. Why are bile salts referred to as detergents?

19.16 In Figure 19.1, a micelle composed of the phospholipid, lecithin is shown. Why is lecithin a good molecule for the formation of micelles?

19.17 Review the information on chylomicrons in Chapter 15. Describe the composition of chylomicrons.

19.18 What is an adipocyte?

Applications

19.19 What is the major storage form of fatty acids?

19.20 What is the outstanding structural feature of an adipocyte?

19.21 What is the major metabolic function of adipose tissue?

19.22 What is the general reaction catalyzed by lipases?

19.23 Why are triglycerides more efficient energy-storage molecules than glycogen?

19.24 a. What are very low density lipoproteins?
   b. Compare the function of VLDLs with that of chylomicrons.

19.25 Describe the stages of lipid digestion.

19.26 Describe the transport of lipids digested in the lumen of the intestines to the cells of the body.

Fatty Acid Degradation

Foundations

19.27 What is the energy source for the activation of a fatty acid in preparation for β-oxidation?

19.28 Which bond in fatty acyl CoA is a high-energy bond?

19.29 What is the product of the hydration of an alkene?

19.30 Which reaction in β-oxidation is a hydration reaction? What is the name of the enzyme that catalyzes this reaction? Write an equation representing this reaction.

Applications

19.31 Calculate the number of ATP molecules produced by complete β-oxidation of the fourteen-carbon saturated fatty acid tetradecanoic acid (common name: myristic acid).

19.32 **a.** Write the sequence of steps that would be followed for one round of β-oxidation of hexanoic acid.
b. Calculate the number of ATP molecules produced by complete β-oxidation of hexanoic acid.

19.33 How many molecules of ATP are produced for each molecule of FADH$_2$ that is generated by β-oxidation?

19.34 How many molecules of ATP are produced for each molecule of NADH generated by β-oxidation?

19.35 What is the fate of the acetyl CoA produced by β-oxidation?

19.36 How many ATP molecules are produced from each acetyl CoA molecule generated in β-oxidation that enters the citric acid cycle?

Ketone Bodies

Foundations

19.37 What are ketone bodies?

19.38 What are the chemical properties of ketone bodies?

19.39 Define *ketoacidosis*.

19.40 In what part of the cell does ketogenesis occur? Be specific.

Applications

19.41 Describe the relationship between the formation of ketone bodies and β-oxidation.

19.42 Why do uncontrolled diabetics produce large amounts of ketone bodies?

19.43 How does the presence of ketone bodies in the blood lead to ketoacidosis?

19.44 When does the heart use ketone bodies? When does the brain use ketone bodies?

Fatty Acid Synthesis

Foundations

19.45 Where in the cell does fatty acid biosynthesis occur?

19.46 What is the acyl group carrier in fatty acid biosynthesis?

Applications

19.47 **a.** What is the role of the phosphopantetheine group in fatty acid biosynthesis?
b. From what molecule is phosphopantetheine made?

19.48 Which molecules involved in fatty acid degradation and fatty acid biosynthesis contain the phosphopantetheine group?

19.49 How does the structure of fatty acid synthase differ from that of the enzymes that carry out β-oxidation?

19.50 In which cellular compartments do fatty acid biosynthesis and β-oxidation occur?

Degradation of Amino Acids

Foundations

19.51 What chemical transformation is carried out by transaminases?

19.52 Write a chemical equation for the transfer of an amino group from alanine to α-ketoglutarate, catalyzed by a transaminase.

19.53 Why is the glutamate family of transaminases so important?

19.54 What biochemical reaction is catalyzed by glutamate dehydrogenase?

Applications

19.55 Into which citric acid cycle intermediate is each of the following amino acids converted?
a. Alanine
b. Glutamate
c. Aspartate
d. Phenylalanine
e. Threonine
f. Arginine

19.56 What is the net ATP yield for degradation of each of the amino acids listed in Problem 19.55?

19.57 Write a balanced equation for the oxidative deamination of glutamate that is mediated by the enzyme glutamate dehydrogenase.

19.58 Write a balanced equation for the transamination of aspartate.

The Urea Cycle

19.59 What metabolic condition is produced if the urea cycle does not function properly?

19.60 What is hyperammonemia? How are mild forms of this disease treated?

19.61 The structure of urea is

a. Which substances are the sources of each of the amino groups in the urea molecule?
b. What substance is the source of the carbonyl group?

19.62 What energy source is used for the urea cycle?

The Effects of Insulin and Glucagon on Cellular Metabolism

Foundations

19.63 In general, what is the effect of insulin on catabolic and anabolic or biosynthetic processes?

19.64 What is the trigger that causes insulin to be secreted into the bloodstream?

19.65 What is meant by the term *target cell*?

19.66 Which are the primary target cells of insulin?

19.67 What is the trigger that causes glucagon to be secreted into the bloodstream?

19.68 Which are the primary target cells of glucagon?

Applications

19.69 Where is insulin produced?

19.70 Where is glucagon produced?

19.71 How does insulin affect carbohydrate metabolism?

19.72 How does glucagon affect carbohydrate metabolism?

19.73 How does insulin affect lipid metabolism?

19.74 How does glucagon affect lipid metabolism?

19.75 Why is it said that diabetes mellitus amounts to starvation in the midst of plenty?

19.76 What is the role of the insulin receptor in controlling blood glucose levels?

FOR FURTHER UNDERSTANDING

1. In birds, arginine is an essential amino acid. Can birds produce urea to remove ammonium ions from the blood? Explain your reasoning.

2. Oil-eating bacteria can oxidize long-chain alkanes. In the first step of the pathway, the enzyme monooxygenase catalyzes a reaction that converts the long-chain alkane into a primary alcohol. Data from research studies indicate that three more reactions are required to allow the primary alcohol to enter the β-oxidation pathway. Propose a pathway that would convert the long-chain alcohol into a product that could enter the β-oxidation pathway.

3. A young woman sought the advice of her physician because she was 30 pounds overweight. The excess weight was in the form of triglycerides carried in adipose tissue. Yet when the woman described her diet, it became obvious that she actually ate very moderate amounts of fatty foods. Most of her caloric intake was in the form of carbohydrates. This included candy, cake, beer, and soft drinks. Explain how the excess calories consumed in the form of carbohydrates ended up being stored as triglycerides in adipose tissue.

4. Olestra is a fat substitute that provides no calories, yet has a creamy, tongue-pleasing consistency. Because it can withstand heating, it can be used to prepare foods such as potato chips and crackers. Recently the Food and Drug Administration approved olestra for use in prepared foods.

Olestra is a sucrose polyester produced by esterification of six-, seven-, or eight-carbon fatty acids to molecules of sucrose. Develop a hypothesis to explain why olestra is not a source of dietary calories.

5. Carnitine is a tertiary amine found in mitochondria that is involved in transporting the acyl groups of fatty acids from the cytoplasm into the mitochondria. The fatty acyl group is transferred from a fatty acyl CoA molecule and esterified to carnitine. Inside mitochondria, the reaction is reversed and the fatty acid enters the β-oxidation pathway.

A seventeen-year-old male went to a university medical center complaining of fatigue and poor exercise tolerance. Muscle biopsies revealed droplets of triglycerides in his muscle cells. Biochemical analysis showed that he had only one-fifth of the normal amount of carnitine in his muscle cells.

What effect will carnitine deficiency have on β-oxidation? What effect will carnitine deficiency have on glucose metabolism?

Glossary

A

accuracy (1.4) the nearness of an experimental value to the true value

acetyl coenzyme A (acetyl CoA) (18.6) a molecule composed of coenzyme A and an acetyl group; the intermediate that provides acetyl groups for complete oxidation by aerobic respiration

acid (8.1) a substance that behaves as a proton (H^+) donor

acid-base reaction (4.3) reaction that involves the transfer of a hydrogen ion (H^+) from one reactant to another

activated complex (5.3) the arrangement of atoms at the top of the potential energy barrier as a reaction proceeds

activation energy (5.3) the threshold energy that must be overcome to produce a chemical reaction

active site (16.10) the cleft in the surface of an enzyme that is the site of substrate binding

active transport (15.6) the movement of molecules across a membrane against a concentration gradient

acyl carrier protein (ACP) (19.4) the protein that forms a thioester linkage with fatty acids during fatty acid synthesis

acyl group (13.2) the functional group found in carboxylic acid derivatives that contains the carbonyl group attached to one alkyl or aryl group:

addition polymer (11.4) polymers prepared by the sequential addition of a monomer

addition reaction (11.4, 12.5) a reaction in which two molecules add together to form a new molecule; often involves the addition of one molecule to a double or triple bond in an unsaturated molecule; e.g., the addition of a water molecule to an alkene to form an alcohol

adenosine triphosphate (ATP) (18.1) a nucleotide composed of the purine adenine, the sugar ribose, and three phosphoryl groups; the primary energy storage and transport molecule used by the cells in cellular metabolism

adipocyte (19.1) a fat cell

adipose tissue (19.1) fatty tissue that stores most of the body lipids

aerobic respiration (18.7) the oxygen-requiring degradation of food molecules and production of ATP

alcohol (12.1) an organic compound that contains a hydroxyl group (—OH) attached to an alkyl group

aldehyde (12.5) a class of organic molecules characterized by a carbonyl group; the carbonyl carbon is bonded to a hydrogen atom and to another hydrogen or an alkyl or aryl group. Aldehydes have the following general structure:

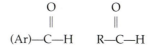

aldose (14.2) a sugar that contains an aldehyde (carbonyl) group

aliphatic hydrocarbon (10.1) any member of the alkanes, alkenes, and alkynes or substituted alkanes, alkenes, and alkynes

alkali metal (2.3) an element within Group IA (1) of the periodic table

alkaline earth metal (2.3) an element within Group IIA (2) of the periodic table

alkane (10.2) a hydrocarbon that contains only carbon and hydrogen and is bonded together through carbon–hydrogen and carbon–carbon single bonds; a saturated hydrocarbon with the general molecular formula C_nH_{2n+2}

alkene (11.1) a hydrocarbon that contains one or more carbon–carbon double bonds; an unsaturated hydrocarbon with the general formula C_nH_{2n}

alkyl group (10.2) a hydrocarbon group that results from the removal of one hydrogen from the original hydrocarbon (e.g., methyl, —CH_3; ethyl, —CH_2CH_3)

alkyl halide (10.4) a substituted hydrocarbon with the general structure R—X, in which R— represents any alkyl group and X = a halogen (F—, Cl—, Br—, or I—)

alkylammonium ion (13.3) the ion formed when the lone pair of electrons of the nitrogen atom of an amine is shared with a proton (H^+) from a water molecule

alkyne (11.1) a hydrocarbon that contains one or more carbon–carbon triple bonds; an unsaturated hydrocarbon with the general formula C_nH_{2n-2}

alpha particle (9.1) a particle consisting of two protons and two neutrons; the alpha particle is identical to a helium nucleus

amide bond (13.4) the bond between the carbonyl carbon of a carboxylic acid and the amino nitrogen of an amine

amides (13.4) the family of organic compounds formed by the reaction between a carboxylic acid derivative and an amine and characterized by the amide group

amines (13.3) the family of organic molecules with the general formula RNH_2, R_2NH, or R_3N (R— can represent either an alkyl or aryl group); they may be viewed as substituted ammonia molecules in which one or more of the ammonia hydrogens has been substituted by a more complex organic group

α-amino acid (16.2) the subunits of proteins composed of an α-carbon bonded to a carboxylate group, a protonated amino group, a hydrogen atom, and a variable R group

aminoacyl tRNA (17.6) the transfer RNA covalently linked to the correct amino acid

aminoacyl tRNA binding site of ribosome (A-site) (17.6) a pocket on the surface of a ribosome that holds the aminoacyl tRNA during translation

aminoacyl tRNA synthetase (17.6) an enzyme that recognizes one tRNA and covalently links the appropriate amino acid to it

amorphous solid (6.3) a solid with no organized, regular structure

amphiprotic (8.1) a substance that can behave either as a Brønsted acid or a Brønsted base

amylopectin (14.6) a highly branched form of amylose; the branches are attached to the C-6 hydroxyl by $\alpha(1 \rightarrow 6)$ glycosidic linkage; a component of starch

amylose (14.6) a linear polymer of α-D-glucose molecules bonded in $\alpha(1 \rightarrow 4)$ glycosidic linkage that is a major component of starch; a polysaccharide storage form

anabolism (18.1) all of the cellular energy-requiring biosynthetic pathways

anaerobic threshold (18.4) the point at which the level of lactate in the exercising muscle inhibits glycolysis and the muscle, deprived of energy, ceases to function

angular structure (3.4) a planar molecule with bond angles other than 180°

anion (2.1) a negatively charged atom or group of atoms

antibodies (16.1) immunoglobulins; specific glycoproteins produced by cells of the immune system in response to invasion by infectious agents

anticodon (17.4) a sequence of three ribonucleotides on a tRNA that are complementary to a codon on the mRNA; codon-anticodon binding results in delivery of the correct amino acid to the site of protein synthesis

antigen (16.1) any substance that is able to stimulate the immune system; generally a protein or large carbohydrate

antiparallel strands (17.2) a term describing the polarities of the two strands of the DNA double helix; on one strand the sugar-phosphate backbone advances in the $5' \rightarrow 3'$ direction; on the opposite, complementary strand the sugar-phosphate backbone advances in the $3' \rightarrow 5'$ direction

apoenzyme (16.10) the protein portion of an enzyme that requires a cofactor to function in catalysis

aqueous solution (7.1) any solution in which the solvent is water

arachidonic acid (15.2) a fatty acid derived from linoleic acid; the precursor of prostaglandins

aromatic hydrocarbon (10.1, 11.5) an organic compound that contains the benzene ring or a derivative of the benzene ring

Arrhenius theory (8.1) a theory that describes an acid as a substance that dissociates to produce H^+ and a base as a substance that dissociates to produce OH^-

artificial radioactivity (9.5) radiation that results from the conversion of a stable nucleus to another, unstable nucleus

atherosclerosis (15.4) deposition of excess plasma cholesterol and other lipids and proteins on the walls of arteries, resulting in decreased artery diameter and increased blood pressure

atom (2.1) the smallest unit of an element that retains the properties of that element

atomic mass (2.1) the mass of an atom expressed in atomic mass units

atomic mass unit (4.1) 1/12 of the mass of a ^{12}C atom, equivalent to 1.661×10^{-24} g

atomic number (2.1) the number of protons in the nucleus of an atom; it is a characteristic identifier of an element

atomic orbital (2.4) a specific region of space where an electron may be found

ATP synthase (18.9) a multiprotein complex within the inner mitochondrial membrane that uses the energy of the proton (H^+) gradient to produce ATP

autoionization (8.1) also known as *self-ionization,* the reaction of a substance, such as water, with itself to produce a positive and a negative ion

Avogadro's law (6.1) a law that states that the volume is directly proportional to the number of moles of gas particles, assuming that the pressure and temperature are constant

Avogadro's number (4.1) 6.022×10^{23} particles of matter contained in 1 mol of a substance

B

background radiation (9.6) the radiation that emanates from natural sources

barometer (6.1) a device for measuring pressure

base (8.1) a substance that behaves as a proton (H^+) acceptor

base pair (17.2) a hydrogen-bonded pair of bases within the DNA double helix; the standard base pairs always involve a purine and a pyrimidine; in particular, adenine always base pairs with thymine and cytosine with guanine

Benedict's reagent (14.4) a buffered solution of Cu^{2+} ions that can be used to test for reducing sugars or to distinguish between aldehydes and ketones

Benedict's test (12.5) a test used to determine the presence of reducing sugars or to distinguish between aldehydes and ketones; it requires a buffered solution of Cu^{2+} ions that are reduced to Cu^+, which precipitates as brick-red Cu_2O

beta particle (9.1) an electron formed in the nucleus by the conversion of a neutron into a proton

bile (19.1) micelles of lecithin, cholesterol, bile salts, protein, inorganic ions, and bile pigments that aid in lipid digestion by emulsifying fat droplets

binding energy (9.3) the energy required to break down the nucleus into its component parts

boiling point (3.3) the temperature at which the vapor pressure of a liquid is equal to the atmospheric pressure

bond energy (3.4) the amount of energy necessary to break a chemical bond

Boyle's law (6.1) a law stating that the volume of a gas varies inversely with the pressure exerted if the temperature and number of moles of gas are constant

breeder reactor (9.4) a nuclear reactor that produces its own fuel in the process of providing electrical energy

Brønsted-Lowry theory (8.1) a theory that describes an acid as a proton donor and a base as a proton acceptor

buffer capacity (8.4) a measure of the ability of a solution to resist large changes in pH when a strong acid or strong base is added

buffer solution (8.4) a solution containing a weak acid or base and its salt (the conjugate base or acid) that is resistant to large changes in pH upon addition of strong acids or bases

buret (8.3) a device calibrated to deliver accurately known volumes of liquid, as in a titration

C

C-terminal amino acid (16.3) the amino acid in a peptide that has a free α-CO_2^- group; the last amino acid in a peptide

calorimetry (5.2) the measurement of heat energy changes during a chemical reaction

cap structure (17.4) a 7-methylguanosine unit covalently bonded to the 5′ end of a mRNA by a 5′–5′ triphosphate bridge

carbinol carbon (12.1) that carbon in an alcohol to which the hydroxyl group is attached

carbohydrate (14.1) generally sugars and polymers of sugars; the primary source of energy for the cell

carbonyl group (12.5) the functional group that contains a carbon-oxygen double bond: —C=O; the functional group found in aldehydes and ketones

carboxyl group (13.1) the —COOH functional group; the functional group found in carboxylic acids

carboxylic acid (13.1) a member of the family of organic compounds that contain the —COOH functional group

carboxylic acid derivative (13.2) any of several families of organic compounds, including the esters and amides, that are derived from carboxylic acids and have the general formula

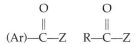

Z = —OR or OAr for the esters, and Z = —NH$_2$ for the amides

carcinogen (17.7) any chemical or physical agent that causes mutations in the DNA that lead to uncontrolled cell growth or cancer

catabolism (18.1) the degradation of fuel molecules and production of ATP for cellular functions

catalyst (5.3) any substance that increases the rate of a chemical reaction (by lowering the activation energy of the reaction) and that is not destroyed in the course of the reaction

cathode rays (2.2) a stream of electrons that is given off by the cathode (negative electrode) in a cathode ray tube

cation (2.1) a positively charged atom or group of atoms

cellulose (14.6) a polymer of β-D-glucose linked by $\beta(1 \rightarrow 4)$ glycosidic bonds

central dogma (17.4) a statement of the directional transfer of the genetic information in cells: DNA → RNA → Protein

chain reaction (9.4) the process in a fission reactor that involves neutron production and causes subsequent reactions accompanied by the production of more neutrons in a continuing process

Charles's law (6.1) a law stating that the volume of a gas is directly proportional to the temperature of the gas, assuming that the pressure and number of moles of the gas are constant

chemical bond (3.1) the attractive force holding two atomic nuclei together in a chemical compound

chemical equation (4.3) a record of chemical change, showing the conversion of reactants to products

chemical formula (4.2) the representation of a compound or ion in which elemental symbols represent types of atoms and subscripts show the relative numbers of atoms

chemical property (1.2) characteristic of a substance that relates to the substance's participation in a chemical reaction

chemical reaction (1.2) a process in which atoms are rearranged to produce new combinations

chemistry (1.1) the study of matter and the changes that matter undergoes

chiral carbon (14.3) a carbon atom bonded to four different atoms or groups of atoms

chiral molecule (14.3) molecule capable of existing in mirror-image forms

cholesterol (15.4) a twenty-seven-carbon steroid ring structure that serves as the precursor of steroid hormones

chylomicron (15.5, 19.1) a plasma lipoprotein (aggregate of protein and triglycerides) that carries triglycerides from the intestine to all body tissues via the bloodstream

citric acid cycle (18.8) a cyclic biochemical pathway that is the final stage of degradation of carbohydrates, fats, and amino acids. It results in the complete oxidation of acetyl groups derived from these dietary fuels

codon (17.4) a group of three ribonucleotides on the mRNA that specifies the addition of a specific amino acid onto the growing peptide chain

coenzyme (16.10) an organic group required by some enzymes; it generally serves as a donor or acceptor of electrons or a functional group in a reaction

coenzyme A (18.6) a molecule derived from ATP and the vitamin pantothenic acid; coenzyme A functions in the transfer of acetyl groups in lipid and carbohydrate metabolism

cofactor (16.10) an inorganic group, usually a metal ion, that must be bound to an apoenzyme to maintain the correct configuration of the active site

colipase (19.1) a protein that aids in lipid digestion by binding to the surface of lipid droplets and facilitating binding of pancreatic lipase

colligative property (7.4) property of a solution that is dependent only on the concentration of solute particles

colloidal suspension (7.1) a heterogeneous mixture of solute particles in a solvent; distribution of solute particles is not uniform because of the size of the particles

combination reaction (4.3) a reaction in which two substances join to form another substance

combined gas law (6.1) an equation that describes the behavior of a gas when volume, pressure, and temperature may change simultaneously

combustion (10.4) the oxidation of hydrocarbons by burning in the presence of air to produce carbon dioxide and water

competitive inhibitor (16.10) a structural analog; a molecule that has a structure very similar to the natural substrate of an enzyme, competes with the natural substrate for binding to the enzyme active site, and inhibits the reaction

complementary strands (17.2) the opposite strands of the double helix are hydrogen-bonded to one another such that adenine and thymine or guanine and cytosine are always paired

complex lipid (15.5) a lipid bonded to other types of molecules

compound (1.2) a substance that is characterized by constant composition and that can be chemically broken down into elements

concentration (5.3, 7.2) a measure of the quantity of a substance contained in a specified volume of solution

concentration gradient (7.4) region where concentration decreases over distance

condensation (6.2) the conversion of a gas to a liquid

condensation polymer (13.2) a polymer, which is a large molecule formed by combination of many small molecules (monomers) that results from joining of monomers in a reaction that forms a small molecule, such as water or an alcohol

condensed formula (10.2) a structural formula showing all of the atoms in a molecule and placing them in a sequential arrangement that details which atoms are bonded to each other; the bonds themselves are not shown

conjugate acid (8.1) substance that has one more proton than the base from which it is derived

conjugate acid-base pair (8.1) two species related to each other through the gain or loss of a proton

conjugate base (8.1) substance that has one less proton than the acid from which it is derived

constitutional isomers (10.2) two molecules having the same molecular formulas, but different chemical structures

Cori Cycle (18.10) a metabolic pathway in which the lactate produced by working muscle is taken up by cells in the liver and converted back to glucose by gluconeogenesis

corrosion (4.3) the unwanted oxidation of a metal

covalent bonding (3.1) a situation in which a pair of electrons is shared between two atoms

covalent solid (6.3) a collection of atoms held together by covalent bonds

cristae (18.5) the folds of the inner membrane of the mitochondria

crystal lattice (3.2) a unit of a solid characterized by a regular arrangement of components

crystalline solid (6.3) a solid having a regular repeating atomic structure

curie (9.7) the quantity of radioactive material that produces 3.7×10^{10} nuclear disintegrations per second

cycloalkane (10.3) a cyclic alkane; a saturated hydrocarbon that has the general formula C_nH_{2n}

D

Dalton's law (6.1) also called the law of partial pressures; states that the total pressure exerted by a gas mixture is the sum of the partial pressures of the component gases

data (1.3) a group of facts resulting from an experiment

decomposition reaction (4.3) the breakdown of a substance into two or more substances

defense proteins (16.1) proteins that defend the body against infectious diseases. Antibodies are defense proteins

degenerate code (17.5) a term used to describe the fact that several triplet codons may be used to specify a single amino acid in the genetic code

dehydration (of alcohols) (12.1) a reaction that involves the loss of a water molecule, in this case the loss of water from an alcohol and the simultaneous formation of an alkene

deletion mutation (17.7) a mutation that results in the loss of one or more nucleotides from a DNA sequence

denaturation (16.9) the process by which the organized structure of a protein is disrupted, resulting in a completely disorganized, nonfunctional form of the protein

density (1.5) mass per unit volume of a substance

deoxyribonucleic acid (DNA) (17.1) the nucleic acid molecule that carries all of the genetic information of an organism; the DNA molecule is a double helix composed of two strands, each of which is composed of phosphate groups, deoxyribose, and the nitrogenous bases thymine, cytosine, adenine, and guanine

deoxyribonucleotide (17.1) a nucleoside phosphate or nucleotide composed of a nitrogenous base in β-*N*-glycosidic linkage to the 1′ carbon of the sugar 2′-deoxyribose and with one, two, or three phosphoryl groups esterified at the hydroxyl of the 5′ carbon

diabetes mellitus (19.3) a disease caused by the production of insufficient levels of insulin and characterized by the appearance of very high levels of glucose in the blood and urine

dialysis (7.6) the removal of waste material via transport across a membrane

diffusion (7.4) net movement of solute or solvent molecules from an area of high concentration to an area of low concentration

diglyceride (15.3) the product of esterification of glycerol at two positions

dipole–dipole interactions (6.2) attractive forces between polar molecules

disaccharide (14.1) a sugar composed of two monosaccharides joined through an oxygen atom bridge

dissociation (3.3) production of positive and negative ions when an ionic compound dissolves in water

disulfide (12.4) an organic compound that contains a disulfide group (—S—S—)

double bond (3.4) a bond in which two pairs of electrons are shared by two atoms

double helix (17.2) the spiral staircase-like structure of the DNA molecule characterized by two sugar-phosphate backbones wound around the outside and nitrogenous bases extending into the center

double-replacement reaction (4.3) a chemical change in which cations and anions "exchange partners"

dynamic equilibrium (5.4) the state that exists when the rate of change in the concentration of products and reactants is equal, resulting in no net concentration change

E

eicosanoid (15.2) any of the derivatives of twenty-carbon fatty acids, including prostaglandins, leukotrienes, and thromboxanes

electrolyte (3.3, 7.1) a material that dissolves in water to produce a solution that conducts an electrical current

electrolytic solution (3.3) a solution composed of an electrolytic solute dissolved in water

electron (2.1) a negatively charged particle outside of the nucleus of an atom

electron affinity (2.6) the energy released when an electron is added to an isolated atom

electron configuration (2.4) the arrangement of electrons around a nucleus of an atom, ion, or a collection of nuclei of a molecule

electron transport system (18.9) the series of electron transport proteins embedded in the inner mitochondrial membrane that accept high-energy electrons from NADH and $FADH_2$ and transfer them in stepwise fashion to molecular oxygen (O_2)

electronegativity (3.1) a measure of the tendency of an atom in a molecule to attract shared electrons

element (1.2) a substance that cannot be decomposed into simpler substances by chemical or physical means

elimination reaction (12.5) a reaction in which a molecule loses atoms or ions from its structure

elongation factors (17.6) proteins that facilitate the elongation phase of translation

emulsifying agent (15.3) a bipolar molecule that aids in the suspension of fats in water

enantiomers (14.3) stereoisomers that are nonsuperimposable mirror images of one another

endothermic reaction (5.1) a chemical or physical change in which energy is absorbed

energy (1.1) the capacity to do work

energy levels (2.4) atomic region where electrons may be found

enthalpy (5.1) a term that represents heat energy

entropy (5.1) a measure of randomness or disorder

enzyme (16.1, 16.10) a protein that serves as a biological catalyst

enzyme-substrate complex (16.10) a molecular aggregate formed when the substrate binds to the active site of the enzyme

equilibrium reaction (5.4) a reaction that is reversible and the rates of the forward and reverse reactions are equal

equivalence point (8.3) the situation in which reactants have been mixed in the molar ratio corresponding to the balanced equation

equivalent (7.4) number of grams of an ion corresponding to Avogadro's number of electrical charges

error (1.4) the difference between the true value and the experimental value for data or results

essential fatty acids (15.2) the fatty acids linolenic and linoleic acids that must be supplied in the diet because they cannot be synthesized by the body

ester (13.2) a carboxylic acid derivative formed by the reaction of a carboxylic acid and an alcohol. Esters have the following general formula:

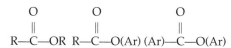

esterification (15.2) the formation of an ester in the reaction of a carboxylic acid and an alcohol

ether (12.3) an organic compound that contains two alkyl and/or aryl groups attached to an oxygen atom; R—O—R, Ar—O—R, and Ar—O—Ar

evaporation (6.2) the conversion of a liquid to a gas below the boiling point of the liquid

exon (17.4) protein-coding sequences of a gene found on the final mature mRNA

exothermic reaction (5.1) a chemical or physical change that releases energy

extensive property (1.2) a property of a substance that depends on the quantity of the substance

F

F_0F_1 complex (18.9) an alternative term for the ATP synthase, the multiprotein complex in the inner mitochondrial membrane that uses the energy of the proton gradient to produce ATP

facilitated diffusion (15.6) movement of a solute across a membrane from an area of high concentration to an area of low concentration through a transmembrane protein, or permease

fatty acid (13.1, 15.2) any member of the family of continuous-chain carboxylic acids that generally contain four to twenty carbon atoms; the most concentrated source of energy used by the cell

fermentation (12.1, 18.4) anaerobic (in the absence of oxygen) catabolic reactions that occur with no net oxidation. Pyruvate or an organic compound produced from pyruvate is reduced as NADH is oxidized

fibrous protein (16.5) a protein composed of peptides arranged in long sheets or fibers

Fischer projection (14.3) a two-dimensional drawing of a molecule, which shows a chiral carbon at the intersection of two lines and horizontal lines representing bonds projecting out of the page and vertical lines representing bonds that project into the page

fission (9.4) the splitting of heavy nuclei into lighter nuclei accompanied by the release of large quantities of energy

fluid mosaic model (15.6) the model of membrane structure that describes the fluid nature of the lipid bilayer and the presence of numerous proteins embedded within the membrane

formula (3.2) the representation of the fundamental compound unit using chemical symbols and numerical subscripts

formula unit (4.2) the smallest collection of atoms from which the formula of a compound can be established

formula weight (4.2) the mass of a formula unit of a compound relative to a standard (carbon-12)

fructose (14.4) a ketohexose that is also called levulose and fruit sugar; the sweetest of all sugars, abundant in honey and fruits

fuel value (5.2) the amount of energy derived from a given mass of material

functional group (10.1) an atom (or group of atoms and their bonds) that imparts specific chemical and physical properties to a molecule

fusion (9.4) the joining of light nuclei to form heavier nuclei, accompanied by the release of large amounts of energy

G

galactose (14.4) an aldohexose that is a component of lactose (milk sugar)

galactosemia (14.5) a human genetic disease caused by the inability to convert galactose to a phosphorylated form of glucose (glucose-1-phosphate) that can be used in cellular metabolic reactions

gamma ray (9.1) a high-energy emission from nuclear processes, traveling at the speed of light; the high-energy region of the electromagnetic spectrum

gaseous state (1.2) a physical state of matter characterized by a lack of fixed shape or volume and ease of compressibility

geometric isomer (11.3) an isomer that differs from another isomer in the placement of substituents on a double bond or a ring

globular protein (16.6) a protein composed of polypeptide chains that are tightly folded into a compact spherical shape

glucagon (18.11, 19.7) a peptide hormone synthesized by the α-cells of the islets of Langerhans in the pancreas and secreted in response to low blood glucose levels; glucagon promotes glycogenolysis and gluconeogenesis and thereby increases the concentration of blood glucose

gluconeogenesis (18.10) the synthesis of glucose from noncarbohydrate precursors

glucose (14.4) an aldohexose, the most abundant monosaccharide; it is a component of many disaccharides, such as lactose and sucrose, and of polysaccharides, such as cellulose, starch, and glycogen

glyceraldehyde (14.3) an aldotriose that is the simplest carbohydrate; phosphorylated forms of glyceraldehyde are important intermediates in cellular metabolic reactions

glyceride (15.3) a lipid that contains glycerol

glycogen (14.6, 18.11) a long, branched polymer of glucose stored in liver and muscles of animals; it consists of a linear backbone of α-D-glucose in α(1 → 4) linkage, with numerous short branches attached to the C-6 hydroxyl group by α(1 → 6) linkage

glycogenesis (18.11) the metabolic pathway that results in the addition of glucose to growing glycogen polymers when blood glucose levels are high

glycogen granule (18.11) a core of glycogen surrounded by enzymes responsible for glycogen synthesis and degradation

glycogenolysis (18.11) the biochemical pathway that results in the removal of glucose molecules from glycogen polymers when blood glucose levels are low

glycolysis (18.3) the enzymatic pathway that converts a glucose molecule into two molecules of pyruvate; this anaerobic process generates a net energy yield of two molecules of ATP and two molecules of NADH

glycoprotein (16.7) a protein bonded to sugar groups

glycosidic bond (14.1) the bond between the hydroxyl group of the C-1 carbon of one sugar and a hydroxyl group of another sugar

group (2.3) any one of eighteen vertical columns of elements; often referred to as a family

H

half-life ($t_{1/2}$) (9.3) the length of time required for one-half of the initial mass of an isotope to decay to products

halogen (2.3) an element found in Group VIIA (17) of the periodic table

halogenation (10.4, 11.4) a reaction in which one of the C—H bonds of a hydrocarbon is replaced with a C—X bond (X = Br or Cl generally)

Haworth projection (14.4) a means of representing the orientation of substituent groups around a cyclic sugar molecule

α-helix (16.5) a right-handed coiled secondary structure maintained by hydrogen bonds between the amide hydrogen of one amino acid and the carbonyl oxygen of an amino acid four residues away

hemoglobin (16.8) the major protein component of red blood cells; the function of this red, iron-containing protein is transport of oxygen

Henry's law (7.1) a law stating that the number of moles of a gas dissolved in a liquid at a given temperature is proportional to the partial pressure of the gas

heterocyclic amine (15.2) a heterocyclic compound that contains nitrogen in at least one position in the ring skeleton

heterocyclic aromatic compound (11.6) cyclic aromatic compound having at least one atom other than carbon in the structure of the aromatic ring

heterogeneous mixture (1.2) a mixture of two or more substances characterized by nonuniform composition

hexose (14.2) a six-carbon monosaccharide

high-density lipoprotein (HDL) (15.5) a plasma lipoprotein that transports cholesterol from peripheral tissue to the liver

holoenzyme (16.10) an active enzyme consisting of an apoenzyme bound to a cofactor

homogeneous mixture (1.2) a mixture of two or more substances characterized by uniform composition

hydrate (4.2) any substance that has water molecules incorporated in its structure

hydration (11.4, 12.1) a reaction in which water is added to a molecule, e.g., the addition of water to an alkene to form an alcohol

hydrocarbon (10.1) a compound composed solely of the elements carbon and hydrogen

hydrogen bonding (6.2) the attractive force between a hydrogen atom covalently bonded to a small, highly electronegative atom and another atom containing an unshared pair of electrons

hydrogenation (11.4, 12.5, 15.2) a reaction in which hydrogen (H_2) is added to a double or a triple bond

hydrolase (16.10) an enzyme that catalyzes hydrolysis reactions

hydrolysis (13.2) a chemical change that involves the reaction of a molecule with water; the process by which molecules are broken into their constituents by addition of water

hydronium ion (8.1) a protonated water molecule, H_3O^+

hydrophilic amino acid (16.2) "water loving"; a polar or ionic amino acid that has a high affinity for water

hydrophobic amino acid (16.2) "water fearing"; a nonpolar amino acid that prefers contact with other nonpolar amino acids to contact with water

hydroxyl group (12.1) the —OH functional group that is characteristic of alcohols

hyperammonemia (19.6) a genetic defect in one of the enzymes of the urea cycle that results in toxic or even fatal elevation of the concentration of ammonium ions in the body

hyperglycemia (18.11) blood glucose levels that are higher than normal

hypertonic solution (7.4) the more concentrated solution of two separated by a semipermeable membrane

hypoglycemia (18.11) blood glucose levels that are lower than normal

hypothesis (1.1) an attempt to explain observations in a commonsense way

hypotonic solution (7.4) the more dilute solution of two separated by a semipermeable membrane

I

ideal gas (6.1) a gas in which the particles do not interact and the volume of the individual gas particles is assumed to be negligible

ideal gas law (6.1) a law stating that for an ideal gas the product of pressure and volume is proportional to the product of the number of moles of the gas and its temperature; the proportionality constant for an ideal gas is symbolized R

indicator (8.3) a solute that shows some condition of a solution (such as acidity or basicity) by its color

induced fit model (16.10) the theory of enzyme-substrate binding that assumes that the enzyme is a flexible molecule and that both the substrate and the enzyme change their shapes to accommodate one another as the enzyme-substrate complex forms

initiation factors (17.6) proteins that are required for formation of the translation initiation complex, which is composed of the large and small ribosomal subunits, the mRNA, and the initiator tRNA, methionyl tRNA

inner mitochondrial membrane (18.5) the highly folded, impermeable membrane within the mitochondrion that is the location of the electron transport system and ATP synthase

insertion mutation (17.7) a mutation that results in the addition of one or more nucleotides to a DNA sequence

insulin (18.11, 19.7) a hormone released from the pancreas in response to high blood glucose levels; insulin stimulates glycogenesis, fat storage, and cellular uptake and storage of glucose from the blood

intensive property (1.2) a property of a substance that is independent of the quantity of the substance

intermembrane space (18.5) the region between the outer and inner mitochondrial membranes, which is the location of the proton (H^+) reservoir that drives ATP synthesis

intermolecular force (3.5) any attractive force that occurs between molecules

intramolecular force (3.5) any attractive force that occurs within molecules

intron (17.4) a noncoding sequence within a eukaryotic gene that must be removed from the primary transcript to produce a functional mRNA

ion (2.1) an electrically charged particle formed by the gain or loss of electrons

ionic bonding (3.1) an electrostatic attractive force between ions resulting from electron transfer

ionic solid (6.3) a solid composed of positive and negative ions in a regular three-dimensional crystalline arrangement

ionization energy (2.6) the energy needed to remove an electron from an atom in the gas phase

ionizing radiation (9.1) radiation that is sufficiently high in energy to cause ion formation upon impact with an atom

ion pair (3.1) the simplest formula unit for an ionic compound

ion product for water (8.1) the product of the hydronium and hydroxide ion concentrations in pure water at a specified temperature; at 25°C, it has a value of 1.0×10^{-14}

irreversible enzyme inhibitor (16.10) a chemical that binds strongly to the R groups of an amino acid in the active site and eliminates enzyme activity

isoelectronic (2.5) atoms, ions, and molecules containing the same number of electrons

isomerase (16.10) an enzyme that catalyzes the conversion of one isomer to another

isotonic solution (7.4) a solution that has the same solute concentration as another solution with which it is being compared; a solution that has the same osmotic pressure as a solution existing within a cell

isotope (2.1) atom of the same element that differs in mass because it contains different numbers of neutrons

I.U.P.A.C. Nomenclature System (10.2) the International Union of Pure and Applied Chemistry (I.U.P.A.C.) standard, universal system for the nomenclature of organic compounds

K

α-keratin (16.5) a member of the family of fibrous proteins that form the covering of most land animals; major components of fur, skin, beaks, and nails

ketoacidosis (19.3) a drop in the pH of the blood caused by elevated levels of ketone bodies

ketone (12.5) a family of organic molecules characterized by a carbonyl group; the carbonyl carbon is bonded to two alkyl groups, two aryl groups, or one alkyl and one aryl group; ketones have the following general structures:

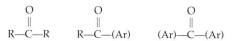

ketone bodies (19.3) acetone, acetoacetone, and β-hydroxybutyrate produced from fatty acids in the liver via acetyl CoA

ketose (14.2) a sugar that contains a ketone (carbonyl) group

ketosis (19.3) an abnormal rise in the level of ketone bodies in the blood

kinetic-molecular theory (6.1) the fundamental model of particle behavior in the gas phase

kinetics (5.3) the study of rates of chemical reactions

L

lactose (14.5) a disaccharide composed of β-D-galactose and either α- or β-D-glucose in β(1 → 4) glycosidic linkage; milk sugar

lactose intolerance (14.5) the inability to produce the digestive enzyme lactase, which degrades lactose to galactose and glucose

law (1.1) a summary of a large quantity of information

law of conservation of mass (4.3) a law stating that, in chemical change, matter cannot be created or destroyed

LeChatelier's principle (5.4) a law stating that when a system at equilibrium is disturbed, the equilibrium shifts in the direction that minimizes the disturbance

lethal dose (LD$_{50}$) (9.7) the quantity of toxic material (such as radiation) that causes the death of 50% of a population of an organism

Lewis symbol (3.1) representation of an atom or ion using the atomic symbol (for the nucleus and core electrons) and dots to represent valence electrons

ligase (16.10) an enzyme that catalyzes the joining of two molecules

linear structure (3.4) the structure of a molecule in which the bond angles about the central atom(s) is (are) 180°

line formula (10.2) the simplest representation of a molecule in which it is assumed that there is a carbon atom at any location where two or more lines intersect, there is a carbon at the end of any line, and each carbon is bonded to the correct number of hydrogen atoms

lipase (19.1) an enzyme that hydrolyzes the ester linkage between glycerol and the fatty acids of triglycerides

lipid (15.1) a member of the group of biological molecules of varying composition that are classified together on the basis of their solubility in nonpolar solvents

liquid state (1.2) a physical state of matter characterized by a fixed volume and the absence of a fixed shape

lock-and-key model (16.10) the theory of enzyme-substrate binding that depicts enzymes as inflexible molecules; the substrate fits into the rigid active site in the same way a key fits into a lock

London forces (6.2) weak attractive forces between molecules that result from short-lived dipoles that occur because of the continuous movement of electrons in the molecules

lone pair (3.4) an electron pair that is not involved in bonding

low-density lipoprotein (LDL) (15.5) a plasma lipoprotein that carries cholesterol to peripheral tissues and helps to regulate cholesterol levels in those tissues

lyase (16.10) an enzyme that catalyzes a reaction involving double bonds

M

maltose (14.5) a disaccharide composed of α-D-glucose and a second glucose molecule in α(1 → 4) glycosidic linkage

Markovnikov's rule (11.4) the rule stating that a hydrogen atom, adding to a carbon–carbon double bond, will add to the carbon having the larger number of hydrogens attached to it

mass (1.5) a quantity of matter

mass number (2.1) the sum of the number of protons and neutrons in an atom

matrix space (18.5) the region of the mitochondrion within the inner membrane; the location of the enzymes that carry out the reactions of the citric acid cycle and β-oxidation of fatty acids

matter (1.1) the material component of the universe

melting point (3.3, 6.3) the temperature at which a solid converts to a liquid

messenger RNA (17.4) an RNA species produced by transcription that specifies the amino acid sequence for a protein

metal (2.3) an element located on the left side of the periodic table (left of the "staircase" boundary)

metallic bond (6.3) a bond that results from the orbital overlap of metal atoms

metallic solid (6.3) a solid composed of metal atoms held together by metallic bonds

metalloid (2.3) an element along the "staircase" boundary between metals and nonmetals; metalloids exhibit both metallic and nonmetallic properties

metastable isotope (9.2) an isotope that will give up some energy to produce a more stable form of the same isotope

micelle (19.1) an aggregation of molecules having nonpolar and polar regions; the nonpolar regions of the molecules aggregate, leaving the polar regions facing the surrounding water

mitochondria (18.5) the cellular "power plants" in which the reactions of the citric acid cycle, the electron transport system, and ATP synthase function to produce ATP

mixture (1.2) a material composed of two or more substances

molar mass (4.1, 4.2) the mass in grams of one mol of a substance

molar volume (6.1) the volume occupied by one mol of a substance

molarity (7.3) the number of moles of solute per liter of solution

mole (4.1) the amount of substance containing Avogadro's number of particles

molecular formula (10.2) a formula that provides the atoms and number of each type of atom in a molecule but gives no information regarding the bonding pattern involved in the structure of the molecule

molecular solid (6.3) a solid in which the molecules are held together by dipole-dipole and London forces (van der Waals forces)

molecule (3.2) a unit in which the atoms of two or more elements are held together by covalent bonds

monatomic ion (3.2) an ion formed by electron gain or loss from a single atom

monoglyceride (15.3) the product of the esterification of glycerol at one position

monomer (11.4) the individual molecules from which a polymer is formed

monosaccharide (14.1) the simplest type of carbohydrate consisting of a single saccharide unit

movement protein (16.1) a protein involved in any aspect of movement in an organism, for instance, actin and myosin in muscle tissue and flagellin that composes bacterial flagella

mutagen (17.7) any chemical or physical agent that causes changes in the nucleotide sequence of a gene

mutation (17.7) any change in the nucleotide sequence of a gene

myoglobin (16.8) the oxygen storage protein found in muscle

N

N-terminal amino acid (16.3) the amino acid in a peptide that has a free α-N$^+$H$_3$ group; the first amino acid of a peptide

natural radioactivity (9.5) the spontaneous decay of a nucleus to produce high-energy particles or rays

neutral glyceride (15.3) the product of the esterification of glycerol at one, two, or three positions

neutralization (8.3) the reaction between an acid and a base

neutron (2.1) an uncharged particle, with the same mass as the proton, in the nucleus of an atom

nicotinamide adenine dinucleotide (NAD$^+$) (18.3) a molecule synthesized from the vitamin niacin and the nucleotide ATP and that serves as a carrier of hydride anions; a coenzyme that is an oxidizing agent used in a variety of metabolic processes

noble gas (2.3) elements in Group VIIIA (18) of the periodic table

nomenclature (3.2) a system for naming chemical compounds

nonelectrolyte (3.3, 7.1) a substance that, when dissolved in water, produces a solution that does not conduct an electrical current

nonessential amino acid (18.11) any amino acid that can be synthesized by the body

nonmetal (2.3) an element located on the right side of the periodic table (right of the "staircase" boundary)

nonreducing sugar (14.5) a sugar that cannot be oxidized by Benedict's or Tollens' reagent

normal boiling point (6.2) the temperature at which a substance will boil at 1 atm pressure

nuclear equation (9.2) a balanced equation accounting for the products and reactants in a nuclear reaction

nuclear imaging (9.5) the generation of images of components of the body (organs, tissues) using techniques based on the measurement of radiation

nuclear medicine (9.5) a field of medicine that uses radioisotopes for diagnostic and therapeutic purposes

nuclear reactor (9.5) a device for conversion of nuclear energy into electrical energy

nucleotide (17.1) a molecule composed of a nitrogenous base, a five-carbon sugar, and one, two, or three phosphoryl groups

nucleus (2.1) the small, dense center of positive charge in the atom

nuclide (9.1) any atom characterized by an atomic number and a mass number

nutritional Calorie (5.2) equivalent to one kilocalorie (1000 calories); also known as a large Calorie

O

octet rule (2.5) a rule predicting that atoms form the most stable molecules or ions when they are surrounded by eight electrons in their highest occupied energy level

oligosaccharide (14.1) an intermediate-sized carbohydrate composed of from three to ten monosaccharides

osmolarity (7.4) the number of moles of solute particles per liter of solution

osmosis (7.4) net flow of a solvent across a semipermeable membrane in response to a concentration gradient

osmotic pressure (7.4) the net force with which water enters a solution through a semipermeable membrane; alternatively, the pressure required to stop net transfer of solvent across a semipermeable membrane

outer mitochondrial membrane (18.5) the membrane that surrounds the mitochondrion and separates it from the contents of the cytoplasm; it is highly permeable to small "food" molecules

β-oxidation (19.2) the biochemical pathway that results in the oxidation of fatty acids and the production of acetyl CoA

oxidation (4.3, 12.1) a loss of electrons; in organic compounds it may be recognized as a loss of hydrogen atoms or the gain of oxygen atoms

oxidative deamination (19.5) an oxidation-reduction reaction in which NAD^+ is reduced and the amino acid is deaminated

oxidative phosphorylation (18.9) production of ATP using the energy of electrons harvested during biological oxidation-reduction reactions

oxidizing agent (4.3) a substance that oxidizes, or removes electrons from, another substance; the oxidizing agent is reduced in the process

oxidoreductase (16.10) an enzyme that catalyzes an oxidation-reduction reaction

P

parent compound or parent chain (10.2) in the I.U.P.A.C. Nomenclature System the parent compound is the longest carbon–carbon chain containing the principal functional group in the molecule that is being named

partial pressure (6.1) the pressure exerted by one component of a gas mixture

particle accelerator (9.5) a device for production of high-energy nuclear particles based on the interaction of charged particles with magnetic and electrical fields

passive transport (15.6) the net movement of a solute from an area of

high concentration to an area of low concentration

pentose (14.2) a five-carbon monosaccharide

peptide bond (16.3) the amide bond between two amino acids in a peptide chain

peptidyl tRNA binding site of ribosome (P-site) (17.6) a pocket on the surface of the ribosome that holds the tRNA bound to the growing peptide chain

percent yield (4.5) the ratio of the actual and theoretical yields of a chemical reaction multiplied by 100%

period (2.3) any one of seven horizontal rows of elements in the periodic table

periodic law (2.3) a law stating that properties of elements are periodic functions of their atomic numbers (Note that Mendeleev's original statement was based on atomic masses.)

peripheral membrane protein (15.6) a protein bound to either the inner or the outer surface of a membrane

phenol (12.2) an organic compound that contains a hydroxyl group (—OH) attached to a benzene ring

phenyl group (11.5) a benzene ring that has had a hydrogen atom removed, C_6H_5—

pH optimum (16.10) the pH at which an enzyme catalyzes the reaction at maximum efficiency

phosphatidate (15.3) a molecule of glycerol with fatty acids esterified to C-1 and C-2 of glycerol and a free phosphoryl group esterified at C-3

phosphodiester bond (17.2) the bond between two sugars (either ribose or deoxyribose) through a bridging phosphate group; these linkages form the sugar-phosphate backbone of DNA and RNA

phosphoglyceride (15.3) a molecule with fatty acids esterified at the C-1 and C-2 positions of glycerol and a phosphoryl group esterified at the C-3 position

phospholipid (15.3) a lipid containing a phosphoryl group

phosphopantetheine (19.4) the portion of coenzyme A and the acyl carrier protein that is derived from the vitamin pantothenic acid

pH scale (8.2) a numerical representation of acidity or basicity of a solution; $pH = -\log [H_3O^+]$

physical change (1.2)   a change in the form of a substance but not in its chemical composition; no chemical bonds are broken in a physical change

physical property (1.2)   a characteristic of a substance that can be observed without the substance undergoing change (examples include color, density, melting and boiling points)

plasma lipoprotein (15.5)   a complex composed of lipid and protein that is responsible for the transport of lipids throughout the body

β-pleated sheet (16.5)   a common secondary structure of a peptide chain that resembles the pleats of an Oriental fan

point mutation (17.7)   the substitution of one nucleotide pair for another within a gene

polar covalent bonding (3.4)   a covalent bond in which the electrons are not equally shared

polar covalent molecule (3.4)   a molecule that has a permanent electric dipole moment resulting from an unsymmetrical electron distribution; a dipolar molecule

poly(A) tail (17.4)   a tract of 100–200 adenosine monophosphate units covalently attached to the 3′ end of eukaryotic messenger RNA molecules

polyatomic ion (3.2)   an ion containing a number of atoms

polymer (11.4)   a very large molecule formed by the combination of many small molecules (called monomers) (e.g., polyamides, nylons)

polyprotic substance (8.3)   a substance that can accept or donate more than one proton per molecule

polysaccharide (14.1)   a large, complex carbohydrate composed of long chains of monosaccharides

polysome (17.6)   complexes of many ribosomes all simultaneously translating a single mRNA

positron (9.2)   particle that has the same mass as an electron but opposite (+) charge

post-transcriptional modification (17.4)   alterations of the primary transcripts produced in eukaryotic cells; these include addition of a poly(A) tail to the 3′ end of the mRNA, addition of the cap structure to the 5′ end of the mRNA, and RNA splicing

precipitate (7.1)   an insoluble substance formed and separated from a solution

precision (1.4)   the degree of agreement among replicate measurements of the same quantity

pressure (6.1)   force per unit area

primary (1°) alcohol (12.1)   an alcohol with the general formula RCH_2OH

primary (1°) amine (13.3)   an amine with the general formula RNH_2

primary (1°) carbon (10.2)   a carbon atom that is bonded to only one other carbon atom

primary structure (of a protein) (16.4)   the linear sequence of amino acids in a protein chain determined by the genetic information of the gene for each protein

primary transcript (17.4)   the RNA product of transcription in eukaryotic cells, before posttranscriptional modifications are carried out

product (4.3, 16.10)   the chemical species that results from a chemical reaction and that appears on the right side of a chemical equation

promoter (17.4)   the sequence of nucleotides immediately before a gene that is recognized by the RNA polymerase and signals the starting point and direction of transcription

properties (1.2)   characteristics of matter

prostaglandins (15.2)   a family of hormonelike substances derived from the twenty-carbon fatty acid, arachidonic acid; produced by many cells of the body, they regulate many body functions

prosthetic group (16.7)   the nonprotein portion of a protein that is essential to the biological activity of the protein; often a complex organic compound

protein (16.1)   a macromolecule whose primary structure is a linear sequence of α-amino acids and whose final structure results from folding of the chain into a specific three-dimensional structure; proteins serve as catalysts, structural components, and nutritional elements for the cell

proton (2.1)   a positively charged particle in the nucleus of an atom

pure substance (1.2)   a substance with constant composition

purine (17.1)   a family of nitrogenous bases (heterocyclic amines) that are components of DNA and RNA and consist of a six-sided ring fused to a five-sided ring; the common purines in nucleic acids are adenine and guanine

pyridoxal phosphate (19.5)   a coenzyme derived from vitamin B_6 that is required for all transamination reactions

pyrimidine (17.1)   a family of nitrogenous bases (heterocyclic amines) that are components of nucleic acids and consist of a single six-sided ring; the common pyrimidines of DNA are cytosine and thymine; the common pyrimidines of RNA are cytosine and uracil

pyrimidine dimer (17.7)   UV-light induced covalent bonding of two adjacent pyrimidine bases in a strand of DNA

pyruvate dehydrogenase complex (18.6)   a complex of all enzymes and coenzymes required for synthesis of CO_2 and acetyl CoA from pyruvate

Q

quantization (2.4)   a characteristic that energy can occur only in discrete units (quanta)

quaternary ammonium salt (13.3)   an amine salt with the general formula $R_4N^+ A^-$ (in which R— can be an alkyl or aryl group or a hydrogen atom and A^- can be any anion)

quaternary (4°) carbon (10.2)   a carbon atom that is bonded to four other carbon atoms

quaternary structure (of a protein) (16.7)   aggregation of more than one folded peptide chain to yield a functional protein

R

rad (9.7)   abbreviation for *radiation absorbed dose*, the absorption of 2.4×10^{-3} calories of energy per kilogram of absorbing tissue

radioactivity (9.1)   the process by which atoms emit high-energy particles or rays; the spontaneous decomposition of a nucleus to produce a different nucleus

Raoult's law (7.4)   a law stating that when a nonvolatile solute is added to a solvent, the vapor pressure of the solvent decreases in proportion to the concentration of the solute

rate of chemical reaction (5.3)   the change in concentration of a reactant or product per unit time

reactant (4.3) starting material for a chemical reaction, appearing on the left side of a chemical equation

reducing agent (4.3) a substance that reduces, or donates electrons to, another substance; the reducing agent is itself oxidized in the process

reducing sugar (14.4) a sugar that can be oxidized by Benedict's or Tollens' reagents; includes all monosaccharides and most disaccharides

reduction (4.3, 12.1) the gain of electrons; in organic compounds it may be recognized as a gain of hydrogen atoms or loss of oxygen atoms

regulatory proteins (16.1) proteins that control cell functions such as metabolism and reproduction

release factor (17.6) a protein that binds to the termination codon in the empty A-site of the ribosome and causes the peptidyl transferase to hydrolyze the bond between the peptide and the peptidyl tRNA

rem (9.7) abbreviation for *roentgen equivalent for man,* the product of rad and RBE

representative element (2.3) member of the groups of the periodic table designated as A

result (1.3) the outcome of a designed experiment, often determined from individual bits of data

reversible, competitive enzyme inhibitor (16.10) a chemical that resembles the structure and charge distribution of the natural substrate and competes with it for the active site of an enzyme

ribonucleic acid (RNA) (17.1) single-stranded nucleic acid molecules that are composed of phosphoryl groups, ribose, and the nitrogenous bases uracil, cytosine, adenine, and guanine

ribonucleotide (17.1) a ribonucleoside phosphate or nucleotide composed of a nitrogenous base in β-*N*-glycosidic linkage to the 1' carbon of the sugar ribose and with one, two, or three phosphoryl groups esterified at the hydroxyl of the 5' carbon of the ribose

ribose (14.4) a five-carbon monosaccharide that is a component of RNA and many coenzymes

ribosomal RNA (rRNA) (17.4) the RNA species that are structural and functional components of the small and large ribosomal subunits

ribosome (17.6) an organelle composed of a large and a small subunit, each of which is made up of ribosomal RNA and proteins; the platform on which translation occurs and that carries the enzymatic activity that forms peptide bonds

RNA polymerase (17.4) the enzyme that catalyzes the synthesis of RNA molecules using DNA as the template

RNA splicing (17.4) removal of portions of the primary transcript that do not encode protein sequences

röentgen (9.7) the dose of radiation producing 2.1×10^9 ions in 1 cm^3 of air at 0°C and 1 atm of pressure

S

saccharide (14.1) a sugar molecule

saponification (13.2, 15.2) a reaction in which a soap is produced; more generally, the hydrolysis of an ester by an aqueous base

saturated fatty acid (15.2) a long-chain monocarboxylic acid in which each carbon of the chain is bonded to the maximum number of hydrogen atoms

saturated hydrocarbon (10.1) an alkane; a hydrocarbon that contains only carbon and hydrogen bonded together through carbon–hydrogen and carbon–carbon single bonds

saturated solution (7.1) one in which undissolved solute is in equilibrium with the solution

scientific method (1.1) the process of studying our surroundings that is based on experimentation

scientific notation (1.4) a system used to represent numbers as powers of ten

secondary (2°) alcohol (12.1) an alcohol with the general formula R_2CHOH

secondary (2°) amine (13.3) an amine with the general formula R_2NH

secondary (2°) carbon (10.2) a carbon atom that is bonded to two other carbon atoms

secondary structure (of a protein) (16.5) folding of the primary structure of a protein into an α-helix or a β-pleated sheet; folding is maintained by hydrogen bonds between the amide hydrogen and the carbonyl oxygen of the peptide bond

Selectively permeable membrane (7.4) a membrane that restricts diffusion of some ions and molecules (based on size and charge) across the membrane

semiconservative DNA replication (17.3) DNA polymerase "reads" each parental strand of DNA and produces a complementary daughter strand; thus, all newly synthesized DNA molecules consist of one parental and one daughter strand

semipermeable membrane (7.4) a membrane permeable to the solvent but not the solute; a material that allows the transport of certain substances from one side of the membrane to the other

shielding (9.6) material used to provide protection from radiation

sickle cell anemia (16.8) a human genetic disease resulting from inheriting mutant hemoglobin genes from both parents

significant figures (1.4) all digits in a number known with certainty and the first uncertain digit

silent mutation (17.7) a mutation that changes the sequence of the DNA but does not alter the amino acid sequence of the protein encoded by the DNA

single bond (3.4) a bond in which one pair of electrons is shared by two atoms

single-replacement reaction (4.3) also called substitution reaction, one in which one atom in a molecule is displaced by another

soap (13.2) any of a variety of the alkali metal salts of fatty acids

solid state (1.2) a physical state of matter characterized by its rigidity and fixed volume and shape

solubility (3.5, 7.1) the amount of a substance that will dissolve in a given volume of solvent at a specified temperature

solute (7.1) a component of a solution that is present in lesser quantity than the solvent

solution (7.1) a homogeneous (uniform) mixture of two or more substances

solvent (7.1) the solution component that is present in the largest quantity

specific gravity (1.5) the ratio of the density of a substance to the density of water at 4°C or any specified temperature

specific heat (5.2) the quantity of heat (calories) required to raise the temperature of one g of a substance one degree Celsius

sphingolipid (15.4) a phospholipid that is derived from the amino alcohol sphingosine rather than from glycerol

sphingomyelin (15.4) a sphingolipid found in abundance in the myelin sheath that surrounds and insulates cells of the central nervous system

standard solution (8.3) a solution whose concentration is accurately known

standard temperature and pressure (STP) (6.1) defined as 273 K and 1 atm

stereochemistry (14.3) the study of the spatial arrangement of atoms in a molecule

stereoisomers (14.3) a pair of molecules having the same structural formula and bonding pattern but differing in the arrangement of the atoms in space

steroid (15.4) a lipid derived from cholesterol and composed of one five-sided ring and three six-sided rings; steroids include sex hormones and anti-inflammatory compounds

storage protein (16.1) a protein that serves as a source of amino acids for embryos or infants

structural analog (16.10) a chemical having a structure and charge distribution very similar to those of a natural enzyme substrate

structural formula (10.2) a formula showing all of the atoms in a molecule and exhibiting all bonds as lines

structural isomers (10.2) molecules having the same molecular formula but different chemical structures

structural protein (16.1) a protein that provides mechanical support for large plants and animals

sublevel (2.4) a set of equal-energy orbitals within a principal energy level

substituted hydrocarbon (10.1) a hydrocarbon in which one or more hydrogen atoms are replaced by another atom or group of atoms

substitution reaction (10.4, 11.5) a reaction that results in the replacement of one group for another

substrate (16.10) the reactant in a chemical reaction that binds to an enzyme active site and is converted to product

substrate-level phosphorylation (18.3) the production of ATP by the transfer of a phosphoryl group from the substrate of a reaction to ADP

sucrose (14.5) a disaccharide composed of α-D-glucose and β-D-fructose in (α1→β2) glycosidic linkage; table sugar

supersaturated solution (7.1) a solution that is more concentrated than a saturated solution (Note that such a solution is not at equilibrium.)

surface tension (6.2) a measure of the strength of the attractive forces at the surface of a liquid

surfactant (6.2) a substance that decreases the surface tension of a liquid

surroundings (5.1) the universe outside of the system

suspension (7.1) a heterogeneous mixture of particles; the suspended particles are larger than those found in a colloidal suspension

system (5.1) the process under study

T

temperature (1.5) a measure of the relative "hotness" or "coldness" of an object

temperature optimum (16.10) the temperature at which an enzyme functions optimally and the rate of reaction is maximal

terminal electron acceptor (18.9) the final electron acceptor in an electron transport system that removes the low-energy electrons from the system; in aerobic organisms the terminal electron acceptor is molecular oxygen

termination codon (17.6) a triplet of ribonucleotides with no corresponding anticodon on a tRNA; as a result, translation will end because there is no amino acid to transfer to the peptide chain

tertiary (3°) alcohol (12.1) an alcohol with the general formula R_3COH

tertiary (3°) amine (13.3) an amine with the general formula R_3N

tertiary (3°) carbon (10.2) a carbon atom that is bonded to three other carbon atoms

tertiary structure (of a protein) (16.6) the globular, three-dimensional structure of a protein that results from folding the regions of secondary structure; this folding occurs spontaneously as a result of interactions of the side chains or R groups of the amino acids

tetrahedral structure (3.4) a molecule consisting of four groups attached to a central atom that occupy the four corners of an imagined regular tetrahedron

tetrose (14.2) a four-carbon monosaccharide

theoretical yield (4.5) the maximum amount of product that can be produced from a given amount of reactant

theory (1.1) a hypothesis supported by extensive testing that explains and predicts facts

thermodynamics (5.1) the branch of science that deals with the relationship between energies of systems, work, and heat

thiol (12.4) an organic compound that contains a thiol group (—SH)

titration (8.3) the process of adding a solution from a buret to a sample until a reaction is complete, at which time the volume is accurately measured and the concentration of the sample is calculated

Tollens' test (12.5) a test reagent (silver nitrate in ammonium hydroxide) used to distinguish aldehydes and ketones; also called the Tollens' silver mirror test

tracer (9.5) a radioisotope that is rapidly and selectively transmitted to the part of the body for which diagnosis is desired

transaminase (19.5) an enzyme that catalyzes the transfer of an amino group from one molecule to another

transamination (19.5) a reaction in which an amino group is transferred from one molecule to another

transcription (17.4) the synthesis of RNA from a DNA template

transferase (16.10) an enzyme that catalyzes the transfer of a functional group from one molecule to another

transfer RNA (tRNA) (17.4) small RNAs that bind to a specific amino acid at the 3' end and mediate its addition at the appropriate site in a growing peptide chain; accomplished by recognition of the correct codon on the mRNA by the complementary anticodon on the tRNA

transition element (2.3) any element located between Groups IIA (2) and IIIA (13) in the long periods of the periodic table

transition state (16.10) the unstable intermediate in catalysis in which the enzyme has altered the form of the substrate so that it now shares properties of both the substrate and the product

translation (17.4) the synthesis of a protein from the genetic code carried on the mRNA

translocation (17.6) movement of the ribosome along the mRNA during translation

transmembrane protein (15.6) a protein that is embedded within a membrane and crosses the lipid bilayer, protruding from the membrane both inside and outside the cell

transport protein (16.1) a protein that transports materials across the cell membrane or throughout the body

triglyceride (15.3, 19.1) triacylglycerol; a molecule composed of glycerol esterified to three fatty acids

trigonal pyramidal molecule (3.4) a nonplanar structure involving three groups bonded to a central atom in which each group is equidistant from the central atom

triose (14.2) a three-carbon monosaccharide

triple bond (3.4) a bond in which three pairs of electrons are shared by two atoms

U

uncertainty (1.4) the degree of doubt in a single measurement

unit (1.3) a determinate quantity (of length, time, etc.) that has been adopted as a standard of measurement

unsaturated fatty acid (15.2) a long-chain monocarboxylic acid having at least one carbon-to-carbon double bond

unsaturated hydrocarbon (10.1, 11.1) a hydrocarbon containing at least one multiple (double or triple) bond

urea cycle (19.6) a cyclic series of reactions that detoxifies ammonium ions by incorporating them into urea, which is excreted from the body

V

valence electron (2.4) electron in the outermost shell (principal quantum level) of an atom

valence shell electron pair repulsion theory (VSEPR) (3.4) a model that predicts molecular geometry using the premise that electron pairs will arrange themselves as far apart as possible, to minimize electron repulsion

van der Waals forces (6.2) a general term for intermolecular forces that include dipole–dipole and London forces

vapor pressure of a liquid (6.2) the pressure exerted by the vapor at the surface of a liquid at equilibrium

very low density lipoprotein (VLDL) (15.5) a plasma lipoprotein that binds triglycerides synthesized by the liver and carries them to adipose tissue for storage

viscosity (6.2) a measure of the resistance to flow of a substance at constant temperature

vitamin (16.10) an organic substance that is required in the diet in small amounts; water-soluble vitamins are used in the synthesis of coenzymes required for the function of cellular enzymes; lipid-soluble vitamins are involved in calcium metabolism, vision, and blood clotting

W

wax (15.4) a collection of lipids that are generally considered to be esters of long-chain alcohols

weight (1.5) the force exerted on an object by gravity

weight/volume percent [% (W/V)] (7.2) the concentration of a solution expressed as a ratio of grams of solute to milliliters of solution multiplied by 100%

weight/weight percent [% (W/W)] (7.2) the concentration of a solution expressed as a ratio of grams of solute to grams of solution multiplied by 100%

Z

Zaitsev's rule (12.1) states that in an elimination reaction, the alkene with the greatest number of alkyl groups on the double-bonded carbon (the more highly substituted alkene) is the major product of the reaction

Answers to Odd-Numbered Problems

Chapter 1

1.1 Organized way of doing science; uses carefully planned experiments to study our environment.

1.3 **a.** Physical property
b. Chemical property
c. Physical property
d. Physical property
e. Physical property

1.5 Intensive property

1.7 **a.** Pure substance
b. Heterogeneous mixture
c. Homogeneous mixture
d. Pure substance

1.9 Data include 5.40 g and 2.00 cm^3; the result is the density, 2.70 g/cm^3.

1.11 0.0775 ton

1.13 **a.** 1.0×10^3 mL
b. 1.0×10^6 μL
c. 1.0×10^{-3} kL
d. 1.0×10^2 cL
e. 1.0×10^{-1} daL

1.15 **a.** 1.3×10^{-2} m
b. 0.71 L
c. 2.00 oz
d. 1.5×10^{-4} m^2

1.17 **a.** Three
b. Three
c. Four
d. Two
e. Three

1.19 **a.** 2.4×10^{-3}
b. 1.80×10^{-2}
c. 2.24×10^2

1.21 **a.** 8.09
b. 5.9
c. 20.19

1.23 **a.** 51
b. 8.0×10^1
c. 1.6×10^2

1.25 **a.** 61.4
b. 6.17
c. 6.65×10^{-2}

1.27 **a.** 0°C
b. 273 K

1.29 23.7 g

1.31 **a.** Chemistry is the study of matter and the changes that matter undergoes.
b. Matter is the material component of the universe.
c. Energy is the ability to do work.

1.33 **a.** Gram (or kilogram)
b. Liter
c. Meter

1.35 Mass is an independent quantity while weight is dependent on gravity.

1.37 Density is mass per volume. Specific gravity is the ratio of the density of a substance to the density of water at 4°C.

1.39 A theory.

1.41 A physical property is a characteristic of a substance that can be observed without the substance undergoing a change in chemical composition.

1.43 Mixtures are composed of two or more substances. A homogeneous mixture has uniform composition while a heterogeneous mixture has non-uniform composition.

1.45 **a.** Chemical reaction
b. Physical change
c. Physical change

1.47 **a.** Physical property
b. Chemical property

1.49 **a.** Pure substance
b. Pure substance
c. Mixture

1.51 **a.** Homogeneous
b. Homogeneous
c. Homogeneous

1.53 **a.** Extensive property
b. Extensive property
c. Intensive property

1.55 **a.** 32 oz
b. 1.0×10^{-3} t
c. 9.1×10^2 g
d. 9.1×10^5 mg
e. 9.1×10^1 da

1.57 **a.** 6.6×10^{-3} lb
b. 1.1×10^{-1} oz
c. 3.0×10^{-3} kg
d. 3.0×10^2 cg
e. 3.0×10^3 mg

1.59 **a.** 10.0°C
b. 283.2 K

1.61 **a.** 293.2 K
b. 68.0°F

1.63 4 L

1.65 101°F

1.67 5 cm is shorter than 5 in.

1.69 a. 3
 b. 3
 c. 3
 d. 4
 e. 4
 f. 3
1.71 a. 3.87×10^{-3}
 b. 5.20×10^{-2}
 c. 2.62×10^{-3}
 d. 2.43×10^{1}
 e. 2.40×10^{2}
 f. 2.41×10^{0}
1.73 a. 1.5×10^{4}
 b. 2.41×10^{-1}
 c. 5.99
 d. 1139.42
 e. 7.21×10^{3}
1.75 a. 1.23×10^{1}
 b. 5.69×10^{-2}
 c. -1.527×10^{3}
 d. 7.89×10^{-7}
 e. 9.2×10^{7}
 f. 5.280×10^{-3}
 g. 1.279×10^{0}
 h. -5.3177×10^{2}
1.77 6.00 g/mL
1.79 1.08×10^{3} g
1.81 1.008 g/mL
1.83 Lead has the lowest density and platinum has the greatest density.
1.85 12.6 mL

Chapter 2

2.1 a. 16 protons, 16 electrons, 16 neutrons
 b. 11 protons, 11 electrons, 12 neutrons
2.3 20.18 amu
2.5 Our understanding of the nucleus is based on the gold foil experiment performed by Geiger and interpreted by Rutherford. In this experiment, Geiger bombarded a piece of gold foil with alpha particles, and observed that some alpha particles passed straight through the foil, others were deflected and some simply bounced back. This led Rutherford to propose that the atom consisted of a small, dense nucleus (alpha particles bounced back), surrounded by a cloud of electrons (some alpha particles were deflected). The size of the nucleus is small when compared to the volume of the atom (most alpha particles were able to pass through the foil).
2.7 a. Zr (zirconium)
 b. 22.99
 c. Cr (chromium)
 d. At (astatine)
2.9 a. Helium, atomic number = 2, mass = 4.00 amu
 b. Fluorine, atomic number = 9, mass = 19.00 amu
 c. Manganese, atomic number = 25, mass = 54.94 amu
2.11 a. Total electrons = 11, valence electrons = 1
 b. Total electrons = 12, valence electrons = 2
 c. Total electrons = 16, valence electrons = 6
 d. Total electrons = 17, valence electrons = 7
 e. Total electrons = 18, valence electrons = 8
2.13 a. Sulfur: $1s^2, 2s^2, 2p^6, 3s^2, 3p^4$
 b. Calcium: $1s^2, 2s^2, 2p^6, 3s^2, 3p^6, 4s^2$

2.15 a. [Ne] $3s^2, 3p^4$
 b. [Ar] $4s^2$
2.17 a. Ca^{2+} and Ar are isoelectronic
 b. Sr^{2+} and Kr are isoelectronic
 c. S^{2-} and Ar are isoelectronic
 d. Mg^{2+} and Ne are isoelectronic
 e. P^{3-} and Ar are isoelectronic
2.19 a. (Smallest) F, N, Be (largest)
 b. (Lowest) Be, N, F (highest)
 c. (Lowest) Be, N, F, (highest)
2.21 a. 8 protons, 8 electrons, 8 neutrons.
 b. 16 neutrons.
2.23

Particle	Mass	Charge
a. electron	5.4×10^{-4} amu	−1
b. proton	1.00 amu	+1
c. neutron	1.00 amu	0

2.25 a. An ion is a charged atom or group of atoms formed by the loss or gain of electrons.
 b. A loss of electrons by a neutral species results in a cation.
 c. A gain of electrons by a neutral species results in an anion.
2.27 From the periodic table, all isotopes of Rn have 86 protons. Isotopes differ in the number of neutrons.
2.29 a. 34
 b. 46
2.31 a. $^{1}_{1}H$
 b. $^{14}_{6}C$
2.33

Atomic Symbol	# Protons	# Neutrons	# Electrons	Charge
a. $^{23}_{11}Na$	11	12	11	0
b. $^{32}_{16}S^{2-}$	16	16	18	2−
c. $^{16}_{8}O$	8	8	8	0
d. $^{24}_{12}Mg^{2+}$	12	12	10	2+
e. $^{39}_{19}K^{+}$	19	20	18	1+

2.35 a. Neutrons
 b. Protons
 c. Protons, neutrons
 d. Ion
 e. Nucleus, negative
2.37 • All matter consists of tiny particles called atoms.
 • Atoms cannot be created, divided, destroyed, or converted to any other type of atom.
 • All atoms of a particular element have identical properties.
 • Atoms of different elements have different properties.
 • Atoms combine in simple whole-number ratios.
 • Chemical change involves joining, separating, or rearranging atoms.
2.39 a. Chadwick—demonstrated the existence of the neutron in 1932.
 b. Goldstein—identified positive charge in the atom.
 c. Crookes—developed the cathode ray tube.
2.41 Negative charge, affected by electric field, affected by magnetic field
2.43 A cathode ray is the negatively charged particle formed in a cathode ray tube.
2.45 a. Sodium
 b. Potassium
 c. Magnesium
 d. Boron
2.47 Group IA (or 1) is known collectively as the alkali metals and consists of lithium, sodium, potassium, rubidium, cesium, and francium.

2.49 Group VIIA (or 17) is known collectively as the halogens and consists of fluorine, chlorine, bromine, iodine, and astatine.

2.51 **a.** True
　　b. True

2.53 **a.** Na, Ni, Al
　　b. Na, Al
　　c. Na, Ni, Al
　　d. Ar

2.55 **a.** One
　　b. One
　　c. Three
　　d. Seven
　　e. Zero (or eight)
　　f. Zero (or two)

2.57 A principal energy level is designated $n = 1, 2, 3$, and so forth. It is similar to Bohr's orbits in concept. A sublevel is a part of a principal energy level and is designated s, p, d, and f.

2.59 The s orbital represents the probability of finding an electron in a region of space surrounding the nucleus.

2.61 Three p orbitals (p_x, p_y, p_z) can exist in a given principal energy level.

2.63 A $3p$ orbital is a higher energy orbital than a $2p$ orbital because it is a part of a higher energy principal energy level.

2.65 $2\,e^-$ for $n = 1$
　　$8\,e^-$ for $n = 2$
　　$18\,e^-$ for $n = 3$

2.67 **a.** $3p$ orbital
　　b. $3s$ orbital
　　c. $3d$ orbital
　　d. $4s$ orbital
　　e. $3d$ orbital
　　f. $3p$ orbital

2.69 **a.** Not possible
　　b. Possible
　　c. Not possible
　　d. Not possible

2.71 **a.** Li^+
　　b. O^{2-}
　　c. Ca^{2+}
　　d. Br^-
　　e. S^{2-}
　　f. Al^{3+}

2.73 **a.** Isoelectronic
　　b. Isoelectronic

2.75 **a.** Na^+
　　b. S^{2-}
　　c. Cl^-

2.77 **a.** $1s^2, 2s^2, 2p^6, 3s^2, 3p^6$
　　b. $1s^2, 2s^2, 2p^6$

2.79 **a.** (Smallest) F, O, N (Largest)
　　b. (Smallest) Li, K, Cs (Largest)
　　c. (Smallest) Cl, Br, I (Largest)

2.81 **a.** (Smallest) O, N, F (Largest)
　　b. (Smallest) Cs, K, Li (Largest)
　　c. (Smallest) I, Br, Cl (Largest)

2.83 A positive ion is always smaller than its parent atom because the positive charge of the nucleus is shared among fewer electrons in the ion. As a result, each electron is pulled closer to the nucleus and the volume of the ion decreases.

2.85 The fluoride ion has a completed octet of electrons and an electron configuration resembling its nearest noble gas.

Chapter 3

3.1 **a.** $\cdot Li \,+\, :\!\overset{\cdot\cdot}{\underset{\cdot\cdot}{Br}}\!\cdot \longrightarrow Li^+ \,+\, :\!\overset{\cdot\cdot}{\underset{\cdot\cdot}{Br}}\!:^-$

　　b. $\cdot Mg \cdot \,+\, 2\,:\!\overset{\cdot\cdot}{\underset{\cdot\cdot}{Cl}}\!\cdot \longrightarrow Mg^{2+} \,+\, 2\,:\!\overset{\cdot\cdot}{\underset{\cdot\cdot}{Cl}}\!:^-$

3.3 Yes; it has a nonzero electronegativity difference.

3.5 **a.** LiBr
　　b. $CaBr_2$
　　c. Ca_3N_2

3.7 **a.** Potassium cyanide
　　b. Magnesium sulfide
　　c. Magnesium acetate

3.9 **a.** $CaCO_3$
　　b. $NaHCO_3$
　　c. Cu_2SO_4

3.11 **a.** Diboron trioxide
　　b. Nitrogen oxide
　　c. Iodine chloride
　　d. Phosphorus trichloride

3.13 **a.** P_2O_5
　　b. SiO_2

3.15 **a.** H:Ö:　　**b.** H:C:H with H below

3.17 **a.** The bonded nuclei are closer together when a double bond exists, in comparison to a single bond.
　　b. The bond strength increases as the bond order increases. Therefore, a double bond is stronger than a single bond.

3.19 **a.** H:P:H with H below (and P structure at right)
　　b. H:Si:H with H below (and Si structure at right)

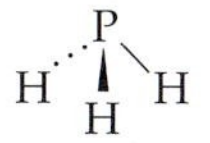

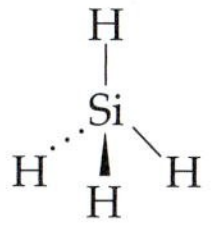

3.21 **a.** Oxygen is more electronegative than sulfur; the bond is polar. The electrons are pulled toward the oxygen atom.
　　b. Nitrogen is more electronegative than carbon; the bond is polar. The electrons are pulled toward the nitrogen atom.
　　c. There is no electronegativity difference between two identical atoms; the bond is nonpolar.
　　d. Chlorine is more electronegative than iodine; the bond is polar. The electrons are pulled toward the chlorine atom.

3.23 **a.** Nonpolar
　　b. Polar
　　c. Polar
　　d. Nonpolar

3.25 **a.** H_2O
　　b. CO
　　c. NH_3
　　d. ICl

3.27 **a.** Ionic
　　b. Covalent
　　c. Covalent
　　d. Covalent

3.29 **a.** Covalent
　　b. Covalent
　　c. Covalent
　　d. Ionic

3.31 **a.** $:\!\overset{\cdot\cdot}{S}\!\cdot \,+\, 2H\cdot \longrightarrow :\!\overset{\cdot\cdot}{S}\!:H$ with H below

　　b. $\cdot\overset{\cdot\cdot}{P}\!\cdot \,+\, 3H\cdot \longrightarrow H:\!\overset{\cdot\cdot}{P}\!:H$ with H below

3.33 He has two valence electrons (electron configuration $1s^2$) and a complete N=1 level. It has a stable electron configuration, with no tendency to gain or lose electrons, and satisfies the octet rule (2 e$^-$ for period 1). Hence, it is nonreactive.
He:

3.35 **a.** Copper (II) ion (cupric ion)
    b. Iron (II) ion (ferrous ion)
    c. Iron (III) ion (ferric ion)

3.37 **a.** K^+
    b. Br^-

3.39 **a.** SO_4^{2-}
    b. NO_3^-

3.41 **a.** NaCl
    b. $MgBr_2$

3.43 **a.** AgCN
    b. NH_4Cl

3.45 **a.** Magnesium chloride
    b. Aluminum chloride

3.47 **a.** Nitrogen dioxide
    b. Sulfur trioxide

3.49 **a.** Al_2O_3
    b. Li_2S

3.51 **a.** CO or CO_2
    b. H_2S

3.53 **a.** Sodium hypochlorite
    b. Sodium chlorite

3.55 Ionic solid state compounds exist in regular, repeating, three-dimensional structures; the crystal lattice. The crystal lattice is made up of positive and negative ions. Solid state covalent compounds are made up of molecules which may be arranged in a regular crystalline pattern or in an irregular (amorphous) structure.

3.57 The boiling points of ionic solids are generally much higher than those of covalent solids.

3.59 KCl would be expected to exist as a solid at room temperature; it is an ionic compound, and ionic compounds are characterized by high melting points.

3.61 Water will have a higher boiling point. Water is a polar molecule with strong intermolecular attractive forces, whereas carbon tetrachloride is a nonpolar molecule with weak intermolecular attractive forces. More energy, hence, a higher temperature is required to overcome the attractive forces among the water molecules.

3.63 **a.** H ·
    b. He :
    c. · C ·
    d. ·N·

3.65 **a.** Li^+
    b. Mg^{2+}
    c. :Cl:$^-$
    d. :P:$^{3-}$

3.67 **a.** :Cl : N : Cl:
          :Cl:

              H
    b. H : C : O : H
              H

    c. :S :: C :: S:

3.69 **a.** :Cl : N : Cl:
           :Cl:
    Pyramidal,
    polar
    water soluble

           H
    b. H : C : O : H
           H
    Tetrahedral around C,
    angular around O
    polar
    water soluble

    c. :S :: C :: S:
    Linear
    nonpolar
    not water soluble

          H H
3.71 H : C : C : O : H
          H H

           H O H
3.73 H : C : C : C : H
          H   H

3.75 **a.** Polar covalent
    b. Polar covalent
    c. Ionic
    d. Ionic
    e. Ionic

3.77 **a.** :O : S :: O
    b. [:Si≡P:]$^-$
    c), d), and e) are ionic compounds.

3.79 A molecule containing no polar bonds *must* be nonpolar. A molecule containing polar bonds may or may not itself be polar. It depends upon the number and arrangement of the bonds.

3.81 Polar compounds have strong intermolecular attractive forces. Higher temperatures are needed to overcome these forces and convert the solid to a liquid; hence, we predict higher melting points for polar compounds when compared to nonpolar compounds.

3.83 Yes

Chapter 4

4.1 26.98 g Al/mol Al

4.3 **a.** 1.51×10^{24} oxygen atoms
    b. 3.01×10^{24} oxygen atoms

4.5 14.0 g He

4.7 **a.** 17.04 g/mol.
    b. 180.18 g/mol.
    c. 237.95 g/mol.

4.9 **a.** DR
    b. SR
    c. DR
    d. D

4.11 **a.** $KCl(aq) + AgNO_3(aq) \rightarrow KNO_3(aq) + AgCl(s)$
    A precipitation reaction occurs.
    b. $CH_3COOK(aq) + AgNO_3(aq) \rightarrow$ no reaction
    No precipitation reaction occurs.

4.13 $Ca \rightarrow Ca^{2+} + 2\,\bar{e}$
    $\underline{S + 2\,\bar{e} \rightarrow S^{2-}}$
    $Ca + S \rightarrow CaS$

4.15 **a.** $4Fe(s) + 3O_2(g) \rightarrow 2Fe_2O_3(s)$
    b. $2C_6H_6(l) + 15O_2(g) \rightarrow 12CO_2(g) + 6H_2O(g)$

4.17 **a.** 90.1 g H_2O
    b. 0.590 mol LiCl

4.19 **a.** 3 mol O_2
    b. 96.00 g O_2

4.21 a. $4Fe(s) + 3O_2(g) \rightarrow 2Fe_2O_3(s)$
   b. 3.50 g Fe
4.23 a. 132.0 g SnF_2
   b. 3.79 % yield
4.25 Examples of other packaging units include a *ream* of paper (500 sheets of paper), a *six-pack* of soft drinks, a *case* of canned goods (24 cans), to name a few.
4.27 a. 28.09 g.
   b. 107.9 g.
4.29 39.95 g.
4.31 40.36 g Ne
4.33 4.00 g He/mol He
4.35 a. 5.00 mol He
   b. 1.7 mol Na
   c. 4.2×10^{-2} mol Cl_2
4.37 A molecule is a single unit comprised of atoms joined by covalent bonds. An ion-pair is composed of positive and negatively charged ions joined by electrostatic attraction, the ionic bond. The ion pairs, unlike the molecule, do not form single units; the electrostatic charge is directed to other ions in a crystal lattice, as well.
4.39 a. 58.44 g/mol
   b. 142.04 g/mol
   c. 357.49 g/mol
4.41 32.00 g/mol O_2
4.43 249.70 grams
4.45 a. 0.257 mol NaCl
   b. 0.106 mol Na_2SO_4
4.47 a. 18.02 g H_2O
   b. 116.9 g NaCl
4.49 a. 40.0 g He
   b. 2.02×10^2 g H
4.51 a. 0.420 mol KBr
   b. 0.415 mol $MgSO_4$
4.53 a. 6.57×10^{-1} mol CS_2
   b. 2.14×10^{-1} mol $Al_2(CO_3)_3$
4.55 The subscript tells us the number of atoms or ions contained in one unit of the compound.
4.57 a. $MgCO_3 (s) \xrightarrow{\Delta} MgO (s) + CO_2 (g)$
   b. $Zn(s) + CuSO_4(aq) \rightarrow ZnSO_4(aq) + Cu(s)$
4.59 $2NaOH(aq) + FeCl_2(aq) \rightarrow Fe(OH)_2(s) + 2NaCl(aq)$
4.61 Heat is necessary for the reaction to occur.
4.63 If we change the subscript we change the identity of the compound.
4.65 Reactants are found on the left side of the reaction arrow.
4.67 a. $2C_2H_6(g) + 7O_2(g) \rightarrow 4CO_2(g) + 6H_2O(g)$
   b. $6K_2O(s) + P_4O_{10}(s) \rightarrow 4K_3PO_4(s)$
   c. $MgBr_2(aq) + H_2SO_4(aq) \rightarrow 2HBr(g) + MgSO_4(aq)$
4.69 a. $Ca(s) + F_2(g) \rightarrow CaF_2(s)$
   b. $2Mg(s) + O_2(g) \rightarrow 2MgO(s)$
   c. $3H_2(g) + N_2(g) \rightarrow 2NH_3(g)$
4.71 a. $2C_4H_{10}(g) + 13O_2(g) \rightarrow 10H_2O(g) + 8CO_2(g)$
   b. $Au_2S_3(s) + 3H_2(g) \rightarrow 2Au(s) + 3H_2S(g)$
   c. $Al(OH)_3(s) + 3HCl(aq) \rightarrow AlCl_3(aq) + 3H_2O(l)$
   d. $(NH_4)_2Cr_2O_7(s) \rightarrow Cr_2O_3(s) + N_2(g) + 4H_2O(g)$
   e. $C_2H_5OH(l) + 3O_2(g) \rightarrow 2CO_2(g) + 3H_2O(g)$
4.73 a. $N_2(g) + 3H_2(g) \rightarrow 2NH_3(g)$
   b. $HCl(aq) + NaOH(aq) \rightarrow NaCl(aq) + H_2O(l)$
   c. $2HNO_3(aq) + Ca(OH)_2(aq) \rightarrow Ca(NO_3)_2(aq) + 2H_2O(l)$
   d. $2C_4H_{10}(g) + 13O_2(g) \rightarrow 10H_2O(g) + 8CO_2(g)$
4.75 50.3 g B_2O_3
4.77 a. $N_2(g) + 3H_2(g) \rightarrow 2NH_3(g)$
   b. Three moles of H_2 will react with one mole of N_2
   c. One mole of N_2 will produce two moles of the product NH_3

   d. 1.50 mol H_2
   e. 17.0 g NH_3
4.79 a. 149.21 g/mol.
   b. 1.20×10^{24} O atoms
   c. 32.00 g O
   d. 10.7 g O
4.81 6.14×10^4 g O_2
4.83 9.13×10^2 g N_2
4.85 6.85×10^2 g N_2

Chapter 5

5.1 Intensive
5.3 Electrical energy is converted to light energy.
5.5 a. Exothermic
   b. Exothermic
   c. Exothermic
5.7 He(g)
5.9 Nonspontaneous
5.11 13°C
5.13 2.7×10^3 J
5.15 $\dfrac{2.1 \times 10^2 \text{ nutritional Cal}}{\text{candy bar}}$
5.17 Heat energy produced by the friction of striking the match provides the activation energy necessary for this combustion process.
5.19 If the enzyme catalyzed a process needed to sustain life, the substance interfering with that enzyme would be classified as a poison.
5.21 At rush hour, approximately the same number of passengers enter and exit the train at any given stop. Throughout the trip, the number of passengers on the train may be essentially unchanged, but the identity of the individual passengers is continually changing.
5.23 Measure the concentrations of products and reactants at a series of times until no further concentration change is observed.
5.25 a. A would decrease
   b. A would increase
   c. A would decrease
   d. A would remain the same
5.27 joule
5.29 An exothermic reaction is one in which energy is released during chemical change.
5.31 The temperature of the water (or solution) is measured in a calorimeter. If the reaction being studied is exothermic, released energy heats the water and the temperature increases. In an endothermic reaction, heat flows from the water to the reaction and the water temperature decreases.
5.33 Double-walled containers, used in calorimeters, provide a small airspace between the part of the calorimeter (inside wall) containing the sample solution and the outside wall, contacting the surroundings. This makes heat transfer more difficult.
5.35 The first law of thermodynamics, the law of conservation of energy, states that the energy of the universe is constant.
5.37 Enthalpy is a measure of heat energy.
5.39 1.20×10^3 cal
5.41 5.02×10^3 J
5.43 a. Entropy increases.
   b. Entropy increases.
5.45 Isopropyl alcohol quickly evaporates (liquid → gas) after being applied to the skin. Conversion of a liquid to a gas requires heat energy. The heat energy is supplied by the skin. When this heat is lost, the skin temperature drops.

5.47 Decomposition of leaves and twigs to produce soil.

5.49 The activated complex is the arrangement of reactants in an unstable transition state as a chemical reaction proceeds. The activated complex must form in order to convert reactants to products.

5.51 A catalyst increases the rate of a reaction without itself undergoing change.

5.53

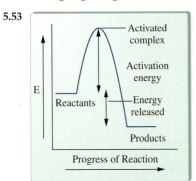

Non-catalyzed reaction
Higher activation energy

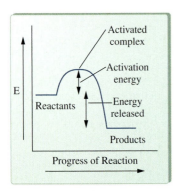

Catalyzed reaction
Lower activation energy

5.55 Enzymes are biological catalysts. The enzyme lysozyme catalyzes a process that results in the destruction of the cell walls of many harmful bacteria This helps to prevent disease in organisms. The breakdown of foods to produce material for construction and repair of body tissue, as well as energy, is catalyzed by a variety of enzymes. For example, amylase begins the hydrolysis of starch in the mouth.

5.57 An increase in concentration of reactants means that there are more molecules in a certain volume. The probability of collision is enhanced because each molecule travels a shorter distance before meeting another molecule. The rate is proportional to the number of collisions per unit time.

5.59 A catalyst speeds up a chemical reaction by facilitating the formation of the activated complex, thus lowering the activation energy, the energy barrier for the reaction.

5.61 Le Chaltelier's principle states that when a system at equilibrium is disturbed, the equilibrium shifts in the direction that minimizes the disturbance.

5.63 A dynamic equilibrium has fixed concentrations of all reactants and products—these concentrations do not change with time. However, the process is dynamic because products and reactants are continuously being formed and consumed. The concentrations do not change because the rates of production and consumption are equal.

5.65 **a.** Equilibrium shifts to the left.
  b. No change
  c. No change

5.67 **a.** False
  b. False

5.69 **a.** PCl_3 increases
  b. PCl_3 decreases
  c. PCl_3 decreases
  d. PCl_3 decreases
  e. PCl_3 remains the same

5.71 Decrease

5.73 False; a catalyst has no effect on the equilibrium position.

5.75 Removing the cap allows CO_2 to escape into the atmosphere. This corresponds to the removal of product (CO_2):
$$CO_2 \,(aq) \rightleftharpoons CO_2 \,(g)$$
The equilibrium shifts to the right, dissolved CO_2 is lost, and the beverage goes "flat."

Chapter 6

6.1 **a.** 0.954 atm
  b. 0.382 atm
  c. 0.730 atm

6.3 **a.** 38 atm
  b. 25 atm

6.5 **a.** 3.76 L
  b. 3.41 L
  c. 2.75 L

6.7 0.200 atm

6.9 4.46 mol H_2

6.11 9.00 L

6.13 0.223 mol N_2

6.15 Intermolecular forces in liquids are considerably stronger than intermolecular forces in gases. Particles are, on average, much closer together in liquids and the strength of attraction is inversely proportional to the distance of separation.

6.17 Evaporation is the conversion of a liquid to a gas at a temperature lower than the boiling point of the liquid. Condensation is the conversion of a gas to a liquid at a temperature lower than the boiling point of the liquid.

6.19 Solids are essentially incompressible because the average distance of separation among particles in the solid state is small. There is literally no space for the particles to crowd closer together.

6.21 In all cases, gas particles are much further apart than similar particles in the liquid or solid state. In most cases, particles in the liquid state are, on average, farther apart than those in the solid state. Water is the exception; liquid water's molecules are closer together than they are in the solid state.

6.23 Pressure is a force/unit area. Gas particles are in continuous, random motion. Collisions with the walls of the container results in a force (mass × acceleration) on the walls of the container. The sum of these collisional forces constitutes the pressure exerted by the gas.

6.25 Gases are easily compressed simply because there is a great deal of space between particles; they can be pushed closer together (compressed) because the space is available.

6.27 Gases exhibit more ideal behavior at low pressures. At low pressures, gas particles are more widely separated and therefore the attractive forces between particles are less. The ideal gas model assumes negligible attractive forces between gas particles.

6.29 The kinetic molecular theory states that the average kinetic energy of the gas particles increases as the temperature increases. Kinetic energy is proportional to (velocity)^2. Therefore, as the temperature increases the gas particle velocity increases and the rate of mixing increases as well.

6.31 The volume of the balloon is directly proportional to the pressure the gas exerts on the inside surface of the balloon.

As the balloon cools, its pressure drops, and the balloon contracts. (Recall that the speed, hence the force exerted by the molecules decreases as the temperature decreases.)

6.33 Volume will decrease according to Boyle's law. Volume is inversely proportional to the pressure exerted on the gas.

6.35 The Kelvin scale is the only scale that is directly proportional to molecular motion, and it is the motion that determines the physical properties of gases.

6.37 • Volume and temperature are *directly* proportional; increasing T *increases* V.
• Volume and pressure are *inversely* proportional; decreasing P *increases* V.
Therefore, both variables work together to *increase* the volume.

6.39 1 atm

6.41 5 L-atm

6.43 5.23 atm

6.45 No. The volume is proportional to the temperature in K, not Celsius.

6.47 0.96 L

6.49 $V_f = \dfrac{P_i V_i T_f}{P_f T_i}$

6.51 1.82×10^{-2} L

6.53 6.00 L

6.55 No. One mole of an ideal gas will occupy exactly 22.4 L; however, there is no completely ideal gas and careful measurement will show a different volume.

6.57 Standard temperature is 273K.

6.59 0.80 mol

6.61 22.4 L

6.63 0.276 mol

6.65 5.94×10^{-2} L

6.67 22.4 L

6.69 Dalton's law states that the total pressure of a mixture of gases is the sum of the partial pressures of the component gases.

6.71 0.74 atm

6.73 The vapor pressure of a liquid increases as the temperature of the liquid increases.

6.75 Viscosity is the resistance to flow caused by intermolecular attractive forces. Complex molecules may become entangled and not slide smoothly across one another.

6.77 All molecules exhibit London forces.

6.79 Only methanol exhibits hydrogen bonding. Methanol has an oxygen atom bonded to a hydrogen atom, a necessary condition for hydrogen bonding.

6.81 **a.** High melting temperature, brittle
b. High melting temperature, hard

6.83 Beryllium.

6.85 Mercury.

Chapter 7

7.1 A chemical analysis must be performed in order to determine the identity of all components, a qualitative analysis. If only one component is found, it is a pure substance; two or more components indicates a true solution.

7.3 After the container of soft drink is opened, CO_2 diffuses into the surrounding atmosphere; consequently the partial pressure of CO_2 over the soft drink decreases and the equilibrium

$$CO_2 (g) \rightleftharpoons CO_2 (aq)$$

shifts to the left, lowering the concentration of CO_2 in the soft drink.

7.5 16.7% NaCl

7.7 7.50% KCl

7.9 2.56×10^{-2} % oxygen

7.11 20.0 % oxygen

7.13 2.00×10^2 ppt and 2.00×10^5 ppm

7.15 0.125 mol HCl

7.17 To prepare the solution, dilute 1.7×10^{-2} L of 12 M HCl with sufficient water to produce 1.0×10^2 mL of total solution.

7.19 Pure water

7.21 Water flows from the cucumber to the surrounding solution.

7.23 Water flows from the cells of the sailor to the fluid surrounding the cells.

7.25 Polar carbon monoxide is more soluble in water.

7.27 1.54×10^{-2} mol Na^+/L

7.29 **a.** 2.00% NaCl
b. 6.60% $C_6H_{12}O_6$

7.31 **a.** 21.0% NaCl
b. 3.75% NaCl

7.33 **a.** 2.25 g NaCl
b. 3.13 g $NaC_2H_3O_2$

7.35 19.5% KNO_3

7.37 1.00 g sugar

7.39 2.0×10^{-3} ppt

7.41 **a.** 0.342 M NaCl
b. 0.367 M $C_6H_{12}O_6$

7.43 **a.** 1.46 g NaCl
b. 9.00 g $C_6H_{12}O_6$

7.45 0.146 M $C_{12}H_{22}O_{11}$

7.47 5.00×10^{-2} L

7.49 20.0 M

7.51 A colligative property is a solution property that depends on the concentration of solute particles rather than the identity of the particles.

7.53 Salt is an ionic substance that dissociates in water to produce positive and negative ions. These ions (or particles) lower the freezing point of water. If the concentration of salt particles is large, the freezing point may be depressed below the surrounding temperature, and the ice would melt.

7.55 Chemical properties depend on the identity of the substance, whereas colligative properties depend on concentration, not identity.

7.57 Raoult's law states that when a solute is added to a solvent, the vapor pressure of the solvent decreases in proportion to the concentration of the solute.

7.59 One mole of $CaCl_2$ produces three moles of particles in solution whereas one mole of NaCl produces two moles of particles in solution. Therefore, a one molar $CaCl_2$ solution contains a greater number of particles than a one molar NaCl solution and will produce a greater freezing-point depression.

7.61 Sucrose

7.63 Sucrose

7.65 **a.** A → B
b. A → B

7.67 **a.** Hypertonic
b. Hypotonic

7.69 Water is often termed the "universal solvent" because it is a polar molecule and will dissolve, at least to some extent, most ionic and polar covalent compounds. The majority of our body mass is water and this water is an important part of the nutrient transport system due to its solvent properties. This is true in other animals and plants as well. Because of its ability to hydrogen bond, water has a high boiling point and a low vapor pressure. Also, water is abundant and easily purified.

7.71 The shelf life is a function of the stability of the ammonia-water solution. The ammonia can react with the water to convert to the extremely soluble and stable ammonium ion. Also, ammonia and water are polar molecules. Polar interactions, particularly hydrogen bonding, are strong and contribute to the long-term solution stability.

7.73

7.75 Polar; like dissolves like (H_2O is polar)

7.77 In dialysis, sodium ions move from a region of high concentration to a region of low concentration. If we wish to remove (transport) sodium ions from the blood, they can move to a region of lower concentration, the dialysis solution.

7.79 Elevated concentrations of sodium ion in the blood may occur whenever large amounts of water are lost. Diarrhea, diabetes, and certain high-protein diets are particularly problematic.

Chapter 8

8.1 **a.** $HF(aq) + H_2O\ (l) \rightleftharpoons H_3O^+\ (aq) + F^-\ (aq)$
  b. $NH_3(aq) + H_2O\ (l) \rightleftharpoons NH_4^+\ (aq) + OH^-(aq)$

8.3 **a.** HF and F^-; H_2O and H_3O^+
  b. NH_3 and NH_4^+; H_2O and OH^-

8.5 **a.** NH_4^+
  b. H_2SO_4

8.7 1.0×10^{-11} M

8.9 12.00

8.11 3.2×10^{-9} M

8.13 0.1000 M NaOH

8.15 $CO_2 + H_2O \rightleftharpoons H_2CO_3 \rightleftharpoons H_3O^+ + HCO_3^-$
An increase in the partial pressure of CO_2 is a stress on the left side of the equilibrium. The equilibrium will shift to the right in an effort to decrease the concentration of CO_2. This will cause the molar concentration of H_2CO_3 to increase.

8.17 In Question 8.15, the equilibrium shifts to the right. Therefore the molar concentration of H_3O^+ should increase.
  In Question 8.16, the equilibrium shifts to the left. Therefore the molar concentration of H_3O^+ should decrease.

8.19 **a.** An Arrhenius acid is a substance that dissociates, producing hydrogen ions.
  b. A Brønsted-Lowry acid is a substance that behaves as a proton donor.

8.21 The Brønsted-Lowry theory provides a broader view of acid-base theory than does the Arrhenius theory. Brønsted-Lowry emphasizes the role of the solvent in the dissociation process.

8.23 **a.** $HNO_2\ (aq) + H_2O\ (l) \rightleftharpoons H_3O^+\ (aq) + NO_2^-\ (aq)$
  b. $HCN\ (aq) + H_2O\ (l) \rightleftharpoons H_3O^+\ (aq) + CN^-\ (aq)$

8.25 **a.** HNO_2 and NO_2^-; H_2O and H_3O^+
  b. HCN and CN^-; H_2O and H_3O^+

8.27 **a.** Weak
  b. Weak
  c. Weak

8.29 **a.** CN^- and HCN; NH_3 and NH_4^+
  b. CO_3^{2-} and HCO_3^-; Cl^- and HCl

8.31 Concentration refers to the quantity of acid or base contained in a specified volume of solvent. Strength refers to the degree of dissociation of the acid or base.

8.33 **a.** Brønsted acid
  b. Brønsted base
  c. Both

8.35 **a.** Brønsted acid
  b. Both
  c. Brønsted base

8.37 HCN

8.39 I^-

8.41 **a.** 1.0×10^{-7} M
  b. 1.0×10^{-11} M

8.43 **a.** Neutral
  b. Basic

8.45 **a.** 7.00
  b. 5.00

8.47 **a.** $[H_3O^+] = 1.0 \times 10^{-1}$ M
    $[OH^-] = 1.0 \times 10^{-13}$ M
  b. $[H_3O^+] = 1.0 \times 10^{-9}$ M
    $[OH^-] = 1.0 \times 10^{-5}$ M

8.49 **a.** $[H_3O^+] = 5.0 \times 10^{-2}$ M
    $[OH^-] = 2.0 \times 10^{-13}$ M
  b. $[H_3O^+] = 2.0 \times 10^{-10}$ M
    $[OH^-] = 5.0 \times 10^{-5}$ M

8.51 A neutralization reaction is one in which an acid and a base react to produce water and a salt (a "neutral" solution).

8.53 **a.** $[H_3O^+] = 1.0 \times 10^{-6}$ M
    $[OH^-] = 1.0 \times 10^{-8}$ M
  b. $[H_3O^+] = 6.3 \times 10^{-6}$ M
    $[OH^-] = 1.6 \times 10^{-9}$ M
  c. $[H_3O^+] = 1.6 \times 10^{-8}$ M
    $[OH^-] = 6.3 \times 10^{-7}$ M

8.55 **a.** 1×10^2
  b. 1×10^4
  c. 1×10^{10}

8.57 **a.** $[H_3O^+] = 1 \times 10^{-5}$
  b. $[H_3O^+] = 1 \times 10^{-12}$
  c. $[H_3O^+] = 3.2 \times 10^{-6}$

8.59 **a.** pH = 6.00
  b. pH = 8.00
  c. pH = 3.25

8.61 pH = 3.12

8.63 pH = 10.74

8.65 **a.** NH_3 and NH_4Cl can form a buffer solution.
  b. HNO_3 and KNO_3 cannot form a buffer solution.

8.67 **a.** A buffer solution contains components (a weak acid and its salt or a weak base and its salt) that enable the solution to resist large changes in pH when acids or bases are added.
  b. Acidosis is a medical condition characterized by higher-than-normal levels of CO_2 in the blood and lower-than-normal blood pH.

8.69 **a.** Addition of strong acid is equivalent to adding H_3O^+. This is a stress on the right side of the equilibrium and the equilibrium will shift to the left. Consequently the $[CH_3COOH]$ increases.
  b. Water, in this case, is a solvent and does not appear in the equilibrium expression. Hence, it does not alter the position of the equilibrium.

Chapter 9

9.1 The nucleus

9.3 **a.** $^{85}_{36}Kr \rightarrow ^{85}_{37}Rb + ^{0}_{-1}e$
  b. $^{226}_{88}Ra \rightarrow ^{4}_{2}He + ^{222}_{86}Rn$

9.5 6.3 ng of sodium-24 remain after 2.5 days

9.7 1/4 of the radioisotope remains after 2 half-lives

9.9 Isotopes with short half-lives release their radiation rapidly. There is much more radiation per unit time observed with short half-life substances; hence, the signal is stronger and the sensitivity of the procedure is enhanced.

9.11 The rem takes into account the relative biological effect of the radiation in addition to the quantity of radiation. This provides a more meaningful estimate of potential radiation damage to human tissue.

9.13 Natural radioactivity is the spontaneous decay of a nucleus to produce high-energy particles or rays.

9.15 Two protons and two neutrons

9.17 An electron with a -1 charge.

9.19 A positron has a positive charge and a beta particle has a negative charge.

9.21 • charge, $\alpha = +2$, $\beta = -1$
• mass, $\alpha = 4$ amu, $\beta = 0.000549$ amu
• velocity, $\alpha = 10\%$ of the speed of light, $\beta = 90\%$ of the speed of light

9.23 Chemical reactions involve joining, separating and rearranging atoms; valence electrons are critically involved. Nuclear reactions involve only changes in nuclear composition.

9.25 ^4_2He

9.27 $^{235}_{92}\text{U}$

9.29 ^1_1H $1 - 1 = 0$ neutrons
^2_1H $2 - 1 = 1$ neutron
^3_1H $3 - 1 = 2$ neutrons

9.31 $^{15}_7\text{N}$

9.33 $^{60}_{27}\text{Co} \rightarrow ^{60}_{28}\text{Ni} + ^{\ 0}_{-1}\beta + \gamma$

9.35 a. $^{23}_{11}\text{Na} + ^2_1\text{H} \rightarrow ^{24}_{11}\text{Na} + ^1_1\text{H}$
b. $^{238}_{92}\text{U} + ^{14}_7\text{N} \rightarrow ^{246}_{99}\text{Np} + 6^1_0\text{n}$
c. $^{24}_{10}\text{Ne} \rightarrow ^{\ 0}_{-1}\beta + ^{24}_{11}\text{Na}$

9.37 $^{209}_{83}\text{Bi} + ^{54}_{24}\text{Cr} \rightarrow ^{262}_{107}\text{Bh} + ^1_0\text{n}$

9.39 $^{27}_{12}\text{Mg} \rightarrow ^{27}_{13}\text{Al} + ^{\ 0}_{-1}\text{e}$

9.41 $^{12}_7\text{N} \rightarrow ^{12}_6\text{C} + ^0_1\text{e}$

9.43 Natural radioactivity is a spontaneous process; artificial radioactivity is nonspontaneous and results from a nuclear reaction that produces an unstable nucleus.

9.45 • Nuclei for light atoms tend to be most stable if their neutron/proton ratio is close to 1.
• Nuclei with more than 84 protons tend to be unstable.
• Isotopes with a "magic number" of protons or neutrons (2, 8, 20, 50, 82, or 126 protons or neutrons) tend to be stable.
• Isotopes with even numbers of protons or neutrons tend to be more stable.

9.47 0.40 mg of iodine-131 remains

9.49 13 mg of iron-59 remains

9.51 201 hours

9.53 Fission

9.55 a. The fission process involves the breaking down of large, unstable nuclei into smaller, more stable nuclei. This process releases energy in the form of heat and/or light.
b. The heat generated during the fission process could be used to generate steam, which is then used to drive a turbine to create electricity.

9.57 $^3_1\text{H} + ^1_1\text{H} \rightarrow ^4_2\text{He} + \text{energy}$

9.59 A "breeder" reactor creates the fuel which can be used by a conventional fission reactor during its fission process.

9.61 The reaction in a fission reactor that involves neutron production and causes subsequent reactions accompanied by the production of more neutrons in a continuing process.

9.63 High operating temperatures

9.65 $^{108}_{47}\text{Ag} + ^4_2\text{He} \rightarrow ^{112}_{49}\text{In}$

9.67 a. Technetium-99m is used to study the heart (cardiac output, size, and shape), kidney (follow-up procedure for kidney transplant), and liver and spleen (size, shape, presence of tumors).
b. Xenon-133 is used to locate regions of reduced ventilation and presence of tumors in the lung.

9.69 Radiation therapy provides sufficient energy to destroy molecules critical to the reproduction of cancer cells.

9.71 Background radiation, radiation from natural sources, is emitted by the sun as cosmic radiation, and from naturally radioactive isotopes found throughout our environment.

9.73 Level decreases

9.75 Positive effect

9.77 Positive effect

9.79 Yes

9.81 A film badge detects gamma radiation by darkening photographic film in proportion to the amount of radiation exposure over time. Badges are periodically collected and evaluated for their level of exposure. This mirrors the level of exposure of the personnel wearing the badges.

9.83 Relative biological effect is a measure of the damage to biological tissue caused by different forms of radiation.

Chapter 10

10.1 The student could test the solubility of the substance in water and in an organic solvent, such as hexane. Solubility in hexane would suggest an organic substance; whereas solubility in water would indicate an inorganic compound. The student could also determine the melting and boiling points of the substance. If the melting and boiling points are very high, an inorganic substance would be suspected.

10.3

Hexane

3-Methylpentane

2-Methylpentane

2,2-Dimethylbutane

2,3-Dimethylbutane

10.5 a.

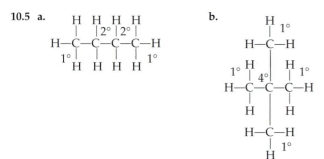

b.

```
        H  1°
        |
   H—C—H
        H |   H  1°
   1° | 4° | 1°
   H—C—C—C—H
        |   |
        H   H
        |
   H—C—H
        H  1°
```

c.

```
       1°  H
           |
      H—C—H
   H  H      H  H
   | 3°| 3°   | 2°|
H—C—C—C—C—C—H
  1° |   |   |  1°
   H   |  H  H  H
      H—C—H
       1° |
          H
```

10.7 a. 2,3-Dimethylbutane
 b. 2,2-Dimethylpentane
 c. Dimethylpropane
 d. 1,2,3-Tribromopropane

10.9 a. The straight chain isomers of molecular formula C_4H_9Br:

```
  H  H  H  H              H  H  Br H
  |  |  |  |              |  |  |  |
H—C—C—C—C—Br         H—C—C—C—C—H
  |  |  |  |              |  |  |  |
  H  H  H  H              H  H  H  H
```

b. The straight chain isomers of molecular formula $C_4H_8Br_2$:

```
  H  H  H  Br             H  H  Br H
  |  |  |  |              |  |  |  |
H—C—C—C—C—Br         H—C—C—C—C—Br
  |  |  |  |              |  |  |  |
  H  H  H  H              H  H  H  H

  H  Br H  H             H  H  H  H
  |  |  |  |             |  |  |  |
H—C—C—C—C—Br        Br—C—C—C—C—Br
  |  |  |  |             |  |  |  |
  H  H  H  H             H  H  H  H

  H  H  Br H             H  Br Br H
  |  |  |  |             |  |  |  |
H—C—C—C—C—H         H—C—C—C—C—H
  |  |  |  |             |  |  |  |
  H  H  Br H             H  H  H  H
```

10.11 a. 1-Bromo-2-ethylcyclobutane
 b. 1,2-Dimethylcyclopropane
 c. Propylcyclohexane

10.13 a.

◇ + 6O_2 ⟶ 4CO_2 + 4H_2O + heat energy

b.

```
  H  H  H                       H  Br H
  |  |  |      Light or heat     |  |  |
H—C—C—C—H + Br₂  ⟶         H—C—C—C—H + HBr
  |  |  |                        |  |  |
  H  H  H                        H  H  H

  H  H  H                       H  H  Br
  |  |  |      Light or heat     |  |  |
H—C—C—C—H + Br₂  ⟶         H—C—C—C—H + HBr
  |  |  |                        |  |  |
  H  H  H                        H  H  H
```

c. $2CH_3CH_3 + 7O_2 \rightarrow 4CO_2 + 6H_2O + energy$

d.

```
  H  H  H  H                          H  H  H  H
  |  |  |  |     Heat or light         |  |  |  |
H—C—C—C—C—H + Cl₂  ⟶            H—C—C—C—C—H + HCl
  |  |  |  |                          |  |  |  |
  H  H  H  H                          H  Cl H  H

  H  H  H  H                          H  H  H  H
  |  |  |  |     Heat or light         |  |  |  |
H—C—C—C—C—H + Cl₂  ⟶            H—C—C—C—C—H + HCl
  |  |  |  |                          |  |  |  |
  H  H  H  H                          Cl H  H  H
```

10.15 The products in the reactions in Problem 10.13b are 1-bromopropane and 2-bromopropane. The products of the reactions in Problem 10.13d are 1-chlorobutane and 2-chlorobutane.

10.17 The number of organic compounds is nearly limitless because carbon forms stable covalent bonds with other carbon atoms in a variety of different patterns. In addition, carbon can form stable bonds with other elements and functional groups, producing many families of organic compounds, including alcohols, aldehydes, ketones, esters, ethers, amines, and amides. Finally, carbon can form double or triple bonds with other carbon atoms to produce organic molecules with different properties.

10.19 The allotropes of carbon include graphite, diamond, and buckminsterfullerene.

10.21 Because ionic substances often form three-dimensional crystals made up of many positive and negative ions, they generally have much higher melting and boiling points than covalent compounds.

10.23 a. $LiCl > H_2O > CH_4$
 b. $NaCl > C_3H_8 > C_2H_6$

10.25 a. LiCl would be a solid; H_2O would be a liquid; and CH_4 would be a gas.
 b. NaCl would be a solid; both C_3H_8 and C_2H_6 would be gases.

10.27 a. Water-soluble inorganic compounds
 b. Inorganic compounds
 c. Organic compounds
 d. Inorganic compounds
 e. Organic compounds

10.29 a.

```
       H           H
       |           |
  H—C—H  H—C—H
       H |  H   |  H
       | |  |   |
  H—C—C——C——C—C—H
       |  |  |   |  |
       H  H  H   H  H
```

b.

```
  H  Br H  H
  |  |  |  |
H—C—C—C—C—H
  |  |  |  |
  H  H  Br H
```

10.31 a. $CH_3CH_2CHCH_3$
          |
          CH_3

b.

CH_3 CH_3
 | |
$CH_3CH_2CCH_2CH_2CHCH_3$
 |
CH_3

10.33 a.   **b.**   **c.**

10.35 a.   **b.**   **c.**
Br

10.37 Structure b is not possible because there are five bonds to carbon-2. Structure d is not possible because there are five bonds to carbon-3. Structure e is not possible because there are five bonds to carbon-2 and carbon-3. Structure f is not possible because there are five bonds to carbon-3.

10.39 a.

```
            H
            |
        H—C—H
            |
    H   H   |   H   H
    |   |   |   |   |
H—C—C—C—C—C—H
    |   |   |   |   |
    H   H   |   H   H
        H—C—H
            |
            H
```

b.

```
    H   H   H   H   H
    |   |   |   |   |
H—C—C—C—C—C—H
    |   |   |   |   |
    H   H   H   H   H
```

c.

```
            H
            |
        H—C—H
            |
        H—C—H
            |
    H   H   |   H   H   H
    |   |   |   |   |   |
H—C—C—C—C—C—C—H
    |   |   |   |   |   |
    H   H   H   H   H   H
```

d.

```
            H
            |
        H—C—H
            |
    H   |   H   H   H   H
    |   |   |   |   |   |
H—C—C—C—C—C—C—H
    |   |   |   |   |   |
    H   H   H   H   H   H
            |
        H—C—H
            |
            H
```

10.41 An alcohol

```
    H   H
    |   |
H—C—C—OH
    |   |
    H   H
```

An aldehyde

```
    H   O
    |   ‖
H—C—C—H
    |
    H
```

A ketone

```
    H   O   H
    |   ‖   |
H—C—C—C—H
    |       |
    H       H
```

A carboxylic acid

```
    H   O
    |   ‖
H—C—C—OH
    |
    H
```

An amine

```
    H   H       H
    |   |      /
H—C—C—N
    |   |      \
    H   H       H
```

10.43 a. C_nH_{2n+2}
b. C_nH_{2n-2}
c. C_nH_{2n}
d. C_nH_{2n}
e. C_nH_{2n-2}

10.45 a. Carboxyl group: —COOH
b. Amino group: —NH$_2$
c. Hydroxyl group: —OH

10.47

Aspirin
Acetylsalicylic acid

10.49 Hydrocarbons are nonpolar molecules, and hence are not soluble in water.

10.51 a. heptane > hexane > butane > ethane
b. $CH_3CH_2CH_2CH_2CH_2CH_2CH_2CH_2CH_3$ > $CH_3CH_2CH_2CH_2CH_3$ > $CH_3CH_2CH_3$

10.53 a. Heptane and hexane would be liquid at room temperature; butane and ethane would be gases.
b. $CH_3CH_2CH_2CH_2CH_2CH_2CH_2CH_2CH_3$ and $CH_3CH_2CH_2CH_2CH_3$ would be liquids at room temperature; $CH_3CH_2CH_3$ would be a gas.

10.55 a.

```
    Br
    |
CH_3CHCH_2CH_3
```

b.

```
        Cl
        |
CH_3—C—CH_3
        |
        CH_3
```

c.

```
        CH_3
        |
CH_3—C—CH_2CH_2CH_2CH_3
        |
        CH_3
```

10.57 a. 2,2-Dibromobutane:

```
    H   Br  H   H
    |   |   |   |
H—C—C—C—C—H
    |   |   |   |
    H   Br  H   H
```

b. 2-Iododecane:

```
    H   I   H   H   H   H   H   H   H   H
    |   |   |   |   |   |   |   |   |   |
H—C—C—C—C—C—C—C—C—C—C—H
    |   |   |   |   |   |   |   |   |   |
    H   H   H   H   H   H   H   H   H   H
```

c. 1,2-Dichloropentane:

```
     H   Cl  H   H   H
     |   |   |   |   |
Cl—C—C—C—C—C—H
     |   |   |   |   |
     H   H   H   H   H
```

d. 1-Bromo-2-methylpentane:

```
            H
            |
        H—C—H
            |
    H   |   H   H   H
    |   |   |   |   |
H—C—C—C—C—C—H
    |   |   |   |   |
    Br  H   H   H   H
```

10.59 a. 3-Methylpentane
b. 2,5-Dimethylhexane
c. 1-Bromoheptane
d. 1-Chloro-3-methylbutane

10.61 **a.** 2-Chloropropane
b. 2-Iodobutane
c. 2,2-Dibromopropane
d. 1-Chloro-2-methylpropane
e. 2-Iodo-2-methylpropane

10.63 **a.** Identical—both are 2-Bromobutane
b. Identical—both are 3-Bromo-5-methylhexane
c. Identical—both are 2,2-Dibromobutane
d. Isomers of molecular formula $C_6H_{12}Br_2$: 1,3-Dibromo-3-methylpentane and 1,4-Dibromo-2-ethylbutane

10.65 Cycloalkanes are a family of molecules having carbon-to-carbon bonds in a ring structure.

10.67 The general formula for a cycloalkane is C_nH_{2n}.

10.69 **a.** Chlorocyclopropane
b. 1,2-Dichlorocyclopropane
c. Bromocyclobutane

10.71 **a.** 1-Bromo-2-methylcyclobutane: **b.** Iodocyclopropane:

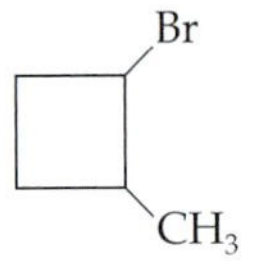

c. 1-Bromo-3-chlorocyclopentane: 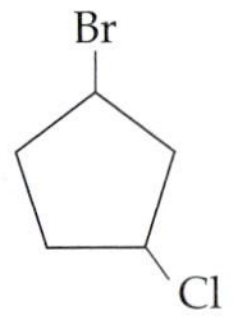 **d.** 1,2-Dibromo-3-methylcyclohexane:

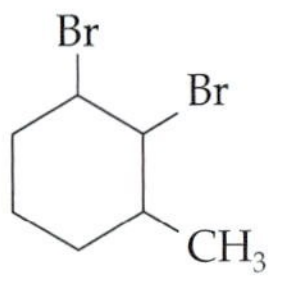

10.73 **a.** $C_3H_8 + 5O_2 \rightarrow 4H_2O + 3CO_2$
b. $C_7H_{16} + 11O_2 \rightarrow 8H_2O + 7CO_2$
c. $C_9H_{20} + 14O_2 \rightarrow 10H_2O + 9CO_2$
d. $2C_{10}H_{22} + 31O_2 \rightarrow 22H_2O + 20CO_2$

10.75 **a.** $8CO_2 + 10H_2O$
b.

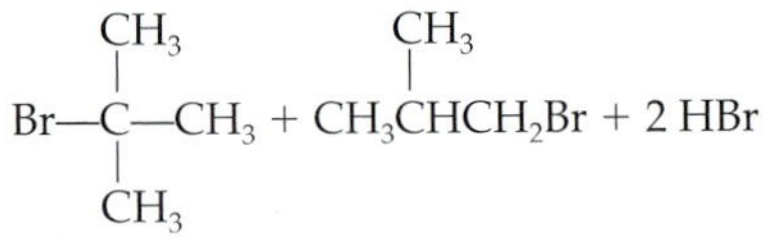

c. Cl_2 + light

10.77 The following molecules are all isomers of C_6H_{14}.

$CH_3CH_2CH_2CH_2CH_2CH_3$ $CH_3CHCH_2CH_2CH_3$

Hexane 2-Methylpentane

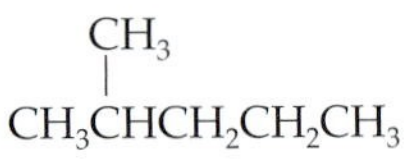

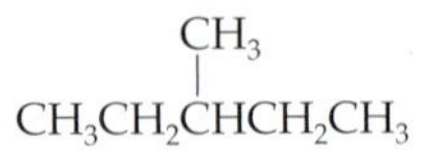

3-Methylpentane 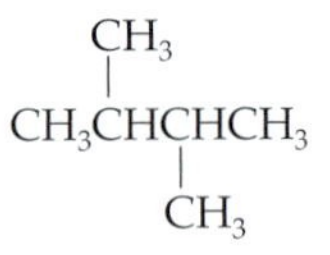 2,3-Dimethylbutane

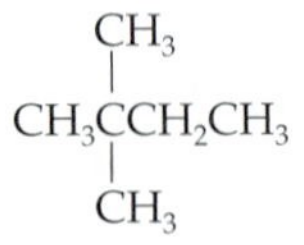

2,2-Dimethylbutane

a. 2,3-Dimethylbutane produces only two monobrominated derivatives: 1-bromo-2,3-dimethylbutane and 2-bromo-2,3-dimethylbutane.
b. Hexane produces three monobrominated products: 1-bromohexane, 2-bromohexane, and 3-bromohexane. 2,2-Dimethylbutane also produces three monobrominated products: 1-bromo-2,2-dimethylbutane, 2-bromo-3,3-dimethylbutane, and 1-bromo-3,3-dimethylbutane.
c. 3-Methylpentane produces four monobrominated products: 1-bromo-3-methylpentane, 2-bromo-3-methylpentane, 3-bromo-3-methylpentane, and 1-bromo-2-ethylbutane.

10.79 The hydrocarbon is cyclooctane, having a molecular formula of C_8H_{16}.

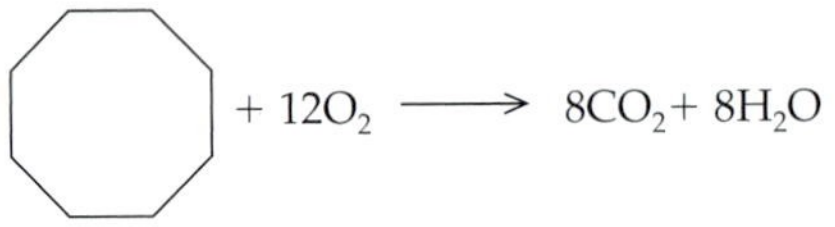

Chapter 11

11.1 **a.**

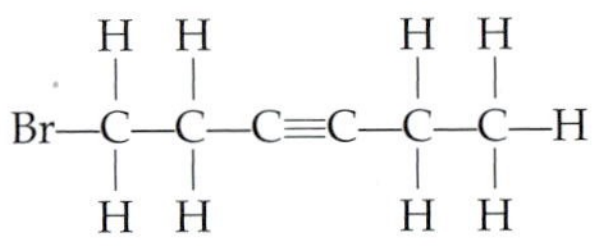

b.

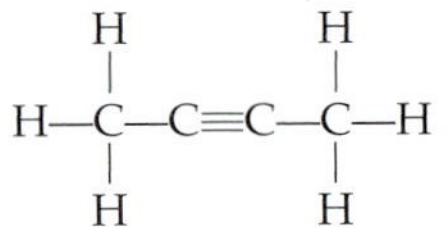

c. $Cl—C\equiv C—Cl$

d.

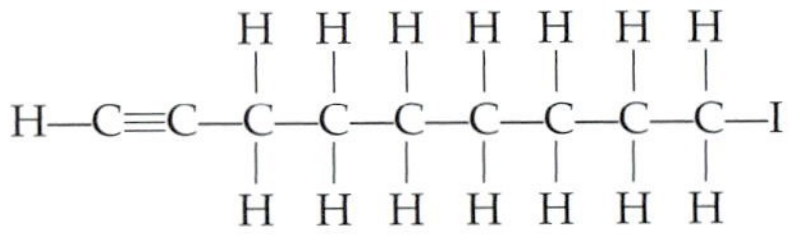

11.3 **a.**

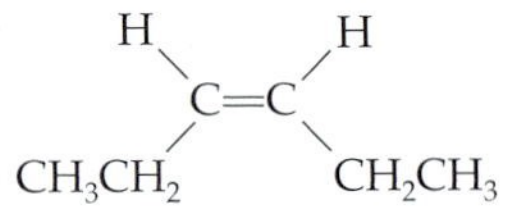

cis-isomer 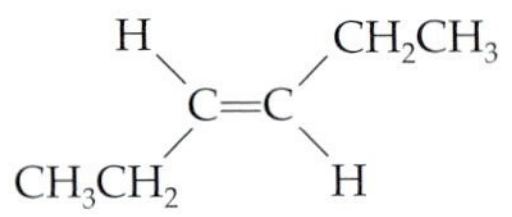 trans-isomer

b.

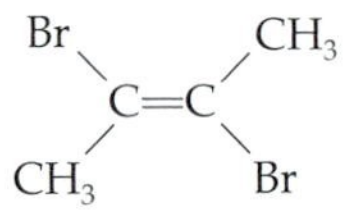

trans-isomer 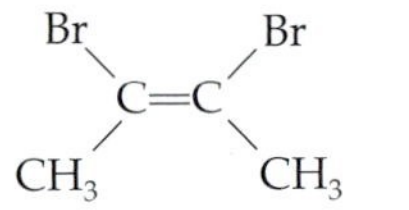 cis-isomer

11.5 Molecule c can exist as cis- and trans-isomers because there are two different groups on each of the carbon atoms attached by the double bond.

11.7 **a.** **b.**

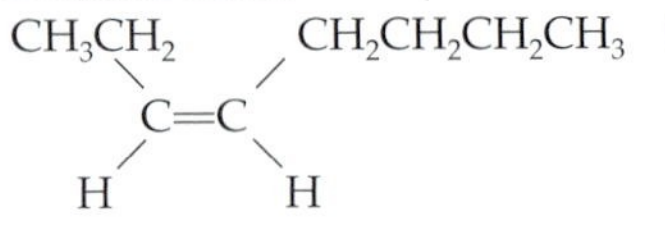

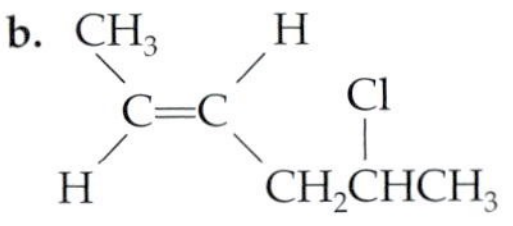

c.

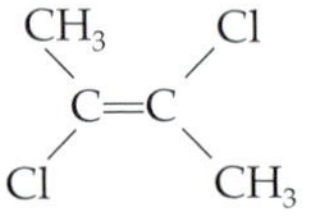

11.9 The hydrogenation of the cis and trans isomers of 2-pentene would produce the same product, pentane.

11.11 a.

$$H_3CC{\equiv}CCH_3 + 2\,H_2 \xrightarrow{\text{Ni}} CH_3CH_2CH_2CH_3$$

b.

$$H_3CC{\equiv}CCH_2CH_3 + 2\,H_2 \xrightarrow{\text{Ni}} CH_3CH_2CH_2CH_2CH_3$$

11.13 a. $CH_3CH{=}CH_2 + Br_2 \longrightarrow$

H Br H
| | |
H—C—C—CH
| | |
H H Br

b. $CH_3CH{=}CHCH_3 + Br_2 \longrightarrow$

H H Br H
| | | |
H—C—C—C—C—H
| | | |
H Br H H

11.15 a. $CH_3C{\equiv}CCH_3 + 2\,Cl_2 \longrightarrow$

H Cl Cl H
| | | |
H—C—C—C—C—H
| | | |
H Cl Cl H

b.

$CH_3C{\equiv}CCH_2CH_3 + 2\,Cl_2 \longrightarrow$

H Cl Cl H H
| | | | |
H—C—C—C—C—C—H
| | | | |
H Cl Cl H H

11.17 a.

$CH_3CH{=}CHCH_3 + H_2O \xrightarrow{H^+} CH_3CHCH_2CH_3$ (only product)
|
OH

b. $CH_2{=}CHCH_2CH_2CHCH_3 + H_2O \xrightarrow{H^+}$
|
CH_3

CH_3CHCH_2CH_2CHCH_3 (major product)
| |
OH CH_3

$CH_2{=}CHCH_2CH_2CHCH_3 + H_2O \xrightarrow{H^+}$
|
CH_3

CH_2CH_2CH_2CH_2CHCH_3 (minor product)
| |
OH CH_3

c. $CH_3CH_2CH_2CH{=}CHCH_2CH_3 + H_2O \xrightarrow{H^+}$

CH_3CH_2CH_2CHCH_2CH_2CH_3
|
OH

$CH_3CH_2CH_2CH{=}CHCH_2CH_3 + H_2O \xrightarrow{H^+}$

CH_3CH_2CH_2CH_2CHCH_2CH_3
|
OH

These products will be formed in approximately equal amounts.

d. $CH_3CHClCH{=}CHCHClCH_3 + H_2O \xrightarrow{H^+}$

CH_3CHClCHCH_2CHClCH_3 (only product)
|
OH

11.19 a.

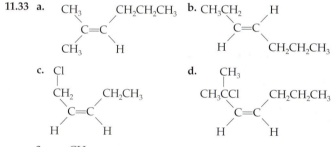

b.

OH
|
Br

Br

c.

NH_2

NO_2

11.21 The longer the carbon chain of an alkene, the higher the boiling point.

11.23 The general formula for an alkane is C_nH_{2n+2}.
The general formula for an alkene is C_nH_{2n}.
The general formula for an alkyne is C_nH_{2n-2}.

11.25 Ethene is a planar molecule. All of the bond angles are 120°.

11.27 a. 2-Pentyne > Propyne > Ethyne
b. 3-Decene > 2-Butene > Ethene

11.29 Identify the longest carbon chain containing the carbon-to-carbon double or triple bond. Replace the –ane suffix of the alkane name with –ene for an alkene or –yne for an alkyne. Number the chain to give the lowest number to the first of the two carbons involved in the double or triple bond. Determine the name and carbon number of each substituent group and place that information as a prefix in front of the name of the parent compound.

11.31 Geometric isomers of alkenes differ from one another in the placement of substituents attached to each of the carbon atoms of the double bond. Of the pair of geometric isomers, the cis-isomer, is the one in which identical groups are on the same side of the double bond.

11.33 a.

CH_3   CH_2CH_2CH_3
  \   /
   C=C
  /   \
CH_3   H

b.

CH_3CH_2   H
      \   /
       C=C
      /   \
     H   CH_2CH_2CH_3

c.

Cl
|
CH_2   CH_2CH_3
  \   /
   C=C
  /   \
H   H

d.

CH_3
|
CH_3CCl   CH_2CH_2CH_3
     \   /
      C=C
     /   \
    H   H

e.

CH_3
|
CH_3CH   H
     \   /
      C=C
     /   \
    H   CH—CHCH_2CH_3
              | |
             Br   CH_3

11.35 a. 3-Methyl-1-pentene
b. 7-Bromo-1-heptene
c. 5-Bromo-3-heptene
d. 1-t-Butyl-4-methylcyclohexene

11.37 a. 2, 3-Dibromobutane could not exist as cis and trans isomers.

b.

CH_3   CH_2CH_2CH_2CH_3
  \   /
   C=C
  /   \
H   H

CH_3   H
  \   /
   C=C
  /   \
H   CH_2CH_2CH_2CH_3

cis-2-Heptene   trans-2-Heptene

c.

cis-2,3-Dibromo-2-butene   trans-2,3-Dibromo-2-butene

d. Propene cannot exist as *cis* and *trans* isomers.

11.39 Alkenes b and c would not exhibit *cis-trans* isomerism.

11.41 Alkenes b and d can exist as both *cis-* and *trans*-isomers.

11.43 $\underset{R}{\overset{R}{>}}C=C\underset{R}{\overset{R}{<}}$ + H$_2$ $\xrightarrow[\text{heat or pressure}]{\text{Pt, Pd, or Ni}}$ R—C—C—R (with H R / R H)

11.45 $\underset{R}{\overset{R}{>}}C=C\underset{R}{\overset{R}{<}}$ + X$_2$ $\longrightarrow$ R—C—C—R (with X R / R X)

11.47 $\underset{R}{\overset{R}{>}}C=C\underset{R}{\overset{R}{<}}$ + H$_2$O $\xrightarrow{\text{H}^+}$ R—C—C—R (with H R / R OH)

11.49 Addition of bromine (Br$_2$) to an alkene results in a color change from red to colorless. If equimolar quantities of Br$_2$ are added to hexene, the reaction mixture will change from red to colorless. This color change will not occur if cyclohexane is used.

11.51 **a.** H$_2$ **d.** $19O_2 \rightarrow 12CO_2 + 14H_2O$
b. H$_2$O   **e.** Cl$_2$
c. Br$_2$ **f.** (cyclopentene structure)

11.53 **a.**

$H_3CC{\equiv}CCH_3$ + 2H$_2$ $\xrightarrow[\text{heat or pressure}]{\text{Pt, Pd, or Ni}}$ H_3C—C—C—CH$_3$ (with H H / H H)

2-Butyne

b.

$CH_3CH_2C{\equiv}CCH_3$ + 2 X$_2$ $\longrightarrow$ CH_3CH_2—C—C—CH$_3$ (with X X / X X)

2-Pentyne

11.55 **a.** Br Br
$CH_3CHCHCH_3$

b.
$CH_3CH_2\overset{OH}{\underset{CH_3}{C}}CH_2CH_2CH_3$ + $CH_3\overset{OH}{C}HCHCH_2CH_2CH_3$ (with CH$_3$)

(major product)   (minor product)

c. (cyclopentane structure)

11.57 A polymer is a macromolecule composed of repeating structural units called *monomers*.

11.59

$n\ \underset{F}{\overset{F}{>}}C=C\underset{F}{\overset{F}{<}}$ $\longrightarrow$ $\left[\begin{array}{c}F\ F\\ -C-C-\\ F\ F\end{array}\right]_n$

Tetrafluoroethene   Teflon

11.61 **a.** $CH_3\overset{H}{\underset{H}{C}}{=}CCH_2CH_3$ + H$_2$O $\xrightarrow{\text{H}^+}$ $CH_3CHCH_2CH_2CH_3$ (with OH)

2-Pentene

+

$CH_3CH_2CHCH_2CH_3$ (with OH)

These products will be formed in approximately equal amounts.

b.

$\underset{Br}{\overset{H\ H}{CH_2C}}{=}\overset{}{C}{-}H$ + H$_2$O $\xrightarrow{\text{H}^+}$ CH_2CHCH_3 (major product) (with Br OH)

3-Bromo-1-propene

+

CH_2CH_2CHOH (minor product) (with Br)

c. (3,4-dimethylcyclohexene) + H$_2$O $\xrightarrow{\text{H}^+}$ (cyclohexanol with OH, CH$_3$, CH$_3$)

3,4-Dimethylcyclohexene

+

(cyclohexanol HO, CH$_3$, CH$_3$)

These products will be formed in approximately equal amounts.

11.63 **a.**

$CH_2{=}CHCH_2\overset{CH_3}{C}HCH_3$ + H$_2$O $\xrightarrow{\text{H}^+}$ $CH_2CH_2CH_2\overset{CH_3}{C}HCH_3$ (with OH)

(This is the minor product of this reaction.)

b.

$CH_3\overset{H}{\underset{H}{C}}{=}CCH_2CH_2CH_3$ + H$_2$O $\longrightarrow$ $CH_3CH_2CHCH_2CH_2CH_3$ (with OH)

OR

$CH_3CH_2\overset{H}{\underset{H}{C}}{=}CCH_2CH_3$ + H$_2$O $\longrightarrow$ $CH_3CH_2CHCH_2CH_2CH_3$ (with OH)

c. (cyclopentene)—CH$_2$CH$_3$ + H$_2$O $\xrightarrow{\text{H}^+}$ (cyclopentanol)—CH$_2$CH$_3$ (with OH)

11.65 **a.**
$CH_2{=}CHCH_2CH{=}CHCH_3$ + 2H$_2$ $\xrightarrow[\text{heat}]{\text{Pt}}$ $CH_3(CH_2)_4CH_3$

1,4-Hexadiene   Hexane

b.

$$CH_3CH=CHCH=CHCH=CHCH_3 + 3H_2 \xrightarrow[\text{heat}]{\text{Ni}}$$

2,4,6-Octatriene

$$CH_3(CH_2)_6CH_3$$

Octane

c.

1,3-Cyclohexadiene   Cyclohexane

d.

1,3,5-Cyclooctatriene   Cyclooctane

11.67 The term aromatic hydrocarbon was first used as a term to describe the pleasant-smelling resins of tropical trees.

11.69 **a.**

b.

c. CH_3CHCH_3

d.

11.71 **a.**

b. $CH_2CH_2CH_3$

c.

d. CH_3

11.73

Pyrimidine

11.75

Purine

Chapter 12

12.1 **a.** 4-Methyl-1-pentanol
b. 4-Methyl-2-hexanol

c. 1, 2, 3-Propanetriol
d. 4-Chloro-3-methyl-1-hexanol

12.3 **a.** Primary
b. Secondary
c. Tertiary

12.5 **a.**

$$CH_3CH=CH_2 + H_2O \xrightarrow{H^+} CH_3CH(OH)CH_3 + CH_3CH_2CH_2OH$$
major product   minor product

b. $CH_2=CH_2 + H_2O \xrightarrow{H^+} CH_3CH_2OH$

c. $CH_3CH_2CH=CHCH_2CH_3 + H_2O \xrightarrow{H^+}$
$CH_3CH_2CH_2CH(OH)CH_2CH_3$

12.7 **a.** The major product is a secondary alcohol (2-propanol) and the minor product is a primary alcohol (1-propanol).
b. The product, ethanol, is a primary alcohol.
c. The product, 3-hexanol, is a secondary alcohol.

12.9 **a.** Ethanol
b. 2-Propanol (major product), 1-Propanol (minor product)
c. 2-Butanol

12.11 **a.** 2-Butanol is the major product. 1-Butanol is the minor product.
b. 2-Methyl-2-propanol is the major product. 2-Methyl-1-propanol is the minor product.

12.13

a.

b. $CH_3CH_2OH \rightarrow CH_3\overset{\overset{\displaystyle O}{\|}}{C}-H$

12.15

a. $CH_3\overset{\overset{\displaystyle OH}{|}}{C}HCH_2CH_3 \rightarrow CH_3\overset{\overset{\displaystyle O}{\|}}{C}CH_2CH_3$

b. $CH_3\overset{\overset{\displaystyle OH}{|}}{C}HCH_2CH_2CH_3 \rightarrow CH_3\overset{\overset{\displaystyle O}{\|}}{C}CH_2CH_2CH_3$

12.17 **a.** I.U.P.A.C.: 1-Ethoxypropane
Common: Ethyl propyl ether
b. I.U.P.A.C.: 1-Methoxypropane
Common: Methyl propyl ether

12.19

$$CH_3CH_2OH + CH_3CH_2OH \xrightarrow{H^+} CH_3CH_2-O-CH_2CH_3 + H_2O$$

Ethanol   Diethyl ether   Water

12.21 **a.**

$$CH_3-\overset{\overset{\displaystyle O}{\|}}{C}-CH_3$$

b.

$$CH_3\overset{\overset{\displaystyle OH}{|}}{C}HCH_2CH_2CH_3$$

12.23 **a.** I.U.P.A.C.: 3, 4-Dimethylpentanal
Common: β, γ-Dimethylvaleraldehyde
b. I.U.P.A.C.: 2-Ethylpentanal
Common: α-Ethylvaleraldehyde

12.25 **a.** 3-Iodobutanone
b. 4-Methyl-2-octanone

12.27

$$CH_3-\overset{\overset{\displaystyle O}{\|}}{C}-H$$

12.29

$$CH_3CH_2CH_2OH \xrightarrow{H_2Cr_2O_7} CH_3CH_2\overset{\overset{\displaystyle O}{\|}}{C}H$$

1-Propanol   Propanal

12.31

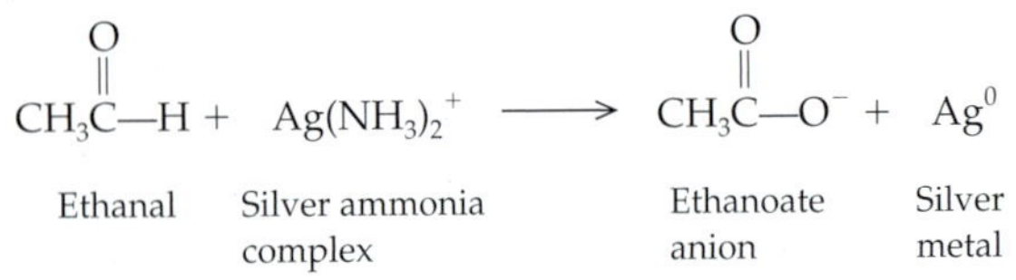

$$CH_3\overset{\overset{\displaystyle O}{\|}}{C}-H + Ag(NH_3)_2^+ \longrightarrow CH_3\overset{\overset{\displaystyle O}{\|}}{C}-O^- + Ag^0$$

Ethanal   Silver ammonia   Ethanoate   Silver
                 complex   anion   metal

12.33 a. Reduction   **d.** Oxidation
b. Reduction   **e.** Reduction
c. Reduction

12.35 The longer the hydrocarbon tail of an alcohol becomes, the less water soluble it will be.

12.37 a < d < c < b

12.39 The I.U.P.A.C. rules for the nomenclature of alcohols require you to name the parent compound, that is the longest continuous carbon chain bonded to the —OH group. Replace the -e ending of the parent alkane with -ol of the alcohol. Number the parent chain so that the carbon bearing the hydroxyl group has the lowest possible number. Name and number all other substituents. If there is more than one hydroxyl group, the -ol ending will be modified to reflect the number. If there are two —OH groups, the suffix -diol is used; if it has three —OH groups, the suffix -triol is used, etc.

12.41 a. 1-Heptanol   **b.** 2-Propanol
c. 2, 2-Dimethylpropanol

12.43 a. 3-Hexanol:

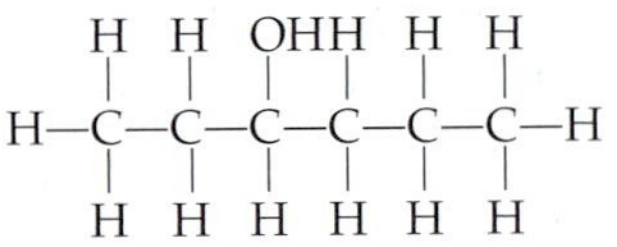

b. 1,2,3-Pentanetriol:

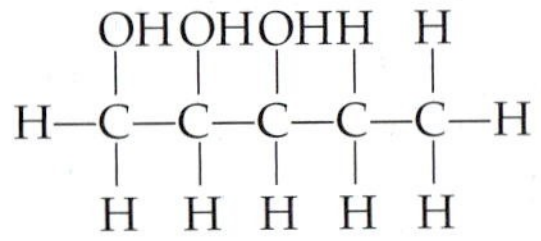

c. 2-Methyl-2-pentanol:

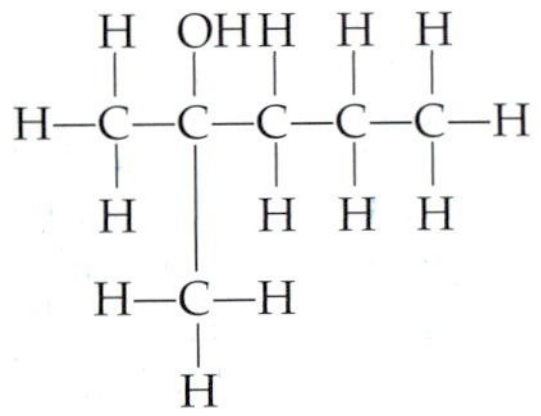

d. Cyclohexanol: 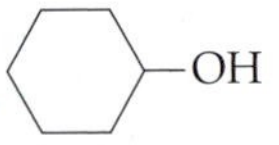   **e.** 3,4-Dimethyl-3-heptanol:

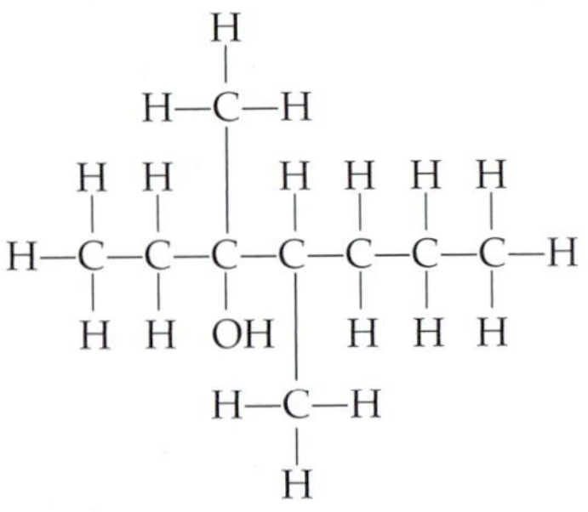

12.45 Methanol is commonly used as a solvent and as a starting material for the synthesis of formaldehyde. Ethanol is also used as a solvent and is the alcohol found in alcoholic beverages. Isopropyl alcohol has been used as rubbing alcohol. Patients with a high fever were given alcohol baths. The rapid evaporation of the alcohol causes cooling of the skin and helps reduce fever. Isopropanol is also used as an antiseptic.

12.47 When the ethanol concentration in a fermentation reaches 12−13%, the yeast producing the ethanol are killed by it. To produce a liquor of higher alcohol concentration, the product of the original fermentation must be distilled.

12.49 The *carbinol carbon* is the one to which the hydroxyl group is bonded.

12.51 a. Primary   **d.** Tertiary
b. Secondary   **e.** Tertiary
c. Tertiary

12.53 a.

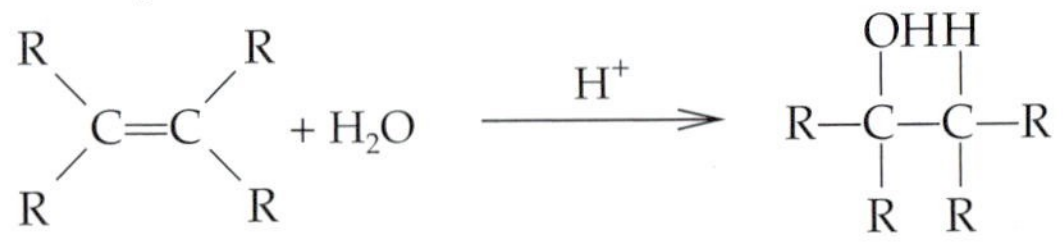

Alkene   Alcohol

b.

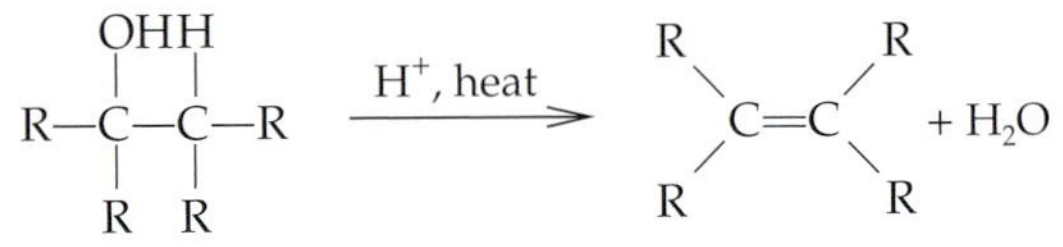

Alcohol   Alkene

12.55 a. 2-Pentanol (major product), 1-pentanol (minor product)
b. 2-Pentanol and 3-pentanol
c. 3-Methyl-2-butanol (major product), 3-methyl-1-butanol (minor product)
d. 3, 3-Dimethyl-2-butanol (major product), 3, 3-dimethyl-1-butanol (minor product)

12.57 a. Butanone
b. N.R.
c. Cyclohexanone
d. N.R.

12.59 Phenols are compounds with an —OH attached to a benzene ring. Like alcohols, phenols are polar compounds because of the polar hydroxyl group. Thus the simpler phenols are somewhat soluble in water.

12.61 Alcohols of molecular formula $C_4H_{10}O$

$CH_3CH_2CH_2CH_2OH,$

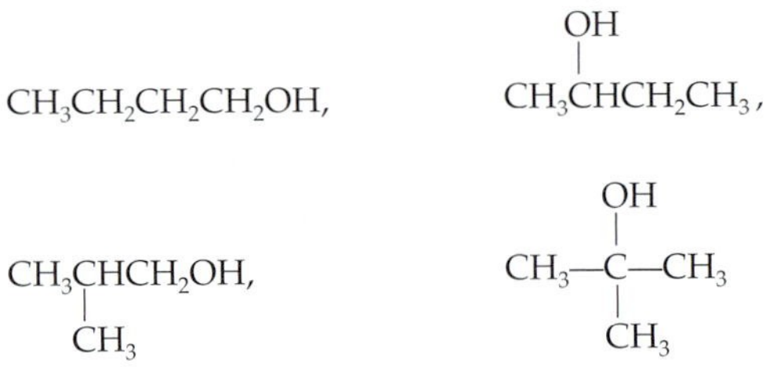

Ethers of molecular formula $C_4H_{10}O$
$CH_3-O-CH_2CH_2CH_3$ $CH_3CH_2-O-CH_2CH_3$
$CH_3-O-CHCH_3$
          $|$
          CH_3

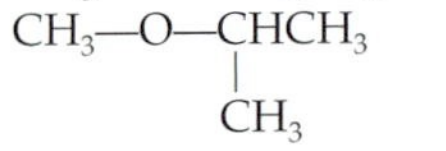

12.63 a. $CH_3CH_2-O-CH_2CH_3 + H_2O$
b. $CH_3CH_2-O-CH_2CH_3 + CH_3-O-CH_3 + CH_3-O-CH_2CH_3 + H_2O$

12.65 Cystine:

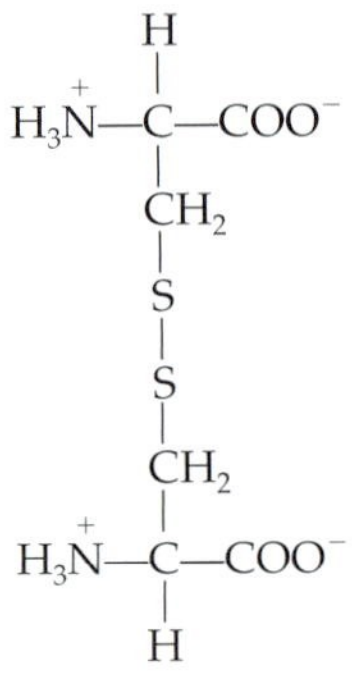

12.67 As the carbon chain length increases, the compounds become less polar and more hydrocarbonlike. As a result, their solubility in water decreases.

12.69

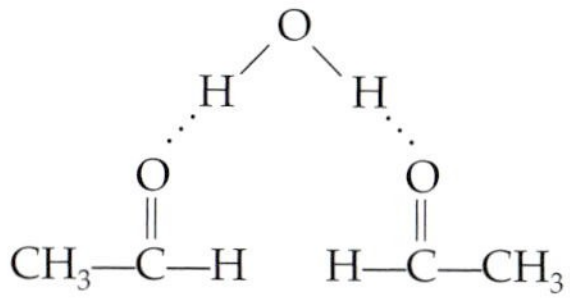

12.71 To name an aldehyde using the I.U.P.A.C. nomenclature system, identify and name the longest carbon chain containing the carbonyl group. Replace the final -e of the alkane name with -al. Number and name all substituents as usual. Remember that the carbonyl carbon is always carbon-1 and does not need to be numbered in the name of the compound.

12.73 **a.** Butanone **c.** Butanal
b. 2-Ethylhexanal **d.** 4-Bromo-4-methylpentanal

12.75 **a.**

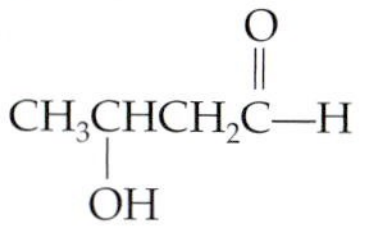

b.

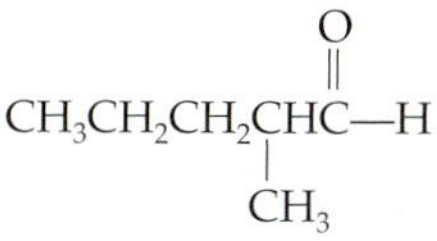

c.

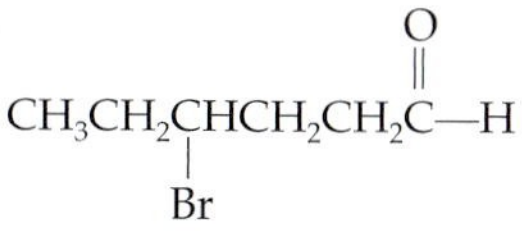

d.

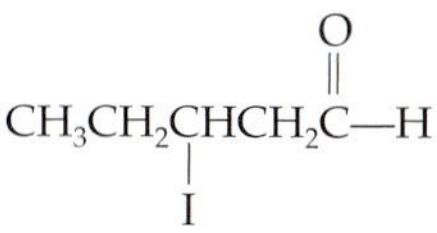

e.

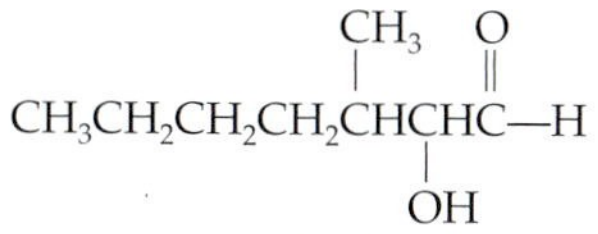

12.77 Acetone is a good solvent because it can dissolve a wide range of compounds. It has both a polar carbonyl group and nonpolar side chains. As a result, it dissolves organic compounds and is also miscible in water.

12.79 **a.** False **c.** False
b. True **d.** False

12.81 **a.**

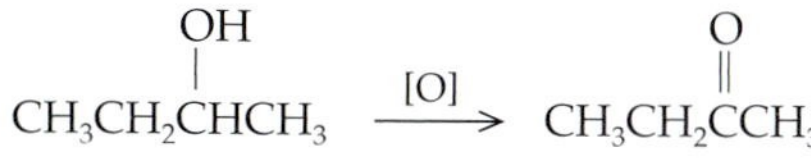

2-Butanol → Butanone

b.

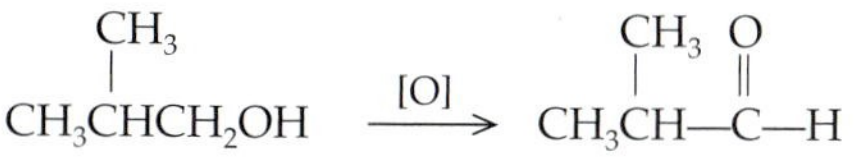

2-Methyl-1-propanol → Methylpropanal

Note that methylpropanal can be further oxidized to methylpropanoic acid.

c.

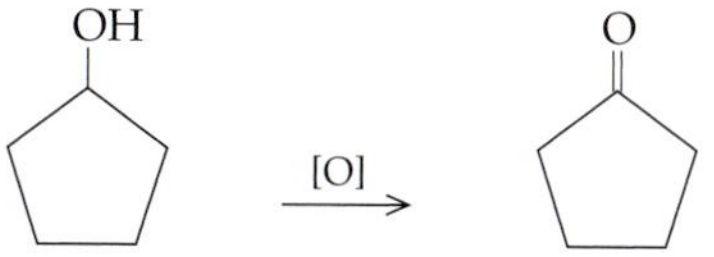

Cyclopentanol → Cyclopentanone

d.

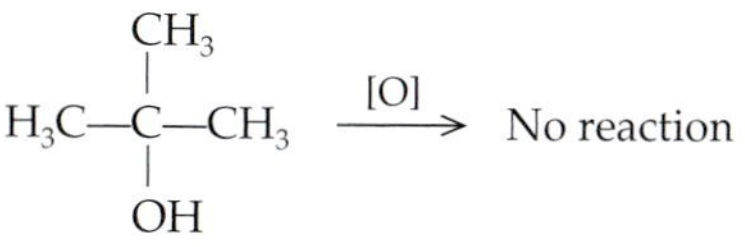

2-Methyl-2-propanol

e.

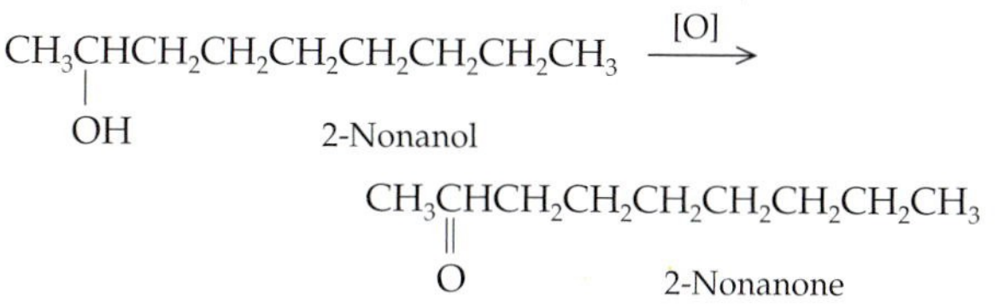

$CH_3CHCH_2CH_2CH_2CH_2CH_2CH_2CH_3$ (OH) 2-Nonanol $\xrightarrow{[O]}$

$CH_3CHCH_2CH_2CH_2CH_2CH_2CH_2CH_3$ (O) 2-Nonanone

f.

$CH_2CH_2CH_2CH_2CH_2CH_2CH_2CH_2CH_2CH_3$ (OH) 1-Decanol $\xrightarrow{[O]}$

$HCCH_2CH_2CH_2CH_2CH_2CH_2CH_2CH_2CH_3$ (O) Decanal

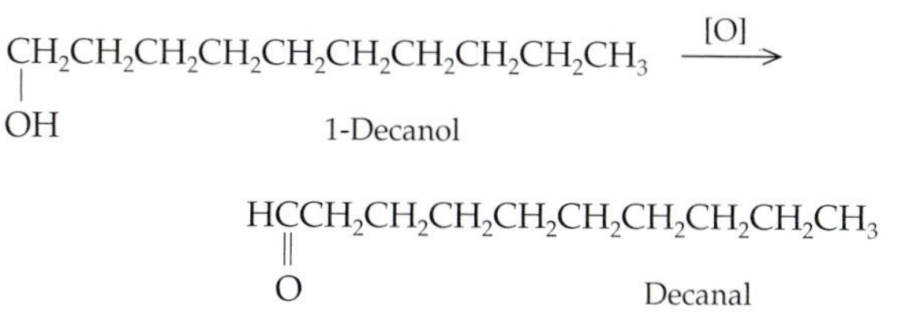

Note that decanal can be further oxidized to decanoic acid.

12.83 **a.**

b.

c.

d.

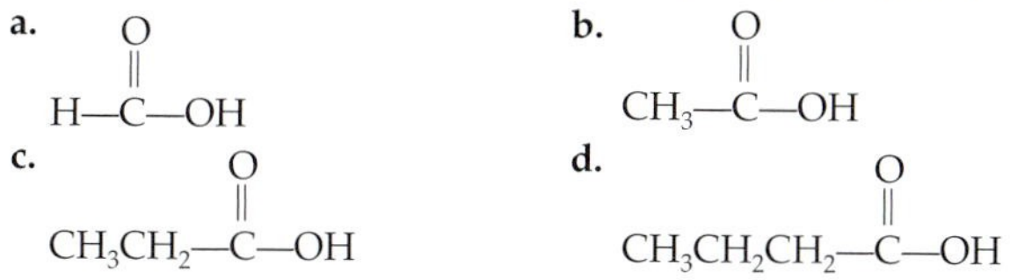

Chapter 13

13.1 **a.** Ketone
b. Ketone
c. Alkane

13.3 The carboxyl group consists of two very polar groups, the carbonyl group and the hydroxyl group. Thus, carboxylic acids are very polar, in addition to which, they can hydrogen bond to one another. Aldehydes are polar, as a result of the carbonyl group, but cannot hydrogen bond to one another. As a result, carboxylic acids have higher boiling points than aldehydes of the same carbon chain length.

13.5 **a.** 2,4-Dimethylpentanoic acid
b. 2,4-Dichlorobutanoic acid

13.7 **a.** α,γ-Dimethylvaleric acid
b. α,γ-Dichlorobutyric acid

13.9 **a.**

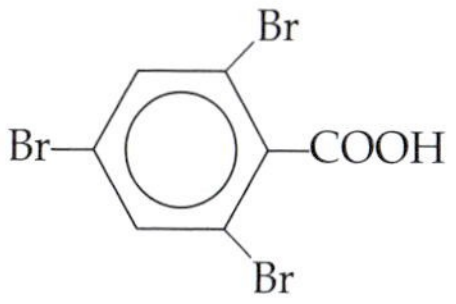

b.

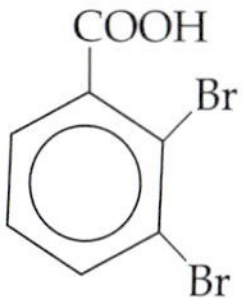

13.11 **a.**

$CH_3CH_2-C-H \longrightarrow CH_3CH_2-C-OH$

Propanal would be the first oxidation product. However, it would quickly be oxidized further to propanoic acid.

b.

$HO-C-CH_2CH_2CH_2CH_3$

13.13 **a.** Propyl butanoate (propyl butyrate)
b. Ethyl butanoate (ethyl butyrate)

13.15 The following reaction between 1-butanol and ethanoic acid produces butyl ethanoate. It requires a trace of acid and heat. It is also reversible.

$$CH_3CH_2CH_2CH_2OH + CH_3COOH \leftrightarrows CH_3C-OCH_2CH_2CH_2CH_3$$

The following reaction between ethanol and propanoic acid produces ethyl propanoate. It requires a trace of acid and heat. It is also reversible.

$$CH_3CH_2OH + CH_3CH_2COOH \leftrightarrows CH_3CH_2C-OCH_2CH_3$$

13.17 a. CH₃COOH + CH₃CH₂CH₂OH
Ethanoic acid   1-Propanol

b. CH₃CH₂CH₂CH₂CH₂COO⁻K⁺ + CH₃CH₂CH₂OH
Potassium hexanoate   1-Propanol

13.19 a. Tertiary
b. Primary
c. Secondary

13.21

$$
\begin{array}{c}
\text{O} \qquad\qquad \text{H}\\
\text{H}\cdots\text{H}\quad\text{H}\cdots\text{O}\\
\text{CH}_3\!-\!\text{N}\qquad\text{N}\!-\!\text{H}\quad\text{H}\\
\text{H}\quad\text{CH}_3\quad\text{CH}_3\ \text{CH}_3
\end{array}
$$

13.23 a. Methanol because the intermolecular hydrogen bonds between alcohol molecules will be stronger.

b. Water because the intermolecular hydrogen bonds between water molecules will be stronger.

c. Ethylamine because it has a higher molecular weight.

d. Propylamine because propylamine molecules can form intermolecular hydrogen bonds while the nonpolar butane cannot do so.

13.25 a.

```
    H  H  H
    |  |  |
H — C— C— C —H
    |  |  |
    H  N  H
       / \
      H   H
```

b.

```
          H     H
           \   /
  H  H  N  H  H  H  H  H
  |  |  |  |  |  |  |  |
H—C— C— C— C— C— C— C— C—H
  |  |  |  |  |  |  |  |
  H  H  H  H  H  H  H  H
```

c.

```
  H  H  H  H  H  H  H
  |  |  |  |  |  |  |
H—C— C— C— C— C— C— C—H
  |  |  |  |  |  |  |
  H  |  H  H  H  H  H
     N—H
      |
    H—C—H
      |
    H—C—H
      |
      H
```

d.

```
        H     H
         \   /
  H  N  H  H  H
  |  |  |  |  |
H—C— C— C— C— C—H
  |  |  |  |  |
  H  |  H  H  H
   H—C—H
     |
     H
```

e.

```
   H     H
    \   /
     N  H  H  H  H  H  H  H
     |  |  |  |  |  |  |  |
H — C— C— C— C— C— C— C— C—H
     |  |  |  |  |  |  |  |
     H  H  H  Cl I  H  H  H  H
```

f.

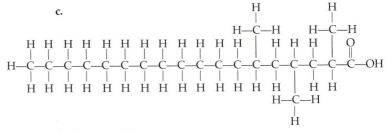

13.27 Aldehydes are polar, as a result of the carbonyl group, but cannot hydrogen bond to one another. Alcohols are polar and can hydrogen bond as a result of the polar hydroxyl group. The carboxyl group of the carboxylic acids consists of both of these groups: the carbonyl group and the hydroxyl group. Thus, carboxylic acids are more polar than either aldehydes or alcohols, in addition to which, they can hydrogen bond to one another. As a result, carboxylic acids have higher boiling points than aldehydes or alcohols of the same carbon chain length.

13.29 a. 3-Hexanone
b. 3-Hexanone
c. Hexane
d. Dipropyl ether
e. Hexanal
f. Ethanol

13.31 Determine the name of the parent compound, that is the longest carbon chain containing the carboxyl group. Change the *-e* ending of the alkane name to *-oic* acid. Number the chain so that the carboxyl carbon is carbon-1. Name and number substituents in the usual way.

13.33 a.

```
    H  H  H  H  O
    |  |  |  |  ‖
H — C— C— C— C— C—OH
    |  |  |  |
    H  H  H  Br
```

b.

```
              H
              |
   H    H—C—H   H  O
   |      |     |  ‖
H—C ———— C ———— C— C—OH
   |      |     |
   H      H     Br
```

c.

```
                                    H              H
                                    |              |
                                  H—C—H          H—C—H
  H  H  H  H  H  H  H  H  H  H  H  H  H  H  H     H  H  H  O
  |  |  |  |  |  |  |  |  |  |  |  |  |  |  |     |  |  |  ‖
H—C— C— C— C— C— C— C— C— C— C— C— C— C— C— C—————C— C— C— C—OH
  |  |  |  |  |  |  |  |  |  |  |  |  |  |  |     |  |  |
  H  H  H  H  H  H  H  H  H  H  H  H  H  H  H   H—C—H H  H
                                                  |
                                                  H
```

d.

```
 H        H
  \      /
   C == C
  /      \
 H        C==O
          |
          OH
```

13.35 a. I.U.P.A.C. name: 2-Hydroxypropanoic acid
Common name: α-Hydroxypropionic acid

b. I.U.P.A.C. name: 3-Hydroxybutanoic acid
Common name: β-Hydroxybutyric acid

c. I.U.P.A.C. name: 4, 4-Dimethylpentanoic acid
Common name: γ, γ-Dimethylvaleric acid

d. I.U.P.A.C. name: 3, 3-Dichloropentanoic acid
Common name: β, β-Dichlorovaleric acid

13.37 Soaps are made from water, a strong base, and natural fats or oils.

13.39 a. (1) H_2CrO_4 (3) HCl

(2) $CH_3\overset{\overset{\displaystyle OH}{|}}{C}HCH_3$ (4) $CH_3CH_2\overset{\overset{\displaystyle O}{\|}}{C}-O^-\ Na^+$

b.

$CH_3\overset{\overset{\displaystyle O}{\|}}{C}-O^-\ Na^+$

c.

$CH_3CH_2CH_2CH_2CH_2\overset{\overset{\displaystyle O}{\|}}{C}-O^-\ Na^+$

d. CH_3COOH

13.41 Esters are mildly polar as a result of the polar carbonyl group within the structure.

13.43 a.

$\overset{\overset{\displaystyle O}{\|}}{C}-OCH_3$ (attached to benzene ring)

b.

$CH_3CH_2CH_2CH_2CH_2CH_2CH_2CH_2CH_2-\overset{\overset{\displaystyle O}{\|}}{C}-O-CH_2CH_2CH_2CH_3$

c.

$CH_3CH_2-\overset{\overset{\displaystyle O}{\|}}{C}-O-CH_3$

d.

$CH_3CH_2-\overset{\overset{\displaystyle O}{\|}}{C}-O-CH_2CH_3$

13.45 The synthesis of an ester is referred to as a dehydration reaction because a water molecule is eliminated during the course of the reaction.

13.47

$\text{(salicylic acid)}\ -\overset{\overset{\displaystyle O}{\|}}{C}-OH + CH_3OH \xrightarrow{H^+}$

b. Salicylic acid

Methyl salicylate

$-\overset{\overset{\displaystyle O}{\|}}{C}-OCH_3 + H_2O$

13.49 a.

$CH_3CH_2-\overset{\overset{\displaystyle O}{\|}}{C}-OCH_2CH_2CH_3 \underset{\longrightarrow}{\overset{H^+,\ heat}{\longleftarrow}}$
Propyl propanoate

$CH_3CH_2-\overset{\overset{\displaystyle O}{\|}}{C}-OH\ +\ CH_3CH_2CH_2OH$
Propanoic acid   1-Propanol

b.

$H-\overset{\overset{\displaystyle O}{\|}}{C}-OCH_2CH_2CH_2CH_3 \underset{\longrightarrow}{\overset{H^+,\ heat}{\longleftarrow}}$
Butyl methanoate

$H-\overset{\overset{\displaystyle O}{\|}}{C}-OH\ +\ CH_3CH_2CH_2CH_2OH$
Methanoic acid   1-Butanol

c.

$H-\overset{\overset{\displaystyle O}{\|}}{C}-OCH_2CH_3 \underset{\longrightarrow}{\overset{H^+,\ heat}{\longleftarrow}}$
Ethyl methanoate

$H-\overset{\overset{\displaystyle O}{\|}}{C}-OH\ +\ CH_3CH_2OH$
Methanoic acid   Ethanol

d.

$CH_3CH_2CH_2CH_2-\overset{\overset{\displaystyle O}{\|}}{C}-OCH_3 \underset{\longrightarrow}{\overset{H^+,\ heat}{\longleftarrow}}$
Methyl pentanoate

$CH_3CH_2CH_2CH_2-\overset{\overset{\displaystyle O}{\|}}{C}-OH\ +\ CH_3OH$
Pentanoic acid   Ethanol

13.51 In systematic nomenclature, primary amines are named by determining the name of the parent compound, the longest continuous carbon chain containing the amine group. The -*e* ending of the alkane chain is replaced with -*amine*. Thus, an alkane becomes an alkanamine. The parent chain is then numbered to give the carbon bearing the amine group the lowest possible number. Finally, all substituents are named and numbered and added as prefixes to the "alkanamine" name.

13.53 Amphetamines elevate blood pressure and pulse rate. They also decrease the appetite.

13.55 a. Ethanol has a higher boiling point than ethanamine because the hydroxyl group oxygen is more electronegative than the amine nitrogen atom. As a result, the intermolecular hydrogen bonds between alcohol molecules are stronger than those between primary amines.

b. 1-Propanamine has a higher boiling point than butane due to the polar $-NH_2$ of 1-propanamine that can form intermolecular hydrogen bonds with other 1-propanamine molecules. Butane is unable to form intermolecular hydrogen bonds.

c. Water has a higher boiling point because the H—OH hydrogen bonds between water molecules are stronger than the H—NH hydrogen bonds between methanamine molecules. This is because oxygen is more electronegative than nitrogen.

d. Ethylmethylamine has a higher boiling point than butane because the polar —NH of ethylmethylamine can form intermolecular hydrogen bonds with other ethylmethylamine molecules. Butane is unable to form intermolecular hydrogen bonds.

13.57 a. 2-Butanamine (2-butylamine)
b. 3-Hexanamine (3-hexylamine)
c. 2-Methyl-2-propanamine (*t*-butylamine)
d. 1-Octanamine (1-octylamine)
e. *N,N*-Dimethylethanamine (dimethylethylamine)

13.59 a. cyclohexane with NH_2

b. cyclopentane with NH_2 and Br

c. $CH_3CH_2\overset{\overset{\displaystyle CH_2CH_3}{|}}{\underset{\underset{\displaystyle CH_2CH_3}{|}}{N^+}}CH_2CH_3I^-$

d.
benzene ring with NH_2 and Br

13.61 a. Primary
  b. Secondary
  c. Primary
  d. Tertiary
13.63 a. H_2O
  b. HBr
  c. $CH_3CH_2CH_2—N^+H_3$
  d. $CH_3CH_2—N^+H_2Cl^-$
        |
        CH_2CH_3
13.65 Drugs containing amine groups are generally administered as ammonium salts because the salt is more soluble in water and, hence, in body fluids.
13.67 Amides have very high boiling points because the amide group consists of two very polar functional groups, the carbonyl group and the amino group. Strong intermolecular hydrogen bonding between the N—H bond of one amide and the C=O group of a second amide results in very high boiling points.
13.69 a. I.U.P.A.C. name: Propanamide
        Common name: Propionamide
  b. I.U.P.A.C. name: Pentanamide
        Common name: Valeramide
  c. I.U.P.A.C. name: N,N-Dimethylethanamide
        Common name: N,N-Dimethylacetamide

13.71 a. $CH_3—\overset{\overset{O}{\|}}{C}—NH_2$

  b. $CH_3CH_2—\overset{\overset{O}{\|}}{C}—NH—CH_3$

  c. (phenyl)$—\overset{\overset{O}{\|}}{C}—\underset{\underset{CH_2CH_3}{|}}{N}—CH_2CH_3$

  d. $CH_3CH_2\underset{\underset{Br}{|}}{\overset{\overset{CH_3}{|}}{CH}}CHCH_2—\overset{\overset{O}{\|}}{C}—NH_2$

  e. $CH_3—\overset{\overset{O}{\|}}{C}—\underset{\underset{CH_3}{|}}{N}—CH_3$

13.73 Amides are not proton acceptors (bases) because the highly electronegative carbonyl oxygen has a strong attraction for the nitrogen lone pair of electrons. As a result they cannot "hold" a proton.

13.75

Amide group
Carboxyl group
COOH
CH₃
O
N
CH₃
S
$CH_3(CH_2)_3SCH_2CONH$

Penicillin BT

13.77 a. $CH_3—\overset{\overset{O}{\|}}{C}—NHCH_3 + H_3O^+ \longrightarrow$

  N-Methylethanamide

  $CH_3COOH + CH_3NH_3^+$

  Ethanoic acid   Methanamine

b. $CH_3CH_2CH_2—\overset{\overset{O}{\|}}{C}—NH—CH_3 + H_3O^+ \longrightarrow$

  N-Methylbutanamide

  $CH_3CH_2CH_2COOH + CH_3NH_3^+$

  Butanoic acid   Methanamine

c. $CH_3\underset{\underset{CH_3}{|}}{CH}CH_2—\overset{\overset{O}{\|}}{C}—NH—CH_2CH_3 + H_3O^+ \longrightarrow$

  N-Ethyl-3-methylbutanamide

  $CH_3\underset{\underset{CH_3}{|}}{CH}CH_2COOH + CH_3CH_2NH_3^+$

  3-Methylbutanoic acid   Ethanamine

Chapter 14

14.1 It is currently recommended that 45–65% of the calories in the diet should be carbohydrates. Of that amount, no more than 10% should be simple sugars.
14.3 An aldose is a sugar with an aldehyde functional group. A ketose is a sugar with a ketone functional group.
14.5 a. Ketose   **d.** Aldose
  b. Aldose   **e.** Ketose
  c. Ketose   **f.** Aldose

14.7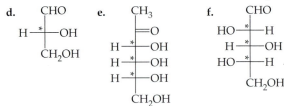

14.9 a. D-   **b.** L-   **c.** D-   **d.** D-   **e.** D-   **f.** L-

14.11

CHO
H——OH
H——OH
H——OH
  CH_2OH

D-Ribose

14.13

CHO
HO——H
HO——H
HO——H
  CH_2OH

L-Ribose

14.15

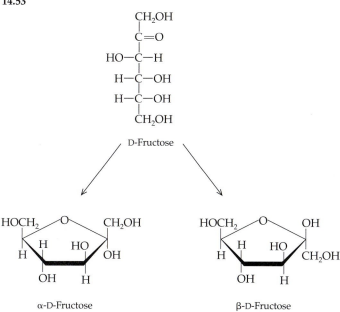

β-D-Galactose   α-D-Galactose

14.17 α-Amylase and β-amylase are digestive enzymes that break down the starch amylose. α-Amylase cleaves glycosidic bonds of the amylose chain at random, producing shorter polysaccharide chains. β-Amylase sequentially cleaves maltose (a disaccharide of glucose) from the reducing end of the polysaccharide chain.

14.19 A monosaccharide is the simplest sugar and consists of a single saccharide unit. A disaccharide is made up of two monosaccharides joined covalently by a glycosidic bond.

14.21 The molecular formula for a simple sugar is $(CH_2O)_n$. Typically n is an integer from 3 to 7.

14.23 Mashed potato flakes, rice, and corn starch contain amylose and amylopectin, both of which are polysaccharides. A candy bar contains sucrose, a disaccharide. Orange juice contains fructose, a monosaccharide. It may also contain sucrose if the label indicates that sugar has been added.

14.25 Four

14.27

D-Galactose (An aldohexose)   D-Fructose (A ketohexose)

14.29 An *aldose* is a sugar that contains an aldehyde (carbonyl) group.

14.31 A ketopentose is a sugar with a five-carbon backbone and containing a ketone (carbonyl) group.

14.33 **a.** β-D-Glucose
b. β-D-Fructose
c. α-D-Galactose

14.35

D-Glyceraldehyde   L-Glyceraldehyde

14.37 Stereoisomers are a pair of molecules that have the same structural formula and bonding pattern but that differ in the arrangement of the atoms in space.

14.39 A *chiral carbon* is one that is bonded to four different chemical groups.

14.41 A polarimeter converts monochromatic light into monochromatic plane-polarized light. This plane-polarized light is passed through a sample and into an analyzer. If the sample is optically active, it will rotate the plane of the light. The degree and angle of rotation are measured by the analyzer.

14.43 A Fischer projection is a two-dimensional drawing of a molecule that shows a chiral carbon at the intersection of two lines. Horizontal lines at the intersection represent bonds projecting out of the page and vertical lines represent bonds that project into the page.

14.45 Dextrose is a common name used for D-glucose.

14.47 D- and L-Glyceraldehyde are a pair of enantiomers, that is, they are nonsuperimposable mirror images of one another.

14.49 **a.**   **b.**   **c.**

14.51 A Haworth projection is a means of representing the orientation of substituent groups around a cyclic sugar molecule.

14.53

D-Fructose

α-D-Fructose   β-D-Fructose

14.55 β-Maltose and α-lactose would give positive Benedict's tests. Glycogen would give only a weak reaction because there are fewer reducing ends for a given mass of the carbohydrate.

14.57 Enantiomers are stereoisomers that are nonsuperimposable mirror images of one another. For instance:

D-Glyceraldehyde   L-Glyceraldehyde

14.59 A glycosidic bond is the bond formed between the hydroxyl group of the C-1 carbon of one sugar and a hydroxyl group of another sugar.

14.61

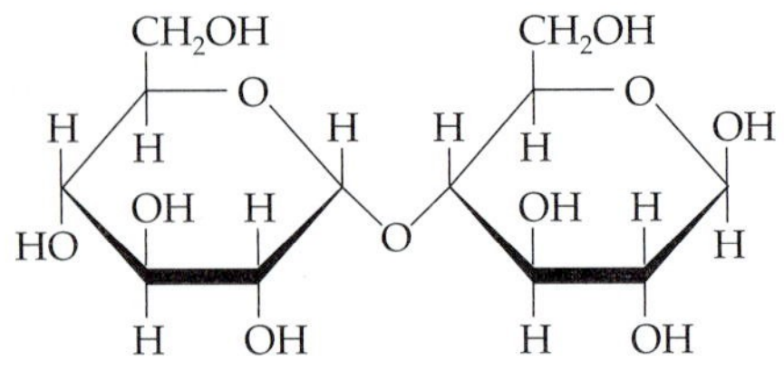

β-Maltose

14.63 Milk

14.65 Untreated galactosemia leads to severe mental retardation, cataracts, and early death.

14.67 A polymer is a very large molecule formed by the combination of many small molecules, called monomers.

14.69 Starch

14.71 The glucose units of amylose are joined by α $(1 \rightarrow 4)$ glycosidic bonds and those of cellulose are bonded together by β $(1 \rightarrow 4)$ glycosidic bonds.

14.73 Glycogen serves as a storage molecule for glucose. Glycogen is stored in the liver and muscle tissue.

Chapter 15

15.1 **a.** $CH_3(CH_2)_7CH=CH(CH_2)_7COOH$
b. $CH_3(CH_2)_{10}COOH$
c. $CH_3(CH_2)_4CH=CH-CH_2-CH=CH(CH_2)_7COOH$
d. $CH_3(CH_2)_{16}COOH$

15.3 $CH_3(CH_2)_{10}COOH \ + \ CH_3CH_2OH \ \xrightarrow{H^+,\ heat}$

Lauric acid   Ethanol
Dodecanoic acid

$$CH_3(CH_2)_{10}-\overset{\overset{\displaystyle O}{\|}}{C}-OCH_2CH_3 \ + \ H_2O$$

Ethyl dodecanoate

15.5

$$CH_3CH_2-\overset{\overset{\displaystyle O}{\|}}{C}-OCH_2CH_2CH_2CH_3 \ + \ H_2O \ \xrightarrow{H^+,\ heat}$$

Butyl propionate

$$CH_3CH_2CH_2CH_2OH \ + \ CH_3CH_2COOH$$

Butyl alcohol   Propionic acid

15.7

$$CH_3CH_2-\overset{\overset{\displaystyle O}{\|}}{C}-OCH_2CH_2CH_2CH_3 \ + \ KOH \ \longrightarrow$$

Butyl propionate

$$CH_3CH_2COO^-K^+ \ + \ CH_3CH_2CH_2CH_2OH$$

Potassium propionate   Butyl alcohol

15.9

$$CH_3(CH_2)_5CH=CH(CH_2)_7COOH \ + \ H_2 \ \xrightarrow{Ni}$$

cis-9-Hexadecenoic acid

$$CH_3(CH_2)_{14}COOH$$

Hexadecanoic acid

15.11

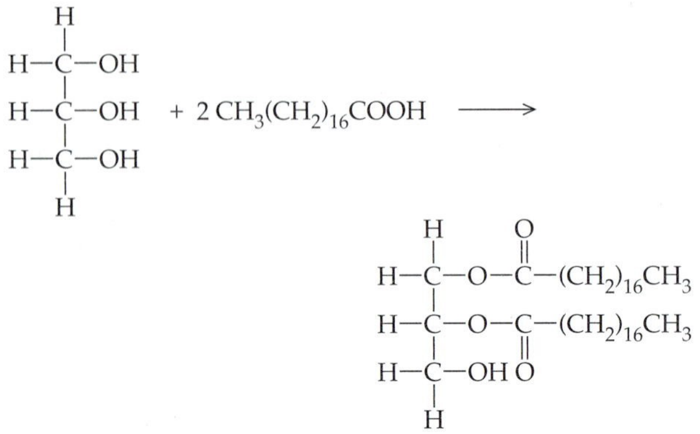

15.13 a. $CH_3(CH_2)_7CH=CH(CH_2)_7\!-\!C\!-\!O\!-\!CH_2$

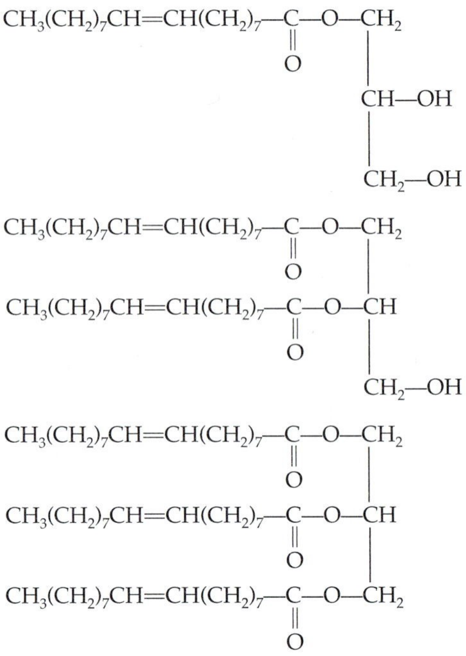

b. $CH_3(CH_2)_8\!-\!C\!-\!O\!-\!CH_2$

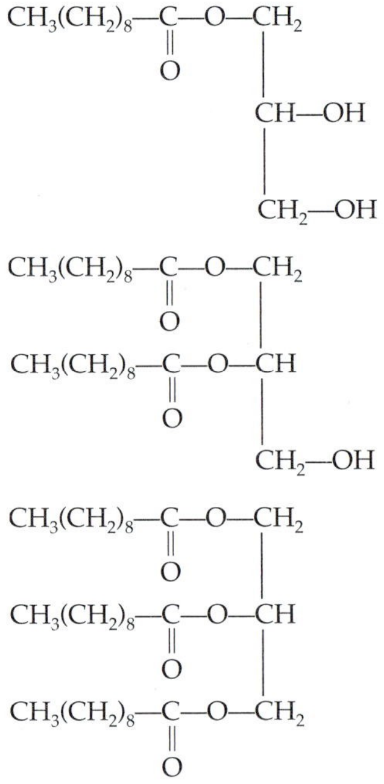

15.15

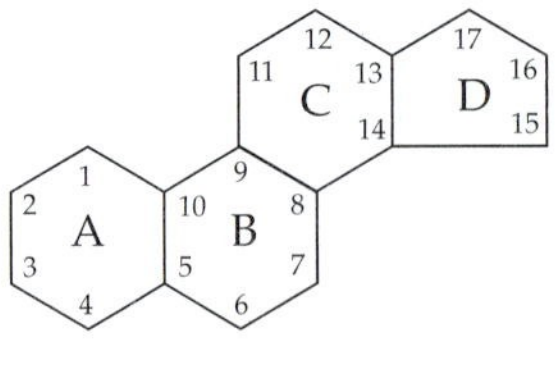

Steroid nucleus

15.17 Diffusion is the net movement of a solute from an area of higher concentration to an area of lower concentration.

15.19 Fatty acids, glycerides, nonglyceride lipids, and complex lipids

15.21 Lipid-soluble vitamins are transported into cells of the small intestine in association with dietary fat molecules. Thus, a diet low in fat reduces the amount of vitamins A, D, E, and K that enters the body.

15.23 A saturated fatty acid is one in which the hydrocarbon tail has only carbon-to-carbon single bonds. An unsaturated fatty acid has at least one carbon-to-carbon double bond. Palmitic acid (hexadecanoic acid) is a saturated fatty acid that has the following structure:

$$CH_3(CH_2)_{14}-\overset{\overset{\displaystyle O}{\|}}{C}-OH$$

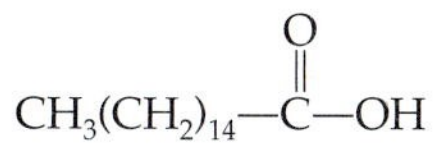

Palmitoleic acid (*cis*-9-hexadecenoic acid) is a monounsaturated fatty acid that has the following structure:

$$CH_3(CH_2)_5CH=CH(CH_2)_7\overset{\overset{\displaystyle O}{\|}}{C}-OH$$

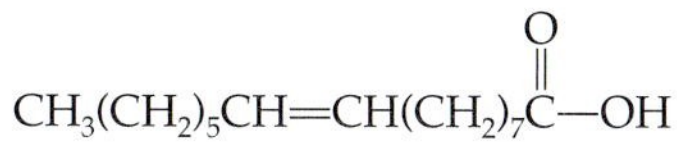

15.25 The melting points of fatty acids increase as the length of the hydrocarbon chains increase. This is because the intermolecular attractive forces, including van der Waals forces, increase as the length of the hydrocarbon chain increases.

15.27 **a.** Decanoic acid
$CH_3(CH_2)_8COOH$
b. Stearic acid
$CH_3(CH_2)_{16}COOH$
c. *trans*-5-Decenoic acid:

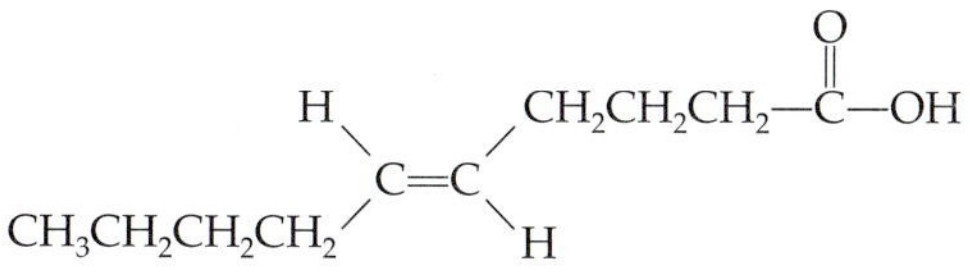

d. *cis*-5-Decenoic acid:

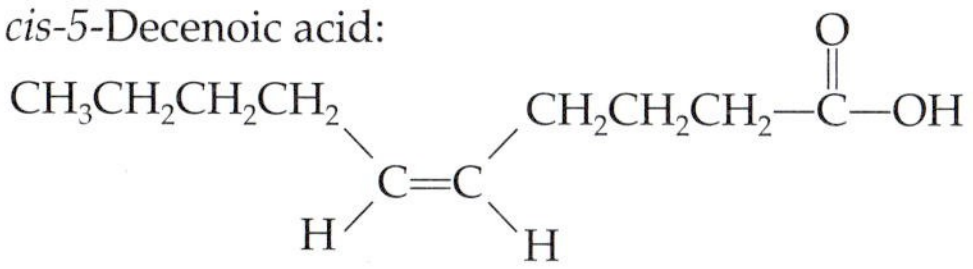

15.29 **a.**

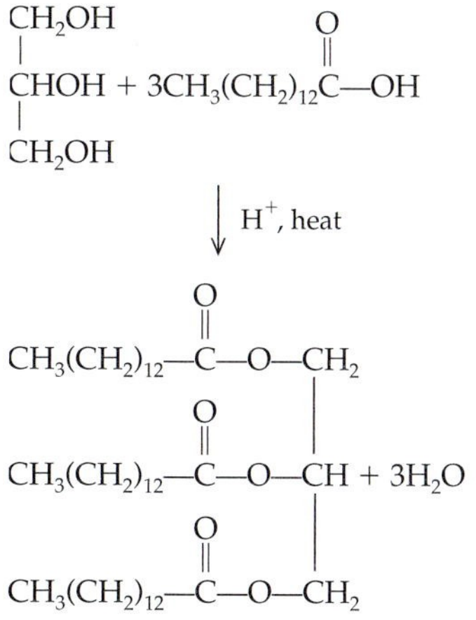

b.

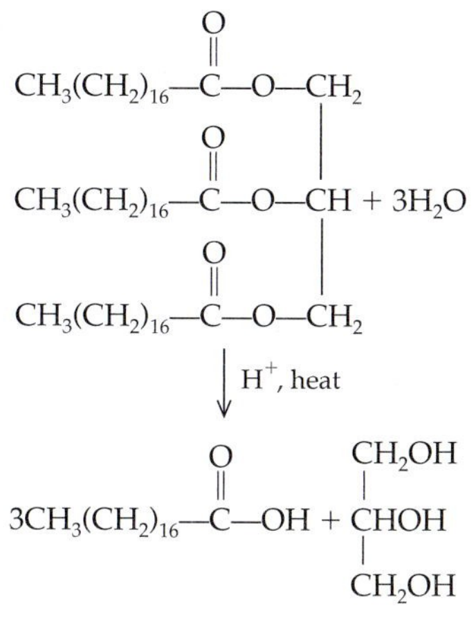

c. $CH_3CH_2CH_2CH_2CH_2CH_2CH_2CH_2CH_2-\overset{\overset{\displaystyle O}{\|}}{C}-OH$

$\downarrow$ KOH

$CH_3CH_2CH_2CH_2CH_2CH_2CH_2CH_2CH_2-\overset{\overset{\displaystyle O}{\|}}{C}-O^-K^+ + H_2O$

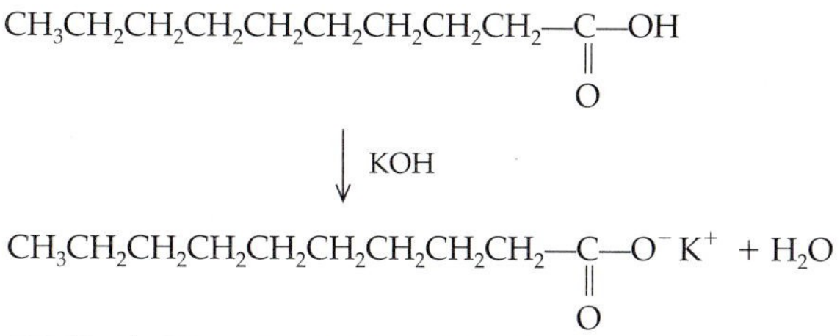

d. $CH_3(CH_2)_4CH=CHCH_2CH=CH(CH_2)_7-\overset{\overset{\displaystyle O}{\|}}{C}-OH + 2H_2$

$\downarrow$ Ni

$CH_3(CH_2)_{16}-\overset{\overset{\displaystyle O}{\|}}{C}-OH$

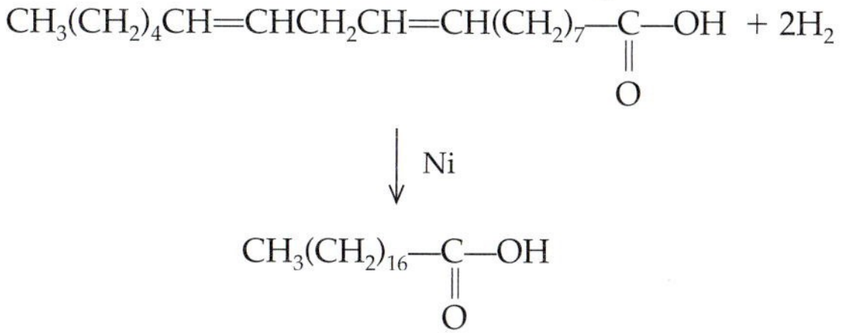

15.31 The essential fatty acid linoleic acid is required for the synthesis of arachidonic acid, a precursor for the synthesis of the prostaglandins, a group of hormonelike molecules.

15.33 Smooth muscle contraction, enhancement of fever and swelling associated with the inflammatory response, bronchial dilation, inhibition of secretion of acid into the stomach.

15.35 An emulsifying agent is a molecule that aids in the suspension of triglycerides in water. They are amphipathic molecules, such as lecithin, that serve as bridges holding together the highly polar water molecules and the nonpolar triglycerides.

15.37 A triglyceride with three saturated fatty acid tails would be a solid at room temperature. The long, straight fatty acid tails would stack with one another because of strong intermolecular and intramolecular attractions.

15.39

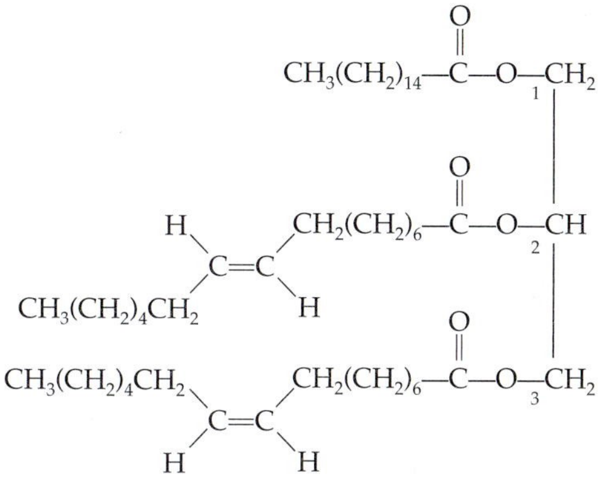

15.41 A sphingolipid is a lipid that is not derived from glycerol, but rather from sphingosine, a long-chain, nitrogen-containing (amino) alcohol. Like phospholipids, sphingolipids are amphipathic. The two major types of sphingolipids are the sphingomyelins and the glycosphingolipids.

15.43 Sphingomyelins are important structural lipid components of nerve cell membranes. They are found in the myelin sheath that surrounds and insulates cells of the central nervous system.

15.45 Cholesterol is readily soluble in the hydrophobic region of biological membranes. It is involved in regulating the fluidity of the membrane.

15.47 Progesterone is the most important hormone associated with pregnancy. Testosterone is needed for development of male secondary sexual characteristics. Estrone is required for proper development of female secondary sexual characteristics.

15.49 Cortisone is used to treat rheumatoid arthritis, asthma, gastrointestinal disorders, and many skin conditions.

15.51 Myricyl palmitate (beeswax) is made up of the fatty acid palmitic acid and the alcohol myricyl alcohol— $CH_3(CH_2)_{28}CH_2OH$.

15.53 Chylomicrons, high-density lipoproteins, low-density lipoproteins, and very low density lipoproteins

15.55 *Chylomicrons* carry dietary lipids from the intestine to other tissues. They are approximately 4% phospholipids, 90% triglycerides, 5% cholesterol, and 1% protein.

Very low density lipoproteins carry triglycerides synthesized in the liver to adipose tissue for storage. They are approximately 18% phospholipids, 60% triglycerides, 14% cholesterol, and 8% protein.

Low density lipoproteins carry cholesterol to peripheral tissues and help regulate cholesterol levels in those tissues. They are approximately 20% phospholipids, 10% triglycerides, 45% cholesterol, and 25% protein.

High density lipoproteins transport cholesterol from peripheral tissues to the liver. They are approximately 30% phospholipids, 5% triglycerides, 20% cholesterol, and 45% protein.

15.57 The basic structure of a biological membrane is a bilayer of phospholipid molecules arranged so that the hydrophobic hydrocarbon tails are packed in the center and the hydrophilic head groups are exposed on the inner and outer surfaces.

15.59 A peripheral membrane protein is bound to only one surface of the membrane, either inside or outside the cell.

15.61 Cholesterol is freely soluble in the hydrophobic layer of a biological membrane. It moderates the fluidity of the membrane by disrupting the stacking of the fatty acid tails of membrane phospholipids.

15.63 If the fatty acid tails of membrane phospholipids are converted from saturated to unsaturated, the fluidity of the membrane will increase.

15.65 In simple diffusion the molecule moves directly across the membrane, whereas in facilitated diffusion a protein channel through the membrane is required.

15.67 Active transport requires an energy input to transport molecules or ions against the gradient (from an area of lower concentration to an area of higher concentration). Facilitated diffusion is a means of passive transport in which molecules or ions pass from regions of higher concentration to regions of lower concentration through a permease protein. No energy is expended by the cell in facilitated diffusion.

15.69 An antiport transport mechanism is one in which one molecule or ion is transported into the cell while a different molecule or ion is transported out of the cell.

15.71 One ATP molecule is hydrolyzed to transport 3 Na^+ out of the cell and 2 K^+ into the cell.

Chapter 16

16.1 a. Glycine (gly):

$$H_3^+N—\overset{\underset{\displaystyle |}{COO^-}}{\underset{\underset{\displaystyle H}{|}}{C}}—H$$

b. Proline (pro):

c. Threonine (thr):

$$H_3^+N—\overset{\underset{\displaystyle |}{COO^-}}{C}—H$$
$$H—\overset{}{C}—OH$$
$$CH_3$$

d. Aspartate (asp):

$$H_3^+N—\overset{\underset{\displaystyle |}{COO^-}}{C}—H$$
$$H—\overset{}{C}—H$$
$$COO^-$$

e. Lysine (lys):

$$H_3^+N—\overset{\underset{\displaystyle |}{COO^-}}{C}—H$$
$$H—\overset{}{C}—H$$
$$H—\overset{}{C}—H$$
$$H—\overset{}{C}—H$$
$$H—\overset{}{C}—H$$
$$N^+H_3$$

16.3 a. Alanyl-phenylalanine:

b. Lysyl-alanine:

c. Phenylalanyl-tyrosyl-leucine:

16.5 The primary structure of a protein is the amino acid sequence of the protein chain. Regular, repeating folding of the peptide chain caused by hydrogen bonding between the amide hydrogens and carbonyl oxygens of the peptide bond is the secondary structure of a protein. The two most common types of secondary structure are the α-helix and the β-pleated sheet. Tertiary structure is the further folding of the regions of α-helix and β-pleated sheet into a compact, spherical structure. Formation and maintenance of the tertiary structure results from weak attractions between amino acid R groups. The binding of two or more peptides to produce a functional protein defines the quaternary structure.

16.7 **a.** Transferase   **d.** Oxidoreductase
b. Transferase   **e.** Hydrolase
c. Isomerase

16.9 The induced fit model assumes that the enzyme is flexible. Both the enzyme and the substrate are able to change shape to form the enzyme-substrate complex. The lock-and-key model assumes that the enzyme is inflexible (the lock) and the substrate (the key) fits into a specific rigid site (the active site) on the enzyme to form the enzyme-substrate complex.

16.11 An enzyme might distort a bond, thereby catalyzing bond breakage. An enzyme could bring two reactants into close proximity and in the proper orientation for the reaction to occur. Finally, an enzyme could alter the pH of the microenvironment of the active site, thereby serving as a transient donor or acceptor of H^+.

16.13 Water-soluble vitamins are required by the body for the synthesis of coenzymes that are required for the function of a variety of enzymes.

16.15 A decrease in pH will change the degree of ionization of the R groups within a peptide chain. This disturbs the weak interactions that maintain the structure of an enzyme, which may denature the enzyme. Less drastic alterations in the charge of R groups in the active site of the enzyme can inhibit enzyme-substrate binding or destroy the catalytic ability of the active site.

16.17 An enzyme is a protein that serves as a biological catalyst, speeding up biological reactions.

16.19 Enzymes speed up reactions that might take days or weeks to occur on their own. They also catalyze reactions that might require very high temperatures or harsh conditions if carried out in the laboratory. In the body, these reactions occur quickly under physiological conditions.

16.21 The general structure of an L-α-amino acid:

$$\overset{\displaystyle COO^-}{\underset{\displaystyle R}{H_3{}^+N-\overset{\displaystyle |}{\underset{\displaystyle |}{C}}-H}}$$

16.23 A chiral carbon is one that has four different atoms or groups of atoms attached to it.

16.25 Interactions between the R groups of the amino acids in a polypeptide chain are important for the formation and maintenance of the tertiary and quaternary structures of proteins.

16.27 A peptide bond is an amide bond between two amino acids in a peptide chain.

16.29 **a.** His-trp-cys:

b. Gly-leu-ser:

c. Arg-ile-val:

16.31 The primary structure of a protein is the sequence of amino acids bonded to one another by peptide bonds.

16.33 The primary structure of a protein determines its three dimensional shape because the location of R groups along the protein chain is determined by the primary structure. The interactions among the R groups, based on their location in the chain, will govern how the protein folds. This, in turn, dictates its three-dimensional structure and biological function.

16.35 The secondary structure of a protein is the folding of the primary structure into an α-helix or β-pleated sheet.

16.37 **a.** α-Helix
b. β-Pleated sheet

16.39 The tertiary structure of a protein is the globular, three-dimensional structure of a protein that results from folding the regions of secondary structure.

16.41 Cystine is formed in an oxidation reaction between two cysteine amino acids which may be in the same peptide chain or in different peptide chains. Thus, cystine forms a covalent bridge between two peptide chains or between two regions of the same peptide chain. In the first case, cystine formation helps maintain the quaternary structure of a protein composed of more than one peptide chain. In the second case, cystine formation helps stabilize the tertiary structure of the protein.

16.43 Quaternary protein structure is the aggregation of two or more folded peptide chains to produce a functional protein.

16.45 A glycoprotein is a protein with covalently attached sugars.

16.47 Hemoglobin is the protein in red blood cells that is responsible for carrying oxygen to the cells of the body. It is composed of four subunits: two α-subunits and two β-subunits. Each of the subunits contains a heme group with an Fe^{2+} which serves as the binding site for the oxygen.

16.49 Because carbon monoxide binds tightly to the heme groups of hemoglobin, it is not easily removed or replaced by oxygen. As a result, the effects of oxygen deprivation (suffocation) occur.

16.51 When sickle cell hemoglobin (HbS) is deoxygenated, the amino acid valine fits into a hydrophobic pocket on the surface of another HbS molecule. Many such sickle cell hemoglobin molecules polymerize into long rods that cause the red blood cell to sickle. In normal hemoglobin, glutamic acid is found in the place of the valine. This negatively charged amino acid will not "fit" into the hydrophobic pocket.

16.53 *Denaturation* is the process by which the organized structure of a protein is disrupted, resulting in a completely disorganized, nonfunctional form of the protein.

16.55 Heat is an effective means of sterilization because it destroys the proteins of microbial life-forms, including fungi, bacteria, and viruses.

16.57 The common name of an enzyme is often derived from the name of the substrate and/or the type of reaction that it catalyzes.

16.59 A substrate is the chemical reactant in a chemical reaction that binds to an enzyme active site and is converted to product.

16.61 An enzyme-substrate complex is a molecular aggregate formed when the substrate binds to the active site of the enzyme.

16.63 A cofactor is an organic group, often a metal ion, that must be bound to an apoenzyme to maintain the correct configuration of the active site.

16.65 **a.** Citrate decarboxylase catalyzes the cleavage of a carboxyl group from citrate.
   b. Adenosine diphosphate phosphorylase catalyzes the addition of a phosphate group to ADP.
   c. Oxalate reductase catalyzes the reduction of oxalate.
   d. Nitrite oxidase catalyzes the oxidation of nitrite.
   e. *cis-trans* Isomerase catalyzes interconversion of *cis* and *trans* isomers.

16.67 The activation energy of a reaction is the energy required for the reaction to occur.

16.69 The lock-and-key model of enzyme-substrate binding was proposed by Emil Fischer in 1894. He thought that the active site was a rigid region of the enzyme into which the substrate fit perfectly. Thus, the model purports that the substrate simply snaps into place within the active site, like two pieces of a jigsaw puzzle fitting together.

16.71 The first step of an enzyme-catalyzed reaction is the formation of the enzyme-substrate complex. In the second step, the transition state is formed. This is the state in which the substrate assumes a form intermediate between the original substrate and the product. In step 3 the substrate is converted to product and the enzyme-product complex is formed. Step 4 involves the release of the product and regeneration of the enzyme in its original form.

16.73 $NAD^+/NADH$ serves an acceptor/donor of hydride anions in biochemical reactions. $NAD^+/NADH$ serves as a coenzyme for oxidoreductases.

16.75 Each of the following answers assumes that the enzyme was purified from an organism with optimal conditions for life near 37°C, pH 7.
   a. Decreasing the temperature from 37°C to 10°C will cause the rate of an enzyme-catalyzed reaction to decrease because the frequency of collisions between enzyme and substrate will decrease as the rate of molecular movement decreases.
   b. Increasing the pH from 7 to 11 will generally cause a decrease in the rate of an enzyme-catalyzed reaction. In fact, most enzymes would be denatured by a pH of 11 and enzyme activity would cease.
   c. Heating an enzyme from 37°C to 100°C will destroy enzyme activity because the enzyme would be denatured by the extreme heat.

16.77 The sulfa drugs are structural analogs of *para*-aminobenzoic acid (PABA). PABA is the substrate of an enzyme involved in the pathway for the biosynthesis of folic acid. The sulfa drugs act as competitive inhibitors of this enzyme.

   Folic acid is a vitamin required for the synthesis of a coenzyme needed to make the amino acid methionine and the purine and pyrimidine nitrogenous bases for DNA and RNA. When a sulfa drug binds to the enzyme, no product is formed, folic acid is not made, and the biosynthesis of methionine and nitrogenous bases ceases. This eventually kills the microorganism. Humans are not harmed by the sulfa drugs because we do not synthesize our own folic acid. It is obtained in the diet.

16.79 The compound would be a competitive inhibitor of the enzyme.

16.81 Creatine phosphokinase (CPK), lactate dehydrogenase (LDH), and aspartate aminotransferase (AST/SGOT)

Chapter 17

17.1 **a.** Adenosine diphosphate:

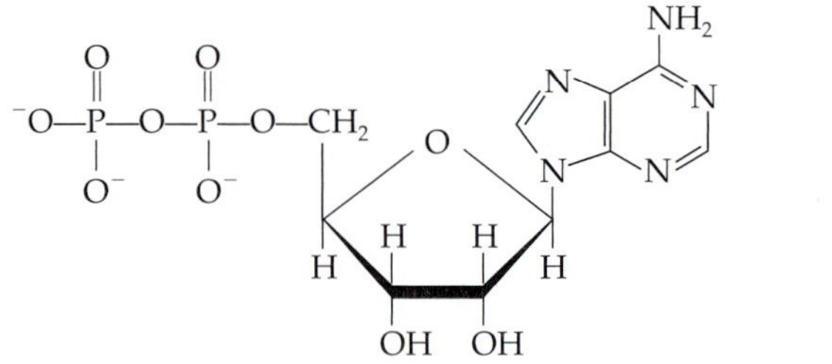

   b. Deoxyguanosine triphosphate:

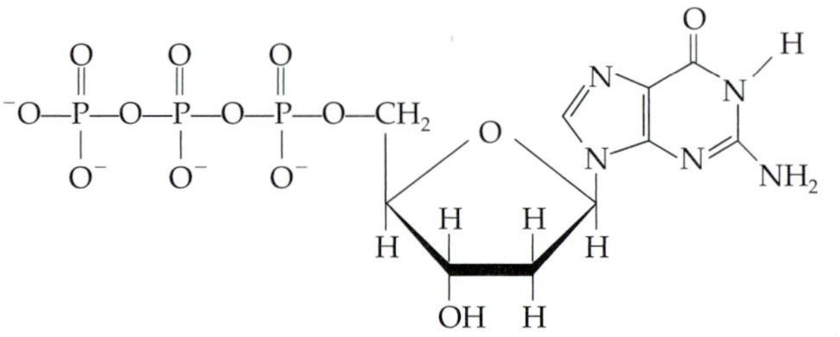

17.3 The RNA polymerase recognizes the promoter site for a gene, separates the strands of DNA, and catalyzes the polymerization of an RNA strand complementary to the DNA strand that carries the genetic code for a protein. It recognizes a termination site at the end of the gene and releases the RNA molecule.

17.5 The genetic code is said to be degenerate because several different triplet codons may serve as code words for a single amino acid.

17.7 The nitrogenous bases of the codons are complementary to those of the anticodons. As a result they are able to hydrogen bond to one another according to the base pairing rules.

17.9 The ribosomal P-site holds the peptidyl tRNA during protein synthesis. The peptidyl tRNA is the tRNA carrying the growing peptide chain. The only exception to this is during initiation of translation when the P-site holds the initiator tRNA.

17.11 The normal mRNA sequence, AUG-CCC-GAC-UUU, would encode the peptide sequence methionine-proline-aspartate-phenylalanine. The mutant mRNA sequence, AUG-CGC-GAC-UUU, would encode the mutant peptide sequence methionine-arginine-aspartate-phenylalanine. This would not be a silent mutation because a hydrophobic amino acid (proline) has been replaced by a positively charged amino acid (arginine).

17.13 It is the N-9 of the purine that forms the *N*-glycosidic bond with C-1 of the five-carbon sugar. The general structure of the purine ring is shown below:

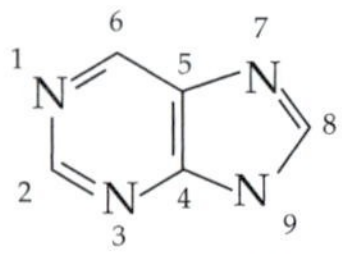

17.15 The ATP nucleotide is composed of the five-carbon sugar ribose, the purine adenine, and a triphosphate group.

17.17 The two strands of DNA in the double helix are said to be *antiparallel* because they run in opposite directions. One strand progresses in the $5' \rightarrow 3'$ direction, and the opposite strand progresses in the $3' \rightarrow 5'$ direction.

17.19 Two

17.21

17.23 The bacterial chromosome is a circular DNA molecule that is supercoiled, that is, the helix is coiled on itself.

17.25 The term *semiconservative DNA replication* refers to the fact that each parental DNA strand serves as the template for the synthesis of a daughter strand. As a result, each of the daughter DNA molecules is made up of one strand of the original parental DNA and one strand of newly synthesized DNA.

17.27 3'-TACGCCGATCTTATAAGGT-5'

17.29 DNA → RNA → Protein

17.31 Anticodons are found on transfer RNA molecules.

17.33 3'-AUGGAUCGAGACCAGUAAUUCCGUCAU-5'.

17.35 *RNA splicing* is the process by which the noncoding sequences (introns) of the primary transcript of a eukaryotic mRNA are removed and the protein coding sequences (exons) are spliced together.

17.37 Messenger RNA, transfer RNA, and ribosomal RNA

17.39 Introns must be removed from a primary transcript because they do not code for protein sequences and would result in the synthesis of a nonfunctional protein.

17.41 The *poly(A) tail* is a stretch of 100−200 adenosine nucleotides polymerized onto the 3' end of a mRNA by the enzyme poly(A) polymerase.

17.43 The *cap structure* is made up of the nucleotide 7-methylguanosine attached to the 5' end of a mRNA by a 5'-5' triphosphate bridge. Generally the first two nucleotides of the mRNA are also methylated.

17.45 Sixty-four

17.47 The reading frame of a gene is the sequential set of triplet codons that carries the genetic code for the primary structure of a protein.

17.49 Methionine and tryptophan

17.51 The codon 5'-UUU-3' encodes the amino acid phenylalanine. The mutant codon 5'-UUA-3' encodes the amino acid leucine. Both leucine and phenylalanine are hydrophobic amino acids, however, leucine has a smaller R group. It is possible that the smaller R group would disrupt the structure of the protein.

17.53 The ribosomes serve as a platform on which protein synthesis can occur. They also carry the enzymatic activity that forms peptide bonds.

17.55 The sequence of DNA nucleotides in a gene is transcribed to produce a complementary sequence of RNA nucleotides in a messenger RNA (mRNA). In the process of translation the sequence of the mRNA is read sequentially in words of three nucleotides (codons) to produce a protein. Each codon calls for the addition of a particular amino acid to the growing peptide chain. Through these processes, the sequence of nucleotides in a gene determines the sequence of amino acids in the primary structure of a protein.

17.57 In the initiation of translation, initiation factors, methionyl tRNA (the initiator tRNA), the mRNA, and the small and large ribosomal subunits form the initiation complex. During the elongation stage of translation, an aminoacyl tRNA binds to the A-site of the ribosome. Peptidyl transferase catalyzes the formation of a peptide bond and the peptide chain is transferred to the tRNA in the A-site. Translocation shifts the peptidyl tRNA from the A-site into the P-site, leaving the A-site available for the next aminoacyl tRNA. In the termination stage of translation, a termination codon is encountered. A release factor binds to the empty A-site and peptidyl transferase catalyzes the hydrolysis of the bond between the peptidyl tRNA and the completed peptide chain.

17.59 An ester bond

17.61 A point mutation is the substitution of one nucleotide pair for another in a gene.

17.63 Some mutations are silent because the change in the nucleotide sequence does not alter the amino acid sequence of the protein. This can happen because there are many amino acids encoded by multiple codons.

17.65 UV light causes the formation of pyrimidine dimers, the covalent bonding of two adjacent pyrimidine bases. Mutations occur when the UV damage repair system makes an error during the repair process. This causes a change in the nucleotide sequence of the DNA.

17.67 A *carcinogen* is a compound that causes cancer. Cancers are caused by mutations in the genes responsible for controlling cell division. Carcinogens cause DNA damage that results in changes in the nucleotide sequence of the gene. Thus, carcinogens are also mutagens.

Chapter 18

18.1 ATP is called the universal energy currency because it is the major molecule used by all organisms to store energy.

18.3 The first stage of catabolism is the digestion (hydrolysis) of dietary macromolecules in the stomach and intestine.

In the second stage of catabolism, monosaccharides, amino acids, fatty acids, and glycerol are converted by metabolic reactions into molecules that can be completely oxidized.

In the third stage of catabolism, the two-carbon acetyl group of acetyl CoA is completely oxidized by the reactions of the citric acid cycle. The energy of the electrons harvested in these oxidation reactions is used to make ATP.

18.5 Substrate level phosphorylation is one way the cell can make ATP. In this reaction, a high-energy phosphoryl group of a substrate in the reaction is transferred to ADP to produce ATP.

18.7 Glycolysis is a pathway involving ten reactions. In reactions 1–3, energy is invested in the beginning substrate, glucose. This is done by transferring high-energy phosphoryl groups from ATP to the intermediates in the pathway. The product is fructose-1,6-bisphosphate. In the energy-harvesting reactions of glycolysis, fructose-1,6-bisphosphate is split into two three-carbon molecules that begin a series of rearrangement, oxidation-reduction, and substrate-level phosphorylation reactions that produce four ATP, two NADH, and two pyruvate molecules. Because of the investment of two ATP in the early steps of glycolysis, the net yield of ATP is two.

18.9 Both the alcohol and lactate fermentations are anaerobic reactions that use the pyruvate and re-oxidize the NADH produced in glycolysis.

18.11

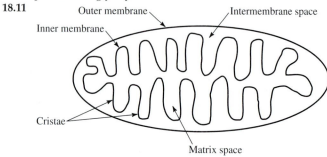

18.13 Pyruvate is converted to acetyl CoA by the pyruvate dehydrogenase complex. This huge enzyme complex requires four coenzymes, each of which is made from a different vitamin. The four coenzymes are thiamine pyrophosphate (made from thiamine), FAD (made from riboflavin), NAD^+ (made from niacin), and coenzyme A (made from the vitamin pantothenic acid). The coenzyme lipoamide is also involved in this reaction.

18.15 *Oxidative phosphorylation* is the process by which the energy of electrons harvested from oxidation of a fuel molecule is used to phosphorylate ADP to produce ATP.

18.17 $NAD^+ + H:^- \rightarrow NADH$

18.19 Glucagon indirectly stimulates glycogen phosphorylase, the first enzyme of glycogenolysis. This speeds up glycogen degradation. Glucagon also inhibits glycogen synthase, the first enzyme in glycogenesis. This inhibits glycogen synthesis.

18.21 ATP

18.23 A high-energy bond is a weak bond that, on breaking, can form a much stronger bond and, in the process, releases energy.

18.25 Carbohydrates

18.27 The following equation represents the hydrolysis of maltose:

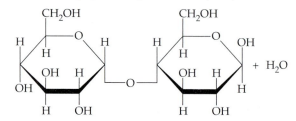

β-Maltose

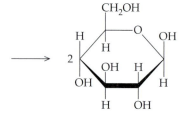

β-D-Glucose

18.29 The hydrolysis of a triglyceride containing oleic acid, stearic acid, and linoleic acid is represented in the following equations:

$$H-\overset{H}{\underset{H}{C}}-O-\overset{O}{\overset{\|}{C}}-(CH_2)_7CH=CH(CH_2)_7CH_3$$
$$H-\overset{}{\underset{}{C}}-O-\overset{O}{\overset{\|}{C}}-(CH_2)_{16}CH_3 \quad + 3H_2O \longrightarrow$$
$$H-\overset{}{\underset{H}{C}}-O-\overset{O}{\overset{\|}{C}}-(CH_2)_7CH=CHCH_2CH=CH(CH_2)_4CH_3$$

$$\begin{array}{c}H-\overset{H}{\underset{|}{C}}-OH\\H-\overset{|}{\underset{|}{C}}-OH\\H-\overset{|}{\underset{|}{C}}-OH\\\overset{|}{H}\end{array}$$
Glycerol

+

$$HO-\overset{O}{\overset{\|}{C}}-(CH_2)_7CH=CH(CH_2)_7CH_3$$
Oleic acid

+

$$HO-\overset{O}{\overset{\|}{C}}-(CH_2)_{16}CH_3$$
Stearic acid

+

$$HO-\overset{O}{\overset{\|}{C}}-(CH_2)_7CH=CHCH_2CH=CH(CH_2)_4CH_3$$
Linoleic acid

18.31 The end products of glycolysis are 2 molecules of pyruvate, 4 molecules of ATP (net production of 2 ATP), and 2 molecules of NADH.

18.33 Glycolysis occurs in the cytoplasm of the cell.

18.35 A kinase transfers a phosphoryl group from one molecule to another.

18.37 NAD^+ is reduced, accepting a hydride anion.

18.39

$$\underset{\text{Acetaldehyde}}{CH_2-\overset{O}{\overset{\|}{C}}-H} \xrightarrow{\text{NADH} \quad \text{NAD}^+} \underset{\text{Ethanol}}{CH_3CH_2OH}$$

18.41 This child must have the enzymes to carry out the alcohol fermentation. When the child exercised hard, there was not enough oxygen in the cells to maintain aerobic respiration. As a result, glycolysis and the alcohol fermentation were responsible for the majority of the ATP production by the child. The accumulation of alcohol (ethanol) in the child caused the symptoms of drunkenness.

18.43 The *mitochondrion* is an organelle that serves as the cellular power plant. The reactions of the citric acid cycle, the electron transport system, and ATP synthase function together within the mitochondrion to harvest ATP energy for the cell.

18.45 The final oxidation of carbohydrates, amino acid carbon skeletons, and fatty acids occur in the mitochondrial matrix. The pathways that carry out these reactions are the citric acid cycle and β-oxidation of fatty acids.

18.47 Coenzyme A is a molecule derived from ATP and the vitamin pantothenic acid. It functions in the transfer of acetyl groups in lipid and carbohydrate metabolism.

18.49 Under aerobic conditions pyruvate is converted to acetyl CoA.

18.51 The coenzymes NAD^+, FAD, thiamine pyrophosphate, and coenzyme A are required by the pyruvate dehydrogenase complex for the conversion of pyruvate to acetyl CoA. These coenzymes are synthesized from the vitamins niacin, riboflavin, thiamine, and pantothenic acid, respectively. If the vitamins are not available, the coenzymes will not be available and pyruvate cannot be converted to acetyl CoA. Because the complete oxidation of the acetyl group of acetyl CoA produces the vast majority of the ATP for the body, ATP production would be severely inhibited by a deficiency of any of these vitamins.

18.53 Isomers are molecules with the same molecular formula but different chemical structures.

18.55 **a.** False **c.** True
b. False **d.** True

18.57 The acetyl group of acetyl CoA is converted into 2 CO_2 during oxidation in the reactions of the citric acid cycle. One ATP molecule, one $FADH_2$, and 3 NADH are produced in the process.

18.59 First, write an equation representing the conversion of pyruvate to acetyl CoA to determine which carbon in acetyl CoA is labeled:

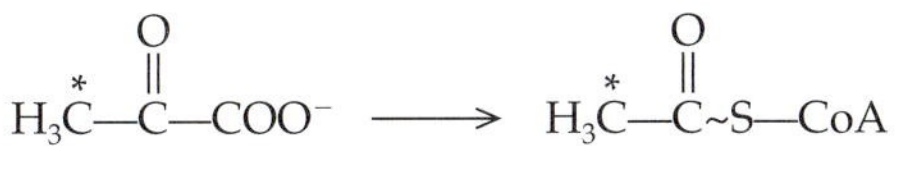

Pyruvate Acetyl CoA

Next, write an equation representing the reaction of acetyl CoA and oxaloacetate to show the location of the radiolabeled carbon in citrate.

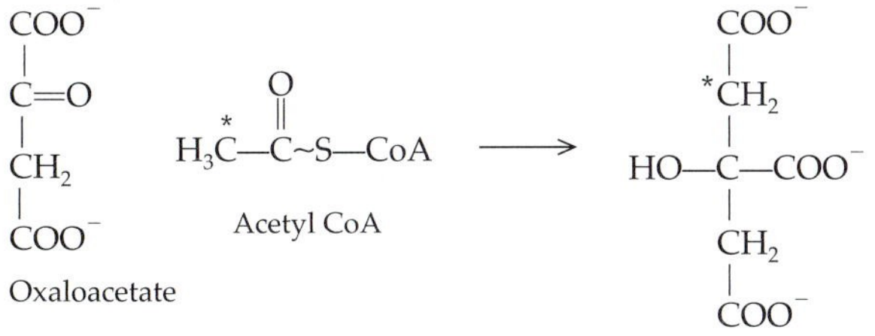

Oxaloacetate Acetyl CoA Citrate

Now draw out the intermediates of the citric acid cycle. Place an asterisk on the radiolabeled carbon and circle the $—COO^-$ groups that are released as CO_2.

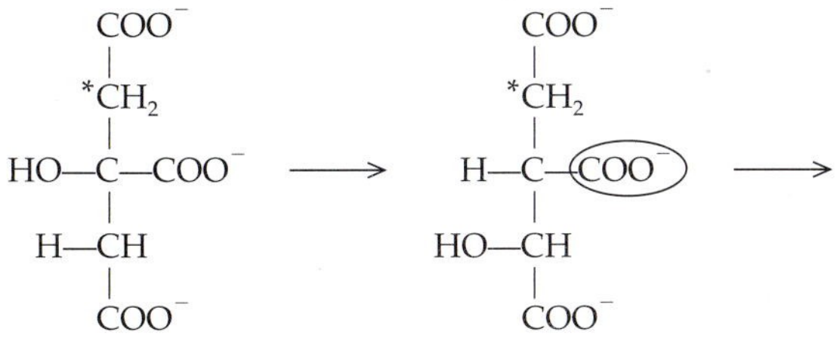

Citrate Isocitrate

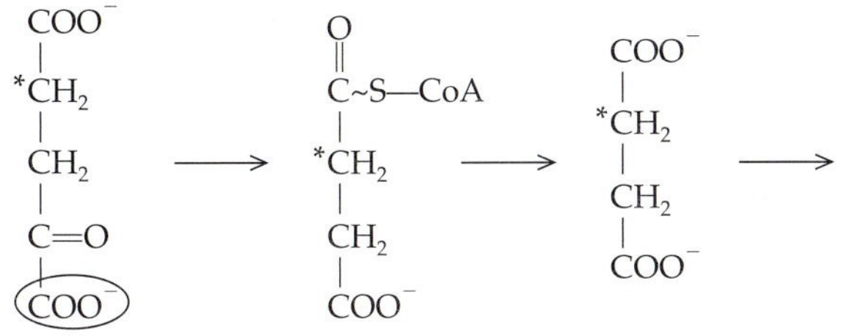

α-Ketoglutarate Succinyl CoA Succinate

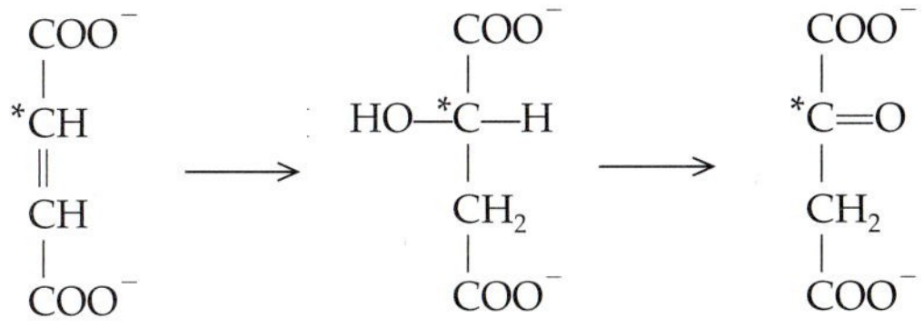

Fumarate Malate Oxaloacetate

18.61 It is a kinase because it transfers a phosphoryl group from one molecule to another. Kinases are a specific type of transferase.

18.63 The electron transport system is series of electron transport proteins embedded in the inner mitochondrial membrane that accept high-energy electrons from NADH and $FADH_2$ and transfer them in stepwise fashion to molecular oxygen (O_2).

18.65 Three ATP

18.67 The oxidation of a variety of fuel molecules, including carbohydrates, the carbon skeletons of amino acids, and fatty acids provides the electrons. The energy of these electrons is used to produce an H^+ reservoir. The energy of this proton reservoir is used for ATP synthesis.

18.69 **a.** Two ATP per glucose (net yield) are produced in glycolysis, whereas the complete oxidation of glucose in aerobic respiration (glycolysis, the citric acid cycle, and oxidative phosphorylation) results in the production of thirty-six ATP per glucose.
b. Thus, aerobic respiration harvests nearly 40% of the potential energy of glucose, and anaerobic glycolysis harvests only about 2% of the potential energy of glucose.

18.71 Gluconeogenesis is production of glucose from noncarbohydrate starting materials. This pathway can provide glucose when starvation or strenuous exercise leads to a depletion of glucose from the body.

18.73 Gluconeogenesis produces glucose from non-carbohydrate molecules in times when blood glucose levels are low, as during strenuous exercise, a high protein/low carbohydrate diet, or starvation. This ensures proper function of brain and red blood cells, which only use glucose as fuel.

18.75 The liver and pancreas

18.77 *Hypoglycemia* is the condition in which blood glucose levels are too low.

18.79 **a.** Insulin stimulates glycogen synthase, the first enzyme in glycogen synthesis. It also stimulates uptake of glucose from the bloodstream into cells and phosphorylation of glucose by the enzyme glucokinase.
b. This traps glucose within liver cells and increases the storage of glucose in the form of glycogen.
c. These processes decrease blood glucose levels.

Chapter 19

19.1 Because dietary lipids are hydrophobic, they arrive in the small intestine as large fat globules. The bile salts emulsify these fat globules into tiny fat droplets. This greatly increases the surface area of the lipids, allowing them to be more accessible to pancreatic lipases and thus more easily digested.

19.3 **a.** Three acetyl CoA, two NADH, and two $FADH_2$
b. Nine acetyl CoA, eight NADH, and eight $FADH_2$

19.5

$$CH_3CH_2CH_2-\overset{\overset{\displaystyle O}{\|}}{C}\sim S-CoA + FAD \longrightarrow$$

$$CH_3CH=CH-\overset{\overset{\displaystyle O}{\|}}{C}\sim S-CoA + FADH_2$$

$\downarrow \curvearrowright H_2O$

$$\longleftarrow \quad CH_3-\overset{\overset{\displaystyle OH}{|}}{\underset{\underset{\displaystyle H}{|}}{C}}-CH_2-\overset{\overset{\displaystyle O}{\|}}{C}\sim S-CoA + NAD^+$$

$$CH_3-\overset{\overset{\displaystyle O}{\|}}{C}-CH_2-\overset{\overset{\displaystyle O}{\|}}{C}\sim S-CoA + NADH$$

$\downarrow \curvearrowright$ Coenzyme A

$$2CH_3-\overset{\overset{\displaystyle O}{\|}}{C}\sim S-CoA$$

19.7 Starvation, a diet low in carbohydrates, and diabetes mellitus are conditions that lead to the production of ketone bodies.

19.9 (1) Fatty acid biosynthesis occurs in the cytoplasm whereas β-oxidation occurs in the mitochondria.
(2) The acyl group carrier in fatty acid biosynthesis is acyl carrier protein while the acyl group carrier in β-oxidation is coenzyme A.
(3) The seven enzymes of fatty acid biosynthesis are associated as a multienzyme complex called *fatty acid synthase*. The enzymes involved in β-oxidation are not physically associated with one another.
(4) NADPH is the reducing agent used in fatty acid biosynthesis. NADH and FADH₂ are produced by β-oxidation.

19.11 The purpose of the urea cycle is to convert toxic ammonium ions to urea, which is excreted in the urine of land animals.

19.13 Insulin stimulates uptake of glucose and amino acids by cells, glycogen and protein synthesis, and storage of lipids. It inhibits glycogenolysis, gluconeogenesis, breakdown of stored triglycerides, and ketogenesis.

19.15 Bile consists of micelles of lecithin, cholesterol, bile salts, proteins, inorganic ions, and bile pigments that aid in lipid digestion by emulsifying fat droplets. Bile salts are detergents because they have polar heads which make them soluble in aqueous solutions and hydrophobic tails that dissolve in nonpolar solvents and, in this case, bind triglycerides.

19.17 A chylomicron is a plasma lipoprotein that carries triglycerides from the intestine to all body tissues via the bloodstream. That function is reflected in the composition of the chylomicron, which is approximately 85% triglycerides, 9% phospholipids, 3% cholesterol esters, 2% protein, and 1% cholesterol.

19.19 Triglycerides

19.21 The major metabolic function of adipose tissue is the storage of triglycerides.

19.23 Triglycerides represent a highly reduced source of fuel. The complete oxidation of fatty acids releases much more energy than the complete oxidation of the same amount of glycogen. Thus, more energy can be stored in a smaller space as triglycerides.

19.25 When dietary lipids in the form of fat globules reach the duodenum, they are emulsified by bile salts. The triglycerides in the resulting tiny fat droplets are hydrolyzed into monoglycerides and fatty acids by the action of pancreatic lipases, assisted by colipase. The monoglycerides and fatty acids are absorbed by cells lining the intestine.

19.27 The energy source for the activation of a fatty acid entering β-oxidation is the hydrolysis of ATP into AMP and PP$_i$ (pyrophosphate group), an energy expense of two high-energy phosphoester bonds.

19.29 An alcohol is the product of the hydration of an alkene.

19.31 112 ATP

19.33 Two ATP

19.35 The acetyl CoA produced by β-oxidation will enter the citric acid cycle.

19.37 Ketone bodies include the compounds acetone, acetoacetone, and β-hydroxybutyrate, which are produced from fatty acids in the liver via acetyl CoA.

19.39 Ketoacidosis is a drop in the pH of the blood caused by elevated concentrations of ketone bodies.

19.41 If β-oxidation is going on at a rapid rate, it will produce large amounts of acetyl CoA. Normally the acetyl CoA would enter the citric acid cycle. However, if there is not enough oxaloacetate to allow all the acetyl CoA to enter the citric acid cycle, it will be used for the synthesis of ketone bodies.

19.43 Ketone bodies are relatively strong acids. If they are present in high concentrations in the blood, they will dissociate to release large amounts of H$^+$. This causes the blood to become more acidic, a condition called ketoacidosis.

19.45 Cytoplasm

19.47 a. The phosphopantetheine group allows formation of a high-energy thioester bond with a fatty acid.
b. It is derived from the vitamin pantothenic acid.

19.49 Fatty acid synthase is a huge multienzyme complex consisting of the seven enzymes involved in fatty acid synthesis. It is found in the cell cytoplasm. The enzymes involved in β-oxidation are not physically associated with one another. They are free in the mitochondrial matrix space.

19.51 Transaminases transfer amino groups from amino acids to ketoacids.

19.53 The glutamate family of transaminases is very important because the ketoacid corresponding to glutamate is α-ketoglutarate, one of the citric acid cycle intermediates. This provides a link between the citric acid cycle and amino acid metabolism. These transaminases provide amino groups for amino acid synthesis and collect amino groups during catabolism of amino acids.

19.55 a. Pyruvate   **d.** Acetyl CoA
b. α-Ketoglutarate   **e.** Succinate
c. Oxaloacetate   **f.** α-Ketoglutarate

19.57

$$\overset{\overset{\displaystyle N^+H_3}{|}}{HC}-COO^-$$
$$\overset{|}{\underset{\underset{\displaystyle COO^-}{|}}{\underset{\underset{\displaystyle CH_2}{|}}{CH_2}}} \quad + NAD^+ + H_2O \longrightarrow$$

$$\overset{\overset{\displaystyle O}{\|}}{C}-COO^-$$
$$\overset{|}{\underset{\underset{\displaystyle COO^-}{|}}{\underset{\underset{\displaystyle CH_2}{|}}{CH_2}}} \quad + NADH$$

Glutamate   α-Ketoglutarate

19.59 Hyperammonemia

19.61 a. The source of one amino group of urea is the ammonium ion and the source of the other is the α-amino group of the amino acid aspartate.

   b. The carbonyl group of urea is derived from CO_2.

19.63 In general, insulin stimulates anabolic processes, including glycogen synthesis, uptake of amino acids and protein synthesis, and triglyceride synthesis. At the same time, catabolic processes such as glycogenolysis are inhibited.

19.65 A target cell is one that has a receptor for a particular hormone.

19.67 Decreased blood glucose levels trigger the secretion of glucagon into the bloodstream.

19.69 Insulin is produced in the β-cells of the islets of Langerhans in the pancreas.

19.71 Insulin stimulates the uptake of glucose from the blood into cells. It enhances glucose storage by stimulating glycogenesis and inhibiting glycogen degradation and gluconeogenesis.

19.73 Insulin stimulates synthesis and storage of triglycerides.

19.75 Untreated diabetes mellitus is starvation in the midst of plenty because blood glucose levels are very high. However, in the absence of insulin, blood glucose can't be taken up into cells. The excess glucose is excreted into the urine while the cells of the body are starved for energy.

Credits

(Flower): © Vol. 101/Corbis; (Caterpillar): © Norm Thomas/PhotoResearchers, Inc.; **13.7a,b:** © Vol. 94/Corbis; **13.7c:** © Gregory G. Dimijian/ Photo Researchers, Inc.

Chapter 14

Opener: © Digital Vision/Getty Images; **14.1:** © USDA; **14.2:** © Vol. 67/PhotoDisc/Getty Images; **p. 414:** © StanleyFlegler/Visuals Unlimited; **p. 415:** © The McGraw-Hill Companies, Inc./ Louis Rosenstock, photographer; **14.3:** © Corbis Royalty Free; **p. 432:** © Jean Claude Levy/ISM/Phototake.

Chapter 15

Opener: © Jane Grushow/Grant Heilman Photography; **p. 449:** Courtesy Special Collec-

tions, Edinburgh University; **p. 443:** Lester Lefkowitz/Corbis; **p. 458:** © Hans Pfletschinger/ Peter Arnold, Inc.

Chapter 16

Opener: Courtesy R. John Collier, Harvard Medical School/NIAID Biodefense Research; **p. 474:** © Vol. 18/PhotoDisc/Getty Images; **p. 480:** © Vol. 270/Corbis; **16.13:** © Meckes/ Ottawa/Photo Researchers, Inc.

Chapter 17

Opener: © Dennis MacDonald/PhotoEdit; **p. 528:** © Dr. Charles S. Helling/USDA; **p. 529:** Courtesy of Orchid Cellmark, Germantown, Maryland.

Chapter 18

Opener: © Dynamics Graphics/Jupiter Images; **p. 550:** © The McGraw Hill Companies, Inc./ Louis Rosenstock, photographer; **18.8:** © CNRI/ Phototake; **p. 553:** Vol. 10/PhotoDisc/Getty Images.

Chapter 19

Opener: © Royalty-Free Corbis; **p. 579:** Vol. 110/PhotoDisc/Getty Images.

Index